Ecosystems and Nature

Environmental Analysis and Economic Policy

Series Editors: Kenneth Button
Professor of Public Policy
The Institute of Public Policy, George Mason University, USA

Peter Nijkamp
Professor in Regional, Urban and Environmental Economics
Free University, Amsterdam, The Netherlands

1. Global Aspects of the Environment (Volumes I and II)
   Robert U. Ayres, Kenneth Button and Peter Nijkamp
2. Environment, Land Use and Urban Policy
   David Banister, Kenneth Button and Peter Nijkamp
3. Environmental Valuation (Volumes I and II)
   Kenneth G. Willis, Kenneth Button and Peter Nijkamp
4. The Environment and Transport
   Yoshitsugu Hayashi, Kenneth Button and Peter Nijkamp
5. Environmental Economics and Development
   J(Hans). B. Opschoor, Kenneth Button and Peter Nijkamp
6. Environmental Instruments and Institutions
   Thomas Tietenberg, Kenneth Button and Peter Nijkamp
7. Ecosystems and Nature
   R. Kerry Turner, Kenneth Button and Peter Nijkamp

Wherever possible, the articles in these volumes have been reproduced as originally published using facsimile reproduction, inclusive of footnotes and pagination to facilitate ease of reference.

For a list of all Edward Elgar published titles visit our site on the World Wide Web at http://www.e-elgar.co.uk

Ecosystems and Nature

Economics, Science and Policy

Edited by

R. Kerry Turner

Director of the Centre for Social and Economic Research on the Global Environment (CSERGE), University of East Anglia and University College London, UK

Kenneth Button

Professor of Public Policy, The Institute of Public Policy, George Mason University, USA

and

Peter Nijkamp

Professor in Regional, Urban and Environmental Economics, Free University, Amsterdam, The Netherlands

ENVIRONMENTAL ANALYSIS AND ECONOMIC POLICY

An Elgar Reference Collection
Cheltenham, UK • Northampton, MA, USA

Published by
Edward Elgar Publishing Ltd
Glensanda House
Montpellier Parade
Cheltenham
Glos GL50 1UA
UK

Edward Elgar Publishing, Inc.
136 West Street
Suite 202
Northampton
Massachusetts 01060
USA

A catalogue record for this book is available from the British Library.

Library of Congress Cataloguing in Publication Data

Ecosystems and nature: economics, science, and policy / edited by R. Kerry Turner, Kenneth Button and Peter Nijkamp.
(Environmental analysis and economic policy: 7)
Includes index.
1. Biological diversity conservation. 2. Biological diversity conservation—Economic aspects. 3. Biological diversity conservation —Government Policy. I. Turner, R. Kerry. II. Button, Kenneth. III. Nijkamp, Peter. IV. Series. V. Series: An Elgar reference collection.
QH75.E3236 1999
333.95'16—dc21 99–27515
CIP

ISBN 1 85898 793 8

Printed and bound in Great Britain by Biddles Ltd, Guildford and King's Lynn

Contents

Acknowledgements

The editors and publishers wish to thank the authors and the following publishers who have kindly given permission for the use of copyright material.

Ambio for articles: Alan Randall (1991), 'The Value of Biodiversity', *Ambio*, **20** (2), April, 64–8; Madhav Gadgil (1992), 'Conserving Biodiversity as if People Matter: A Case Study from India', *Ambio*, **21** (3), May, 266–70; W. Neil Adger, Katrina Brown, Raffaello Cervigni and Dominic Moran (1995), 'Total Economic Value of Forests in Mexico', *Ambio*, **24** (5), August, 286–96; Priya Shyamsundar and Randall Kramer (1997), 'Biodiversity Conservation – At What Cost? A Study of Households in the Vicinity of Madagascar's Mantadia National Park', *Ambio*, **26** (3), May, 180–84.

American Association for the Advancement of Science for articles: Kenneth Arrow, Bert Bolin, Robert Costanza, Partha Dasgupta, Carl Folke, C.S. Holling, Bengt-Owe Jansson, Simon Levin, Karl-Göran Mäler, Charles Perrings and David Pimentel (1995), 'Economic Growth, Carrying Capacity, and the Environment', *Science*, **268**, 28 April, 520–21; Stuart L. Pimm, Gareth J. Russell, John L. Gittleman and Thomas M. Brooks (1995), 'The Future of Biodiversity', *Science*, **269**, 21 July, 347–50.

Annual Reviews, Inc. for article: C.S. Holling (1973), 'Resilience and Stability of Ecological Systems', *Annual Review of Ecology and Systematics*, **4**, 1–23.

Ecological Applications for article: Carl Folke, C.S. Holling and Charles Perrings (1996), 'Biological Diversity, Ecosystems, and the Human Scale', *Ecological Applications*, **6** (4), November, 1018–24.

Elsevier Science BV for articles: Robert Costanza, Stephen C. Farber and Judith Maxwell (1989), 'Valuation and Management of Wetland Ecosystems', *Ecological Economics*, **1** (4), December, 335–61; H. Jack Ruitenbeek (1992), 'The Rainforest Supply Price: A Tool for Evaluating Rainforest Conservation Expenditures', *Ecological Economics*, **6** (1), July, 57–78; H. Jack Ruitenbeek (1994), 'Modelling Economy–Ecology Linkages in Mangroves: Economic Evidence for Promoting Conservation in Bintuni Bay, Indonesia', *Ecological Economics*, **10** (3), August, 233–47; James R. Kahn and Judith A. McDonald (1995), 'Third-World Debt and Tropical Deforestation', *Ecological Economics*, **12** (2), February, 107–23; Bryan G. Norton (1995), 'Evaluating Ecosystem States: Two Competing Paradigms', *Ecological Economics*, **14** (2), August, 113–27; Gail Bingham, Richard Bishop, Michael Brody, Daniel Bromley, Edwin (Toby) Clark, William Cooper, Robert Costanza, Thomas Hale, Gregory Hayden, Stephen Kellert, Richard Norgaard, Bryan Norton, John Payne, Clifford Russell and Glenn Suter (1995), 'Issues in Ecosystem Valuation: Improving Information for Decision Making', *Ecological Economics*, **14** (2), August, 73–90; N. Bockstael, R. Costanza,

I. Strand, W. Boynton, K. Bell and L. Wainger (1995), 'Ecological Economic Modeling and Valuation of Ecosystems', *Ecological Economics*, **14** (2), August, 143–59; Talbot Page (1995), 'Harmony and Pathology', *Ecological Economics*, **15** (2), November, 141–4; Charles Perrings and Brian Walker (1997), 'Biodiversity, Resilience and the Control of Ecological-Economic Systems: The Case of Fire-Driven Rangelands', *Ecological Economics*, **22** (1), July, 73–83.

Elsevier Science Ltd for articles: Dominic Moran, David Pearce and Anouk Wendelaar (1996), 'Global Biodiversity Priorities: A Cost-Effectiveness Index for Investments', *Global Environmental Change*, **6** (2), June, 103–19; Kris H. Johnson, Kristiina A. Vogt, Heidi J. Clark, Oswald J. Schmitz and Daniel J. Vogt (1996), 'Biodiversity and the Productivity and Stability of Ecosystems', *Trends in Ecology and Evolution*, **11** (9), September, 372–7.

Kluwer Academic Publishers BV for articles: Andreas Hohl and Clement A. Tisdell (1993), 'How Useful are Environmental Safety Standards in Economics? – The Example of Safe Minimum Standards for Protection of Species', *Biodiversity and Conservation*, **2** (2), April, 168–81; Charles Perrings and David Pearce (1994), 'Threshold Effects and Incentives for the Conservation of Biodiversity', *Environmental and Resource Economics*, **4** (1), February, 13–28; Brent M. Swallow and Daniel W. Bromley (1995), 'Institutions, Governance and Incentives in Common Property Regimes for African Rangelands', *Environmental and Resource Economics*, **6**, 99–118; Katrina Brown (1997), 'Plain Tales from the Grasslands: Extraction, Value and Utilization of Biomass in Royal Bardia National Park, Nepal', *Biodiversity and Conservation*, **6** (1), January, 59–74.

Marine Resource Economics for articles: John B. Loomis and Douglas M. Larson (1994), 'Total Economic Values of Increasing Gray Whale Populations: Results from a Contingent Valuation Survey of Visitors and Households', *Marine Resource Economics*, **9**, 275–86; Stephen K. Swallow (1994), 'Renewable and Nonrenewable Resource Theory Applied to Coastal Agriculture, Forest, Wetland, and Fishery Linkages', *Marine Resource Economics*, **9** (4), Winter, 291–310.

New York Botanical Garden for article: Ricardo Godoy, Ruben Lubowski and Anil Markandya (1993), 'A Method for the Economic Valuation of Non-Timber Tropical Forest Products', *Economic Botany*, **47** (3), 220–33.

Oxford University Press for article: Timothy M. Swanson (1994), 'The Economics of Extinction Revisited and Revised: A Generalised Framework for the Analysis of the Problems of Endangered Species and Biodiversity Losses', *Oxford Economic Papers*, **46**, Supplementary Issue, October, 800–821.

Holmes Rolston III and *Environmental Ethics* for article: Holmes Rolston, III (1985), 'Valuing Wildlands', *Environmental Ethics*, **7**, Spring, 23–48.

Springer-Verlag GmbH & Co. KG for article: R.K. Turner, S. Subak and W.N. Adger (1996), 'Pressures, Trends, and Impacts in Coastal Zones: Interactions Between Socioeconomic and Natural Systems', *Environmental Management*, **20** (2), 159–73.

University of Wisconsin Press for articles: Thomas D. Crocker (1985), 'On the Value of the Condition of a Forest Stock', *Land Economics*, **61** (3), August, 244–54; Sandra S. Batie and Carl C. Mabbs-Zeno (1985), 'Opportunity Costs of Preserving Coastal Wetlands: A Case Study of a Recreational Housing Development', *Land Economics*, **61** (1), February, 1–9; Edward B. Barbier (1994), 'Valuing Environmental Functions: Tropical Wetlands', *Land Economics*, **70** (2), May, 155–73; Martin Whitby and Caroline Saunders (1996), 'Estimating the Supply of Conservation Goods in Britain: A Comparison of the Financial Efficiency of Two Policy Instruments', *Land Economics*, **72** (3), August, 313–25; John M. Gowdy (1997), 'The Value of Biodiversity: Markets, Society, and Ecosystems', *Land Economics*, **73** (1), February, 25–41.

Walter de Gruyter & Co. for article: Lars Drake (1992), 'The Non-Market Value of the Swedish Agricultural Landscape', *European Review of Agricultural Economics*, **19**, 351–64.

Every effort has been made to trace all the copyright holders but if any have been inadvertently overlooked the publishers will be pleased to make the necessary arrangement at the first opportunity.

In addition the publishers wish to thank the Library of the London School of Economics and Political Science and the Marshall Library of Economics, Cambridge University for their assistance in obtaining these articles.

Series Preface

Kenneth Button and Peter Nijkamp

The environment has been a dominant theme for research and public policy during the latter part of the twentieth century and there is no sign that this will change as we move to the new millennium. The global interest in environmental matters in part stems from the increased pressures that a mounting population and increased production puts on the planet's natural resource base. Additionally, as personal incomes rise and leisure time becomes more freely available in the developed world, concern with more immediate human needs give way to an interest in preservation and conservation for future generations and for others. Science has also contributed by both highlighting emerging environmental problems and discovering the importance of many components of the natural environment for medical and other uses.

Historically, the interest in the environment can be traced through several phases, each with its own particular emphasis. The interest is also far from new and in antiquity there were often laws governing such things as the removal of night soil and the use of noisy wagons into towns. Traditionally, also the natural environment was seen as a source of the necessities of life and was controlled through farming techniques to ensure that food was adequate to maintain the population. Land use was controlled and water resources were often regulated at the local and regional levels.

With the Industrial Revolution came population growth, urbanization and the accelerated exploitation of natural resources. Malthus advanced ideas of over population while others, such as Jevons, became concerned about the finite nature of such resources as coal. The new technologies contaminated water courses and the widespread use of carbon fuels led to atmospheric pollution. In cities, dirt, noise and urban decay led to new ideas of town planning and ideal cities.

Much of the interest in the latter part of the last century centred around public health. As knowledge of how disease was spread – urban authorities in particular sought to improve the local environment by such measures as sewage control and clean water supply to reduce the spread of germs and infection. This trend, much later, spread to policies embodied in various pieces of clean air legislation to reduce local atmospheric pollution that causes smog and other harmful effects.

Wealthy societies, and the better off within poorer societies, with the time and resources to expand, became concerned with the built environment and in shaping nature in ways they found aesthetically pleasing. Over the centuries this has led to landscaping of the countryside and the provision of parks and gardens in urban areas.

The twentieth century has seen many of these trends continue although some of the problems, such as depletion of natural resources, initially appeared less severe as new sources were found in Africa, South America and Asia. Difficulties in feeding and otherwise sustaining the burgeoning populations of these regions, however, resurrected Malthusian concerns over whether the planet could sustain itself. Increasing reliance on fossil fuels for energy and the geographical concentration of the main reserves of oil in politically unstable areas resurrected concern over natural resource reserves in the post Second World War period.

More fundamental to modern debates has been the global orientation of recent concerns. This has been accompanied by a much more popular interest in the state of the natural environment. Kenneth Boulding's development of the idea of 'spaceship earth' attracted much attention, as did the rather doom ladened prophesies of the Club of Rome, but this was subsequently overshadowed by the publication of the Brundtland Report, *Our Common Future*, with its emphasis on the need to maintain resources for future generations. With the report came the much touted, but universally ambiguous, concept of sustainable development.

Just as the academic interest in the environment has evolved so has the institutional setting in which the topic is debated and researched. In a very general way as our understanding of the environment has progressed and as technology has developed so there has been a movement out from micro-policy making in the hands of city councils, lords of the manor and the like to macro-policy making, involving global agencies such as the United Nations and World Bank. Indeed, one of the pressing issues of contemporary environmental debate is the appropriate jurisdiction for different levels of government.

For those interested in researching environmental topics, either from the standpoint of the beginner or of that of an expert wishing to expand an established basis of knowledge, there are practical problems. Environmental analysis extends across virtually all academic disciplines although many such as economics and engineering have specific sub-areas of interest specifically devoted to environmental issues. The multidisciplinary and transdisciplinary nature of environmental research makes it difficult to keep track of the key literature. To those new to the field there is the added problem of tracing back to find classic papers or research results. There are now a large number of specialist journals that deal with environmental topics but this is often adding to the problem of collecting information rather than reducing it. While these journals and periodicals provide explicitly environmental research findings, other general journals in economics, sociology, geography, engineering, medicine, biology and so on still continue to publish important findings.

The aim of this series of edited books is to provide both a set of reference volumes for those already in the field and to facilitate newcomers with accessible core material. The aim is not to offer readings on the purely scientific issues but rather to limit the papers to those concerned with social science aspects of the environment. This does not mean that we have excluded papers with a natural science orientation but rather the focus is on the softer debates and research. There are seven volumes in the series. The papers have all been previously published but do not constitute a set of traditional classics. While many of the articles have an established pedigree, equally there are less often referenced pieces. The aim of each volume is to paint a picture of how a topic area has developed, provide an indication as to some of the key contributions to this development but also look at some of the more recent literature that covers policy considerations.

The books cover a wide range of topic areas. At one extreme there are the macro-oriented volumes edited with Bob Ayres looking at the *Global Aspects of the Environment* while, in conjunction with Kerry Turner, there is a collection concerned with *Ecosystems and Nature*. Also at the aggregate level, edited with Hans Opschoor, there is a volume which focuses on *Environmental Economics and Development*. At the other extreme there are volumes concerned with micro-issues such as that along with Ken Willis that deals with *Environmental*

Valuation and with David Banister on *The Environment, Land Use and Urban Policy*. Transport's impact on the environment is multifaceted and can be severe hence there is a volume edited with Yoshi Hayashi, *The Environment and Transport* dealing exclusively with that subject. Finally, edited with Tom Tietenberg there is a collection that looks more generally at *Environmental Instruments and Institutions*.

We accept that there are many omissions from the collections and that others may have made different selections or have structured the material differently. Space limits what can be included and there is much more that could have been justified in a larger exercise (although this would add further to deforestation). We have also had to omit work by many eminent and important scholars for the same reason. As for structuring the material, a number of possible cuts are possible, the ones selected here just seem logical to us and would seem to fit with the needs of both researchers and, possibly, teachers.

Introduction

R. Kerry Turner, Kenneth Button and Peter Nijkamp

Scoping Framework and Function-based Conceptual Model

Given the relentless and cumulative process of global environmental change (GEC), driven by, among others, population growth, urbanization, industrial development, trade and capital flow liberalization, trans-national corporative activity and lifestyle and attitudinal changes, the world's ecosystems and their services have come under increasingly severe pressure. The consequences of this process manifest themselves across a range of spatial and temporal scales. They now pose a significant threat to the Earth's biodiversity. This is the world's natural biological capital with the capacity, when sustainably managed, to provide extensive opportunities for wealth creation and the maintenance/enhancement of the quality of life for all nations. The planet's biodiversity is also a storehouse for a number of ethical, aesthetic, spiritual, cultural and religious symbols and values. The loss of biodiversity may impose negative impacts on the functioning and adaptability of ecological systems themselves and their goods and services provision. It also seems to be the case that contemporary rates of biodiversity loss are unprecedentedly high. Human appropriation of the life-supporting products of photosynthesis (measured scientifically as net primary productivity) has already reached between 25 and 39 per cent of the total available for terrestrial systems indicating approaching possible 'limits' to the process of economic growth.

Before exploring some of these issues in more detail, the four part organizational framework for the volume as a whole will be outlined. In order to scope the myriad of issues, problems and arguments surrounding the scientific analysis, valuation and management of biodiversity and its linkages to the abiotic environment, a simplified organizational and auditing framework will be adopted. This is the pressure-state-impact-response (P-S-I-R) approach, which although simple is flexible enough to be conceptually valid across a range of spatial scales, from individual habitats to drainage-basin and adjacent coastal waters. It also highlights the dynamic characteristics of ecosystem and socio-economic system changes, involving multiple feedbacks within a possible co-evolutionary process (Turner, Perrings and Folke, 1997). Environmental pressure builds up via socio-economic driving forces – demographic, economic, institutional and technological – which cause changes in environmental systems 'states'. These changes include increased nutrient, sediment and water fluxes across drainage basins and into the marine environment; conversion loss, fragmentation and degradation of habitats; pollution of soil, water and atmosphere; and climate alteration. The processing and functioning capabilities of ecosystems will be affected and this results in impacts on human welfare via productivity, health, amenity and other value changes. The impacts impose social welfare gains and losses across a spectrum of different stakeholders, depending on the spatial, socio-economic, political and cultural setting. Policy response mechanisms will then be triggered within this continuous feedback process (see Turner, Subak and Adger, Chapter 17). Parts I and II of this volume deal with pressures, state changes and impacts related to biodiversity loss, while Parts III and IV are devoted to policy response and related ethical and social issues.

At the core of this interdisciplinary analytical framework is a conceptual model which links ecosystem processes and functions with outputs of goods and services, which can then be assigned monetary economic and/or other values, see Barbier (Chapter 14). While it is customary to distinguish between genetic diversity (differences between the individuals of a species which provide the diversity that is found between species), organizational diversity and ecological diversity, a fourth concept, functional diversity is also important for the methodology used in this volume. Functional diversity can be defined as the variety of different responses to environmental change, in particular the variety of spatial and temporal scales with which organisms react to each other and to the environment (Steele, 1991). Marine and terrestrial ecosystems differ significantly in their functional responses to environmental change and this will have practical implications for management strategies. Thus although marine systems may be much more sensitive to changes in their environments, they may also be much more resilient (that is, more adaptable in terms of their recovery responses to stress and shock). The concentration of scientific attention on species extinction rates, the abundance of taxa and the preservation of genetic information has led to a conservation strategy dominated by the establishment of protected reserves in the megadiversity ('hot spots') regions of the world. The functional diversity concept encourages analysts to take a wider perspective and examine biological diversity in large-scale ecological processes, together with the relevant socio-economic driving forces causing biodiversity loss. The focus is then on the ability of interdependent ecological–economic systems to maintain functionality under a range of stress and shock conditions.

Economic development these days is supposed to be constrained by the goal of sustainability. But although an economy can be both efficient and sustainable, economic efficiency does not in itself guarantee sustainability (Pearce and Turner, 1990). The 'scale' of socio-economic activity must be kept within biophysical limits and carrying capacity, a concept related to resilience (see Arrow, Chapter 5). Keeping the scale of socio-economic activities within the resilience of their underpinning resource base, locally, regionally and globally, is the main challenge for sustainable development policy. Biodiversity loss may be related to a loss of system resilience, but the limits are not static and appropriate behaviour, management systems and institutions can serve to delay or postpone their onset. The sustainable utilization of resources principle will be a key component of any future biodiversity conservation strategy. A recent collection of articles published in the journal *Science* has highlighted the fact that while agroforestry does reduce biodiversity, it can also act as an effective buffer to forest clearance and conversion to other land uses which are potentially more destructive to forested ecosystems. In agriculture the use of ecologically based management strategies can increase the sustainability of production while minimizing negative externalities. As far as global fisheries are concerned a precautionary approach is now an urgent requirement as the fish catch approaches its upper limit. Harvest rates must be cut back despite the socio-economic and political pressures working in the opposite direction (Nobler and Dirzo, 1997; Botsford, Castilla and Peterson, 1997; and Matson *et al.*, 1997).

The real challenge will be to demonstrate in practical ways both the economic value of biodiversity (use and non-use values) and capture mechanisms by which local people can participate in an equitable sharing of income and assets connected to the conservation of biodiversity. The rate of return derived from conservation-motivated activities (the 'capture' rate of return) must be greater than the rate of return available from alternative options (the opportunity cost of conservation), allowing for the prior correction of any existing market and institutional failures.

A wide variety of measures can be used to conserve biodiversity including both *in situ* and *ex situ* methods (Watson *et al.* 1995). *Ex situ* conservation will be especially important for any future habitat restoration and rehabilitation programmes, since they can provide the biological material for the breeding of domestic plants and animals. But no management prescriptions can possibly anticipate all the vagaries of the environmental change process. Flexibility and adaptability will be necessary characteristics for any future successful management strategy. Nevertheless, the fundamental uncertainty of the effects of biodiversity loss will remain. There is generally considerable uncertainty about the threshold values of either populations of organisms or biogeochemical cycles for many of the most important ecosystems. Further, if a threshold is breached the implications are completely unknown (see Perrings and Pearce, Chapter 6). Some judgement about the socially acceptable margin of safety (the 'precautionary principle') in the exploitation of biodiversity will be required. This is essentially an ethical judgement and there are a number of ways in which such an ethical dimension can be reflected in the decision-making process. It may be possible to discern a hierarchy of economic and ethical considerations. Perhaps the most important task is to clarify where conventional economic values are sufficient for sustainability decisions and where broader human values – including non-monetary values – and criteria are more appropriate (see Bingham *et al.*, Chapter 8).

The contents of this volume have therefore been set out in the following way:

- **Part I** contains readings which explore basic scientific and economic definitions, principles and methodological problems surrounding biodiversity. It sets out the fundamentals of the causes of the pressures leading to the different forms of biodiversity loss. The Holling paper (Chapter 1) is now a classic scientific treatment of meaning of functional diversity in ecosystems and the much more recent paper by Folke, Holling and Perrings (Chapter 2) draws out the economic and policy implications of the functional diversity concept. Chapter 3 by Johnston *et al.* and Chapter 4 by Pimm *et al.* are up-to-date authoritative surveys of the science of biodiversity and species loss. This part of the volume is completed by three papers by Arrow *et al.* (Chapter 5), Perrings and Pearce (Chapter 6) and Swanson (Chapter 7) which highlight the economic implications and problems concerned with concepts such as carrying capacity, threshold effects and the causes of biodiversity loss. Swanson generalizes the economic analysis of biodiversity by including land-based resources together with marine resources which were the sole focus of attention in the bioeconomic literature prior to the publication of Swanson's work.
- **Part II** explores the consequences of biodiversity loss and evaluates their significance in terms of social welfare impacts; it examines both the methods and techniques available for the demonstration of biodiversity conservation value and the results of

empirical studies drawn from the main categories of resource systems. Chapters 8, 9 and 10 by Bingham *et al.*, Bockstael *et al.* and Gowdy respectively provide an overview of the issues and problems surrounding the valuation of ecosystem functions, goods and services. Gowdy includes a critique of conventional economic valuation approaches, methods and techniques. The following suite of papers have been selected both for their expositional clarity and also coverage of some of the main ecosystem types – forests, wetlands, marine and grasslands and agriculture – and complexities posed by the valuation procedure. This coverage is picked up again in Part III Section B in Chapter 28 by Perrings and Walker, which links science and management issues.

- **Part III** deals with value capture mechanisms and the practicalities of their implementation in a number of different socio-economic and cultural settings. Kahn and McDonald (Chapter 22) provide a rigorous analysis of the relationship between debt and ecosystem degradation in the topics. Chapter 24 by Swallow and Bromley and Chapter 25 by Gadgil have been included because they serve to highlight the importance of institutions, property rights, history and culture at the local level. All of these factors are vital to the success of practical conservation measures. Chapter 23 by Ruitenbeek is conceptually important and introduces the concept of supply price (linked to opportunity costs) into conservation policy. Batie and Mabbs-Zeno (Chapter 31) and Shyamsundar and Kramer Chapter 30 take the opportunity costs calculations further in a developed and developing country context respectively, while Chapter 29 by Hohl and Tisdell questions the usefulness of the safe minimum standards/ precautionary approach to biodiversity conservation. Whitby and Saunders (Chapter 26) explore the cost-effectiveness of two primary conservation mechanisms in developed countries, zoning and landowner management agreements. Finally, Chapter 27 by Moran *et al.* has been included because it trys to develop a practical diversity index which can play a 'second-best' role in the absence of precise scientific data on taxonomic categories and difference measures. This practical index includes the available scientific data together with socio-economic data on environmental pressures.
- **Part IV** presents an overview of the ethical questions and answers that any biodiversity conservation policy will stimulate. The conventional economic and philosophical viewpoint expounded by Randall is contrasted with the much more ecocentric positions favoured in the papers by Rolston, Norton and Page.

Biodiversity: Definitions and Scientific Concepts

Biodiversity is defined by the UN Convention on Biological Diversity as 'the variability among living organisms from all sources, including terrestrial, marine and other aquatic ecosystems and the ecological complexes of which they are part; this includes diversity within species, between species and for ecosystems'. According to the recent Global Biodiversity Assessment Report (Heywood and Watson, 1995), put simply biodiversity is the variety of the world's organisms, including their genetic make-up and the communities they form. Biodiversity is dynamic because the genetic composition of species changes over time in response to natural and human-induced selection pressures; and the occurrence and relative abundance of species in ecological communities changes as a result of ecological and physical factors.

It is possible to distinguish between genetic diversity, organismal species diversity, ecological diversity and functional diversity (see Table 1). The loss of such diversity may have significant consequences for the functioning of ecological systems and human welfare. But the choice of diversity measure has implications for conservation management options. The first two measures tend to bias management actions towards *in situ* and *ex situ* methods designed to support protected reserves in the mega-diversity regions of the world. The functional diversity approach implies that biodiversity conservation to assure the resilience of ecosystems is required for all systems (not just protected areas) no matter how heavily impacted they are. Secondly, that policy needs to encompass and deal with the socio-economic and cultural driving forces that lie at the core of the loss process. This requires the correction of existing market and policy failures which under-price biodiversity and a recognition of the asymmetry between private and social values of biodiversity. New institutions will also be required to enable a sustainable utilization of biological resources (see Holling, Chapter 1 and Folke, Holling and Perrings, Chapter 2).

Table 1 Forms of biological diversity

Genetic diversity	Populations, individuals, chromosomes, genes, nucleotides
Organismal diversity	Kingdom, phyla, families, genera, species, subspecies, populations, individuals
Ecological diversity	Biomes, bioregions, landscapes, ecosystems, habitats, populations
Functional diversity	Based on a functional relationship between organisms diversity and sets of ecological services and affected by the differential space and time scales with which organisms react to each other and to the environment; keystone process species, resilience-related 'natural insurance capital' species
Cultural diversity	Human interactions at all levels

Sources: Heywood and Watson (1995); Steele (1991); Folke, Holling and Perrings (Chapter 2).

Only about one million species have been identified and described but there are anything from ten to 50 million species in existence (see Pimm *et al.*, Chapter 4). Problems of estimating the total number of species are formidable and most species are as yet undescribed in every species rich group. Calculations suggest (but cannot prove) that recent extinction rates of around 20 to 200 species per one million species-years are unprecedented, that is, are substantially above background rates in the absence of humanity (Pimm *et al.*, Chapter 4). The species most prone to current extinction are rare and local. Past extinctions have been concentrated in small endemic-rich areas ('hot spots') because by chance alone, small endemic-rich areas will contribute disproportionately to the total number of extinctions. However, there is still great uncertainty since we know the geographical ranges of only a small proportion of the already small proportion of species for which we have names. As yet there is no comprehensive understanding of the geographical patterns of species richness and so our current knowledge of endemism is insufficient for us to know the future of biodiversity with much precision.

A long standing (since the 1950s) theoretical axiom predicts that species diversity enhances the productivity and stability of ecosystems. But the results of some field studies in the 1960s and 1970s began to challenge the universality of this paradigm. There are now four competing hypotheses about the functional role of species diversity in ecosystems: (1) the original diversity–stability hypothesis; (2) the rivet hypothesis, which suggests that the ability of ecosystems to absorb changes in abundance of some species and maintain performance, can decline as species are lost and that the collapse can be sudden and the consequences drastic as the lost species ('rivets popping') go beyond some threshold number; (3) the redundancy hypothesis proposes that some species may be expendable in terms of ecosystems maintenance if any extant species can take the lost species' functional place in the community; and (4) the idiosyncratic hypothesis, which acknowledges that there may be no pattern or an indeterminate relationship between species diversity and ecosystem function (see, Johnson *et al.*, Chapter 3). Empirical testing of these competing hypotheses so far has not resulted in conclusive results. Species diversity enhances productivity and stability in some ecosystems, but not in others. It seems clear that for any ecosystem function to be sustained, a minimum composition of organisms is required and that this composition varies with the environmental conditions in which the system operates. What we do not know is the critical thresholds of diversity associated with different environmental conditions at different temporal and spatial scales (Perrings and Pearce, Chapter 6).

Studies of terrestrial ecosystems suggest that individual species population dynamics are more sensitive to stress than are ecosystem processes. The system's functional robustness may be linked to the prevalence of a relatively few processes, driven by a limited number of organisms and groups of organisms ('keystone process species') (Folke, Holling and Perrings, Chapter 2). But keystone process species alone will not guarantee the continuation of the ecosystem in question, the management of the system in order to protect its 'natural insurance capital' is also crucial.

Economic activities are sustainable only if the life-support ecosystems on which they depend are resilient (that is able to absorb some level of stress and shock without flipping to a new locally sustainable equilibrium). Loss of ecosystem resilience may be associated with a sudden loss of biological productivity and consequent human welfare loss. It may also lead to the irreversible loss of options for both current and future generations of humans (Arrow *et al.*, Chapter 5). From a management point of view, the situation is made more complex and uncertain because the ecosystems often 'fail to signal' the long term consequences of loss of resilience and will continue to function 'as normal' in the short run. Further, current socio-political and economic market institutions in society do not necessarily respond very efficiently to environmental feedbacks – interrelated market and institutional failure phenomena are commonplace (Turner, 1991).

The stability and sustainability of joint ecological–economic system will therefore be determined by a complex set of factors: ecosystem structure; types of stress and shock experienced, the resource usage and management regime that is in place and the feedback mechanisms that exist (or not) to signal the need for adjustments as environmental conditions change, and in particular as thresholds (carrying or assimilative capacities) are approached.

Causes of Biodiversity Loss

The underlying fundamental causes of biodiversity loss are (Pearce and Moran, 1997/98):

- human population growth which pressurizes resource systems via multiple and competing usages;
- market failures, many of the goods and services provided by conserved ecosystems – climate regulation, water regulation, nutrient cycling, waste treatment and so on are under-priced or zero-priced and thus fail to compete in the economic development process with alternative uses for the resource system in which biodiversity resides and contributes to; the allocation of biological resources on the basis of these market signals is both inefficient and inequitable, since the poor of today and all future generations will bear an unjust cost burden (Costanza *et al.* 1997);
- intervention failures, inefficient and/or uncoordinated regulations, policies and institutions which exacerbate unsustainable resource usages. Subsidies, for example, are extensively deployed in most economies and often result in biodiversity loss and other environmental damage. Over fishing is directly linked to government subsidy policy, since to catch $70 billion worth of fish, the global fishing industry recently incurred annual costs of $12 billion (Safina, 1995). The high level of subsidy has also encouraged over-capitalization. Between 1970 and 1990 the world's industrial fishing fleet grew at twice the rate of the global catch. As profits declined the asset value of ships also fell, forcing owners to continue to maximize fishing effort to service loan repayments on their boats.

Economic Theory of Species Extinction

Economic explanations of species extinction have their origins in the theory of the fishery (Gordon, 1954; Clark 1973 and 1990). The Gordon–Clark models emphasize the threat to economically significant resources from inappropriate property rights and the relative rates of return over time to conservation versus exploitation. For conservation to be the chosen option the actual ('capture') rate of return must be greater than the rates of return available from alternative resource usage options (the opportunity cost of conservation). The property rights regime that is in force is also crucial, with open access having the highest risk of extinction for the species and common property having a much reduced but still non-zero level of extinction risk (see Swallow and Bromley, Chapter 24). Advances in technology which reduce effort-costs increase extinction risks. During the 1950s and 1960s, for example, the fishing industry adapted various military technologies such as radar, long range navigation and satellite positioning systems to improve catch rates (Safina, 1995).

In the single owner, profit-maximizing context, extinction can be the result of deliberate planning. If the growth rate of the species stock does not equal the single owner's discount rate, extinction is a risk. The growth rate of the species stock is effectively the rate of return on the resource. If this is less than the rate of return the single owner can get by investing elsewhere the resource will be run down to zero. Vessel owners who may be caught in a debt repayment trap could counter the loss of preferred fish by switching to species of lesser value, currently those positioned lower in the food web. The wider environmental consequences of this practice may be very damaging as it robs larger fish, marine mammals and sea birds of a food source. Owners who re-invest in aquaculture do not necessarily contribute to any reduction in the overall pressure on wild fish populations. Biomass fishing practice is often used to feed farmed shrimp populations and the infrastructure such as pens and other constructions often displace natural resources such as mangroves with their multiple function benefits.

The combination of uncertainty about threshold fish population sizes and open access property rights is a particularly troublesome one and slow growth species will automatically be at risk of extinction. In order to attempt to protect some fisheries, government regulations have purposefully been promoted in an attempt to reduce catch efficiency. Regulations on catch quotas, net mesh size, type of boat allowed and so on have all ben instituted in order to lower efficiency, while at the same time meeting wider and social and ecological goals.

Swanson (Chapter 7) has extended the Gordon–Clark model in order to incorporate land-based resources. The core of his argument is the fact that land-based biological resources are typically subject to more extensive human management than marine resources. The resource system is also subject to a wider range of competing land uses. Considering all uses as part of a potential asset portfolio, the issue becomes one of relative returns to the scarce factor. Fisheries tend to occupy space that has relatively low opportunity costs – fewer competing direct human uses for the open sea. On the other hand, land-based biodiversity conservation has to compete with multiple alternative uses – agriculture, housing, industrial development and so on. Thus the Swanson model concentrates on the 'base resource', that is the environmental space occupied by the resource. Marine and land-based resources also differ in terms of carrying capacity. The former generally have a fixed carrying capacity, while the latter has a variable carrying capacity determined by decisions about the level of base resource allocated to biodiversity conservation. In coastal zones where fisheries compete directly with alternative uses, the relevance of positive opportunity cost makes the analysis closer to the land-based model.

The Swanson model serves to highlight the fact that it is not enough to take the property-rights regime as given, as in the Gordon–Clark models, because it is a matter of choice and management. Under population growth and development pressure, biodiversity is competing with alternative uses of land and their rates of return. Correction of market and policy failures together with mechanisms for the equitable sharing of benefits provides some grounds for optimism about future successful conservation. But such measures do not provide a guarantee of sustainability since the overall life-support function value of biodiversity is not necessarily captured (Turner and Pearce, 1993).

Valuation of Biodiversity

Because biodiversity conservation has to compete with economic values of converted land, values which are expressed in market prices, resource and environmental economists have concentrated on the search for economic values of conservation which bear a direct comparison with the economic values of development (Pearce and Moran, 1997/98). While this approach is still relevant and important it is a relatively narrow perspective. Ecological economics has been seeking a wider systems-focused perspective and related analysis of value. (Turner, Perrings and Folke, 1997). In this systems view, biodiversity is assigned a 'primary' or 'glue' or 'infrastructural' value, linked to the overall life-support function. This dimension of value is not readily capturable by existing economic valuation methods and techniques.

The problem of valuation has been the cause of extensive controversy and debate in both academic and policy circles. In environmental economics, a preference-based value system operates in which the benefits of environmental gain (or the damages) are measured by social opportunity cost or total economic value. Within this approach distinctions are made between use values (direct and indirect use values), option value and quasi-option value and non-use values (bequest and existence value). While the meaning of use value (direct and indirect) is straightforward, option and non-use values are somewhat more obscure in both literatures. Option value is essentially the willingness of an individual to pay for an 'insurance premium' to safeguard future resource use and appreciation opportunities. Quasi-option value is a reflection of the conditional expected value of information and is not commensurate with other value components. Bequest value is motivated by a desire to ensure that opportunities are kept available for future generations. Existence value covers a number of motivations individuals might hold not in the context of resource use, but relating to the mere knowledge that certain resources are conserved and will continue to be so.

There are two main economic approaches to estimating the value of non-marketed environmental effects of economic activity:

- the use of surrogate markets to reveal preferences of resource users; and
- the use of simulated markets to get resource users to state preferences.

Table 2 summarizes the environmental change impact categories to be expected and the valuation methods appropriate for the different welfare gains/losses. The empirical studies reprinted in Part II Sections B, C, D and E cover a variety of ecosystems and valuation methods.

Debate continues over the precise boundaries between these different components of economic value, but the conventionally accepted approach to the valuation of environmental resources is based on the assumption that households maximise utility deriving from these different sources of value subject to an income constraint; and that their private willingness-to-pay is a function of prices, income and household tastes (including environmental attitudes), together with conditioning variables such as household size and so on. The social value of environmental resources committed to some use is then defined as the aggregation of private values.

Ecological economics also uses these value categories, but emphasis on the limiting role of ecological processes and the importance of ecological resilience means that it places much greater emphasis on the 'health' of overall systems and life support functions served by ecological processes and biogeochemical cycles. From this perspective it is argued that the aggregation of private values does not adequately capture the 'true' social value of ecosystems and their interrelationships. One formulation of this argument is that total economic value captures only 'secondary' values (associated with useful functions and/or services of an ecosystem) but not 'primary' values (the value of an overall healthy and

Table 2 Environmental change impacts and valuation methods

Effects categories	Valuation method options
Productivity such as fisheries, agriculture, tourism, water resources, industrial production, marine transport, storm buffering and coastal protection.	Market valuation via prices or surrogates Preventive expenditure Replacement cost/shadow projects/cost-effectiveness analysis Defensive expenditure
Health	Human capital or cost of illness Contingent valuation Preventive expenditure
Amenity Coastal ecosystems, wetlands, dunes, beaches, and so on, and some landscapes, including cultural assets and structures.	Contingent valuation/ranking Travel cost Hedonic property method
Existence values Ecosystems; cultural assets	Contingent valuation

evolving ecosystem, necessary to support the continuing provision of a range of secondary values) (Turner and Pearce, 1993; Gren *et al.* 1994). System integrity needs to be conserved by ensuring that sufficiently large 'chunks' of ecosystems are protected so as to retain species diversity and complexity of relationships.

Some positions within ecological economics give greater emphasis to 'intrinsic' values, or values 'in' things compared to purely 'instrumental' values, or values 'of' things. This has led to questions being raised about the nature of the conventional distinction between use and non-use value and therefore whether there is such a thing as environmental existence value that can be measured. There are variations in the extent to which such wider values are encompassed, partly depending on the practicality of entering intrinsic values into decision making and the problems of choosing a meta-ethical principle that enables trade-offs between intrinsic and instrumental values.

It seems reasonable to argue that while there are limits to the economic calculus, that is, not everything is amenable to meaningful monetary valuation, economic valuation methods and techniques can and should play a significant role in the project, programme and policy appraisal process which leads to the setting of relative values (including environmental assets values). Costanza *et al.* (1997) controversially estimated the current economic value of 17 ecosystem services on a biosphere-wide basis at between US\$16–54 trillion (10^{12}) per year, giving an average annual value some 1.8 times the current global Gross National Product. The rationale behind this valuation exercise could be based on a number of arguments:

- Due to a lack of adequate market price data (or absence of data), together with inadequate (or absent) property rights regimes which ensure that resource values can be practicably appropriated, ecosystem services are assigned too little or zero value and weight in policy decisions.

- Some important environmental science research and debate, together with related policymaking (that is international agreements and conventions) necessarily takes place at the global scale. There is need therefore for social science research to 'engage' science and policy at this scale. But such an engagement must, in our view, encompass analysis which will show clearly why globally aggregated social science data/estimates are often not meaningful, if the objective is to move beyond mere dialogue towards a more rational policy process.
- It is important to prove how valuable ecosystem services really are and to formulate mechanisms by which such function-based values can be realistically captured. Such capture must be relevant for everyday socio-economic and political activity and decision taking, through national income/resource accounting and project cost-benefit appraisal, down to the grass roots level in developed and developing countries.

Costanza *et al.*'s study goes some of the way towards meeting the arguments set out above. Their paper has engaged environmental scientists and policy makers, but the global biome scale economic value calculations risk ridicule from both scientists and economists. On the basis of the data and methods cited in the article and supporting inventory, the conclusion that the value of the biosphere services really is, on average US$33 trillion per year, is not supportable. Apart from raising policy-maker, scientist and citizen awareness about the environment's economic value and the possible significance of the loss of that value over time, the global value calculations do not serve to advance meaningful policy debate in efficiency and equity terms, in practical conservation versus development contexts. Such calculations with their 'single number' outcomes shroud a number of fundamental 'scaling' problems to do with valuation contexts, that is the temporal, spatial and cultural specificity of economic value estimates. Such values can also only meaningfully be assigned to relatively small ('marginal') changes in ecosystem capabilities (functions/services.) The practical problem is that determining precisely what is and what is not a discrete and marginal change in complex ecological systems is not straightforward.

The specificity problem (relative scarcity basis of value) is generic and serves to constrain the transfer of site-based function and/or system services economic values across time and geographical and cultural space. It is not being argued that all such 'benefits transfer' is invalid, but that such procedures must be handled with extreme caution and have real limits. Many value estimates will not be amenable to legitimate aggregation beyond local to 'regional' (defined biogeographically and including cross-national boundaries where necessary) scales. Further research to more precisely define these limits and to formulate a robust validity and reliability testing protocol is an urgent requirement (Turner, Adger and Brouwer, 1998).

Economic valuation based on a functional approach properly appreciates the links between ecosystem structure, the characteristics of which provide society directly with extractive and/or non-extractive benefits (goods and services) and ecosystem processes providing indirect benefits. This appreciation of the system's underlying dynamics is essential in order to sort out the stocks and flows involved and to ensure against possible double counting. Both stock and flow value estimates have, for example, been calculated and reported in forest services valuation studies, and researchers seeking to obtain aggregate figures have incorrectly summed both types of numbers. The often quoted world median value for non-timber forest products (NTFPs) of US$50 per ha per yr is an example of such an error (Batagoda and Turner, 1997). There is a need therefore for researchers to carefully unpick the existing value data work, in order to separate out individual function stock and flow estimates (see Godoy, Lubowski and Markandya, Chapter 11). For NTFPs, four types of value exist but only the last three are economically and policy relevant: (1) the total inventory stock value; (2) the total extractable stock value (based on some sustainability rule(s)), (3) the potential extractable flow value (linked to physical accessibility factors and the non-uniformity of species distribution across a forest area); and (4) the actual extracted flow value (conditioned by market accessibility, market demand and cultural value factors). It may then be legitimate to estimate a mean value for each of the last three value categories, as long as they are further 'regionally' differentiated across the major forest types, that is tropical rainforests, tropical dry forest and temperate forests (Batagoda and Turner, 1997).

Aggregation across different functions provided by a given ecosystem should be constrained by the danger of double counting. Exploitation of one function may preclude another, while some keystone processes and related functions may underpin others and the system infrastructure. There may also be possible incompatibilities between different valuation measures (such as opportunity costs, consumer surplus, market prices) as they are applied to different functions. Within a wetland, the exploitation of a particular function service such as wastewater cycling, for example, will preclude, or limit, the provision of other services, such as recreation. The global value of mangroves cited in Costanza *et al.* (1997) of $9990 per ha, for example, is in fact made up of substitution cost estimates for the coastal protection function, replacement cost estimates for the nutrient cycling service, market price value of food production and a travel cost estimate of the recreation service.

The issue of double counting and the extent of control necessary in economic valuation studies of complex ecosystems, to account for their characteristics and functional values, has been addressed in a recent meta-analysis of wetland valuation studies (Brouwer *et al.*, 1997). In this study, a distinction is made between economic use and non-use values associated with the ecological, biogeochemical and hydrological functions provided by wetlands. The latter were shown to have a significant impact on stated preferences for a large number of wetland conservation projects or programmes in different countries. Average values were derived for different wetland functions after controlling for issues like population sample and procedural valuation characteristics.

Scaling up values of single functions of an ecosystem, or even more ambitiously, aggregate systems value to biogeographical and global scales, on a per hectare basis, increases the difficulties by an order of magnitude. If a single ecosystem service, in economic welfare terms, is relatively independent of location and its socio-cultural context, then the scaling up procedure can be regarded as more or less valid. Thus the economic value of carbon sequestration provided by coastal or terrestrial ecosystems has equal global social value (postponement of possible global climate change and its impacts) wherever it is provided. Or, in the case of coral reefs, the recreation value of reef visits and diving may be conditioned by a reasonably common set of site characteristics, regardless of precise location, and excluding totally inaccessible sites cases.

On the other hand, the value of many services are primarily determined by the locational factors. Although some criteria for 'benefits transfer' have been discussed in the literature a comprehensive testing protocol is not in prospect (Brouwer and Spaninks, 1997). More research is required on these scaling possibilities and limits. It seems likely that the biogeographical regional scale will be the limit for a large number of service values.

Finally from a systems perspective, the aggregation and scaling up of individual ecosystem services value estimates does not lead to the quantification of a total ecosystem value. An evolving 'healthy' ecosystem presumably requires some, currently unknown, minimum configuration of structure, processes and functions to retain its resilience property. Economic valuation studies estimate the value of the flow of services assuming the sustainability of ecosystems and their interrelationships between themselves and the abiotic environment.

Option Value and Biodiversity Conservation Priority Setting

According to Pearce and Moran (1997/98) although option value (interpreted as flexibility value and it maximization) is the most frequently cited rationale for conservation and the focus of common interest in economics and biology, convincing quantitative assessments of option value are lacking. The quantification and valuation of direct and indirect usages of ecosystems or their components would provide a lower bound on option value – see the studies in Part II of this volume. But analysts have also been exploring more precise biological diversity measures as a potential basis for ranking relative diversity 'value' of species and sets of species. Such measures offer some insight on the total system value in terms of its diversity or flexibility dimensions and can also inform conservation prioritization. Since not all biodiversity can be saved some priority setting is inevitable and will require some relative valuations. Value may derive from any number of criteria, ranging from basic diversity, reflected in phylogenetic tree pattern to functional diversity and resilience links.

All concepts of diversity imply some measure of distance between the elements of a set, but the particular measure of distance judged to be relevant depends on the significance of diversity. Some ecological estimates of the value of biodiversity assume that the relative value of each species depends directly on the phylgenetic distance between species, but in the context of functional diversity two different concepts have to be combined, that is the diversity of the ecological functions performed by different species and the diversity of species performing a given ecological function. The measure requires the number of species required for a given process and the number of species supporting the structuring set of processes under different environmental conditions. The latter component is vital in the relation between diversity and ecosystem resilience.

The value of functional diversity in general is not readily observable, but in some cases the value of the loss of functional diversity can be obtained by indirect methods. In the case of semi-arid rangelands, range livestock managers face a problem in terms of the large inter-annual variation in fodder production in response to variation in rainfall. Economic decisions with respect to stocking and offtake levels have resulted in a significant change in the functional diversity and hence resilience of many rangelands. These decisions lead to changes in species, as perennial grasses are replaced by annuals; phenological diversity is also lost in the grass sward because grazing pressure reduces early-season palatable species which are replaced by later-growing species. In years of mostly early-season rains, production will be less than expected. Perrings and Walker (Chapter 28) discuss the management problems of rangelands.

Some recent empirical evidence derived from microbial microcosm studies seems to demonstrate that a simple measure of biodiversity, the number of species (S) per functional or trophic group (F), or S/F, serves as a valuable index of ecosystem reliability. These results support the hypothetical 'insurance' value of biodiversity, that is insurance against the failure of ecosystems to provide goods and services (Folke, Holling and Perrings, Chapter 2). The mechanistic basis for such insurance is compensatory growth, a process widely observed in many ecosystems. Greater diversity should therefore enhance the ecosystem's capacity to continue to provide a fixed level of functioning over a given unit of time.

More generally, the search for sustainability indicators can be guided by a theory of sustainable development (defined as non-declining per capita human well-being) and the two interpretations of the conditions necessary to achieve this: strong sustainability (SS) and weak sustainability (WS). Depending on the state of belief about these conditions a different set of indicators may be deemed important. The package currently on offer includes 'green' national income accounts and genuine savings measures (net saving which accounts for the depletion of natural resources and the accumulation of pollution) and some less well defined ecological indicators linked to resilience, biodiversity, carrying capacity and 'distance to goal' (that is cost of attaining some pre-emptive sustainability 'targets') concepts.

Pearce and Moran (1997) note that the choice of species and areas for conservation as determined via phylogenetic data, may be different to the choice that would result if the identifiable uses approach to option value alone was deployed. However, it is unlikely that science will be able to generate sufficient phylogenetic data in the short run to allow the comparison of national diversities. Surrogate information is required and the use of species-richness information, for example, may prove to be another practicable option (see Moran, Pearce and Wendelaar, Chapter 27) which will take the issue of priorities beyond the current focus on 'hot spots', that is areas where species are under most threat. Assuming data on a species exist in a geographic location, the problem becomes one of maximization of species richness in a set of reserves subject to a given financial budget.

Mounting concern over the continuing loss of tropical forests and other habitats and concerted international efforts to implement the UN's Biodiversity Convention, has focused much attention on the determination of global priorities, beyond mere 'hot spot' delineation. Rankings based on threat indices (pressures) such as deforestation rates, population growth and the extent of existing protected areas, combined with cost criteria might prove useful indexes. Further, while the national scale may be appropriate for decision making, analysis at this level provides a less than comprehensive picture of the dynamics of biodiversity conservation or loss at local and regional levels. The priority rank index does not obviate the need for a project by project approach irrespective of national boundaries. The issue of whose priorities we are actually setting matters. Global priority setting cannot be divorced from a proper understanding of local priorities, demands and unfulfilled needs (see Gadgil, Chapter 25). There is evidence that the mainly poor and rural communities close to national parks and protected areas generally pay a heavy price in terms of uncompensated forgone economic opportunities. The consequence is often illegal and/or 'open access' exploitation of resources, and increased conflicts between local people and the state. Some moves towards community-based conservation have been made utilizing buffer zones around park boundaries and fostering more local participation. The impact of these integrated conservation development projects has so far been limited (Wells and Brandon, 1993). Evidence from South America is even more pessimistic about the prospects for such efforts (Southgate and Clark, 1993). The links between the local and international levels are explored in a little more detail below.

Value Capture Mechanisms

A sustainable biological resources utilization and management strategy should be focused on the maintenance overtime of ecosystems resilience. It will need to be buttressed by created economic incentive mechanisms that reduce the differences between the value of biological resources to the private individual and their value to the wider community, up to the global scale. A precondition for this is the valuation of resources, but this is a necessary not a sufficient requirement. What is also needed is the development of effective enabling institutions. To be effective, given the uncertainties surrounding biodiversity science and policy, institutions will have to be flexible. They need to be able to respond and adapt to environmental feedbacks before the resilience of systems is compromised and significant welfare losses are risked.

It is now becoming increasingly clear that much of the existing literature on the financial and economic returns from sustainable utilization of biodiversity was somewhat over-optimistic. Taking the case of NTFPs derived from tropical forests, as multiple site data is accumulated the estimated returns that can be expected from such harvesting have fallen significantly. They are highly sensitive to regional forest type variation, the full costs and frequency of harvesting and forest access and marketability (Batagoda and Turner, 1997). Any economic justification of conservation will therefore have to go beyond these local direct use values if it is to stack up against forest clearance for plantation or other development options.

Equally clear is the fact that tropical deforestation stimulates more than local ecological and economic well-being impacts. It also gives rise to significant effects outside of any particular country (usually a developing country). Biodiversity loss may mean significant potential loss of internationally important food, medicinal and material products, together with regional climate disruption, global warming contribution and loss of international existence value. All of these factors have contributed to an increase in international efforts to conserve tropical forests focused in particular on the carbon storage values of tropical forests and the more general global existence value possessed by biodiversity. But if local value capture is insufficient to justify conservation, there will be a need for significant transfers of resources to developing countries to ensure that pressures for deforestation are countered.

The concept of a rainforest supply price (RSP) has been suggested in order to provide a means by which the international community can evaluate in monetary terms whether it will be beneficial to make resource transfers to a particular developing country in recognition of explicit conservation programmes (see Ruitenbeek, Chapter 23). The RSP is the amount that a developing country must capture through direct transfers from the rest of the world in order for it to save a particular rainforest. Based on a common denominator (one hectare of rainforest), the RSP evaluates rain forest conservation expenditures on a cross-country basis. The RSP is not, however, capable of identifying the optimum amount of rainforest since it excludes global benefits of forest conservation. It does give donor agencies and international funding bodies an objective assessment of how much must be deployed in any given area to save a rainforest. In essence RSP is the financial payment that a developing country must be given by the rest of the world to induce it to put through reforms required to conserve forests.

An application of the RSP concept within a cost–benefit study of the Korup National Park in Cameroon, demonstrated that the park project was not in Cameroon's interest if it were required to pay for the entire project. Cameroon would require compensatory transfers of 1000 ECU per km^2 per year. Transfers of twice this level would be necessary were it not for selected local environmental benefits derived from forest conservation. Between 1987 and 1989 the international community had sanctioned forest conservation transfer funding in the range of 15 to 1575 ECU per km^2 per year. So the transfer cost for the Korup project is within these limits (Ruitenbeek, Chapter 23).

The evidence of high local opportunity costs incurred because of the Korup conservation project is mirrored in other countries' experiences. Thus the opportunity costs of biodiversity conservation in Kenya also seem to be high. At the national level, agriculture and livestock production in the parks, reserves and forests of Kenya could support 4.2 million Kenyans and generate gross annual revenues of $565 million and net returns of $203 million. These forgone net returns of $203 million, some 2.8 per cent of Gross Domestic Product, represent the opportunity cost to Kenya of biodiversity conservation. The current combined net revenues of $42 million from wildlife tourism and forestry are clearly well below the opportunity costs to land (Norton-Griffiths and Southey, 1995). The impact of Mantadia National Park in Madagascar on local communities has been very costly. The locals have lost access to more that 800 ha of agricultural lands and to a significant amount of forest products as a result of the park. The net present value of costs to the average household because of protection is US$419, which amounts to 18 per cent of their total gross income in 1991. If all the households in the immediate vicinity of the Park are impacted, the aggregate net present value resulting from conservation is US$305 590. The analysis undertaken also shows that these costs vary according to geographical location. Of course the analysis excludes benefits to these households from forest conservation such as flood control. The cost estimates represent a minimum value of the benefits required to make the park into an economically viable enterprise. They also represent the benefits required to make the park into a protected area that is acceptable to local residents (see Shyamsundar and Kramer, Chapter 30).

The sign of biodiversity conservation opportunity costs is not, however, uniformly highly positive. In Europe and North America, the opportunity costs of protecting resources such as wetlands have often been low or negative. In the USA, for example, alternatives uses of some wetlands such as agriculture and recreational housing yield low net social values (see Batie and Mabbs-Zeno, Chapter 31). Wetland conversions in the UK in the 1970s and 1980s actually generated net economic costs and only occurred because of subsidies provided by the EU's Common Agricultural Policy (Turner, Dent and Hey, 1983). The policy response in North America and Europe has been a combination of zoning, standards and incentive approaches. Thus in the USA, for example, there is a national 'no net loss' standard in force for wetlands. While in the UK sensitive areas are offered some measure of protection by landowners management agreements, or environmentally sensitive area designation which prescribes the intensity of land use (see Whitby and Saunders, Chapter 26).

A number of measures such as subsidy/tax break elimination, improved local access to credit, environmental performance bonds, forest compacts, transferable development rights, debt-for-nature swaps and so on have been designed to generate biodiversity conservation incentives while at the same time discouraging perverse incentives that work against conservation. Some combination of these measures should be deployed as a package at both the local and international levels.

The links between Third-World debt and tropical deforestation have been much debated. It seems to be the case that debt can lead to myopic behaviour, the manifestation of which is unsustainable logging practices and forest clearance. Kahn and McDonald (1995, Chapter 22) have conducted a detailed study which shows that debt is significantly correlated with deforestation across tropical countries. Debt-for-nature swaps (that is the writing off of debt repayments in return for specified conservation actions) may have a dual effect on deforestation. The contractual agreement itself is designed to conserve forests but the debt reduction may provide an indirect pressure reduction effect, although its magnitude is small. Transferable development rights (TDRs) on the other hand, enable a country or private land owner to sell the 'right' to convert a natural habitat for a price that fully covers the opportunity cost. Except for a minimum level of legislation to establish the TDRs and some pump-priming to stimulate their demand, the system relies on the market to determine their value and to achieve the optimal level of biodiversity conservation (Panayotou, 1994).

Policy Implications and Ethics

Policy guided only by economic efficiency with valuation linked to current individual preferences and utility maximization is not a sufficient response to the problems of biodiversity loss. Rather policy needs to address multiple goals – sustainability, equity and efficiency – and the inevitable trade-offs between them. Social preferences, technology and the distribution of income and assets all influence perceptions of value (see Randall, Chapter 35). The debate about biodiversity conservation necessarily exposes a hierarchy of economic and ethical considerations, from immediate self-interest, through possible self-interest (risk avoidance), to care for future generations (Bingham *et al.*, Chapter 8; and Gowdy, Chapter 10).

All future values are particularly sensitive to threshold effects and the fundamental uncertainty associated with such effects. Under conditions of fundamental uncertainty it is not possible to equate marginal net private benefit with marginal external cost in order to derive standard economic efficiency solutions. Instruments designed to protect thresholds will have to rely on non-economic criteria, which somehow define the socially acceptable margin of safety in the exploitation of biodiversity. This is essentially an ethical judgement about the appropriate way to respond to collective ignorance about the future effects of current resource utilization. Environmental standards are fairly commonplace at the micro-economic level (fishing permits, pollution emissions permits, zoning, technology-based pollution limits and so on). They try to combine efficiency in allocation with the safety of the limit. But macro-economic environmental standards to address the 'scale' issue are more controversial (Daly and Cobb, 1989). The policy implications are that there is a role for the precautionary principle and the 'safe minimum standards' notion (combined with social opportunity cost assessments). Meanwhile, science works to reduce some aspects of the uncertainty; for example, is it species preservation, or populations, or something else, that should be the basis for conservation policy? Social scientists are also still required to provide valuation data in order to make more transparent the choices involved in socio-economic terms. Such choices are made in a number of institutional settings and at and across various levels. Biodiversity conservation policy will be determined just as much by socio-economic gains and losses and their distribution across humans, as by concerns about the loss of evolutionary history *per se*.

Biodiversity should be assigned 'critical' status implying limited substitutability in a strong sustainability sense. The result is that policy options involve ethical judgements and competing world views about socially responsible or precautionary safety margins.

Deciding whether or not the social costs of imposing environmental conservation/preservation standards are 'unacceptable' will be a broad social decision, conditioned by the specifications of a given resource conflict situation, including local circumstances. Thus the safe minimum standards (SMS) approach 'is deliberately "fuzzy" because it does not rely on a single criterion for making discrete choices' (see Hohl and Tisdell, 1993, Chapter 29). Non-economic criteria, such as an ethical assessment of our obligation to future generations, will also influence the final policy formulation. By acting as a constraint on economic efficiency objectives, SMS and its requirement that economic gains be forfeited unless the costs are unacceptable, can provide a framework for addressing the notion that individuals may foster both consumer and citizen preferences.

Social and not individual preference values will be part of the SMS setting process and economic analysis can play a constructive role alongside other political and ethical considerations. Consumer preferences can be measured via economic willingness to pay criteria, from which the social costs of imposing SMS can be assessed. As Crowards (1995) argues a social/political procedure will then need to be instituted to reflect citizen preferences about the acceptability of paying costs now to avoid exposure to uncertain future losses. The pivotal question is then, how are these citizen preferences to be aired and acknowledged in a manner that is relevant to each specific location and 'conflict' scenario? Some degree of community participation will be required in the decision-making process wherever environmental resources are threatened with irreversible loss.

Ecological economists believe that traditional ethical reasoning is faced with a number of challenges in the context of the sustainable development debate. Their adoption of a systems perspective leads to the need for an ethical re-evaluation of the role and rights of present individual humans, as against the system's survival and therefore the welfare of future human generations and present and future non-human species. The welfare of the current generation will include the existence of poverty and income inequalities, as well as the institutions that perpetuate such conditions (see Norton, Chapter 33; and Page, Chapter 34).

So 'concern for others' is an important ethical issue in the debate. Given that in certain circumstances individuals are, to a greater or lesser extent, self-interested and greedy, ecological economists have explored the extent to which such behaviour could be modified and how to achieve the modification. Thus a significant number of ecological economists support a Rawlsian rather than a utilitarian approach to intergenerational equity. Sustainable development is seen as a condition imposed on the present generation in order to satisfy some notion of intergenerational equity. The present generation can assure the welfare of future generations by underwriting the health, diversity, resilience and productivity of natural systems. The widely held view in ecological economics that the requirements of the system sometimes outweighs those of its individual components contradicts the principle of consumer sovereignty by recognizing the bounds within which sovereignty may be exercised. It also lies behind arguments for a new environmental ethic or morality. Various candidate ethical codes have been put forward in the 'bioethics' literature, ranging from human-centred stewardship through to 'deep ecology'; which in the latter's case wishes to ascribe interests and rights to non-human species and even inanimate (see Rolston, Chapter 32).

The environmental economics valuation literature has addressed the interests of future generations via an investigation of irreversible choices (development versus conservation) and the contradictions between discounting and obligations to the distant future. How societies value the environment is part and parcel of the sustainability question. But incorporating environmental values into decision making will not in itself guarantee sustainability values unless each generation is committed to transferring to the next sufficient reproducible and natural capital to make development sustainable (Howarth and Norgaard, 1992). Howarth (1995) has recently shown that a 'Kantian' approach to intergenerational equity requires the prior imposition of sustainability criteria, as a moral constraint on the maximization of social preferences concerning the distribution of welfare between generations. Such an approach precludes actions that yield present benefits but impose the risk of irreversible future losses when scientific research would permit the effective resolution of uncertainty over generational time.

An important long term goal in ecological economics is the provision of support for environmental sustainability, decision making based on the principles of economic efficiency, optimal scale, distributional fairness and justice, participation and legitimacy and precaution. In the face of an increasing degree of environmental risk, uncertainty and ignorance, the distributional incidence of the risks, costs and benefits becomes a key issue. Public consultation and consequent acceptance of the inevitable trade-offs involved will be an important requirement in any legitimization process. The challenge for a sustainable biodiversity utilization strategy is not to limit itself to the protection (irrespective of opportunity costs) of all wild resources in some designated areas, but to ensure that economic development does not threaten critical biodiversity thresholds in all areas. Protection of functional diversity and ecosystem resilience is, however difficult, necessary to mitigate the formidable risks posed by the process of global environmental change, but it also offers new and extensive opportunities for wealth creation and redistribution, together with an enhanced quality of life.

References

Batagoda, S. and Turner, R.K. (1997), 'Towards policy relevant ecosystem services and natural capital values: rainforest', Centre for Social and Economic Research on the Global Environment, Mimeo, University of East Anglia, Norwich and University College London, UK.
Botsford, L.W., Castilla, J.C. and Peterson, C.H. (1997), 'The management of fisheries and marine ecosystems', *Science*, **277**, 504–9.
Brouwer, R. and Spaninks, F.A. (1997), 'The validity of transferring environmental benefits: further empirical testing'. Global Environmental Change Working Paper GEC 97-07. Centre for Social and Economic Research on the Global Environment, University of East Anglia and University College London, UK.
Brouwer, R., Langford, I.H., Bateman, I.J., Crowards, T.C. and Turner, R.K. (1997), 'A meta-analysis of wetland contingent valuation studies', Global Environmental Change Working Paper 97–20, Centre for Social and Economic Research on the Global Environment, University of East Anglia and University College London, UK.
Clark, C. (1973), 'The economics of overexploitation', *Science*, **181**, 630–4.
Clarke, D. (1990), *Mathematical Bioeconomics*, 2nd edn, New York: Wiley.
Costanza, R. *et al.* (1997), 'The value of the world's ecosystem services and natural capital', *Nature*, **387**, 253–60.
Crowards, T.M. (1995), 'Safe minimum standards: costs and opportunities', Centre for Social and Economic Research on the Global Environment, Mimeo, University of East Anglia and University College London, UK.
Daly, H. and Cobb, J.B. (1989), *For the Common Good*, Boston: Beacon Press.
Gordon, H.S. (1954), 'The economic theory of a common property reserve: the fishery', *Journal of Political Economy*, **62**, 124–42.
Gren, I.M., Folke, C., Turner, R.K. and Bateman, I. (1994), 'Primary and secondary values of wetland ecosystems', *Environmental and Resources Economics*, **4**, 55–74.
Heywood, V. and Watson, R.T. (eds) (1995), *Global Biodiversity Assessment*, Cambridge: Cambridge University Press.
Howarth, R.B. (1995), 'Sustainability under uncertainty: a deontological approach', *Land Economics*, **71**, 417–27.
Howarth, R.B. and Norgaard, R.B. (1992), 'Environmental valuation under sustainable development', *American Economics Association Papers and Proceedings*, **82**, 473–77.
Matson, P.A., Parton, W.J., Power, A.G. and Swift, M.J. (1997), 'Agricultural intensification and ecosystem properties, *Science*, **277**, 504–9.
Noble, I.R. and Dirzo, R. (1997), 'Forests as human-diminated ecosystems', *Science*, **277**, 522–5.
Norton-Griffiths, M. and Southey, C. (1995), 'The opportunity costs of biodiversity conservation in Kenya', *Ecological Economics*, **12**, 125–40.
Panayotou, T. (1994), 'Conservation of biodiversity and economic development: the concept of transferable development rights', *Environmental and Resource Economics*, **4**, 95–110.
Pearce, D.W. and Moran, D. (1997/98), 'The economics of biodiversity', in H. Folmer and T. Tietenberg (eds), *The International Yearbook of Environmental and Resource Economics*, Cheltenham: Edward Elgar Publishing.
Pearce, D.W. and Turner, R.K. (1990), *The Economics of Natural Resouces and the Environment*, Hemel Hempstead: Harvester Wheatsheaf.
Safina, C. (1995), 'The world's imperilled fish', *Scientific American*, November, 46–53.
Steele, J.H. (1991), 'Marine functional diversity', *BioScience*, **41**, 470–74.
Southgate, D. and Clarke, H.L. (1993), 'Can conservation projects save biodiversity in South America?', *Ambio*, **22**, 146–63.
Turner, R.K. (1991), 'Economics and wetland management', *Ambio*, **20**, 59–63.
Turner, R.K. and Pearce, D.W. (1993), 'Sustainable economic development: economic and ethical principles', in E.B. Barbier (ed.), *Economics and Ecology*, London: Chapman and Hall, 177–94.

Turner, R.K., Dent, D. and Hey, R.D. (1983), 'Valuation of the environmental impact of wetland flood protection and drainage schemes', *Environment and Planning A*, **15**, 871–88.

Turner, R.K., Perrings, C. and Folke, C. (1997), 'Ecological economics: paradigm or perspective', in J. van den Bergh and J. van der Straaten (eds), *Economy and Ecosystems in Change*, Cheltenham: Edward Elgar Publishing, 25–49.

Turner, R.K., Adger, W.N. and Brouwer, R. (1998), 'Ecosystem services value, research needs and policy relevance: a commentary', *Ecological Economics*, **25**, 61–5.

Watson, R.T., Heywood, V.H., Baste, I., Dias, B. Gámez, R., Janetos, T., Reid, W. and Ruark, G. (1995), *Global Biodiversity Assessment: Summary for Policy Makers*, Cambridge: Cambridge University Press.

Wells, M.P. and Brandon, K.E. (1993), 'The principles and practices of buffer zones and local participation in biodiversity conservation, *Ambio*, **22**, 157–72.

Part I
Biodiversity:
Basic Science and Economics

[1]

RESILIENCE AND STABILITY OF ECOLOGICAL SYSTEMS

C. S. Holling

Institute of Resource Ecology, University of British Columbia, Vancouver, Canada

INTRODUCTION

Individuals die, populations disappear, and species become extinct. That is one view of the world. But another view of the world concentrates not so much on presence or absence as upon the numbers of organisms and the degree of constancy of their numbers. These are two very different ways of viewing the behavior of systems and the usefulness of the view depends very much on the properties of the system concerned. If we are examining a particular device designed by the engineer to perform specific tasks under a rather narrow range of predictable external conditions, we are likely to be more concerned with consistent nonvariable performance in which slight departures from the performance goal are immediately counteracted. A quantitative view of the behavior of the system is, therefore, essential. With attention focused upon achieving constancy, the critical events seem to be the amplitude and frequency of oscillations. But if we are dealing with a system profoundly affected by changes external to it, and continually confronted by the unexpected, the constancy of its behavior becomes less important than the persistence of the relationships. Attention shifts, therefore, to the qualitative and to questions of existence or not.

Our traditions of analysis in theoretical and empirical ecology have been largely inherited from developments in classical physics and its applied variants. Inevitably, there has been a tendency to emphasize the quantitative rather than the qualitative, for it is important in this tradition to know not just that a quantity is larger than another quantity, but precisely how much larger. It is similarly important, if a quantity fluctuates, to know its amplitude and period of fluctuation. But this orientation may simply reflect an analytic approach developed in one area because it was useful and then transferred to another where it may not be.

Our traditional view of natural systems, therefore, might well be less a meaningful reality than a perceptual convenience. There can in some years be more owls and fewer mice and in others, the reverse. Fish populations wax and wane as a natural condition, and insect populations can range over extremes that only logarithmic transformations can easily illustrate. Moreover, over distinct areas, during long or short periods of time, species can completely disappear and then reappear. Different and useful insight might be obtained, therefore, by viewing the behavior of ecological systems in terms of the probability of extinction of their elements, and by shifting emphasis from the equilibrium states to the conditions for persistence.

An equilibrium centered view is essentially static and provides little insight into the transient behavior of systems that are not near the equilibrium. Natural, undisturbed systems are likely to be continually in a transient state; they will be equally so under the influence of man. As man's numbers and economic demands increase, his use of resources shifts equilibrium states and moves populations away from equilibria. The present concerns for pollution and endangered species are specific signals that the well-being of the world is not adequately described by concentrating on equilibria and conditions near them. Moreover, strategies based upon these two different views of the world might well be antagonistic. It is at least conceivable that the effective and responsible effort to provide a maximum sustained yield from a fish population or a nonfluctuating supply of water from a watershed (both equilibrium-centered views) might paradoxically increase the chance for extinctions.

The purpose of this review is to explore both ecological theory and the behavior of natural systems to see if different perspectives of their behavior can yield different insights useful for both theory and practice.

Some Theory

Let us first consider the behavior of two interacting populations: a predator and its prey, a herbivore and its resource, or two competitors. If the interrelations are at all regulated we might expect a disturbance of one or both populations in a constant environment to be followed by fluctuations that gradually decrease in amplitude. They might be represented as in Figure 1, where the fluctuations of each population over time are shown as the sides of a box. In this example the two populations in some sense are regulating each other, but the lags in the response generate a series of oscillations whose amplitude gradually reduces to a constant and sustained value for each population. But if we are also concerned with persistence we would like to know not just how the populations behave from one particular pair of starting values, but from all possible pairs since there might well be combinations of starting populations for which ultimately the fate of one or other of the populations is extinction. It becomes very difficult on time plots to show the full variety of responses possible, and it proves convenient to plot a trajectory in a phase plane. This is shown by the end of the box in Figure 1 where the two axes represent the density of the two populations.

The trajectory shown on that plane represents the sequential change of the two populations at constant time intervals. Each point represents the unique density of each population at a particular point in time and the arrows indicate the direction of change over time. If oscillations are damped, as in the case shown, then the trajectory is represented as a closed spiral that eventually reaches a stable equilibrium.

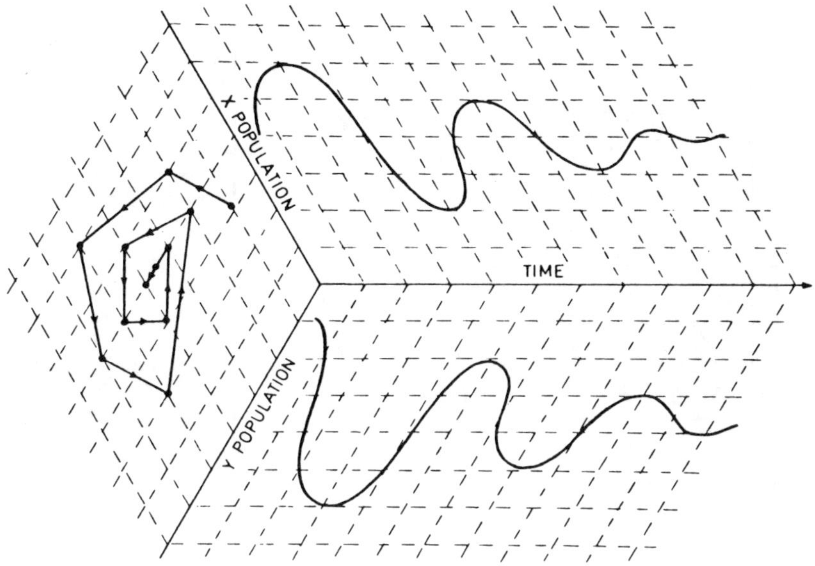

Figure 1 Derivation of a phase plane showing the changes in numbers of two populations over time.

We can imagine a number of different forms for trajectories in the phase plane (Figure 2). Figure 2a shows an open spiral which would represent situations where fluctuations gradually increase in amplitude. The small arrows are added to suggest that this condition holds no matter what combination of populations initiates the trajectory. In Figure 2b the trajectories are closed and given any starting point eventually return to that point. It is particularly significant that each starting point generates a unique cycle and there is no tendency for points to converge to a single cycle or point. This can be termed "neutral stability" and it is the kind of stability achieved by an imaginary frictionless pendulum.

Figure 2c represents a stable system similar to that of Figure 1, in which all possible trajectories in the phase plane spiral into an equilibrium. These three examples are relatively simple and, however relevant for classical stability analysis, may well be theoretical curiosities in ecology. Figures 2d–2f add some complexities. In a sense Figure 2d represents a combination of a and c, with a region in the center of the phase plane within which all possible trajectories spiral inwards to equilibrium. Those outside this region spiral outwards and lead eventually to extinction of one or the other populations. This is an example of local stability in contrast to the global stability of Figure 2c. I designate the region within which stability occurs as the domain of attraction, and the line that contains this domain as the boundary of the attraction domain.

The trajectories in Figure 2e behave in just the opposite way. There is an internal region within which the trajectories spiral out to a stable limit cycle and beyond

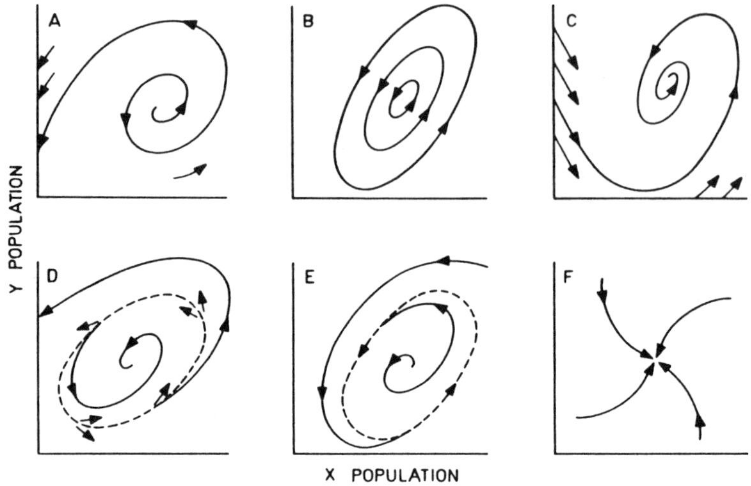

Figure 2 Examples of possible behaviors of systems in a phase plane; (a) unstable equilibrium, (b) neutrally stable cycles, (c) stable equilibrium, (d) domain of attraction, (e) stable limit cycle, (f) stable node.

which they spiral inwards to it. Finally, a stable node is shown in Figure 2f in which there are no oscillations and the trajectories approach the node monotonically. These six figures could be combined in an almost infinite variety of ways to produce several domains of attraction within which there could be a stable equilibrium, a stable limit cycle, a stable node, or even neutrally stable orbits. Although I have presumed a constant world throughout, in the presence of random fluctuations of parameters or of driving variables (Walters 39), any one trajectory could wander with only its general form approaching the shape of the trajectory shown. These added complications are explored later when we consider real systems. For the moment, however, let us review theoretical treatments in the light of the possibilities suggested in Figure 2.

The present status of ecological stability theory is very well summarized in a number of analyses of classical models, particularly May's (23–25) insightful analyses of the Lotka-Volterra model and its expansions, the graphical stability analyses of Rosenzweig (33, 34), and the methodological review of Lewontin (20).

May (24) reviews the large class of coupled differential equations expressing the rate of change of two populations as continuous functions of both. The behavior of these models results from the interplay between (*a*) stabilizing negative feedback or density-dependent responses to resources and predation, and (*b*) the destabilizing effects produced by the way individual predators attack and predator numbers respond to prey density [termed the functional and numerical responses, as in Holling (11)]. Various forms have been given to these terms; the familiar Lotka-Volterra model includes the simplest and least realistic, in which death of prey is caused only by predation, predation is a linear function of the product of prey and predator populations, and growth of the predator population is linearly proportional to the same product. This model generates neutral stability as in Figure 2b, but the assumptions are very unrealistic since very few components are included, there are no explicit lags or spatial elements, and thresholds, limits, and nonlinearities are missing.

These features have all been shown to be essential properties of the predation process (Holling 12, 13) and the effect of adding some of them has been analyzed by May (24). He points out that traditional ways of analyzing the stability properties of models using analytical or graphical means (Rosenzweig & MacArthur 33, Rosenzweig 34, 35) concentrate about the immediate neighborhood of the equilibrium. By doing this, linear techniques of analysis can be applied that are analytically tractable. Such analyses show that with certain defined sets of parameters stable equilibrium points or nodes exist (such as Figure 2c), while for other sets they do not, and in such cases the system is, by default, presumed to be unstable, as in Figure 2a. May (24), however, invokes a little-used theorem of Kolmogorov (Minorksy 26) to show that all these models have either a stable equilibrium point or a stable limit cycle (as in Figure 2e). Hence he concludes that the conditions presumed by linear analysis are unstable, and in fact must lead to stable limit cycles. In every instance, however, the models are globally rather than locally stable, limiting their behavior to that shown in either Figures 2c or 2e.

There is another tradition of models that recognizes the basically discontinuous features of ecological systems and incorporates explicit lags. Nicholson and Bailey initiated this tradition when they developed a model using the output of attacks and survivals within one generation as the input for the next (29). The introduction of this explicit lag generates oscillations that increase in amplitude until one or other of the species becomes extinct (Figure 2a). Their assumptions are as unrealistically simple as Lotka's and Volterra's; the instability results because the number of attacking predators at any moment is so much a consequence of events in the previous generation that there are "too many" when prey are declining and "too few" when prey are increasing. If a lag is introduced into the Lotka-Volterra formulation (Wangersky & Cunningham 40) the same instability results.

The sense one gains, then, of the behavior of the traditional models is that they are either globally unstable or globally stable, that neutral stability is very unlikely, and that when the models are stable a limit cycle is a likely consequence.

Many, but not all, of the simplifying assumptions have been relaxed in simulation models, and there is one example (Holling & Ewing 14) that joins the two traditions initiated by Lotka-Volterra and Nicholson and Bailey and, further, includes more realism in the operation of the stabilizing and destabilizing forces. These modifications are described in more detail later; the important features accounting for the difference in behavior result from the introduction of explicit lags, a functional response of predators that rises monotonically to a plateau, a nonrandom (or contagious) attack by predators, and a minimum prey density below which reproduction does not occur. With these changes a very different pattern emerges that conforms most closely to Figure 2d. That is, there exists a domain of attraction within which there is a stable equilibrium; beyond that domain the prey population becomes extinct. Unlike the Nicholson and Bailey model, the stability becomes possible, although in a limited region, because of contagious attack. [Contagious attack implies that for one reason or another some prey have a greater probability of being attacked than others, a condition that is common in nature (Griffiths & Holling 9).] The influence of contagious attack becomes significant whenever predators become abundant in relation to the prey, for then the susceptible prey receive the burden of attention, allowing more prey to escape than would be expected by random contact. This "inefficiency" of the predator allows the system to counteract the destabilizing effects of the lag.

If this were the only difference the system would be globally stable, much as Figure 2c. The inability of the prey to reproduce at low densities, however, allows some of the trajectories to cut this reproduction threshold, and the prey become extinct. This introduces a lower prey density boundary to the attraction domain and, at the same time, a higher prey density boundary above which the amplitudes of the oscillations inevitably carry the population below the reproduction threshold. The other modifications in the model, some of which have been touched on above, alter this picture in degree only. The essential point is that a more realistic representation of the behavior of interacting populations indicates the existence of at least one domain of attraction. It is quite possible, within this domain, to imagine stable equilibrium points, stable nodes, or stable limit cycles. Whatever the detailed configuration, the existence of discrete domains of attraction immediately suggests important consequences for the persistence of the system and the probability of its extinction.

Such models, however complex, are still so simple that they should not be viewed in a definitive and quantitative way. They are more powerfully used as a starting point to organize and guide understanding. It becomes valuable, therefore, to ask what the models leave out and whether such omissions make isolated domains of attraction more or less likely.

Theoretical models generally have not done well in simultaneously incorporating realistic behavior of the processes involved, randomness, spatial heterogeneity, and an adequate number of dimensions or state variables. This situation is changing very rapidly as theory and empirical studies develop a closer technical partnership. In what follows I refer to real world examples to determine how the four elements that tend to be left out might further affect the behavior of ecological systems.

SOME REAL WORLD EXAMPLES

Self-Contained Ecosystems

In the broadest sense, the closest approximation we could make of a real world example that did not grossly depart from the assumptions of the theoretical models would be a self-contained system that was fairly homogenous and in which climatic fluctuations were reasonably small. If such systems could be discovered they would reveal how the more realistic interaction of real world processes could modify the patterns of systems behavior described above. Very close approximations to any of these conditions are not likely to be found, but if any exist, they are apt to be fresh water aquatic ones. Fresh water lakes are reasonably contained systems, at least within their watersheds; the fish show considerable mobility throughout, and the properties of the water buffer the more extreme effects of climate. Moreover, there have been enough documented man-made disturbances to liken them to perturbed systems in which either the parameter values or the levels of the constituent populations are changed. In a crude way, then, the lake studies can be likened to a partial exploration of a phase space of the sorts shown in Figure 2. Two major classes of disturbances have occurred: first, the impact of nutrient enrichment from man's domestic and industrial wastes, and second, changes in fish populations by harvesting.

The paleolimnologists have been remarkably successful in tracing the impact of man's activities on lake systems over surprisingly long periods. For example, Hutchinson (17) has reconstructed the series of events occurring in a small crater lake in Italy from the last glacial period in the Alps (2000 to 1800 BC) to the present. Between the beginning of the record and Roman times the lake had established a trophic equilibrium with a low level of productivity which persisted in spite of dramatic changes in surroundings from *Artemesia* steppe, through grassland, to fir and mixed oak forest. Then suddenly the whole aquatic system altered. This alteration towards eutrophication seems to have been initiated by the construction of the Via Cassia about 171 BC, which caused a subtle change in the hydrographic regime. The whole sequence of environmental changes can be viewed as changes in parameters or driving variables, and the long persistence in the face of these major changes suggests that natural systems have a high capacity to absorb change without dramatically altering. But this resilient character has its limits, and when the limits are passed, as by the construction of the Roman highway, the system rapidly changes to another condition.

More recently the activities of man have accelerated and limnologists have recorded some of the responses to these changes. The most dramatic change consists of blooms of algae in surface waters, an extraordinary growth triggered, in most instances, by nutrient additions from agricultural and domestic sources.

While such instances of nutrient addition provide some of the few examples available of perturbation effects in nature, there are no controls and the perturbations are exceedingly difficult to document. Nevertheless, the qualitative pattern seems consistent, particularly in those lakes (Edmundson 4, Hasler 10) to which sewage has been added for a time and then diverted elsewhere. This pulse of disturbance characteristically triggers periodic algal blooms, low oxygen conditions, the sudden disappearance of some plankton species, and appearance of others. As only one example, the nutrient changes in Lake Michigan (Beeton 2) have been accompanied by the replacement of the cladoceran *Bosmina coregoni* by *B. Longirostris, Diaptomus oregonensis* has become an important copepod species, and a brackish water copepod *Eurytemora affinis* is a new addition to the zooplankton.

In Lake Erie, which has been particularly affected because of its shallowness and intensity of use, the mayfly *Hexagenia,* which originally dominated the benthic community, has been almost totally replaced by oligochetes. There have been blooms of the diatom *Melosira binderana,* which had never been reported from the

United States until 1961 but now comprises as much as 99% of the total phytoplankton around certain islands. In those cases where sewage has been subsequently diverted there is a gradual return to less extreme conditions, the slowness of the return related to the accumulation of nutrients in sediments.

The overall pattern emerging from these examples is the sudden appearance or disappearance of populations, a wide amplitude of fluctuations, and the establishment of new domains of attraction.

The history of the Great Lakes provides not only some particularly good information on responses to man-made enrichment, but also on responses of fish populations to fishing pressure. The eutrophication experience touched on above can be viewed as an example of systems changes in driving variables and parameters, whereas the fishing example is more an experiment in changing state variables. The fisheries of the Great Lakes have always selectively concentrated on abundant species that are in high demand. Prior to 1930, before eutrophication complicated the story, the lake sturgeon in all the Great Lakes, the lake herring in Lake Erie, and the lake whitefish in Lake Huron were intensively fished (Smith 37). In each case the pattern was similar: a period of intense exploitation during which there was a prolonged high level harvest, followed by a sudden and precipitous drop in populations. Most significantly, even though fishing pressure was then relaxed, none of these populations showed any sign of returning to their previous levels of abundance. This is not unexpected for sturgeon because of their slow growth and late maturity, but it is unexpected for herring and whitefish. The maintenance of these low populations in recent times might be attributed to the increasingly unfavorable chemical or biological environment, but in the case of the herring, at least, the declines took place in the early 1920s before the major deterioration in environment occurred. It is as if the population had been shifted by fishing pressure from a domain with a high equilibrium to one with a lower one. This is clearly not a condition of neutral stability as suggested in Figure 2b since once the populations were lowered to a certain point the decline continued even though fishing pressure was relaxed. It can be better interpreted as a variant of Figure 2d where populations have been moved from one domain of attraction to another.

Since 1940 there has been a series of similar catastrophic changes in the Great Lakes that has led to major changes in the fish stocks. Beeton (2) provides graphs summarizing the catch statistics in the lakes for many species since 1900. Lake trout, whitefish, herring, walleye, sauger, and blue pike have experienced precipitous declines of populations to very low values in all of the lakes. The changes generally conform to the same pattern. After sustained but fluctuating levels of harvest the catch dropped dramatically in a span of a very few years, covering a range of from one to four orders of magnitude. In a number of examples particularly high catches were obtained just before the drop. Although catch statistics inevitably exaggerate the step-like character of the pattern, populations must have generally behaved in the way described.

The explanations for these changes have been explored in part, and involve various combinations of intense fishing pressure, changes in the physical and chemical environment, and the appearance of a foreign predator (the sea lamprey) and foreign competitors (the alewife and carp). For our purpose the specific cause is of less interest than the inferences that can be drawn concerning the resilience of these systems and their stability behavior. The events in Lake Michigan provide a typical example of the pattern in other lakes (Smith 37). The catch of lake trout was high, but fluctuated at around six million pounds annually from 1898 to 1940. For four years catches increased noticeably and then suddenly collapsed to near extinction by the 1950s due to a complete failure of natural reproduction. Lake herring and whitefish followed a similar pattern (Beeton 2: Figure 7). Smith (37) argues that the trigger for the lake trout collapse was the appearance of the sea lamprey that had spread through the Great Lakes after the construction of the Welland Canal. Although lamprey populations were extremely small at the time of the collapse, Smith argues that even a small mortality, added to a commercial harvest that was probably at the maximum for sustained yield, was sufficient to cause the collapse. Moreover, Ricker (31) has shown that fishing pressure shifts the age structure of fish populations towards younger ages. He demonstrates that a point can come where only slight increases in mortality can trigger a collapse of the kind noted for lake trout. In addition, the lake trout was coupled in a network of competitive and predatory interconnections with other species, and pressures on these might have contributed as well.

Whatever the specific causes, it is clear that the precondition for the collapse was set by the harvesting of fish, even though during a long period there were no obvious signs of problems. The fishing activity, however, progressively reduced the resilience of the system so that when the inevitable unexpected event occurred, the populations collapsed. If it had not been the lamprey, it would have been something else: a change in climate as part of the normal pattern of fluctuation, a change in the chemical or physical environment, or a change in competitors or predators. These examples again suggest distinct domains of attraction in which the populations forced close to the boundary of the domain can then flip over it.

The above examples are not isolated ones. In 1939 an experimental fishery was started in Lake Windermere to improve stocks of salmonids by reducing the abundance of perch (a competitor) and pike (a predator). Perch populations were particularly affected by trapping and the populations fell drastically in the first three years. Most significantly, although no perch have been removed from the North Basin since 1947, populations have still not shown any tendency to return to their previous level (Le Cren et al 19).

The same patterns have even been suggested for terrestrial systems. Many of the arid cattle grazing lands of the western United States have gradually become invaded and dominated by shrubs and trees like mesquite and cholla. In some instances grazing and the reduced incidence of fire through fire prevention programs allowed invasion and establishment of shrubs and trees at the expense of grass. Nevertheless, Glendening (8) has demonstrated, from data collected in a 17-year experiment in which intensity of grazing was manipulated, that once the trees have gained sufficient size and density to completely utilize or materially reduce the moisture supply, elimination of grazing will not result in the grassland reestablishing itself. In short, there is a level of the state variable "trees" that, once achieved, moves the system from one domain of attraction to another. Return to the original domain can only be made by an explicit reduction of the trees and shrubs.

These examples point to one or more distinct domains of attraction in which the important point is not so much how stable they are within the domain, but how likely it is for the system to move from one domain into another and so persist in a changed configuration.

This sampling of examples is inevitably biased. There are few cases well documented over a long period of time, and certainly some systems that have been greatly disturbed have fully recovered their original state once the disturbance was removed. But the recovery in most instances is in open systems in which reinvasion is the key ingredient. These cases are discussed below in connection with the effects of spatial heterogeneity. For the moment I conclude that distinct domains of attraction are not uncommon within closed systems. If such is the case, then further confirmation should be found from empirical evidence of the way processes which link organisms operate, for it is these processes that are the cause of the behavior observed.

Process Analysis

One way to represent the combined effects of processes like fecundity, predation, and competition is by using Ricker's (30) reproduction curves. These simply represent the population in one generation as a function of the population in the previous generation, and examples are shown in Figures 3a, c, and e. In the simplest form, and the one most used in practical fisheries management (Figure 3a), the reproduction curve is dome-shaped. When it crosses a line with slope 1 (the straight line in the figures) an equilibrium condition is possible, for at such cross-overs the popula-

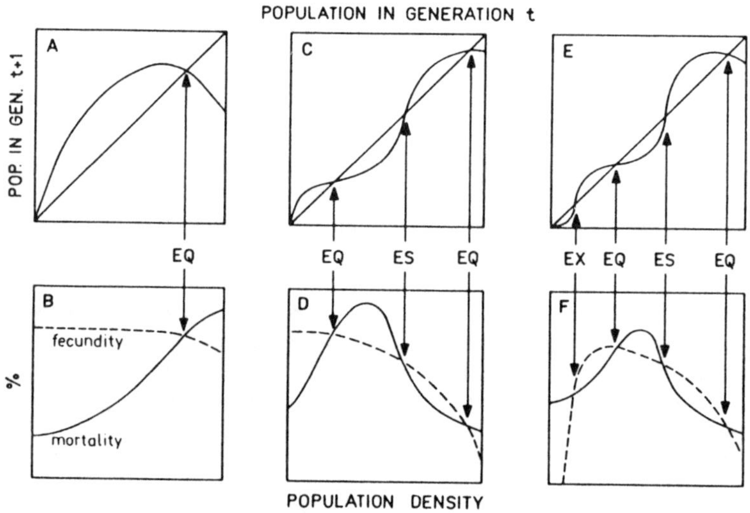

Figure 3 Examples of various reproduction curves (a, c, and e) and their derivation from the contributions of fecundity and mortality (b, d, and f).

tion in one generation will produce the same number in the next. It is extremely difficult to detect the precise form of such curves in nature, however; variability is high, typically data are only available for parts of any one curve, and the treatment really only applies to situations where there are no lags. It is possible to deduce various forms of reproduction curves, however, by disaggregating the contributions of fecundity and mortality. The three lower graphs in Figure 3b, 3d, and 3f represent this disaggregation of their counterpart reproduction curves. The simplest types of reproduction curve (Figure 3a) can arise from a mortality that regularly increases with density and either a constant fecundity or a declining one. With fecundity expressed as the percentage mortality necessary to just balance reproduction, the cross-over point of the curves represents the equilibrium condition. But we know that the effects of density on fecundity and mortality can be very much more complicated.

Mortality from predation, for example, has been shown to take a number of classic forms (Holling 11, 13). The individual attack by predators as a function of prey density (the functional response to prey density) can increase with a linear rise to a plateau (type 1), a concave or negatively accelerated rise to a plateau (type 2), or an S-shaped rise to a plateau (type 3). The resulting contribution to mortality from these responses can therefore show ranges of prey density in which there is direct density dependence (negative feedback from the positively accelerated portions of the type 3 response), density independence (the straight line rise of type 1), and inverse dependence (the positive feedback from the negatively accelerated and plateau portions of the curves). There are, in addition, various numerical responses generated by changes in the number of predators as the density of their prey increases. Even for those predators whose populations respond by increasing, there often will be a limit to the increase set by other conditions in the environment. When populations are increasing they tend to augment the negative feedback features (although with a delay), but when populations are constant, despite increasing prey density, the percent mortality will inevitably decline since individual attack eventually saturates at complete satiation (the plateaux of all three functional responses). In Figures 3d and 3f the mortality curves shown summarize a common type. The rising or direct density-dependent limb of the curve is induced by increasing predator populations and by the reduced intensity of attack at low densities, shown by the initial positively accelerated portion of the S-shaped type 3 response. Such a condition is common for predators with alternate prey, both vertebrates (Holling 14) and at least some invertebrates (Steele 38). The declining inverse density-dependent limb is induced by satiation of the predator and a numerical response that has been reduced or stopped.

Fecundity curves that decline regularly over a very wide range of increasing population densities (as in Figure 3d) are common and have been referred to as *Drosophila*-type curves (Fujita 6). This decline in fecundity is caused by increased competition for oviposition sites, interference with mating, and increased sterility. The interaction between a dome-shaped mortality curve and a monotonically decreasing fecundity curve can generate equilibrium conditions (Figure 3d). Two stable equilibria are possible, but between these two is a transient equilibrium designated as the escape threshold (ES in Figure 3). Effects of random changes on populations or parameters could readily shift densities from around the lower equilibrium to above this escape threshold, and in these circumstances populations would inevitably increase to the higher equilibrium.

The fecundity curves are likely to be more complex, however, since it seems inevitable that at some very low densities fecundity will decline because of difficulties in finding mates and the reduced effect of a variety of social facilitation behaviors. We might even logically conclude that for many species there is a minimum density below which fecundity is zero. A fecundity curve of this Allee-type (Fujita 6) has been empirically demonstrated for a number of insects (Watt 42) and is shown in Figure 3f. Its interaction with the dome-shaped mortality curve can add another transient equilibrium, the extinction threshold (EX in Figure 3f). With this addition there is a lower density such that if populations slip below it they will proceed inexorably to extinction. The extinction threshold is particularly likely since it has been shown mathematically that each of the three functional response curves will intersect with the ordinate of percent predation at a value above zero (Holling 13).

Empirical evidence, therefore, suggests that realistic forms to fecundity and mortality curves will generate sinuous reproduction curves like those in Figures 3c and 3e with the possibility of a number of equilibrium states, some transient and some stable. These are precisely the conditions that will generate domains of attraction, with each domain separated from others by the extinction and escape thresholds. This analysis of process hence adds support to the field observations discussed earlier.

The behavior of systems in phase space cannot be completely understood by the graphical representations presented above. These graphs are appropriate only when effects are immediate; in the face of the lags that generate cyclic behavior the reproduction curve should really produce two values for the population in generation $t + 1$ for each value of the population in generation t. The graphical treatment of Rosenzweig & MacArthur (33) to a degree can accommodate these lags and cyclic behavior. In their treatment they divide phase planes of the kind shown in Figure 2 into various regions of increasing and decreasing x and y populations. The regions are separated by two lines, one representing the collection of points at which the prey population does not change in density ($dx/dt = 0$, the prey isocline) and one in which the predator population does not so change ($dy/dt = 0$, the predator isocline). They deduce that the prey isocline will be dome-shaped for much the same reason as described for the fecundity curves of Figure 3f. The predator isocline, in the simplest condition, is presumed to be vertical, assuming that only one fixed level of prey is necessary to just maintain the predator population at a zero instantaneous rate of change.

Intersection of the two isoclines indicates a point where both populations are at equilibrium. Using traditional linear stability analysis one can infer whether these equilibrium states are stable (Figure 2c) or not (Figure 2a). Considerable importance is attached to whether the predator isocline intersects the rising or falling portion of the prey isocline. As mentioned earlier these techniques are only appropriate near equilibrium (May 24), and the presumed unstable conditions in fact generate stable limit cycles (Figure 2e). Moreover, it is unlikely that the predator isocline is a vertical one in the real world, since competition between predators at high predator densities would so interfere with the attack process that a larger number of prey would be required for stable predator populations. It is precisely this condition that was demonstrated by Griffiths & Holling (9) when they showed that a large number of species of parasites distribute their attacks contagiously. The result is a "squabbling predator behavior" (Rosenzweig 34, 35) that decreases the efficiency of predation at high predator/prey ratios. This converts an unstable system (Figure 2a) to a stable one (Figure 2c); it is likely that stability is the rule, rather than the exception, irrespective of where the two isoclines cross.

The empirical evidence described above shows that realistic fecundity and mortality (particularly predation) processes will generate forms that the theorists might tend to identify as special subsets of more general conditions. But it is just these special subsets that separate the real world from all possible ones, and these more realistic forms will modify the general conclusions of simpler theory. The ascending limb of the Allee-type fecundity curve will establish, through interaction with mortality, a minimum density below which prey will become extinct. This can at the same time establish an upper prey density above which prey will become extinct because the amplitude of prey fluctuations will eventually carry the population over the extinction threshold, as shown in the outer trajectory of Figure 2d. These conditions alone are sufficient to establish a domain of attraction, although the boundaries of this domain need not be closed. Within the domain the contagious attack by predators can produce a stable equilibrium or a stable node. Other behaviors of the mortality agents, however, could result in stable limit cycles.

More realistic forms of functional response change this pattern in degree only. For example, a negatively accelerated type of functional response would tend to make the domain of attraction somewhat smaller, and an S-shaped one larger. Limitations in the predator's numerical response and thresholds for reproduction of predators, similar to those for prey, could further change the form of the domain. Moreover, the behaviors that produce the sinuous reproduction curves of Figures 3c and 3e can add additional domains. The essential point, however, is that these systems are not globally stable but can have distinct domains of attraction. So long as the populations remain within one domain they have a consistent and regular form of behavior. If populations pass a boundary to the domain by chance or through intervention of man, then the behavior suddenly changes in much the way suggested from the field examples discussed earlier.

The Random World

To this point, I have argued as if the world were completely deterministic. In fact, the behavior of ecological systems is profoundly affected by random events. It is important, therefore, to add another level of realism at this point to determine how the above arguments may be modified. Again, it is applied ecology that tends to supply the best information from field studies since it is only in such situations that data have been collected in a sufficiently intensive and extensive manner. As one example, for 28 years there has been a major and intensive study of the spruce budworm and its interaction with the spruce-fir forests of eastern Canada (Morris

27). There have been six outbreaks of the spruce budworm since the early 1700s (Baskerville 1) and between these outbreaks the budworm has been an exceedingly rare species. When the outbreaks occur there is major destruction of balsam fir in all the mature forests, leaving only the less susceptible spruce, the nonsusceptible white birch, and a dense regeneration of fir and spruce. The more immature stands suffer less damage and more fir survives. Between outbreaks the young balsam grow, together with spruce and birch, to form dense stands in which the spruce and birch, in particular, suffer from crowding. This process evolves to produce stands of mature and overmature trees with fir a predominant feature.

This is a necessary, but not sufficient, condition for the appearance of an outbreak; outbreaks occur only when there is also a sequence of unusually dry years (Wellington 43). Until this sequence occurs, it is argued (Morris 27) that various natural enemies with limited numerical responses maintain the budworm populations around a low equilibrium. If a sequence of dry years occurs when there are mature stand of fir, the budworm populations rapidly increase and escape the control by predators and parasites. Their continued increase eventually causes enough tree mortality to force a collapse of the populations and the reinstatement of control around the lower equilibrium. The reproduction curves therefore would be similar to those in Figures 3c or 3e.

In brief, between outbreaks the fir tends to be favored in its competition with spruce and birch, whereas during an outbreak spruce and birch are favored because they are less susceptible to budworm attack. This interplay with the budworm thus maintains the spruce and birch which otherwise would be excluded through competition. The fir persists because of its regenerative powers and the interplay of forest growth rates and climatic conditions that determine the timing of budworm outbreaks.

This behavior could be viewed as a stable limit cycle with large amplitude, but it can be more accurately represented by a distinct domain of attraction determined by the interaction between budworm and its associated natural enemies, which is periodically exceeded through the chance consequence of climatic conditions. If we view the budworm only in relation to its associated predators and parasites we might argue that it is highly unstable in the sense that populations fluctuate widely. But these very fluctuations are essential features that maintain persistence of the budworm, together with its natural enemies and its host and associated trees. By so fluctuating, successive generations of forests are replaced, assuring a continued food supply for future generations of budworm and the persistence of the system.

Until now I have avoided formal identification of different kinds of behavior of ecological systems. The more realistic situations like budworm, however, make it necessary to begin to give more formal definition to their behavior. It is useful to distinguish two kinds of behavior. One can be termed stability, which represents the ability of a system to return to an equilibrium state after a temporary disturbance; the more rapidly it returns and the less it fluctuates, the more stable it would be. But there is another property, termed resilience, that is a measure of the persistence of systems and of their ability to absorb change and disturbance and still maintain the same relationships between populations or state variables. In this sense, the budworm forest community is highly unstable and it is because of this instability that it has an enormous resilience. I return to this view frequently throughout the remainder of this paper.

The influence of random events on systems with domains of attraction is found in aquatic systems as well. For example, pink salmon populations can become stabilized for several years at very different levels, the new levels being reached by sudden steps rather than by gradual transition (Neave 28). The explanation is very much the same as that proposed for the budworm, involving an interrelation between negative and positive feedback mortality of the kinds described in Figures 3d and 3f, and random effects unrelated to density. The same pattern has been described by Larkin (18) in his simulation model of the Adams River sockeye salmon. This particular run of salmon has been characterized by a regular four-year periodicity since 1922, with one large or dominant year, one small or subdominant, and two years with very small populations. The same explanation as described above has been proposed with the added reality of a lag. Essentially, during the dominant year limited numerical responses produce an inverse density-dependent response as in the descending limb of the mortality curves of Figure 3d and 3f. The abundance of the prey in that year is nevertheless sufficient to establish populations of predators that have a major impact on the three succeeding low years. Buffering of predation by the smolts of the dominant year accounts for the larger size of the subdominant. These effects have been simulated (Larkin 18), and when random influences are imposed in order to simulate climatic variations the system has a distinct probability of flipping into another stable configuration that is actually reproduced in nature by sockeye salmon runs in other rivers. When subdominant escapement reaches a critical level there is about an equal chance that it may become the same size as the dominant one or shrivel to a very small size.

Random events, of course, are not exclusively climatic. The impact of fires on terrestrial ecosystems is particularly illuminating (Cooper 3) and the periodic appearance of fires has played a decisive role in the persistence of grasslands as well as certain forest communities. As an example, the random perturbation caused by fires in Wisconsin forests (Loucks 21) has resulted in a sequence of transient changes that move forest communities from one domain of attraction to another. The apparent instability of this forest community is best viewed not as an unstable condition alone, but as one that produces a highly resilient system capable of repeating itself and persisting over time until a disturbance restarts the sequence.

In summary, these examples of the influence of random events upon natural systems further confirm the existence of domains of attraction. Most importantly they suggest that instability, in the sense of large fluctuations, may introduce a resilience and a capacity to persist. It points out the very different view of the world that can be obtained if we concentrate on the boundaries to the domain of attraction rather than on equilibrium states. Although the equilibrium-centered view is analytically more tractable, it does not always provide a realistic understanding of the systems' behavior. Moreover, if this perspective is used as the exclusive guide to the management activities of man, exactly the reverse behavior and result can be produced than is expected.

The Spatial Mosaic

To this point, I have proceeded in a series of steps to gradually add more and more reality. I started with self-contained closed systems, proceeded to a more detailed explanation of how ecological processes operate, and then considered the influence of random events, which introduced heterogeneity over time.

The final step is now to recognize that the natural world is not very homogeneous over space, as well, but consists of a mosaic of spatial elements with distinct biological, physical, and chemical characteristics that are linked by mechanisms of biological and physical transport. The role of spatial heterogeneity has not been well explored in ecology because of the enormous logistic difficulties. Its importance, however, was revealed in a classic experiment that involved the interaction between a predatory mite, its phytophagous mite prey, and the prey's food source (Huffaker et al 15). Briefly, in the relatively small enclosures used, when there was unimpeded movement throughout the experimental universe, the system was unstable and oscillations increased in amplitude. When barriers were introduced to impede dispersal between parts of the universe, however, the interaction persisted. Thus populations in one small locale that suffer chance extinctions could be reestablished by invasion from other populations having high numbers—a conclusion that is confirmed by Roff's mathematical analysis of spatial heterogeneity (32).

There is one study that has been largely neglected that is, in a sense, a much more realistic example of the effects of both temporal and spatial heterogeneity of a population in nature (Wellington 44, 45). There is a peninsula on Vancouver Island in which the topography and climate combine to make a mosaic of favorable locales for the tent caterpillar. From year to year the size of these locales enlarges or contracts depending on climate; Wellington was able to use the easily observed changes in cloud patterns in any year to define these areas. The tent caterpillar, to add a further element of realism, has identifiable behavioral types that are determined not by genetics but by the nutritional history of the parents. These types represent a range from sluggish to very active, and the proportion of types affects the shape of the easily visible web the tent caterpillars spin. By combining these defined differences of behavior with observations on changing numbers, shape of webs, and changing cloud patterns, an elegant story of systems behavior emerges. In a favorable year locales that previously could not support tent caterpillars now can, and populations are established through invasion by the vigorous dispersers from other locales. In these new areas they tend to produce another generation with a high proportion of vigorous behavioral types. Because of their high dispersal behavior and the small area of the locale in relation to its periphery, they then tend to leave in greater numbers than they arrive. The result is a gradual increase in the proportion of more sluggish types to the point where the local population collapses. But, although its fluctuations are considerable, even under the most unfavorable conditions there are always enclaves suitable for the insect. It is an example of a population with high fluctuations that can take advantage of transient periods of favorable conditions and that has, because of this variability, a high degree of resilience and capacity to persist.

A further embellishment has been added in a study of natural insect populations by Gilbert & Hughes (7). They combined an insightful field study of the interaction between aphids and their parasites with a simulation model, concentrating upon a specific locale and the events within it under different conditions of immigration from other locales. Again the important focus was upon persistence rather than degree of fluctuation. They found that specific features of the parasite-host interaction allowed the parasite to make full use of its aphid resources just short of driving the host to extinction. It is particularly intriguing that the parasite and its host were introduced into Australia from Europe and in the short period that the parasite has been present in Australia there have been dramatic changes in its developmental rate and fecundity. The other major difference between conditions in Europe and Australia is that the immigration rate of the host in England is considerably higher than in Australia. If the immigration rate in Australia increased to the English level, then, according to the model the parasite should increase its fecundity from the Australian level to the English to make the most of its opportunity short of extinction. This study provides, therefore, a remarkable example of a parasite and its host evolving together to permit persistence, and further confirms the importance of systems resilience as distinct from systems stability.

SYNTHESIS

Some Definitions

Traditionally, discussion and analyses of stability have essentially equated stability to systems behavior. In ecology, at least, this has caused confusion since, in mathematical analyses, stability has tended to assume definitions that relate to conditions very near equilibrium points. This is a simple convenience dictated by the enormous analytical difficulties of treating the behavior of nonlinear systems at some distance from equilibrium. On the other hand, more general treatments have touched on questions of persistence and the probability of extinction, defining these measures as aspects of stability as well. To avoid this confusion I propose that the behavior of ecological systems could well be defined by two distinct properties: resilience and stability.

Resilience determines the persistence of relationships within a system and is a measure of the ability of these systems to absorb changes of state variables, driving variables, and parameters, and still persist. In this definition resilience is the property of the system and persistence or probability of extinction is the result. Stability, on the other hand, is the ability of a system to return to an equilibrium state after a temporary disturbance. The more rapidly it returns, and with the least fluctuation, the more stable it is. In this definition stability is the property of the system and the degree of fluctuation around specific states the result.

Resilience versus Stability

With these definitions in mind a system can be very resilient and still fluctuate greatly, i.e. have low stability. I have touched above on examples like the spruce budworm forest community in which the very fact of low stability seems to introduce high resilience. Nor are such cases isolated ones, as Watt (41) has shown in his analysis of thirty years of data collected for every major forest insect throughout Canada by the Insect Survey program of the Canada Department of the Environment. This statistical analysis shows that in those areas subjected to extreme climatic conditions the populations fluctuate widely but have a high capability of absorbing periodic extremes of fluctuation. They are, therefore, unstable using the restricted definition above, but highly resilient. In more benign, less variable climatic regions the populations are much less able to absorb chance climatic extremes even though the populations tend to be more constant. These situations show a high degree of stability and a lower resilience. The balance between resilience and stability is clearly a product of the evolutionary history of these systems in the face of the range of random fluctuations they have experienced.

In Slobodkin's terms (36) evolution is like a game, but a distinctive one in which the only payoff is to stay in the game. Therefore, a major strategy selected is not one maximizing either efficiency or a particular reward, but one which allows persistence by maintaining flexibility above all else. A population responds to any environmental change by the initiation of a series of physiological, behavioral, ecological, and genetic changes that restore its ability to respond to subsequent unpredictable environmental changes. Variability over space and time results in variability in numbers, and with this variability the population can simultaneously retain genetic and behavioral types that can maintain their existence in low populations together with others that can capitalize on chance opportunities for dramatic increase. The more homogeneous the environment in space and time, the more likely is the system to have low fluctuations and low resilience. It is not surprising, therefore, that the commerical fishery systems of the Great Lakes have provided a vivid example of the sensitivity of ecological systems to disruption by man, for they represent climatically buffered, fairly homogeneous and self-contained systems with relatively low variability and hence high stability and low resilience. Moreover, the goal of producing a maximum sustained yield may result in a more stable system of reduced resilience.

Nor is it surprising that however readily fish stocks in lakes can be driven to extinction, it has been extremely difficult to do the same to insect pests of man's crops. Pest systems are highly variable in space and time; as open systems they are much affected by dispersal and therefore have a high resilience. Similarly, some Arctic ecosystems thought of as fragile may be highly resilient, although unstable. Certainly this is not true for some subsystems in the Arctic, such as Arctic frozen soil, self-contained Arctic lakes, and cohesive social populations like caribou, but these might be exceptions to a general rule.

The notion of an interplay between resilience and stabilty might also resolve the conflicting views of the role of diversity and stability of ecological communities. Elton (5) and MacArthur (22) have argued cogently from empirical and theoretical points of view that stability is roughly proportional to the number of links between species in a trophic web. In essence, if there are a variety of trophic links the same flow of energy and nutrients will be maintained through alternate links when a species becomes rare. However, May's (23) recent mathematical analyses of models of a large number of interacting populations shows that this relation between increased diversity and stability is not a mathematical truism. He shows that randomly assembled complex systems are in general less stable, and never more stable, than less complex ones. He points out that ecological systems are likely to have evolved to a very small subset of all possible sets and that MacArthur's conclusions, therefore, might still apply in the real world. The definition of stability used, however, is the equilibrium-centered one. What May has shown is that complex systems might fluctuate more than less complex ones. But if there is more than one domain of attraction, then the increased variability could simply move the system from one domain to another. Also, the more species there are, the more equilibria there may be and, although numbers may thereby fluctuate considerably, the overall persistence might be enhanced. It would be useful to explore the possibility that instability in numbers can result in more diversity of species and in spatial patchiness, and hence in increased resilience.

Measurement

If there is a worthwhile distinction between resilience and stability it is important that both be measurable. In a theoretical world such measurements could be developed from the behavior of model systems in phase space. Just as it was useful to disaggregate the reproduction curves into their constituent components of mortality and fecundity, so it is useful to disaggregate the information in a phase plane. There are two components that are important: one that concerns the cyclic behavior and its frequency and amplitude, and one that concerns the configuration of forces caused by the positive and negative feedback relations.

To separate the two we need to imagine first the appearance of a phase space in which there are no such forces operating. This would produce a referent trajectory containing only the cyclic properties of the system. If the forces were operating, departure from this referent trajectory would be a measure of the intensity of the forces. The referent trajectories that would seem to be most useful would be the neutrally stable orbits of Figure 2b, for we can arbitrarily imagine these trajectories as moving on a flat plane. At least for more realistic models parameter values can be discovered that do generate neutrally stable orbits. In the complex predator-prey model of Holling (14), if a range of parameters is chosen to explore the effects of different degrees of contagion of attack, the interaction is unstable when attack is random and stable when it is contagious. We have recently shown that there is a critical level of contagion between these extremes that generates neutrally stable orbits. These orbits, then, have a certain frequency and amplitude and the departure of more realistic trajectories from these referent ones should allow the computation of the vector of forces. If these were integrated a potential field would be represented with peaks and valleys. If the whole potential field were a shallow bowl the system would be globally stable and all trajectories would spiral to the bottom of the bowl, the equilibrium point. But if, at a minimum, there were a lower extinction threshold for prey then, in effect, the bowl would have a slice taken out of one side, as suggested in Figure 4. Trajectories that initiated far up on the side of the bowl would have amplitude that would carry the trajectory over the slice cut out of it. Only those

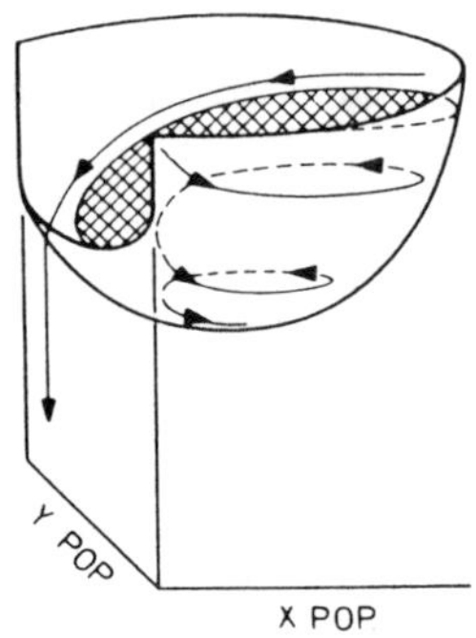

Figure 4 Diagramatic representation showing the feedback forces as a potential field upon which trajectories move. The shaded portion is the domain of attraction.

trajectories that just avoided the lowest point of the gap formed by the slice would spiral in to the bowl's bottom. If we termed the bowl the basin of attraction (Lewontin 20) then the domain of attraction would be determined by both the cyclic behavior and the configuration of forces. It would be confined to a smaller portion of the bottom of the bowl, and one edge would touch the bottom portion of the slice taken out of the basin.

This approach, then, suggests ways to measure relative amounts of resilience and stability. There are two resilience measures: Since resilience is concerned with probabilities of extinction, firstly, the overall area of the domain of attraction will in part determine whether chance shifts in state variables will move trajectories outside the domain. Secondly, the height of the lowest point of the basin of attraction (e.g. the bottom of the slice described above) above equilibrium will be a measure of how much the forces have to be changed before all trajectories move to extinction of one or more of the state variables.

The measures of stability would be designed in just the opposite way from those that measure resilience. They would be centered on the equilibrium rather than on the boundary of the domain, and could be represented by a frequency distribution of the slopes of the potential field and by the velocity of the neutral orbits around the equilibrium.

But such measures require an immeanse amount of knowledge of a system and it is unlikely that we will often have all that is necessary. Hughes & Gilbert (16), however, have suggested a promising approach to measuring probabilities of extinction and hence of resilience. They were able to show in a stochastic model that the distribution of surviving population sizes at any given time does not differ significantly from a negative binomial. This of course is just a description, but it does provide a way to estimate the very small probability of zero, i.e. of extinction, from the observed mean and variance. The configuration of the potential field and the cyclic behavior will determine the number and form of the domains of attraction, and these will in turn affect the parameter values of the negative binomial or of any other distribution function that seems appropriate. Changes in the zero class of the distribution, that is, in the probability of extinction, will be caused by these parameter values, which can then be viewed as the relative measures of resilience. It will be important to explore this technique first with a number of theoretical models so that the appropriate distributions and their behavior can be identified. It will then be quite feasible, in the field, to sample populations in defined areas, apply the appropriate distribution, and use the parameter values as measures of the degree of resilience.

APPLICATION

The resilience and stability viewpoints of the behavior of ecological systems can yield very different approaches to the management of resources. The stability view emphasizes the equilibrium, the maintenance of a predictable world, and the harvesting of nature's excess production with as little fluctuation as possible. The resilience view emphasizes domains of attraction and the need for persistence. But extinction is not purely a random event; it results from the interaction of random events with those deterministic forces that define the shape, size, and characteristics of the domain of attraction. The very approach, therefore, that assures a stable maximum sustained yield of a renewable resource might so change these deterministic conditions that the resilience is lost or reduced so that a chance and rare event that previously could be absorbed can trigger a sudden dramatic change and loss of structural integrity of the system.

A management approach based on resilience, on the other hand, would emphasize the need to keep options open, the need to view events in a regional rather than a local context, and the need to emphasize heterogeneity. Flowing from this would be not the presumption of sufficient knowledge, but the recognition of our ignorance; not the assumption that future events are expected, but that they will be unexpected. The resilience framework can accommodate this shift of perspective, for it does not require a precise capacity to predict the future, but only a qualitative capacity to devise systems that can absorb and accommodate future events in whatever unexpected form they may take.

Literature Cited

1. Baskerville, G. L. 1971. *The Fir-Spruce-Birch Forest and the Budworm*. Forestry Service, Canada Dept. Environ., Fredericton, N. B. Unpublished
2. Beeton, A. D. 1969. Changes in the environment and biota of the Great Lakes. *Eutrophication: Causes, Consequences, Correctives*. Washington DC: Nat. Acad. Sci.
3. Cooper, C. F. 1961. The ecology of fire. *Sci. Am.* 204:150–6, 158, 160
4. Edmondson, W. T. 1961. Changes in Lake Washington following increase in nutrient income. *Verh. Int. Ver. Limnol.* 14:167–75
5. Elton, C. S. 1958. *The Ecology of Invasions by Animals and Plants*. London: Methuen
6. Fujita, H. 1954. An interpretation of the changes in type of the population density effect upon the oviposition rate. *Ecology* 35:253–7
7. Gilbert, N., Hughes, R. D. 1971. A model of an aphid population—three adventures. *J. Anim. Ecol.* 40:525–34

8. Glendening, G. 1952. Some quantitative data on the increase of mesquite and cactus on a desert grassland range in southern Arizona. *Ecology* 33:319–28
9. Griffiths, K. J., Holling, C. S. 1969. A competition submodel for parasites and predators. *Can. Entomol.* 101:785–818
10. Hasler, A. D. 1947. Eutrophication of lakes by domestic sewage. *Ecology* 28: 383–95
11. Holling, C. S. 1961. Principles of insect predation. *Ann. Rev. Entomol.* 6:163–82
12. Holling, C. S. 1966. The functional response of invertebrate predators to prey density. *Mem. Entomol. Soc. Can.* 48: 1–86
13. Holling, C. S. 1965. The functional response of predators to prey density and its role in mimicry and population regulations. *Mem. Entomol. Soc. Can.* 45: 1–60
14. Holling, C. S., Ewing, S. 1971. Blind man's buff: exploring the response space generated by realistic ecological simulation models. *Proc. Int. Symp. Statist. Ecol.* New Haven, Conn.: Yale Univ. Press 2:207–29
15. Huffaker, C. D., Shea, K. P., Herman, S. S. 1963. Experimental studies on predation. Complex dispersion and levels of food in an acarine predator-prey interaction. *Hilgardia* 34:305–30
16. Hughes, R. D., Gilbert, N. 1968. A model of an aphid population—a general statement. *J. Anim. Ecol.* 40:525–34
17. Hutchinson, G. E. 1970. Ianula: an account of the history and development of the Lago di Monterosi, Latium, Italy. *Trans. Am. Phil. Soc.* 60:1–178
18. Larkin, P. A. 1971. Simulation studies of the Adams River Sockeye Salmon *(Oncarhynchus nerka). J. Fish. Res. Bd. Can.* 28:1493–1502
19. Le Cren, E. D., Kipling, C., McCormack, J. C. 1972. Windermere: effects of exploitation and eutrophication on the salmonid community. *J. Fish. Res. Bd. Can.* 29:819–32
20. Lewontin, R. C. 1969. The meaning of stability. *Diversity and Stability of Ecological Systems, Brookhaven Symp. Biol.* 22:13–24
21. Loucks, O. L. 1970. Evolution of diversity, efficiency and community stability. *Am. Zool.* 10:17–25
22. MacArthur, R. 1955. Fluctuations of animal populations and a measure of community stability. *Ecology* 36:533–6
23. May, R. M. 1971. Stability in multi-species community models. *Math. Biosci.* 12:59–79
24. May, R. M. 1972. Limit cycles in predator-prey communities. *Science* 177: 900–2
25. May, R. M. 1972. Will a large complex system be stable? *Nature* 238:413–14
26. Minorsky, N. 1962. *Nonlinear Oscillations.* Princeton, N.J.: Van Nostrand
27. Morris, R. F. 1963. The dynamics of epidemic spruce budworm populations. *Mem. Entomol. Soc. Can.* 31:1–332
28. Neave, F. 1953. Principles affecting the size of pink and chum salmon populations in British Columbia. *J. Fish. Res. Bd. Can.* 9:450–91
29. Nicholson, A. J., Bailey, V. A. 1935. The balance of animal populations—Part I. *Proc. Zool. Soc. London* 1935: 551–98
30. Ricker, W. E. 1954. Stock and recruitment. *J. Fish. Res. Bd. Can.* 11:559–623
31. Ricker, W. E. 1963. Big effects from small causes: two examples from fish population dynamics. *J. Fish. Res. Bd. Can.* 20:257–84
32. Roff, D. A. 1973. Spatial heterogeneity and the persistence of populations. *J. Theor. Pop. Biol.* In press
33. Rosenzweig, M. L., MacArthur, R. H. 1963. Graphical representation and stability condition of predator-prey interactions. *Am. Natur.* 97:209–23
34. Rosenzweig, M. L. 1971. Paradox of enrichment: destabilization of exploitation ecosystems in ecological time. *Science* 171:385–7
35. Rosenzweig, M. L. 1972. Stability of enriched aquatic ecosystems. *Science* 175: 564–5
36. Slobodkin, L. B. 1964. The strategy of evolution. *Am. Sci.* 52:342–57
37. Smith, S. H. 1968. Species succession and fishery exploitation in the Great Lakes. *J. Fish. Res. Bd. Can.* 25:667–93
38. Steele, J. H. 1971. Factors controlling marine ecosystems. *The Changing Chemistry of the Oceans,* ed. D. Dryssen, D. Jaquer, 209–21. Nobel Symposium 20, New York: Wiley
39. Walters, C. J. 1971. Systems ecology: the systems approach and mathematical models in ecology. *Fundamentals of Ecology,* ed. E. P. Odum. Philadelphia: Saunders. 3rd ed.
40. Wangersky, P. J., Cunningham, W. J. 1957. Time lag in prey-predator population models. *Ecology* 38:136–9
41. Watt, K. E. F. 1968. A computer approach to analysis of data on weather, population fluctuations, and disease. *Biometeorology, 1967 Biology Colloquium,* ed. W. P. Lowry. Corvallis, Oregon: Oregon State Univ. Press

42. Watt, K. E. F. 1960. The effect of population density on fecundity in insects. *Can. Entomol.* 92:674–95
43. Wellington, W. G. 1952. Air mass climatology of Ontario north of Lake Huron and Lake Superior before outbreaks of the spruce budworm and the forest tree caterpillar. *Can. J. Zool.* 30: 114–27
44. Wellington, W. G. 1964. Qualitative changes in populations in unstable environments. *Can. Entomol.* 96:436–51
45. Wellington, W. G. 1965. The use of cloud patterns to outline areas with different climates during population studies. *Can. Entomol.* 97:617–31

[2]

Ecological Applications, 6(4), 1996, pp. 1018–1024

BIOLOGICAL DIVERSITY, ECOSYSTEMS, AND THE HUMAN SCALE[1,2]

CARL FOLKE
The Beijer International Institute of Ecological Economics, The Royal Swedish Academy of Sciences, P.O. Box 50005, S-10405 Stockholm, Sweden, and Department of Systems Ecology, Stockholm University, Stockholm, Sweden

C. S. HOLLING
Department of Zoology, University of Florida, Gainesville, Florida 32611, USA

CHARLES PERRINGS
Department of Environmental Economics and Environmental Management, University of York, Heslington, York YO1 5DD, United Kingdom

Abstract. This paper considers the significance of biological diversity in relation to large-scale processes in complex and dynamic ecological–economic systems. It focuses on functional diversity, and its relation to production and maintenance of ecological services that underpin human societies. Within functional groups of organisms two important categories of species are identified: keystone process species and those essential for ecosystem resilience. The latter group represents "natural insurance capital." In addition to basic research on the interplay among biological diversity, functional performance, and resilience in complex self-organizing systems, we suggest that a functional approach has two main implications for a strategy for biodiversity conservation: (1) Biodiversity conservation to assure the resilience of ecosystems is required for all systems, no matter how heavily impacted they are. It should not be limited to protected areas. (2) The social, cultural, and economic driving forces in society that cause biodiversity loss need to be addressed directly. Specifically, (a) differences between the value of biological diversity to the private individual and its fundamental value to society as a whole need to be removed; (b) social and economic policies that encourage biodiversity loss should be reformed, especially where there is a risk of irreversible damage to ecosystems and diversity; and (c) institutions that are adaptive and work in synergy with ecosystem processes and functions are critical and should be created at all levels.

Key words: biodiversity; biological diversity, conservation of; critical ecosystem processes; disturbance, capacity to buffer; ecological services, maintenance of; ecosystem function and resilience; functional diversity; market externalities; multiple equilibria; nature reserves; transboundary effects.

INTRODUCTION

It is generally accepted that the significance of biodiversity includes much more than the mix of species. Wilson (1992:393) defines it to include ". . . the variety of ecosystems, which comprise both the communities of organisms within particular habitats and the physical conditions under which they live." This is also the position taken by the Convention on Biological Diversity of the United Nations Conference on Environment and Development in Rio de Janeiro 1992, signed by more than 150 nations (UNEP 1992).

Despite this, much research and policy on biodiversity conservation assumes that what matters is the number of genetically distinct organisms. There has been a tendency for biologists to estimate extinction rates, stress the importance of the abundance of taxa, and discuss the potential for preservation of genetic information (Olney et al. 1993, Prendergast et al. 1993, Smith et al. 1993). This approach has led to a conservation strategy dominated by the establishment of protected reserves in the mega-diversity regions of the world.

We strongly support the ethical argument for conserving the uniqueness and diversity of life. However, we question whether the focus on genotypic diversity and megadiversity hot-spots is the most appropriate way to conceptualize, analyze, or respond to biodiversity loss. Nor is it clear to us that it actually satisfies the ethical goal. We are suggesting an approach that complements the conventional approach, but which has a different focus. We will discuss biological diversity in large-scale ecological processes, and the intensifying human driving forces behind its loss. Specifically, we will analyze biodiversity loss in terms of its impact on the ability of interdependent ecological–economic systems to maintain functionality under a range of environmental conditions. Human demographic, social, cultural, and economic trends are not seen as external to

[1] Manuscript received 20 December 1994; revised 27 May 1995; accepted 21 June 1995; final version received 8 November 1995.

[2] For reprints of this group of papers, see footnote 1 on p. 975.

ecosystems, but as parts of the biogeochemical and hydrological flows of the ecosphere. That is, we take an ecological–economics approach.

Whether we like it or not the growing human impact on the planet is a fact. "Keeping humans out of nature" through a protected-area strategy may buy time, but it does not address the factors in society driving the loss of biodiversity. This paper stresses the importance of creating incentives for people and economies to act more in harmony than in conflict with essential processes that control the dynamics and structure of ecosystems, and of which biological diversity is a crucial part. Incentives need to be created to conserve biodiversity not just in protected areas but everywhere.

The approach requires that the functional relationship between the diversity of organisms and the set of ecological services on which humanity depends is addressed (Ehrlich and Mooney 1983, Perrings et al. 1992, 1995a). In addition to the fact that they house the genetic library, organisms help to sustain a flow of ecological services that are prerequisites for economic activities. These include photosynthesis, provision of food and other renewable resources, soil generation and preservation, pollination of crops, recycling of nutrients, filtering of pollutants and waste assimilation, flood control, climate moderation, operation of the hydrological cycle, and maintenance of the gaseous composition of the atmosphere. These functions sustain and protect human activities, and so human well being (Folke 1991, de Groot 1992, Ehrlich and Ehrlich 1992). In economic terms, ecosystems are fundamental "factors of production"—factors that are becoming increasingly scarce as a consequence of the rapid human population growth, and human behavior towards the natural capital base (Barbier et al. 1994, Jansson et al. 1994).

Increasing globalization of human activities and large-scale movements of people mean that humankind is in an era of novel co-evolution of ecological and socio-economic systems at regional and even planetary scales (Holling 1994). Ensuring the capacity of ecosystems to continue to generate ecological services on which the well being of human societies depends is a major challenge. Developing a feasible and useful strategy for biodiversity conservation is a central component of this challenge.

Functional Diversity in Large-scale Ecosystems

The critical aspect of the diversity of organisms and their environments in this context is functional diversity (Schulze and Mooney 1993). We know that for any ecosystem function to be sustained, a minimum composition of organisms is required to develop the relations between primary producers, consumers, and decomposers that mediate the flow of energy, the cycling of elements, and spatial and temporal patterns of vegetation. We also know that this composition varies with the environmental conditions in which the system operates. What is currently missing is detailed knowledge of the critical thresholds of diversity associated with different environmental conditions at different temporal and spatial scales. It is not clear, for example, how specific sets of genes, genotypes, species, populations, and communities influence ecosystem functions over the existing range of environmental conditions, or what the critical levels of diversity in communities and ecosystems and the factors that control them are (Solbrig 1991, Schulze and Mooney 1993). Nor is it clear how things would change with a change in environmental conditions. To a large extent this is a consequence of the fact that organisms and their environments are connected by a complex web of interrelations and feedbacks that are non-linear, and contain lags and discontinuities, thresholds, and limits (Kay 1991, Costanza et al. 1993).

Recent small-scale, multi-species experiments indicate that species deletion under given environmental conditions may lead to loss of function (Naeem et al. 1994, Tilman and Downing 1994). Such experiments have been undertaken in 1–4 m^2 plots or enclosures and contain species that live their life at those scales. It is not self-evident that results from small assemblages of species on small scales are valid for large-scale ecosystems.

Empirical findings from large-scale ecosystem studies of lakes, forests, marine and savanna ecosystems indicate that there are differencies in the link between species and functional diversity and the generation of ecosystem services. Studies of natural, managed, and disturbed or impacted ecosystems have shown that individual species population dynamics are more sensitive to stress than are ecosystem processes (Schindler 1990, Vitousek 1990). This implies that an ecosystem under stress may be expected to keep more of its functional performance than its species composition (Holling et al. 1995). This functional robustness is based on evidence that a relatively few processes, having distinct frequencies in space and time, structure ecosystems and set the rhythm of ecosystem dynamics (Holling 1992), a pattern that seems to be particularly true for terrestrial ecosystems (Holling et al. 1995). A limited number of organisms and groups of organisms seem to drive or control the critical processes necessary for ecosystem functioning, while the remaining organisms exist in the niches formed by these keystone process species. Such organisms modify, maintain, and create habitats. Jones et al. (1994) refer to them as "ecosystem engineers." (We prefer the term "keystone process species," to avoid confusion with the field in applied ecology called "ecological engineering" [e.g., Mitsch and Jorgensen 1989].) Nor is the set of such species necessarily constant over time, since it depends on environmental conditions (Lawton and Brown 1993, Holling et al. 1995).

As ecosystems are complex self-organizing systems, they are characterized by multiple locally stable equilibria or persistent states, each of which may correspond to a distinct set of environmental conditions, and may be controlled by a distinct set of keystone process species (Schneider and Kay 1994, Perrings et al. 1995*a*). But keystone process species alone will not guarantee the continuation of the ecosystem in question. It is in this context that the concept of ecosystem resilience becomes crucial in biodiversity conservation. "Resilience," as the term is conventionally used in ecology, refers to resistance to disturbance and speed of return to a stable equilibrium state (Pimm 1984)—what might be termed "efficiency of function." Following the work of Holling (1973), we use resilience to capture the existence of function in systems that, at any given moment, will be away from any one of a number of locally stable equilibrium states. Resilience in this sense is a measure of the perturbation that can be absorbed before an ecosystem in the attractor domain of one equilibrium state is dislodged into that of another equilibrium state. It is the capacity of the system to buffer disturbance. As an example of multiple equilibrial states, forest land cleared for agriculture in Amazonia, once abandoned, tends to develop into grassland or savanna and may not revert to the original forest. Different sets of interacting physical and biological processes and organisms control ecosystem functioning in each of the equilibrium states. Such phenomena have been documented in savanna ecosystems, and are also observed in boreal forests, fisheries, and agriculture (Regier and Baskerville 1986, Westoby et al. 1989, Trenbath et al. 1990, Gunderson et al. 1995).

The significance of functional diversity involving the controlling species in such cases may be illustrated by the semi-arid grasslands of eastern and southern Africa. There are (at least) two equilibrium states in these ecosystems, one dominated by grasses, and one by shrubs. The grassland state is controlled by two groups of grasses, each of which has different functions. One group contains species that are tolerant to grazing and drought, with the capacity to hold soil and water because of deep roots. The second group is more productive in terms of plant biomass than the first group, which makes it attractive to grazers, but is less drought tolerant. The species of the second group have a competitive advantage over the first group during periods when grazing is less intensive, and when rainfall is higher. The reverse occurs during transient but intense grazing pulses of migrating herbivores like zebras or antelopes. In this way, a diversity of both types of grass species is maintained in a manner that assures that the ecological functions underpinning ecosystem productivity are preserved, over a range of climatic conditions.

When fixed management rules are applied, such as the stocking of ranched cattle at a moderate but "sustained" level, the functional diversity of the system is gradually lost. This is due to a shift from the natural intense pulses of grazing to more modest but persistent grazing pressure. Continuous moderate grazing shifts the competitive advantage to the productive group of grass species, at the expense of the drought-resistant grasses. As a consequence, it slowly reduces diversity to one type of function, with the consequence that the system loses its capacity to function under as wide a range of climatic conditions. That is, the resilience contracts. An episode of drought that previously could be absorbed can flip the ecosystem into a state that is dominated and controlled by woody shrubs of low value for cattle ranching (Perrings and Walker 1995).

This sort of behavior is also observed in other managed systems (Ludwig et al. 1993, Gunderson et al. 1995). Many ecosystems evolve through management to become more spatially uniform, less functionally diverse, and more sensitive to disturbances that otherwise could have been absorbed. They have lost resilience.

Biological Diversity as Insurance

The problem to be addressed in developing a strategy for biodiversity conservation is that current institutions in society, including markets, do not respond to environmental feedbacks (Berkes and Folke 1994). That is, many of the most important environmental effects of human behavior are not recognized in the set of market prices. They are external to the market. The implication of this is that individual users of biological resources will not take the true cost of their actions into account. The problem is frequently exacerbated by governmental policies that, by subsidizing users of biological resources, deepen the wedge between the private and social cost of their behavior. Indeed, the failure of markets and the inability of government policies to correct the failure of markets can be seen as the prime driving force behind the loss of biodiversity.

The problem is also exacerbated by the fact that ecosystems themselves often "fail to signal" the long-term consequences of loss of resilience, continuing to function in the short term even as resilience declines. That is, large-scale ecosystems continue to function even when the composition and number of organisms is reduced. It seems like ecosystems frequently signal loss of resilience only at the point at which external shocks, previously absorbed by a diversity of organisms with overlapping influences, now flip those systems into some other regime of behavior, as in the grassland case above (Holling et al. 1996).

An important ecological objective of a biodiversity conservation strategy to avoid such flips is the protection of those organisms that establish and maintain the niches formed by keystone process species. These species provide a buffer to rare and extreme events. The vulnerability of key structuring processes is a function of the number of organisms that can take over and run such processes when the system is perturbed (Holling et al. 1995). Such organisms are essential for ecosystem resilience. Their loss implies a reduction in ecosystem plasticity and capacity for self-organization and evolution, which, in turn, threatens the capacity of the system to produce valuable ecological services that are needed for human existence. These organisms can therefore be viewed as "natural insurance capital" for securing the generation of ecological services, both at present and in the future (Barbier et al. 1994).

The insurance function of biodiversity certainly includes its role in the genetic evolution of microbial, plant, animal, and human life. But it also includes the capacity of the ecosystems in which those reservoirs are contained to function under current and future environmental conditions. When loss of biodiversity reduces ecosystem resilience, it threatens the functions of that system, and hence the foundation for economic activity and human welfare (Perrings et al. 1995*a*). In most cases, however, the reduction in the "insurance value" of biodiversity is not signalled in the incentive structures of human society, including the price mechanism. Nor is it within the reach of the policies of any one government.

The problem may be illustrated by the link between migratory insectivorous bird populations and changes in insect outbreaks in boreal regions of Canada. A set of insectivorous birds is one of the controlling factors of the forest renewal patterns produced by budworm population cycles. Their existence contributes to the resilience of the boreal forest. Simulations based on long-term studies of budworm–forest systems dynamics indicate that the total bird population would have to be reduced by ≈75% before the system would flip to a different pattern of behavior (Holling 1988). A large proportion of the bird species spend the winter in Central America and parts of South America. Radar images of flights of migratory birds across the Gulf of Mexico over a roughly 20-yr period have revealed that the frequency of trans-Gulf flights has declined by almost 50% (Gauthreaux 1992), approaching the range of uncertainty in the simulation estimate above. Hence, in addition to regional forest fragmentation and its negative effects on nesting success of migratory birds (Robinson et al. 1995), Canadian boreal forests and the economic activities dependent upon their functioning seem to be threatened by increasing land-use pressures in neo-tropical countries and along the migration paths of insectivorous birds (Holling 1994).

Similarly, the widespread cutting of mangrove ecosystems in Southeast Asia and South America for shrimp farming causes the loss of resilience of these coastal ecosystems to provide spawning and nursery grounds for fish and shellfish. In this case the degradation takes place in the coastal area of one country, but can cause reduced or lost yields of adult fish harvested in feeding grounds that belong to other countries. In neither case is it possible for the government of the country that is most affected to address the problem directly (Barbier et al. 1994).

Biodiversity loss through ecosystem modification is a new form of international environmental impact, a new type of "transboundary pollution." Many of the ecosystem modifications and the environmental effects that they cause, which today are regarded as of only local or national concern, are in fact of regional and ultimately global concern. The web of connections linking one ecosystem with the next is intensifying across all scales in both space and time. Local human influences on air, land, and oceans slowly accumulate to trigger sudden abrupt changes when thresholds are reached, directly affecting the generation of ecological services and the vitality of human societies elsewhere. Local environmental problems may have their cause half a world away, as illustrated by the above examples. Everyone is now in everyone else's backyard.

Implications for Biodiversity Conservation

What implications does this have for a strategy of biodiversity conservation? In addition to basic research to increase our understanding of the interplay among biological diversity, functional performance, and resilience in complex and nested self-organizing systems, we suggest that there are two main implications.

First, preserving biodiversity through nature reserves and other protected areas is an important short-term step, which we endorse, but it is not sufficient to solve the problem of biodiversity loss. Nature reserves are embedded in their larger environments. Most reserves cannot alone deal with ecological attributes that cover larger scales, such as those discussed in this article. For example, wading birds in the Everglades National Park depend on areas more than 10 times the size of the park. Conservation efforts should be planned at the scale of the regional landscape. Small reserves will lose their distinctive species if they are surrounded by a hostile landscape (Askins 1995). A hostile landscape is, generally, the result of increasing and intensifying human activities. Human influences at scales from the local ecosystem to the planet as a whole are becoming so extensive and pervasive that conditions that justified locating a protected area in one place originally may disappear over decades and move elsewhere. A biodiversity-conservation strategy that only focuses on preserving as much of the genetic library as possible in a few small areas may not be effective even in its own terms.

Second, we have argued for a strategy that aims at conserving the capacity of ecosystems to continue to deliver life-support and other ecological services to humanity under a wide range of environmental conditions. The role of biological diversity in the functioning of ecosystem performance is not limited to protected areas. Hence the conservation of biodiversity should be addressed everywhere. From this perspective, the heavily debated concept of "sustainability" might be interpreted as the maintenance of a level of biological diversity that will guarantee the resilience of ecosystems that sustain human societies. The goal of a conservation strategy should be to protect not all biodiversity in some areas, but biodiversity thresholds in all areas. Conserving essential self-organizing processes and the resilient patterns that they produce is the foundation for preservation of the biodiversity heritage (Holling et al. 1996). This approach has several implications.

1) To do this, it is necessary to identify the major social and economic forces that are currently driving the loss of functional diversity and to create incentives to redirect those forces. This includes both the proximate and the underlying forces (Perrings et al. 1992). The proximate forces refer to the direct reduction of biodiversity because of over-exploitation of species, land-use changes, and landscape fragmentation. The underlying forces include inappropriate government policies, the structure of property rights, pressure of human population growth and poverty, patterns of consumption and production, and the values of society. It is the underlying forces that need to be addressed in a strategy for biodiversity conservation (Barbier et al. 1994, Perrings et al. 1995*b*).

2) Following (1), it is necessary to create economic incentives that reduce the differences between the value of biological diversity to the private individual and its value to society as a whole. That is, it is necessary to internalize the external costs of biodiversity loss (in the language of economists). Internalizing only a few of the external costs of biodiversity loss is often enough to motivate the conservation of biodiversity.

3) A precondition for this is the development of effective institutions for biodiversity conservation. Institutions are the humanly devised constraints that shape human interaction. They structure incentives in human exchange, whether political, social, or economic, and shape the way societies evolve through time (North 1990). Institutions provide the framework for human actions, but to be effective they have to be adaptive (Gunderson et al. 1995). Being adaptive means, among other things, being able to respond to environmental feedbacks before those effects challenge the resilience of the entire resource base and the economic activities that depend on it. That is, it is necessary to frame the level of economic activity in a way that minimizes the risk of irreversible damage to the systems on which human activity depends (Perrings and Opschoor 1994). While many recent institutional innovations tend to do just the opposite, there are success stories from both traditional and contemporary societies from which we can learn (Feeny et al. 1990, Ostrom 1990, Berkes et al. 1995).

4) Rather than focusing attention on areas with the highest count of genetically distinct organisms, we should be stimulating the development of institutions, policies, and patterns of human consumption and production that work in synergy with ecosystem functions and processes. We should be reversing the trend towards large-scale and intensive monoculture. These simplified ecosystems are characterized by very low levels of diversity and even lower levels of resilience. They need to be redirected to mimic functional ecosystems (Jordan et al. 1987, Mitsch and Jörgensen 1989, Soulé and Piper 1992). In this way biodiversity may be conserved and even enhanced, while ecosystem resilience and the insurance value of biological resources may be increased. Nature reserves in both terrestrial and aquatic environments become important in this context, as a source of immigrant organisms for redevelopment and restoration of degraded areas.

Conclusion

Biodiversity loss includes not just the extinction of species, but any change in the mix of species and ecosystems that results from human activity and compromises the structuring processes upon which human well being and survival depend (Perrings et al. 1992, 1995*a*, *b*). This approach to biodiversity conservation supports a strategy that reorients biodiversity research and policy away from genetic information and the preservation of species for tourism and recreation in nature reserves, and towards conservation for the protection of ecosystem function and resilience. This complementary view of biodiversity implies that the benefits derived from its conservation are much wider and more fundamental than previously conceived both in science and in policy. It illuminates the fact that people depend on biological diversity for their well being and survival. The emphasis in research and conservation work should not be as closely tied, as it currently is, to protected areas in the mega-diversity zones. Biodiversity loss is a matter of consequence in any ecosystem in which resilience is threatened by the deletion of populations, irrespective of the species head-count in such systems. Moreover, the most powerful instrument for biodiversity conservation is not the park fence in isolation, but policies and reforms that make conservation a matter of private as well as social interest.

Acknowledgments

This article is one of the products of the research program on the Ecology and Economics of Biodiversity Loss, of the Beijer International Institute of Ecological Economics, The Royal Swedish Academy of Sciences. Perrings has served as the Director of the program. The article has been discussed in the Swedish Scientific Committee on Biological Diversity, and at courses in Systems Ecology and Ecological Economics. We are grateful to constructive criticisms we have received during those discussions, and to the reviewers for their input. Folke's work was in part financed by the Swedish Council for Forestry and Agricultural Research (SJFR).

Literature Cited

Askins, R. A. 1995. Hostile landscapes and the decline of migratory songbirds. Science **267**:1956–1957.

Barbier, E. B., J. Burgess, and C. Folke. 1994. Paradise lost? The ecological economics of biodiversity. Earthscan, London, England.

Berkes, F., and C. Folke. 1994. Investing in cultural capital for a sustainable use of natural capital. Pages 128–149 *in* A. M. Jansson, M. Hammer, C. Folke, and R. Costanza, editors. Investing in natural capital: the ecological economics approach to sustainability. Island Press, Washington, D.C., USA.

Berkes, F., C. Folke, and M. Gadgil. 1995. Traditional ecological knowledge, biodiversity, resilience and sustainability. Pages 281–299 *in* C. A. Perrings, K.-G. Mäler, C. Folke, C. S. Holling, and B.-O. Jansson, editors. Biodiversity conservation: problems and policies. Kluwer Academic Publishers, Dordrecht, The Netherlands.

Costanza, R., L. Waigner, C. Folke, and K.-G. Mäler. 1993. Modeling complex ecological economic systems: toward an evolutionary, dynamic understanding of people and nature. BioScience **43**:545–555.

de Groot, R. S. 1992. Functions of nature. Wolters Noordhoff BV, Groningen, The Netherlands.

Ehrlich, P. R., and A. E. Ehrlich. 1992. The value of biodiversity. Ambio **21**:219–226.

Ehrlich, P. R., and H. A. Mooney. 1983. Extinction, substitution and ecosystem services. BioScience **33**:248–254.

Feeny, D., F. Berkes, B. J. McCay, and J. M. Acheson. 1990. The tragedy of the commons: twenty-two years later. Human Ecology **18**:1–19.

Folke, C. 1991. Socioeconomic dependence on the life-supporting environment. Pages 77–94 *in* C. Folke and T. Kåberger, editors. Linking the natural environment and the economy: essays from the Eco-Eco group. Kluwer Academic Publishers, Dordrecht, The Netherlands.

Gauthreaux, S. A. 1992. The use of weather radar to monitor long-term patterns of trans-Gulf migration in spring. Pages 96–100 *in* J. M. Hagen III and D. W. Johnston, editors. Ecology and conservation of neotropical migrant landbirds. Smithsonian Institution Press, Washington, D.C., USA.

Gunderson, L., C. S. Holling, and S. Light, editors. 1995. Barriers and bridges to the renewal of ecosystems and institutions. Columbia University Press, New York, New York, USA.

Holling, C. S. 1973. Resilience and stability of ecological systems. Annual Review of Ecology and Systematics **4**:1–23.

———. 1988. Temperate forest insect outbreaks, tropical deforestation and migratory birds. Memoirs of the Entomological Society of Canada **146**:21–32.

———. 1992. Cross-scale morphology, geometry, and dynamics of ecosystems. Ecological Monographs **62**:447–502.

———. 1994. An ecologists view of the Malthusian conflict. Pages 79–103 *in* K. Lindahl-Kiessling and H. Landberg, editors. Population, economic development, and the environment. Oxford University Press, Oxford, UK.

Holling, C. S., G. Peterson, P. Marples, J. Sendzimir, K. Redford, L. Gunderson, and D. Lampert. 1996. Self-organization in ecosystems. Pages 346–384 *in* B. H. Walker and W. L. Steffen, editors. Global change and terrestrial ecosystem. Cambridge University Press, Cambridge, England.

Holling, C. S., D. W. Schindler, B. H. Walker, and J. Roughgarden. 1995. Biodiversity in the functioning of ecosystems: an ecological primer and synthesis. Pages 44-83 *in* C. A. Perrings, K.-G. Mäler, C. Folke, C. S. Holling, and B.-O. Jansson, editors. Biodiversity loss: ecological and economic issues. Cambridge University Press, Cambridge, England.

Jansson, A. M., M. Hammer, C. Folke, and R. Costanza, editors. 1994. Investing in natural capital: the ecological economics approach to sustainability. Island Press, Washington, D.C., USA.

Jones, C. G., J. H. Lawton, and M. Shachak. 1994. Organisms as ecosystem engineers. Oikos **69**:373–386.

Jordan, W. R., M. E. Gilpin, and J. D. Aber. 1987. Restoration ecology: a synthetic approach to ecological research. Island Press, Washington, D.C., USA.

Kay, J. J. 1991. A nonequilibrium thermodynamic framework for discussing ecosystem integrity. Environmental Management **15**:483–495.

Lawton, J. H., and V. K. Brown. 1993. Redundancy in ecosystems. Pages 255–270 *in* E.-D. Schulze and H. A. Mooney, editors. Biodiversity and ecosystem function. Springer, New York, New York, USA.

Ludwig, D., R. Hilborn, and C. Walters. 1993. Uncertainty, resource exploitation, and conservation: lessons from history. Science **260**:17, 36.

Mitsch, W. J., and S. E. Jörgensen, editors. 1989. Ecological engineering: an introduction to ecotechnology. John Wiley & Sons, New York, New York, USA.

Naeem, S., L. J. Thompson, S. P. Lawler, J. H. Lawton, and R. M. Woodfin. 1994. Declining biodiversity can alter the performance of ecosystems. Nature **368**:734–737.

North, D. C. 1990. Institutions, institutional change and economic performance. Cambridge University Press, Cambridge, England.

Olney, P. J. S., G. M. Mace, and A. T. C. Feistner, editors. 1993. Creative conservation. Chapman & Hall, London, England.

Ostrom, E. 1990. Governing the Commons: the evolution of institutions for collective action. Cambridge University Press, Cambridge, England.

Perrings, C. A., C. Folke, and K.-G. Mäler. 1992. The ecology and economics of biodiversity loss: the research agenda. Ambio **21**:201–211.

Perrings, C. A., and H. Opschoor. 1994. The loss of biological diversity: some policy implications. Environmental and Resource Economics **4**:1–11.

Perrings, C. A., K.-G. Mäler, C. Folke, C. S. Holling, and B.-O. Jansson, editors. 1995*a*. Biodiversity loss: ecological and economic issues. Cambridge University Press, Cambridge, England.

Perrings, C. A., K.-G. Mäler, C. Folke, C. S. Holling, and B.-O. Jansson, editors. 1995*b*. Biodiversity conservation: problems and policies. Kluwer Academic Publishers, Dordrecht, The Netherlands.

Perrings, C., and B. H. Walker. 1995. Biodiversity loss and the economics of discontinuous change in semi-arid rangelands. Pages 190–210 *in* C. A. Perrings, K.-G. Mäler, C. Folke, C. S. Holling, and B.-O. Jansson, editors. Biodiversity loss: ecological and economic issues. Cambridge University Press, Cambridge, England.

Pimm, S. L. 1984. The complexity and stability of ecosystems. Nature **307**:321–326.

Prendergast, J. R., R. M. Quinn, J. H. Lawton, B. C. Eversham, and D. W. Gibbons. 1993. Rare species, the coincidence of diversity hotspots and conservation strategies. Nature **365**:335–337.

Regier, H. A., and G. L. Baskerville. 1986. Sustainable redevelopment of regional ecosystems degraded by exploitive development. Pages 75–101 *in* W. C. Clark and R. E. Munn, editors. Sustainable development of the biosphere. Cambridge University Press, Cambridge, England.

Robinson, S. K., F. R. Thompson III, T. M. Donovan, D. R. Whitehead, and J. Faaborg. 1995. Regional forest fragmentation and the nesting success of migratory birds. Science **267**:1987–1990.

Schindler, D. W. 1990. Experimental perturbations of whole lakes as tests of hypotheses concerning ecosystem structure and function. Oikos **57**:25–41

Schneider, E., and J. J. Kay. 1994. Complexity and thermodynamics: towards a new ecology. Futures **24**:626–647.

Schulze, E.-D., and H. A. Mooney, editors. 1993. Biodiversity and ecosystem function. Springer, New York, New York, USA.

Smith, F. D. M., R. M. May, P. Pellew. T. H. Johnson, and K. S. Walter. 1993. Estimating extinction rates. Nature **364**:494–496.

Solbrig, O. T. 1991. The origin and function of biodiversity. Environment **33**:16–38.

Soulé, J. D., and J. K. Piper. 1992. Farming in nature's image: an ecological approach to agriculture. Island Press, Washington D.C., USA.

Tilman, D., and J. A. Downing. 1994. Biodiversity and stability in grasslands. Nature **367**:363–365.

Trenbath, B., G. Conway, and I. Craig. 1990. Threats to sustainability in intensified agricultural systems: analysis and implications for management. Pages 337–370 *in* S. R. Gliessman, editor. Agroecology. Springer, New York, New York, USA.

Vitousek, P. M. 1990. Biological invasions and ecosystem processes: towards an integration of population biology and ecosystem studies. Oikos **57**:7–13.

Westoby, M., B. H. Walker, and I. Noy-Meir. 1989. Opportunistic management for rangelands not at equilibrium. Journal of Rangeland Management **42**:266–274.

Wilson, E. O. 1992. The diversity of life. Belknap, Cambridge, Massachusetts, USA.

UNEP [United Nations Environment Programme]. 1992. Convention on biological diversity. United Nations Environment Programme, Nairobi, Kenya.

[3]

Biodiversity and the productivity and stability of ecosystems

Kris H. Johnson, Kristiina A. Vogt, Heidi J. Clark, Oswald J. Schmitz and Daniel J. Vogt

Attempts to unveil the relationships between the taxonomic diversity, productivity and stability of ecosystems continue to generate inconclusive, contradictory and controversial conclusions. New insights from recent studies support the hypothesis that species diversity enhances productivity and stability in some ecosystems, but not in others. Appreciation is growing for the ways that particular ecosystem features, such as environmental variability and nutrient stress, can influence biotic interactions. Alternatives to the diversity–stability hypothesis have been proposed, and experimental approaches are starting to evolve to test these hypotheses and to elucidate the mechanisms underlying the functional role of species diversity.

Kris Johnson, Kristiina Vogt, Heidi Clark, Oswald Schmitz and Daniel Vogt are at the Yale School of Forestry and Environmental Studies, 205 Prospect Street, New Haven, CT 06511, USA (kjohnson@minerva.cis.yale.edu).

Resolution of the relationships between the diversity of life forms occupying ecosystems and the behavior of those systems is a prime directive of ecological research. The problem ultimately encompasses all questions about how species coexist and how communities of populations influence ecosystem performance. Early theoretical discussions established the axiom that diverse, complex ecological communities are the most stable[1]. Results of some field studies in the 1960s and 1970s began to challenge the universality of this paradigm. Theoretical advances based on mathematical modelling[1] indicated that the nature of species interactions, rather than species number alone, determines the stability of ecological systems. Subsequently, several hypotheses about diversity and ecosystem function relationships have been proposed. Recently, growing concern over the loss of biodiversity and new empirical evidence has prompted revisitation of the idea that species diversity enhances the productivity and stability of ecosystems[2].

Hypotheses and theoretical foundations

Four prevalent hypotheses (in addition to the null model) are summarized in Box 1: the diversity–stability, rivet or rivet-popper, redundancy, and idiosyncratic response hypotheses.

The diversity–stability hypothesis[3] introduced the idea that increasing the number of trophically interacting species in an ecological community should increase the collective ability of member populations to maintain their abundances after disturbance. In his presentation of the hypothesis, MacArthur[3] implicitly recognized the transfer of energy from one trophic level to another as a quintessential ecosystem function. Biomass accumulation is the intuitively sensible, practical measure of energy assimilation, and field studies that followed used some measure of biomass as the ecosystem function response variable. The formalization of the diversity–stability hypothesis served as the original impetus for framing ecological questions in terms of the relationship between diversity and stability. The hypothesis came to be popularly acclaimed as a law of nature in spite of the fact that results of some field studies failed to support it[1].

The remaining hypotheses were introduced as alternatives to the diversity–stability hypothesis. The rivet hypothesis[4] suggests that ecosystem resistance – the ability of a system to absorb changes in abundances of some species without drastically changing ecosystem performance (e.g. biomass production)[5] – can decline as species are deleted even if system performance appears outwardly unaffected,

  PII: S0169-5347(96)10040-9 TREE vol. 11, no. 9 September 1996 with potentially sudden and drastic consequences. Ehrlich and Ehrlich[4] presumed that it may not be known ahead of time which, or how many, species deletions may cause ecosystem collapse (Box 1). The redundancy hypothesis, as presented by Walker[6], proposes that some species may be expendable in terms of ecosystem maintenance if any extant species can take the extinct species' functional place in the community. This idea originated with Ehrlich and Ehrlich[4] as part of the rivet analogy; Lawton[7] recently advocated a useful distinction between these two hypotheses (Box 1). The redundancy hypothesis can be viewed as a redirection of the rivet hypothesis towards explicit consideration of which 'rivets' are likely to be functionally expendable within a given scenario of previous deletions[6]. Most recently, Lawton[7] introduced the idiosyncratic hypothesis, which acknowledges that there may be no pattern or an indeterminate relationship between species diversity and ecosystem function. Results of experiments designed to identify the functional significance of species diversity can be interpreted to support one or more of these hypotheses.

The development of the alternative hypotheses traces back to the diversity–stability hypothesis by association. MacArthur[3] was primarily concerned with the temporal constancy of population numbers in the event of disturbance. The contemporary definition of ecological stability refers to the ability of a community to (1) resist change in its ability to maintain an ecosystem function (resistance), and (2) recover to normal levels of function after disturbance (resilience)[5]. This refinement of the meaning of ecological stability occurred partly through debate following MacArthur's proposal[1]. Walker[6] referenced the historic mathematical challenges to MacArthur's hypothesis[1] in justifying his articulation of the redundancy hypothesis. Lawton[7] considered the redundancy[6], rivet[4] and idiosyncratic[7] hypotheses together to address rates of ecosystem processes. McNaughton[8,9] and his colleagues extrapolated the diversity–stability hypothesis to the ecosystem level by treating primary production rate as the response variable in species diversity experiments. Tilman and Downing[10] explicitly state that the redundancy hypothesis is an alternative to the diversity–stability hypothesis. All of the hypotheses in Box 1 are, therefore, phylogenetically related, and have come to be accepted as contextually homologous by investigators of the functional significance of biodiversity[2,10–12]. Indeed, ecosystem resilience could very well serve as the 'process' for all of the graphs in Box 1.

Patterns in empirical evidence

Support for the idea that species diversity is crucial for maintaining ecosystem functions has come primarily from a microcosm experiment in a controlled laboratory facility called the Ecotron[11,13], and from plant community responses in field experiments conducted in herb-dominated systems in seasonal environments[8–10,14,15]. Examinations of other systems[16,17] have generated support for some of the alternative hypotheses (Box 1), suggesting that intrinsic characteristics of particular ecosystems influence the functional roles of their resident species and, therefore, of species diversity. Among field experiments, findings about the functional role of species diversity vary somewhat, depending on the choice of state variables, the way that those variables are measured, and functional distinction between the species involved[7,8,18].

Table 1 summarizes experimental approaches and the results of some classical investigations into the influence of species diversity on the productivity and stability of ecosystems, focusing mainly on terrestrial systems for brevity.

Box 1. Hypotheses for the functional role of species diversity in ecosystems

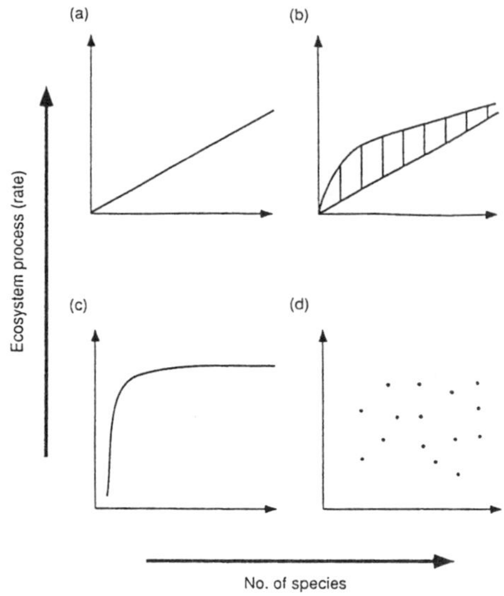

A number of hypotheses about the functional role of species diversity within ecological communities have been proposed.

Diversity–stability hypothesis: this hypothesis [represented in Fig. (a)] predicts that ecological communities will increase in energetic efficiency (productivity), and in ability to recover from disturbance, as the number of species in the system increases. According to this concept, deletion of any species from a food web will increase the susceptibility of the system to disruption via perturbation. MacArthur[3] proposed this hypothesis in 1955 under the premise that energy flow in complex food webs (meaning those featuring the greatest number of interspecific links or 'connectance') will be least disrupted by disturbance because alternative pathways for energy flow will be used. Elton's restatement of this hypothesis[42] is based on the same principle and is also frequently cited as the origin of the idea. Insofar as the diversity–stability hypothesis has been accepted as an alternative to any of the other hypotheses graphically portrayed here [Figs (b)–(d)][2,6,10–12], our graphical caricature [Fig. (a)] is logically derived. In its original form[3], the diversity–stability hypothesis did not include assertions of linearity in the effect of species richness on ecosystem function.

Rivet hypothesis: introduced by Ehrlich and Ehrlich[4] in 1981, this hypothesis [Fig. (b)] likens species in an ecosystem to rivets holding an airplane together – the removal of rivets beyond some threshold number may cause the airplane, or the ecosystem, to suddenly and catastrophically collapse. Explicit in their presentation is the appreciation that a few extinctions may go unnoticed in terms of system performance because some species may be redundant[4], generating a nonlinear relationship between species richness and ecosystem function.

Redundancy hypothesis: advanced by Walker[6] in 1992, the redundancy hypothesis [Fig. (c)] proposes that certain species have some ability to expand their 'jobs' in ecosystems in order to compensate for neighbor species that go extinct, *sensu* Ehrlich and Ehrlich[4]. At the heart of this concept is the idea that species may be segregated into functional groups; those species within the same functional group are predicted to be more expendable in terms of ecosystem function relative to one another than species without functional analogs.

Idiosyncratic hypothesis: Lawton[7] included this model [Fig. (d)] to address the possibility of a null or indeterminate relationship between species composition and ecosystem function. This relationship would be expected, for example, in communities featuring higher-order interactions.

All figures, except for part (a), are redrawn, with permission, from Ref. 7.

In most cases, species diversity increased and rate of biomass production decreased with successional time[9,10,19–21]. In the first of two experiments by Tilman and colleagues in Minnesota grasslands[10], nitrogen (N) fertilization ultimately resulted in a decline in species richness, so the plots that accumulated the greatest biomass (standing crop) also contained the fewest species. In a second experiment specifically designed to test for species diversity effects, Tilman *et al.* seeded plots with various numbers of grasses randomly selected from a pool of native species, and observed a positive diversity–productivity relationship[15]. In a novel experiment employing the Ecotron, Naeem *et al.* examined a suite of ecosystem functions relative to differences in randomly assigned species assemblages. One of their findings was that primary productivity, measured as light absorbed by plants and as carbon dioxide (CO_2) fixation, increased with plant species diversity[13].

Analyses of community stability also produced mixed results. The stability of plant productivity increased with species diversity in most studies[10,20,21], as did retention of mineral N in the most recent study from Minnesota[15]. However, California annual grasslands displayed no effect of species diversity on fluctuations in standing crop between years[9], while persistence of community types (measured as proportions of grasses, forbs and other plants) in British hay meadows declined with increasing diversity[20,21].

Tilman and Downing's report of diversity–stability relations in Minnesota grasslands[10] has received considerable attention[11,12,22,23]. This experiment was not originally designed to examine the role of species diversity in maintaining ecosystem function, and there was no control system for comparison[22]. However, the observation that aboveground biomass in species-rich plots changed less and recovered to initial levels faster than in species-poor plots (Table 1) was taken as support for the diversity–stability hypothesis. Since the theoretical mechanism behind the diversity–stability hypothesis (Box 1) is that additional species have different niches and consequently contribute to productivity in some additive fashion[8,10], it is interesting that overall biomass declined with increasing diversity in their study[10]. Such discrepancy between the mechanistic theory underlying the diversity–stability hypothesis and the presumably supportive empirical evidence testifies to the challenge posed by the pursuit of reliable knowledge of diversity and ecosystem function relationships. The disparity in the findings about diversity–productivity relationships between the Tilman and Downing study[10] and the more recent experiments from Minnesota[15] and the Ecotron[13] (Table 1) has not been addressed.

McNaughton[9] recently re-examined results of some early studies of New York old fields that included observations on the responses of consumer communities, as well as those of herbaceous plants, to pulsed nutrient enrichment (Table 1). Communities of arthropod herbivores and carnivores appeared, like plants, to increase in number and equitability (evenness) of species with successional time. Herbivores, unlike plants and carnivores, exhibited positive relationships between productivity and species diversity. In contrast to plants, the most diverse consumer communities were most sensitive to, and slowest to recover from, the environmental changes caused by fertilization. This illustrates that the nature of the perturbation and the selection of species for study can affect the outcome of diversity–stability experiments.

While the discrepancies apparent between theory and empirical evidence have led some toward skepticism of ardently theoretical dialogues about the functional role of diversity (see discussions by McNaughton[8] and Hairston[24]), it may be neglect of mechanistic thinking about species interactions relative to ecosystem function that fosters confusion. The basis upon which MacArthur[3] formed the diversity–stability hypothesis was implicitly a purely mechanistic blend of food web ecology and ecosystem ecology; as direct (consumer versus producer) links in a community became more numerous, interspecific compensatory responses (a type of indirect interaction) became more likely. Walker[6] considered such interpopulation compensation as the hallmark of redundancy. Frost *et al.*[25] have begun to formalize this concept mathematically, explicitly identifying negative interactions related to resource competition as likely indicators of compensatory potential. In the absence of such compensatory mechanisms, one or another of the alternative hypotheses (Box 1) is mechanistically more plausible.

Table 1. Summary of selected diversity/ecosystem function studies[a]

Ecosystem	Disturbance type	Diversity measure	Productivity measure	Stability measure	Diversity/ productivity relationship	Diversity/ stability relationship
California annual grassland[9]	annual variation	S[a]	SC[a] (aboveground)	SC(NPP)$^{-1}$	plants (–)	plants (0)
New York old fields[9]	annual variation; N–P–K fertilizer	S/$\log_2$ N; S and H′	NPP[a] (above- and belowground)	ΔNPP	plants (–) herbivores (–) carnivores (–),(0)	plants (+) herbivores (–) carnivores (–)
Serengeti[8]	Δ precipitation; grazing	H′	NPP (aboveground)	ΔSC	NA[a]	plants (+)
Yellowstone grasslands[14]	drought; grazing	H′	SC (aboveground)	Δ relative abundances	NA	plants (+)
British grasslands[20,21]	fertilization; mowing; Δ precipitation	S	SC (aboveground)	Δ vegetation; composition[20] ΔSC[20,21]	plants (–)	plants (–)[20], (+)[20,21]
Minnesota grasslands[10,15]	drought[10]; ΔS[15]	S	SC (aboveground)	ΔSC; recovery rate[b]; NA[15]	plants (–)[10], (+)[15]	plants (+)[10] NA[15]
Costa Rican tropical forest[19]	annual variation	S	soil fertility[c]	Δ soil fertility	NA	plants (+)
Ecotron[13]	S	S	NPP	NA	plants (+)	NA

[a]S, species richness; SC, standing crop; NPP, net primary productivity; N, total community biomass; Δ, change in parameter; H′, Shannon index; NA, not assessed; +, positive association; –, negative association; 0, no relationship.
[b]Resistance was measured as $\ln(\partial b/B\partial t)$ and resilience was measured as $\ln(b/B)$, where b is biomass at time t after disturbance and B is biomass before disturbance.
[c]Components of soil fertility included: available nitrogen (N), phosphorous (P) and potassium (K) content; and pH and cation exchange capacity.

The question of how important species diversity is to ecosystem function boils down to consideration of how species in that system interact[6,25]. The relatively consistent finding from the studies of old fields and grasslands summarized in Table 1 that stability increases with plant species diversity reflects influences of seasonality on species interactions. Herbaceous species in seasonal environments exhibit phenology reflecting seasonality in availability of limiting resources[26], especially water, N and phosphorous (P). Competitive coexistence is achieved with temporal resource partitioning through phenological differences that result in a sort of relay from one species to another in dominance of the vegetation throughout a season[8,26]. Periodic disturbances can also be important since they can 'reset' a seral system before competitive dominants extinguish competitor populations[16]. Finally, in systems regulated by consumers, the ability of herbivores to change their diets in response to changes in relative plant species abundances and antiherbivore defenses also influences community dynamics[27]. In this manner, we see that the intrinsic characteristics of an ecosystem influence the feedback interactions among its resident species. And, in the case of most seasonal herb communities, a positive correlation between species diversity and functional stability is predictable.

Observations about other ecosystems

Because the living and abiotic elements interacting together necessarily vary between ecosystems, the nature of the interplay of those elements must also vary between systems[28]. The functions of species in ecosystems are therefore expected to be system dependent. While this concept seems self evident, and has long been acknowledged in discussions of other classes of ecological experiments[29,30], it has largely escaped mention in recent reports of diversity/productivity/stability studies. This has resulted in liberal extrapolation of results from some of the studies mentioned above to the functional role of species diversity in general[11]. However, research from other systems suggests that ecosystem functions may not be universally dependent on species diversity.

For example, Cowling *et al.*[16] have contrasted the growth-form redundancy of plant species occupying South Africa's fire-frequented Cape shrublands, called fynbos, with that of adjacent forests. Fynbos vegetation is characteristically flammable owing to soil infertility[16], and the associated disturbance regime profoundly affects the way that life forms interact with one another and with the environment. Conversely, the adjacent forest ecosystems characteristically have a longer disturbance frequency and tend to achieve a greater degree of successional equilibrium. Cowling *et al.*[16] classified plants into 11 functional groups based on growth form, and compared the growth-form richness and species richness for the two ecosystem types. These groupings can be interpreted as surrogates for measures of trophic niche width under the assumption that growth-form is related to resource acquisition strategy.

The nutrient poverty of fynbos should set the stage for intense exploitative competition for resources between co-occurring species with overlapping resource needs. Selection would result either in niche partitioning and, ultimately, adaptive radiation, or competitive exclusion, if the disturbance frequency was long enough to allow communities to approach equilibrium. This scenario would result in phenotypically distinctive, functionally unique species with either exclusive fundamental or realized niches[31]. However, the disturbance frequency mediates exploitative competition between plants in fynbos[16], rendering some prominent plant species functionally redundant. Thus, Cowling *et al.*[16] argued that functional redundancy apparently increased as disturbances became more frequent. Perturbation experiments for explicitly testing the fynbos redundancy hypothesis with regard to primary productivity and nutrient cycling have not been performed, but are certainly feasible following the approach of McNaughton[8].

Environmental factors other than disturbance also can influence the functional redundancy of different species. Vitt and Chee[17] found that vascular flora do not drive ecosystem processes in Canadian boreal peat bogs. Rather, vascular-plant communities are downstream by-products of geochemical processes largely independent of biotic activity within the bogs. Plant communities assemble along gradients of pH (bryophytes) and organic macronutrients (vascular plants). Nutrient supply depends on the chemistry of the water flowing into the peat bogs[17]. Nitrogen and P have been found to be particularly limiting to plant productivity in these systems[17].

Vitt and Chee[17] classified bogs into three types: low, moderate and high mineral (elemental cations) content. Plant species richness was greatest for the moderate class bogs, which were also the most temporally and spatially variable with regard to water chemistry, especially N concentration. Thus, plant species richness did not correspond to ecosystem dynamics in a manner supportive of the diversity–stability hypothesis. An idiosyncratic or null relationship between species diversity and ecosystem performance, especially with regard to nutrient fluxes, better describes such bog vascular-plant communities.

Are functional distinctions strongest at the species level?

Central to questioning the role of species diversity in maintaining ecosystem function is whether or not species designations best distinguish functional groupings. The important problem lies in determining where in taxonomic hierarchy functionally distinct phenotypes exist. Phenotypes have a strong genetic determinant, and it is phenotype that translates into functional role. Taxonomy itself has continually evolved to embrace new technology and conventions in classifying biota. With recent advances in protein, chromosome and DNA isolation, genetics has become increasingly important in species determination. However, explicit functional distinction has not yet become an important taxonomic criterion for species identification[32]. Some species designations continue to be controversial and dynamic. For this reason alone, species names seem unreliable as a basis for differentiating functional groups.

Examples have been described where either genetic or functional differences are prominent between taxonomic levels other than species. Cossalter and Boyle[33] have found that populations of the tropical legume *Pterocarpus macrocarpus* differ from one another in resistance to drought stress. As another example, limber pine (*Pinus flexilis*) along the eastern foothills of the central Rocky Mountains exhibits a multi-trunk growth form and, until recently, each cluster was considered a single, distinct individual. Carsey and Tomback[34], however, discovered that the trunk clump actually develops from caches of variably related seeds constructed by the bird, Clark's nutcracker (*Nucifraga columbiana*). In this case, unusually high genetic variability occurs within and between individual tree clumps. Additionally, Duarte *et al.*[32] recently pointed out that plants of different ancestry above the species level can be functionally alike. These examples from plant ecology show that the genetically bounded phenotypic distinctions that determine functional uniqueness may not conform simply to current systematic distinctions at the species level.

MacGillivray and Grime[5] are developing a systematic approach for functional classification of plants. Their technique employs principle component analysis of morphological traits in order to group species into functional types. The traits reflect life history strategies related to stress tolerance and ability to compete for nutrients. To test the notion that resistance and resilience of community productivity to various disturbances should be predictable based on their classification criteria, they subjected communities of grass species in various functional groups to an array of realistic disturbances (drought, frost and fire). Results from their field trials (not included in Table 1) generally confirmed predictions that resistance would increase and resilience would decrease as community types changed from ruderals to slow-growing, long-lived competitive dominants, as expected during succession[5].

Effects of phenotypic plasticity and mycorrhizae on plant species redundancy

Phenotypic plasticity refers to the pronounced influence of environment over expression of the genes that determine growth-form or behavior. The presence of phenotypic plasticity can blur functional distinctions between competing species, especially those closely related, rendering them functionally redundant[32]. Plasticity, which appears to be inversely related to genetic heterozygosity[35], tends to be selectively advantageous, and therefore most common, where the environment varies unpredictably[33]. Thus, where competition might result in competitive exclusion or adaptive radiation into functionally unique life forms, phenotypic plasticity provides a plausible mechanism for long-term persistence of redundant species. This is a pattern that might be expected between generalists with overlapping resource requirements.

Mycorrhizal associations also muddy the functional distinctions between species because they buffer plants from environmental stresses, especially nutrient deficiencies, that might otherwise instigate intense competition. For instance, different kinds of trees in mixed-species stands host many of the same generalist species of mycorrhizal fungi, some of which help each host tree to acquire N and P (Ref. 36). The fungal symbioses ease interspecific competition by homogenizing the distribution of limiting nutrients, and by equalizing nutrient acquisition capabilities. Such effects certainly increase functional redundancy (in terms of primary productivity) of tree species, and appear strongest for systems in nutrient poor, seasonably variable environments[36,37]. Coexistence of plant species under such conditions may be attributable to processes other than divergence of trophic niches. In such cases, species redundancy should be relatively high, so tests of the species diversity–stability hypothesis would probably produce different results to those reported for Minnesota grasslands[10]. Further study of the mediation of influence of species diversity on ecosystem attributes by mycorrhizae and other mechanisms begetting functional plasticity can contribute immensely to empirical resolution of patterns of diversity/productivity/stability relationships.

Directions for the future

It is noteworthy, given the importance of belowground processes to ecosystem function[5,38] – especially in grasslands[19] – that roots have received little attention in the studies reported in Table 1. Only in New York old fields[9] was belowground production monitored. Focusing on belowground responses to perturbation can help ecologists detect ecosystem-level changes because root morphology and biomass are particularly sensitive indicators of environmental stress[38].

Furthermore, scientists focusing on interannual variation in ecosystem attributes between experimental communities under natural field conditions are resigned somewhat to descriptive analyses because climatic effects are largely uncontrollable. Explicit tests of diversity–ecosystem stability hypotheses, like all experiments, require that observations about ecosystem function be made relative to variation in independent variables, and that controls exist for those independent variables. Interannual change in environment and ecosystem attributes can occur for a system independently of deviations from ground state[18]. Assessment of deviation from ground state can be made only by comparing observed responses in treatments with those in real controls[18]. The Ecotron offers unique potential to examine ecosystem attributes under carefully controlled conditions, and the most recent experiment by Tilman *et al.*[15] exemplifies a controlled design for examining the effects of species diversity manipulations under field conditions.

Finally, it is clear that identification of functionally distinct groups of organisms is necessary if ecologists are to generate a priori predictions about ecosystem sensitivity to extinctions[5,6]. For nutrient-related ecosystem attributes, such as primary productivity, functional groupings will reflect resource acquisition strategies of member species. Studies of food webs have sometimes blackboxed species into functional groups based on trophic levels[39–41]. However, this approach may lead to underestimating functional differences between species within a trophic level, as might be seen with a keystone predator. In terms of ecosystem functions involving the processing of nutrients, it is likely that functionally unique species will be trophic specialists and will exhibit relatively non-plastic phenotypes. Plasticity, which should be an evolutionarily stable strategy where resource abundances fluctuate unpredictably, can be genetically programmed, or achieved through symbiotic interactions between individuals of different species. Functional redundancy between species is likely to be highest where plasticity or other mechanisms facilitate coexistence of species that otherwise compete for consumable resources.

Reliable knowledge about the functional role of species diversity in ecosystems can only come from cause–effect resolution through meaningful hypothesis-testing in controlled experiments[15,18,24]. The difficulty of designing and implementing controlled experiments for hypothesis testing is daunting. Most studies to date, with some exceptions[8,13,15,19], have generated only correlative conclusions about diversity–ecosystem function relations because species diversity has rarely been manipulated independently of ecosystem attributes like succession and total nutrient abundance. The most compelling evidence to date that species diversity can improve productivity under certain conditions comes from the innovatively designed experiments of Naeem *et al.* in the Ecotron[13] and Tilman *et al.* in grasslands[15]. These conclusions will not apply to all ecosystems[13]. The challenge posed by the task of identifying functionally distinct groups[6] for a given ecosystem is formidable in itself, but is also necessary to undertake in order to generate meaningful predictions about the effect of particular species deletions on system performance[5]. Experiments designed to determine the types and strengths of interspecific interactions are suggested to be the best path to take in determining which species have the potential to compensate at the population level for declines in or extinctions of other species for specific ecosystems[6,25].

The array of hypotheses in Box 1 illustrates a growing appreciation that species diversity can influence the stability and productivity of ecosystems in a variety of ways. Our understanding of these relationships will continue to progress as ecologists identify the mechanisms determining the degree to which taxa perform unique and critical functions. Identifying the taxonomic level at which functional distinctions might exist for a given ecosystem remains a key consideration in this pursuit. The approach to functional ordination based on morphological traits currently being developed by MacGillivray and Grime[5] offers promise as a means of predicting the functional importance of diversity within the context of a specific ecosystem.

Acknowledgements

This manuscript was greatly improved over earlier drafts by following suggestions made by John Lawton, Maria Uriarte, Andrew Beckerman and an anonymous reviewer. This review was supported in part by National Science Foundation grants to Yale University, USA under the Terrestrial Ecology Program with the University of Puerto Rico and the International Institute of Tropical Forestry, US Forest Service, as part of the Luquillo Experimental Forest Long-Term Ecological Research funding to K.A.V. and D.J.V., and by grant NSF-DEB-9508604 to O.J.S.

References

1 Shugart, H.H. and O'Neill, R.V., eds (1979) **Editors' comments on papers 12–15** in *Systems Ecology* (Benchmark Papers in Ecology, Vol. 9), pp. 167–168, Dowden, Hutchinson and Ross

2 Baskin, Y. (1995) **Ecosystem function of biodiversity,** *BioScience* 44, 657–660

3 MacArthur, R. (1955) **Fluctuations of animal populations and a measure of community stability,** *Ecology* 36, 533–536

4 Ehrlich, P.R. and Ehrlich, A.H. (1981) *Extinction. The Causes and Consequences of the Disappearance of Species*, Random House

5 MacGillivray, C.W. and Grime, J.P. (1995) **Testing predictions of the resistance and resilience of vegetation subjected to extreme events,** *Funct. Ecol.* 9, 640–649

6 Walker, B. (1992) **Biodiversity and ecological redundancy,** *Conserv. Biol.* 6, 18–23

7 Lawton, J. (1994) **What do species do in ecosystems,** *Oikos* 71, 367–374

8 McNaughton, S.J. (1977) **Diversity and stability of ecological communities: a comment on the role of empiricism in ecology,** *Am. Nat.* 111, 515–525

9 McNaughton, S.J. (1993) **Biodiversity and function of grazing ecosystems** in *Biodiversity and Ecosystem Function* (Ecological Studies, Vol. 99) (Schulze, E-D. and Mooney, H.A., eds), pp. 361–383, Springer-Verlag

10 Tilman, D. and Downing, J.A. (1994) **Biodiversity and stability in grasslands,** *Nature* 367, 363–365

11 Kareiva, P. (1994) **Diversity begets productivity,** *Nature* 368, 686–687

12 Collins, S. (1995) **The measurement of stability in grasslands,** *Trends Ecol. Evol.* 10, 95–96

13 Naeem, S. *et al.* (1994) **Declining biodiversity can alter the performance of ecosystems,** *Nature* 368, 734–737

14 Frank, D.A. and McNaughton, S.J. (1991) **Stability increases with diversity in plant communities: empirical evidence from the 1988 Yellowstone drought,** *Oikos* 62, 360–362

15 Tilman, D., Wedin, D. and Knops, J. (1996) **Productivity and sustainability influenced by biodiversity in grassland ecosystems,** *Nature* 379, 718–720

16 Cowling, R.M. *et al.* (1994) **Species diversity; functional diversity and functional redundancy in fynbos communities,** *South Afr. J. Sci.* 90, 333–337

17 Vitt, D.H. and Chee, W-L. (1990) **The relationship of vegetation to surface water chemistry and peat chemistry in fens of Alberta, Canada,** *Vegetatio* 89, 87–106

18 Smedes, G.W. and Hurd, L.E. (1981) **An empirical test of community stability: resistance of a fouling community to a biological patch-forming disturbance,** *Ecology* 62, 1561–1572

19 Ewel, J.J., Mazzarino, M.J. and Berish, C.W. (1991) **Tropical soil fertility changes under monocultures and successional communities of different structure,** *Ecol. Appl.* 1, 289–302

20 Silvertown, J. *et al.* (1994) **Rainfall, biomass variation, and community composition in the Park Grass Experiment,** *Ecology* 75, 2430–2437

21 Dodd, M. *et al.* (1994) **Community stability: a 60-year record of trends and outbreaks in the occurrence of species in the Park Grass Experiment,** *J. Ecol.* 83, 277–285

22 Givnish, T.J. (1994) **Does diversity beget stability?** *Nature* 371, 113–114

23 Walker, B. (1995) **Conserving biological diversity through ecosystem resilience,** *Conserv. Biol.* 9, 747–752

24 Hairston, N.G. (1989) **Ecological problems and how they are approached** in *Ecological Experiments: Purpose, Design, and Execution*, pp. 1–22, Cambridge University Press

25 Frost, T.M. *et al.* (1995) **Species compensation and complementarity in ecosystem function** in *Linking Species and Ecosystems* (Jones, C.G. and Lawton, J.H., eds), pp. 224–239, Chapman & Hall

26 Sarmiento, G. (1984) **The seasonal rhythms of the savanna species** in *The Ecology of Neotropical Savannas*, pp. 43–74, Harvard University Press

27 Huntly, N. (1991) **Herbivores and the dynamics of communities and ecosystems,** *Annu. Rev. Ecol. Syst.* 22, 477–503

28 Steele, J.H. (1991) **Marine functional diversity: ocean and land ecosystems may have different time scales for their responses to change,** *BioScience* 41, 470–474

29 Oksanen, L. *et al.* (1981) **Exploitation ecosystems in gradients of primary productivity,** *Am. Nat.* 118, 240–261

30 Steele, J.H. (1985) **Comparison of marine and terrestrial ecological systems,** *Nature* 313, 355–358

31 Grant, R. and Grant, P. (1993) **Evolution of Darwin's finches caused by a rare climatic event,** *Proc. R. Soc. London Ser. B* 251, 111–117

32 Duarte, C.M. *et al.* (1995) **Comparative functional plant ecology: rationale and potentials,** *Trends Ecol. Evol.* 10, 418–421

33 Cossalter, C. and Boyle, T. in *Management of Soil, Water and Nutrients in Tropical Plantation Forest,* (Nambiar, S., Brown, A. and Cossalter, C., eds), Australian Center for International Agricultural Research (in press)

34 Carsey, K.S. and Tomback, D.F. (1994) **Growth form distribution and genetic relationships in the tree cluster of *Pinus flexilis*, a bird-dispersed pine,** *Oecologia* 98, 402–411

35 Knowles, P. and Grant, M.C. (1981) **Genetic patterns associated with growth variability in Ponderosa pine,** *Am. J. Bot.* 68, 942–946

36 Perry, D.A., Bell, T. and Amaranthus, M.P. (1992) **Mycorrhizal fungi in mixed-species forests and other tales of positive feedback, redundancy and stability** in *The Ecology of Mixed-Species Stands of Trees* (Cannell, M.G.R., Malcom, D.C. and Robertson, P.A., eds), pp. 151–179, Blackwell

37 Harley, J.L. and Smith, S.E. (1983) *Mycorrhizal Symbiosis*, Academic Press

38 Vogt, K. *et al.* (1993) **Belowground responses as indicators of environmental change,** *Environ. Exp. Bot.* 33, 189–205

39 Havens, K.E. (1994) **Experimental perturbation of a freshwater plankton community: a test of hypotheses regarding the effects of stress,** *Oikos* 69, 147–153

40 Grime, J.P. (1995) **Functional types: a comment on Steneck and Dethier,** *Oikos* 73, 120–121

41 Steneck, R.S. and Dethier, M.N. (1995) **A functional group approach to the structure of algal-dominated communities,** *Oikos* 69, 476–498

42 Elton, C.S. (1958) **The reasons for conservation** in *The Ecology of Invasions by Animals and Plants,* pp. 143–153, Chapman & Hall

[4]

The Future of Biodiversity

Stuart L. Pimm,* Gareth J. Russell, John L. Gittleman, Thomas M. Brooks

Recent extinction rates are 100 to 1000 times their pre-human levels in well-known, but taxonomically diverse groups from widely different environments. If all species currently deemed "threatened" become extinct in the next century, then future extinction rates will be 10 times recent rates. Some threatened species will survive the century, but many species not now threatened will succumb. Regions rich in species found only within them (endemics) dominate the global patterns of extinction. Although new technology provides details of habitat losses, estimates of future extinctions are hampered by our limited knowledge of which areas are rich in endemics.

Debates about the consequences of human population growth are not new. Our numbers have increased dramatically since Malthus but so has our technology (*1*). Will technical ingenuity keep pace with increasing population problems? Ingenuity can replace a whale-oil lamp with an electric light bulb, but not the whales we may hunt to extinction. Species matter to us (*2*). How fast we drive them to extinction is a matter of our future. Critics consider high estimates of current and future extinction rates to be "doomsday myths," contending that it is the "facts, not the species" that are endangered (*3*). Here, we review these estimates.

Extinctions have always been a part of Earth's history. So what is the background rate of extinction: how fast did species disappear in the absence of humanity (*4*)? A summary of 11 studies of marine invertebrates suggests that fossil species last from 10^6 to 10^7 years (*5*). For ease of comparison, we use the number of extinctions (*E*) per 10^6 species-years (MSY) or *E*/MSY. If species last from 10^6 to 10^7 years, then their rate of extinction is 1 to 0.1 *E*/MSY.

These estimates derive from the abundant and widespread species that dominate the fossil record. The species most prone to current extinction are rare and local. Moreover, we emphasize terrestrial vertebrates in our discussions of current extinctions. There are only two studies of their fossils (*5*), and these suggest high background rates (~1 *E*/MSY). Interestingly, we can supplement these estimates from our knowledge of speciation rates. These could not be much less than the extinction rates, or the groups would not be here for us to study.

Molecular phylogenies are now produced rapidly and extensively. There is one for 1700 bird species (*6*). Using the relative time axis of molecular distances, we can elucidate the patterns of species formation. Models in which every lineage has the same, constant probability of giving birth to a new lineage (speciation) or going extinct (death) permit estimation of the rate parameters (*7*). The rich details of this approach offer hope in testing for important factors controlling the relative rates of background speciation and extinction. Obviously, absolute rates require accurately dated events, such as the first appearance of a species or genus in the fossil record. There are genetic distance and paleontological estimates of divergence times for 72 carnivore and 14 primate species or subspecies (*8*). Given their importance as a benchmark against which to compare modern extinction rates, we plead for more absolutely timed accounts.

How Many Species Are There?

Any absolute estimate of extinction rate requires that we know how many species there are. In fact, we do not. May (*9*) shows that the problems of estimating their numbers are formidable. Only ~10^6 species are described and <10^5—terrestrial vertebrates, some flowering plants, and invertebrates with pretty shells or wings—are popular enough to be known well. Birds are exceptional in that differences in taxonomic opinion [~8500 to 9500 species (*6*)] far exceed the annual descriptions of new species (~1). Most species are as yet undescribed in every species-rich group (Fig. 1). Major uncertainties lie in those groups in which we have scant or conflicting evidence of very high diversity. There are ~10^6 described insects, yet estimates range from ~10^7 to nearly 10^8 species. Some potentially rich communities, such as the deep-sea benthos, have been sparsely sampled.

How can we be confident in our extrapolations of extinction rates from the <10^5 well-known species to the ~10^6 described, or to the conservative grand total of ~10^7 (*5*)? If extinction rates in diverse taxa and regions are broadly similar, then they are

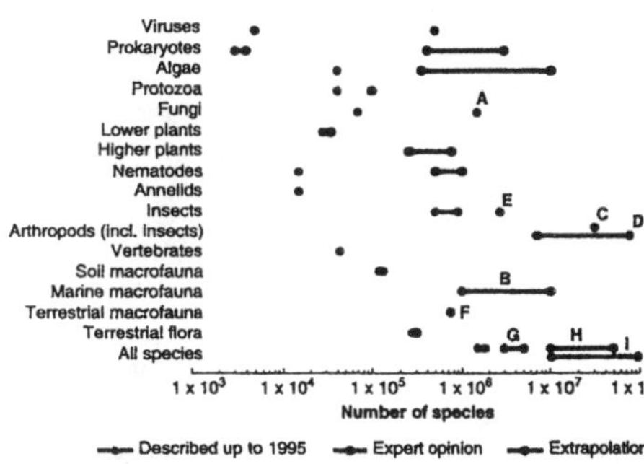

Fig. 1. Numbers of described species and estimates of species numbers, including expert opinions of taxonomic specialists (*12*) and various extrapolations (*26*). **(A)** The British ratio of 6 fungi species:1 plant species and a world total of 2.5 × 10^5 plant species suggests ~1.5 million species of fungi worldwide. **(B)** A world total of 10^6 to 10^7 species of marine macrofauna comes from the accumulation of new species along sample transects. **(C)** A large sample of canopy-dwelling beetles from one species of tropical trees had 163 species specific to it. There are 5 × 10^4 tree species, and so 163 × 5 × 10^4 ≈ 8 × 10^6 species of canopy beetles. Because 40% of described insects are beetles, the total number of canopy insects is 2 × 10^7. Adding half that number for arthropod species on the ground gives a grand total of 3 × 10^7. **(D)** If only 20% of canopy insects are beetles, but there are at least as many ground as canopy species, then the grand total is 8 × 10^7. **(E)** Some 63% of the 1690 species on ~500 Indonesian tree species were previously unrecorded. The ~10^6 described insect species thus suggest a total of 2.7 × 10^6 species. **(F)** Across many food webs, there are roughly three times as many herbivores and carnivores combined as there are plants. This resulting estimate of terrestrial animal species, ~7.5 × 10^5, is certainly too low, because published food webs omit many species. **(G)** There are about two tropical bird and mammal species for each temperate or boreal species. Yet, of the ~1.5 × 10^6 described species, about one-third is tropical. The prediction of 3 × 10^6 species is an underestimate, because not all temperate species are described. **(H)** There is a linear increase in species numbers with decreasing body size. Below a threshold level, however, the numbers drops, perhaps because of sampling bias. If the true pattern remained linear, there would be 1 × 10^7 to 5 × 10^7 species. **(I)** We added the more detailed estimates for the numbers of species in the largest groups.

The authors are in the Department of Ecology and Evolutionary Biology, University of Tennessee, Knoxville, TN 37996, USA.

*To whom correspondence should be addressed.

likely to be representative. If we understand the underlying mechanisms, we may find they operate universally.

The Past as a Guide to the Future

Unambiguous evidence of human impact on extinction comes from before-and-after comparisons of floras and faunas (*10*). Polynesians reached the planet's last habitable areas—Pacific islands—within the last 1000 to 4000 years. The bones of many bird species persist into, but not through, archaeological zones that show the presence of humans. No species disappeared in the longer intervals before the first human contact. Adding known and inferred extinctions, it seems that with only Stone Age technology, the Polynesians exterminated >2000 bird species, some ~15% of the world total.

We must infer extinctions, because we will not find the bones of every now-extinct species. From the overlap in species known from bones and those survivors seen by naturalists, sampling theory infers that ~50% of the species are still missing (*10*). Faunal reconstruction affords a second inference. For example, Steadman (*10*) contends that every one of ~800 Pacific islands should have had at least one unique species of rail. A few remote islands still have rails. Others lost theirs to introduced rats in the last century. Large volcanic islands typically lost several species of rails. Accessible islands lost their rails earlier, for every survey of bones from islands now rail-free has found species that did not survive human contact.

High extinction rates also followed the Pacific's colonization by Europeans. Since 1778, the Hawaiian islands have lost 18 species of birds; the fate of 12 more is unknown (*10*). Nor are birds unusual. Of 980 native Hawaiian plants, 84 are extinct and 133 have wild populations of <100 individuals (*11*). Across the Pacific, a predatory snail introduced to control another introduced snail ate to extinction hundreds of local varieties of land snails (*12*).

Nor are Pacific islands unusual: of 60 mammalian extinctions worldwide, 19 are from Caribbean islands (*12*). In the last 300 years, Mauritius, Rodrigues, and Réunion in the Indian Ocean lost 33 species of birds, including the dodo, 30 species of land snails, and 11 reptiles. St. Helena and Madeira in the Atlantic Ocean have lost 36 species of land snails (*12*).

Importantly, extinction centers are not necessarily on islands nor only in terrestrial environments. The fynbos, a floral region in southern Africa, has lost 36 plant species (of ~8500); 618 more are threatened with extinction (*12*). Extinctions of 18 (of 282) species of Australian mammals rival those from the Caribbean; 43 more are threatened (*12*). In the last century, North American freshwater environments lost 21 of 297 mussel and clam species (120 are threatened) and 40 of ~950 fish species (*12*).

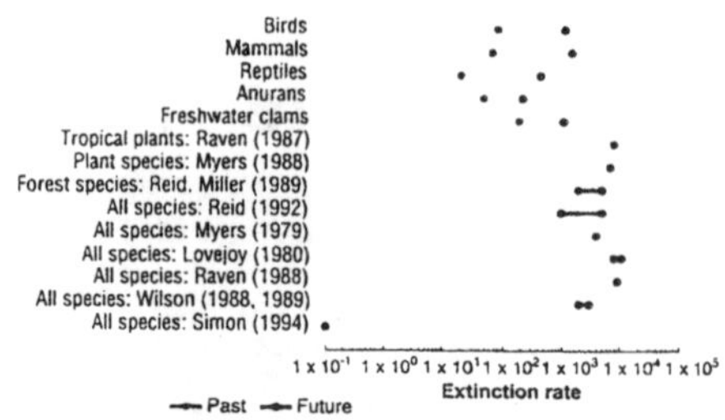

Fig. 2. Estimates of extinction rates expressed as extinctions per million species-years. For birds through clams, we derive past rates from known extinctions in the last 100 years; we derive future rates by assuming that all currently threatened species will be extinct in 100 years. The latter rates are much higher than the former but are still far too low. The remaining estimates are previously published (*1, 16, 27*). Myers (1979) (*27*) assumes an exponential increase in the number of extinctions. Myers (1988) (*16*) assumes the loss of a small number of areas rich in endemics. With the exception of Simon, the rest are estimates based on the relation between habitat loss and species loss. Simon's claims (*1*) of one (or a few) species per year (out of a conservative total of 10^7 species) are not scientifically credible.

This world tour of extinction centers has remarkable features (*12*). Recent extinction rates are 20 to 200 *E*/MSY (Fig. 2)—a small range given, among other things, the uncertainties of whether to average rates over a century or a shorter interval that reflects more recent human impacts. We find high rates in mainlands and islands, in arid lands and rivers, and for both plants and animals. Although we know less about invertebrates, high rates characterize bivalves of continental rivers and island land snails. There is nothing intrinsic to the diverse life histories of these species to predict their being unusually prone to extinction.

What obvious features unite extinction centers? We know the species and places well—as did naturalists a century ago. Importantly, each area holds a high proportion of species restricted to it. Such endemics constituted 90% of Hawaiian plants, 100% of Hawaiian land birds, ~70% of fynbos plants, and 74% of Australian mammals (*12*). In contrast, only ~1% of Britain's birds and plants are endemics (*12*). Remote islands are typically rich in endemics, but so are many areas within continents (*13*). Past extinctions are so concentrated in small, endemic-rich areas that the analysis of global extinction is effectively the study of extinctions in one or a few extinction centers (*12*). Why should this be?

Random extinction is the simplest model. Some species groups and some places will suffer more extinctions than others, but generally the more species present, the more there will be to lose. This model does a poor job of predicting global patterns. If island birds were intrinsically vulnerable to extinction, then Hawai'i and Britain with roughly the same number of species of breeding land birds (~135) would have suffered equally. Hawai'i had >100 extinctions, Britain only 3 (*12*). Nor is the number of species an area houses a good predictor of the total extinctions. Islands house few species and suffer many extinctions.

Imagine a cookie-cutter model where some cause destroys (cuts out) a randomly selected area. Species also found elsewhere survive, for they can recolonize. But some of the endemics go extinct, the proportion depending on the extent of the destruction. We do not assume that island biotas are intrinsically more vulnerable than mainlands. For random species ranges, the number of extinctions correlates weakly with the area's total number of species, but strongly with the number of its endemics. By chance alone, small endemic-rich areas will contribute disproportionately to the total number of extinctions.

This model is consistent with known mechanisms of extinction. Habitat destruction cuts out areas, as the model implies. Introduced species also destroy species regionally. Species need not be entirely within the area destroyed to succumb to extinction: The populations outside may be too small to persist (*14*). Moreover, across many taxa, range-restricted species have lower local densities than widespread species (*15*). The former are not only more likely to be cut in the first place, but their surviving populations will have lower densities and thus higher risks of extinction than widespread species. This entirely self-evident model emphasizes the localization of endemics—Myers' "hot spots" (*16*)—as the key variable in understanding global patterns of recent and future extinctions.

Predicting Future Rates of Extinction

Projecting past extinction rates into the future is absurd for no other reason than that the ultimate cause of these extinctions—the human population—is increasing exponentially. For vertebrates, we have worldwide surveys of threatened species

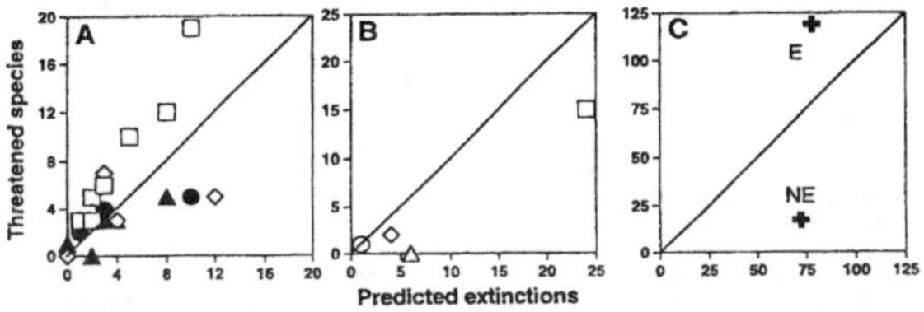

Fig. 3. The numbers of species currently threatened with extinction plotted against the numbers of species predicted to become extinct. Squares, Philippines; circles, North Wallacea; diamonds, Greater Sundas; triangles, Lesser Sundas. The predictions use satellite estimates of remaining forest cover and the relation between species numbers and area. **(A)** For the subset of species endemic to single islands, these numbers match; they straddle the graph's diagonal. **(B)** For the subset of species that are endemic to several islands within each region, the predicted extinctions consistently exceed the numbers of threatened species. **(C)** When we combine the species found on one or several islands (E) across the entire region, the predictions based on forest losses (78) are slightly smaller than the number of threatened species (119). The remaining subset comprises widely occurring species (NE). Their predicted extinctions (72) far exceed the few (17) actually threatened with extinction. Predictions of extinctions from habitat loss must be scaled to the number of endemics the area contains (*23*).

(*12*). Is it reasonable to assume that all these species will be extinct in <100 years, thus making future rates 200 to 1500 *E*/MSY (Fig. 2)?

Some threatened species are declining rapidly and will soon be extinct. Others, not so obviously doomed, have small numbers ($<10^2$). They risk the demographic vagaries of sex (all the young of a generation being of the same sex) and death (all the individuals dying in the same year from independent causes). For these, both models and empirical, long-term studies of island populations suggest times to extinction on the order of decades (*17*). Population fluctuations, and the environmental vagaries that cause them, drive the extinction of larger populations ($>10^2$) (*14*). Over 20 years, bird densities can vary 10-fold, and insect densities 10,000-fold (*14*). Ecologists have been slow to combine models and data. Yet even in the absence of a formal analysis, such fluctuations can obviously doom even quite large populations.

Our predictions may err because some threatened species will survive the century (*18*). The more serious problem with our predictions is that species not now threatened will become extinct. For birds—the one group for which we have detailed lists of the causes of threats—limited habitat is the most frequently cited factor, implicated in ~75% of threatened species (*18*). Increasingly well documented studies (*19*) show that habitat destruction is continuing and perhaps accelerating. Some now-common species will lose their habitats within decades.

Interestingly, accidentally or deliberately introduced species are blamed for only 6% of currently threatened birds (*18*). Yet introduced species, and the predation, competition, disease, and habitat modification they cause, are the most frequently cited factors in all the extinction centers we discussed above (*12*). Undoubtedly, many species will be lost to introduced species in ways that we cannot now anticipate. For example, no one considered the birds on the island of Guam to be in danger 30 years ago, but an introduced snake has eliminated all the island's birds since then (*14*). Were this predator to reach Hawai'i, all its birds would be at risk.

Calibrating Species Loss from Habitat Loss: A Tale of Two Forests

So far, we have sampled well-known, but disparate species whose high extinction rates probably typify the unknown majority. We now consider a typical mechanism of extinction: habitat loss. Can we predict species losses from estimates of habitat losses? The function $S = cA^z$ relates the number of species counted (S) to the area surveyed (A); c and z are constants (*20*). If the original habitat area, A_o, is reduced to A_n, we expect the original number of species, S_o, to decline eventually to S_n. Now $S_n/S_o = cA_n^z/cA_o^z$ or $(A_n/A_o)^z$— an expression that is independent of c. Across different situations, z varies from 0.1 to 1.0, but it is often taken to be ~¼ (*20*). This value is typical of islands isolated by sea-level changes, a process that may be the best model for large habitat fragments isolated by deforestation (*20, 21*).

This recipe forms the basis of the predictions of 1000 to 10,000 *E*/MSY shown in Fig. 2. To challenge these estimates, critics point to the few bird extinctions after the clearing of North America's eastern forests (*3*). Is the recipe flawed? Only if interpreted naïvely are these results a poor model for what happens elsewhere.

An extinction "cold spot." European colonists cut >95% of the eastern forests of North America, but not simultaneously. Locally, forests reclaimed abandoned fields, and regionally forests recovered in the Northeast as settlers moved westward. Of the region's 2.87×10^6 km^2 area, forests always covered >50% (*21*). So 16% (= $0.5^{0.25}$)—or 26 of the ~160 forest species—should have gone extinct. Only 4 did so (*21*). Yet, such predictions are naïve. Not enough time may have elapsed for the extinctions to occur. However, all but 28 of these species occur widely across North America. They would have survived elsewhere even if all the forest had been permanently cleared. The cookie-cutter model restricts the analysis to the region's 28 endemics, whence the predicted and observed number of extinctions correspond (4 ≈ 16% of 28).

Simply, this region has very few endemics and so few species to lose. In contrast, tropical moist forests may hold two-thirds of all species on Earth (*22*). Despite inevitable differences in their definition, satellite imaging yields detailed and rapidly changing estimates showing their rapid depletion (*19*). The forests' global extent is variously estimated at 8×10^6 to 12.8×10^6 km^2 and their rate of clearing as 1.2×10^6 to 1.4×10^6 km^2 per decade (*19*).

An extinction "hot spot." The 1.47×10^6 km^2 of forests in the Philippines and Indonesia (excluding Irian Jaya) hold 545 endemic bird species—20 times the number in America's eastern forest in half the area (*23*). Only 0.91×10^6 km^2 of forest remains, and ~10% of the original area is cleared per decade. Using current satellite-based estimates of forest cover, the species-area recipe adequately predicts the number of species endemic to single islands that are currently threatened (Fig. 3A). The recipe, however, overestimates the numbers of currently threatened species that are found on several islands (Fig. 3B) and greatly overestimates the number of currently threatened species that are widespread (Fig. 3C).

Estimates of extinctions from habitat losses (Fig. 2) use an area's total number of species, not its smaller number of endemics. Does this reliance on such totals inflate these rates? In general, it does not, because many tropical areas are unusually rich in endemics (*13*). For example, 18 areas worldwide are so rich in endemics as to encompass ~20% of the known species of flowering plants in a total area of 0.74×10^6 km^2 (*16*). A larger area than this was cleared from the eastern American forests in the 19th century. The fate of these areas obviously dominates the calculations of future extinction rates. Details of land use changes in these areas are critical, but the details are not sufficient in themselves. We also need the detailed patterns of endemism.

Unfortunately, we know the geographical ranges of only a small proportion of the already small proportion of species for which we have names. We do have a comprehensive understanding of the geographical patterns of species richness (*20*). Its lessons are not encouraging. First, we cannot extrapolate from one species group to the next. For instance, across a continent species richness in frogs may not correlate with the species richness in birds (*24*). Worse, the direction of the correlation—positive or negative—may differ between continents (*24*). Second, areas rich in species are not always rich in endemics (*24*). Simply, our understanding of endemism is insufficient for us to know the future of biodiversity with precision (*25*).

REFERENCES AND NOTES

1. N. Myers and J. L. Simon, *Scarcity or Abundance* (Norton, New York, 1994).
2. P. R. Ehrlich and A. H. Ehrlich, *Extinction: The Causes and Consequences of the Disappearance of Species* (Random House, New York, 1981).
3. J. L. Simon and A. Wildavsky, *New York Times*, 13 May 1993, p. A23; S. Budiansky, *U.S. News World Rep.*, 13 December 1993, p. 81; *Nature* **370**, 105 (1994).
4. For extinction to occur, there need not be a background rate—a slow, characteristic winking out of species—punctuated by extreme events such as the one that eliminated the dinosaurs. In the geological record, the number of extinctions per time interval scales continuously as 1/frequency. Thus, there is no mean rate, for the rate depends on the interval over which it is measured. See D. Jablonski, in *Dynamics of Extinction*, D. K. Elliott, Ed. (Wiley, New York, NY, 1986), pp. 193–229.
5. R. M. May, J. H. Lawton, N. E. Stork, in *Extinction Rates*, J. H. Lawton and R. M. May, Eds. (Oxford Univ. Press, Oxford, 1995), pp. 1–24.
6. C. G. Sibley and J. E. Ahlquist, *Phylogeny and Classification of the Birds* (Yale Univ. Press, New Haven, CT, 1990).
7. S. Nee, A. O. Mooers, P. H. Harvey, *Proc. Natl. Acad. Sci. U.S.A.* **89**, 8322 (1992); S. Nee, E. C. Holmes, R. M. May, P. H. Harvey, *Proc. R. Soc. London Ser. B* **344**, 77 (1994).
8. R. K. Wayne, R. E. Benveniste, D. N. Janczewski, S. J. O'Brien, in *Carnivore Behavior, Ecology, and Evolution*, J. L. Gittleman, Ed. (Cornell Univ. Press, Ithaca, NY, 1989), pp. 465–494; R. K. Wayne, B. Van Valkenburgh, S. J. O'Brien, *Mol. Biol. Evol.* **8**, 297 (1991).
9. R. M. May, *Philos. Trans. R. Soc. London Ser. B* **330**, 293 (1990).
10. D. W. Steadman, *Science* **267**, 1123 (1995); S. L. Pimm, M. P. Moulton, L. J. Justice, *Philos. Trans. R. Soc. London Ser. B* **344**, 27 (1994).
11. S. Sohmer, in *Biodiversity and Terrestrial Ecosystems*, C.-I. Peng and C. H. Chou, Eds. (Academia Sinica Monograph Series 14, Taipei, 1994), pp. 43–51.
12. World Conservation Monitoring Centre, *Global Biodiversity: Status of the Earth's Living Resources* (Chapman & Hall, London, 1992); M. P. Nott, E. Rogers, S. L. Pimm, *Curr. Biol.* **5**, 14 (1995); J. Parslow, *Breeding Birds of Britain and Ireland: A Historical Survey* (Poyser, Berkhamsted, 1973).
13. International Council for Bird Preservation (ICBP), *Putting Biodiversity on the Map: Priority Areas for Global Conservation* (ICBP, Cambridge, 1992).
14. S. L. Pimm, *The Balance of Nature? Ecological Issues in the Conservation of Species and Communities* (Univ. of Chicago Press, Chicago, IL, 1992).
15. K. J. Gaston, *Rarity* (Chapman & Hall, London, 1994).
16. N. Myers, *Environmentalist* **8**, 187 (1988); *ibid.* **10**, 243 (1990).
17. S. L. Pimm, J. M. Diamond, T. M. Reed, G. J. Russell, J. M. Verner, *Proc. Natl. Acad. Sci. U.S.A.* **90**, 10871 (1993).
18. N. J. Collar, M. J. Crosby, A. J. Stattersfield, *Birds to Watch 2: The World List of Threatened Birds* (BirdLife International, Cambridge, 1994).
19. D. Skole and C. Tucker, *Science* **260**, 1905 (1993); Food and Agricultural Organization, *The Forest Resources of the Tropical Zone by Main Ecological Regions* (Food and Agricultural Organization, Rome, 1992); N. Myers, in *The Causes of Tropical Deforestation*, K. Brown and D. W. Pearce, Eds. (University College Press, London, 1994), pp. 27–40.
20. M. L. Rosenzweig, *Species Diversity in Space and Time* (Cambridge Univ. Press, Cambridge, 1995).
21. S. L. Pimm and R. A. Askins, *Proc. Natl. Acad. Sci. U.S.A.*, in press.
22. P. H. Raven, in *Biodiversity*, E. O. Wilson, Ed. (National Academy Press, Washington, DC, 1988), pp. 119–122.
23. T. M. Brooks, S. L. Pimm, N. J. Collar, unpublished results.
24. J. J. Schall and E. R. Pianka, *Science* **201**, 679 (1978); J. R. Prendergast, R. M. Quinn, J. H. Lawton, B. C. Eversham, D. W. Gibbons, *Nature* **365**, 335 (1993); J. Curnutt, J. Lockwood, H.-K. Luh, P. Nott, G. Russell, *ibid.* **367**, 326 (1994).
25. S. L. Pimm and J. L. Gittleman, *Science* **255**, 940 (1992).
26. D. L. Hawksworth, *Mycol. Res.* **95**, 441 (1991); J. F. Grassle and N. J. Maciolek, *Am. Nat.* **139**, 313 (1992); T. L. Erwin, *Coleopt. Bull.* **36**, 74 (1982); N. E. Stork, *Biol. J. Linn. Soc.* **35**, 321 (1988); I. D. Hodkinson and D. Casson, *ibid.* **43**, 101 (1990); S. L. Pimm, J. H. Lawton, J. E. Cohen, *Nature* **350**, 669 (1991); P. H. Raven, *Futurist* **19**, 8 (1985); R. M. May, *Science* **241**, 1441 (1988).
27. M. V. Reid, in *Tropical Deforestation and Species Extinction*, T. C. Whitmore and J. A. Sayer, Eds. (Chapman & Hall, London, 1992), pp. 55–73; P. H. Raven, in *Botanic Gardens and the World Conservation Strategy*, D. Bramwell, O. Hamann, V. Heywood, H. Synge, Eds. (Academic Press, London, 1987), pp. 19–29; W. V. Reid and K. R. Miller, *Keeping Options Alive: The Scientific Basis for Conserving Biodiversity* (World Resources Institute, Washington, DC, 1989); N. Myers, *The Sinking Ark: A New Look at the Problem of Disappearing Species* (Pergamon Press, Oxford, 1979); T. E. Lovejoy, *The Global 2000 Report to the President*, vol. 2 (Council on Environmental Quality, Washington, DC, 1980), p. 328; P. H. Raven, in *Biodiversity*, E. O. Wilson and F. M. Peter, Eds. (National Academy Press, Washington, DC, 1988), pp. 1190–1222; E. O. Wilson, in *Biodiversity*, E. O. Wilson and F. M. Peter, Eds. (National Academy Press, Washington, DC, 1988), pp. 3–18; *Sci. Am.* **261**, 108 (September 1989).

28. We thank H.-K. Luh, K. Norris, P. Nott, J. Tobias, and three anonymous reviewers for help and comments. S.L.P. is supported by a Pew Scholarship in Conservation and the Environment.

[5]

■ POLICY FORUM

Economic Growth, Carrying Capacity, and the Environment

Kenneth Arrow, Bert Bolin, Robert Costanza, Partha Dasgupta, Carl Folke, C. S. Holling, Bengt-Owe Jansson, Simon Levin, Karl-Göran Mäler, Charles Perrings, David Pimentel

National and international economic policy has usually ignored the environment. In areas where the environment is beginning to impinge on policy, as in the General Agreement on Tariffs and Trade (GATT) and the North American Free Trade Agreement (NAFTA), it remains a tangential concern, and the presumption is often made that economic growth and economic liberalization (including the liberalization of international trade) are, in some sense, good for the environment. This notion has meant that economy-wide policy reforms designed to promote growth and liberalization have been encouraged with little regard to their environmental consequences, presumably on the assumption that these consequences would either take care of themselves or could be dealt with separately.

In this article we discuss the relation between economic growth and environmental quality, and the link between economic activity and the carrying capacity and resilience of the environment (*1*).

Economic Growth, Institutions, and the Environment

The general proposition that economic growth is good for the environment has been justified by the claim that there exists an empirical relation between per capita income and some measures of environmental quality. It has been observed that as income goes up there is increasing environmental degradation up to a point, after which environmental quality improves. (The relation has an "inverted-U" shape.)

One explanation of this finding is that people in poor countries cannot afford to emphasize amenities over material well-being. Consequently, in the earlier stages of economic development, increased pollution is regarded as an acceptable side effect of economic growth. However, when a country has attained a sufficiently high standard of living, people give greater attention to environmental amenities. This leads to environmental legislation, new institutions for the protection of the environment, and so forth.

The above argument does not, however, pertain to the environmental resource basis of material well-being, a matter we shall return to subsequently.

So far the inverted U-shaped curve has been shown to apply to a selected set of pollutants only (*2*, *3*). However, because it is consistent with the notion that people spend proportionately more on environmental quality as their income rises, economists have conjectured that the curve applies to environmental quality generally (*4*). But it is important to be clear about the conclusions that can be drawn from these empirical findings. While they do indicate that economic growth may be associated with improvements in some environmental indicators, they imply neither that economic growth is sufficient to induce environmental improvement in general, nor that the environmental effects of growth may be ignored, nor, indeed, that the Earth's resource base is capable of supporting indefinite economic growth. In fact, if this base were to be irreversibly degraded, economic activity itself could be at risk (*5*).

There are other reasons for caution in interpreting these inverted U-shaped curves. First, the relation has been shown to be valid for pollutants involving local short-term costs (for example sulfur, particulates, and fecal coliforms), not for the accumulation of stocks of waste or for pollutants involving long-term and more dispersed costs (such as CO_2), which are often increasing functions of income (*6*).

Second, the inverted-U relations have been uncovered for emissions of pollutants, not resource stocks. The relation is less likely to hold wherever the feedback effects of resource stocks are significant, such as those involving soil and its cover, forests, and other ecosystems.

Third, the inverted-U curves, as they have been estimated, say nothing about the system-wide consequences of emission reductions. For example, reductions in one pollutant in one country may involve increases in other pollutants in the same country or transfers of pollutants to other countries (*7*).

And fourth, in most cases where emissions have declined with rising income, the reductions have been due to local institutional reforms, such as environmental legislation and market-based incentives to reduce environmental impacts. But such reforms often ignore international and intergenerational consequences. Where the environmental costs of economic activity are borne by the poor, by future generations, or by other countries, the incentives to correct the problem are likely to be weak. The environmental consequences of growing economic activity may, accordingly, be very mixed.

The solution to environmental degradation lies in such institutional reforms as would compel private users of environmental resources to take account of the social costs of their actions (*8*). The inverted-U relation is evidence that this has happened in some cases. It does not constitute evidence that it will happen in all cases or that it will happen in time to avert the important and irreversible global consequences of growth.

Carrying Capacity and Ecosystem Resilience

The environmental resource base upon which all economic activity ultimately depends includes ecological systems that produce a wide variety of services. This resource base is finite. Furthermore, imprudent use of the environmental resource base may irreversibly reduce the capacity for generating material production in the future. All of this implies that there are limits to the carrying capacity of the planet. It is, of course, possible that improvements in the management of resource systems, accompanied by resource-conserving structural changes in the economy, would enable economic and population growth to take place despite the finiteness of the environmental resource base, at least for some period of time. However, for that to be even conceivable, signals that effectively reflect increasing scarcities of the resource base need to be generated within the economic system.

Carrying capacities in nature are not fixed, static, or simple relations. They are contingent on technology, preferences, and

K. Arrow is in the Department of Economics, Stanford University, Stanford, CA 94305, USA. B. Bolin is in the Department of Meteorology, University of Stockholm, 106 91 Stockholm, Sweden. R. Costanza is director of the Maryland International Institute for Ecological Economics, University of Maryland, Box 38, Solomons, MD 20688, USA. P. Dasgupta is on the Faculty of Economics, Cambridge University, Cambridge CB3 9DD, UK. C. Folke is at the Beijer International Institute of Ecological Economics, Royal Swedish Academy of Sciences, Box 50005, S-104 05 Stockholm, Sweden. C. S. Holling is in the Department of Zoology, University of Florida, Gainesville, FL 32611, USA. B.-O. Jansson is in the Department of Systems Ecology, University of Stockholm, S-106 91 Stockholm, Sweden. S. Levin is in the Department of Ecology and Evolutionary Biology, Princeton University, Princeton, NJ 08544, USA. K.-G. Mäler is director of the Beijer International Institute of Ecological Economics, Royal Swedish Academy of Sciences, Box 50005, S-104 05 Stockholm, Sweden. C. Perrings is head of the Department of Environmental Economics and Environmental Management, University of York, York Y01 5DD, UK. D. Pimentel is in the Department of Entomology and Section of Ecology and Systematics, Cornell University, Ithaca, NY 14853, USA.

the structure of production and consumption. They are also contingent on the ever-changing state of interactions between the physical and biotic environment. A single number for human carrying capacity would be meaningless because the consequences of both human innovation and biological evolution are inherently unknowable. Nevertheless, a general index of the current scale or intensity of the human economy in relation to that of the biosphere is still useful. For example, Vitousek *et al.* (*9*) calculated that the total net terrestrial primary production of the biosphere currently being appropriated for human consumption is around 40%. This does put the scale of the human presence on the planet in perspective.

A more useful index of environmental sustainability is ecosystem resilience. One way of thinking about resilience is to focus on ecosystem dynamics where there are multiple (locally) stable equilibria (*10*). Resilience in this sense is a measure of the magnitude of disturbances that can be absorbed before a system centered on one locally stable equilibrium flips tc another (*11*). Economic activities are sustainable only if the life-support ecosystems on which they depend are resilient. Even though ecological resilience is difficult to measure and even though it varies from system to system and from one kind of disturbance to another, it may be possible to identify indicators and early-warning signals of environmental stress. For example, the diversity of organisms or the heterogeneity of ecological functions have been suggested as signals of ecosystem resilience. But ultimately, the resilience of systems may only be tested by intelligently perturbing them and observing the response with what has been called "adaptive management" (*12*).

The loss of ecosystem resilience is potentially important for at least three reasons. First, the discontinuous change in ecosystem functions as the system flips from one equilibrium to another could be associated with a sudden loss of biological productivity, and so to a reduced capacity to support human life. Second, it may imply an irreversible change in the set of options open both to present and future generations (examples include soil erosion, depletion of groundwater reservoirs, desertification, and loss of biodiversity). Third, discontinuous and irreversible changes from familiar to unfamiliar states increase the uncertainties associated with the environmental effects of economic activities.

If human activities are to be sustainable, we need to ensure that the ecological systems on which our economies depend are resilient. The problem involved in devising environmental policies is to ensure that resilience is maintained, even though the limits on the nature and scale of economic activities thus required are necessarily uncertain.

Economic Growth and Environmental Policy

We conclude that economic liberalization and other policies that promote gross national product growth are not substitutes for environmental policy. On the contrary, it may well be desirable that they are accompanied by stricter policy reforms. Of particular importance is the need for reforms that would improve the signals that are received by resource users. Environmental damages, including loss of ecological resilience, often occur abruptly. They are frequently not reversible. But abrupt changes can seldom be anticipated from systems of signals that are typically received by decision-makers in the world today. Moreover, the signals that do exist are often not observed, or are wrongly interpreted, or are not part of the incentive structure of societies. This is due to ignorance about the dynamic effects of changes in ecosystem variables (for example, thresholds, buffering capacity, and loss of resilience) and to the presence of institutional impediments, such as lack of well-defined property rights. The development of appropriate institutions depends, among other things, on understanding ecosystem dynamics and on relying on appropriate indicators of change. Above all, given the fundamental uncertainties about the nature of ecosystem dynamics and the dramatic consequences we would face if we were to guess wrong, it is necessary that we act in a precautionary way so as to maintain the diversity and resilience of ecosystems.

Economic growth is not a panacea for environmental quality; indeed, it is not even the main issue. What matters is the content of growth—the composition of inputs (including environmental resources) and outputs (including waste products). This content is determined by, among other things, the economic institutions within which human activities are conducted. These institutions need to be designed so that they provide the right incentives for protecting the resilience of ecological systems. Such measures will not only promote greater efficiency in the allocation of environmental resources at all income levels, but they would also assure a sustainable scale of economic activity within the ecological life-support system. Protecting the capacity of ecological systems to sustain welfare is of as much importance to poor countries as it is to those that are rich.

REFERENCES AND NOTES

1. This is a report of the Second Askö Meeting, held 31 August to 2 September 1994, in the archipelago outside Stockholm, Sweden. The meeting was organized by the Beijer International Institute of Ecological Economics, Royal Swedish Academy of Sciences, Box 50005, S10405 Stockholm, Sweden. The aim of the meeting was to establish a substantive dialogue among a small group of ecologists and economists to gauge whether an interdisciplinary consensus exists on the issues of economic growth, carrying capacity, and the environment and to determine what could be said about the joint development of economic and environmental policy.
2. Phenomena for which the relation holds most clearly include poor sanitation, impure water supplies, suspended particulates, SO_2, NO_x, and CO (*3*) [G. M. Grossman and A. B. Krueger, in *The U.S. Mexico Free Trade Agreement*, P. Garber, Ed. (MIT Press, Cambridge, MA, 1993), pp. 165–177].
3. N. Shafik and S. Bandyopadhay, "Economic growth and environmental quality: time series and cross country evidence" (background paper for the World Development Report, World Bank, Washington, DC, 1992).
4. W. Beckerman, *World Dev.* **20**, 481 (1992).
5. A. M. Jansson, M. Hammer, C. Folke, R. Costanza, Eds, *Investing in Natural Capital: the Ecological Economics Approach to Sustainability* (Island Press, Washington, DC, 1994).
6. N. Shafik and S. Bandyopadhay (*3*) find that CO_2 emissions are an increasing function of per capita income. W. Moomaw and M. Tullis [in *Industrial Ecology and Global Change*, R. Socolow, C. Andrews, F. Berkhout, V. Thomas, Eds. (Cambridge Univ. Press, Cambridge, 1994), pp. 157–172] show that while cross-sectionally CO_2 emissions per capita rise with income per capita, the experience in individual countries over time is highly variable, depending on the structure of individual economies. They conclude that there are many different development paths, some of which can de-link economic growth from CO_2 emissions, but adequate institutions are required and it is by no means an automatic consequence of growth.
7. R. Lopez, *J. Enivron. Econ. Manage.* **27**, 163 (1994); T. M. Selden and D. Song; *ibid.*, p. 147; K. Anderson and R. Blackhurst, Eds, *The Greening of World Trade Issues* (Harvester Wheatsheaf, Hemel Hempstead, Herts, UK, 1992).
8. A related issue is the measurement of economic growth at the macro level. The conventional measure, GNP, is far from adequate as a measure of true economic performance. We need a more comprehensive index that includes the flow of environmental services as well as the value of net changes in the stocks of natural capital so that the true social costs of growth can be internalized. With this improved index, many of the apparent conflicts between "growth" and the environment would disappear because environmental degradation would be subtracted from the index. [See, for example, P. Dasgupta and K.-G. Mäler, *The Environment and Emerging Development Issues*, Proceedings of the Annual Bank Conference on Development Economics, 1990 (Supplement to the World Bank Economic Review, 1991).
9. P. M. Vitousek, P. R. Ehrlich, A. H. Ehrlich, P. A. Matson, *BioScience* **36**, 368 (1986).
10. Another way is to focus on ecosystem dynamics in the neighborhood of an equilibrium. The object here is to measure the resistance of a system to disturbances and to determine the speed with which it returns to equilibrium. [See S. L. Pimm, *Nature* **307**, 321 (1984).]
11. C. S. Holling, *Annu. Rev. Ecol. System.* **4**, 123 (1973).
12. Adaptive management views regional development policy and management as "experiments," where interventions at several scales are made to achieve understanding, to produce a social or economic product, and to identify options. [See, for example, C. S. Holling, Ed., *Adaptive Environmental Assessment and Management* (Wiley, London, 1978); C. J. Walters, *Adaptive Management of Renewable Resources* (Macmillan, New York, 1986); K. Lee, *Compass and the Gyroscope* (Island Press, Washington, DC, 1993).
13. We thank the two anonymous referees for helpful comments.

[6]

Threshold Effects and Incentives for the Conservation of Biodiversity

CHARLES PERRINGS[1] and DAVID PEARCE[2]
[1] *Beijer Institute and University of York, U.K.;* [2] *CSERGE and University College, London, U.K.*

Abstract. Biological diversity is a central component of the stock of natural capital on which all economic development is based. Other things being equal, loss of biologically diversity implies loss of development potential, and its conservation through sustainable use or outright protection implies the protection of that potential. One characteristics of biodiversity loss of especial importance is that biodiversity loss, more than any other current environmental problem, is associated with ecological threshold effects. The paper explores the implication of this characteristics for the properties of a biodiversity conservation strategy.

Key words.

1. Introduction

The intractability of the problem of biodiversity loss is closely related to the irreversibility of many of the social costs of species deletion. The irreversibility of species loss is one facet of a more general characteristic of change in complex ecological systems: namely that such change tends to be discontinuous around system thresholds. Ecological systems are highly non-linear, non-convex dissipative systems, the dynamics of which are discontinuous in the neighbourhood of system thresholds. These thresholds are defined by certain critical values for either populations of organisms or biogeochemical cycles. At such values the ecosystem loses resilience and so self-organization. In thermodynamic terms, they are the values at which the system 'flips' from one thermodynamic branch to another, or from one basin of attraction to another. The implication of this is that marginal change in the population of some organism in the neighbourhood of a threshold may be associated with enormous social costs, if that marginal change causes the system to flip from one state to another.

The existence of threshold effects complicates the valuation of the opportunity cost of biodiversity-depleting activities because of the uncertainty it engenders. This makes the development of economic instruments to protect thresholds problematic. There is generally considerable uncertainty about the threshold values of either populations of organisms or biogeochemical cycles for many of the most important ecosystem types, and there is often fundamental ignorance about the implications of crossing a threshold.

Environmental and Resource Economics **4**: 13–28, 1994.

This paper considers the policy implications of threshold effects. It is organized in five sections. The following section offers a preliminary discussion of the optimal exploitation of biological resources with and without threshold effects. In section three, we offer a more formal discussion of the properties of an optimal biodiversity conservation policy, paying special attention to the distinction between use and non-use values. In addition, the section offers some examples of policies which reflect these properties. Section four extends the treatment to cover threshold effects. A final section offers some concluding remarks.

2. Optimal Biodiversity Loss with Threshold Effects

The value of biological resources committed to some use is equal to the sum of the instrumental value of those resources to private users, plus the value of the intratemporal and intertemporal external effects associated with that use. External effects, in this context, refers to effects of the use of a resource on third parties that are not taken into account by the user. These include (but are not restricted to) effects on third parties that are not mediated by the market. Intratemporal external costs are the subject of a large literature both in environmental and public economics, and though ignorance about the structure of ecosystems might lead us to doubt whether we can generate precise estimates of the costs involved, the associated policy options are reasonably well defined. Intertemporal external costs, on the other hand, is more problematic in that it invokes a further criterion for optimal resource use — intergenerational equity.

Both inter- and intra-temporal externalities involve values which have been categorized as non-use (Randall, 1991). This type of value has been conceptualized in a number of different ways in the literature. Non-use values may be 'bequest' value, involving a future potential use (Krutilla, 1967); as 'existence' value, unassociated with any use now or in the future (Pearce and Turner, 1990); as currently non-consumptive but future use vlaues — 'option' value (Weisbrod, 1964); as 'scientific' use value (Krutilla, 1967); and as 'quasi-option' value (Arrow and Fisher, 1974). Bequest, option and existence values all relate to the idea of biodiversity loss as an erosion of the opportunity set. Quasi-option and scientific values refer very specifically to the value that biological assets have as sources of yet-to-be-uncovered information, and so concern only a subset of the opportunities foregone as a result of biodiversity loss. Since these are all future values, however, they are especially sensitive to threshold effects and to the fundamental uncertainty associated with such effects.

To approach the policy issues raised by externalities it is convenient, for the moment, to ignore dynamic effects, and to consider the most basic definition of the total economic cost of a resource committed to some particular use:

$$TEC = C + E \tag{2.1}$$

where $C = C(w, q)$ denotes private cost, and $E = E(r(q))$ denotes external cost, w being a strictly positive vector of market input costs, r a non-negative vector of (unpriced) biological resources, and q output. The function $C(\cdot)$ is assumed to be continuous, differentiable and increasing in both q, and w. $E(\cdot)$ is assumed, in the general case, only to be increasing in the level of output. That is, it may be neither differentiable nor continuous. If $E(\cdot)$ is in fact a continuous, differentiable function, this corresponds to the standard optimal externality problem.

Denoting private revenue $R = R(p, q)$, p being the price of output, then in this particular case the privately optimal level of output is simply that at which

$$\partial R/\partial q = \partial C/\partial q \tag{2.2}$$

and the socially optimal level of output is that at which

$$(\partial \Pi_p/\partial q) = (\partial E/\partial r)(\partial r/\partial q) \tag{2.3}$$

Π_p being marginal net private benefit (private profit), or private revenue minus private cost. This is the point at which marginal net private benefit is equal to marginal external cost. The two conditions are satisfied at the levels of output q_p^* and q_s^* in Fig. 1. That is, if there is no uncertainty about either private cost or environmental damage functions, and if both are continuous and differentiable, the first order conditions for a social optimum are satisfied by equality between marginal net private benefit and marginal external cost.

This is the simplest possible case, and it is well understood that its main value is heuristic. If the output of some economic activity that makes use of biological resources is excessive because the users are not confronted with the costs of the resources, then the solution lies in some mechanism which does confront them with those costs. This is the familiar basis of the Pigovian taxes and subsidies that are a primary part of the armoury of environmental economists, and the oft-cited 'user' or 'polluter pays' principle. The case is deceptive in its simplicity, but it is useful as a baseline to show how threshold effects change our perception of the problem.

Consider, once again, the external cost function. We have already remarked that the existence of threshold values for various parameters of the ecological systems on which economic activity depends implies that there are limits to the ability of such systems to withstand the stress imposed by depletion or pollution. If an ecosystem is stressed beyond these limits it will collapse. The thresholds accordingly imply discontinuities in or at least the non-smoothness of the external cost function. Taking the more severe disruption for illustration, it is immediate that the first order conditions will not hold wherever a discontinuity occurs at a point in the external cost function where $\partial E/\partial q < \partial \Pi/\partial q$. Such a case is shown in Fig. 2.

In this case the first order conditions cease to hold, and maximization of net social benefit implies that the socially optimal level of private output will

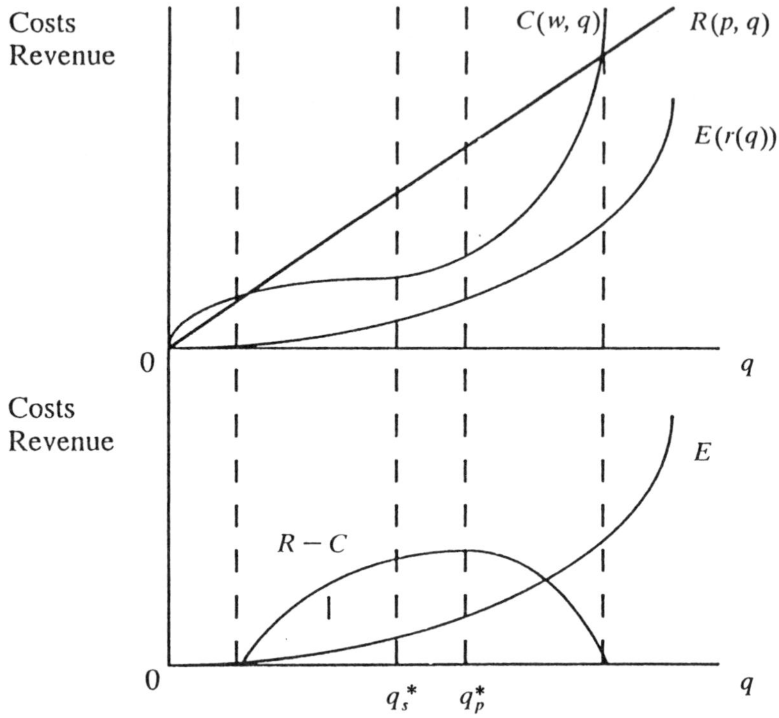

Fig. 1. Private and social optima: continuous external cost function.

be at the point of ecological discontinuity. This last point turns out to be of considerable importance from a policy perspective. It suggests that wherever output is constrained by the resilience of ecological systems, it will be optimal to 'stress' those systems up to the limits of their resilience. This is analogous to the historic practice of treating the maximum sustainable yield of a natural resource as the optimal level of output (irrespective of the set of relative prices or the rate of discount). It is particularly problematic in this case, however, because of uncertainty — both as to the precise threshold values and as to the costs of a potentially irreversible, unprecedented change in the self-organization of an ecosystem. For this reason economists and ecologists alike have argued for some physical restriction on the change in the stock of biological resources wherever they are believed to be important to the health of the ecosystems in which they are found, and wherever there is uncertainty about the resilience of those ecosystems in the neighbourhood of thresholds. We consider the problem of ecological discontinuities and the meaning to be attached to 'physical restrictions' around such discontinuities

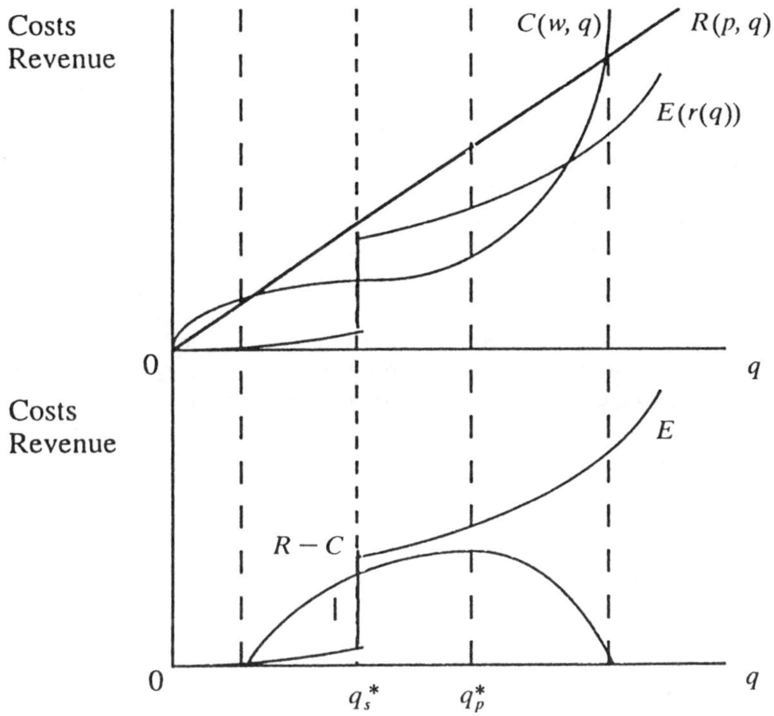

Fig. 2. Private and social optima: discontinuous external cost function.

below. What is important here is that the concern that resources lost in the reorganization of ecological systems may not only be irreplaceable, but may also be vital to the life support system, has translated as a call for the preservation of at least some natural assets.

3. Biodiversity Conservation: A Characterization of Optimal Policy

The most immediate implication of our discussion of the cost of biological resources to this point is that a sufficient condition for the private decisions of resource users to be socially optimal is the elimination of the gap between the private and social cost of those resources. That is, a sufficient condition for the maximization of net social benefit is the internalization of environmental externalities. However, internalization of environmental externalities is especially problematic in the case of externalities that involve the crossing of ecological thresholds. To ease discussion of this case, this section offers a brief characterization of optimal policy where there are dynamic feedbacks between biological resources and economic output, but in the absence of threshold effects and the uncertainty they bring.

The social problem in the general case is to maximize social net benefit over some meaningful (finite) time horizon by parametric control of the level of economic output, subject to the effect of the level of economic activity on the biological resources on which economic activity depends. Parametric control implies that the levers available to society are indirect. That is, the level of economic output is influenced through some set of incentives — in practice a set of market based instruments. Since the problem is dynamic we are able to include all of the elements in the total economic value of resources. To do this, we assume that there are two components to the social net benefit of economic activity. The first is a stream of income deriving from that activity, denoted by the integral function $Y[\cdot]$. The second is the value of the stock of biological resources that remains at the terminal time, T. It is denoted by an algebraic function $W[\cdot]$. The first of these approximates the sum of the private use values of biological resources. The second approximates the bequest, existence or (Weisbrod) option value of those resources. The external effects associated with the depletion of biological resources, are registered in the equations describing the evolution of the biological resource base: the equations of motion of the system.

Formally the problem is to maximize the sum of the two components of value, Π_s, through choice of the structure of incentives, k, where

$$\Pi_s = W[r(T), T]e^{-\delta T} + \int_0^T Y[r(t), q(t), t]e^{-\delta t}\, dt, \tag{3.1}$$

subject to

$$\dot{r}(t) = f[r(t), q(t), t], \quad 0 \leqslant t \leqslant T, \tag{3.2}$$

$$r(0) = r_0, \tag{3.3}$$

with

$$q(t) = q[k, w(t), t] \qquad 0 \leqslant t \leqslant T, \tag{3.4}$$

$$w(t) = w[q(t), t] \qquad 0 \leqslant t \leqslant T. \tag{3.5}$$

The vectors r and w have already been defined. r is a non-negative vector of biological resources in the public domain. In the context of the control problem, it describes the state variables of the system. Equations (3.2), the equations of motion, describe the evolution of the state variables as a function of both their natural rate of regeneration and the level of economic activity. q is a vector of economic outputs, and in this context is the vector of control variables — reached through the control parameters, or the set of economic incentives k. δ is the social rate of discount, and $t = 0 \leqslant t \leqslant T$ denotes time.

The first order conditions for the maximization of net social benefit provide sufficient insight into the general properties of an optimal policy for our purposes. These conditions include the following:

$$\lambda'(t) = -H_r[\cdot] = -Y_r(t)e^{-\delta t} - \lambda'(t)F_r(t), \tag{3.6}$$

$$\lambda'(T) = W_r(T)e^{-\delta T}, \tag{3.7}$$

$$0 = H_k[\cdot] = Y_k e^{-\delta t} + \lambda'(t)F_k(t), \tag{3.8}$$

where

$$H[r(t), q(t), \lambda(t), t] = Y[r(t), q(t), t]e^{-\delta t} + \lambda'(t)f[r(t), q(t), t], \tag{3.9}$$

is the Hamiltonian for the problem, and λ' is a vector of Lagrangian multipliers adjoining the equations of motion to the benefit function. Formally, the first order conditions give the necessary conditions on $\lambda(t)$, $H_k[\cdot]$ and $H_r[\cdot]$ if all parts of the first variation of the benefit function with respect to the control parameters are to be equal to zero in the neighbourhood of an optimal trajectory.

To interpret these conditions, note that the first two, (3.6) and (3.7), concern the social value of biological resources under an optimal policy. The adjoint vector, $\lambda'(T)$, measures the sensitivity of the benefit function to perturbations in the state variables along an optimal trajectory at time t. Hence $\lambda_i(t)$ is the marginal benefit of an increase in the availability of the ith biological resource: the social value or social opportunity cost of that resource. (3.6) states that under an optimal policy the rate of change in the value of any biological resource at time t, $(\lambda'(t) + \lambda'(t)F_r(t))$, is equal to the negative of the discounted change in net social benefit associated with its use. This last is the discounted value of the difference between marginal net private benefit and the intratemporal external cost of the biological resources used in production. If a resource is being depleted its value will be rising, and vice versa. The transversality condition, (3.7), states that under an optimal policy the social opportunity cost of the vector of biological resources at the terminal time T is the discounted terminal (bequest/existence/option) value of those resources.

The maximum condition, (3.8), then requires that controls be applied up to the point at which the marginal discounted net social benefit derived from the exploitation of biological resources, $Y_k e^{-\delta t}$, is equal to the marginal social cost of the biological resources foregone, $-\lambda'(t)F_k(t)$. In other words, private resource use decisions should be modified through the system of incentives so as to equate the social costs and benefits of the use of biological resources at the margin — given the rate at which future costs and benefits are discounted, and given the initial stock of biological resources.

Sustainability, in this formulation of the problem, is satisfied through the terminal benefit function, $W[\cdot]$. The social value of the stock of biological resources at the terminal time, T, is determined on a different basis from its value during the course of the policy. For $t < T$ the value of biological resources is assessed on the basis of the flow of goods and services it yields. We have suggested that $W[\cdot]$, on the other hand, is an increasing function of the set of opportunities kept open through the conservation of biological resources. In the absence of information on future preferences and technologies this might be approximated by the number of components of $r(T)$ that are strictly positive *and* of a value greater than their threshold value.[1] It may be helpful to think of $W[\cdot]$ as an ethically driven function, concerned with the impact of biodiversity on the choices available to generations yet to be born, and to think of $Y[\cdot]$ as a standard welfare function concerned with the benefits to existing generations of the exploitation of the biological resource base.

The inclusion of $W[\cdot]$ aside, the formulation corresponds to the simplest case discussed in section 2, and the requirements of an optimal policy match the requirements of that case reasonably closely. That is, under an optimal policy the structure of incentives should drive the privately optimal level of activity to the point at which the marginal social benefit of biological resource use is equal to the marginal social cost. Moreover, as in the simplest case, these general requirements say nothing about the properties of the incentives to be used in particular circumstances. These properties will depend on a range of things: the source of the wedge between private and social cost, institutional constraints on the choice of instrument, factors in the responsiveness of resource users to a change in the effective price of resources (income/output and substitution effects), and so on.

4. Biodiversity Conservation and Ecological Sustainability

Now consider the problem of threshold effects. The existence of thresholds or points of discontinuity in the evolution of ecological systems, and the uncertainty associated with threshold effects, creates a particularly intractable problem for policy. The cost of exceeding thresholds is fundamentally uncertain. Since an optimal policy depends on the assessment of the social costs and benefits of the use of biological resources, this is a real point of difficulty. It implies that the quality of the information available on the outcomes of a policy is highly sensitive to whether or not the policy threatens the thresholds. Policies that assure a balance between biological resources that are consistent with the resilience of ecosystems (that stay within the system thresholds) will have statistically predictable outcomes. Policies that change the balance between biological resources in the neighbourhood of thresholds will not. That is, they will have dynamic general equilibrium effects which cannot be estimated ex ante. In this section we consider the implications of this problem for the sustainability of development policy, and for the characteristics of the appropriate policy instruments. More particularly, we consider both the properties of an optimal policy under these conditions, and the significance of 'quantity' restrictions in the choice of instruments.

The aspect of the biodiversity problem that concerns us here is, therefore, the impact of a change in biodiversity on the resilience of ecosystems. In terms of the general formulation of the policy problem offered in section 3, the implication of loss of resilience is that any policy which compromises the resilience of ecosystems will have uncontrolled effects. Under such a policy, the system will accordingly be neither observable nor controllable. A small change in policy may have dramatic but unforseeable effects. This does not mean that there is no basis on which to make decisions, but it does mean that the properties of an optimal policy differ from those described in section 3. The existence of discontinuities and threshold effects compromises the controllability of the system. In these circumstances the problem is to maximize the social net benefits of the exploitation of biological resources subject both to the ecological dynamics of the system (the equations of motion) and to the protection of system stability or resilience.

Such a constraint implies that the level of economic activity be restricted in such a way as to preclude the possibility of the collapse of components of the system through overexploitation, *where there is reason to believe that the social costs of system collapse are major.* It is similar, in intent, to arguments by a variety of authors (Page, 1977; Pearce, 1987; Daly and Cobb, 1989) for the limitation of either the level of economic activity, or the form of economic activity. It is, however, differently motivated. The motivation in this case derives from the unpredictability of losses due to the crossing of ecosystem thresholds (Common and Perrings, 1992; Perrings, 1991a). Formally, it implies a problem with the same structure as that outlined in equations 3.1 to 3.5, but including what may be thought of as a stability or ecological sustainability constraint:

$$0 \geqslant h[r(t), q(t), t] \quad 0 \leqslant t \leqslant T. \tag{4.1}$$

This vector imposes a restriction on the admissible values of the set of biological resources, $r(t)$. If $h_i[\cdot] < 0$, the constraint is non-binding, implying that the biological resource concerned is not close to its threshold value. If $h_i[\cdot] = 0$ the constraint is binding, and the biological resource is in the neighbourhood of its threshold value. Adjoining this to the social net benefit function (3.1) by a non-negative vector of Lagrange multipliers, η, yields the augmented benefit function:

$$\Pi_s = W[\cdot]e^{-\delta T} + \int_0^T \{Y[\cdot]e^{-\delta t} + \lambda'(t)\,[f[\cdot] - \dot{x}(t)] + \\ + \eta'(t)\,h[r(t), q(t), t]\}\,\mathrm{d}t \tag{4.2}$$

in which

$$\eta'(t) = 0 \text{ if } h[\cdot] < 0$$

$$\eta'(t) \geqslant 0 \text{ if } h[\cdot] = 0$$

and the first order conditions for the maximization of net social benefit include:

$$\lambda'(t) = \begin{cases} -Y_r(t)e^{-\delta t} - \lambda'(t)F_r(t) & h[\cdot] < 0 \\ -Y_r(t)e^{-\delta t} - \lambda'(t)F_r(t) - \eta'(t)H_r(t) & h[\cdot] = 0 \end{cases} \tag{4.3}$$

$$\lambda'(T) = W_r(t)e^{-\delta T} \tag{4.4}$$

$$0 = \begin{cases} Y_k(t)e^{-\delta t} + \lambda_0(t)F_k(t) & h[\cdot] < 0 \\ Y_k(t)e^{-\delta t} + \lambda_0(t)F_k(t) + \eta'(t)H_k(t), & h[\cdot] = 0 \end{cases} \tag{4.5}$$

where

$$H[\cdot] = Y[\cdot]e^{-\delta t} + \lambda'(t)f[\cdot] + \eta'(t)c[\cdot]$$

is the Hamiltonian for the problem. The two forms of the adjoint and the maximum condition correspond to the case where the sustainability constraint is/is not effective along an optimal trajectory. If the constraint is non-binding, the corresponding multiplier has a zero value. If it is binding, the corresponding multiplier is positive and is in principle equal to the marginal benefit of the relaxation of the constraint. It implies that access to the resource is restricted.

The policy problem, in this case, is the protection of the thresholds. In our formulation of the problem this implies selection of an appropriate set of incentives, *k*, and this requires some explanation. In cases of this kind, it is generally argued that the appropriate instruments involve quantitative controls — examples being tradable or marketable permits, direct controls, and a variety of safe minimum standards. Indeed, instruments of this kind, particularly marketable pollution or emission permits and transferable quota, have long been applied to problems of both pollution and extraction (Tietenberg, 1990). But consider where quantitative restrictions enter such instruments. It is true that they all imply the existence of an underlying set of safe minimum standards (SMS), since it is these which determine the limits imposed on the aggregate harvest or level of emissions. But none implies direct control over the resource allocation decisions of the agents concerned. Despite the fact that a comparison of the welfare costs of 'price' and 'quantity' instruments is a major part of the conventional treatment of environmental policy (Baumol and Oates, 1988), what is ultimately at issue in all instruments is the form of the private cost function. The significance of quantitative 'limits' or 'standards' lies in their implications for the shape of the private cost function. The 'enforcement' of fisheries quota, emission standards or any other quantitative restriction, is nothing other than the imposition of penalties on resource users who exceed the standards. The effect of such penalties is that expected private marginal costs increase at the standard by the penalty multiplied by the probability that it will be 'enforced'. Notice that compliance with standards in the absence of penalties implies that those standards are a non-binding constraint on private activity, and so are irrelevant.

The significance of this point for the structure of incentives in the neighbourhood of thresholds is immediate. The standard result due to Weitzman (1974), that 'quantitative' restrictions dominate price incentives under uncertainty if the marginal social benefits are greater than the marginal private costs of reducing environmentally damaging activity, may be interpreted in this way. The welfare costs of an incentives policy administered under uncertainty will be greater, the greater the difference between the marginal private costs (net of incentives) and marginal social costs of environmentally damaging activity. In the neighbourhood of thresholds where the marginal social benefits of reducing environmentally damaging activity are very high, the welfare costs of error in estimating the marginal private benefits of environmentally damaging activity will be smaller, the more closely the marginal private cost of such activity, net of incentives, approximates the marginal social cost.

Aside from the problem of measurement errors, there are two sources of uncertainty to be addressed in the determination of incentives, both reflecting errors in the specification of the system dynamics. The first is uncertainty as to the threshold values for the state variables, the second is uncertainty as to the costs of exceeding thresholds. Whereas measurement error may be filtered out in many instances, error in the specification of system dynamics may not. Such errors exist because the available information set does not include a sufficient profile of the statistical properties of the unavailable information set to predict its conditioning effect on the future behaviour of the system. We cannot know what the effect of crossing various ecological thresholds will be. Leaving uncertainty as to the threshold values of biological resources to one side for the moment, if there is fundamental uncertainty about the net costs of exceeding thresholds the appropriate point of reference is the marginal private net benefit function. Supposing that the conjectured minimum net social cost of exceeding thresholds is positive, the penalties should then be greater than the maximum private net benefit of exceeding the threshold.

It is beyond the scope of this paper to determine an optimal penalty function for the case of fundamental uncertainty, but we can at least illustrate the important properties of a penalty function that will assure a socially optimal level of activity. Consider the example shown in Fig. 3.

The external costs of biological resource use are discontinuous at some level of economic activity, q_s^*, and are thought to jump at that point to some

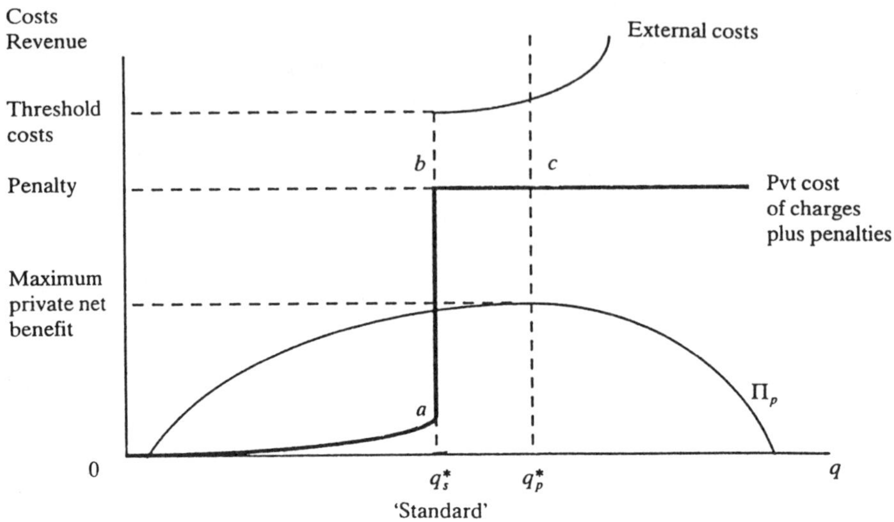

Fig. 3. Penalty function: thresholds known with certainty.

much higher level, and then to continue to increase with the level of economic activity. However, insufficient data exist to predict the distribution of such costs beyond the threshold.

The level of activity q_s^* is accordingly the level at which some ecological threshold is crossed with adverse but unknown effects on either the stock of the resource concerned or the flow of ecological services to which that resource contributes. If the threshold costs — the social costs of exceeding the threshold — are greater than the maximum private net benefit of exceeding the threshold, any arbitrary penalty between the two will result in the socially optimal level of activity. The private cost function illustrated in Fig. 3, described by the line 0abc, combines user charges for the expected social costs of biological resource use below the threshold level of activity (0a), and a penalty above the threshold level of activity (bc). Since any penalty greater than or equal to the maximum private net benefit of biological resource use would have the same effect, the difference between the penalty and the maximum private net benefit may be thought of as providing a margin for safety.

In this example q_s^* has been defined as the level at which some ecological threshold is crossed. If this is known with certainty, and if the penalty is enforced (is incurred with probability one), it will be sufficient to protect society against the costs of exceeding the threshold. However, the rationale for safe minimum standards lies in the fact that the threshold values for many species and ecosystems — just as the breaking point of physical materials —

is not known with certainty. In this case, the standard which triggers the penalty would be such as to reduce the 'risk' of overshooting ecological thresholds and the consequential losses to society to acceptable levels (Ciriacy-Wantrup, 1952). Notice that there is no necessary relationship between the use of standards in these cases and the use of minimax strategies — as was argued by Bishop (1982). Although wherever standards are set conservatively relative to the ecological thresholds they are designed to protect, it does indicate a degree of 'risk' aversion. In the general formulation of the problem of thresholds offered here, the SMSs refer to the minimum stocks of biological resources consistent with the resilience of the ecosystems of interest, and are captured in the constraints $0 \geqslant h[r(t), q(t), t]$. In the example considered above, this implies that the standard which triggers the penalty will be associated with a lower level of activity than that which actually breaches the ecological threshold. That is, the penalty function 0abc will lie to the left of the discontinuity at q_s^*. This is illustrated in Fig. 4.

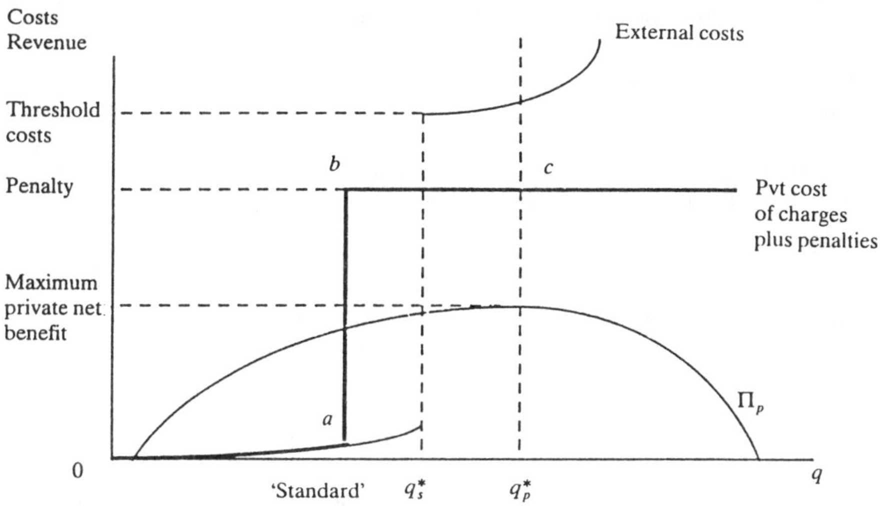

Fig. 4. Penalty function: thresholds not known with certainty.

The existence of ecological thresholds associated with the size and mix of biotic species is a major, but still neglected, aspect of the biodiversity problem. The development of threshold instruments has been largely restricted to species that are 'endangered' as a consequence of commercial exploitation. Quota for the harvesting of fish, shellfish, terrestrial mammals and so on are common examples. But comparatively little attention has been directed to threshold instruments designed to protect the resilience of ecosystems, where this is sensitive to the mix of species, and where loss of resilience due to a change in biodiversity has potentially severe and irreversible consequences. Part of the reason for this lies in the fact that the ecological services of biodiversity are in the nature of a 'layered' public good. The external costs of biodiversity loss are not only reciprocal but also highly diffused — occurring in multiple jurisdictions, and over multiple time periods. But while this complicates matters, it does not prevent the development of appropriate instruments (Pearce and Perrings, 1993).

5. Concluding remarks

Although threshold effects are not the only source of special difficulty in the conservation of biodiversity, they are one of the least tractable. This is partly because the fundamental uncertainty with which they are associated demands a different approach to decision-making from that conventionally adopted in both theoretical and applied economics, and partly because they invalidate normal tests for efficiency in the allocation of resources. It is no longer possible to equate marginal net private benefit with marginal external cost. The result is that the economic instruments required to protect thresholds cannot be motivated by conventional economic objectives, such as the maximization of expected utility or welfare, but must rely on non-economic criteria. Whether these instruments are defined by reference to the physical restrictions they imply (safe minimum standards) or to the cost functions associated with the enforcement of those restrictions (environmental levies or fines), they must be motivated by a judgement about the socially acceptable margin of safety in the exploitation of the natural environment. This is essentially an ethical judgement. The much-discussed Precautionary Principle implies an ethical statement about the appropriate way to respond to collective ignorance about the future effects of economic activity (Perrings, 1991b).

Standards are in fact the oldest and most extensive of all biodiversity conservation instruments. Taboos on the exploitation of particular species or habitats, harvest limits, closed hunting and fishing seasons are all examples of conservation instruments that have a very long history in most societies. Moreover, at the microeconomic level, they are re-emerging as favoured instruments amongst environmental economists. Since they can be used to set only the aggregate limit on the exploitation of species or habitats, leaving the allocation within that limit to the market, they combine efficiency in allocation with safety of the limit. Marketable fishing or emission permits are modern examples of a very old instrument. There is not nearly the same level of support for standards at the macroeconomic level. Daly's (1993) arguments for the restriction of the scale of aggregate economic activity — for the adoption of an economic equivalent to the plimsoll line — have much less support amongst environmental economists, and it has to be said that if the appropriate restrictions are in place at the microeconomic level, it is not clear what the advantage of an aggregate restriction would be. The problem of biodiversity loss derives from the fact that individual private users of resources have no incentive to take the social cost of their actions into account. It is therefore at the level of the individual user that conservation instruments should work.

All standards have the function of restricting the sphere of economic activity, but the motivation behind them differs. Quota in fisheries tend to be motivated by reference to the need to regulate access to common property resources. In this paper, standards of the same sort are motivated by reference to the existence of threshold effects and the fundamental uncertainty with which they are often associated. Of course the two need not be mutually exclusive. Threshold effects exist in the exploitation of common and private property resources alike. In either case, resource users ignore the social costs of their actions and the solution to the problem lies beyond the play of free markets. The socially responsible strategy is therefore one that bounds the sphere of economic activity. Though the costs and benefits of imposing limits cannot be computed precisely (in the nature of the problem), they will be appropriate wherever there is sufficient reason to believe that the net present value of the imposition of the restrictions is positive.

Acknowledgment

David Pearce acknowledges the support of the UK Economic and Social Research Council (UK ESRC). CSERGE is a designated research centre of ESRC.

Note

[1] If one was interested only in the preservation of examples of all species, it would not matter if the numbers of each were below the critical minimum required for the resilience of the ecosystems of which they were a part. In this case, $W(\cdot)$ would be an increasing function of the number of strictly positive components of $r(T)$, and would not depend on the value of those components.

References

Arrow, K. J. and A. C. Fisher (1974), 'Environmental Preservation, Uncertainty, and Irreversibility', *Quarterly Journal of Economics* **88**(2), 312–319.

Baumol, W. and W. Oates (1988), *The Theory of Environmental Policy*, Cambridge University Press, Cambridge.

Bishop, R. C. (1982), 'Option Values: An Exposition and Extension', *Land Economics* **58**, 1–15.

Bromley, D. (1991), *Environment and Economy: Property Rights and Public Policy*, Blackwell, Oxford.

Ciriacy-Wantrup, S. V. (1952), *Resource Conservation: Economics and Policies*, University of California Press, Berkeley.

Common, M. and C. Perrings (1992), 'Towards an Ecological Economics of Sustainability', *Ecological Economics* **6**(1), 7—34.

Daly, H. and W. Cobb (1990), *For the Common Good: Redirecting the Economy Toward Community, the Environment and a Sustainable Future*, Beacon Press, Boston.

Dasgupta, P. (1991), 'The Environment as a Commodity', in D. Helm, ed., *Economic Policy Towards the Environment*, Blackwell, Oxford, pp. 25—51.

Krutilla, J. V. (1967), 'Conservation Reconsidered', *American Economic Review* **57**(4), 778—786.

Page, T. (1977), *Conservation and Economic Efficiency*, Johns Hopkins University Press, Baltimore.

Pearce, D. W. (1987), 'Foundations of an Ecological Economics', *Ecological Modelling* **38**, 9—18.

Pearce, D. W. and C. Perrings (1993), 'Biodiversity Conservation and Economic Development: Local and Global Dimensions', in C. Perrings, K. G. Mäler, C. S. Holling, B. O. Jansson, and C. Folke, eds., *Biodiversity Loss: Ecological and Economic Issues*, Cambridge University Press, New York, forthcoming.

Pearce, D. W. and R. K. Turner (1990), *Economics of Natural Resources and the Environment*, Harvester-Wheatsheaf, London.

Perrings, C. (1991a), 'Ecological Sustainability and Environmental Control', *Structural Change and Economic Dynamics* **2**, 275—295.

Perrings, C. (1991b), 'Reserved Rationality and the Precautionary Principle: Technological Change, Time and Uncertainty in Environmental Decision-Making', in R. Costanza, ed., *The Ecological Economics: The Science and Management of Sustainability*, Columbia University Press, New York.

Randall, A. (1991), 'Total and Nonuse Values', in J. B. Braden and C. D. Kolstad, eds., *Measuring The Demand for Environmental Quality*, Amsterdam: North Holland, 303—322.

Tietenberg, T. H. (1990), 'Economic Instruments for Environmental Regulation', *Oxford Review of Economic Policy* **6**(1), 17—33.

Turner, R. K. and D. W. Pearce (1992), 'The Ethical Foundations of Sustainable Development', *Advances in Human Ecology* **1**, 177—195.

Weisbrod, B. (1964), 'Collective Consumption Services of Individual Consumption Goods', *Quarterly Journal of Economics* **77**, 71—77.

Weitzman, M. L. (1991), 'On Diversity', *Environmental and Resource Economics* (forthcoming).

Weitzman, M. L. (1974), 'Prices vs Quantities', *Review of Economics Studies* **41**, 477—491.

[7]

Oxford Economic Papers 46 (1994), 800–821

THE ECONOMICS OF EXTINCTION REVISITED AND REVISED: A GENERALISED FRAMEWORK FOR THE ANALYSIS OF THE PROBLEMS OF ENDANGERED SPECIES AND BIODIVERSITY LOSSES

By TIMOTHY M. SWANSON

Faculty of Economics, University of Cambridge, Austin Robinson Building, Sidgwick Avenue, Cambridge CB3 9DD, and CSERGE, University College London

1. Introduction

THE ECONOMICS of species extinction was initially modelled by Clark (1973). (See also Clark and Munro 1978; Clark *et al.* 1979.) It built upon Gordon's (1954) model of fishery over-exploitation to demonstrate the conditions under which open access regimes might generate non-viable stock levels of a given species. His work provided the framework for most of the analysis that has followed since. Given that the past 20 years have seen the first attempts at policies to address the problem of species extinction, the analysis of extinction has achieved some amount of policy relevance. Many of the ideas contained within the Clark and Gordon models of 'open access overexploitation' serve as the basis for the continuing discussions concerning the optimal policies for the regulation of endangered species.

However, during this period, the perception of the nature of the problem of species extinctions has altered quite a lot, from well-focused concerns about the endangerment of individual well-recognised species to a much broader concern that includes potential losses of millions of virtually unknown life forms. The latter sort of problems are usually categorised as 'biodiversity losses'. This paper constructs a general framework for addressing species extinction and biodiversity losses, by analysing the more fundamental forces affecting species and their habitats across the whole earth. This is a general problem that has usually been discussed in many different contexts, in regard to concerns about rainforests, individual species (such as the African elephant), or general genetic diversity.

The generalisation of the analysis of extinction requires that it be brought back 'on-shore', within the context of the myriad of conflicts and relationships that this implies. The original model placed the analysis of species extinction squarely within the previous literature on fishery economics. Extrapolation from this context was problematic because there are far fewer human uses competing for oceanic resources than there are for terrestrial ones. Extinction is an economic problem precisely because it is a resource allocation problem: it derives from the human allocation of the various life-sustaining resources between competing species. A more generalised analysis of extinction must take into consideration all of the various resources that are required to sustain a particular form of life within the context of the environment in which it lives.

Finally, it must be stated that the entirety of this paper is positive, rather than normative, in motivation. The object of the enquiry is to ascertain the nature of the forces that drive extinctions so that they might better be controlled; why they should be controlled is outside the scope of this paper but covered by me elsewhere (Swanson 1991, 1992, 1993, 1994).

Therefore, the general object of this paper is to create a more generalised framework for the analysis of both endangered species and biodiversity problems. Specifically first, to integrate the problems of endangered species and biodiversity losses; second, to bring the analysis of species extinction back 'onshore' by incorporating the other resources required for survival within the model; and, third, to illustrate the dramatically different policy implications that derive from a more generalised analysis. The paper will start in Section 2 by giving an illustration of the Clark model, and its policy implications, in the context of the African elephant. Section 3 will then integrate the problems of well-known species extinctions such as the elephant with the processes afflicting hundreds of thousands of lesser known species—the biodiversity problem. Here, the economics of extinction will be treated within a generalised framework of choice regarding productive biological assets. Then, in Section 4, this investment-based model is revised to allow the incorporation of all the various resource constraints that operate upon a terrestrial species. Section 5 discusses the implications of these revisions to the economic model of extinctions. It demonstrates that the different types of endangerment (over-exploitation, conversions, diversity losses) are merely different routes to the same ultimate result, driven by the same fundamental forces. This indicates that the policies for addressing all forms of endangerment, extinction, and diversity losses must be created in a fashion that will address this more fundamental problem. These reforms will often require policies much different from those that have been created in order to address the proximate (as opposed to the fundamental) cause of endangerment.

2. The economics of extinction revisited: the framework and its implications

Extinction problems are usually thought of in the context of a few very noticeable examples of endangered species, such as the African elephant. In fact, the decline of the African elephant during the decade of the 1980s is a classic example of the ostensibly good fit between the Clark model and actual species decline, as well as the policy implications that derive from that analysis. It will be used as a case study to illustrate the message of this paper.

During the past decade it is generally estimated that the population of the African elephant declined by about half, from 1,343,340 in 1979 to 609,000 in 1989 (Douglas-Hamilton 1989). In addition, the trade in ivory, the principal product derived from this species, also doubled between the early 1970s and the early 1980s—from an average of about 550 tonnes to an average of about 1,000 tonnes (Barbier *et al.* 1990). On account of these trends, population modellers predicted the imminent extinction of the species, over a period of about 20 years (Caughley and Goddard 1975; Renewable Resources Assessment Group 1989).

How does the Clark model explain the decline and possible extinction of an individual species, such as the African elephant? It does so in relation to three factors:

(i) open access to the resource;
(ii) relative price to cost of harvesting the resource; and
(iii) relative growth rate of the resource.

These three factors are sufficient to produce a bioeconomic model showing that the optimal stock of the resource goes to zero in the steady-state. In essence, these conditions create incentives for continued harvesting of the resource, even to an extent incompatible with the capacity of the resource to regenerate itself.

2.1. *A demonstration of the Clark model*

A simple model will illustrate how these conditions combine to provide the conditions for extinction. Let x represent the stock of a diverse biological resource, where y is the corresponding flow variable (the periodic harvest). The cost of harvesting the resource is then negatively related to the stock of the resource, while positively related to the flow, $c(y, x)$. Taking an example from Dasgupta (1982), we may say that the cost function is of the following generalised form, representing the costs of harvest.

$$C(y, x) = By^{\alpha}x^{-\gamma} \tag{1}$$

Using the assumption (i) above, open access conditions of harvest imply that the harvest will continue until the average cost of the harvest of another unit of the resource is equal to its price, i.e. $P = AC$ (Gordon 1954). This is the open access equilibrium.

$$P = By^{\alpha-1}x^{-\gamma} \tag{2}$$

We also know the annual change in stock over time, x, will be equal to the intrinsic growth rate of the resource, $H(x)$, less that year's harvest, y.

$$\dot{x} = H(x) - y \tag{3}$$

Combining (2) and (3) gives an equation that demonstrates that the annual change in the resource stock is: (i) a positive function of the growth rate of the resource; and (ii) a negative function of the 'harvesting effort' (which is positively related to the price-cost ratio of harvests). Bioeconomic equilibrium results where these two forces are in balance.

$$\dot{x} = H(x) - (P/B)^{1/(\alpha-1)}x^{+\gamma/(\alpha-1)} \tag{4}$$

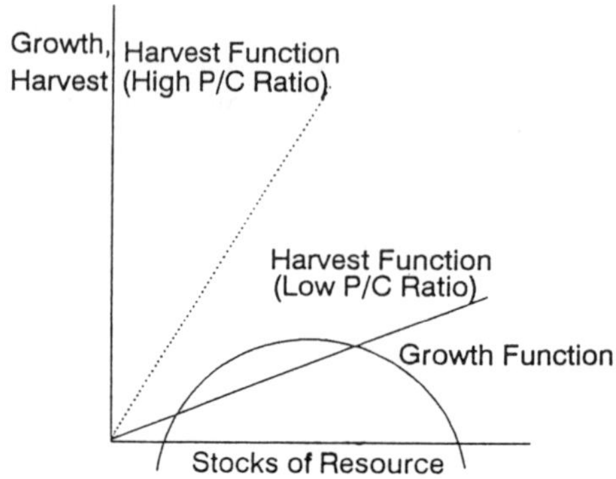

FIG. 1. Clark bio-economic model of extinction

In Fig. 1, the intersection of these two curves occurs at the stock level that will exist in the steady state equilibrium. Bioeconomic equilibrium represents the equation between the forces of biological regeneration and societal exploitation.

The dashed line in Fig. 1 indicates the nature of the possibility of 'economic extinction'. In the case of this harvesting function, there is no intersection between the growth function and the harvest function, at any stock level. This means that the incentives to harvest the resource exist at all stock levels, and the growth of the resource is insufficient to maintain a population in the face of these pressures. The bioeconomic equilibrium is extinction. The basic principle derived from this analysis is: low growth rates combined with relatively high price–cost ratios, under the assumptions of the Clark model, result in forces pointing to the ultimate extinction of the species.

2.2. *The policy implications of the original model*

The policy implications of this model are straightforward. Extinctions result from low growth rates and high price–cost ratios, in the context of poorly managed access. Since it is assumed there is little to be done about affecting resource growth rates or habitat conditions, the implication is then to work through the price–cost ratio. In the context of Fig. 1, an endangered species would be found initially in the situation of a harvest function represented by the dashed line. Any act that would either decrease the price of the resulting products, or increase the costs of the production process, would shift the harvest function downwards (to the unbroken line in Fig. 1) creating the possibility of a stable bioeconomic equilibrium.

The application of this paradigm leads to some fairly straightforward policy conclusions. A quick fix for the endangered species within this framework is the enforced criminalisation of the production process. This is believed to be effective because it should have two important impacts on the prevailing incentive structure: first, criminalisation should eliminate from the market some demand for the products; and second, it should increase the cost of accessing and marketing the resource. Therefore, the policy of enforced criminalisation is intended to have positive impacts working on both sides of the price–cost ratio, with, it is hoped, the resulting downward shift in the harvest function restoring bioeconomic equilibrium at a stock greater than zero.

This is precisely the policy shift that occurred as a result of the decline of the African elephant in the 1980s. The international community has acted to impose an international 'ban' on the trade in ivory effective from 1990. Acting within the context of the Convention on International Trade in Endangered Species (CITES), the parties passed a resolution to list the African elephant on Appendix I of the treaty, effectively disallowing all further trade in that species' products and requiring all member states to enforce this ban (Barbier *et al.* 1990). The ivory trade had been criminalised in order to save the elephant. Was this the correct analysis of and approach to the problem of species endangerment (Swanson 1993)?

3. The economics of extinction generalised: species extinction and biodiversity losses

Despite the existence of these international and national policies to address the problem of species endangerment, the natural world is nevertheless heading toward a period of severe mass extinctions. Within the fossil record, there is evidence of about five occasions on which 50% or more of the then-existing species became extinct over a short period of time. Several of the word's most eminent scientists project a mass extinction rivalling these that is set to occur over the next century (Ehrlich and Ehrlich 1981; Lovejoy 1980). Even if the projections of a 'doomsday scenario' are inaccurate, the current rate of extinction is extraordinary. Averaged over the 4.5 billion years of biological history, the long term rate of extinction is approximately nine species per million per annum. Scientists estimate that the current rate of extinction is three or four orders of magnitude above this background rate (Wilson 1988).

Most of these extinctions are not so newsworthy as the elephant's decline, and so they go largely unnoticed. In fact, many if not most of these disappearing species are being destroyed without ever having been classified. This is known as the problem of biodiversity losses, i.e. the narrowing of the gene pool through the loss of many species as well as the loss of many of the varieties within any given species. It is the problem of the general homogenisation of the natural world, and the extinctions that are resulting from this process (Swanson 1994). Within the framework of this paper these two problems can be considered as cases of the same general problem, only applied to more and less noticeable species.

3.1. *The problem of endangered species generalised: societal asset portfolios*

The problems of specific endangered species and general biodiversity losses have their sources in the same fundamental economic problem, *viz.* the human choice of the set of biological assets upon which society relies. Human societies must select a portfolio of assets from which they then derive a flow of benefits, and one important part of this portfolio is the range of biological resources upon which we depend. These resources are assets (stocks which generate flows) simply by virtue of being biological in nature.

One would expect in the context of a developing society, large changes in that society's portfolio to occur. One important aspect of the development process is the accumulation of new forms of capital. Given capital constraints it will sometimes be optimal to disinvest in one asset and invest the receipts in another asset; that is, it will be socially optimal to engage in conversions between assets in order to equilibrate returns. Sometimes a natural asset will be converted to a human-made asset (Solow 1974). That is, human society will disinvest in the natural resource in order to invest in human-made capital items, such as machinery. Conversions occur between biological assets as well; a biological asset will be converted to a different form of biological asset, such as the clearing of a forest for a cattle ranch (Swanson 1991).

It is this process of disinvestment that lies at the base of the decline of any species. Although there are various manners in which the conversion process might be manifested (the mining of stocks, overexploitation, land use conversions), the fundamental force driving species' decline is always the relative rate of investment by the human species. It is the human choice of another asset, over a given biological asset, that results in the inevitable decline of that species. Extinctions, whether of specific breeds or of general diversity, are the result of their non-inclusion in the human asset portfolio. The remainder of this paper demonstrates how this is the case.

3.2. *Species as productive assets—the model generalised*

The generalised model of biological extinction focuses on the investment potential of different varieties of biological resources. It is based on both the Clark model, and the Solow model of asset substitution (Solow 1974). It analyses extinction as a process of human choice of which productive assets to retain in the natural portfolio. The general dynamic model, building upon that found in the fisheries literature, may be set up as follows (see e.g. Conrad and Clark 1987). The following variables must be defined, each assumed to be subscripted for time. Also, the variables will all be assumed to apply to a specific diverse resource within a given country. This terminology will be retained throughout the remainder of the paper.

- $p(y)$ the inverse demand curve—a function of the aggregate harvest, y
- y_i the harvest of the ith agent accessing the resource
- $c(x)$ the (constant) average cost of harvest—a function of stock level
- $H(x)$ the growth of the resource—a function of the stock of the resource
- r the marginal cost of capital within a country
- λ the 'shadow price' of a unit of resource stock (equates with resource 'rent' in equilibrium)

At this juncture it is necessary to interject a few stylised facts in order to define the appropriate societal objective regarding its diverse resource stocks. First, when discussing the problem of global diversity, it is necessary for the discussion to focus on a few states in the developing world, as the vast majority of the world's biological diversity resides in a mere handful of states (McNeely *et al.* 1990). Secondly, these states are some of the poorest in the world, and they are therefore justifiably interested in the maximisation of the appropriable values from their diverse resources (Swanson, 1991). Thirdly, the greatest values that are attached to diverse resources generally flow from the developed world, where these forms of resources are relatively more scarce. These assumptions imply that the state with significant quantities of diverse resources is primarily concerned with the capture of the maximum appropriable producer surplus to be derived from its natural assets. This is the societal objective depicted in eq. (5).

$$\max_{y} \int_{0}^{\infty} [p(y)y - c(x)y]\, e^{-rt}\, dt \qquad (5)$$

$$\text{s.t.} \quad \dot{x} = H(x) - y$$

This society will sove this problem in order to determine the optimal stock levels for each of its diverse resources. From the perspective of this producer/state, species are differentially valuable as productive assets depending upon the appropriable values that they generate. Given scarce investment capital, the decision as to which species to maintain within the societal portfolio of assets will depend upon their relative productivity. For purposes of later comparisons, the optimal flow (y^*) is shown in eq. (6), then the optimal stock level (x^*) is depicted in (7).

$$y^*\colon \lambda = p\left[\frac{1}{\varepsilon_d} + 1\right] - c \qquad (6)$$

where ε_d is the elasticity of demand for the diverse resource.

$$x^*\colon \frac{\dot{\lambda}}{\lambda} - \frac{c'(x)y}{\lambda} + H'(x) = r \qquad (7)$$

We are concerned here with the steady-state stock levels (i.e. for $\dot{\lambda} = 0$, $\dot{x} = 0$), because these are indicative of the societally targeted stock levels for this particular biological asset. Of course, this target might take quite a long time to reach, but the analysis of extinction must necessarily be concerned with the very long run; it means little to speak of a long transient phase prior to extinction, unless that phase persists for at least hundreds of thousands of years. Equation (7) implies that the resource would be maintained at a stock level that equates the return from that asset with the return from other assets. The return to the stock of a given diverse resource is dependent upon two factors: (i) the relative growth rate (which slows as the stock level reaches carrying capacity); and (ii) the cost of access—'search costs' (which decline with

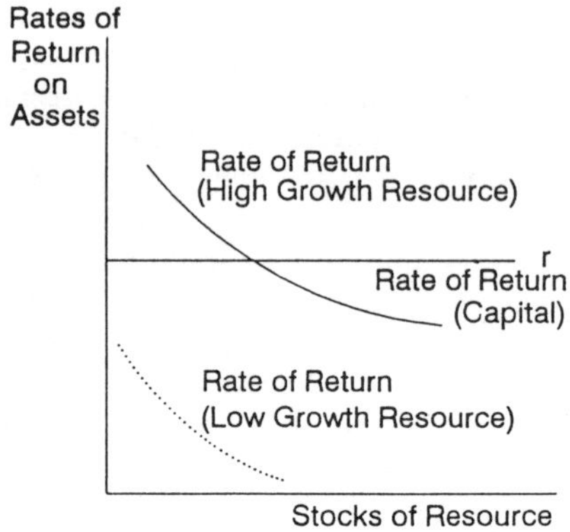

FIG. 2. Investment-based mode of extinction

increasing stocks). Society will target the stock level (through investments or disinvestments) that generates a competitive return (*r*).

Figure 2 demonstrates this point. An investment-based model indicates that the stock level for each biological resource would be targeted at the level that equilibrates its returns with all other productive assets in the economy. Since the marginal return on further investments in the stocks of a given biological asset will decline, as the carrying capacity of its habitat is reached, there will always exist an equilibrium stock level for a given species. It is clear, within this framework, that the nature of an extinct resource is that of a noncompetitive asset. The dashed line in Fig. 2 represents a very slow-growing resource which fails to reach the prevailing marginal rate of return from assets within the economy. There are incentives for the conversion of the entire stock of this asset to other, more productive forms of capital in order to achieve a better return.

Therefore, slow growth relative to other assets in the economy is in and of itself a route to species extinction. Resources, even biological resources, must be competitive as productive assets if there is to be a force for their retention in a world of scarce resources. This point was first discussed in the context of the debate concerning the 'optimal harvest strategy' applicable to the blue whale; certain very slow-growing biological resources are subject to pressures for their complete disinvestment (Clark 1976; Spence 1975; Dasgupta 1982). It was further analysed by Clark and co-authors (Clark and Munro 1978) and several others in a series of articles (Cropper *et al.* 1979; Cropper 1988; Swallow 1990; Swallow *et al.* 1990).

3.3. *Open access regimes and investment incentives*

The results derived in the previous section are true even from the perspective of a perfectly well-managed biological resource; non-competitive growth rates are sufficient conditions for asset disinvestment. Slow growth is not the only route to extinction however. An alternative route to the creation of a force for disinvestment is the operation of an open-access regime. Open-access regimes can render otherwise viable biological resources largely valueless as economic assets.

The impact of uncontrolled access on a resource can be demonstrated within the generalised model set forth above, in the context of a decentralised group of resource harvesters. In this case, each of a large number of competing harvesters attempts to maximise its own profits from individual harvesting activity (y_i). The maximisation problem for each individual harvester is then as follows

$$\max_{y_i} \int_0^\infty [p(y)y_i - c(x)y_i]\, e^{-rt}\, dt \tag{8}$$

$$\text{s.t.} \quad \dot{x} = H(x) - y$$

Open-loop Nash equilibrium conjectures will be used to determine how the competing harvesters will use the open access resource. The use of open-loop conjectures implies that harvesters do not take account of stock effects in their decision making. Nash conjectures imply that there is no strategic interaction between harvesters; each harvester takes the harvests of others as a given. Open-loop Nash equilibrium conjectures are plausible in the context of an open-access regime, precisely because open access usually implies a large number of competing harvesters. We will return in the next section to the precise meaning of an open-access regime, but at this point the sole implication is a large number of competing harvesters.

Under these assumptions, the first order conditions for an optimal solution to each of the harvesters' decision problems are as follows for respectively optimal flow (for competitive entry) and optimal stock.

$$y_i^*: \lambda = p - c \tag{9}$$

$$x^*: \frac{\dot{\lambda}}{\lambda} - \frac{c'(x)y}{\lambda N} + H'(x) = r \tag{10}$$

$$\text{for } \dot{\lambda} = 0 \quad \text{or} \quad \lambda = \frac{c'(x)y}{(r - H'(x))N}$$

Open access operates to remove all incentives for investment in the diverse resource, via the long-known process of 'rent dissipation' (Gordon 1954; Cheung 1970). That is, under open access, the shadow price of the resource stock goes to zero. With a value of zero, there are no incentives for the asset to remain invested in that particular form. Equations (9) and (10) together demonstrate this process of rent dissipation occurring in the context of open access management. From (10), as $N \to \infty$, $\lambda \to 0$. From (9), this implies that $p = c$; that is, the price received from the flow of the diverse resource is equal to the costs (of the other factors) involved in its production. There is no rent accruing to the resource itself. Therefore, under open-access management regimes the resource rental value is driven to zero through competitive access.

This means that the use of an open-access regime for the harvesting of any resource is equivalent to deeming the resource as one suitable only for conversion. Irrespective of the competitiveness of the natural asset, the application of an open access regime over-rides its inherent characteristics and renders it an unworthy investment.

3.4. *Conclusion—extinctions as the result of conversion*

All biological resources are by their nature productive assets. Assets are kept in a developing society's portfolio on account of the appropriable flows that they return. Nature (the evolutionary process) initially endowed this world with positive stocks of all biological assets; it performed the role of allocating niches between competing life forms. Now this role has been unsurped by the human species. For whatever reason, it is now within the capacity of human societies to determine the portfolio of life forms that will continue to exist on earth; once this capacity exists, these life forms are subject to the same forces that determine the stock levels of all the other forms of productive assets under human jurisdiction. The general nature of these forces is equivalent to portfolio balancing, and the decline of individual species is the result of disinvestment. This process of conversion between assets threatens all biological assets (i.e. species) with depletion in accord with their relative productivities. The general nature of the extinction threat to all species, both the more and the less noticeable, is the same. The fundamental cause of extinction is that a species will be seen as an inferior asset, and thus be excluded from the human portfolio of assets.

4. The economics of extinction revised

Exclusion from the human portfolio of biological assets is a sufficient condition for biological extinction, but it is not the force that acts directly upon the species to bring about its decline. Exclusion from the human portfolio is only the most fundamental force at work; it is the 'targeting' of a zero stock level for a given species. This brings into play the forces for conversion, and then the local decision makers may select any one of a number of different routes to the ultimate destination. These are the more consequential actions, resulting from this more fundamental force, that actually bring about the decline of the species. It is the confusion between the consequential and the fundamental that gives rise to inappropriate policymaking, and thus it is important to clearly distinguish between the two.

There are three distinct forms of consequential forces that actually bring about species decline. That is, the fundamental decision not to include a specific biological asset within the human portfolio may ultimately affect the stock level of that asset via:

(i) stock disinvestments—the removal of the stocks of the asset for sale and allocation of receipts to the acquisition of other, more competitive assets;
(ii) base resource re-allocation—the refusal of an allocation of base resources (land, water, foods) to the asset, and allocation of these base resources to other, more competitive assets; or
(iii) management services re-allocations—the refusal of an allocation of management services to the asset, and allocation of these services to other, more competitive assets.

All of these routes have the same observable effect (i.e. species decline); however, it is more important to show that each has the same fundamental source (i.e. non-inclusion in the portfolio of assets).

4.1. *Investments in resource stocks*

The requirement of asset competitiveness for investment in resource stocks was discussed in the previous section. As indicated in Fig. 2, if a resource is unable to generate a competitive return, there will be the incentive to convert it into another asset form. This is the scenario of 'optimal exinction' developed by Clark (1973).

Although investment in stocks is necessary in order to avoid extinction, it is not often the case that a species is eradicated by reason of this form of stock disinvestment. For example, if there is no demand for the species' products, then there is no incentive for stock level disinvestments of this type. Therefore, inadequate growth rates do little to explain the nature of the 'biodiversity problem', i.e. the loss of millions of unknown and unclassified species.

Also, there is mounting empirical evidence that the decline of many of the well-known and well-documented species also does not fit this pattern. This is because, although there are clear and substantial markets for these species' products, their values are not being realised by their producers. For example, studies have shown that little, if any, of the rental value of the resource is being appropriated by the harvesters of many natural resources, e.g. elephants, rhinos, and live birds (Swanson and Barbier 1992; Swanson 1992d; Repetto 1988).

Therefore, the non-competitiveness of the stocks of a resource in terms of growth certainly does generate incentives for direct disinvestment, but this does not explain a very substantial proportion of the current (and projected) extinctions. In order to explain most species endangerment and general diversity depletion, it is necessary to appeal to other considerations.

4.2. *Investments in base resources*

The resources required for the continuing survival of an individual species are more than one. Most obviously, in order for the species to be able to regenerate itself, there must be some non-zero stock of the species retained (see above). Second, in order for that stock to sustain itself within the biosphere, there must be an allocation of a flow of biological services, or base resources.

All living organisms are dependent upon various biological necessities for their sustenance (food, light, air, water, etc.). Since human societies now have control of these base natural resources (by virtue of their control over land-use decisions), biological assets must compete for an allocation of these resources as well. Initially, nature itself (i.e. the evolutionary process) determined the allocation of base resources between life forms, in accordance with relative fitness. Now this role has been usurped by the human species. The natural phenomenon of extinction has been translated into the socio-economic sphere because the fates of individual species are now determined within the realm of socially optimal resource allocation: the allocation by society of resources between competing life forms (Swanson 1994).

This implies two fundamental revisions of the basic bioeconomic model. First, the bioeconomic model of extinction must be revised in order to focus on the fundamental importance of investments in these base resources for the survival of terrestrial species. What is required is the jettisoning of an implicit assumption derived from fishery economics, with the effect of bringing the economics of extinction 'on shore'. Specifically, it is important to revise the model in order to provide for the possibility of competing uses for the resources upon which a species relies, and therefore to provide for the necessity of realising a competitive return for the species' use of these resources.

The growth function is the focal point for this revision of the pre-existing model. In general, a logistic form of growth function has been applied for the analysis of population dynamics for most of the past 150 years. This function relates the flow from a stock, $H(x)$, to the level of the existing stock, x, and to the extent to which that level is less than the available carrying capacity, R. In essence, the logistic growth function implies that if there exists a vacant niche to be filled, the individuals of a species will jointly place more of their aggregate energy into reproduction in order to fill the niche. However, if the species has previously expanded to fill its niche, then most of the energy used by the species will be absorbed in maintenance rather than growth. Therefore, growth rates are linked to the existing stock of a species relative to its available niche. This provides the growth function of the general form, $H(x) = x(R - x)$, with its distinctive logistic shape.

Since the bioeconomic model was developed within the literature of fishery economics, it has always been assumed that the carrying capacity for a given species was a fixed, exogenously-given characteristic of the seas it inhabited. And, since humans had few other competing uses for this habitat, the opportunity cost of using it for fish production was assumed to be near zero. These assumptions are wildly off the mark in regard to terrestrial resources. On land, the amount of habitat available to a given species is endogenous to the process of determining how many individuals of that species will continue to exist, and it is probably the single most important factor determining species viability in the short and medium run. The decision concerning how many base resources to allocate to a species shifts the growth function for that species, determining (in part) the rate of productivity of that species. The nature of that

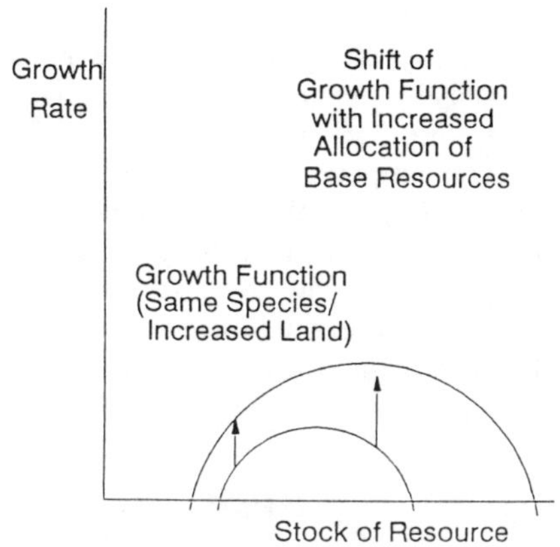

FIG. 3. Endogenising rates of growth

shift is indicated by the two different logistic functions shown in Fig. 3, each corresponding to the same species, but with different base resource allotments.

Therefore, the 'natural' growth rate of a biological resource is affected by the allotment of natural habitat. However, the cost of providing that habitat is not without a price. The total opportunity cost of the habitat equals the flow of benefits expected to be received from the best alternative use of the land, together with the potential returns available from mining the existing stock of natural resources on that land and investing the proceeds elsewhere in the economy. The cost of these implicit investments must be incorporated into the basic model determining the feasibility of investments in diverse resources (see Swallow, 1990 for related analyses).

The changes in the extinction model that these alterations imply result in the revised societal objective function.

$$\max_{y,R} \int_0^\infty [p(y)y - c(x)y - r\rho_R R]\,e^{-rt}\,dt \qquad (11)$$

$$\text{s.t.} \quad \dot{x} = H(x; R) - y$$

where ρ_R is the price of a unit of the base resource R (land).

These alterations indicate that, if this species is to continue to have the use of the resources (R) on which it depends, then it must be able to afford a competitive return (r) on those. The relevant return accords with the marginal rate of return available on investments in this society. The implicit assumption in previous bioeconomic models was that biological resources were naturally 'free goods' that did not require investment. It was believed that biological product resulted from natural processes of growth deriving from the biosphere's capture of photosynthetic energy. This is definitely true in an aggregate sense.

The earth does produce a continuous flow of biological product as the result of natural photosynthetic processes. However, the particular form that the biosphere will take is crucially dependent upon relative investment rates, i.e. on the rate at which base resources are allocated to individual species.

A determination of noncompetitiveness renders not only investments in stocks of that species inefficient, but also renders inefficient any investments in the base resources required for the sustenance of that species. Without that natural resource base the species will be undercut, resulting in its inevitable decline. This alternative route to extinction may also be illustrated within the context of this model. When the additional decision variable (R) is included in this analysis, there is an additional first order condition that must obtain in the steady state.

$$R^*: \frac{\lambda \cdot H_R}{\rho_R} = r \tag{12}$$

This condition states that a particular species will receive allocations of base resources (approximated by their surrogate-land allocations) only to the extent that the species is able to generate a competitive rate of return from this use. The particular life forms inhabiting a parcel of land are viewed by biologists as an outgrowth from the energy gradients represented by that site, using that spot on earth for the appropriation of the sun's energy for regeneration and growth (Futuyma 1986). Equation (12) implies that life forms must now compete for their 'place in the sun', subject to human decisions concerning land use. Alternatively, a species must not only be capable of generating a competitive return on its own stock values, but it must also be able to earn a competitive return on the ancillary resources that it requires for its sustenance. This condition is illustrated in Fig. 4.

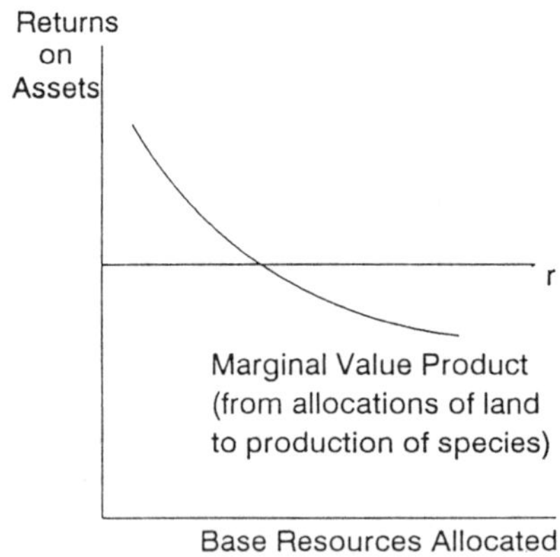

FIG. 4. Optimal allocation of base resources (land)

Therefore, a noncompetitive biological asset is subject to more than one form of disinvestment. It might be removed through an active disinvestment programme (mining of stocks for rent appropriation and asset conversion), but this does not appear empirically to be the most pervasive threat. Most of the extinctions and endangerment today are occurring without notice and those that are well-publicised usually occur without the appropriation of many rents. In most cases, extinctions are the direct result of passive 'undercutting'. That is, for most threatened life forms, the problem is an unwillingness by humans to invest in the required ancillary resources (such as the base resources required for biological survival).

4.3. *Investment in management services for diverse resources*

Although the passive 'undercutting' of species through base resource reallocations probably explains most species endangerment and extinctions, it does not explain them all. In fact, there is a sub-set of endangerment that fits neither the 'mining' paradigm nor the 'undercutting' paradigm discussed in the preceding sections. This is the case with regard to the endangerment of those well-recognised species (rhinos, birds, elephants, etc.) whose decline appears to be most closely related to the 'over-exploitation model' discussed at the beginning of the paper. This is the model developed by Gordon (1954) and Clark (1976) in the context of the fishery, and in which the decline of a species is explained by reference to the existence of an 'open access' regime. In the over-exploitation model, species decline is believed to be 'caused' by open access forms of resource management. This is because the existence of an open access regime renders most investment uneconomic. The nature of the regime rather than the resource determines its competitiveness as an asset.

In that case, it is important to ask why productive assets potentially worthy of investment would be subjected to such a regime. The answer within this framework is straightforward. From the perspective of the generalised investment model, it is more likely that open access regimes are caused by decisions not to invest in diverse resources, rather than a cause of such decisions. Management services are another form of ancillary resource requirement necessary for the survival of many forms of life, especially the more noticeable and charismatic species. These services can be just as critical to the survival of a particular species as are the services that flow from base resources such as land and water. If there are demand pressures that are resource-specific, open-access conditions doom such a resource to disinvestment. The failure to provide management resources for such a resource is equivalent to a decision not to provide an essential requirement for its survival; such a decision is determined by more fundamental considerations.

This point may also be generated within the analytical framework of the investment model of extinction. All that is required is a more complete listing of the essential requirements for the sustenance of a species; these are now seen to be both biological and institutional in nature. This could be considered within the context of the above model by including another input, M, representing the management services allocated to the resource. These services might be considered as a 'stock', in the sense of investments in 'institution-building': the creation of a management regime that will yield a flow over the course of many years of resource management. Thus the societal objective regarding diverse resource is

$$\max_{y,R,M} \int_0^\infty e^{-rt}[S(y; R, M) - c(x)y - r\rho_M M - r\rho_R R]\,dt \qquad (13)$$

$$\text{s.t.} \quad \dot{x} = H(x) - y$$

where $S(y; M, R)$ is the flow of social benefits from y given the quality of resource management generated by the level of investment in management services (M).

Optimal investment in management institutions is given by

$$M^*: \frac{S'_M}{\rho_M} = r \qquad (14)$$

Equation (14) states that the decision whether to invest in institution-building with regard to a given diverse resource depends upon the perceived benefits that will flow from such an investment. As with all other resources, those available for resource managment are scarce and competitively allocated. A resource will only receive an allocation of management services, as in Fig. 5, if it is able to afford a competitive return to these.

This indicates the true nature of the open access regime. It is not an exogenous 'state of nature', but rather an endogenous 'notion of the state'. That is, open access regimes are caused by a state-level decision to invest at the lowest possible level for the management of certain resources, i.e. $M = 0$. Under open access,

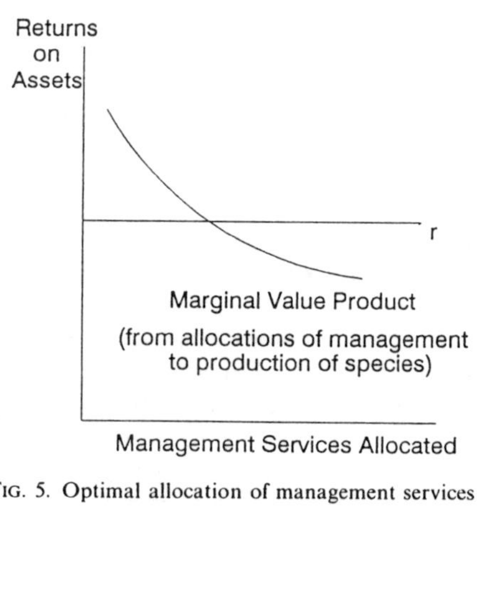

FIG. 5. Optimal allocation of management services

the 'quality' of resource management is at its lowest possible threshold. When this occurs, we are returned to the world of the Clark model (in which 'open access' was the assumed state of nature). In this state of the world, there are no incentives to invest in any way that is associated with the unmanaged species, as the prospects for the return of future benefit flows are severely discounted. Then, all that stands between a resource and its extinction is the net costliness of its harvesting (i.e. its price/cost ratio), as is described in the Clark model.

A positive level of institutional investment (i.e. $M > 0$) is indeed a precondition to all other investment. However, open access regimes are not base forces for extinction, but merely links in the chain of causation between the base forces and observed extinctions. That is, the same fundamental forces that generate the general incentives for underinvestment (low relative returns) cause underinvestment in management institutions. Therefore, open access is an 'effect' rather than a 'cause' of the fundamental forces driving the extinction process, and even over-exploitation is the result of underinvestment.

5. Implications of the revised model of extinction

There are two sets of implications to be derived from this revised model of the extinction process. The first concerns the multiple routes to extinction engendered by diverse resource noncompetitiveness, and the integration of the various issues of resource degradation within a single framework. The second concerns the dramatically different policy implications of this revised model.

5.1. *Integration of the alternative routes to extinction*

There are three alternative routes to extinction for terrestrial species. These three routes are:

5.1.1. *Stock disinvestment* ('resource mining'). These are resources with relatively high values but low growth. In this context, there are incentives to harvest the entirety of the resource (for its high value) and to invest the funds in other assets (for their greater growth rates) (Clark 1973). This is termed 'resource mining' within the literature.

An example of this force in action may be the deforestation of the tropical hardwood forests. These trees represent substantial amounts of standing value, but they have very low growth potential. Thus it is economically rational to 'cash in' the hardwoods and invest the returns in other, more productive assets.

It has also been used to explain why it is that there may be no non-zero optimal stock level for some species, such as the blue whale (Spence 1975). This raises the problem of 'optimal extinction': a society's need to decide whether it will mine a species into oblivion.

5.1.2. *Management resources diversion* ('overexploitation'). These are resources of 'medium' value but relatively low growth. Since they are slow growing resources, they make little sense as assets; society has no incentive to invest in their growth capacity. In addition, on account of their relatively low value, they do not justify a commitment of substantial amounts of national resources for the management of the exploitation process. Then the nation may allow these resources to be depleted through unmanaged overexploitation.

Examples of this process include the depletion of many forms of wildlife: the African elephant and rhinoceros, the wild birds in trade, and many other diverse life forms. In almost every instance, the species of concern go almost entirely unmanaged in their state of origin. This is the result of a combination of demand pressures and state neglect or criminalisation. For example, the trading in wildlife or its products is prohibited by law in almost all South American countries, resulting in the maintenance of a trade under unmanageable conditions. Similarly the taking of ivory has been illegal throughout Sub-Saharan Africa in most countries for the past 20 years. Such a legal structure renders the creation of effective management regimes difficult in the context of open forests and savannahs. Combined with underfunding, the prohibition amounts to the erection of a tacit open access regime.

5.1.3. *Base resource conversion* (*'biodiversity depletion'*). These are resources that are of little or no known individual value to humans. The own-stock value of such resources does not place any survival constraint upon the species, but the costs of the other resources they require become an almost insurmountable hurdle. No living species can persist without an allocation of the ancillary resources they require, and the competition for these resources is the one that is resulting in the removal of the majority of life forms endangered today.

An example of this process is the depletion of many types of virtually unknown life forms (e.g. plants and insects) when land is deforested and converted to other forms of use, such as cattle ranching. This branch of the force for extinction is generally termed the 'biodiversity problem'.

Hence, it can be seen that three of the most-discussed issues in the literature on environment and development (resource mining, overexploitation, and biodiversity depletion) are all alternative outcomes from the same fundamental social process. Each is a possible outcome of the development process as applied to living resources with various combinations of innate characteristics (relative value, relative growth). If a resource has a high relative value and growth rate, then it attracts investment and is unthreatened by the development process. However, if a resource has either low value or low growth, then it is subject to disinvestment pressures *via* one of the routes outlined above. At base, however, it is the nature of the development process itself (the rebalancing of the national portfolio of assets) that is driving the depletion and endangerment of biological diversity.

5.2. *Rent appropriation and investment incentives*

Although there are these three alternative routes to extinction, they are all consequential not fundamental causes of extinction. As was shown in Section 4, each of these forces is the consequence of the more fundamental determination that the biological asset is unworthy of investment. The crucial difference between these models for policy purposes is the comparative statistics analysis regarding the impact of the price–cost ratio on harvesting incentives. In the Clark model, as set out in Section 2, the impact of an increased price–cost ratio on the resource was unambiguous. It generated enhanced incentives for harvesting, without a concomitant increase in the resource flow or stock levels (as the growth function was exogenous to the model). This resulted in an unambiguous increase in the threat to the viability of the species.

In this model, the implications of an increased price–cost ratio can be very different. It all depends upon the inherent characteristics of the species. If the species is perceived to be of high or medium value, but very low growth, then a change in the price of the species' products cannot alter its ultimate fate; it will be targeted toward extinction in any event. However, if the species has a significant rate of growth, then an increase in its value may have the effect of making investments in its ancillary resources economically worthwhile. Then an increased price might actually generate the conditions under which the species might avoid extinction.

For example, in the context of eq. (12), the effect is unambiguous. An increase in rent capture (e.g. resulting in an increase in the equilibrium level of λ in eq. (12)) increases the relative rate of return for this species (i.e. the right-hand side of eq. (12)). More lands may be made available to the species because this species is now better able to 'pay its way'. This analysis applies equally to the allocation of management services to the diverse resource. If social benefits are defined to be equal to at least the sum of all individual harvests of the species, then an increase in the per unit appropriable value will increase the incentives to invest in this factor as well. In general, enhanced rent appropriation will enhance the incentives to invest in the ancillary resources required for the survival of this species. This may also be depicted by means of a 'shifting out' of the rate of return curves in Figs 4 and 5. In essence, the increase in rent value has increased the investment-worthiness of the diverse resource with respect to the ancillary resources it requires, and hence the ultimate survivability of the species. Therefore

$$\frac{\partial R^*}{\partial \lambda} > 0 \tag{15}$$

Of course, this result does depend upon the initial conditions in which the diverse resource is found and the magnitude of the change in price. If the resource is initially subject to disinvestment by reason of the diversion of management services and it is perceived to be unworthy of investment, then it will already be subject to an 'open-access' regime. Under the circumstances of an open-access regime, a marginally increased rental value will only increase the rate of species decline, while maintaining the inefficient management regime in place. However, this does not mean that the policy implications of this model are context dependent, only that the reforms required may be more drastic under some circumstances than in others. It may be necessary to both subsidise the start-up costs with regard to institution-building in some contexts, while simultaneously enhancing the rental value of the resource to create incentives for ongoing investment.

In any event, it is clear that a species that is subject to an open access regime cannot be saved through a policy of 'rent destruction' in the long run, it can only be shifted between the various avenues to extinction. If a species is perceived as being unworthy of investment, then a reduction in its marketed value will simply render it more likely to be lost by reason of base resource conversion and less likely to be lost by reason of management services diversions. This change in policy cannot save any terrestrial species, only change the proximate cause of its decline.

The only policies that can alter the ultimate fate accorded by human society to any particular species are those that address the fundamental cause of decline: perceived investment-unworthiness. It was previously mentioned that portfolio reconstruction occurs as a matter of fact within the development process, and this might have been taken to imply that these declines in diversity are unavoidable. This is not the case, simple because there are alternative paths available for development; not every country need base its portfolio on the same set of assets. Opting for the same development path as used by the previous states is the safe option, but not necessarily the optimal route, even for the developing state itself. What is required is some manner of international institution that will attract developing countries toward diverse development paths, enabling them to disavow the customary routes (Swanson 1994). This implies that diverse resources must be accorded very substantial values, including market values, if they are to receive the investments that they require for their survival. The policy of 'demand destruction' is at odds with the fundamental solution concept to the problem of endangered species and biodiversity decline.

7. Conclusion

Even biological resources are now economic resources, in the sense that human societies have control over the terrestrial biosphere and human decisions on resource allocations will determine which life forms will continue to exist into the future. Given that humans have usurped this power from nature, the purpose of this paper has been to identify the characteristics of the resources that will be in decline under the circumstances and the nature of the policies that might alter their fate. The answer to both questions lies in the capacity of the resource (present and future) to attract human investment.

Although the fundamental force causing extinctions is disinvestment, this may occur via a number of different routes: 'resource mining', 'land use conversions', or 'over-exploitation'. These are all alternative routes that lie between the society's determination of investment-unworthiness and the ultimate extinction of the resource. The choice between the routes depends only upon the relative value and growth rate of the resource. In any event, it is only the high-value/high-growth resources that will ultimately survive. For this reason policies based on reducing the value of a resource can only have the effect of altering its route to extinction.

All of these environmental problems have their source in the developmental process. Developing countries are those which are currently engaging in the most dramatic rebalancing of their national portfolios of natural assets. Those resources which are under threat are those which are perceived to be unworthy of significant investment for purposes of development. The answer is for the peoples of other countries who are concerned about these resources to create the conditions under which the resource is viewed more favourably as an asset. This will allow some countries to consider the option of developing down paths built upon a diversity of assets.

ACKNOWLEDGEMENTS

I would like to thank Professors David Pearce, Bob Rowthorn, and Nick Stern as well as numerous seminar participants for their comments on this paper.

REFERENCES

BARBIER, E., BURGESS, J., SWANSON, T., and PEARCE, D. (1990). *Elephants, Economics and Ivory*, Earthscan, London.

CAUGHLEY, G. and GODDARD, J. (1975). 'Abundance and Distribution of Elephants in Luangwa Valley, Zambia', *African Journal of Ecology*, **26**, 323–7.

CHEUNG, S. (1970). 'The Structure of a Contract and the Theory of a Non-Exclusive Resource', *Journal of Law and Economics*, **13**, 49–70.

CLARK, C. (1976). *Mathematical Bioeconomics: The Optimal Management of Renewable Resources*, John Wiley, New York.

CLARK, C. (1973). 'Profit Maximisation and the Extinction of Animal Species', Journal of Political Economy, **81**, 950–61.

CLARK, C. and MUNRO, G. (1978). 'Renewable Resources and Extinction: Note', *Journal of Environmental Economics and Management*, **5**, 23–9.

CLARK, C., CLARK, F., and MUNRO, G. (1979). 'The Optimal Exploitation of Renewable Resource Stocks: Problems of Irreversible Investment', *Econometrica*, **47**, 25–49.

CONRAD, J. and CLARK, C. (1987). *Nature Resource Economics*. Cambridge University Press, Cambridge.

CROPPER, M. (1988). 'A Note on the Extinction of Renewable Resources', *Journal of Environmental Economics and Management*, **15**, 25–30.

CROPPER, M., LEE, D., and PANNU, S. (1979). 'The Optimal Extinction of a Renewable Natural Resource', Journal of Environmental Economics and Management, **6**, 49–55.

DASGUPTA, P. (1982). *The Conrad of Resources*, Blackwell, Oxford.

DOUGLAS-HAMILTON, I. (1989). 'Overview of Status and Trends of the African Elephant', in S. Cobb (ed.), *The Ivory Trade and the Future of the African Elephant*, Report of the Ivory Trade Review Group to the CITES Secretariat.

FUTUYMA, D. (1986). *Evolutionary Biology*, Sinauer, Sunderland.

GORDON, H. S. (1954). 'The Economic Theory of a Common-Property Resource: The Fishery', *Journal of Political Economy*, **62**, 124–42.

LOVEJOY, T. (1980). 'A Projection of Species Extinctions', in G. Barney (ed.), *The Global 2000 Report to the President*. Washington, Council on Environmental Quality.

MCNEELY, J., MILLER, K., REID, W., MITTERMEIER, R., and WERNER, T. (1990). *Conserving the World's Biological Diversity*, IUCN, Gland, Switzerland.

RENEWABLE RESOURCES ASSESSMENT GROUP (1989). 'The Impact of the Ivory Trade on the African Elephant Population', in S. Cobb (ed.), *The Ivory Trade and the Future of the African Elephant*, Report of the Ivory Trade Review Group to the CITES Secretariat.

REPETTO, R. and GILLIS, M. (1988). *Public Policies and the Misuse of Forest Resources*, Cambridge University Press, Cambridge.

SOLOW, R. (1974). 'The Economics of Resources or the Resources of Economics', *American Economic Review*, **64**.

SPENCE, M. (1975). 'Blue Whales and Applied Control Theory', in H. Gottinger (ed.), *System Approaches and Environmental Problems*, Vandenhoeck, Göttingen.

SWALLOW, S. (1990), 'Depletion of the Environmental Bias for Renewable Resources', *Journal of Environmental Economics and Management*, **19**, 281–96.

SWALLOW, S., BROOKS, P., and WEAR, D. (1990). 'Policy-Relevant Nonconvexities in the Production of Multiple Forest Benefits', *Journal of Environmental Economics and Management*, **19**, 276–80.

SWANSON, T. (1991). 'Conserving Biological Diversity', in D. Pearce (ed.), *Blueprint 2: Greening the World Economy*, Earthscan, London.

SWANSON, T. (1992a). 'Policies for the Conservation of Biological Diversity', in T. Swanson and E. Barbier (eds), *Economics for the Wilds: Wildlands, Wildlife, Diversity and Development*, Earthscan, London.

SWANSON, T. (1992b). 'The Economics of a Biodiversity Convention', *Ambio*, **21**, 250–7.

SWANSON, T. (1992c). 'The Evolving Trade Mechanisms in CITES', *Review of European Community and International Environmental Law*, **1**, 57–63.

SWANSON, T. (1992d). 'Animal Welfare and Economics: The Case of the Live Bird Trade', in S. Edwards and J. Thomsen (eds), *Conservation and Management of Wild Birds in Trade*, Report to the Conference of the Parties to CITES, Kyoto, Japan.

SWANSON, T. (1993). 'Regulating Endangered Species', *Economic Policy*, **11**, 187.

SWANSON, T. (1994). *The International Regulation of Extinction*, Macmillan, London.

SWANSON, T. and BARBIER, E. (eds) (1992). *Economics for the Wilds: Wildlands, Wildlife, Diversity and Development*, Earthscan, London.

WILSON, E. (1988). *Biodiversity*, National Academy Press, Washington.

Part II
Valuing Ecosystem Functions and Services: Demonstrating the Value of Natural Biological Capital

A Overviews

[8]

ELSEVIER

Ecological Economics 14 (1995) 73–90

ECOLOGICAL ECONOMICS

Analysis

Issues in ecosystem valuation: improving information for decision making

Gail Bingham [a], Richard Bishop [b], Michael Brody [c,*], Daniel Bromley [b], Edwin (Toby) Clark [d], William Cooper [e], Robert Costanza [f], Thomas Hale [c,1], Gregory Hayden [g], Stephen Kellert [h], Richard Norgaard [i], Bryan Norton [j], John Payne [k], Clifford Russell [l], Glenn Suter [m]

[a] *Resolve, Washington, DC, USA*
[b] *University of Wisconsin, Madison, WI, USA*
[c] *Environmental Protection Agency, Washington, DC, USA*
[d] *Delaware Department of Natural Resources, Dover, DE, USA*
[e] *Michigan State University, East Lansing, MI, USA*
[f] *University of Maryland, Solomons, MD, USA*
[g] *University of Nebraska, Lincoln, NE, USA*
[h] *Yale University, New Haven, CT, USA*
[i] *University of California, Berkeley, CA, USA*
[j] *Georgia Institute of Technology, Atlanta, GA, USA*
[k] *Duke University, Durham, NC, USA*
[l] *Vanderbilt University, Nashville, TN, USA*
[m] *Oak Ridge National Laboratory, Oak Ridge, TN, USA*

Received 6 April 1994; accepted 10 April 1995

Abstract

In Spring 1991, the U.S. Environmental Protection Agency convened an expert group of ecologists, economists and other social scientists for the purpose of advancing the state of the art of ecosystem valuation methods. This Ecosystem Valuation Forum was organized as a dialogue because it has been clear from the outset that agreement even on the meaning of the term "ecosystem valuation" could not be taken for granted. Individuals from diverse disciplines, and from industry, environmental groups and government agencies disagree about what information about ecosystem services is needed, how it should be used and, therefore, what would constitute an advance in the methods that analysts should employ. The Forum discussed the varied ways in which experts from different disciplines approach valuation, what ecosystem attributes or services are important to value, and the factors that complicate the task of assigning values to ecosystem attributes. The Forum placed particular importance on approaching the problem of ecosystem valuation from the perspective of decision makers. Therefore, members discussed the variety of decision makers who might need valuation information, the controversy over where balancing decisions about costs and benefits should be made, and the implications for what information is needed

* Corresponding author.
[1] Current address: Louis Berger, Inc., Washington, DC, USA.

0921-8009/95/$09.50
SSDI 0921-8009(95)00021-6

within different institutional constraints. In addition, agency decision makers operate under real time and resource constraints. Thus, the Forum discussed the need to develop protocols that would guide analysts in a search for decisive information. The Forum concluded that the time is ripe for making new progress in solving some of these problems, while acknowledging that it may not be possible to develop a single unifying definition of value. Instead, the goal would be to understand how various concepts of value are structured, how they relate to each other, and how they can guide us toward a more integrated valuation process. The Forum recommended that next steps in addressing these issues be organized around case studies, particularly those that would enable researchers to improve linkages between ecological and economic methods and to develop improved protocols for valuation studies.

Keywords: Ecosystem valuation; Decision makers

1. The problem

Public and private decision makers want and need better information about the values of ecosystems in weighing the advantages and disadvantages of human actions that may affect ecosystems. The level of public interest in environmental protection has never been higher, nor has the investment of public and private dollars. Although environmental and business interests disagree about when and how information about the economic costs of achieving environmental objectives should be weighed, all sides are concerned about improving the availability and use of information about ecosystem values in making policy decisions.

However, information often is lacking about: (1) the physical changes to ecosystems and the socio-economic consequences that might result from alternative courses of action; and (2) the "value" of those changes. Information is inadequate because the capacity to value alterations in ecosystem attributes is not well developed, particularly for ecosystem functions and processes. While some progress has been made in valuing certain aspects of ecosystems as commodities—recreation, for instance—much work remains before satisfactory methods for valuing all the services and attributes of ecosystems will be available.

One of the limits to providing sufficient ecosystem valuation information to decision makers is that it is extremely difficult to measure fully the functions and processes of an ecological system or to predict the ecological impacts of disturbances to those complex systems. Furthermore, even where relatively simple ecosystems are fairly well defined, it is difficult to determine the causal relationships between human actions and ecosystem functions and processes. Much needs to be done before the consequences of human alterations to ecosystems will be well understood or predictable.

Early in the first meeting of the Forum, it became clear that even the phrase "ecosystem valuation" presumed a level of agreement among the individuals and disciplines that could not be taken for granted. Because experts from within and across disciplines currently approach the question of ecosystem valuation from such very different perspectives, it has been critical to explore the problem in valuing ecosystem services and attributes and to construct an interdisciplinary agenda for discussion.

It is desirable to place this effort to value ecosystem services within some historical context that describes how concerns about valuing the services of the natural environment become a part of the information set used in policy decisions. In 1958, a subcommittee of the Inter-agency Committee on Water Resources issued a report on Proposed Practices for Economic Analysis of River Basin Projects [2] (known to many as the "green book") which had a profound and lasting effect on the principles and practice of economic analysis of public projects. Many of the problems in fully valuing ecosystem attributes still faced today were anticipated in that report:

[2] U.S. Federal Inter-agency River Basin Committee. Subcommittee on Standards and Evaluation. "Proposed Practices for Economic Analysis of River Basin Projects" a report to the Inter-Agency Committee on Water Resources, Washington DC, 1958 (revised).

"The public policies governing the development of the Nation's water and related land resources are not necessarily determined solely on the basis of economic considerations." (p. 3)

"... from the standpoint of society as a whole there may be beneficial or adverse effects that would escape consideration in a summation of individual evaluations, as for example,... value of resource conservation to future generations..." (p. 6)

"The problem of evaluating, from a public viewpoint, the extent to which a project [policy] accomplishes the aim of satisfying human needs and desires presents a major difficulty at the outset, because there are no common terms in which all effects of a project [policy] are normally expressed." (p. 6)

The members of the Forum are conscious of the pertinence of these observations today, some three decades later. However, it is important to recognize the evolutionary nature of our knowledge base, and the need to be aware that the ability to characterize and estimate the economic values of environmental services, (available largely outside organized markets) changes with the advance of the scientific and the economic disciplines involved. Perhaps natural and social sciences have advanced sufficiently that needed progress can now be made.

The need to be aware that the ability to characterize and estimate monetary values of environmental services has grown tremendously since the "Green Book" appeared. Of course, a great many problems remain unresolved, and further progress on tools to establish monetary values is needed. Beyond that, three problem areas persist that have been less well recognized.

First, to a considerable degree, ecosystem valuation could be improved if currently available tools for monetary valuation could be competently applied more often. Additional resources for economic studies would help, but resources, even if augmented, will likely always be inadequate to assess every situation in which ecosystems are damaged or at risk. Such studies could well be a waste of resources as well. The challenge is to define what sorts of valuation information is likely to be decisive: i.e., that is really needed to make decisions. Ways to build on existing information rather than to rely on expensive primary data gathering need to be developed.

Second, the fact that good valuation studies depend on sound ecological results is not adequately appreciated. Ecosystem services can be valued in dollars only after those services are well understood, yet predicting how ecosystem service flows will change as a result of human intervention is often difficult or impossible. As ecology itself evolves to better understand the services ecosystems provide and to better predict how they will change as a result of human actions, existing economic tools can be applied to better evaluate ecosystems. In the meantime, decision criteria to adequately address ecological uncertainties need to be devised, and such criteria are likely to involve human values beyond monetary values.

Third, it would be a mistake to think that if ecosystem services were perfectly understood and evaluated in monetary units according to accepted economic precepts, the problems of ecosystem valuation would be completely solved. Many would question whether monetary valuation alone adequately captures what decision makers need to know to confront irreversible ecosystem modifications that could have serious long-term economic and social repercussions. Perhaps the most important task is to clarify where conventional economic values are sufficient for decisions and where broader human values—including non-monetary values—and criteria for decision making are more appropriate.

1.1. How do we value ecosystems?

A discussion of what one means by "value" is inescapable if people from diverse disciplines and perspectives are to find a common language for dialogue. Value to the average citizen is not a confusing word. It means the general importance or desirability of something. As often happens with words, more precise definitions of value have evolved in different disciplines to meet different needs, but that greater precision sometimes limits interdisciplinary inquiry.

We acknowledge that the Forum may not be able to develop a single unifying definition of value. Instead, our goal is to understand how various concepts of value are structured, how they relate to each other, and how they can guide us toward a more integrated valuation process.

Ecosystems have important attributes, both structural and functional, which influence an ecologist's perspective about how the value of ecosystems should be understood. To some economists, ecosystem values correspond to what people will pay to maintain or restore that system or some of its attributes; to others, it may mean societal value as determined through the political process; to still others it means valuing the consequences to the community. To philosophers the term "value" is a verb encompassing utilitarian, aesthetic and moral assessments. To social-psychologists, ecosystem values may mean characterizing the reasons, feelings and beliefs people express for preferring some ecosystem attributes over others.

Values are by definition anthropogenic (a forest cannot tell us whether it prefers to be old growth or young and vigorously growing), however, ecosystem values need not derive from human use of the systems or their components. That is, ecosystems may be valuable to people as ecosystems as well as producers of timber and clean water.

The Forum recognizes the need to separate issues of semantics from issues central to improving ecosystem valuation. Thus, although public discussions often include references to "intrinsic" values for species and ecosystems as opposed to their instrumental values to humans, the Forum will not employ this dichotomy, in part because its current use obscures important distinctions. The broad variety of values derived from ecosystems fall upon a continuum ranging from easily priced tangible benefits (such as food and pharmaceuticals); through the values associated with less easily priced services, aesthetic experiences and bequest values; all the way to moral and spiritual values.

Separating these many different ways in which ecosystems are valuable to people is a significant first step, particularly because the methods for measurement will vary. Our goal, therefore, is not to classify these by locating them within or outside the process of human valuation, nor is it to separate goals from the means by which they are achieved. Instead, we intend to recognize the entire range of human values and to seek ways to integrate these multiple values in an improved decision framework.

It also is important to understand the various roles that individuals play and how these roles affect the notion of "valuing" something. It can be observed empirically that the same individual, whether lay citizen or expert, will value ecosystem attributes differently when expressing personal values and when serving in some advisory or decision-making role for a public institution. People also reason differently when acting as decision makers or advisors (for example, when they vote to institute strict land-use regulations) than they do as private agents (when they might, by contrast, use their own land to the maximum intensity allowed by current law). Similarly, people seem to emphasize quite different issues and decision criteria when their concern shifts from short-term considerations relevant over months or years to concerns about sustaining processes over several generations.

Thus, some ecosystem valuation experts question which of these valuation contexts should be used to determine people's "real" values. To many economists, the answer is to use the values people express as individual households or consumers, because the generally accepted norm is to attempt to act (and therefore to value outcomes) based on the perceived preferences or goals of the group being represented. Thus, these multiple concepts of value may be distinct and not necessarily contradictory.

However, as the science of ecology has matured, mankind's knowledge about the interconnectedness of ecosystem processes and structures has grown. As a result, individual citizens have shown increasing concern about the effects of human actions on ecosystems. An important attribute of this concern, though, is the awareness of many citizens that they do not understand all of the attributes of ecosystems that are necessary to support the ability of the natural environment to produce those services that they benefit from directly, and that much more remains to be learned. This lack of information affects people's abilities to place a value on ecosystem attributes.

This raises new questions about whether there is a prior step to choosing whose values to count. Might people be willing to allow scientific advisors—and the ways of valuing inherent to their disciplines—to play a role in shaping the valuation information used in collective decision-making processes? This could take various forms. People might want experts to inform them about the interconnectedness of ecosystem processes and structures and let them, as citizens, express their values through democratic processes. Or, people might want experts to develop other kinds of methods for valuing certain ecosystem attributes determined by scientists to be crucial to the viability of ecosystems, but which are not directly useful to people.

We cannot calculate ultimate values for ecosystems. But, by emphasizing the varied ways in which ecosystems are valued, we propose to offer a variety of guides to improved decision making with an accompanying set of criteria or "rules of thumb" indicating which guides to emphasize in particular contexts.

1.2. What is to be valued?

Ecosystems are dynamic—populations of species rise and fall, one species may substitute for another species, physical processes change. Although such changes occur naturally, human actions often cause more rapid or unanticipated changes. The effects of such changes are not necessarily ecosystem collapse; alternate ecological states are possible. However, some ecosystems are more desirable to people than others. Some ecosystems provide more habitat for threatened or endangered species than others; some ecosystems provide more water purification services than others; and some ecosystems provide more recreation opportunities than others.

Experts continue to struggle with the fundamental question of what services to value. The choice of which ecosystem attributes to value is itself a valuation decision, and a challenging one. The very term "ecosystem" is a multi-scale concept, referring to such widely different circumstances as a rotting log, a prairie, and the earth itself. Even individual attributes of concern occur at widely different temporal and spatial scales. One needs to know what is important to measure, and why.

The notion of importance is central to any coherent approach to valuation. After all, if there is no agreement regarding the ends, there can be no agreement regarding valuation. Put somewhat differently, "values" in ecosystem valuation are derived from criteria whereby particular courses of action are regarded as good or bad. Ecologists may think in terms of ensuring the resilience of a particular ecosystem, or of maintaining its productivity. Economists may think in terms of a monetary expression of the goals derived from preferences. Other social scientists will think in terms of other expressions of human preferences. Regardless of the form taken, valuation of ecosystems cannot logically proceed in isolation from a clear idea about what ends are being sought.

An identification and categorization of various ecosystem attributes also is necessary in organizing an inquiry to improve methods for valuing those attributes.

Ecosystems have many different attributes and offer many different services, the value of some or all of which decision makers may wish to consider in weighing the advantages and disadvantages of a course of action. A partial list could include, among many others:

- food (e.g., oceans)
- sources of wild medicinal plants (e.g., forests)
- water purification (e.g., wetlands)
- flood control (e.g., wetlands)
- erosion control (e.g., forests, wetlands)
- carbon sequestration (e.g., forests, oceans)
- habitat for wildlife (most ecosystems)
- reservoir of biological diversity
- nutrient recycling
- detoxification of chemicals
- recreation and outdoor adventure
- aesthetic enjoyment, solitude, and spiritual fulfillment

Several factors complicate the task of assigning values to ecosystem attributes. For example, many important structures or functions of ecosystems do not directly benefit people, but are necessary for the ecosystem as a whole to provide the services that people do use. For example, attributes that are more directly useful to humans include food, medicine, recreation and aesthetics, while those that help sustain viable ecosystems include such attributes as habitat, nutrient recycling and genetic diversity. Changes in these latter services could, in theory, be translated into effects on valued goods and services more directly useful to people. In practice, however, the time, data, or methods for that translation often are not available. Although the task presents a difficult set of conceptual problems, progress in improving the valuation information available to decision makers must begin to consider the value of those ecosystem attributes that help sustain the ecosystem itself. [3]

Ecosystem services also have present and future dimensions. The patterns in which they are used today will affect their ability to continue to "produce" the outputs that people care about and, thus, people's "option" to use such goods and services in the future.

In addition, human knowledge about ecosystems is very incomplete. Thus, we may not anticipate all the goods and services that an ecosystem does or could provide. For example, the Pacific Yew—previously considered a "weed" species—has recently been found to have medicinal value in the treatment of certain cancers. There also may be unknown future values in ecosystem goods and services other than those that have commodity value. For example, 30 years ago, neutralizing acid deposition and sequestering carbon were not recognized to be valuable ecosystem services.

Furthermore, ecological impairment may not just be a matter of damage to various life-support and ecosystem functions, but also to some very important notions of a meaningful quality of life. Often the environmental harm of concern to people is more a matter of a less satisfying, uglier and diminished existence that, despite increased material benefits, people describe as less beneficial in intangible terms than the nurturing landscape that has been lost.

Categorizing the services that ecosystems provide may help in deciding what to measure or in selecting among different valuation methods for different services. Also, different methods may be based on different categorizations. If more than one valuation method is determined to be useful in making a decision, then an understanding of the categories assumed will allow an assessment of whether certain attributes or services are being double counted.

As an example, the following categories were modified from a list developed by Steven Kellert (Kellert and Clark, 1991):

Naturalistic / outdoor recreational services e.g., observation of wildlife in a natural environment or fishing on a secluded lake.

Ecological services e.g., nutrient cycling.

Existence services e.g., knowing a species or ecosystem (e.g., Yellowstone) exists even though you never intend to see it.

Scientific services e.g., the potential of species and systems to increase human knowledge about the natural world.

Aesthetic services e.g., the beauty of natural systems and individual species.

Utilitarian goods or services e.g., direct resource commodities such as fish production, medicines, agricultural products, etc.

Cultural, symbolic, moral and historic services e.g., the role a species can play in a particular society, such as the bald eagle in some American Indian cultures.

[3] There is another type of characterization that arises frequently in economic modeling. This one is defined from the perspective of whether the good or service once "produced" by the ecosystem enters preferences directly or if it requires some type of transformations before it is of direct value. Some goods and services of an ecosystem may be recognized by people, but they may or may not have direct value to such individuals. For example, timber is harvested from forests because it is useful to people, but in meeting these needs individuals generally do not purchase trees. Instead, they buy milled lumber, furniture or houses made from wood products. Similarly, it may be the case that some people do not value ground water other than as a water supply. In this situation the ground water may be a final output of the aquifer, but an intermediate input to the production activities required to extract it, assure it is safe drinking water, and deliver it to the people wishing to use it. The ultimate drinking water available for consumption at each person's house is the final product.

G. Bingham et al. / Ecological Economics 14 (1995) 73–90 79

The Forum plans to emphasize the values resulting from ecosystem functions. This does not imply that ecosystem structures are unimportant. Rather, the commodity values of fish and timber, the recreational values of lakes and forests, and other values of ecosystem structures are relatively well understood. In contrast, the services provided by ecosystem functions, such as climate moderation and water purification, are poorly appreciated by decision makers and the public, poorly estimated by environmental scientists, and incompletely valued by current valuation methods. The values of these poorly characterized services are potentially quite large. Also, in contrast to some other poorly characterized services, such as aesthetics or existence services, those provided by ecosystem functions may be more amenable to assessment and quantitative valuation.

The development of a categorization system is identified in (see Section 5) as a topic for future work.

2. Information needs

2.1. The choice problem

The debate about how choices should be made, and by whom, forms the backdrop to any effort to improve valuation methods. At the most basic level, choice implies value. Whether the decision maker chooses to do something or not, the act of choice implicitly reveals a threshold for the value the decision maker has assigned to what is at stake.

In the United States, decisions are made by individuals, executive branch agencies, legislative bodies, or the courts within limits set by existing law and the constitution. People differ, however, about which decisions are appropriately made by individuals and which by collective processes. Within the category of collective processes for social choice, the relationship between legislative, executive and judicial bodies becomes important in shaping different views about where balancing decisions should be made and how.

A key factor is the degree of specificity of desired outcomes set in the legislative process. In

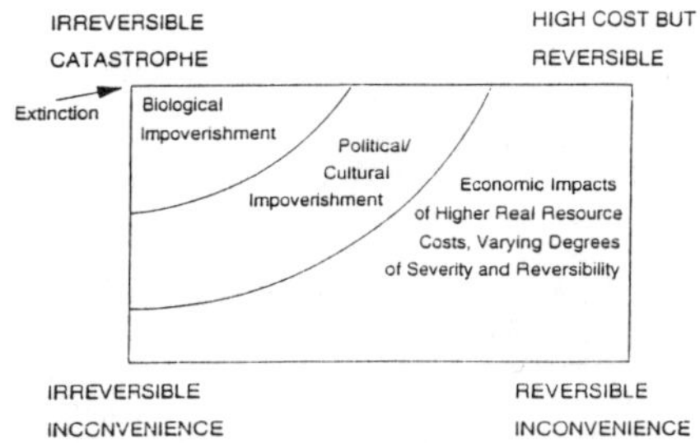

Fig. 1. Potential impacts on the natural environment: severity and reversibility.

some statutes, Congress establishes general goals. Under these circumstances, agencies weigh costs and benefits in the rulemaking process to specify desired outcomes. In other cases, Congress establishes the desired outcome more specifically, weighing costs and benefits in the legislative process, and asks the Executive Branch to determine the best means to achieve these pre-established ends. In actual practice, of course, these distinctions are blurred, particularly because ends set by Congress usually can be achieved to different degrees. [4]

These general scenarios play a significant role in determining the place of cost-effectiveness and benefit-cost analysis in particular (see Section 2.2 for examples). The specific directives set by Congress also have implications for what types of information will be most useful (see Section 2.3 distinguishing information about ecosystem effects and values). Different methods for providing information also may be useful depending on whether the information is needed to describe, predict or value alterations to ecosystem attributes, and on how easily or accurately the ecosystem attribute being assessed can be measured in monetary terms (see Section 4.2 on methods and Section 4.3.4 on units of measure and the question of how to deal with incommensurate terms.)

[4] Disagreements about the limits of discretion of the executive branch are frequently decided through judicial action.

The perceived social cost of a "wrong" decision is important in understanding the debate over whether to make particular decisions in one "locus" or another. There are several ways to consider the costs of a wrong decision, which need to be explored further. As one example, Fig. 1 presents two dimensions for mapping the outcomes of decisions facing society—the severity of effect and the degree of irreversibility. (Other dimensions of concern might also be appropriate, and more work is needed in defining the scale for any of these measures: e.g., "severity" could be measured by recovery time, spatial scale, numbers of species, etc.).

Decisions about the "locus" of choice can be overlaid on this map. Decisions in the lower right corner involving low severity and complete reversibility in the ecosystem from the perspective of the collective society—are often made by individuals. Decisions in the upper left corner—involving potentially catastrophic and irreversible effects—are often made through some collective choice process, often set or constrained by legislative decision. For these latter decisions, it may be necessary to organize information differently and focus on a "safe minimum standard" and cost-effective ways of realizing it. The decisions in the middle also are often made collectively in executive branch agencies where conventional benefit-cost analysis may be used. Then the question for debate is where to draw the lines between these groups and what information is most useful for decision makers particularly at these important boundary areas.

The use of benefit-cost analysis for making collective decisions is strongly supported by some and strongly questioned by others. The appropriate, and broadly accepted use of this and other economic tools is a matter worth continuing consideration.

It is important to recognize the limits of monetary valuation in making decisions. According to conventional economic theory, monetary terms can be used to analyze the efficiency of resource use. However, depending on the assignment of rights, economies can operate efficiently in very different ways. Many environmental problems routinely involve equity issues as well as efficiency considerations, not only in assigning rights to resource use between present and future generations, but also between those having disparate power or resources within present populations.

Bromley (1976) states the point as follows:

"As the empirical vehicle for welfare economics, benefit-cost analysis ostensibly guides society toward [a] Pareto optimal point. Unfortunately, this move is often confused—by virtually all 'policy makers' and not a few economists—with being analogous to socially optimal. We seem to require constant reminding that a move to a Pareto optimal point may not involve a Pareto improvement, since some individuals will be made worse off, and others better off. The essence of public choice is the shifting of comparative advantages; it is the restructuring of rights and the exposure to the rights of others."

Thus, redistributing rights to natural resources and environmental services is a collective choice which, in turn, determines a new efficient allocation of resources and services. Such a shift in how resources and services are allocated results in new prices, including new implicit values for the resources and services themselves (Norgaard, 1991). These collective choices must be made before valuation can occur. For example, determining the efficient use of child labor in the United States was made moot by the collective decision that child labor—irrespective of potential economic benefits—is morally unacceptable.

2.2. Implications for economic analysis

As stated previously, the choice to weigh costs and benefits in legislative versus executive bodies determines the context in which ecosystem valuation methods will be used—in particular, whether valuation information will contribute to a cost-effectiveness or a benefit-cost analysis when the decision reaches an executive branch agency.

A fundamental difference exists between those decisions in which a complete, monetary benefit-cost analysis, as an economist might understand that concept, is employed and a decision in which a different balancing of advantages and disadvantages is employed. We do not take a position on which approach is the more correct, but rather argue that both approaches must be understood as equally legitimate processes for choice depending on the context.

Although this section will focus on distinctions between these two contexts for applying monetary methods for resource valuation, we recognize that additionally relevant valuation strategies might involve assessments based upon ecological and/or social-psychological methods. This would certainly be the case for many resource values, such as existence or aesthetic values, which are not expressed or measured well monetarily.

Cost-effectiveness analysis

Under some circumstances, decision makers analyze the relative costs of alternative means of achieving previously set objectives. This is an analysis of how to accomplish what one has already decided to do. For environmental regulations, often the first step is to set "safe minimum standards" which can help ensure that only alternatives that achieve stated objectives are compared.

For example, states set water quality standards using EPA criteria. If surface water downstream from a permitted outfall, say from a municipal wastewater treatment plant, fails to meet state water quality standards, the treatment plant will be required to take action. A benefit-cost analysis of whether to improve the water quality downstream is not undertaken, but an analysis of the cost effectiveness of various options for correcting the problem might still be very appropriate for minimizing both economic and environmental costs of corrective actions.

Similarly, the Endangered Species Act, although it does allow for exceptions under rare circumstances, embodies the objective that no species should become extinct as a result of human actions. Once a species is listed as threatened or endangered, a plan is developed to protect the species. This plan may include an analysis of the economic impact of different options. Though not a formal cost-effectiveness analysis, such an impact assessment may serve the same functions.

Superfund requires that hazardous sites be cleaned up. Sites are added to the National Priority List on the basis of a hazard ranking system, not whether the monetary benefits of clean-up exceed the monetary costs. After a site is placed on the National Priority List, a remedial investigation and feasibility study is conducted, which includes a risk assessment to determine if the degree of risk warrants remedy. After options that provide an acceptable level of clean-up are developed, costs are compared.

Benefit-cost analysis

Under other circumstances, decision makers let prices, or other indicators of relative worth, determine choices. This calls for a comparison of benefits and costs to determine whether to do something.

To comply with Executive Order 12291, EPA must demonstrate that the benefits outweigh the economic costs of new controls before promulgating many, if not most of its regulations dealing with ecological effects. [5]

For example, if EPA were to promulgate additional regulations under Subtitle D of the Resource Conservation and Recovery Act to prevent adverse environmental effects from gold or other metals mining wastes, decision makers would conduct a benefit-cost analysis.

Benefits of regulation might include reducing on-site bird and mammal mortality from exposure to cyanide leaching solution, or reducing the loss of fish production caused by increased acidity or heavy metals on freshwater streams, but methods for quantifying these benefits are insufficient. Although current valuation methods can measure changes in the recreation value of affected areas, more work is needed to more fully value ecosystem attributes that are affected, before ecologists and economists can agree that information about ecosystem values is adequate for decision making. Crucial to this goal will be the determination of

[5] In most instances, Congress has precluded consideration of the costs of controls to protect human health. Although the use of benefit-cost analysis and regulatory impact assessments as part of the process for defining standards to protect human health is debated, the emphasis is on obtaining information to define thresholds where no significant health effect will occur.

production functions—defined broadly to include structures and processes that transform matter and energy inputs into ecosystem services—that are either directly valued by people or are important in supporting features of those ecological systems that are valued.

2.3. Ecosystem effects and ecosystem values

In addition to the choices about where and how costs and benefits will be weighed in making decisions, it also is useful to distinguish between two different types of information—environmental effects and value. [6] Information about effects and information about value needs to be better linked, however. Quantitative information about the increase or decrease in an ecosystem service can be a necessary. prerequisite to valuation, but values may also determine (implicitly or explicitly) which effects one chooses to measure.

Decision makers need information about the effects of human actions on ecosystems. Often this requires the ability to make predictions. In the case of deposition of hazardous air pollutants EPA has not yet decided which environmental effects (often called "endpoints") are significant to assess and, therefore, does not yet have information about either the nature or magnitude of the effects.

In another example, EPA is required periodically to review its ambient air quality standards for ground level ozone. To conduct such a review, decision makers need information about what changes in ecosystems, including forest ecosystems, would occur at different concentrations of ozone under different climatic conditions. Effects can be physical, chemical or biological. Improving information about environmental effects requires a systems perspective, a focus on appropriate models, and clear thinking about relevant measurement endpoints as well as appropriate temporal and spatial scales.

The identification of ecosystem values has clear and important implications for ecological risk assessment. The endpoints of risk assessments are the important ecosystem attributes that are believed to be potentially susceptible to the hazardous agent. Therefore, risk assessors must develop methods to estimate the likelihood of changes in those attributes that are to be valued. For example, if detoxification of chemicals is deemed to be a valued service of wetlands, then the capacity for detoxification is an assessment endpoint, and methods should be developed to estimate changes in detoxification rates in response to increased toxicant loading, dredging, filling, or changes in the hydrologic regime.

Decision makers also need information to determine the value to society of avoiding adverse environmental effects in at least two broad categories of decisions.

Executive Order 12291 requires that agency decision makers conduct a benefit-cost analysis for all new, and for any revisions to, major regulations, except where legislation or judicial decisions expressly forbid it. EPA has issued guidelines for conducting such analysis, which acknowledge that "estimating the benefits (or damages averted) of environmental regulations that affect ecosystems is perhaps the most complex problem in benefits analysis" (USEPA, 1983).

Executive branch agencies also have the responsibility under several statutes to assess damages for certain actions that have caused harm to natural resources. Liability for natural resource damages can be traced to the Trans-Alaska Pipeline Act, but it has gained much greater influence with CERCLA (and its re-authorizing amendments SARA) and the Oil Pollution Act of 1990. This legislation acknowledges that natural resources are assets and that the consumptive and nonconsumption services they provide must be considered in determining their value for the

[6] Decision makers also need information about the sources of stress to ecosystems in order to design effective environmental protection strategies. For example, Congress in its recent amendments to the Clean Air Act directed EPA to assess and perhaps control deposition of hazardous air pollutants on the Great Lakes and other large water bodies. EPA is currently measuring the total loading of pollutants to the Great Lakes and assessing how much of it comes from the air. Sources can be of different types, from point and non-point locations, directly from human actions or indirectly from other parts of the environment. However, the need for information about sources is not directly applicable to the questions of valuation before the Forum.

purpose of assessing damages. Regulations setting guidelines for placing a value on natural resources for the purpose of damage assessment have been repromulgated by the Department of the Interior after litigation over an earlier version. The National Oceanographic and Atmospheric Administration also is preparing its own regulations under different statutory authority.

Clearly, more needs to be done to improve the information available to decision makers about the nature and magnitude of environmental effects (particularly predictive information) and about the value of those effects. For specific decision points, decision makers may only need ecological data or only need economic analysis, but the point here is that, for broad advancement in ecosystem valuation methods, improvements in both are needed. Progress in understanding how ecosystems function and how they are affected by human activities provides a necessary but not sufficient basis for most choices. Conversely, improving our ability to value ecosystem functions and services will have limited utility unless we also improve our understanding of how ecological systems respond to perturbations.

3. The search for decisive information

We are reminded of two other aspects of choice. First, there is the problem of structuring the choice process so that related phenomena with possibly similar implications are regarded similarly. Deriving reasonably standardized decision protocols can be useful in achieving consistency across choices. Such decision protocols can also be very practical in economizing on decision costs. After all, choices require information acquisition, processing, interpretation, and synthesis. These actions require staff time, and hence imply considerable expense.

The second aspect of choice regards what we shall call decisive information. By decisive information we mean that information which is necessary and just sufficient to allow choice. This concept of decisive information, as well as the concept of decision protocols, warrant further clarification.

3.1. On decision protocols

As suggested above, decision protocols economize on decision costs by offering systematic rules for describing how information should be made available for each of the decisions involved in any action. Indeed, one may regard a decision tree as a decision protocol. Such maps guide the decision process in a logical and consistent search for the point at which enough is known about ecosystem effects to stop studying the problem and make a decision.

Determining the critical pathway of physical effects or exposure routes may be a useful step in developing a decision tree for valuation protocols. For example, the procedures for conducting valuation studies might suggest the circumstances under which physical data about possible exposures should be collected first.

It might be useful periodically to review mandated valuation information tasks, such as those under E.O. 12291, to determine whether the information that must be generated on a routine basis is normally relevant to the decision. In addition, protocols for conducting valuation studies ought also to provide rules of thumb for when a full study should be undertaken, de novo, and when extrapolation from previous studies is valid. This would require clear criteria for determining which classes of decisions or problems are similar.

As an example, the DOI natural resource damage regulations begin to describe valuation protocols. While crude, their Type A computation model for marine environments illustrates one approach for defining systematic procedures for developing these estimates. Often labeled as benefit transfers, most valuation estimates must re-organize existing information (often for a different resource or at best the same resource under different conditions) to the needs of a particular policy analysis. This Forum's development of a consistent framework for integrating ecological and economic "values" will not change this feature of the problem. So we must consider the implications of that resolution for the existing practices of benefit transfer and how the usually implicit protocols should be re-written.

3.2. On decisive information

Information is costly to obtain, and decision makers have limited amounts of time and money. Thus, it is important to identify in advance what information is relevant. One test is to ask what information, if known, has the potential to affect the decision one way or another. Identifying "decisive" information in advance could substantially reduce information costs of public decisions.

Perfect information will never be available on the systems affected by human decisions—private or collective. Thus, uncertainty will be an inherent feature of all important decisions. The uncertainty can be treated as part of the system, a randomness in outcomes, and/or our ability to observe it. Based on those observations, we must characterize what we know about these systems and use summaries of those observations to inform decisions. The value structure adopted influences how we collect information to inform decisions. Learning from that information is likely to change aspects of the value structure. Decisive information is that information that would permit this implicit question to be answered.

The collection of decisive information about ecosystems is especially difficult. The Environmental Protection Agency is striving to incorporate more holistic ways of understanding complex ecosystems to respond to the rising public consciousness of ecosystems as a whole. However, existing data and ways of knowing reflect already well-established values and disciplines. This creates two problems. First, disciplinary ways of knowing and their associated data inherently value those things they consider and dismiss those things they do not. Second, it is frequently impossible to synthesize or weight the incongruent data and disparate, conventional ways of knowing with new ways of knowing and recently collected data. A more holistic understanding quite frequently entails a difficult process of discourse as experts of different backgrounds strive to think in new ways to comprehend whole ecosystems.

The identification of decisive information is related to the estimation of uncertainty. A decision cannot be made on the basis of benefit-cost analysis if the confidence interval on the estimate of costs significantly overlaps the confidence interval on the estimate of benefits (estimated subjectively or by uncertainty analysis). From this perspective, decisive information is information that is feasible to obtain and will reduce uncertainty so that the confidence intervals no longer overlap significantly.

Finally, where only partial information about values is available, analysts may benefit from guidance about the appropriate use or conclusions to draw from such information.

For example, a decision rule for evaluating actions that involve irreversible transformations of unique natural environments could suggest that we monetize both the benefits and the costs of all aspects of the policy involving the ecosystem and consider their difference, recognizing that there may be many services that are not included. Then, we could ask how large the value of the omitted services would have to be in order for the policy maker to be indifferent between taking the action or not. Essentially, this is asking how large an unknown has to be to change a decision.

4. Agenda for future

Progress on improving methods for providing information to decision makers about the value of ecosystem attributes will require an interdisciplinary dialogue. Establishing an interdisciplinary agenda is no easy task, however.

During a significant fraction of their time together, Forum members have struggled to identify the reasons for their very different notions of what it means to value an ecological system. An understanding of the terms of that struggle will help to place in perspective this report about information needs, and the limits of current methods for producing useful information.

Some members of the Forum were comfortable from the beginning with the notion that changes in the service flows from ecological systems to human society ought to be valued in monetary terms to the extent supportable by available data and techniques. They felt that this would bring such services into policy discussions in terms commensurate with marketed goods and services. As a general matter this would, they felt, improve the efficiency with which society used its environmental resources. One reason for this view is the belief that such pricing would encourage preservation by making explicit the opportunity cost of development and other economic activities.

Other Forum members expressed serious reservations about this view. Their hesitation arose from the limited knowledge about the behavior of both large and small-scale ecological systems, the concern that current prices (or monetary values) do not and cannot reflect the tastes of and technologies available to future generations, and the possibility that irreversible ecological damage may result from large-scale, long-term human intervention in the environment.

In considering these differences, the Ecosystem Valuation Forum has formed the following preliminary conclusions.

4.1. Contextual issues

Information needs

Clarity about the uses of valuation information is crucial if efforts to improve valuation methods are to be productively focussed. Focusing on linkages is one key. It is essential to link the information produced by both ecological studies and valuation methods to the needs of policy makers. The information produced by ecological studies and the information needed to implement valuation methods also needs to be linked. Currently, none of these linkages are adequate.

In addition, time and resource constraints of decision makers must be taken into account so that the information collected will be useful in making a decision.

However, care must be taken to consider the implications of the personal, normative values that experts bring to the information-collection process. Success in providing improved information to decision makers will require frequent reminders about (1) the different patterns of thinking in different disciplines, (2) differences between the needs of users of information and the personal views of scientists attempting to provide it, (3) the complex interaction between natural and social systems, and (4) the difficulty (and potentially irreversible impacts) of environmental "experiments" (Norgaard, 1992).

Terminology

Semantic difficulties incurrent valuation terminology are a barrier to progress in developing improved ecosystem valuation methods. Some words in common usage, such as "benefit", "value" and "function" have special and different meanings across different disciplines.

In addition, the same terms sometimes are used to mean different things, and different terms are used to mean the same thing. Examples of the former include the word benefit, which in different disciplines might refer to avoided ecosystem impacts, economic development, or individual satisfaction. Examples of the use of different terms to mean the same thing, sometimes incorrectly, include existence value, option value, and intrinsic value; functions, processes, services, and attributes. Important issues will be more difficult to resolve as long as critical terms are being used differently.

It may be difficult, however, to create a simple, cross-disciplinary "glossary" that links terms from one discipline with those of another via straightforward translation rules. As is emphasized in the currently popular idea of disciplinary "paradigms", the meanings of scientific terms are suffused with the assumptions, methods, goals and values that give identity to the disciplines themselves.

4.2. Methodological issues

Ecological methods

Improving information about the value of ecosystem attributes will require more than improving valuation methods. The capability of ecological methods to describe and quantify ecosystem attributes, and to predict the consequences of human actions on those attributes, also must be improved. Better information is needed about how various services of ecosystems are affected as a result of human actions.

The place to start is with service flows—what are the services that an ecosystem provides, and how is the production of those services likely to change under alternative courses of action? With this causal nexus established we can begin to develop methods for estimating the value of those changes for the decision maker who faces choices among several options. The capacity of ecological methods to predict environmental effects is highly variable, however.

The fields of ecology and toxicology have advanced to a relatively high degree of accuracy in predicting the effect of some actions on particular ecosystem attributes. For example, toxicologists can predict what percentage of certain fish populations will remain at different levels of contamination much more accurately than they can predict levels of terrestrial wildlife populations.

In many other important areas, however, a great deal of uncertainty remains. Ecosystems are complex, and a high potential exists for non-additive and synergistic effects. Equally important, however, insufficient consensus exists regarding which features of ecosystems are essential to maintain.

In addition, some ecosystem attributes, such as primary productivity, support the overall viability of the system. These aspects are "services" to the ecosystem, but are not direct services to people. This does not make them less important. To the contrary, they are sometimes the most important attributes of the ecosystem from an ecological perspective because the loss of one or more of these attributes could result in the loss of many, if not all, of the other services the ecosystem provides. Thus, services that may be intermediate to the ecosystem will have a value because of their contribution to the "production" of the services that people do value.

The complexity of ecosystems poses enormous challenges for predicting alterations resulting from human actions. The variety of ecosystems and their attributes have not been completely catalogued, and not all causal relationships are known. Thus, the analyst is presented with difficult questions concerning the selection of relevant attributes as well as how to present data in a form that can be used for valuation purposes.

Valuation methods

Valuation methods may originate from economics, ecology, social-psychology, philosophy, or other disciplines.

Several methods already exist for assessing the monetary value of ecosystem attributes. Contingent valuation, travel cost and hedonic pricing are three examples of methods for monetary valuation given an existing system of rights and a high degree of knowledge about the ecosystem attribute by the user.

These methods fall into two categories—one relying on observable choices, and a second relying on the responses people make to proposed choices. The observable choice or revealed preference methods (travel cost and hedonic) are preferred by some economists on the grounds that an actual choice demonstrates the commodity (or service) to be valued has been selected by those whose monetary values are being measured. However, it is important to acknowledge that what is usually observed is the selection of a good or service that is linked to the environmental service, not the environmental service itself. Thus, the analyst's judgments can influence the monetary estimates that are inferred from these choices. The survey or contingent valuation approach assumes that stated preferences accurately represent what peoples' preferences would be if they had the choices proposed to them. The description or framing of what is to be valued has been found to be a central element in the reliability of the method.

Restoration cost and replacement cost approaches for valuing ecosystem attributes also provide monetary measures of ecosystem values. Restoration cost sets the value of a system as the cost of restoring it to its pre-damaged condition. One definition of replacement cost calculates the necessary expense if the naturally provided service didn't exist (e.g., for soil productivity, flood control or water purification). These methods rely on defining the set of attributes to be restored or replaced. Subtle changes in descriptions of how the acts of restoration or replacement are judged have significant implications for the criteria used to value the ecosystem. Moreover, these changes also influence the relationship between these approaches to valuation and the economic, monetary measures. For example, if restoration is defined as the act of restoring the predamaged or baseline condition of the ecosystem instead of its functions, the definition implies a very different mix of activities. The same qualification applies to replacement. What is involved is defining and measuring the elements, attributes, or services designated to comprise the important features of the ecosystem and evaluating whether different approaches to obtaining them in fact restore or replace what has been damaged or lost.

The concept of "putting the resource back the way it was" presents at least two challenges. First, it may be technically impossible to do, thus the cost of doing so may be impossible to determine. In addition, restoration cost may not be cost-effective: i.e., the full resource allocation implications may be negative.

Economic methods should be used with an understanding of their limitations. Even as we improve our ability to monetize natural resources, concerns are likely to remain because of uncertainties in scientific ability to predict the effects of human actions on ecosystems, particularly when the recovery time for environmental effects is very long. In addition, although improved economic methods can lead to improved environmental decisions, economics cannot substitute for collective political decisions about distribution issues, including rights to resource use to future generations or within the present generation. Thus, for some types of decisions, the issues of sustainability and ecosystem values ultimately will require collective choices within the political process.

Interdisciplinary approaches may overcome some limitations of existing economic methods. Ecological economic modeling constructs detailed dynamic simulation models of linked ecological economic systems which, after being calibrated to real world situations, can be exercised to determine the linkages and values of the ecological system.

Other disciplines may provide yet additional information relevant to valuation. Social psychologists have used a variety of survey methods to assess human preferences for varying environmental and ecosystem characteristics. For example, attitudinal scales have been developed to quantify aesthetic, ecological, naturalistic, scientific, and even ethical valuations of ecosystems and landscape features. Additionally, researchers in the field of landscape architecture have pioneered techniques for estimating the aesthetic value of varying ecological systems. These techniques have involved both expert estimations and public surveys. However, these methods do not yet capture all of the values that contribute to what people describe as a meaningful quality of life.

Although some ecosystem attributes can be valued using existing methods, current methods do not now value all the services of ecosystems that ecologists or the general public believe are important—each has strengths and weaknesses. Thus, an assessment of what can and can't be valued, and what problems remain after existing methods are improved, is important.

4.3. Challenges

In addition, there are issues related to the complexity of environmental decisions rather than to specific valuation techniques. These include dealing with uncertainty, accounting for irreversibility, and considering the rights of future generations.

Uncertainty

While some level of uncertainty is present in most decisions, the level of uncertainty about ecosystem effects can be extremely large. Decision makers must consider uncertainty both in the likelihood that a particular outcome might occur, and in the severity of the effects, should it occur.

For conceptual purposes, it is helpful to distinguish among and describe the various ways uncertainty can influence the information available to decision makers. In both the natural and social sciences, some processes often are described as having random components, which together with the systematic factors hypothesized to influence these processes contribute to the outcomes (or states) we observe. We might designate these components as "inherent randomness" simply to distinguish them from the variation that arises from incomplete or imperfect observation of these same processes. Imperfect observation is sometimes argued to cause uncertainty in the estimates or summaries of available information. In practice, both types of randomness are simultaneously present in the information available to decision makers because inherent randomness contributes to the randomness caused by incomplete observation. Both affect the summaries developed from data on the processes of interest.

Uncertainty has an added and important influence on the valuation process because people in their own personal decisions—and society in its collective decisions—respond differently to judgments that must be made for processes that are relatively more certain than those that are uncertain (i.e., are characterized as having more of one or both types of randomness). These responses to uncertainty depend upon what is at risk, how much is known about the processes that exhibit the randomness, and whether people can individually or collectively control the ways the outcomes might affect them. At the simplest level, economists describe these reactions as part of people's preferences and characterize as risk averse those who prefer to avoid uncertainty beyond simply acknowledging that the average (or expected values) of the outcomes should be the focus of attention. Retaining this simple mode, economic models attempt to relate individuals' personal attitudes toward uncertainty to the implications of their personal risk preferences for society as a whole.

These descriptions are most relevant to processes where the randomness is reasonably well understood, the outcomes largely monetary in character, and people have acquired experience in dealing with them. For most environmental risks (and especially those associated with ecosystems), this characterization is not appropriate. So the knowledge, risk perception, sense of control, range of outcomes, and opportunities for response to risks will influence both the individual's and society's valuation of changes in uncertainty. Developing methods for responding to these needs will be an important aspect of the Forum's activities during its second phase of work.

One approach follows from an analogy between decision making under uncertainty and purchasing insurance. If a proposed regulation having known economic costs will reduce the likelihood of unknown future environmental impacts, a decision to regulate still might be made when the known benefits are less than the known costs. Basically, the difference between the costs and benefits would be the cost of an "insurance policy" against future impacts. One reason to purchase such "insurance" (often expressed by the desire to be conservative where predictions about future effects are uncertain) is the concern about the limits of science to portray all possible effects—some people would "rather be safe than sorry". Of course, this does not answer the question of how much "insurance" people are willing to buy for unknown future benefits.

Irreversibility

Many changes to ecosystems can be reversed over varying lengths of time. In many other cases, however, changes can only partially be reversed, if at all. Decision makers do not need sophisticated benefit-cost analysis to justify actions to prevent irreversible environmental effects of large magnitude. Under such circumstances, analysis of the cost effectiveness of different policies or actions for preventing such consequences is appropriate. The question, however, remains how large is too large?

Temporal scale

Environmental effects may have a very long recovery time. Thus, the benefits of resource use may accrue to one generation while the costs are borne by multiple generations.

Issues of an intergenerational dimension—ensuring a sustainable flow of ecosystem services—and the impacts of discounting must be addressed. Discounting is commonly used to express future costs and benefits in terms of present monetary value, assuming that a value received now is worth more than the same value provided at some future date. Some critics regard discounting as inappropriately devaluing those ecosystem attributes that take a long time to be realized. Thus, discounting results in greater resource exploitation or use of natural capital now at the expense of future environmental health or availability of natural capital. Uncertainties include the inability of current generations to fully reflect the values and preferences of future generations, the inability to account for future technological change, and the applicability of discounting the value of renewable resources when exploitation would decrease sustainable yield.

Units of measure

For decisions affecting ecosystems, conventional benefit-cost analysis is incomplete because many costs and benefits are difficult to quantify and, even if quantifiable, may be difficult to measure in monetary terms. Common terms certainly make the summation of benefits and costs easier, but many worry that environmental values can never be adequately measured monetarily. Thus, some experts question whether the analysis should be done solely in monetary terms.

Clearly, decision makers often face choices involving values that are measured in incommensurate terms. For example, in deciding whether to purchase land adjacent to the Manassas Battlefield National Park, Congress weighed the monetary cost of the land against the benefit (or value) of maintaining the integrity of the historic site for visitors. In making pesticide registration decisions under FIFRA, EPA may be faced with a decision to ban a pesticide that is known to cause human health effects, for which the only practicable alternative would cause adverse ecological impacts.

In other cases, one might find common, non-monetary units of measure. For example, in deciding whether to ban a pesticide, EPA evaluates alternative products to assess whether the replacement pesticide would cause more or less harm. Number of deaths or years of life lost are generally the common terms for health effects. Presumably, similar comparisons could be made using common terms for ecological effects.

5. Recommendations and conclusions

5.1. Case studies

The Forum recommends that the approach for addressing the research agenda outlined above should be to identify and explore the real challenges of ecosystem valuation through practical case studies, for only through the crucible of real experience will methods be found that will make a useful and realistic contribution to public decision making. In particular, such case studies should enable researchers to improve linkages between ecological and economic methods and to develop improved protocols for valuation studies. EPA staff involved in the cases selected can also expect to benefit directly in their work. Criteria for selection of case studies include:

- variety of issues,
- variety of temporal and spacial scales,
- decision-focused (a choice must be made),
- ecosystem values will/did play a significant role in the decision,
- illustrates one or more of the challenges identified in Phase I, and
- data is available.

5.2. Assumptions

That good valuation studies depend on sound ecological information is not adequately appreciated. Better linkages are needed between ecological and economic methods. For example, existing methods for valuing ecosystem services often rely on predictions about how ecosystem service flows will change as a result of human intervention. Advances in ecological methods themselves will help, as will efforts to integrate methods across disciplines. Forum members recommend joint modelling exercises to enhance the ability to apply existing valuation methods to a greater number of ecosystem attributes and to develop clearer, logical steps from ecosystem damage assessments to valuation.

Joint modelling efforts might be designated to:

- review what information existing ecological model(s) accept and in what form predictive information is generated,
- begin to explore ways the ecological model(s) need to be adapted to serve policy or valuation purposes and/or what economists could do with the information generated, and then
- organize a valuation study given previous work on linkages with the ecological model.

Ecosystem valuation could be improved if currently available tools for monetary valuation could be competently applied more often. Additional resources for economic studies would help, but resources, even if augmented, will likely always be inadequate to assess every situation in which ecosystems are damaged or at risk. Thus, guidance is needed on how to define the valuation information that is most likely to be decisive, on criteria for choosing among the available valuation methods, and on ways to build on existing information rather than to rely on expensive primary data gathering. Here, too, better linkages between ecological and economic methods may shed light on the problem, with decision trees relying on qualitative ecological information a possibility.

Forum members anticipate exploring specific decision scenarios as case studies for developing improved protocols for policy analysts. Possible candidates include the EPA regulations on the land application of sludge from pulp and paper mills or the re-registration of granular carbofuran. The latter case is ripe for a retrospective review of valuation protocols because it has a significant amount of data available in the ecological risk assessment, there is a clear effect on individual birds, and there are interesting valuation issues with respect to endangered species, but with the dilemma of not having been able to establish population or ecosystem effects.

Finally, work is needed to clarify where existing ecosystem valuation methods and monetary measures of value are sufficient for decisions and where new methods, assumptions, or criteria for decision making are needed. Even if ecosystem services were perfectly understood and evaluated in monetary units according to accepted economic percepts, the problems of ecosystem valuation would not be completely solved. The problems of uncertainty (both statistical and scientific), irreversibility, and large disparities in temporal and spatial scale remain.

For example, the policy for classifying wetlands as having high, medium or low value is just such an example; there may be other cases that would allow members to deal with varying temporal and spatial scale, with aesthetic and scientific values, and with multiple units of measure. Forum members recommend addressing these issues of the limits and appropriate use of ecosystem valuation methods in a more generic manner as well.

The challenge of improving ecosystem valuation methods presents an opportunity for partnership — partnership between ecologists, economists and other social scientists and partnership between the research and policy communities. Interdisciplinary dialogue between ecologists, economists, and other social scientists is essential to the task of developing improved methods for valuing ecosystem attributes.

Acknowledgement

Dr. V. Kerry Smith, North Carolina State University, contributed to earlier drafts of this report.

References

Bromley, 1976. Economics and public decisions: roles of the state and issues in economic evaluation. J. Econ. Issues, 10(4): 812.

Kellert, S. and Clark, T., 1991. The theory and application of a wildlife policy framework. In: W.R. Mangun and S.S. Nagel (Editors), Public Policy and Wildlife Conservation. Greenwood, New York, NY.

Norgaard, R., 1992. Environmental science as a social process. J. Environ. Monitor. Assess., 20: 95–110.

Norgaard, R., 1991. Sustainability as Intergenerational Equity: The Challenge to Economic Thought and Practice. Internal Discussion Paper, Asia Regional Series, Report No. IDP 97, The World Bank, June 1991.

USEPA, 1983. Guidelines for Performing Regulatory Impact Analysis. Office of Policy, Planning and Evaluation, December, 1983 (reprinted March, 1991).

[9]

ELSEVIER

Ecological Economics 14 (1995) 143–159

ECOLOGICAL ECONOMICS

Analysis

Ecological economic modeling and valuation of ecosystems

N. Bockstael [a,b,*], R. Costanza [b,c], I. Strand [a,b], W. Boynton [b,c], K. Bell [a], L. Wainger [b,c]

[a] *Department of Agricultural and Resource Economics, University of Maryland, College Park, MD 20742, USA*

[b] *Maryland International Institute for Ecological Economics, Center for Environmental and Estuarine Studies, University of Maryland, Box 38, Solomons, MD 20688-0038, USA*

[c] *Chesapeake Biological Laboratory, Center for Environmental and Estuarine Studies, University of Maryland, Box 38, Solomons, MD 20688-0038, USA*

Received 6 April 1994; accepted 10 August 1994

Abstract

We are attempting to integrate ecological and economic modeling and analysis in order to improve our understanding of regional systems, assess potential future impacts of various land-use, development, and agricultural policy options, and to better assess the value of ecological systems. Starting with an existing spatially articulated ecosystem model of the Patuxent River drainage basin in Maryland, we are adding modules to endogenize the agricultural components of the system (especially the impacts of agricultural practices and crop choice) and the process of land-use decision making. The integrated model will allow us to evaluate the indirect effects over long time horizons of current policy options. These effects are almost always ignored in partial analyses, although they may be very significant and may reverse many long-held assumptions and policy predictions. This paper is a progress report on this modeling effort, indicating our motivations, ideas, and plans for completion.

Keywords: Ecological-economic modeling; Ecosystem models; Patuxent River basin; Spatial modeling; Land use

1. Introduction

In its report, Reducing Risk: Setting Priorities and Strategies for Environmental Protection (September 1990), the U.S. Environmental Protection Agency (EPA)'s Science Advisory Board recommended that "EPA should develop improved methods to value natural resources and to account for long-term environmental effects in its economic analysis" (EPA, 1990). In spring 1991, EPA convened a group of ecologists, economists, and other social scientists at an "Ecosystem Valuation Forum" for the purpose of advancing the state of the art in ecosystem valuation methods. The Forum's report on the concepts and problems in ecosystem valuation (Bingham, 1994) identified two broad areas of research in need of improvement:

1. Understanding how ecosystems function and how they are affected by human activity—for example, what determines human uses and human intervention into ecosystems, and how is this affected, among other things, by the ecosystem's characteristics and regulatory paradigms?

* Corresponding author

Elsevier Science B.V.
SSDI 0921-8009(95)00026-7

2. Improving methods for ecosystem valuation—for example, what services provided by an ecosystem are of value to society, what methods should be used to measure this value, and in what terms should it be expressed?

One of the outcomes of the forum was to commission case studies to begin to address these research needs. We have begun a project to model the interactions of ecological and economic systems to address the above questions. This paper represents a preliminary report on our progress in this project. The project team consists of two economists and two ecologists who are interested in breaking down barriers to communication and developing integrated ecological economic models to address the issues of human interactions with ecosystems and ecosystem valuation.

2. Ecological economic modeling

Ecosystems are extremely complex systems whose functions and processes are not easily characterized. A particularly troublesome issue involves the description of factors affecting the reaction of ecosystems to natural and human stresses. There is some agreement regarding which ecosystem characteristics can be used to evaluate stress, such as food web structures, complexity, and densities. However, there is considerable uncertainty regarding their relative significance. The importance of various attributes and measures of system stocks and flows in conveying the fitness of an ecosystem is poorly understood, and existing knowledge often limits the scope of analysis to a few species and resources, as only a few functional relationships have been quantified. Equally troublesome is our lack of understanding of the effects of ecosystems' states on human well-being. Some characteristics of ecosystems have obvious and immediate economic and/or human significance, while others appear important in a longer term or more global sense. There is considerable uncertainty surrounding the effects of human decisions on natural systems, and vice versa, as well as what features of one prompt responses from the other.

Considerable work over the past decades has attempted to address aspects of the first of these issues, but for the most part this work has concentrated on *either* ecological *or* economic modeling approaches to the problem. The tradition of independent research has resulted in a body of environmental work completed by both sciences that overlooks the fundamental problems associated with their interactions. Namely, the studies are completed using different terms and units of measurement, and the studies neglect to account for the interaction between the processes and populations each seeks to model. For the most part, the ecological models have tended to ignore humans, except to the extent that they introduce apparently exogenous shocks into the ecological system, and economic models have taken stylized versions of "ecological facts" or relationships to connect human actions with the state of the ecosystem.

Both sciences recognize the problematic nature of these tendencies, and there have been some efforts to revise tradition. The literature contains some attempts to grapple with the interactions between ecological functions and economic actions (e.g., Tschirhart and Crocker, 1987; Costanza et al., 1993), but these are largely conceptual papers. In addition to early work such as that reported by Russell (1993), a recent Resource Policy Symposium noted only four integrated ecological economic models developed for analyzing agriculture (Bouzaher, 1993; Hoogervorst, 1993) and natural resource-based tourism (Dixon and Scura, 1993; van den Bergh, 1993). Attempts at interdisciplinary modeling have often produced larger and larger "black boxes", where the level of resolution is reduced to the lowest common denominator and the integrity of each component of the model is lost. This project will take a parallel modeling approach and will utilize advances in computer modeling and spatial data availability to address the transdisciplinary modeling task in new ways.

3. Ecosystem valuation

The second recommendation of the Forum reflects concerns with current ecosystem valuation methods. These concerns include the narrow focus on tangible, on-site resource commodities and the disregard for ecosystem functioning. By evaluating only those components of the ecosystem that have immediate value to individuals, and focusing on short-term changes in the ecosystem, this practice ignores the fact that changes in ecosystems play out over time and space and may indeed be irreversible.

Much of the ecosystem valuation work has focused on wetlands because of current regulatory interest in these ecosystems. This literature illustrates the above problems. Recognizing the difficulty of identifying and measuring the contributions of wetlands, the majority of valuation papers focus on specific well-defined services wetlands provide, such as storm protection, recreation, fisheries, timber, or water purification (Gupta and Foster, 1975; Lynne et al., 1981; Farber, 1988; Costanza et al., 1989; Bergstrom et al., 1990). Attempts have been made to finesse the problem (Shabman et al., 1979; Shabman, 1987) by valuing the resource at its alleged upper bound: the cost of providing substitute services. But this approach begs the fundamental question of definition and measurement of those services: i.e., what is it about wetlands that really matters? (Scodari, 1992).

Another drawback of the existing attempts to value ecosystems is that they tend to be spatially either too general or too specific. Studies either examine aggregate national data or they pursue site-specific case studies, such as those referenced above. Of course the ecology will differ from prairie potholes to the everglades to the tundra to the coastal zone, but the economics will vary as well. Even the "same" wetland in different locations will have different value. And the alternative uses of the wetland will vary with location. For example, two lakes differing only by spatial location (i.e., near an airport versus near a national forest) will provide entirely different vectors of ecosystem services. Conversely, the value of a wetland cannot adequately be captured by local analyses whose scale diminishes the importance of spatial location. Human intervention may set in motion both ecological and economic processes that have far-reaching consequences.

The ultimate purpose of this project is to shed light on ecosystem valuation, but as is clear from the above, this result depends on good ecological and economic modeling that allows for feedback between the human and natural systems as well as spatial and temporal disaggregation. Interestingly, part of that feedback comes in the form of changing private market values of land in different uses. Human decisions invoke changes in ecosystems that play out over time and space and may result in a reconfiguration of the landscape. Reconfiguration results in a new array of market values for land parcels within the landscape and a new array of non-market values emanating from public good amenities of the landscape. The change in private land values signals a new round of human land-use decisions. As has long been recognized, however, the predominance of externalities in land use prevents even the effect of human decisions on market values from being internalized into individuals' decisions. For this reason, as well as the obvious effect of land use on public goods such as recreation, the public sector has imposed an imperfect and ever-shifting pattern of land-use regulations.

In this project, we will model economic and ecological dynamics in a landscape framework which incorporates regulations and human decision-making processes so that more can be learned about the ultimate consequences for ecosystem and land use configuration. The analysis will reveal how market land values are reconfigured and it should shed light on the effects of landscape reconfiguration on non-marketed values, including those that take time to emerge.

4. Conflicts inherent in ecological economic modeling

While the ultimate intent of this project may be to gain insight into the dilemma of valuing ecosystem configurations and the services from those configurations, the more immediate objective is to model the interaction of the ecosystem and human activity, illustrating how humans intervene in the ecosystem and how different ecosystem configurations contribute to human welfare. It is at least conceivable that what appear to be deep-rooted ethical differences between ecologists and economists with regard to the more general goal can be reconciled if we understand more fully the specific nature of humans' interactions with ecosystems and how these interactions induce subsequent reactions as they unfold spatially and temporally. A major criticism by ecologists is that economists are too narrow and anthropocentric in viewing the role, functions, and indeed, importance of ecological systems, since they tend to focus on immediate and on-site impacts rather than including future and indirect implications to humans and ecosystem integrity. Economists are impatient with ecologists for their unwillingness to evaluate the relative contributions of various characteristics of ecosystems to values and their disregard of some human preferences when considering value. Economists are presumably taking the pragmatist's view, disregarding those things that they don't understand or can't measure, no matter how important. Ecologists are taking the purist's view: any valuation scheme that fails to take into account the most important aspects of a problem is unacceptable.

Structurally, the disciplines of economics and ecology have much in common. Both analyze and predict the behavior of complex, interrelated systems in which the behavior of individual agents and flows of energy and matter are central, and its dynamics are governed by the allocation of scarce resources among competing agents (Bernstein, 1981). However, despite these basic similarities, there are significant distinctions in modeling practices. Broad explanations of such differences include the distinctions in their units of measurement, populations of interest, handling of risk and uncertainty, and paradigms of analysis (Bernstein, 1981; Norgaard, 1989).

The dynamics of ecosystems and markets are extremely different, and these differences are reflected in the time steps and time horizons commonly employed in models of ecological and economic systems. Many of the distinctions in timing follow directly from the availability of data employed to model the economic and ecological systems. Data on economic decisions is rarely available for a periodicity any less than one month, and generally only available annually. In contrast, because scientists have studied the growth and development of organisms, hydrological dynamics, etc., ecosystems functions can frequently be modeled at a very high level of temporal and spatial resolution.

While ecologists frequently adopt high time resolution models, they also are interested in long-time horizons, and it is with regard to long-term implications of human actions that ecologists are frequently concerned. Economists have tended to ignore the very long run, believing that future perturbations to the system are virtually impossible to predict.

Other types of boundary conditions differ as well. When faced with limiting the amount of activity to be modeled, ecologists employ geographical or physical bounds, and economists rely on the extent of markets to define their study areas. There is no reason for ecosystem and market boundaries to coincide. As a rule, neither definition nor bounding mechanism subsumes the other. Ecologists think in terms of physical flows such as water, biomass, and energy. Many of these connections have spatial characteristics, and in recent years ecologists have made great advances in the development of spatial modeling techniques (Costanza et al., 1990). For example, the Patuxent Landscape Model (PLM) (DeBellevue et al., 1993), adapted for this project, utilizes significant spatial dynamics and in so doing can model the succession of complex ecosystems over long periods of time. In contrast, economists tend to think in terms of informational flows such as money and prices and use these informational flows as the basis of changes when modeling human behavior. Economists have much experience with time dynamics, but little with spatial considerations, as markets are not typically spatially defined and may have little to do with geographical dimensions. One significant exception is the land market. Although a considerable amount of research has been completed on the land market, economists have struggled somewhat unsuccessfully to fully model its spatial component.

Many differences in the approaches of ecolo-

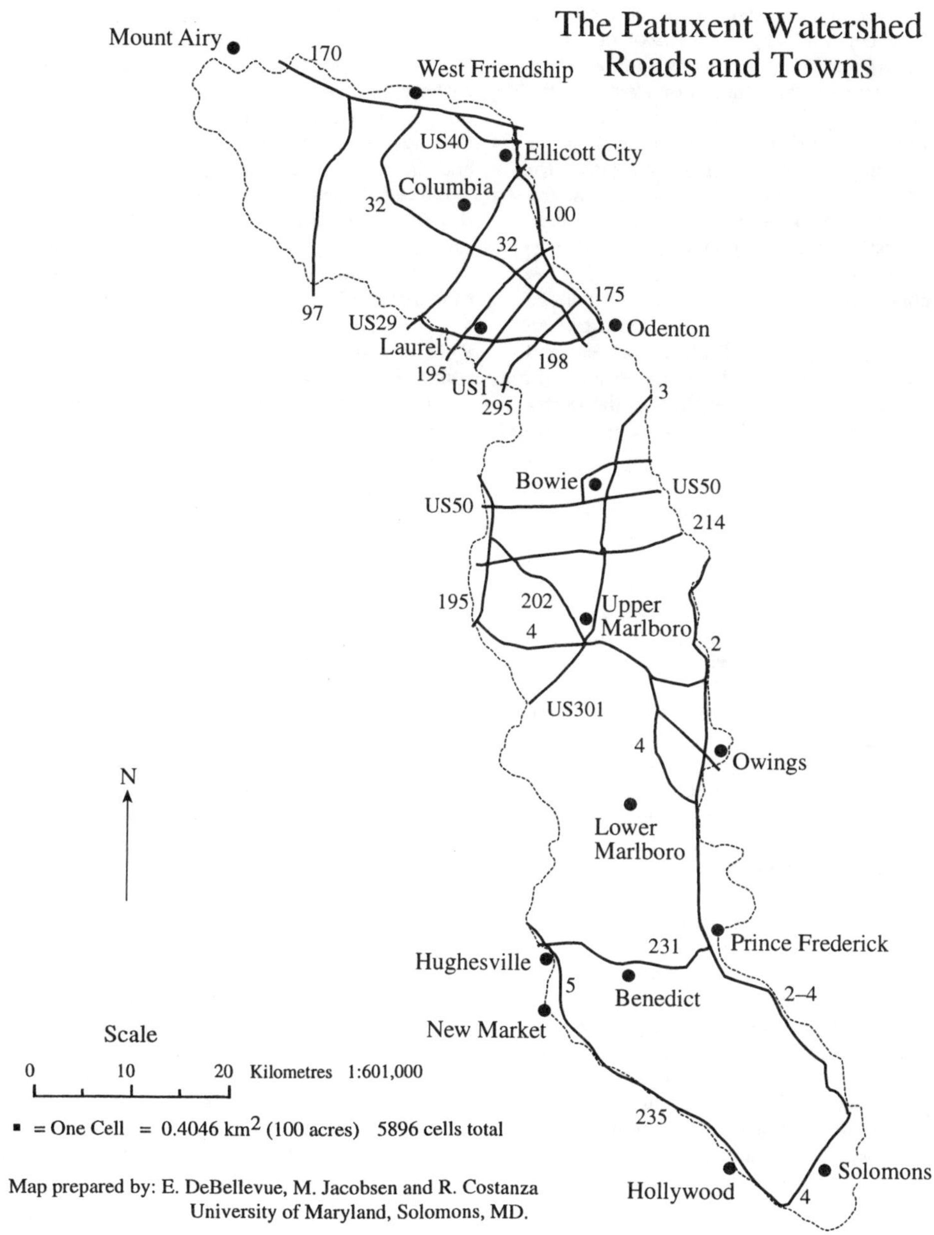
The Patuxent Watershed
Roads and Towns
Mount Airy
170
West Friendship
US40
Ellicott City
Columbia
32
100
32
175
Odenton
97
US29
Laurel
198
195
US1
295
3
Bowie
US50
US50
214
195
202
Upper
Marlboro
4
2
US301
4
Owings
N
Lower
Marlboro
Prince Frederick
231
Hughesville
5
Benedict
2–4
New Market
Scale
0
10
20
Kilometres 1:601,000
▪ = One Cell = 0.4046 km2 (100 acres) 5896 cells total
235
Hollywood
Solomons
4
Map prepared by: E. DeBellevue, M. Jacobsen and R. Costanza
University of Maryland, Solomons, MD.

gists and economists occur because the factors in the environment that are the focus of detail for one discipline are often hopelessly aggregated or abstracted from by the other. For example, the ecological model might not distinguish between agricultural crops such as soybeans and wheat, or among wildlife species on recreational lands, for this distinction may not be important in modeling the ecosystem. Yet, in addressing human behavior and assessing values to humans, these distinctions are significant.

An aim of this research will be to acknowledge and reconcile these conflicts in the context of an integrated modeling effort. Both the process and the modeling exercise will be valuable and will serve an important role in furthering the objectives of the Ecosystem Valuation Forum (Bingham, 1994).

5. An overview of the modeling project

Recognizing the need for improvements in understanding how ecosystems function, how humans interact with ecosystems, and how ecosystems are valued, the following objectives were identified for this project:

1. to increase our understanding of and our ability to model the interactions between human behavior and ecological processes;
2. to increase our understanding of and our ability to model the effect of the regulatory environment on human behavior; and
3. to develop through the above exercises a better understanding of the role of ecological processes in society's well-being, thus improving our methods for valuing ecosystem configurations and our ability to assess the benefits and costs of regulatory consequences.

To fulfill these three objectives, a pilot ecological-economic model is being developed to examine functions and processes of ecosystems, clarify relationships between human actions and ecosystems, and provide detailed information that will enable improved ecosystem valuation.

The model consists of interrelated ecological and economic models that employ a landscape perspective because this perspective captures the spatial and temporal distributions of the services and functions of the natural system and human-related phenomena, such as surrounding land-use patterns and population distributions. Configuration and reconfiguration of the landscape occurs as a result of ecological and economic factors, and these factors are closely intertwined.

The ecological part of the model is based on the Patuxent Landscape Model (PLM), one of several landscape-level spatial simulation models currently under development by Costanza and others at the University of Maryland (Costanza et al., 1990; Fitz et al., 1993; Fitz et al., 1995). The PLM is capable of simulating the succession of complex ecological systems using a landscape perspective. Economic models are being developed to reflect human behavior and economic influences. The effects of human intervention result directly from the conversion of land from one use to another (e.g., wetlands conversion, residential development, power plant siting) or from changes in the practices that take place within specific land uses (e.g., adoption of agricultural best management practices, intensification of congestion and automobile emissions, change in urban water and sewer use, and storm run-off).

The Maryland Patuxent River Watershed, which includes portions of Anne Arundel, Calvert, Charles, Howard, Montgomery, Prince George's, and St. Mary's counties, will serve as the study area (Fig. 1). The ecological model is currently being calibrated for this area. In this phase of the project the economic models will characterize land use and agricultural decisions, and capture the effects on these decisions of institutional influences such as environmental, zoning, transportation, and agricultural policies. The integration of the two models provides a framework for regulatory analysis in the context of risk assessment, non-point source pollution control, wetlands mitigation/restoration, etc.

As part of this exercise, the ease (and cost-effectiveness) of generalizing the model to other regions will be assessed. It is expected that the modeling approaches will remain fairly constant over applications, although the particular behavioral models and ecological processes emphasized, as well as the actual parameters of the models, would differ. Because the models are designed to depict specific processes and ecosystems and to evaluate specific management strategies, they require a significant amount of site-specific data. However, the integrated model will illustrate the circumstances under which human actions set in motion processes that can have far-reaching effects that play out over a broad geographical range and over time. These illustrations will help policymakers define the relevant questions to be asked and effects to be anticipated in other regional settings.

An overview of the components of the integrated modeling system is depicted in Fig. 2. The major components of the system are the spatially explicit, dynamic models for the ecological system and the economic decisions, and the mechanism for exchange of information between the two. The results of the interrelated processes produce a new landscape and a new array of values associated with the landscape. Some of these become signals to private individuals and public agents in future decisions.

6. The ecosystem model

The starting point of the modeling effort, the Patuxent Landscape Model (PLM), is an outgrowth of the coastal ecological landscape spatial simulation (CELSS) model developed by Costanza et al. (1990). This modeling approach has been applied in two previous studies. The model was first implemented in the Atchafalaya Delta Area of coastal Louisiana, where it was developed to model spatial ecosystem processes, succession, and land loss problems, and was used to evaluate the impacts of management strategies and specific projects designed to alleviate coastal erosion problems (Costanza et al., 1990). A more sophisticated version of the model is currently being implemented in the Water Conservation Areas

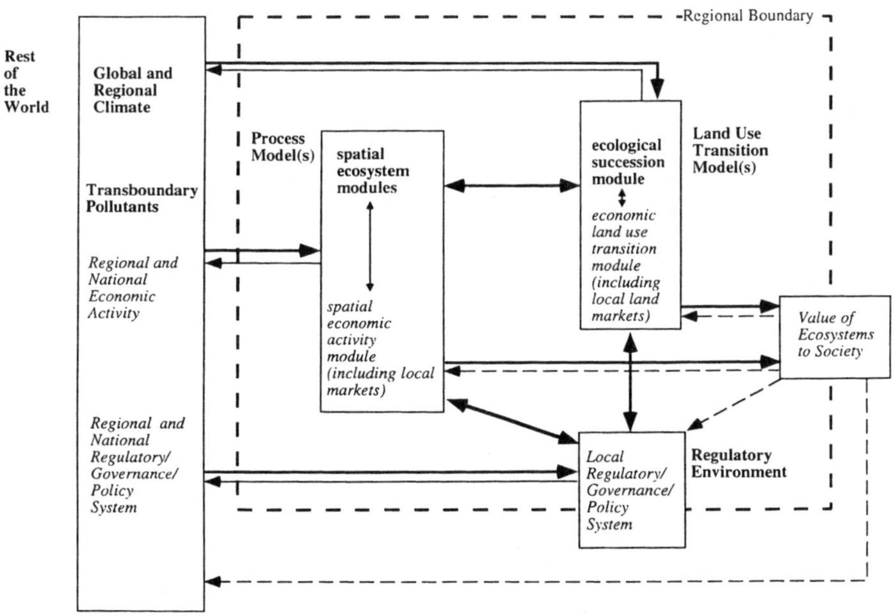

Fig. 2. Integrated ecological economic modeling and valuation framework. Modules in **bold** are already included in the Patuxent Landscape Model (PLM). Modules in *italics* are to be added as part of our ongoing project.

and Everglades National Park in Florida to examine the repercussions of management strategies on elements of the ecosystem such as water levels, nutrients, and plant successional patterns (Fitz et al., 1993).

The PLM uses an integrated, spatial simulation approach to model ecological processes. Central to the PLM is a General Ecosystem Model (GEM) (Fitz et al., 1995; see Fig. 3) that is replicated in each of the grid cells that compose the study area (Fig. 4). The GEM unit models are linked together and to spatial Geographical Information System (GIS) data using the Spatial Modeling Workstation (SMW) (Maxwell and Costanza, 1994; see Fig. 5). In addition to linking flows between cells, the spatial configuration superimposes the river network on the grid of cells allowing better simulation of river flow processes across the landscape.

The GEM unit model simulates fundamental ecological processes with hydrology as its core. The hydrologic sector of the model simulates the availability of water and its vertical and horizontal movements based on topography, geology, land use, and other characteristics within each cell. Characteristics of each cell's habitat type translate to parameters of physical, chemical, and biological equations which are used to simulate such processes as primary production, nutrient fluxes, organic/inorganic sediment suspension and deposition, basic "consumer" dynamics, and decomposition. The dynamics of various ecological processes are expressed as the interaction between state variables (stocks) and flows of material, energy, and information. After the vertical or within-cell dynamics have been simulated, the results of the unit model are processed by the spatial modeling program. In linking the stocks and flows, the model predicts the exchange of matter between cells (horizontal fluxes) and simulates the resulting temporal changes in water availability, water quality, and habitat/ecosystem type. The landscape model employs an algorithm for determining the habitat type of each cell at all times during the simulation. The habitat switching algorithm redefines the habitat/ecosystem type of cells as conditions change, and selects parameter sets as necessary.

Data is collected to calibrate or initialize the model for a given set of ecosystems and a particular landscape. Effort is then devoted to simulating past behavior of the landscape. This allows

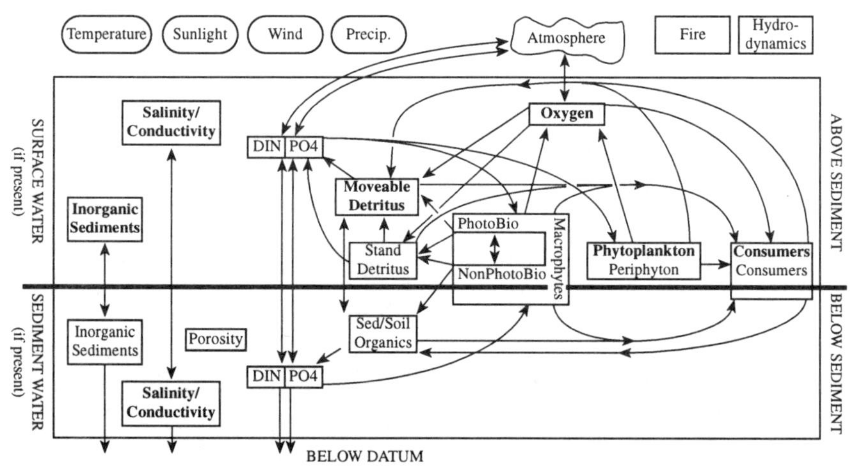

Fig. 3. Summary diagram of the Patuxent Landscape Model's (PLM) unit model (GEM). The state variables and mass fluxes of the unit model, excluding hydrology. The model is run within one cell and parameterized for the cell's habitat. Hydrology is important in the vertical fluxes shown and the unshown horizontal fluxes of materials from cell to cell. State variables are enclosed within rectangles, and those that can flux vertically and horizontally with water movement are shown in outline type.

for calibration of the functions of the model against actual historical data. For the Patuxent project, historical land use and geographic data (GIS) have been acquired from the Maryland Department of Natural Resources (DNR), Maryland Department of Environment (MDE), and the U.S. Geological Survey (USGS). Maps indicating land use estimates for 1973, 1981, and 1985 (with 1991 now available), as well as mappings of the watershed, streams, soils, slopes, contaminants, highways, marinas, railroads, water quality, and wetlands have been acquired. The PLM contains about 6000 spatial cells each of about 100 acres and each containing a GEM with 21 state variables (see Table 1 for a list of these). The model will be calibrated with land use, stream flow, water quality, and other data from 1973 and 1985. The intent is to run scenarios to the year 2020.

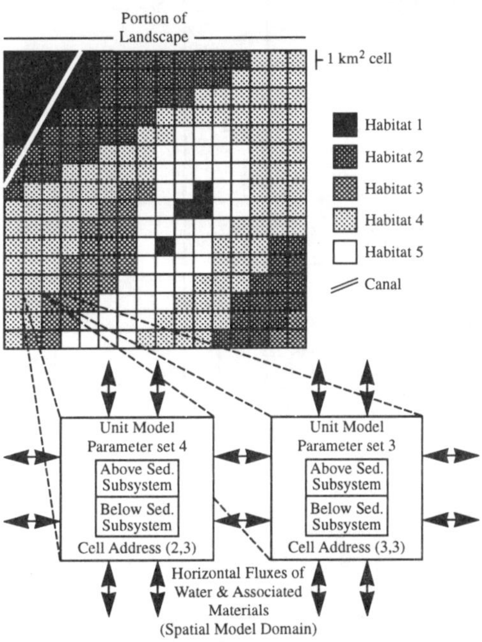

Fig. 4. The cellular structure of the PLM. Each cell has a (variable) habitat type which is used to parameterize the unit model for that cell. The unit model simulates ecosystem dynamics for that cell in the above-sediment and below-sediment subsystems. Nutrients and suspended materials in the surface water and saturated sediment water are fluxed between cells in the domain of the spatial model.

The spatial resolution of the PLM allows evaluation of impacts from changes in land use type or practices for particular groups of cells in the watershed. Our aim is to be able to estimate selected indicator variables which will describe processes of interest in the Patuxent River watershed. For example, fertilizer applications and mechanical soil manipulation (e.g., tillage) influence the total loadings of nutrients and sediment which are carried by surface and sub-surface waterflow off a given land parcel. The flow path of that water over various land types will determine its nutrient and sediment load when it reaches a wetland or stream. The combined effect of multiple land parcels then influences vegetative growth processes in wetlands within the Patuxent estuary. As plant biomass changes, nutrient uptake, evapotranspiration, and sediment retention rates change. The simulation of these multiple interrelated processes allows the PLM to estimate general effects of such practices as vegetative buffers and retention ponds on urban and agriculture lands on total loadings to the estuary and ecosystem types of undeveloped land.

Future additions to the ecological model include adding estimates of fish and wildlife populations based on the pattern and productivity of the landscape elements. At this stage these estimates will not feed back to affect the ecosystem dynamics. Rather, they will be more on the order of "habitat suitability", indicating the type and quantity of various species that could be expected in a given landscape rather than modeling the populations endogenously.

7. Economic models of human decisions

The two notable aspects of the ecological modeling approach—its dynamic nature and its ability to spatially disaggregate—are also important dimensions for the economic sectors that interact with the ecosystem. The human activities of importance form an interrelated, dynamic, spatially distributed system in their own right. Economic models that reflect these dimensions are being conceptualized and estimated. Ensuring the necessary feedback between the two systems will require an integration characterized by exchange of information between the models at an appropriate time and space scale.

Human decisions that determine land-use designation, as well as those that affect flows/extractions from the system conditioned on the land-use designation, will affect the configuration of the landscape. The existing PLM, like other ecological models, ignores the effect of the landscape on humans. In addition, human effects on the landscape, whether through changes in the land use of the cells or through the emissions assumed for any land use, are imposed exogenously rather than modeled as part of the process. The economic considerations are important because hu-

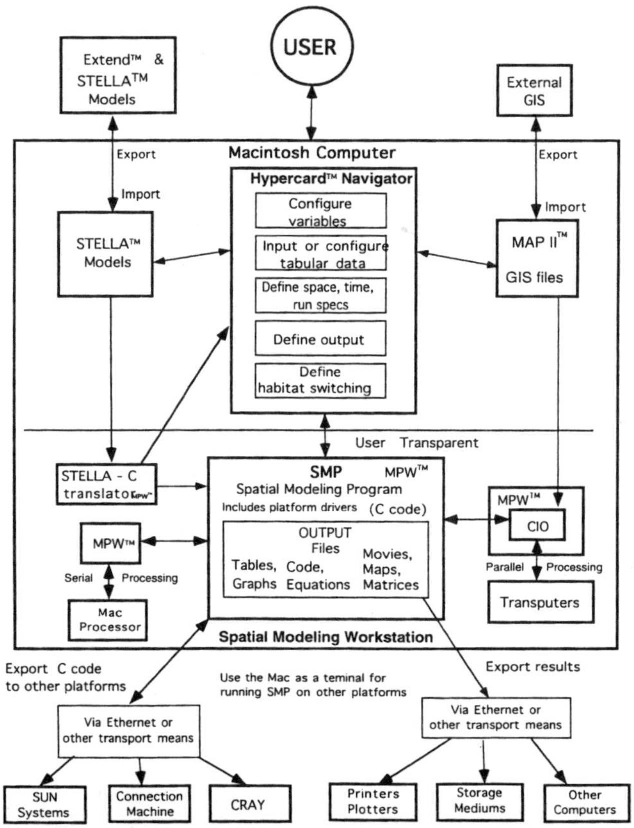

Fig. 5. Diagram of the Patuxent Landscape Model (PLM) implementation architecture. The PLM is programmed and simulated in the Spatial Modeling Workstation (SMW). The SMW consists of: STELLA unit models for execution of ecosystem dynamics; MAP II GIS program for storage and manipulation of GIS data files; the Spatial Modeling Program (SMP) that executes the interchange of flows among unit models on the computers, and imports and exports spatial (GIS) data; the Hypercard Navigator (under development) that is used to guide the user through configuration of models, data sets, execution platforms, and output files; and an Apple Macintosh computer with transputers installed for simulation or as a user terminal interface to other computer platforms.

Table 1
Sectors and state variables in the General Ecosystem Model (GEM) (sectors are in italics, state variables are in uppercase)

Sector and STATE VARIABLE names	Number
Hydrology sector	
SURFACE WATER	1
UNSATURATED WATER	2
SATURATED WATER	3
Hydrodynamic sector	
Horizontal flow, vertical flow, and waves	
Inorganic sediments sector	
SUSPENDED INORGANIC SEDIMENTS	4
DEPOSITED INORGANIC SEDIMENTS	5
PORE SPACE	6
Salt (NaCl) sector (conductivity)	
SALT in SURFACE WATER	7
SALT in SEDIMENT WATER	8
SALT CRYSTALS	9
Dissolved phosphorus sector	
PO_4 in SURFACE WATER	10
PO_4 in SEDIMENT WATER	11
Dissolved nitrogen sector (NO_2, NO_3, NH_4^+)	
DIN in SURFACE WATER	12
DIN in SEDIMENT WATER	13
Dissolved Oxygen Sector	
DISSOLVED OXYGEN in SURFACE WATER	14
Non-macrophyte sector	
ALGAE (phytoplankton and/or periphyton)	15
Macrophyte sector	
MACROPHYTE PHOTOSYNTHETIC BIOMASS	16
MACROPHYTE NON-PHOTOSYNTHETIC BIOMASS	17
Above-sediment organic matter and detritus sector	
STANDING DETRITUS	18
ABOVE-SOIL ORGANIC MATTER AND DETRITUS (Transportable by surface water flow. Includes weight of microscopic decomposers. Includes suspended POC and DOC when systems are wet.)	19
Organic sediments / Soil sector	
SEDIMENT/SOIL ORGANIC MATTER	20
Consumer sector	
CONSUMERS (All fauna except microscopic decomposers)	21
Fire sector	
Fire ignition, fire propagation	
Global inputs sector	
Atmospheric inputs, geographical information, etc.	

man decisions are influenced by the characteristics of the ecosystem landscape, and the landscape changes as a result of the decisions. For ecological reasons, the effects of a single action taken at a given point in time will continue to play out over time and will affect more than just

Table 2
The habitat or land-use designation types relative land use for the Patuxent watershed for 1973 and 1985

Land use description	% 1973 area	% 1985 area	% change
Urban	10.7	14.9	4.2
Low density residential	3.7	6.2	2.5
Medium density residential	2.9	3.5	0.6
High density residential	0.7	1.0	0.3
Commercial	1.1	1.5	0.4
Industrial	0.2	0.2	0.0
Institutional	1.2	1.2	0.0
Extractive	0.3	0.3	0.0
Open urban land	0.6	1.0	0.4
Agricultural	34.2	31.6	−2.6
Cropland	29.6	26.6	−3.0
Pasture	4.6	4.5	−0.1
Orchard	0.2	0.1	−0.1
Feeding operations	0.0	0.1	0.1
Row and garden crops	0.2	0.3	0.1
Forest	48.1	46.0	−2.1
Deciduous forest	31.4	27.6	−3.8
Evergreen forest	1.4	1.6	0.2
Mixed forest	14.3	14.8	0.5
Brush	1.0	2.0	1.0
Water	5.3	5.5	0.2
Fresh open water (plankton)			
Fresh open water (benthic)			
Brackish open water (Plankton)			
Brackish open water (benthic)			
Wetlands	1.2	1.4	0.2
Salt marsh			
Fresh marsh			
Swamp			
Barren land	0.1	1.2	1.1
Beaches	0.0	0.4	0.4
Bare exposed rock	0.0	0.0	0.0
Bare ground	0.1	0.8	0.7
Total	100.0	100.0	0.0

the cell where the action took place. Additionally, that action will have been influenced by the characteristics of the given cell, the characteristics of surrounding cells, and the expectations of changes in those cells.

Perhaps the most important type of economic behavior that requires modeling is land conversion. Since the ecological processes that are modeled in the generic ecosystem model are conditioned on land use, scenarios are driven by the land-use designation of the cells in the landscape at each period of time. The land use conversion model will endogenize those human decisions.

Conditioned on a cell's land-use designation, the cell is programmed to undergo certain processes. But human decisions will affect these processes and may not be predetermined. In fact, they may be affected by feedback from the ecological system, and most surely will be affected by changing policy environments. In the first phase of this project one such behavioral model conditioned on land use will be developed—a model of agricultural practices. Additional behavioral models conditioned on land use will be added in the future. An obvious candidate is a transportation model.

8. Land-use conversion

Economists have tried to explain land values and land-use decisions for some time, but generally in a static context and rarely incorporating

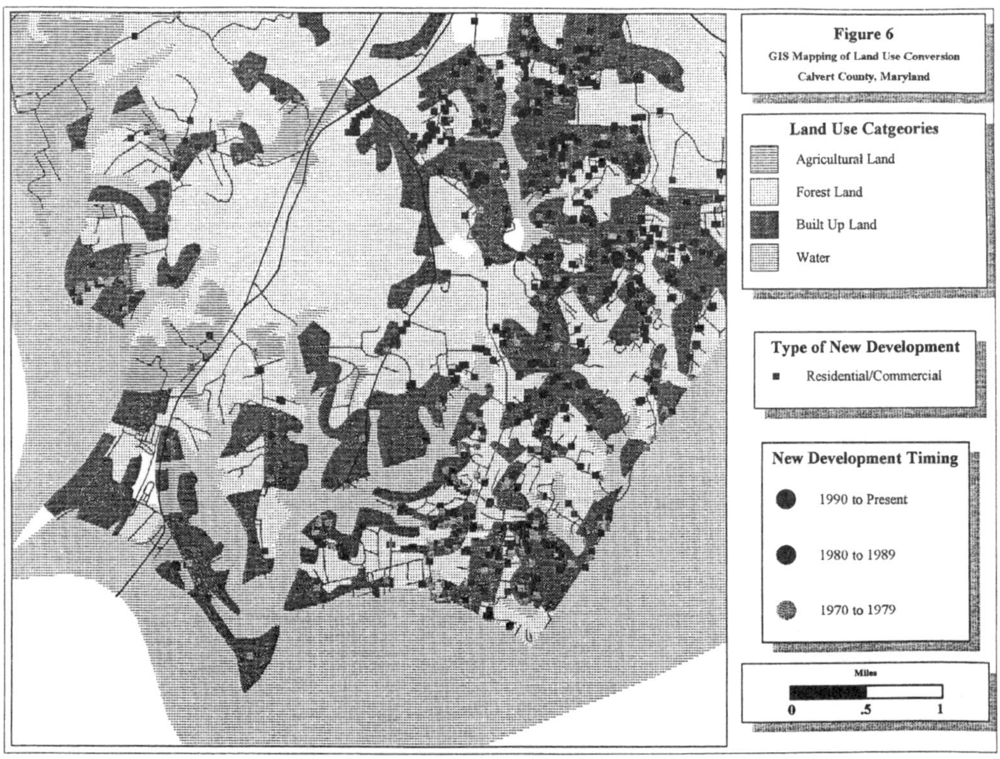

Fig. 6. GIS Mapping of Land Use Conversion, Calvert County, MD.

features of the local ecosystems. Most frequent in this literature are hedonic models of residential transactions prices as functions of local amenities and housing characteristics (e.g., Diamond, 1980; Beaton, 1991). Additionally, some models seek to explain conversion of agricultural land to residential and urban use as functions of profitability of alternative land uses(e.g., Alig and Healy, 1987; Fischel, 1982; McMillan, 1989). Data limitations have restricted these applications to cross-sectional analysis at one point in time.

Our proposed approach relies on spatially disaggregated data from digitized maps and allows dynamic analysis of micro-level changes that can be observed over time. Because the data reside in a geographical information system, it is easy to relate activities on individual parcels of land with local ecological and economic features of the landscape that might be driving these activities.

The land-use conversion decision is modeled probabilistically. This involves estimating the transition probabilities that a given parcel of land, set in a landscape configuration of given ecological/economic description, will be converted from its current state to any other state (including no change). As surrounding landscape conversions take place, the probability of conversion of a given parcel changes.

Here we are interested in considering the discrete and finite set of alternative land uses. The set of uses in the Patuxent watershed for which we have data are listed in Table 2. Depending on the state of the parcel at the time of the decision, some uses may be precluded, at least for realization within a limited time period (of a year or shorter). For example, if the land is already in residential development, it is probably infeasible to return it to forest in one year. Federal laws, such as wetland and coastal zone management, will also affect the likelihood that land parcels switch uses since legislation often prohibits certain land uses and/or makes these uses significantly more costly. Again, the GIS system is useful because many of the land-use restrictions and potential scenarios have been digitized in GIS maps and can be merged with the ecological and economic digitized geographic information (Fig. 6).

This type of probabilistic model is appealing because it allows for variations over individual decision makers facing the same factors, and variation of the factors affecting the decision with spatial location and proximity to centers of economic activity, transportation networks, and ecological features. Thus it predicts the likelihood of each type of transition rather than predicting deterministically one type of conversion for all land of a certain ecological/economic description. These transition probabilities can be interpreted as the percentage of land of a given type that is converted to each alternative land use (including the percentage that remains unconverted) and are functionally equivalent to the transition probabilities in a Markov analysis.

Because the differential in economic returns among commercial, residential, agricultural, and open land uses differs depending on the state of the regional economy, the explanatory model will reflect these exogenous forces and the scenarios will incorporate predictions of these cycles. An existing county-based regional model for Maryland will provide predictions of the "macro-economic" exogenous factors to be used in the simulations.

9. Agricultural practices

The first step of the PLM modeling dynamics is to designate a cell as a given land use. Together with prior information, the economic land-use conversion model and the ecosystem succession model, these designations are made for each "round" of the model's calibrations. However, the land-use designations are broad categories with agricultural uses specified as either cropland, pasture or orchards. It is important to know more about the agricultural land use in order to understand the impact this land use has on the ecosystem. The agricultural practices model attempts this, and in so doing serves as an example of micro-level economic behavior models which capture human decisions that influence the environment, conditioned on a given land use. This economic model will accept information from the "outside world" (including existing policy) and the ecological world (such as topographical information) and will attempt to predict the cropping/livestock patterns, as well as the farming practices that serve as parameters in the model for the subsequent cropping season.

As with the land-use conversion model, the economic model of farming behavior will be probabilistic. That is, it is the probabilities of various alternative decisions that will be generated by the model. These probabilities will need to be interpreted as predicted proportions of similar farming land that is used for given crops/livestock and, conditioned on that choice, is subject to different BMP's.

Models of agricultural crop/livestock choice are numerous, but few are conceived at as "micro" a scale as this project should be able to achieve. Profitability from different farming choices will drive the decisions, but profitability is conditioned, not only on relative output and input prices, but factors of the landscape that affect the production function for different crops (e.g., hydrology, slopes, soil types, etc.). In the Patuxent watershed there is an interesting added consideration. In Maryland, farmers often sell hunting rights on their lands. These rents can be a major source of income for farmers and affect not only how much land stays in agriculture, but also crop and livestock decisions. Depending both on farmers' decisions and the ecological characteristics of the land, the amount and composition of wildlife will differ, endowing the land with differing hunting value.

The choice by farmers among farming practices that could alter their pollution emissions, conditioned on crop/livestock choice, is a difficult one to model, but one that has considerable importance for policy formation. Through provision of technical and financial (cost-sharing) assistance, federal and state agencies have attempted to address these problems by subsidizing the use of "best management practices" (BMPs), defined as farming practices that reduce soil and nutrient losses at reasonable cost (U.S. Department of Agriculture). Adoption of new agricultural technologies depends on a variety of economic and behavioral factors, including scale and scope of operations, topography and other natural factors, human capital or skill considerations, relative prices, attitudes toward risk, and perceptions about the extent of erosion or leaching problems and about returns to adoption (for a survey, see Feder et al., 1985). These factors will be pursued in the agricultural decision model.

In addition to models of crop choice, numerous behavioral studies of best management practice (BMP) adoption in the United States have examined the influence of factors ranging from topography and erodibility to human capital factors (e.g., age and experience) to farm operation characteristics (e.g., cropping patterns, scale and scope and tenure status). A few generalizations can be drawn from the findings of these empirical studies. Adoption is more likely and expenditures on BMPs are greater on more erodible land and on larger operations. Older, more educated and more experienced farmers are more likely to adopt BMPs. Farmers with more off-farm income and greater risk aversion are less likely to adopt. Farmers are more likely to adopt if they perceive erosion to be more of a problem, attach greater importance to conservation and environmental quality or have conservation plans. Cash grain farmers are less likely to adopt BMPs. (Ervin and Ervin, 1982; Lee and Stewart, 1983; Saliba and Bromley, 1986; Shortle and Miranowski, 1986; Norris and Batie, 1987; Lynne et al., 1988; Nielsen et al., 1989). Although these generalizations are revealing, there is no study that comprehensively examines the adoption of numerous, but specific BMPs in a spatially explicit context linked to ecosystem characteristics.

10. Integration of components and resolution of modeling inconsistencies

At present, the ecological model runs fourteen sectors, tracks twenty-one state variables, estimates horizontal and vertical fluxes of water, nutrients, sediments, and biomass, and simulates ecological succession over the landscape. For the purposes of the economic model, two types of interactions between humans and natural resources will be of significance: (1) decisions that determine land-use designation of a cell, and (2) decisions that affect flows/extractions from the system conditioned on the land use type of the cell.

An interrelated model would capture how the distribution of human activities (farming, electric power generation, commercial and residential development, recreation, wastewater treatment, highway construction, fishing) affect the ecosystem. Such a model would also capture the effect of the ecosystem landscape on the quality and value of goods and services (e.g., recreation, wildlife enjoyment, water quantity and quality, housing, environmental aesthetics, etc.) and therefore, on human decisions. The model must also be able to reflect how human activity and its impact on the ecosystem may differ under different regulatory regimes and under alternative regulations.

Our approach will *not* be to meld both ecological and economic models into one "supermodel" because this approach would force each model to adopt the lowest level of resolution in the other. Instead, the two types of models will exist in parallel and will proceed at their own levels of specificity, but will exchange information on ecological and economic elements generated by the other. Because appropriate choices on time step, geographical scale, and level of aggregation differ in the ecological and economic models, there are several potential inconsistencies that may arise in this information exchange.

By design, the PLM model establishes physical boundaries. Activity is modeled in the area bounded by the top of the trees, the drainage basin boundaries, and the bottom of the groundwater. The choice of physical boundaries encompasses the dynamics of the ecosystem. The physical definition of the boundary of the Patuxent River Basin region will coincide with relevant economic boundaries, if the region is a price-taker in markets that extend beyond the boundaries of the region. That is, for markets that extend beyond the Patuxent River Basin Region, the region must not significantly influence the actions of others (i.e., it does not produce or demand a significant amount relative to other trading partners). For the Patuxent watershed, this is likely to be approximately correct and thus no serious inconsistencies with respect to "boundaries" are likely to arise. However, in other applications of this modeling approach, the definition of boundaries could well be conflicting.

The model is structured to simulate a desired level of ecological complexity and resolution. Modeling with greater levels of disaggregation is possible, but this would be costly and require a significant amount of additional research, data, and programming. The behavioral interactions will largely involve the water, nutrients, macrophyte, and consumer sectors. Although these state variables are linked with habitat types, much finer distinctions of the landscape will be necessary for the economic model, as human behavior is likely to depend on more detailed information. In our approach, the details will be introduced exogenously by distinguishing different subgroups of macrophytes and consumers and associating these different subgroups with different land uses. The model will include markers and detailed rules for species loss, treating habitat evaluation as a side calculation. Clearly, there are qualitative distinctions between different composition and phyla. A function will be developed that will indicate the likelihood of game species and other forms of wildlife in particular habitat types. This function allows for the provision of additional complementary information without increasing the level of disaggregation of the unit model.

The ecosystem model operates on a time scale of a day or less. Yet, given data availability, the economic models are most easily estimated on an annual basis. The intra-year timing of the agricultural decisions will be easy to predict since these are governed by seasons. Land conversions pose a different problem, however. They can be predicted only on an annual basis, but their timing is important because construction costs can have different immediate ecosystem impacts depending on the season. Independent data on building starts and construction durations will be used to forecast seasonal impacts of construction.

The geographical scale of resolution between the two models will also differ. Land parcels modeled in the land-use conversion model will not correspond to cell boundaries in the ecosystem model. The latter divides the study area into

5,896 cells with each cell representing approximately 0.364 km^2 or 90 acres. The cell size in the ecological model was selected to balance the desired complexity and resolution of the ecological landscape. The fact that the economic model will produce probabilistic estimates may actually help to resolve the differences in geographical scale, as the probabilities can be interpreted as percentages of land of a given description that is put to a given land use. However, the transition probabilities may still "fracture" cells. The PLM model has the ability to assign to cells split land-use designations, allowing heterogeneity within a cell. Devising weights to monitor what is happening in cells, thresholds can be set so that cells could go from homogeneous to heterogeneous units and vice versa. This additional detail will also allow the model to make inferences on a wider variety of land use restrictions (i.e., agricultural policies,. zoning policies, and environmental protection policies).

In resolving this array of modeling inconsistencies, sensitivity tests will be used to assess the importance of level of detail in both models. Sensitivity games are used now to accommodate uncertainty over elements such as rainfall patterns over time. By generating several patterns for the future and running these different patterns as distinct scenarios, the significance of the uncertainty can be revealed.

Acknowledgements

This project was supported by a cooperative agreement with the USEPA, No. 07-5-25196-3734 titled: Ecological Economic Modelling and Valuation of Ecosystems. Cooperative Agreement with EPA. Michael Brody and MaryJo Kealy were the project officers and provided valuable input on the project direction and scope. The project is a direct spin-off of the "Ecosystem Valuation Forum" sponsored by EPA, which is described in the introductory paper in this special issue. We thank Ed DeBellevue, Carl Fitz, Marla Markowski, and Enrique Reyes for their contributions to the research effort, and Marjan van den Belt for helpful comments on earlier drafts.

References

Alig, R.J. and Healy, R.G., 1987. Urban and built-up land area changes in the United States: an empirical investigation of determinants. Land Econ., 63(3): 216–226.

Beaton, W.P., 1991. The impact of regional land-use controls on property values: the case of the New Jersey Pinelands. Land Econ., 67(2): 172–194.

Bergstrom, J.C., Stoll, J.R., Titre, J.P. and Wright, V.L., 1990. Economic value of wetlands-based recreation. Ecol. Econ., 2: 129–147.

Bernstein, B.B., 1981. Ecology and economics: complex systems in changing environments. Annu. Rev. Ecol. System., 12: 309–330.

Bouzaher, A. et al., 1993. CEEPES: an evolving system for agroenvironmental policy. Paper presented at: The Integrating Economic and Ecological Indicators Symposium, Resource Policy Consortium, World Bank, Washington, DC, May.

Bingham, G., 1994. Issues in ecosystem valuation: improving information for decision making. Ecol. Econ., in press.

Costanza, R., Farber, S.C. and Maxwell, J., 1989. The valuation and management of wetland ecosystems. Ecol. Econ., 1: 335–361.

Costanza, R., Sklar, F.H. and White, M.L., 1990. Modeling coastal landscape dynamics. BioScience, 40(2): 91–107.

Costanza, R., Wainger, L., Folke, C. and Maler, K.-G., 1993. Modeling complex ecological economic systems: toward an evolutionary, dynamic understanding of people and nature. BioScience, 43: 545–555.

DeBellevue, E.B., Maxwell, T., Costanza, R. and Jacobsen, M., 1993. Development of a landscape model for the Patuxent River watershed. Discussion paper #10, Maryland International Institute for Ecol. Econ., Solomons, MD.

Diamond, D.B., 1980. The relationship between amenities and urban land prices. Land Econ., 56(1): 21–32.

Dixon, J.A. and Scura, L.F., 1993. Ecology and economics as joint products: marine parks in the Caribbean. Paper presented at: The Integrating Economic and Ecological Indicators Symposium, Resource Policy Consortium, World Bank, Washington, DC, May.

EPA Science Advisory Board (SAB), 1990. Reducing risk: setting priorities and strategies for environmental protection. SAB-EC-90-021. Washington, DC: EPA.

Ervin C.A. and Ervin, D.E., 1982. Factors affecting the use of soil conservation practices: hypotheses, evidence and implications. Land Econ., 58: 277–292.

Farber, S. 1988. The value of coastal wetlands for recreation: an application of travel cost and contingent valuation methodologies. J. Environ. Manage., 26: 299–312.

Feder, G., Just, R.E. and Zilberman. D., 1985. Adoption of agricultural innovations in developing countries: a survey. Econ. Dev. Cult. Change, 33: 255–298.

Fischel, W.A., 1982. The urbanization of agricultural land: a review of the National Agricultural Land Study. Land Econ., 58(2): 236–259.

Fitz, H.C., Costanza, R. and Reyes, E., 1993. The Everglades Landscape Model (ELM): Summary Report of Task 2, Model Development. Report to the South Florida Water Management District, Everglades Research Division.

Fitz, H.C., DeBellevue, E.B., Costanza, R., Boumans, R., Maxwell, T., Wainger, L. and Sklar, F., 1995 Development of a general ecosystem model (GEM) for a range of scales and ecosystems. Ecol. Model. (in press).

Gupta, T.R. and Foster, J.H., 1975. Economic criteria for freshwater wetland policy in Massachusetts. Am. J. Agric. Econ., 57: 40–45.

Hoogervorst, N.J.P., 1993. Integration of economic and ecological modeling for agriculture in the Netherlands. Paper presented at: The Integrating Economic and Ecological Indicators Symposium, Resource Policy Consortium, World Bank, Washington, DC, May.

Lee, L.K. and Stewart, W.H., 1983. Landownership and the adoption of minimum tillage. Am. J. Agric. Econ., 65: 256–264.

Lynne, G.D., Shonkwiler, J.S. and Rola, L.R., 1988. Attitudes and farmer conservation behavior. Am. J. Agric. Econ., 70: 12–19.

Lynne, G.D., Conroy, P. and Prochaska, F.J., 1981. Economic valuation of marsh areas for marine production processes. J. Environ. Econ. Manage., 8: 175–186.

Maxwell, T. and Costanza, R., 1994. Distributed modular spatial ecosystem modeling. Int. J. Computer Simul., (in press).

McMillan, D.P., 1989. An empirical model of urban fringe land use. Land Econ., 65(2): 138–145.

Nielsen, E.G., Miranowski, J.A. and Morehart, M.J., 1989. Investments in soil conservation and land improvements: factors explaining farmers' decisions. Agric. Econ. Rep., 601, Economic Research Service, U.S. Department of Agriculture, Washington, DC.

Norgaard, R.B., 1989. The case for methodological pluralism. Ecol. Econ., 1: 37–57.

Norris, P.E. and Batie, S.S., 1987. Virginia farmers' soil conservation decisions: an application of Tobit analysis. South. J. Agric. Econ., 19: 79–90.

Russell, C.S., 1993. Old lessons and new contexts in economic-ecological modeling. Resource Policy Consortium, World Bank, Washington, DC.

Saliba, C., Bromley, B. and Bromley, D.W., 1986. Soil management decisions—how they should be compared and what factors influence them. North Central J. Agric. Econ., 8: 305–317.

Scodari, P.F., 1992. Wetlands management benefits. Report prepared under Cooperative Agreement CR 817553, for Economic Analysis and Research Branch, OPPE, EPA, Washington, DC.

Shabman, L.A., 1987. Mitigating damages from coastal wetlands development: policy, economics and financing. Mar. Resource Econ., 4: 248–277.

Shabman, L.A., Batie, S.S. and Mabbs-Zeno, C.C., 1979. The economics of wetlands preservation in Virginia. J. Northeastern Agric. Econ. Council, 8: 101–115.

Shortle, J.S. and Miranowski, J.A., 1986. Effects of risk perception and other characteristics of farmers and farm operations on the adoption of conservation tillage. Appl. Agric. Res., 1: 85–90.

Tschirhart, J. and Crocker, T.D., 1987. Economic valuation of ecosystems. Trans. Am. Fish. Soc., 116: 469–478.

van den Bergh, J., 1993. Dynamic analysis of economic development and natural environment on the Greek Sporades Islands. Paper presented at: The Integrating Economic and Ecological Indicators Symposium, Resource Policy Consortium, World Bank, Washington, DC, May.

[10]

The Value of Biodiversity: Markets, Society, and Ecosystems

John M. Gowdy

ABSTRACT. *The value of biodiversity is discussed at different levels including market value, nonmarket values to humans, and the value of biodiversity to ecosystems. The main conclusion is that, although market exchange values of environmental services may be used to justify biodiversity protection measures, it must be stressed that exchange value constitutes a small portion of total biodiversity value. The total value of existing biodiversity is largely unknown but indications are that it is essential to human existence. The various levels of biodiversity value point to the need for a hierarchical and pluralistic methodology to determine appropriate policies for its preservation.* (JEL Q21)

I. INTRODUCTION

Among ecologists, there is a general consensus that biodiversity[1] is of critical importance to the health of ecosystems and even for the long-term survival of the human species. There is also a consensus that biodiversity is being lost at a rate which is a cause for concern.[2] Economists, on the other hand, generally view biodiversity as just another good which is to be placed in the basket of market choices just as any other. The discussion of the specifics of biodiversity policy has been unsatisfactory partly because of the different meanings of the word "value" used by economists and ecologists. Many ecologists fail to understand the logic of market allocation and why biological resources are used in seemingly irrational ways. Many economists fail to appreciate the narrowness of the concept of economic value as indicated by relative prices determined by market exchange. Following Anderson (1966) and Brown (1984), the discussion below considers economic measures of value to be species of the genus *assigned value* which belongs to the family *value*.

Many biologists and paleontologists believe that we are at a critical point in the history of the human species, and perhaps even in the 600-million-year history of complex multicelled life on earth. If the paleontologists and biologists who study the phenomenon of mass extinction are correct, the current human-induced mass extinction may be of the same order of magnitude as the five other major extinction episodes which destroyed between 20 and 96 percent of existing species on the planet (Ehrlich and Ehrlich 1992; Ward 1994; Wilson 1992). We are living in a truly remarkable period in Earth's $4\frac{1}{2}$-billion-year history. The valuation decisions our species has made in the recent past and will make during the next few decades will determine the fate of life on Earth for the next tens of millions of years. It is critical to clarify the language and concepts we use to estimate the value of biodiversity, and thus policies leading to its destruction or preservation.

The importance attached to the issue of biodiversity preservation necessarily involves ethical judgments about duty to future generations and responsibility toward the nonhuman natural world. Although individuals hold a variety of conflicting beliefs about human responsibility to the natural world, this does not mean that policy choices

Department of Economics, Rensselaer Polytechnic Institute.

The author would like to thank Michael Kosz, Carl McDaniel, David Orr, George Robinson, Sabine O'Hara, and two anonymous reviewers for comments on an earlier draft.

[1] The term biodiversity encompasses all of the species that currently exist on Earth, the variations that exist within each species, and all of the interactions that exist among all of these organisms and their biotic and abiotic environments as well as the integrity of these interactions.

[2] In one of the most thorough surveys to date, Pimm et al. (1995) estimate that current extinction rates are 100 to 1,000 times higher than their pre-human levels. This rate is expected to increase tenfold in the next century. Economist Julian Simon's (in Meyer and Simon 1994) claim that there has been no increase in extinction rates is "not scientifically credible" (Pimm et al. 1995, 348).

about biodiversity projection can or should be divorced from ethical considerations. Advocating economic efficiency as the prime policy criteria is itself an ethical judgment. Arguments for preserving biodiversity necessarily move through a hierarchy of economic and ethical considerations, from immediate economic self-interest, to possible self-interest (risk avoidance), to responsibility toward the future (Bingham et al. 1995).

As discussed below and elsewhere (Gowdy 1994, 1996; Gowdy and McDaniel 1995), there are fundamental conflicts between the rules by which markets allocate resources and the rules which govern ecosystems. The question is how, when, and in what form will these conflicts be resolved. Although our present socioeconomic system cannot continue to expand indefinitely by destroying biodiversity, it is quite possible that economic growth can continue for decades or perhaps even centuries. That is the essential point in a provocative article by Sagoff (1995).[3] If biodiversity loss and all the other forms of environmental degradation will not appreciably affect economic activity in the immediate or even medium-term future, why should we bother protecting it? The thrust of Sagoff's argument is that ecologically minded economists need to address this question directly, and the remainder of this paper is devoted to that task.

II. MARKETS AND BIODIVERSITY

Contemporary microeconomic theory is a theory of allocation. Market-based values are at bottom trade-off ratios (Freeman 1991). Microeconomic theory explains the rules by which markets allocate goods and services among consumers, and productive inputs among producers. For consumers, for any given amount and any given distribution of goods, and assuming that (1) participants in the market have all the relevant information about the objects exchanged in markets, (2) each participant buys or sells a very small portion of the total amount being traded, and (3) there are no barriers to entering and leaving the market for any particular good, unhindered market exchange will result in Pareto optimality; a situation in which no further trading can make one person better of without making someone else worse off. In a real or hypothetical market framework, the value of biodiversity is determined by adding up the market prices times the quantities traded of all the various attributes of biodiversity. Marshallian measures of consumer surplus may be adjusted according to the suggestion of Hicks, taking into account changes in real income as prices go up or down (Hanley and Spash 1993; Mitchell and Carson 1989, 24–30).

Neoclassical economists recognize market imperfections and know that prevailing market prices may not reflect the true value of a particular good. They have developed sophisticated techniques such as contingent valuation and hedonic pricing to capture economic values which are not fully reflected in market prices (Cummings, Brookshire, and Schultze 1986; Hanley and Spash 1993; Mitchell and Carson 1989). The theoretical validity of these widely used valuation methods depends directly on the validity of the indifference curve analysis of the neoclassical model of exchange (Freeman 1991). Economists recognize that externalities might exist, that information may be imperfect, and that markets may not be fully competitive. Many economists, however, fail to recognize the limitations of basing values entirely on the preferences of isolated individuals acting as consumers at a specific point in time, although others, such as Sen (1967) and Marglin (1963), have noted the difference between decision making in a private market context and a social (citizen) context.

[3] Sagoff overstates the case for the ability of markets to overcome resource scarcity. Among other things, he equates a declining resource use per unit of output with declining absolute use, ignores Georgescu-Roegen's subtle distinctions between stocks and funds, relies on Solow's outdated and no longer accepted analysis claiming that pure technological change is responsible for most economic growth, and unquestioningly accepts the neoclassical concept of weak sustainability. Sagoff's optimistic conclusions also rely on markets and technologies which, over time, erode the moral and social basis for further progress (Homer-Dixon 1995). See Daly (1995) for a brief reply to Sagoff's article.

The preferred neoclassical method for estimating the nonmarket values of environmental attributes, contingent valuation, actually captures a hodgepodge of market values and broader values and forces them into the indifference curve framework of market exchange. Moreover, the convention of using willingness-to-pay over willingness-to-accept measures has no coherent theoretical justification, and results in an undervaluation of environmental features (Bromley 1995). Difficulties with establishing total value based on market determined prices include (1) discounting the future (Arrow et al. 1995; Bishop 1992; Price 1993), (2) the incongruity problem, that is, reducing all the complex attributes of something to a single, one-dimensional metric (utility or money) (Georgescu-Roegen 1954; Vatn and Bromley 1994), and (3) the existence of pure uncertainty (Bishop 1978, 1979; Bromley 1989; Ciriacy-Wantrup 1968; Gowdy and McDaniel 1995).

Biologists such as Ehrenfield (1988), Ehrlich and Ehrlich (1992), and Wilson (1992) argue that biodiversity has values that are difficult or impossible to measure in market prices. Ecosystems are complex, indivisible entities which operate on timescales outside the range of individual human experience or perception. For example, the health of some forest ecosystems depends on periodic fires that may occur only once in several decades. Short-term ecosystem instability may be an essential part of long-term stability (Pimm 1991). Wildlife managers are beginning to incorporate natural levels of change into planning policies (Pickett, Parker, and Fiedler 1992). Ecosystems may also be seen as hierarchical layers of processes operating at different levels of complexity, each level operating on vastly different spatial and time scales (O'Neill et al. 1986).

For many biologists, the total value of biodiversity is essentially infinite; it is essential to the sustainability of life on Earth including human life. Ehrlich and Ehrlich (1992, 225) write:

> The ravaging of biodiversity, in our view, is the most serious single environmental peril facing civilization. Biodiversity is a resource for which there is absolutely no substitute; its loss is irreversible on any timescale of interest to society.

To most economists, even to most environmental economists, biodiversity is just another commodity, subject to trade-offs and substitution at the margin, just as any other market good. Solow (1993, 181) writes:

> [H]istory tells us an important fact, namely, that goods and services can be substituted for one another. If you don't eat one species of fish, you can eat another species of fish. Resources are, to use a favorite word of economists, fungible in a certain sense. They can take the place of each other. That is extremely important because it suggests that we do not owe to the future any particular thing. There is no specific object that the goal of sustainability, the obligation of sustainability, requires us to leave untouched . . . . Sustainability doesn't require that any *particular* species of fish or any *particular* tract of forest be preserved. (Italics in original.)

Solow, like most neoclassical economists, does not recognize that species are unique, or that if particular species or ecosystems are "traded" for monetary gain they cannot be replaced. In the standard economic view "utility" derived from enjoying a rainforest is the same as "utility" derived from the sale of lumber from clear-cutting and destroying that rainforest. The value of anything can be expressed in some common denominator, "utils" or whatever, and one util is as good as another. Casting the argument in terms of marginal utility or marginal value does not resolve the issue. Marginal value is the change in total value; if we cannot estimate total value we cannot estimate marginal value.[4] Furthermore, when we talk about ecosystems, the concept of the

[4]Some may object that we only need to rank utilities (and thus values if we make the assumption that utility is measured in the value of actual or contingent consumer goods) not measure them cardinally. This issue has not been resolved in spite of the universal adoption of ordinal ranking in almost every microeconomic text. Georgescu-Roegen (1968) points out that diminishing marginal utility has no meaning without some notion of cardinality. He argues that we need at least *weak cardinality*, that is, we need to be able to not only rank commodity bundles but also to rank the differences between them.

marginal value of biodiversity is problematic. Removing one species will affect all the others in the system. Environmental features are characterized by "functional transparency" (Vatn and Bromley 1994), that is, contribution of one feature of an ecosystem cannot be known until that feature is absent. As discussed below, the fact that precise marginal values cannot be placed on biodiversity does not mean that substitution and trade-offs are not relevant to public policy regarding biodiversity protection. If, however, policy choices are restricted to market trade-offs, higher-order aspects of biodiversity value will be missed.

A useful exercise to examine the conflicting views of the value of biodiversity is to consider the value of biodiversity at different levels in a hierarchical framework, beginning with value in market exchange, through total economically calculable value and total potential value to humans (including uncertainty and the need for precaution), to its value to the stability of the entire biogeophysical configuration of planet Earth. Such an exercise shows the danger of basing biodiversity protection policy on the narrow criterion of value in market exchange.

III. THE MARKET VALUE OF BIODIVERSITY

Market value is the value of a commodity relative to other market commodities available to individual consumers at a specific point in time (Gowdy 1996; Price 1993). The determination of value through the process of exchange is eloquently described by neoclassical microeconomic theory.[5] Numerous studies have estimated the value of biodiversity in the context of market exchange. Specific case studies show that biodiversity can have a substantial market value. The following examples are by no means exhaustive and are reported only to show the range of market values provided by biodiversity and the various ways economists have attempted to capture these market values.

The Commercial Value of Biological Resources as Raw Material

Peters, Gentry, and Mendelson (1989) estimated that the economic value of a hectare of forest in Mishana, Rio Nanay, in the Peruvian Amazon, based on the yield of various forest products (12 products) was $422 per year after deducting transport costs and harvesting cost. The value of timber on delivery to the sawmill was $1,000. Using a discount rate of 5 percent they estimated the net present value of the sustainable harvest at $6,330 per hectare.[6] A frequently cited example of the value of natural products is the rosy periwinkle (*Catharanthus roseus*) of Madagascar which produces two alkaloids that have been used to manufacture drugs that cure Hodgkin's disease and lymphocytic leukemia. The income from the sale of these drugs is estimated to be more than $180 million per year (Wilson 1992, 283). Many plants, for example, the alpine pennycress (*Thlaspi caerulescens*), have been used to remove toxic metals from contaminated soils. Tropical plants seem to be particularly adept at removing toxic metals, possibly because metal accumulation is a defense against harmful insects and microbes (Moffat 1995). It is even possible to

[5]Neoclassical economics does an excellent job of explaining exactly how consumers make market choices. As Bishop (1992) points out, the problem is that neoclassical economists have been lazy in exploring the implications of their analysis of market exchange. As illustrated by the downward sloping indifference curve, markets essentially force all goods to be substitutes. Every introductory text, of course, discusses complementarity, such as that between tennis balls and tennis rackets. Using appropriate techniques testing for separability, these goods can be lumped together as one good, which can be tossed back into the market world of substitutes.

[6]This assumes that 25 percent of the crop is left intact for regeneration ($316.5/.05 = $6,330). At a discount rate higher than 31.6 percent the economically rational decision would be to cut down the trees. Such a discount rate may seem very high but may not be unusual for poor people in the Amazon basin. There are probably additional costs involved in participating in 12 separate markets for forest products as opposed to participating in a single well-established market for lumber. See the extensive discussion of the value of natural products in Oldfield (1988).

burn the plants and then recover valuable metals such as copper and nickel.

A well-known example of the value of preserving biodiversity is the agreement between the Merck & Co. Inc. and Instituto Nacional de Bioversidad de Costa Rica (INBio). Merck has agreed to pay INBio $1 million over two years for the right to prospect for commercially valuable biological material in plants, insects, and soils (Gershon 1992). So far, results have been disappointing for Merck and the INBio contract may not be renewed when it runs out in 1996. Critics of the INBio agreement point out that many of the rainforest products in Costa Rica are also found in neighboring countries which will not share the proceeds of the agreement and that without further conservation policies the Merck agreement will do little to halt biodiversity loss (Martinez-Alier 1994).

In general, it must be said that the total market value of drugs from tropical forests is not overwhelmingly high. Mendelsohn and Balick (1995) estimated the potential value of all yet-to-be discovered rainforest drugs to be between $3 and $4 billion to a private pharmaceutical company and $147 billion to society as a whole, figures far below previous estimates because they (correctly) used net revenues instead of gross revenues to calculate market value. Their net revenue calculations took into account the costs of research and development, taxes, manufacturing, and marketing. Of course, market value is only one part of total value of biodiversity and the fact that the market value of drugs from plants is relatively small does not mean that this contribution of biodiversity is unimportant to medical research. A letter to the journal *Science* (Torrence 1995) representing 138 biomedical research scientists makes this point: "We believe that the progress of biomedical research and disease treatment depends on the maintenance of the greatest possible diversity in nature."

Ecotourism

There are very large direct and indirect economic values of nature through ecotourism. Part of the value of ecotourism arises from geological features such as the Grand Canyon, but much of it is due to biodiversity. Geist (1994) estimates that the direct economic benefit of Wyoming's big game animals, from tourism and hunting is about $1 billion or about $1,000 per large animal. He estimates that the total economic benefit resulting from the protection of wildlife in all North American national parks for tourism is more than $70 billion. In a study of the Greater Yellowstone area, Power (1991) found that recreation, not mining, timber and ranching, is by far the largest generator of income and jobs. Power concluded that not only is preservation of the ecological and biodiversity integrity of the Greater Yellowstone area not in conflict with local economic well-being, it is in fact essential to it.

Costa Rica has been one of the most successful promoters of ecotourism. More than 7 percent of Costa Rica's tourism income comes from ecotourism. In spite of efforts to protect the remaining natural areas, population pressure continues to threaten the integrity of the public and private reserves in Costa Rica and there are few corridors connecting them. The size of existing reserves is too small to support large species such as jaguars. As shown in the economic data for the Greater Yellowstone area, income from recreational use can justify preserving natural areas and biodiversity. In the process, it can also create an economic and political lobby against extractive uses such as mining and logging. There are also potential conflicts between ecotourism and biodiversity preservation. Close contact with some species may interfere with animal behavior patterns. Most areas set aside for ecotourism are rather small, fragmented and vulnerable, a problem acutely illustrated by the plight of the mountain gorillas caught in the civil strife in Rwanda. There are also frequent conflicts between local residents and tourists, particularly in Third World countries.

Ranching and Farming of Wild Species

Ranching and farming of wild species has also been advocated as a way to translate their preservation into market value. Ranches are dependent upon turtle eggs taken from the wild and farms are completely self-contained and use eggs from their own breeding turtles. Ranching of Green Iguanas and Pacas (a large nocturnal rodent related to guinea pigs) has been tried with some success in Central America (Ocana et al. 1988). An economically intractable problem is the low growth rate of many species. In a study of Dugong (Manatees) and Green Turtles, two species hunted for their meat, Tisdell (1986) found that the rapidly maturing turtles are profitable to farm, but such a strategy is not economically feasible for the slow-to-mature Dugong. Dugong do not reach sexual maturity until age 9–10 years, pregnancy rates are very low, and the interval between births is from three to six years. Dugong are a beautiful and peaceful species but the existence of pure time preference, expressed in a high and positive discount rate, makes them uneconomical to preserve.

Farming and ranching may damage vulnerable ecosystems, for example, by displacing other competing species or overgrazing by the animals being farmed. However, it might be the lesser of two evils in tropical forests. Even severely damaged ecosystems might eventually repair themselves if all the pieces are still present. Once species are extinct, they will never return. Farming may be the only alternative to extinction in many cases.

There is little doubt that specific natural products can have great value. However, market-based arguments for the value of biodiversity in general are more problematic. Although there are certainly large direct market benefits from exploiting biodiversity, it should be recognized that these values are a very small part of the world's economic national product, that many of the potential economic uses of biodiversity depend upon tenuous and risky markets, and that many of the proposals to exploit specific species may cause great damage to fragile tropical ecosystems. Those who rush to praise markets as the great protector of biodiversity should remember that one of the greatest "contributions" of wildlife to economic profit is the illegal trade in endangered species. The international trade in wildlife goods such as Siberian tiger parts, rhino horns, and elephant tusks may amount to as much as $5 billion to $8 billion a year (Associated Press 1994).

Many attributes of biodiversity have real and sometimes very high market values. It is, however, quite possible that basing biodiversity preservation decisions on market values could result in their destruction for economic gain. The very narrow and ephemeral framework of the market-based valuation of biodiversity should be explicitly recognized. Much of the neoclassical theory that was developed to analyze the static exchange of a fixed amount of reproducible goods is inappropriate as a basis to formulate policies to allocate irreproducible and absolutely essential biological entities. This point is implicitly recognized by every human society, including our own, which have elaborate and well-defined rules of behavior to protect the greater good from the whims of isolated individual self-interest. Vatn and Bromley (1994), Douglas and Wildavsky (1982), Kneese and Schultze (1985), among many others, discuss how narrowing the range of possible choices can distort perceptions of value.

The neoclassical explanation of biodiversity loss is succinctly stated by Herb Gintis[7]:

> The price of biological diversity is zero because of a market failure (there is no market in biodiversity). Therefore people will use as much as they please of it in the service of other ends (e.g., profit). Therefore it will be used up more rapidly than socially desirable.

This is a good statement of the neoclassical position and it lays bare the narrow conception of value of neoclassical economics and of the market economy neoclassical theory describes. Among the assumptions implicit in this quotation are the following, (1) a meaningful price can be put on biodiversity,

[7]This quote is from an e-mail response to Alan McGowen in the ecological economics forum, August 9, 1994.

(2) biodiversity is substitutable market good just like any other, and (3) it is socially desirable to use up biodiversity (albeit at a slower rate than the present one) in the pursuit of market goals. What are the limitations of this characterization of value and how may we categorize values at levels in the hierarchy above market exchange?

IV. THE TOTAL VALUE OF BIODIVERSITY TO THE HUMAN SPECIES

In the 19th century historian Thomas Arnold referred to economics as "the one-eyed" endeavor and according to Georgescu-Roegen (1984, 21):

> It is a judicious characterization because it has retained its currency ever since. Standard economists . . . have indeed refused to see that economic value extends beyond the market mechanism.

An example by Sagoff (1988) illustrates the difference between market values and the total value people place on the natural world. In the late 1960s Walt Disney Enterprises proposed to develop a ski resort in the Mineral King valley in Sequoia National Park in California. Sagoff, in an informal survey of his students, found a striking difference between the choices people make as consumers and the choices they make as citizens. Very few students expressed a willingness to visit the area if it remained a wilderness; the vast majority indicated a strong interest in visiting Mineral King if it were developed as a ski resort. Would these students choose to develop the area if the choice were up to them?

> The response was nearly unanimous. The students believed that the Disney plan was loathsome and despicable, that the Forest Service had violated a public trust by approving it, and that the values for which we stand as a nation compel us to preserve the little wilderness we have for its own sake and as a heritage for future generations. On these ethical and cultural grounds, and in spite of their consumer preferences, the students opposed the Disney plan to develop Mineral King. (Sagoff 1988, 51)

This example calls into question a basic tenet of welfare economics, that it is possible to construct a social welfare function based on a ranking of individual utility functions as revealed by consumer choices. Economists usually avoid dealing with the philosophical difficulties involved in the distinction between the choices people make as consumers and as citizens. The approach of Varian (1992, 333) is typical:

> The most reasonable interpretation of such a [social welfare] function is that it represents a social decision maker's preferences about how to trade off the utilities of different individuals. We will refrain from making philosophical comments here and just postulate that some such function exists.

Neoclassical economists argue that they are taking the ethical high ground by asserting that individual preferences are sacrosanct. Randall (1988, 217) writes:

> The mainstream economic approach is doggedly nonjudgmental about people's preferences: what the individual wants is presumed to be good for that individual.

The problem is, as the Mineral King case illustrates, that what is good for the individual as indicated by market signals is frequently an inadequate reflection of individual choice as indicated in a more universal (and more realistic) context. The neoclassical social preference function is based only on individual choices made in markets. And, it should be added, these individual choices are assumed to have no adverse effect on other individuals, there is no "impossibility theorem" (Arrow 1951) to contend with, and no collective good apart from that revealed by the sum of market-based, self-interested-individual utilities.

The basic argument here is a simple one made by many other economists and political theorists, namely, that individual preferences cannot be fully expressed in the narrow realm of market exchange (see Daly and Cobb 1989; Elster 1983; Marglin 1963; Mishan 1980; Sen 1977; Sagoff 1988). In the case of biodiversity loss we are clearly confronted with absolute scarcity and irrevocable loss; a loss which may affect the long-run survival of our species. This represents a new problem of valuation and one in which the valuation methodology used cannot pretend to be independent of ethical judgments (Funtowicz and Ravetz 1994, 198).

Economists attempt to capture nonmarket values by invoking the concepts of option, existence, and bequest values to individuals. Option value is the price an individual would be willing to pay to preserve something for future use. Existence value reflects the utility to an individual of preserving something for its own sake whether or not it is used in the future. Bequest values reflect the utility gained from leaving something of value to future generations. Forcing bequest, option, and existence values into a market choice framework ignores the intractable conflicts between market-constrained choice and citizen valuation. The result is a neither fish-nor-fowl mix of the two. Private decisions about available market choices cannot capture collective choices.

Many mainstream economists who recognize the limits of markets to reveal the value of biodiversity argue that an expanded cost-benefit framework can capture all relevant values. Randall (1988, 219) writes:

> Most issues involved with biodiversity violate the special case where market price is a valid indicator of economic value. Nevertheless, the general theory of economic value encompasses these broader concerns.

Other economists argue that attempts to expand the scope of market valuation through techniques such as the contingent valuation method (CVM) "contribute minimally—if at all—to the relevation of values" (Vatn and Bromley 1994, 130). The expressed purpose of CVM is to take "public goods" and put them in a hypothetical market context where they can be evaluated in an indifference curve framework in order to establish a "price" relative to goods traded in the market. Such a method confuses social and private choices. As the Mineral King example illustrates, there is ample evidence that these kinds of choices are different.

The response of many neoclassical economists is to criticize hypothetical surveys for not requiring actual monetary payment. The standard economic view is that any values other than those expressed through market payments are not "real." Neill et al. (1994) criticize open-ended hypothetical surveys for not accurately eliciting "real economic commitments." The thrust of their paper is that any commitment not expressed in terms of market exchange is not "truthful." This view is expressed clearly by Branden, Kolstad, and Miltz (1991, 12):

> Many economists are loath to base economic values—values that will be used to allocate real resources—on information that does not grow out of real economic commitments.

By this view, nonmarket decisions about the value of something, which are not made in the everything-is-substitutable, everything-has-a-price world of neoclassical theory are not "real." In fact, a study by Spash and Hanley (1995) found that valuation methods which elicit bids for biodiversity preservation fail as measures of welfare changes due to the prevalence of lexicographic preferences.[8] In lexicographic cases individuals are unwilling to make trades; one thing is absolutely preferred over another. They found that a significant number of respondents refuse to make trade-offs between biodiversity and market goods. They also found that knowledge about biodiversity and its importance is extremely limited. Both findings call into question the very basis of cost-benefit calculations. In a CVM study estimating the value of four wildlife species in New England (bald eagle, Atlantic salmon, wild turkey, and coyote) Stevens et al. (1991) found evidence for lexicographic preference ordering. Forty-four percent of all respon-

[8] The problem of lexicographic ordering of preferences was raised decades ago by Georgescu-Roegen (1936, 1954). The relevance of his early work in utility theory to environmental economics has been neglected not only by neoclassical economists but also by those sympathetic to his later work about the entropy law and the economic process.

dents agreed with the statement that "preservation of wildlife should not be determined by how much money can be spent." Sixty-seven percent of all respondents agreed with the statement that "As much wildlife as possible should be preserved no matter what the cost." Stevens et al. (1991, 398–99) write:

> A more fundamental issue concerns whether or not respondents made meaningful tradeoffs, and our results suggest that the majority of respondents who were willing to pay used decision-making processes inconsistent with the neoclassical paradigm of tradeoffs between money and wildlife.

Panel study results reported by Stevens, More, and Glass (1994) and by Kahneman and Knetsch (1992) suggest that contingent valuation questions may be capturing things like moral satisfaction and feelings of civic duty to contribute to a worthy cause. Neoclassical economists call this the "embedding effect" and they have devised elaborate techniques to isolate this aberration from standard theory.

An intractable conflict between markets and ecosystems is the necessity for discounting monetary values. The incompatibility between the pure time preference of individuals making decisions at a point in time and the social value of environmental sustainability has been widely discussed and will not be explored in depth here (see Price 1993). Amin (1992, 525) puts it succinctly:

> The practice of discounting the future appears to be an obvious psychological law. However, one cannot extend such discounting to society at large, in any sphere . . . because with a rate of only 4% over 30 years, any value is reduced to two-thirds of its initial value, leaving a margin of almost absolute uncertainty about the validity of choices. But if 30 years can appear a respectable period in the scale of human life, what does it represent in the time frame of the history of nations and of humanity?

Many mainstream economists recognize the discounting problem but fail to come to grips with its intractability in standard economic valuation. Randall (1988, 219–20) correctly points out that discounting could justify the complete collapse of living systems in only a few hundred years from now if it is offset by minor economic gains in the present. Yet he asserts that there is an "almost adequate conceptual basis for economic valuation of biodiversity"; the "almost" referring to the discounting problem.

Another value of biodiversity that transcends market value is called the "biophilia hypothesis" (Wilson 1984). According to this hypothesis, humans are fundamentally complex mammals that seek variety and new stimuli in the context of the biological world. Some regular contact with nature and other species is essential to human mental health and well-being. Although the study of biophilia as a scientific concept is only in its infancy, some research has attempted to test the hypothesis. Ulrich (1993) discusses a number of studies that found a positive relationship between measures of emotional well-being and natural (non-built) environments. According to him, "this aesthetic appreciation for nature may be universally expressed across human cultures" (Ulrich 1993, 49). For example, a study by Yi (1992) comparing Texans and Koreans, found that positive feelings about specific types of landscapes were not culturally specific; urban dwellers and farmers from both cultures responded positively to the same types of natural objects and settings. To the extent that biophilia is not reflected in market values, or is even thwarted by economic activity, this reflects another aspect of biodiversity above and beyond that captured by market or pseudo-market valuation. Orr (1994, 149) writes:

> Biophilia can be suffocated, for example, by the demands of an economy oriented toward accumulation, speed, sensation, and death. But economists have not written much about how an economy encourages or discourages love generally or biophilia in particular.

Biophilia may be part of the reason for the lexicographic ordering of preferences for environmental protection that regularly show up in CVM surveys. Biodiversity may be an essential, non-substitutable, factor for the psychological well-being of humans. It may be part of a hierarchy of essential needs along with food, shelter, sex, and companionship necessary for the enjoyment of life. Some of the founders of modern utility theory (Georgescu-Roegen 1936, 1950; Menger 1950) argued for a hierarchical representation of human needs. Georgescu-Roegen (1968, 263) argued that it is the principle of the irreducibility of wants, not the postulate of indifference, that should be at the base of a realistic theory of choice. If the biophilia hypothesis proves to be true, it could provide the strongest argument of all for the preservation of biodiversity. Biodiversity may not be just another item in the consumer's utility array. It may be essential for our psychological well-being.

Recent theories of social conflict suggest yet another value of biodiversity not captured in market prices. Dasgupta (1995), drawing on empirical studies from a number of disciplines, finds self-reinforcing links between population growth, poverty, and the degradation of local environments including biodiversity loss. A comprehensive study by Homer-Dixon, Boutwell, and Rathjens (1993) concludes that increasing natural resource scarcity can strengthen the hand of the social elite and exacerbate an unequal distribution of resources. The loss of wild biodiversity in agricultural systems can contribute to increased vulnerability and increase the likelihood of social instability (Gowdy 1997). The role that biodiversity plays in maintaining social order is another positive "value," and another attribute that is impossible to measure.

The apparent positive relationship between social stability and natural resource preservation raises fundamental questions about individual choices, democracy, and the social contract. Should policies to protect biodiversity be based upon some democratic representation of the preferences of humans? This is an extremely difficult philosophical problem. One might argue that all choices are necessarily based on individual preferences. The context of these preferences, however, makes a difference (Bromley 1990, 1991; Sagoff 1994; Vatn and Bromley 1994). In the next valuation section below, describing the ecosystem value of biodiversity, an essential point is that there is a real world outside the human mind.[9] Our preferences can be whatever we choose, but if our choices, individual, social or otherwise, are not consistent with biophysical reality, our complex technological culture and perhaps our species will join those other cultures and species that have become extinct. Humans have negative feelings about organisms which are essential parts of ecosystems. Many unappreciated groups of organisms, such as the insects, algae and fungi, are essential for the existence of other forms of life, including humans. Yet these organisms rank low in esteem for humans and there is little chance of improvement in their image (Kellert 1993). Choices based on human preferences, no matter how broadly defined, are bound to undervalue the importance of preserving these creatures. Vatn and Bromley (1994, 138):

> It seems as if some have come to regard the natural environment as a large zoological garden from which we can select for policy attention those parts that happen to hold our momentary affection. But first we have to value it to reassure ourselves that the attention is warranted, or that the attention is efficient.

Ecosystems obey some basic rules that have been established over the $3\frac{1}{2}$-billion-year history of life on planet Earth. The highest hierarchy of the value of biodiversity is its value in stabilizing the life support system which makes human existence possible.

V. THE ECOSYSTEM VALUE OF BIODIVERSITY

For many decades ecologists has suspected that biodiversity plays a positive role in the health of ecosystems. However, until very recently there has been no direct ex-

[9] Some economists argue otherwise. According to Gilder (1981): "Because economies are governed by thought, they reflect not the laws of matter but the laws of the mind." Simon would have us believe that "in the end, copper and oil come out of our minds" (quoted in Daly 1985). For a peculiar mix of radical environmentalism and neoclassical technological utopianism see Sagoff (1995).

perimental evidence for this positive role. Tilman and Downing (1994) found that biodiversity plays a crucial role in maintaining the resilience of ecosystems to environmental shocks. In a study of Minnesota grasslands they found that primary productivity in more diverse plant communities is more resistant to drought and recovers faster.[10] Alarmingly, in view of the predicted sharp increase in extinction rates, they found that each additional species lost in the grasslands they studied had an increasingly greater negative impact on drought resistance. In another controlled experiment Naeem et al. (1994) also found that declining biodiversity adversely affected the performance of ecosystems in terms of plant productivity, nutrient retention, water retention, and decomposition. Tilman et al. (1994) argue that, primarily because of habitat fragmentation, the stage is set for a sharp increase in extinction rates in the near future. They write:

> Even moderate habitat destruction is predicted to cause time-delayed but deterministic extinction of the dominant competitor in remnant patches. Further species are predicted to become extinct, in order from the best to the poorest competitors as habitat destruction increases. Moreover, the more fragmented a habitat already is, the greater is the number of extinctions caused by added destruction. Because such extinctions occur generations after fragmentation, they represent a debt—a future ecological cost of current habitat destruction. (Tilman et al. 1994, 65)

Perhaps the most important ecological value of diverse ecosystems is the preservation of evolutionary potential. Species diversity, as well as genetic diversity within species, allows species and ecosystems to adapt to environmental changes. Natural systems are not in a "state of equilibrium" as economists use the term equilibrium. Economic theory refers to an equilibrium system as one which returns to its initial state after a disturbance. When ecosystems are disturbed, depending on the nature and extent of the disturbance, they do not necessarily revert to their initial state when the disturbance is removed because the initial state is no longer the same. This flexibility to adapt to different environmental conditions has evolved over eons. Geological records show that the Earth is a constantly changing, even chaotic system. Mountains are formed and eroded, ice sheets expand and contract, and volcanoes disrupt climate patterns. Genetic diversity, as well as higher orders of biological diversity, gives ecosystems the ability to adapt to these changes. When we reduce the variability within the biological world through habitat destruction, ecosystem modification, extinction, and genetic erosion we limit the possible responses to future environmental change. Evolutionary potential has obvious value to ecosystems, but we cannot quantify its value to the human species. There are some examples of evolutionary changes which positively affected human prospects for overcoming adverse changes. Paleolithic human cultures in Europe, whose economies were based on big game hunting, were adversely affected after these animals disappeared with the retreat of the glaciers some 15 thousand years ago. The warmer climate also apparently led to the evolution of certain kinds of plants with larger seeds suitable for agriculture which humans were able to exploit. This kind of evolutionary response depends on nature having a variety of genetic material to work with and illustrates the importance of preserving species *in situ*. We know that human activity is significantly changing climatic conditions on earth at both the local and global levels. With the projected changes in the Earth's biosphere as a result of greenhouse warming (Peters and Lovejoy 1992) we may be destroying the ability of the biosphere to adapt to change at one of those critical

[10] The relationship between ecosystem productivity and diversity is a complicated one. While it is generally true that "diversity begets productivity" it also seems to be true that diversity begins to decline when productivity is very high. Apparently, both high and low levels of productivity are characterized by limitational factors (Tilman 1982, 1986). Robinson et al. (1992) found complex patterns of change resulting from habitat fragmentation. The Tilman and Robinson et al. studies point to the need for detailed population analyses of the specific effects of habitat fragmentation.

times in the Earth's history when adaptability is needed most.

Frequently occurring natural agents such as hurricanes, storms, and fires are essential parts of ecosystem health (Solbrig 1991). Such seemingly destructive events have "value" to ecosystems, but impose large economic costs on human society. For example, the fires on Long Island, New York, during the summer of 1995 caused extensive loss of property but were beneficial to the long-term health of the forest ecosystem. According to Wilson (1992) such events have been "coded for" in the genes of the organisms making up ecosystems. The long-standing policy of fire suppression in national parks and forests is currently being hotly debated and shows clearly the spatial and time scale conflicts between the human economy and ecosystem health.

The total ecosystem value of biodiversity may be the value of the existence of humans and all other species. Norton (1988, 205) puts it nicely:

> The value of biodiversity is the value of everything there is. It is the summed value of all the GNPs of all countries from now until the end of the world. We know that, because our very lives, and our economies are dependent upon biodiversity. If biodiversity is reduced sufficiently, and we do not know the disaster point, there will no longer be any conscious beings. With them go all value—economic and otherwise.

VI. POLICY IMPLICATIONS

Humans are part of the natural world and any damage humans do to that world has the potential to increase the risk to ourselves. Many economists counter that humans are fundamentally different from other species. We are unique and not subject to the same laws of nature as other species. It is hard to answer this because it is not a testable scientific argument, merely a statement of optimism; "we'll think of something." To counter this argument one can only point to past and present cases where human society did not think of anything clever to save themselves in the face of self-inflicted environmental disaster. There is mounting evidence that almost every civilization since the adoption of agriculture some 10,000 years ago has overshot and collapsed, most of them because of the misuse of their natural resource base. These include the civilization of Easter Island, the Maya, and the Sumerians (Bahn and Flenley 1992; Ponting 1991; Tainter 1988; Weiskel 1989). Easter Islanders degraded the biodiversity of their island by cutting down all the large trees (two species of palm tree and the toromiro) to use as skids for transporting the large statues for which the island is famous. Without these trees, the islanders could not build boats or wooden houses, as they did formerly. Deforestation also caused soil erosion. After the loss of these trees came a precipitous decline in the standard of living for the Easter Islanders (Bahn and Flenley 1992). Likewise, the Maya cut down vast tracts of forest to obtain wood for houses and for firewood to use in heating limestone to produce the white stucco facade for their temples. This deforestation evidently caused soil erosion and contributed to the demise of that civilization (Dobson 1996, 254–55). Of course the cause of the collapse of past civilizations is a complex mix of resource degradation, including biodiversity loss, social instability, and unresponsive political institutions. In view of the history of our species, the argument that humans do not need to conserve biodiversity because we are so resourceful and intelligent seems imprudent at best.

As long as the "ideology of efficiency" (Bromley 1990) continues to dominate the political discussion in the United States and Europe, policy efforts to preserve biodiversity must confront cost-benefit analysis. The thrust of the various arguments presented above is that even if one argues that individual choices are sacrosanct, these choices cannot be fully expressed in the context of market exchange. Economic valuation studies that attempt to capture social values in a hypothetical market situation should be treated with suspicion. If market values, including extended cost-benefit analysis, are not sufficient as a basis for biodiversity protection policy, what should we do? I suggest the following basic considerations.

1. Devising a meaningful single measure, monetary or otherwise, of the value of biodiversity, is impossible. Ecologists have developed numerous indices of different aspects of biodiversity such as species richness and species abundance (Magurran 1988). The consensus among ecologists is that such measures are useful base numbers to show changes within a particular ecosystem but are not comparable across ecosystems. Not everything can be quantified, pure uncertainty is pervasive, and the time-scales involved are beyond the experience of individual humans. We can, however, use a pluralistic methodology (Norgaard 1989) to construct multiple assessments of the various aspects of biodiversity value. As outlined above, a useful procedure is to group the various value concepts according to the following hierarchy. First estimate the present discounted market value of biodiversity, and where appropriate and using various scenarios, reduce uncertainty to risk. Second, estimate the potential nonmonetary value of biodiversity based on decisions humans make as social creatures, not isolated individuals. Finally, use the best scientific information available to assess the function and importance of biodiversity to specific ecosystems and to the long-run viability of the human species. What is the importance of biodiversity to long-run human survival? If we follow Georgescu-Roegen's admonishment "Love thy species as thyself" there is no conflict between biodiversity preservations and the long-run interests of *Homo sapiens*.[11]

2. Incorporate a broad geographic and ecological perspective in policies for biodiversity protection. Five fundamental policy objectives have emerged in the burgeoning literature on ecosystems protection (Noss 1991): (i) In a protected area all native ecosystem types should be represented in their various natural ranges of variation (Pulliam 1995); (ii) all native species in these protected areas should be represented in their natural patterns of abundance and distribution; (iii) ecological and evolutionary processes should be maintained, including everything from predation patterns to hydrological cycles (Holling 1995); (iv) the potential for evolutionary change should be maintained, that is, the response to long-term and short-term disturbances such as fire and long-term climate variation; and (v) the coevolutionary (Norgaard 1984, 1988) relationship between human and natural systems should be explicitly recognized. New federal, state, and nongovernmental organization (NGO) policies are encouraging. The U.S. Department of Interior has officially adopted an ecosystem approach to endangered species management, the State of Florida has made impressive efforts to protect and restore the Everglades based on coevolutionary ecosystem management objectives (Mann 1995), and the Nature Conservancy, in its "Last Great Places" program, has based its land acquisition and management policies on the ecosystem concept.

3. Emphasize biodiversity strategies that broaden the choice set beyond simple cost-benefit calculations. For example, the U.S. Department of Interior is also pioneering the concept of conservation offsets. When a development project is proposed that might threaten an ecologically valuable area, it may be allowed if the developer agrees to restore or expand another threatened area. This kind of ecological swap can be done without calculating economic "values" and

[11]A reviewer raised the question, "Is the case for preserving all biodiversity at the current margin so strong that it should 'trump' all other social objectives?" I do not argue that the value of *any* biodiversity to humans is always infinite. To say, however, that the marginal value of biodiversity is finite, should not be taken to mean that we have an idea about how to measure it. We know the existence of our species depends on other life forms on the planet, therefore its total value is infinite; we cannot continue to destroy it and survive. Furthermore, we have only a vague idea about how the pieces of biodiversity fit together. On the other hand, if we take a broad view of sacrificing biodiversity at the margin, *and* if we have an idea of what an environmentally sustainable society should be, *and* if we are moving toward that, *maybe* we can sacrifice some biodiversity on the way toward that sustainable path. The important point to remember, I think, is that there are limits to the substitutability between "natural capital" and "manufactured capital," and thus limits to the standard economic notion of weak sustainability (see Gowdy and O'Hara 1996).

without quantifying costs and benefits using a common metric. These sorts of policies cannot be undertaken without collective discussion of choices based on some commonly adhered to notion of democratic discourse (O'Hara 1996; Vatn and Bromley 1994).

4. Expand the safe minimum standard (SMS) approach to biodiversity protection to include the preservation of evolutionary potential. The SMS recognizes that species are unique and that the future repercussions of destroying biodiversity are uncertain. This approach, as stated by Ciriacy-Wantrup (1968) and modified by Bishop (1978, 1979) states that the SMS should be adopted unless the "social costs of doing so are unacceptably large" (Bishop 1978, 10). Although Bishop thought of social cost as the cost of preservation borne by current and near-term generations, it should be recognized that reducing the ability of future generations to adapt to new environmental or social conditions is also a social cost of reducing biodiversity. One might argue that we can leave future generations better off by leaving them less biodiversity and more "capital" including scientific and technical knowledge. This argument can be turned around to say that we should leave future generations more biodiversity because they might have developed the scientific and technical knowledge to use species which we now consider to be economically unimportant.

As Vatn and Bromley (1994, 143) point out, the SMS is a fairly crude measure which was developed for decisions involving the critical harvest levels for renewable natural resources. The basis message of the SMS approach is that we should err on the side of caution. In terms of ecosystem exploitation, we should leave them alone if we are not sure of the side effects. With the increasing knowledge of the complexity of evolutionary processes and ecosystems, this leave-it-along strategy should be extended beyond its original formulation to include preserving the evolutionary potential of ecosystems and genetic diversity (Gowdy 1993; Myers 1993).

VII. CONCLUSION

The problems with the standard economic concept of value as expressed in relative market prices are becoming increasingly clear. Difficulties include problems in cognition, the incongruity problem, indivisibilities and complementarities of ecosystem functions, and pure time preference. Attempts to duplicate market values for environmental attributes through contingent valuation suffer from these same difficulties. It is clear both from theoretical considerations and scientific evidence about the nature of biodiversity, that economists need to broaden their concept of value beyond that determined by market exchange. Such a broader concept of value leads inevitably to the need for a broader array of policy options for the protection of biodiversity beyond those based on simple cost benefit calculations.

The difficulties with the standard economic approach to the valuation of biodiversity raise long-neglected issues in neoclassical utility theory. Fundamental questions were raised decades ago about the meaning of "utility" and were once the subject of a lively debate among leading economists (Georgescu-Roegen 1968). Perhaps a positive spillover effect from the ongoing discussion of "value" among resource economists will be that it will serve as a stimulus to reopen a debate among other economists about the general validity of the concepts of indifference and substitutability which lie at the base of neoclassical theory and policy.

References

Amin, S. 1992. "Can Environmental Problems be Subject to Economic Calculations?" *World Development* 20 (4):523–30.

Anderson, B. 1966. *Social Value*. New York: Augustus M. Kelley.

Arrow, K. 1951. *Social Choice and Individual Values*. New York: Wiley.

Arrow, K., et al. 1995. "Economic Growth, Carrying Capacity, and the Environment." *Science* 268 (28 Apr.):520–21.

Associated Press. 1994. Untitled. America Online News Service.

Bahn, P., and J. Flenley. 1992. *Easter Island, Earth Island*. New York: Thames and Hudson.

Bingham, G., et al. 1995. "Issues in Ecosystem Valuation: Improving Information for Decision Making." *Ecological Economics* 14 (2):73–90.

Bishop, R. C. 1978. "Endangered Species and Uncertainty: The Economics of a Safe Minimum Standard." *American Journal of Agricultural Economics* 60 (1):10–18.

———. 1979. "Endangered Species, Irreversibility, and Uncertainty: A Reply." *American Journal of Agricultural Economics* 61 (2):376–79.

———. 1992. "Economic Efficiency, Sustainability, and Biodiversity." *Ambio* 22 (2–3):69–73.

Brandon, J., C. Kolstad, and D. Miltz. 1991. "Introduction." In *Measuring the Demand for Environmental Quality*, eds. J. Braden and C. Kolstad. Amsterdam: North-Holland.

Bromley, D. W. 1989. "Entitlements, Missing Markets, and Environmental Uncertainty." *Journal of Environmental Economics and Management* 17 (2):181–94.

———. 1990. "The Ideology of Efficiency: Searching for a Theory of Policy Analysis." *Journal of Environmental Economics and Management* 19 (1):86–107.

———. 1991. "Entitlements, Missing Markets, and Environmental Uncertainty: Reply." *Journal of Environmental Economics and Management* 20 (3):297–302.

———. 1995. "Property Rights and Natural Resource Damage Assessments." *Ecological Economics* 14 (2):129–36.

Brown, T. 1984. "The Concept of Value in Resource Allocation." *Land Economics* 60 (Aug.):231–46.

Ciriacy-Wantrup, S. V. 1968. *Resource Conservation: Economics and Policies*. 3rd ed. Berkeley: University of California Press.

Cummings, R., D. Brookshire, and W. Schultze. 1986. *Valuing Environmental Goods: An Assessment of the Contingent Valuation Method*. Totowa, NJ: Roman and Allanheld.

Daly, H. 1985. "The Circular Flow of Exchange Value and the Linear Throughput of Matter-Energy." *Review of Social Economy* 44 (4):279–97.

———. 1995. "Reply to Mark Sagoff's 'Carrying Capacity and Ecological Economics'." *BioScience* 45:620–24.

Daly, H., and J. Cobb. 1989. *For the Common Good*. Boston: Beacon Press.

Dasgupta, P. 1995. "Population, Poverty and the Local Environment." *Scientific American* 272 (Feb.):40–45.

Dobson, A. 1996. *Conservation and Biodiversity*. New York: W. H. Freeman.

Douglas, M., and A. Wildavsky. 1982. *Risk and Culture: An Essay on the Selection of Technical and Environmental Dangers*. Berkeley: University of California Press.

Ehrenfield, D. 1988. "Why Put a Value on Biodiversity?" In *Biodiversity*, ed. E. O. Wilson. Washington, DC: National Academy Press.

Ehrlich, P., and A. Ehrlich. 1992. "The Value of Biodiversity." *Ambio* 21 (3):219–26.

Elster, J. 1983. *Ulysses and the Sirens*. Cambridge: Cambridge University Press.

Freeman, M., III. 1991. "Factorial Survey Methods and Willingness to Pay for Housing Characteristics: A Comment." *Journal of Environmental Economics and Management* 20 (1):92–96.

Funtowicz, S., and J. Ravetz. 1994. "The Worth of a Songbird: Ecological Economics as a Post-Normal Science." *Ecological Economics* 10 (3):197–208.

Geist, V. 1994. "Wildlife Conservation as Wealth." *Nature* 368 (7 Apr.):491–92.

Georgescu-Roegen, N. 1936. "The Pure Theory of Consumer's Behavior." *Quarterly Journal of Economics* 50 (3):545–93.

———. 1950. "The Theory of Choice and the Constancy of Economic Laws." *Quarterly Journal of Economics* 64 (1):125–38.

———. 1954. "Choice, Expectations, and Measurability." *Quarterly Journal of Economics* 68 (2):503–34.

———. 1968. "Utility." In *The International Encyclopedia of the Social Sciences*. New York: Macmillan.

———. 1984. "Feasible Recipes versus Viable Technologies." *Atlantic Economic Journal* 12 (1):21–30.

Gershon, D. 1992. "If Biological Diversity Has a Price, Who Sets it and Who Should Benefit?" *Nature* 359 (15 Oct.):565.

Gilder, G. 1981. *Wealth and Poverty*. New York: Basic Books.

Gowdy, J. M. 1993. "Economic and Biological Aspects of Genetic Diversity." *Society and Natural Resources* 6 (1):1–16.

———. 1994. *Coevolutionary Economics: Economics, the Economy, and the Environment*. Boston: Kluwer Academic Press.

———. 1996. "Society and Ecosystems: Discounting and the Social Aspects of Biodiversity Protection." *International Journal of Social Economics* 23 (4/5/6):49–63. Special issue in honor of Clem Tisdell.

———. 1997. "Biophysical Limits to Industrialization: Prospects for the 21st Century." In *The Coming Age of Scarcity*, eds. M. Dobkowski and I. Walliman. Syracuse: Syracuse University Press.

Gowdy, J. M., and C. M. McDaniel." 1995. "One World, One Experiment: Addressing the Biology-Economics Conflict." *Ecological Economics* 14 (3):180–92.

Gowdy, J. M., and S. O'Hara. 1996. "Weak Sustainability and Viable Technologies." *Ecological Economics*. Special issue in honor of Nicholas Georgescu-Roegen, forthcoming.

Hanley, N., and C. Spash. 1993. *Cost Benefit Analysis and the Environment*. Brookfield, VT: Edward Elgar.

Holling, C. S. 1995. "Sustainability: The Cross-Scale Dimension." In *Defining and Measuring Sustainability: The Biogeophysical Dimensions*, eds. M. Munasinghe and W. Shearer. Washington, DC: The United Nations University and the World Bank.

Homer-Dixon, T. 1995. "The Ingenuity Gap: Can Poor Countries Adapt to Resource Scarcity?" *Population and Development Review* 21: 587–612.

Homer-Dixon, T., J. Boutwell, and G. Rathjens. 1993. "Environmental Change and Violent Conflict." *Scientific American* 268 (Feb.): 38–45.

Kahneman, D., and Knetsch. 1992. "Valuing Public Goods: The Purchase of Moral Satisfaction." *Journal of Environmental Economics and Management* 22 (1):57–70.

Kellert, S. 1993. "Values and Perceptions of Invertebrates." *Conservation Biology* 7 (4): 845–55.

Kneese, A. V., and W. D. Schultze. 1985. "Ethics and Environmental Economics." In *Handbook of Natural Resource and Energy Economics*, eds. A. V. Kneese and W. D. Schultze. New York: North-Holland.

Magurran, A. 1988. *Ecological Diversity and Its Measurement*. Princeton: Princeton University Press.

Mann, C. 1995. "Filling in Florida's Gaps: Species Protection Done Right?" *Science* 269 (21 July):318–20.

Marglin, S. 1963. "The Social Rate of Discount and the Optimal Rate of Investment." *Quarterly Journal of Economics* 77 (1):95–111.

Martinez-Alier, J. 1994. "The Merchandising of Biodiversity." *Ethnoecologica* 2 (1):69–85.

Menger, K. 1950. *Principles of Economics*. Glencoe, IL: Free Press.

Mendelsohn, R., and M. Balick. 1995. "The Value of Undiscovered Pharmaceuticals in Tropical Forests." *Economic Botany* 49 (2):223–28.

Mishan, E. J. 1980. "How Valid are Economic Evaluations of Allocative Changes?" *Journal of Economic Issues* 14 (1):143–61.

Mitchell, R., and R. Carson. 1989. *Using Surveys to Value Public Goods: The Contingent Valuation Method*. Washington, DC: Resources for the Future.

Moffatt, A. S. 1995. "Plants Proving Their Worth in Toxic Metal Cleanup." *Science* 269 (21 July):302–3.

Myers, N. 1993. "Biodiversity and the Precautionary Principle." *Ambio* 22 (2–3):74–79.

Myers, N., and J. Simon. 1994. *Scarcity or Abundance?* New York: Norton.

Naeem, S., L. Thompson, S. Lawler, J. Lawton, and R. Woodfin. 1994. "Declining Biodiversity Can Alter the Performance of Ecosystems." *Nature* 368 (21 Apr.):734–37.

Neill, H., R. Cummings, P. Ganderton, G. Harrison, and T. McGuckin. 1994. "Hypothetical Surveys and Real Economic Commitments." *Land Economics* 70 (May):145–54.

Norgaard, R. 1984. "Coevolutionary Development Potential." *Land Economics* 60 (May): 160–73.

———. 1988. "Sustainable Development: A Coevolutionary View." *Futures* (Dec.):606–20.

———. 1989. "The Case for Methodological Pluralism." *Ecological Economics* 1 (1):37–58.

Norton, B. 1988. "Commodity, Amenity, and Morality: The Limits of Quantification in Valuing Biodiversity." In *Biodiversity*, ed. E. O. Wilson. Washington, DC: National Academy Press.

Noss, R. 1991. "Sustainability and Wilderness." *Conservation Biology* 5 (1):120–22.

Ocana, G., I. Rubinoff, N. Smythe, and D. Werner. 1988. "Alternatives to Destruction." In *Biodiversity*, ed. E. O. Wilson. Washington, DC: National Academy Press.

O'Hara, S. 1996. "Discursive Ethics as a Tool in Ecosystems Valuation and Environmental Policy." *Ecological Economics* 16 (2):95–107.

Oldfield, M. 1988. *The Value of Conserving Genetic Resources*. Sunderland, MA: Sinauer.

O'Neill, R., D. DeAngelis, J. Waide, and T. Allen. 1986. *A Hierarchical Concept of Ecosystems*. Princeton: Princeton University Press.

Orr, D. 1994. *Earth in Mind*. Washington, DC: Island Press.

Peters, C., A. Gentry, and R. Mendelsohn. 1989. "Valuation of an Amazonian Rainforest." *Nature* 339 (29 June):655–56.

Peters, R., and T. Lovejoy. 1992. *Global Warming and Biological Diversity*. New Haven: Yale University Press.

Pickett, S. T., C. T. Parker, and P. L. Fiedler. 1992. "The New Paradigm in Ecology: Implications for Conservation Biology Above the Species Level." In *Conservation Biology*, eds. P. L. Fiedler and S. K. Jain. New York: Chapman and Hall.

Pimm, S. 1991. *The Balance of Nature?* Chicago: University of Chicago Press.

Pimm, S., G. Russell, J. Gittleman, and T. Brooks. 1995. "The Future of Biodiversity." *Science* 269 (21 June):347–50.

Ponting, C. 1991. *A Green History of the World*. New York: Penguin Books.

Power, T. 1991: "Ecosystem Preservation and the Economy in the Greater Yellowstone Area." *Conservation Biology* 5 (3):395–404.

Price, C. 1993. *Time, Discounting and Value*. Cambridge, MA: Basil Blackwell.

Pulliam, H. R. 1995. "Managing Landscapes for Sustainable Biodiversity." In *Defining and Measuring Sustainability: The Biogeophysical Dimensions*, eds. M. Munasinghe and W. Shearer. Washington, DC: The United Nations University and the World Bank.

Randall, A. 1988. "What Mainstream Economists Have to Say about the Value of Biodiversity." In *Biodiversity*, ed. E. O. Wilson. Washington, DC: National Academy Press.

Robinson, G., R. Holt, M. Gaines, S. Hamburg, M. Johnson, H. Fitch, and E. Martinko. 1992. "Diverse and Contrasting Effects of Habitat Fragmentation." *Science* 257 (24 July):524–25.

Sagoff, M. 1988. *The Economy of the Earth*. Cambridge: Cambridge University Press.

———. 1994. "Should Preferences Count?" *Land Economics* 70 (May):127–44.

———. 1995. "Carrying Capacity and Ecological Economics." *BioScience* 45 (9):610–20.

Sen. A. 1967. "Isolation, Assurance, and the Social Rate of Discount." *Quarterly Journal of Economics* 81 (1):112–24.

———. 1977. "Rational Fools: A Critique of the Behavioral Foundations of Economic Theory." *Philosophy and Public Affairs* 6:317–44.

Solbrig, O. 1991. "The Origin and Function of Biodiversity." *Environment* 33:16–20, 34–38.

Solow, R. 1993. "Sustainability: An Economist's Perspective." In *Economics of the Environment*, eds. R. Dorfman and N. Dorfman. New York: Norton.

Spash, C., and N. Hanley. 1995. "Preferences, Information, and Biodiversity Preservation." *Ecological Economics* 12 (3):191–208.

Stevens, T., J. Echeverria, R. Glass, T. Hager, and T. More. 1991. "Measuring the Existence Value of Wildlife: What do CVM Estimates Really Show?" *Land Economics* 67 (Nov.): 390–400.

Stevens, T., T. More, and R. Glass. 1994. "Interpretation and Temporal Stability of CV Bids for Wildlife Existence." *Land Economics* 70 (Nov.):355–63.

Tainter, J. A. 1988. *The Collapse of Complex Societies*. Cambridge: Cambridge University Press.

Tilman, D. 1982. *Resource Competition and Community Structure*. Princeton: Princeton University Press.

———. 1986. "Resources, Competition, and the Dynamics of Plant Communities." In *Plant Ecology*, ed. M. J. Crawley. Oxford: Blackwell.

Tilman, D., and J. Downing. 1994. "Biodiversity and Stability in Grasslands." *Nature* 367 (27 Jan.):363–65.

Tilman, D., R. May, C. Lehman, and M. Nowak. 1994. "Habitat Destruction and the Extinction Debt." *Nature* 371 (1 Sept.):65–66.

Tisdell, C. 1986. "Conflicts about Living Marine Resources in Southeast Asian and Australian Waters: Turtles and Dugong as Cases." *Marine Resource Economics* 3 (1):89–109.

Torrence, P. 1995. "The Endangered Species Act." *Science* 269 (29 Sept.):1803–4.

Ulrich, R. 1993. "Biophilia, Biophobia, and Natural Landscapes." In *The Biophilia Hypothesis*, eds. S. Kellert and E. O. Wilson. Washington, DC: Island Press.

Varian, H. 1992. *Microeconomic Analysis*. 3rd. ed. New York: Norton.

Vatn, A., and D. W. Bromley. 1994. "Choices without Prices without Apologies." *Journal of Environmental Economics and Management* 26 (2):129–48.

Ward, P. 1994. *The End of Evolution*. New York: Bantam Books.

Wiskel, T. 1989. "The Ecological Lessons of the Past: An Anthropology of Environmental Decline." *The Ecologist* 19:104–9.

Wilson, E. O. 1984. *Biophilia*. Cambridge: Harvard University Press.

———. 1992. *The Diversity of Life*. Cambridge: Harvard University Press.

Yi, Y. 1992. "Affect and Cognition in Aesthetic Experiences of Landscapes." Ph.D. diss., Texas A&M University.

B Tropical and Temperate Forests

[11]

A Method for the Economic Valuation of Non-timber Tropical Forest Products[1]

Ricardo Godoy, Ruben Lubowski, and Anil Markandya

Godoy, Ricardo, Ruben Lubowski, *and* **Anil Markandya** *(Harvard Institute for International Development, Harvard University, Cambridge, MA 02138, U.S.A.).* A Method for the Economic Valuation of Non-timber Tropical Forest Products. *Economic Botany 47(3):220–233. 1993. By drawing on quantitative studies in social anthropology, zoology, ethnobotany, and economics we present a method for conducting an economic valuation of non-timber forest products. A review of 24 studies suggests that the median value for non-timber forest products is about $50/ha/year. We discuss problems with past studies and suggest ways to get better estimates of output quantities, marginal costs, and prices.*

Key Words: non-timber forest products; tropical forests; economic valuation; methods; sustainability.

Policy-makers often assume tropical forests have no economic value unless they are logged or farmed (Dove 1983; Hecht, Anderson, and May 1988). Besides timber, tropical forests produce food (Arnold 1991; FAO 1989), construction materials, medicinal plants, fodder, and firewood, all of which villagers use daily (Panayotou and Ashton 1992). As a proxy for the opportunity cost of the forest—the maximum value derivable from the forest before it is put to new uses (Godoy 1992)—the economic value of non-timber tropical forest products is conservative because the forest also produces other benefits, such as biological diversity and environmental services (Panayotou, pers. comm.). Policy-makers and development organizations need an accurate estimate of the opportunity cost of the forest to evaluate proposed projects and filter out economically disadvantageous ones. Under some conditions, leaving the forest unlogged and using it to get non-timber forest goods and environmental services may be socially and economically optimal (Panayotou, pers. comm.).

In this article we review studies estimating the economic value of non-timber tropical forest products. Drawing on anthropology and economics, we present a method for future valuation studies. We limit our discussion to non-timber forest goods because methods for valuing other goods and services from the tropical forest (e.g., timber, biodiversity) are outlined elsewhere (Godoy 1992; Panayotou, pers. comm.) and because the valuation of non-timber products is the area of tropical forestry in which we are likely to make the greatest contribution. Ethnographers are familiar with the social uses of many forest goods through their research on hunting and gathering and swidden agriculture. Ethnobotanists have provided detailed inventories of useful forest plants. Social anthropologists have the classic ethnographic fieldwork tools, such as participant observation, time allocation methods, household diaries, direct records of the goods extracted from the forest, and long-term field research, that give them a comparative advantage in studies of forest products.

We focus on tropical forests for brevity, though the valuation method outlined here could be applied to any forest type. Lastly, we focus on valuation at the forest gate, the point at which villagers first sell or consume the goods, rather than assessing the value of forest goods to the regional, national, or world economy. We omit studies that present too little information to permit calculation of yearly profits per hectare (e.g., Fearnside 1989; Hart 1978; Hecht, Anderson, and May 1988; Jessup and Vayda 1989; Peluso and Pof-

[1] Received 8 April 1993; accepted 5 May 1993.

Economic Botany 47(3) pp. 220–233. 1993

fenberger 1989). In assessing the studies and producing our methods we draw on the anthropological literature on hunters and foragers, surveys of methods in social anthropology (e.g., Bernard et al. 1984, 1986; Borgerhoff and Caro 1985; Gregory and Altman 1989; Gross 1984), conversations with some of the leading researchers in the study of tropical-forest plant and animal use, and our own fieldwork experience in Asia and Latin America.

Until recently few scholars had studied the economic value of non-timber tropical-forest products (e.g., Dunn 1975). In 1988 the International Tropical Timber Organization published a report calling for the rigorous study of these products (Panayotou and Ashton 1992). Since then many researchers have done such valuations (e.g., de Beer and McDermott 1989; Godoy and Feaw 1989; Padoch and de Jong 1989; Peters, Gentry, and Mendelsohn 1989; Plotkin and Famolare 1992; Schwartzman 1989). Table 1 contains a summary of these studies and points up four problems with them: incompatibility of results, a tendency to examine flora (mainly) or fauna but not both, a lack of attention to sustainability, and a disproportionate attention to Latin America.

The net values vary widely, ranging from about $1 to about $420/ha/year. The variation can be explained by the biological and economic diversity of the different study sites sampled, the difference in the methods and assumptions used, and the different products studied. Even when studying the same goods, independent valuations conducted at nearly the same time have produced different results (e.g., Padoch and de Jong 1989; Peters, Gentry, and Mendelsohn 1989). Even a single plot may yield different values depending on the valuation technique used (Abeygunawardena and Wickramasinghe 1992). Future valuations should use a common method and reporting procedure to facilitate the comparison of results (Borgerhoff and Caro 1985; Johnson 1978; NSF/USAID 1989:12; Redford and Robinson 1987).

Though hunting is reputed to produce benefits comparable to those from converting forest to farmland (Panayotou and Ashton 1988; Paucar and Gardner 1981), few researchers have estimated the economic value of the fauna extracted, and, as far as we know, no one has measured the combined economic value of plants and animals. A complete economic valuation of non-timber tropical-forest products must include both fauna and flora. Ethnographies of hunting and fishing are of limited use in estimating the value of wildlife per hectare because their authors do not report both the size of the catchment area and the monetary value of the game caught.

Few researchers discuss the extent to which extractive practices are sustainable (Schwartzman 1989). Lastly, most of the studies come from Latin America, mainly the Amazon basin. We need studies from the tropical forests of Africa and Asia, where different harvesting methods, forest types, societies, and economies may yield different values (Anderson 1990, citing Nair 1985). Besides addressing these issues, researchers must discuss the reasons for their choice of study site, the sample of people studied, and the representativeness of the study period so other researchers can assess the typicality of their findings. They will need to measure the quantity, location, price, and marginal cost of the goods brought from the forest to determine their net value.

For non-marginal use or extraction rates, where extraction is sustainable, the value of the non-timber products may be measured as $\Sigma_{i=0}^{n} Q_i(P_i - C_i)$, where Q_i = the quantity of good extracted, P_i = the forest gate price of the good (which may be equal to its price under competitive market conditions with no externalities), C_i = cost of extraction and i = set of non-timber tropical products. If the extraction rates are so large as to change the prices of the products concerned, or supply a significant share of the market, then allowance needs to be made for the intra-marginal values, by measuring the consumer and producer surplus that is created by the supply of the goods. Furthermore, if the extraction rates are non-sustainable, adjustment should be made for the eventual depletion of the products by adding to C_i, a depletion premium based on the expected date of extraction. Markandya and Pearce (1987) discuss how to calculate this premium. In this context, researchers should evaluate directly the sustainability of plant and animal extraction, rather than assume that local people necessarily use the forest in an ecologically sound manner.

Even after allowing for all these corrections, this method of valuation underestimates the value of non-timber tropical forest products because it does not fully capture the benefits of its services in lean years (Arnold 1991; FAO 1989).

The Context

In few of the studies surveyed do the researchers discuss the biological representativeness of their research sites, so it remains unclear to what extent the sites resemble the forest around them. Since tropical forests are heterogeneous, different study sites will yield different information. Unless researchers discuss the biophysical characteristics of their sites, it is difficult to generalize or to compare the results of different studies. To make sure that the study site represents the surrounding forest and to get a measure of variance one should where possible choose a sample of study sites. Researchers may often be unable to select representative forest sites, but they should always discuss the reasons for their choices to enable other scholars to compare and evaluate their results.

Proper sampling of the population is necessary for a study to be generalizable (Gregory and Altman 1989; Gross 1984). The main attributes of forest-product extractors which influence harvesting should first be identified. Important attributes may include age, technology, and income: men under 17 years of age or over 50 may hunt less frequently (Hill and Hawkes 1983), teenage hunters may be less skilled (Yost and Kelley 1983), shotguns are better than blowguns and bows for certain types of kills (Stearman 1990; Yost and Kelley 1983), and richer villagers may rely less on forest goods (FAO 1989). After identifying the major variables, the researcher should conduct a census, noting the distribution of the attributes of interest, and then select a stratified random sample to ensure that the study group mirrors the entire population in these attributes (Borgerhoff and Caro 1985). Most anthropologists working with extractors of forest-products do not use stratified random samples because these often include people who are difficult to reach. Anthropological samples are usually too small for conventional statistical testing (Behrens 1990). Reviewing time-allocation studies, Minge-Klevana (1980) finds a median sample of about 50 households or persons.

Unless the researcher collects information for 12 consecutive months, sampling must cover the seasonal variations in a year. Hurtado and Hill (1990), Henley (1982:47–49), and Nietschmann (1972, 1973) show that the amount of non-timber tropical forest products collected varies with the season because of climatic changes, government regulations (Hart 1978), or seasonal work in agriculture. To avoid skewed results, researchers should randomly select for observation the same number of weeks within each month of the year and the same number of days within each week (Gross 1984). Even a continuous yearlong study may not reflect the typical composition of goods removed from the forest because the amounts of game and plants vary from year to year (Ashton 1989; Leighton and Leighton 1982; Vickers 1979). Contraction or expansion of the larger economy or a crisis in the household may also influence the degree to which people depend on forest resources. Redford and Robinson (1987) show that only 3 of the 21 studies they review on hunting in the neotropics lasted more than a year. Researchers can begin to decide whether the natural supply of non-timber tropical forest products during their study period is representative by comparing regional climatic information during the period of the study with information from past years.

Quantity

Researchers should distinguish between two types of quantities: the inventory (the stock quantity in the forest) and the flow (the quantity actually used by people). Some researchers value the inventory (e.g., Caballero, Toledo, and Argueta 1978; Peters, Gentry, and Mendelsohn 1989; Prance et al. 1987; Toledo, Caballero, and Argueta 1978), others the flow (e.g., Godoy and Feaw 1989; Schwartzman 1989), and still others value both (e.g., Padoch and de Jong 1989). For most purposes, the value of the inventory is a meaningless concept related neither to present or to sustainable use. The difference in value between the flow and the inventory can be large. Padoch and de Jong (1989) measured the potential and the actual yearly profits of non-timber tropical forest products per hectare in two separate plots in Iquitos, Peru, and found that the value of the goods extracted was only 2.5–3.5% of the value of the inventory. If we assume this relation between flow and inventory, the value reported by Peters and his coworkers for Iquitos would correspond to about $15/ha/year for flow, comparable to the flow value reported by Padoch and de Jong (1989) for a neighboring site.

The most accurate method of valuing the products extracted is to identify, count, weigh, and measure them as they enter the village each day (e.g., Bergman 1986; Dufour 1983; Hill and

Hawkes 1983; Stearman 1990; Vickers 1980; Wilkie and Curran 1991). This technique may be difficult if extractors live scattered in different villages, if they consume goods in the forest (Nietschmann 1973:150), or if they extract goods at night or before dawn. To address the problem of scattered users, researchers must observe a random sample of villages and households. To estimate the type and amount of goods consumed in the forest, they can either ask extractors or randomly follow them and record their consumption. We know of only one scholar who has recorded extraction of products from the forest at night (Scaglione 1986).

In weighing goods, researchers should use scales or balances suited to the different types of goods collected (Hart 1978; Hurtado and Hill 1987, 1990). Weighing the entire daily yield may be time-consuming, but informants can be trained to help. If it is impossible to weigh each animal, the average weight for the species can be determined and used instead. For instance, Yost and Kelley (1983) approximated the weight of adult animals hunted using the average weight for the species but reduced the weight by 60% if hunters classified the animal as young. Redford and Robinson (1990) recommended that researchers weigh a sample of at least five adults to establish the average weight of a species. If such a sample is weighed only once, it may not reflect the seasonal variation in the weight of the animals during a year. Again, if the researcher cannot count the entire catch, average weights for aggregate quantities (e.g., a basket of fish) should be estimated (Gregory and Altman 1989:148). If extractors eat part of the animal in the forest, the weight of the animal should only be estimated for the parts brought back to the village. We should not value what is eaten because that is part of the cost of hunting. When researchers do not see the animal, they can estimate its weight from hunters' reports of its size (Gregory and Altman 1989:146).

When counting and weighing extracted goods, researchers must also identify the species involved. Redford and Robinson (1987) report that the irregular use of scientific names hinders the comparison of hunting studies in the Amazon. Since species identification may be difficult, they recommended elsewhere (1990) that researchers take photographs and collect skulls from which, in combination with information on the weight and the location of the kill, taxonomists can later identify them. Researchers should also note the sex and the age of the animals and the number of fetuses of pregnant ones; such information helps in assessing the sustainability of hunting and fishing (Redford and Robinson 1990).

Through direct observations researchers can measure the extraction area. They can map the terrain covered by extractors using aerial photographs (Bergman 1980:32) or pedometers (Stearman 1990). Vickers (1980) calculated distance indirectly by measuring the time hunters were out in the forest; he found that hunters on average traveled 2–3 km per hour. Researchers can also follow extractors as they forage or question them about the location of their activities, preferably at the end of each day (Vickers 1983: 457). They can also ask them to pinpoint the locations of their kills on maps; Vickers found an informant who gave him a detailed map of local hunting trails. A new device for measuring the size of the extraction area is the Global Positioning System (GPS) receiver, which picks up satellite signals to give its users their ground coordinates (Baksh 1991; Behrens and Sever 1991).

Researchers are limited in the number of persons they can directly observe, but they can increase the amount of information they can gather by questioning extractors about their activities and by training them to help gather information. With literate populations, researchers can train extractors to keep diaries on the results and locations of extractive expeditions and on the allocation of their time in the forest. White (1976), Bergman (1986), Alcorn (1981), Dufour (1983), Padoch and de Jong (1989), and Stephen Beckerman (pers. comm.) trained informants to keep diaries, and extractors have been taught to weigh and measure the daily catch (Bergman 1986:32; Nietschmann 1973:21; Stocks 1983; Wilkie 1989).

Although extractors may give accurate information, one should avoid relying too heavily on their accounts. Distortions arise because people may be careless in their record keeping, lie to the researcher (Nacham 1984), or forget past events. The accuracy of recall information drops when people are asked to remember events far in the past (Bernard et al. 1984). Besides errors from carelessness, lying, and faulty memory, people bias information by reporting culturally significant events or information they believe interviewers want to hear. For instance, Dufour (pers. comm.) says that the Tucano of Colombia do not report all the game and fish caught because they eat large animals in secrecy to avoid customary food-sharing norms. Several researchers (Ayres and Ayres 1979; Redford and Stearman 1990; Smith 1976) say that hunters under-report kills of smaller animals (e.g., birds) because they consider them unimportant. Extractors may also under-report plants or animals for use in the household rather than for sale because the market is seen as more important. Through training and special interviews, researchers can increase the quality of the information reported (Vickers 1984; Yost and Kelley 1983).

Incentives to enhance recall information may introduce distortions. Wilkie (1989) paid informants a wage to show him their catch that was "sufficient to encourage reporting of animals killed but not so much as to promote hunting purely for the reward." He checked hunters' reports by questioning other villagers but concluded that having hunters keep records probably produced selective reporting and underreporting of many animals commonly eaten (Wilkie 1989). After reviewing the accuracy of informant recall, Bernard and his coworkers concluded that "on average, about half of what informants report is probably incorrect in some way" (Bernard et al. 1984). Other researchers echo this conclusion (Bernard et al. 1986; Borgerhoff and Caro 1985; Gross 1984).

When using oral or written reports from extractors, researchers should check the information to ensure that it is reliable. Diaries and other written records should be checked regularly (e.g., Bergman 1986:32; Stocks 1983). Besides failures to complete such records, Bergman (1986:32) identified errors in time-allocation diaries from his memory of what he had seen people do. Yost and Kelley (1983) trained assistants to collect information on hunting activities from all villagers and then cross-referenced extractors' reports with each other and with their own report. Researchers can also randomly select informants for observation in the forest and compare their own records with the extractors' diaries or with the information they provide in interviews at the end of the day.

Methods to correct for errors in extractors' reports are less developed, but several deserve attention. Filling diaries throughout the day reduces misinformation. Engle (pers. comm.) used three techniques to enhance recall from interviews: eliciting the entire story, reviewing it with the informant, and presenting the informant with stories from other individuals (e.g., "What did you do this morning? Here is what X did"). Recall enhancement techniques increased the number of activities recalled from 40% to 63% and the amount of time accurately recalled from 58% to 70% (Engle and Lumpkin, pers. comm.). Similarly, Bernard et al. (1984) describe studies showing that memory enhancement techniques increase accuracy by 5–10%. Their studies also reveal that groups recall details of events better than individuals (pp. 510–511). Devices for enhancing recall might also be used. For example, extractors might be asked at the end of each day to select cards, each stamped with the name or with a photograph of a non-timber product. Pointing to the card, the researcher could prompt the collector: "Do you remember how much of this you took out today?"

Whether researchers get information on quantities of forest products extracted from direct observations or from extractors, they should adjust this information by the losses from spoilage during storage in the village and transport to the market.

Price

The most difficult part of valuation is assigning the products a monetary value. Optimal-foraging theorists use expenditures of labor or energy as a measure of value in nonmarket societies (Smith 1983) but these two measures can produce different results (Gregory and Altman 1989:150–151). Moreover, neither is consistent with modern valuation theory as applied in this area.

The best method of valuation is to make use of the prices that exist of the commodity concerned, or that prevail in related markets. For example, a non-timber product may have no market but may be bartered for a product that does. In that case it is easy to impute a price for the product in which we are interested. Another method could be to use the value of a close substitute with a price. This will require establishing a relative price between the priced and unpriced products, which can be done on the basis of product characteristics. If neither of these methods is feasible, the researcher can resort to a contingent valuation approach where users of the product are asked what is their willingness to pay for the product. Contingent valuation methods have been widely tested in developed countries with strong market traditions and their use in situations where familiarity with, and acceptance of, markets is much more limited should be carried out with great care (Mitchell and Carson 1989). Fortunately it is rarely necessary to resort to contingent valuation methods for most forest products.

Few researchers have measured the shares of non-timber tropical forest products going to the household and to the market (Behrens 1986; Nietschmann 1973). The distinction is important because goods for the market and for the home should be valued differently. Products consumed at home or exchanged with kin should be valued at their retail purchasing price in the village; goods sold on the market should be valued at their selling price (Chibnik 1978; Painter 1986). If the population is literate, a random sample of households can be selected to keep log books of daily income, expenses, and amounts of non-timber tropical forest products sold and consumed at home. In a Philippine village, Hayami (1978) selected 11 households to keep daily records of income, expenses, and uses of production; the records were checked twice a week.

Since the purpose of the exercise is to make a judgement of the valuation of the tropical forest products from a national viewpoint, the analyst must always try to estimate the total economic value of the products. This means adjusting for any taxes and subsidies that may cause the price to deviate from the opportunity cost of the resource. Such adjustments are referred to as shadow prices. For most standard resources (e.g., minerals, capital, labor) these shadow prices are estimated for most countries, and revised periodically. The full set of adjustments that have to be made will depend at which stage of the product cycle the valuation is being carried out. If it is at the forest gate then only those adjustments that are relevant to that stage need to be made (Godoy and Bennett 1991; Godoy and Feaw 1989). If the valuation is being carried out where the product is nationally sold, then a number of additional adjustments may have to be made. As Fearnside (1989) says, shadow pricing is important in providing an economic rationale for extractive resources, which may not be financially profitable. For a detailed discussion of shadow pricing in the appraisal of projects see Squire and van der Tak (1975).

Besides correcting for taxes and subsidies, a full exercise in shadow pricing also requires estimating the externalities of extracting non-timber forest products. In no study have researchers valued the environmental externalities of extracting non-timber tropical forest products. By failing to quantify these factors, conventional valuations, such as those in Table 1, underestimate the economic benefits of these products to society.

Marginal Costs

To get an accurate measure of the marginal costs of extracting and processing non-timber tropical forest products, the researcher must calculate the cost of the materials (e.g., bullets, arrows) used, the labor time directly associated with finding, extracting, processing, and transporting the goods from the forest to the village or to the market, and the temporal cost of resources—the benefits forgone by delaying the sale or the use of the good.

Researchers have not often calculated the cost of the materials used in extraction, and this cost can be high (Bailey and Aunger 1989; Moran 1991; Nietschmann 1972, 1973:153–154; Ross 1978). In studying the Waorani hunters of Ecuador, Yost and Kelley (1983) found that the cost of making goods to trade for shotgun shells and travelling to buy them was three hours per shell.

Studies based on optimal-foraging theories give detailed information on the time spent searching for, pursuing, killing, and processing game (Hawkes and O'Connell 1981; Hill and Hawkes 1983; Smith 1983), but students of plant collection have not included comparable information. To calculate the gross time spent in the forest, researchers can log departures from and returns to the village (Beckerman 1980; Hill and Hawkes 1983; Hill and Kaplan 1989; Hurtado and Hill 1990) or train villagers to do this (Hill and Kaplan 1989; Hurtado and Hill 1990). These methods overestimate the time spent procuring goods because people use part of the time in the forest to farm, eat, or rest (Henley 1982:69; Hill and Hawkes 1983). To gather information on time allocated for extraction and processing, the researcher must either rely on recall information or follow a random sample of villagers in the forest for direct observation.

Researchers can interview extractors at the end of the day to decide how they spent their time in the forest (Gross 1984; O'Connell and Hawkes 1984; Vickers 1980; Yost and Kelley 1983). Extractors can also be taught to record how much time they spent at different activities in the forest,

TABLE 1. SURVEY OF OPPORTUNITY COST OF FOREST FROM FORAGING.

Location	Net value ($/ha/year)	Comments	Source
Venezuela	0.75	Experimental caiman harvest.	Thorbjarnarson (1991)
Korup National Park, Cameroon	1.06	Net present value of forgone gross benefits from hunting = $2.7 m at a 5% discount rate; area = 126 000 ha; hunting said to be non-sustainable.	Ruitenbeek (1988) and Infield (1988)
Mudumalal Sanctuary, South India	3.0	0.02 elephants/ha at $1500 per elephant. Excludes costs of domestication and training. Price refers to a domesticated animal. Assume a 10% discount rate.	Sukumar (1989) and personal communication
Ituri Forest, Zaire	0.50–3.18	318 kg of game/km^2 of primary forest or 50 kg/km^2 in climax forest at $1/kg. Estimate leaves out costs. Price is for prized meats.	Wilkie (1989) and Wilkie and Curran (1991)
Amazon, Brazil	4.8	Estimate is gross return/ha/year; only values flora.	Schwartzman (1989)
Sarawak, East Malaysia	8.00	Values wildlife in one square kilometer.	Caldecott (1987)
Maya Biosphere Reserve, Guatemala	10.00	550 000 hectares produce about $5.5 million/year in exports of chicle, xate palm, and allspice; gross value.	Nations (1992)
Western Amazonia	5–16	Gross value varies by the size of the extraction area (150 to 300 hectares).	Hecht (1992)
Cross River National Park	16.50	Yearly income from hunting gathering, trapping = $826, Naira/person = $108; population = 38 300; area = 250 000 hectares. Gross value.	Ruitenbeek (1989)
Iquitos, Peru	16–22	Based in part on villager's diaries.	Padoch and de Jong (1989)
Iquitos, Peru	20	Potential value of about six species of latex and fruits.	Pinedo-Vasques, Zarin and Jipp (1992)
Belize	36–166	Medicinal plants; 5% discount rate.	Balick and Mendelsohn (1992)
Hantana, Sri Lanka	50	50 randomly chosen household surveyed in three villages; used contingent valuation and opportunity cost approach; estimate leaves out cost of extraction.	Abeygunawardena and Wickramasinghe (1992)
Kalimantan, Indonesia	53	Net present value of cultivated rattan = $529/ha over 25 years with a real discount rate of 10%.	Godoy and Feaw (1989)
Amazon, Brazil	59	Includes kernel, charcoal, and feed meal of babassu palm. Unclear whether returns are net or gross.	Anderson, May and Balick (1991)

TABLE 1. CONTINUED.

Location	Net value ($/ha/year)	Comments	Source
Combú Island, Guamá River, Brazil	79	$3171/year/family; average of 5 families in 1984–1988. Estimate is gross value and only includes semi-wild cacao, açaí, and rubber. We assume one family worked 40 hectares of forest.	Anderson and Ioris (1992)
Tamilnadu, India	80	Fuel and fodder.	Appasamy (1993)
Brazil	97	Value of Brazil nuts only ($/hectare): Collector's price = 97; Exporter's price = 176; Retail price = 1059.	Mori (1992)
Pará, Brazil	110	Value after selective thinning of competitors and pruning of açaí palm.	Anderson (1990b)
Veracruz, Mexico	116	Estimate leaves out lumber and coffee.	Alcorn (1989)
Amazon, Ecuador	120	Values wildlife over 500 square kilometers.	Paucar and Gardner (1981)
India	117–144	Gross benefits; includes fruits, herbs, and medicinal plants.	Chopra (1993)
Jenaro Herrera, Peru	167	Values only harvest of wild *camu camu*; unclear if estimate is gross or net.	Peters quoted in Vásquez and Gentry (1989)
Iquitos, Peru	420	Values the inventory in one hectare; only includes plants.	Peters, Gentry and Mendelsohn (1989)

either at the end of the day or in the course of it. In a study of fishing among the Cocamilla, Stocks (1983) gave out watches and trained local people to record each day the time spent travelling and working; he checked and copied the records every day. Bergman (1986:30) gave Shipibo men diaries and wristwatches and instructed them to keep track of their time. Never having owned watches, most of these men had to learn to tell time, and some of them had to find someone else to write for them. Bergman reviewed and copied the diaries twice a week.

To measure the allocation of time directly, researchers generally use instantaneous sampling. Pioneered by Johnson (1975), the method requires visiting randomly chosen subjects at random times of the day and recording their activity at the instant they are first seen. By collecting many spot observations it is possible to gather a great deal of information in a short time and to estimate the proportion of time people devote to different activities. Instantaneous sampling is not as useful for studying time allocation in the forest. Hames (1980) used instantaneous sampling to study the Ye'Kwana and Yanomamo of Amazonian Venezuela but found it impossible to observe individuals who were out of the village; instead he asked relatives who had stayed behind about the activities of those who had left (see also Colfer 1981). Behrens (1986) says he could not observe individuals out of the village and therefore had to rely on recalls, hearsay, and a larger sample.

Rather than using instantaneous sampling to study the uses of time in the forest, researchers can use focal subject sampling. With this method, the researcher spends a full day following one randomly chosen individual, noting the amount of time spent in different activities (Hill and Hawkes 1983; Hill and Kaplan 1989; Hurtado and Hill 1987). Following extractors in the forest also yields information on the amount of forest products consumed in the forest that will allow correction of the estimates of quantities weighed in the village (Hill and Kaplan 1989). Focal subject sampling will allow the researcher to estimate the time spent by the focal individual in different activities, but if an extractor works in a group it will not permit a reliable estimate of total labor cost. In such cases, the unit of observation should be the group, and the researcher should record the activities of everyone in it. If the group is large or if it splits up, the researcher will have to observe as many individuals as possible and learn about the others from diaries or from interviews at the end of the day.

Whether one follows extractors or relies on diaries and interviews, one should decide on the categories for classifying labor activities in the forest to ensure that the results of the study will be comparable to those of others. The following categories may be useful for classifying extractive activities: travelling, resting, searching, harvesting, pursuing, and processing (Hill and Hawkes 1983).

Having measured the extractors' allocation of time in the forest, researchers must also assign a monetary value to the labor time. Some researchers (e.g., Peters, Gentry, and Mendelsohn 1989) use the country's official wage as an estimate of the unprotected rural wage, but this should not be done uncritically. Instead, researchers should use what people actually pay each other. Such payments can take many forms and can be complex, including not only money wage but also gifts of food (Godoy and Feaw 1989; Godoy and Bennett 1991). Although rural wages vary by season, age, sex, and type of work, most of the studies in Table 1 use only one wage. In future valuation studies, the flow of goods from the forest over time should be matched to the wage prevailing at the time the good is extracted.

In calculating the marginal costs of non-timber tropical-forest products extraction, the cost of capital is seldom measured. Discounting is important when extraction and consumption or sale occur at very different times (Timmer, Falcon, and Pearson 1983). Though private time preferences vary by age, length of waiting period, size of reward, and income (Thaler 1981; Rogers, pers. comm.; Lawrence 1991), the most appropriate practice at the present time is to use an estimate of the social discount rate. The literature on how to estimate this rate is complex (Markandya and Pearce 1991), but the consensus is that the social discount rate is much lower than the market rate of discount in developing countries. The latter can be as high as 15–20%, whereas the social rate of discount is more likely to be around 3–6%. For many countries some estimates of the social discount rate have been made. Where no such estimates exist, the researcher can take a value of 4–5% as representative of the range of numbers that have been established.

Sustainability

Estimating the economic value of non-timber tropical forest products at one point in time leaves unanswered the question of to what extent the estimated value is sustainable. Scholars are divided into at least three camps on this issue. Some say that indigenous people manage non-timber tropical forest products sustainably; others say that they do not, and a third camp says that sustainability is the result of special conditions that must be identified in each case.

Anthropologists who say that indigenous people manage the forest sustainably (e.g., Dufour 1990; Posey, Frechione, and Edding 1984; Vickers 1984) point to practices reportedly designed to enhance production, such as manipulating forest fallow to speed the growth of desirable plants (Irvine 1989; Padoch et al. 1985; Posey 1985, 1989) and animals (Stocks 1987), establishing corridors and scattered gardens and spacing human settlements to stimulate reproduction of wildlife (Linares 1976; Lovejoy and Schubart 1980; Nations and Nigh 1980; Redford and Robinson 1987), mandating the hunting of one animal species to relieve hunting pressure on others (Balée 1985), tabooing the killing of threatened animals, and forbidding hunting and fishing in depleted zones (Ross 1978; McDonald 1977). Other researchers (e.g., Hart 1978; Henley 1982: 50–52; Jessup and Peluso 1986; Johnson 1989; Redford and Robinson 1985, 1987) argue that depletion of forest resources may have started before contact and continues to increase as markets envelop indigenous societies (Gross et al. 1979; Nietschmann 1972, 1973; Roosevelt 1980; Smith 1983). Johnson (1989), Winterhalder (1983), and Hames (1987) suggest that low population density, localized resources, effective social sanctions, and limited possibilities for expansion encourage conservation or intensification of resource use. Redford and Robinson (1985) say that sustainability must be assessed directly by counting animals and plants in the forest.

Hall and Bawa (1993) discuss methods for assessing the biological sustainability of plant extraction. To measure the sustainability of hunting, anthropologists have used indirect measures, such as increases in the distance, frequency, and duration of hunts, increases in the capture of smaller, less desirable animals, movement of human settlements, and people's recall of yields over time (e.g., Stearman 1990; Stocks 1983). The only way to get direct measures of hunting sustainability is to compare rates of game killed in hunting with rates of natural reproduction and mortality from censuses of animal populations in the wild (Redford and Robinson 1985, 1990). As far as we know, researchers have not yet done this analysis.

Where use patterns are not sustainable, the valuation of current extraction rates should allow for that by lowering the value of the non-forest products. The appropriate adjustment is a depletion premium, which will depend on how long the present extraction can continue, and on the discount rate.

Conclusions

Valuation of non-timber tropical forest products is plagued by inadequate measurement of costs, quantities extracted, and prices. Although researchers have produced important case studies, the results of different studies cannot be directly compared because different methods have been used. Most such valuations seem to have been an afterthought. Researchers must pay more attention to problems of method if future valuation studies are to produce generalizable results.

Although there is no substitute for direct observation over several years, the best method for an economic valuation study will depend on the cultural background of the people studied, the level of literacy, and the settlement pattern. The resources and the time available to undertake the study also influence the choice of method. Mindful that diaries and interviews can produce inaccurate information, researchers must use them because they can save time and money. Diaries, if feasible, are potentially more accurate than interviews, but interviews are still important to check on them as well as to give researchers greater flexibility in questioning extractors.

Few scholars have directly measured the sustainability of non-timber tropical forest product extraction (e.g., Hall and Bawa 1993; Reining et al. 1991). Future valuations should include censuses of the plants and animals removed by humans to assess the effects of harvesting on the availability of these goods. Scholars should not assume or infer from direct information that local people manage the forest sustainably.

Most valuation studies have been done by researchers interested in the flora rather than in the fauna of the forest. Ethnobotanists have calculated returns per hectare; students of hunting have calculated returns per unit of labor. Flora and fauna per unit of area must be valued together, by a single method. Valuation of non-timber tropical forest products should begin to pull together scholars from different disciplines. Future valuations will need to combine the skills of, at least, a natural-resource economist or economic anthropologist to price the goods and to measure the flow of goods removed from the forest and a zoologist and a botanist to determine the sustainability of extraction. Indigenous people and local scholars must be recruited to assist in the assessment.

Acknowledgments

This research was partially financed by grants to Harvard University from the Rockefeller Foundation, the John Merck Fund, and the W. Alton Jones Foundation. We thank the following for providing information on their research methods: Paul Appasamy, Stephen Beckerman, Kanchan Chopra, Wade Davis, Patrice Engle, Daniel Gross, Kristen Hawkes, Kim Hill, Stephen Hubbell, Allen Johnson, and William Vickers.

Literature Cited

Abeygunawardena, P., and W. A. R. Wickramasinghe. 1992. An economic evaluation of non-timber products of Hantana forest. Pages 183–190 *in* H. P. M. Gunagina, ed., Multi-purpose tree species in Sri Lanka: research and development. Proceedings of the Second Regional Workshop, Organized by National Research Committee in Multi Purpose Trees Species. Winrock International, F/FRED. Bangkok, Thailand, April 5th to 7th 1992.

Alcorn, J. 1981. Huastec noncrop resource management: implications for prehistoric rain forest management. Human Ecology 9:395–417.

———. 1989. An economic analysis of Huastec Mayan agroforestry management. Pages 182–206 *in* J. G. Browder, ed., Fragile lands of Latin America: strategies for sustainable development. Westview Press, Boulder, CO.

Anderson, A. B. 1990. Extraction and forest management by rural inhabitants in the Amazon estuary. Pages 65–85 *in* A. B. Anderson, ed., Alternatives to deforestation: steps toward sustainable use of the Amazon rain forest. Columbia University Press, New York.

———**, and E. M. Ioris.** 1992. The logic of extraction: resource management and income generation by extractive producers in the Amazon estuary. Pages 175–199 *in* K. H. Redford and C. Padoch, eds.,

Conservation of neotropical forests: working from traditional resource use. Columbia University Press, New York.

———, **P. May, and M. Balick.** 1991. The subsidy from nature: palms, forests, peasantry, and development on an Amazon Frontier. Columbia University Press, New York.

Appasamy, Paul. 1993. Role of non-timber forest products in a subsistence economy: the case of a joint forestry project in India. Economic Botany 47:258–267.

Arnold, J. E. M. 1991. Policy issues related to the role of trees in rural income and welfare security. *In* Forestry and agroforestry policy research needs. International Food Policy Research Institute, Washington, DC.

Ashton, P. A. 1989. Dipterocarp reproductive biology. Pages 219–240 *in* H. Lieth and M. J. A. Werger, eds., Tropical rain forest ecosystems. Elsevier Science Publishers, Amsterdam.

Ayres, J. M., and C. Ayres. 1979. Aspectos de caça no alto Rio Aripuaná. Acta Amazonica 9:287–298.

Bailey, R. C., and R. Aunger, Jr. 1989. Subsistence strategies among BaMbuti Pygmies: net hunters and archers. Human Ecology 17:273–297.

Baksh, M. 1991. Applications of a hand-held GPS receiver in South America rain forests. Pages 227–236 *in* C. A. Behrens and T. L. Sever, eds., Applications of space-age technology in anthropology: proceedings of a conference held at the John C. Stennis Space Center, Mississippi, November 28th 1990.

Balée, W. 1985. Ka'apor ritual hunting. Human Ecology 13:485–510.

Balick, M., and R. O. Mendelsohn. 1992. Assessing the economic value of traditional medicines from tropical forests. Conservation Biology 6:128–130.

Beckerman, S. 1980. Fishing and hunting by the Bari of Colombia. Pages 67–110 *in* R. Hames, ed., Studies of hunting and foraging in the neotropics. Bennington College Working Papers on South American Indians 2, Bennington, Vermont.

Behrens, C. 1986. The cultural ecology of dietary change accompanying changing activity patterns among the Shipibo. Human Ecology 14:367–396.

———. 1990. Qualitative and quantitative approaches to the analysis of anthropological data: a new synthesis. Journal of Quantitative Anthropology 2:305–328.

———, **and T. L. Sever, eds.** 1991. Applicators of space-age technology in anthropology: proceedings of a conference held at the John C. Stennis Space Center, Mississippi, November 28th 1990.

Bergman, R. W. 1986. Amazon economics: the simplicity of Shipibo Indian wealth. University Microfilms International, Ann Arbor.

Bernard, H. R., P. Killworth, D. Kronenfeld, and L. Sailer. 1984. The problem of informant accuracy: the validity of retrospective data. Annual Review of Anthropology 13:495–517.

———, **Pertti J. Pelto, Oswald Werner, James Boster, A. Kimball Romney, Allen Johnson, Carol R. Ember, and Alice Kasakoff.** 1986. The construction of primary data in cultural anthropology. Current Anthropology 27:382–396.

Borgerhoff, M., and T. M. Caro. 1985. The use of quantitative observational techniques in anthropology. Current Anthropology 26:323–335.

Caballero, J., V. M. Toledo, and A. Argueta. 1978. Flora útil o el uso tradicional de plantas. Biótica 3:103–104.

Caldecott, J. 1987. Hunting and wildlife management in Sarawak, Malaysia. World Wildlife Fund, Washington, DC.

Chibnik, M. 1978. The value of subsistence production. Journal of Anthropological Research 34:561–576.

Chopra, K. 1993. The value of non-timber forest products: an estimate from India. Economic Botany 47:251–257.

Colfer, C. J. 1981. Women, men, and time in the forest of East Kalimantan. Borneo Research Bulletin 13:75–83.

de Beer, J. H., and M. J. McDermott. 1989. Economic value of non-timber forest products in Southeast Asia. Council for the International Union of the Conservation of Nature, The Hague.

Dove, M. 1983. Swidden agriculture, or the political economy of ignorance. Agroforestry Systems I:85–99.

Dufour, D. L. 1983. Nutrition in the northwest Amazon: household dietary intake and time-energy expenditure. Pages 329–356 *in* Raymond B. Hames and William T. Vickers, eds., Adaptive responses of native Amazonians. Academic Press, New York.

———. 1990. Use of tropical rainforest by native Amazonians. BioScience 40:652–659.

Dunn, F. L. 1975. Rainforest collectors and traders: a study of resource utilization in modern and ancient Malaya. Monographs of the Malayasian Branch Royal Asiatic Society 5.

FAO. 1989. Household, food security, and forestry: an analysis of socio-economic issues. FAO, Rome.

Fearnside, P. M. 1989. Extractive reserves in Brazilian Amazonia. Bioscience 39:387–393.

Godoy, R. 1992. Organizing principles in the valuation of tropical forest. Forest Ecology and Management 50:171–180.

———, **and C. P. A. Bennett.** 1991. The economics of monocropping and intercropping by smallholders: the case of coconuts in Indonesia. Human Ecology 19:83–98.

———, **and T. C. Feaw.** 1989. The profitability of smallholder rattan cultivation in central Borneo. Human Ecology 7:397–420.

Gregory, C. A., and J. C. Altman. 1989. Observing the economy. Routledge, London.

Gross, D. R. 1984. Time allocation: a tool for the study of cultural behavior. Annual Reviews of Anthropology 13:519–558.

———, **E. George, N. M. Flowers, F. M. Leoi, M. L. Ritter, and D. W. Werner.** 1979. Ecology and acculturation among native peoples of central Brazil. Science 204:1043–1050.

Hall, P., and K. Bawa. 1993. Methods to assess the impact of extraction of non-timber tropical forest products on plant populations. Economic Botany 47:234–247.

Hames, R. 1980. Game depletion and hunting zone rotation among the Ye'Kwana and Yanomamo of Amazonia, Venezuela. Pages 42–68 *in* R. Hames, ed., Studies of hunting and foraging in the neotropics. Bennington College Working Papers on South American Indians 2, Bennington, Vermont.

———. 1987. Game conservation or efficient hunting? Pages 92–102 *in* B. J. MacCay and J. H. Acheson, eds., The question of the commons: the culture and ecology of communal resources. University of Arizona Press, Tucson.

Hart, J. 1978. From subsistence to market: a case study of the Mbuti net hunters. Human Ecology 6:325–55.

Hawkes, K., and J. F. O'Connell. 1981. Affluent hunters? Some comments in light of the Alyawara case. American Anthropologist 83:622–26.

Hayami, Y. 1978. Anatomy of a peasant economy: a rice village in the Philippines. International Rice Research Institute, Manila.

Hecht, S., A. Anderson, and P. May. 1988. The subsidy from nature: shifting cultivation, successional palm forests, and rural development. Human Organization 47:25–35.

———. 1992. Valuing land use in Amazonia: colonist agriculture, cattle, and petty extraction in comparative perspective. Pages 379–399 *in* K. H. Redford and C. Padoch, eds., Conservation of neotropical forests: working from traditional resource use. Columbia University Press, New York.

Henley, P. 1982. The Panare: tradition and change in an Amazon frontier. Yale University Press, New Haven.

Hill, K., and H. Kaplan. 1989. Population and dry subsistence strategies of the recently contacted Yura of Peru. National Geographic Research 5:317–334.

———, **and K. Hawkes.** 1983. Neotropical hunting among the Aché of eastern Paraguay. Pages 139–188 *in* R. B. Hames and W. T. Vickers, eds., Adaptive responses of native Amazonians. Academic Press, New York.

Hurtado, M. A., and K. R. Hill. 1987. Early dry season subsistence ecology of Cuiva (Hiwi) foragers of Venezuela. Human Ecology 15:163–183.

———, **and** ———. 1990. Seasonality in a foraging society: variation in diet, work effort, fertility, and sexual division of labor among the Hiwi of Venezuela. Journal of Anthropological Research 46:293–315.

Infield, M. 1988. Hunting, trapping, and fishing in villages within and on the periphery of the Korup national park. World Wide Fund for Nature Publication 3206/A9.6, London.

Irvine, D. 1989. Succession management and resource distribution in an Amazonian rain forest. Advances in Economic Botany 7:223–237.

Jessup, T. C., and N. L. Peluso. 1986. Minor forest products or common property resources in East Kalimantan, Indonesia. Pages 505–532 *in* Proceedings of the Conference on Common Property Resource Management. National Academy Press, Washington, D.C.

———, **and A. P. Vayda.** 1989. Dayaks and forests of interior Borneo. Expedition 30:5–17.

Johnson, A. 1975. Time allocation in a Machiguenga community. Ethnology 14:301–310.

———. 1978. Research methods in social anthropology. Edward Arnold, London.

———. 1989. How the Machiguenga manage resources: conservation or exploitation of nature? Advances in Economic Botany 7:213–222.

Lawrence, E. C. 1991. Poverty and the rate of time preference: evidence from panel data. Journal of Political Economy 99:54–77.

Leighton, M., and D. R. Leighton. 1982. Vertebrate response to fruiting seasonality within a Bornean rain forest. Pages 181–196 *in* E. C. Leigh, Jr., ed., The ecology of a tropical forest. Smithsonian Institution Press, Washington, DC.

Linares, O. 1976. Garden hunting in the American tropics. Human Ecology 4:331–349.

Lovejoy, T. E., and H. O. R. Schubart. 1980. The ecology of Amazonian development. Pages 21–26 in F. Barbira-Scazzocchio, ed., Land, people, and planning in contemporary Amazonia. Cambridge University Press, Cambridge.

Markandya, A., and D. W. Pearce. 1987. Marginal opportunity cost as a planning concept in natural resource management. Annals of Regional Science 21:18–32.

———, **and** ———. 1991. Development, environment, and the rate of discount. World Bank Research Observer 6:137–152.

McDonald, D. 1977. Food taboos: a primitive environmental protection agency in South America. Anthropos 72:734–748.

Minge-Klevana, W. 1980. Does labor time decline with industrialization? A survey of time-allocation studies. Current Anthropology 21:279–290.

Mitchell, R. C., and R. T. Carson. 1989. Using survey to value public goods: the contingent valuation method. Resources for the Future, Washington, DC.

Moran, E. 1991. Human adaptive strategies in Amazonian backwater frontier. American Anthropologist 93:361–382.

Mori, S. A. 1992. The Brazil nut industry: past, present, and future. Pages 241–251 in M. J. Plotkin and L. M. Famolare, eds., Sustainable harvest and marketing of rain forest products. Island Press, Washington, DC.

Nacham, S. R. 1984. Lies my information told me. Journal of Anthropological Research 40:536–553.

Nair, P. K. R. 1985. Classification of agroforestry systems. Agroforestry Systems 3:97–128.

Nations, J. D. 1992. Xateros, chicleros, and pimenteros: harvesting renewable tropical forest resources in the Guatemalan Petén. Pages 208–219 *in* K. H. Redford and C. Padoch, eds., Conservation of neotropical forests: working from traditional resource use. Columbia University Press, New York.

———**, and R. B. Nigh.** 1980. The evolutionary potential of Lacandon Maya sustained yield tropical forest agriculture. Journal of Anthropological Research 36:1–30.

NSF/USAID. 1989. Funding priorities for research toward effective sustainable management of biodiversity research in tropical Asia: report of a workshop sponsored by the National Science Foundation and the United States Agency for Development, Bangkok, March 27–30, 1989.

Nietschmann, Bernard. 1972. Hunting and fishing focus among the Miskito Indians of eastern Nicaragua. Human Ecology 1:41–67.

———. 1973. Between land and water: the subsistence ecology of the Miskito Indians of eastern Nicaragua. Seminar Press, New York.

O'Connell, J. F. O., and K. Hawkes. 1984. Food choice and foraging sites among the Alyawara. Journal of Anthropological Research 40:504–535.

Padoch, C., and W. de Jong. 1989. Production and profit in agroforestry: an example from the Peruvian Amazon. Pages 102–113 in J. G. Browder, ed., Fragile lands of Latin America: strategies for sustainable development. Westview Press, Boulder, CO.

———**, J. Chota, W. de Jong, and J. Unruh.** 1985. Amazonian agroforestry: a market-oriented system in Peru. Agroforestry Systems 3:47–58.

Painter, M. 1986. The value of peasant labor power in a prolonged transition to capitalism. Journal of Peasant Studies 13:221–239.

Panayotou, T., and P. Ashton. 1988. The case for multiple use management of tropical forests. International Tropical Timber Organization, Yokohama.

———**, and** ———. 1992. Not by timber alone: sustaining tropical forests through multiple use management. Island Press, Covelo, CA.

Paucar, A., and A. Gardner. 1981. Establishment of a scientific research station in the Yasuni national park of the republic of Ecuador. National Zoo, Washington, DC.

Peluso, N., and M. Poffenberger. 1989. Social forestry in Java: reorienting management systems. Human Organization 48:333–344.

Peters, C., A. H. Gentry, and R. O. Mendelsohn. 1989. Valuation of an Amazonian rainforest. Nature 339: 655–656.

Pinedo-Vásquez, M., D. Zarin, and P. Jipp. 1992. Economic returns from forest conversion in the Peruvian Amazon. Ecological Economics 6:163–173.

Plotkin, M. J., and L. M. Famolare, eds. 1992. Sustainable harvest and marketing of rain forest products. Island Press, Washington, DC.

Posey, D. A. 1985. Native and indigenous guidelines for new Amazonian development: understanding biodiversity through ethnoecology. Pages 156–181 *in* J. Hemming, ed., Change in the Amazon Basin, Vol. I: Man's impact on forest and rivers. Manchester University Press, Manchester.

———. 1989. A preliminary report on diversified management of tropical forest by the Kayapo Indians of the Brazilian Amazon. Advances in Economic Botany 7:112–126.

———**, J. Frechione, and J. Edding.** 1984. Ethnoecology as applied anthropology in Amazonian development. Human Organization 43:95–106.

Prance, G., W. Balée, B. M. Boom, and R. L. Carneiro. 1987. Quantitative ethnobotany and the case for conservation in Amazonia. Conservation Biology 1:296–310.

Redford, K., and J. G. Robinson. 1985. Hunting by indigenous peoples and conservation of their game species. Cultural Survival 9:41–44.

———**, and** ———. 1987. The game of choice: patterns of Indian and colonist hunting in the neotropics. American Anthropologist 89:650–667.

———**, and** ———. 1990. A research agenda for studies of subsistence hunting in the neotropics. Florida Journal of Anthropology 6:117–120.

———**, and A. Stearman.** 1990. Local peoples and the Beni Biosphere Reserve, Bolivia. Vida Silvestre Neotropical 2:49–56.

Reining, C. C. S., R. M. Heinzman, M. C. Madrid, S. López, and A. Solórzano. 1991. The non-timber forest product industries of the Maya Biosphere Reserve, Guatemala: results of ecological and socio-economic surveys. Unpublished report. Conservation International Foundation, Washington, DC.

Roosevelt, A. 1980. Parmana: prehistoric maize and manioc subsistence along the Amazon and Orinoco. Academic Press, New York.

Ross, E. B. 1978. Food taboos, diet and hunting strategy: the adaptation to animals in Amazonian cultural ecology. Current Anthropology 19:1–16.

Ruitenbeek, H. J. 1988. Social cost-benefit analysis of the Korup project, Cameroon. World Wide Fund for Nature Publication 3206/A14.1, London.

———. 1989. Economic analysis of issues and projects relating to the establishment of the proposed cross river national park (Oban Division) and support zone. World Wide Fund for Nature, Oban Feasibility Study. London, n.d.

Scaglione, R. 1986. The importance of night time observations in time allocation studies. American Ethnologist 13:537–545.

Schwartzman, S. 1989. Extractive reserves in the Amazon. Pages 150–163 *in* J. G. Browder, ed., Fragile lands of Latin America: strategies for sustainable development. Westview Press, Boulder, CO.

Smith, E. A. 1983. Anthropological applications of optimal foraging theory: a critical review. Current Anthropology 24:625–651.

Smith, N. J. H. 1976. Utililzation of game along Brazil's Trans-Amazon Highway. Acta Amazónica 6:455–466.

Squire, L., and H. G. van der Tak. 1975. Economic analysis of projects. Johns Hopkins University Press, Baltimore.

Stearman, A. 1990. The effects of settler incursion on fish and game resources of the Yuqui, a native Amazonian society of eastern Bolivia. Human Organization 49:373–385.

Stocks, A. 1983. Cocamilla fishing: patch modification and environmental buffering in the Amazon Varzea. Pages 239–268 *in* R. B. Hames and W. T. Vickers, eds., Adaptive responses of native Amazonians. Academic Press, New York.

———. 1987. Resource management in an Amazon Varzea Lake ecosystem: the Cocamilla case. Pages 108–120 *in* B. J. MacCay and J. H. Acheson, eds., The question of the commons: the culture and ecology of communal resources. University of Arizona Press, Tucson.

Sukumar, R. 1989. The Asian elephant: ecology and management. Cambridge University Press, Cambridge.

Thaler, R. 1981. Some empirical evidence on dynamic inconsistency. Economic Newsletter 8:201–207.

Thorbjarnarson, J. H. 1991. An analysis of the spectacled caiman (*Caiman crocodilus*) harvest program in Venezuela. Pages 216–235 in K. H. Redford and J. G. Robinson, eds., Neotropical wildlife use and conservation. University of Chicago Press, Chicago.

Timmer, P., W. Falcon, and S. Pearson. 1983. Food policy analysis. Johns Hopkins University Press, Baltimore.

Toledo, V. M., J. Caballero, and A. Argueta. 1978. El uso múltiple de la selva basado en el conocimiento tradicional. Biótica 3:85–101.

Vásquez, R., and A. H. Gentry. 1989. Use and misuse of forest-harvested fruits in the Iquitos area. Conservation Biology 3:350–361.

Vickers, W. 1979. Native Amazonian subsistence in diverse habitats: the Siona-Secoya of Ecuador. Studies in Third World Societies 7:6–36.

———. 1980. An analysis of Amazonian hunting yields as a function of settlement age. Pages 7–29 *in* R. Hames, ed., Studies of hunting and foraging in the neotropics. Bennington College Working Papers on South American Indians 2, Bennington, Vermont.

———. 1983. The territorial dimensions of Siona Secoya and Encabellado adaptation. Pages 451–478 *in* R. B. Hames and W. T. Vickers, eds., Adaptive responses of native Amazonians. Academic Press, New York.

———. 1984. The faunal components of lowland Amazonian hunting kills. Interciencia 9:366–376.

White, B. N. F. 1976. Production and reproduction in a Javanese village. Ph.D. Dissertation, Columbia University, New York, n.d.

Wilkie, D. S. 1989. Impact of roadside agriculture on subsistence in the Ituri forest of northeastern Zaire. American Journal of Physical Anthropology 78:485–94.

———. 1991. Central Africa global climate change project, biodiversity support program, evaluation of commercially available GPS receivers. Department of Environmental Management, Tufts University, Somerville, MA.

———, and B. Curran. 1991. Why do Mbuti hunters use nets? Ungulate hunting efficiency of archers and net hunters in the Ituri rain forest. American Anthropologist 93:680–689.

Winterhalder, B. 1983. The boreal forest, Cree-Ojibwa foraging, and adaptive management. Pages 331–345 *in* Resources and dynamics of the Boreal zone: proceedings of the conference held at Thunder Bay, Ontario, August 1982. Association of Canadian Universities, Ottawa.

Yost, J. A., and P. M. Kelley. 1983. Shotguns, blowguns, and spears: the analysis of technological efficiency. Pages 189–224 *in* R. B. Hames and W. T. Vickers, eds., Adaptive responses of native Amazonians. Academic Press, New York.

[12]

Article W. Neil Adger, Katrina Brown, Raffaello Cervigni and Dominic Moran

Total Economic Value of Forests in Mexico

Failure to account for the numerous functions and economic uses of forests have led to patterns of global forest use with many detrimental environmental consequences. This study demonstrates the economic techniques for estimating the Total Economic Value (TEV) of forests. For the Mexican forest estate, the results show an annual lower bound value of the services of the total forest area to be in the order of USD 4 billion. This aggregate value stems from the nonmarketed services provided by nonconsumptive use; from future potential uses of the genetic resources and from pure existence values; and the largest proportion of economic value coming from the functional values of hydrological and carbon cycling. However, only a proportion of this value can feasibly be 'captured' within Mexico: much of the benefit of Mexico's forests falls outside the country's borders, and is therefore not considered by forest users or national policy makers.

INTRODUCTION

Forest resources are recognized to have a number of values, including intrinsic, economic, ecological, cultural and aesthetic values. Forests have economic value in so far as they are scarce and capable of generating human welfare. Decisions on logging, management or conversion of forest lands are most frequently determined on economic criteria, such as the demand for timber, for agricultural land or the need to export forest products to generate foreign exhange. Yet, it is widely acknowledged that market transactions provide an incomplete picture of total economic value. Those forest benefits which are not normally exchanged in markets are generally ignored in decision-making. Undervaluation of these welfare enhancing services introduces inherent distortions to efficient resource allocation. In order to improve the efficiency of resource allocation two steps are necessary. First, the estimation of the appropriate (shadow) prices for nonmarketed or partially marketed forest functions is required. Second, appropriate mechanisms need to be developed to capture the estimated economic values.

Consideration of nonmarket valuation illuminates the scale of economic distortions due to undervaluation and is a necessary condition for rational policy making. Social cost-benefit analysis is however, also concerned with equity issues and how benefits accrue to different sectors of society. Indeed, valuation by itself is of little interest to a country owning environmental assets unless they can be turned into revenue flows. This paper offers estimates of the main components of nontimber values for the Mexican forestry sector. We indicate how these benefits accrue spatially (locally, nationally, and globally) and provide recommendations on how global values can be internalized within the borders of Mexico.

TOTAL ECONOMIC VALUE

The continued existence of forests provides a stream of goods and services including timber and nontimber products, recreation, climate regulation, carbon sequestration, existence value and so forth. The Total Economic Value (TEV) of these forests is defined as the amount of resources, expressed in common units of money that society would be worse off if the forests were lost. TEV is given by the sum of a number of components (1, 2).

Total Economic Value (TEV) = Direct-use value + Indirect-use value + Option value + Existence value.

Direct-use values include revenues from timber and values of nontimber forest products. While timber values are not the main focus of interest of this paper, the sector is considered to have considerable commercial potential given appropriate management regimes. Indirect-use values or "functional" values relate to the ecological functions performed by forests, such as global biogeochemical cycling, the protection of soils, and the regulation of watersheds.

Option (or quasi-option; 3, 4) value is the expected value of the information on the benefits of an asset, conditional on its preservation enabling an increase in the stock of knowledge relevant to the utilisation of the asset. A frequently evoked example of quasi-option value is associated with genetic resources; for example, future pharmaceuticals developed from plant materials. Existence value relates to the value of environmental assets irrespective of current or optional uses. Empirical measures of existence values based on donations to conservation organizations, or on the contingent valuation method suggest these can be a significant element in total economic value, especially in contexts where the asset has unique characteristics or cultural significance.

Although intuitively appealing, applications of total economic value to particular environmental resources are few because of the problems of nonadditivity of the elements and of quantification of the components. In the forest context, studies of the individual components of value have been undertaken including estimates of the economic consumer surplus associated with ecotourism in tropical forests (5, 6); nontimber forest products (7); carbon sequestration functions (8); and recreation, landscape and aesthetic values (9). The study presented here attempts to take these studies further by incorporating all the components of value, calibrated for the Mexico forest types, and is to our knowledge the first attempt at providing such an estimate. This paper concentrates on deriving values for tourism; nontimber forest products; the value of carbon storage; water quality; option value associated with the pharmaceutical uses of genetic material; and existence value.

Since an exhaustive coverage of all forest types is beyond the scope of investigation, an implicit assumption is that benefits estimated for one forest site are able to be transferred or extrapolated across the Mexican forest area. Thus, damage estimates from soil erosion at one site may be extrapolated to similar sites where site-specific data are lacking. Although the legitimacy of benefit transfer procedures are still the subject of academic debate (10, 11) we nevertheless make use of them here, as the aim is to present broad orders of magnitude to be used as input to policy making.

Loss of forests reduce soil conservation functions. First year of maize cultivation following clearance, near Laguna Belgica, Chiapas, Mexico. Photo: J. Barkham.

MEXICO'S FOREST SECTOR

Mexico possesses more than 50 mill. ha of forests, covering a quarter of the total 195 mill. ha of the national territory. Of total forest area, half is mixed broadleaved tropical forest, and half temperate forest. About 34 mill. ha are regarded as potential permanent productive forests (12). Despite the extent of commercially-valuable forest resources, the sector's contribution to Mexican Gross Domestic Product has only ranged between 1.8% and 2% in the last decade (12). Domestic production of wood products has been increasingly unable to meet demand, so that Mexico is currently a net importer of all industrial forest product categories, with the exception of wood-based panels (13). Although there is no agreement as to the exact magnitude and causes of forest loss, the rate of loss is thought to have increased in recent years, with estimates ranging between 400 000 ha^{-1} yr^{-1} and 1.5 mill. ha^{-1} yr^{-1} for the mid-1980s (13).

Approximately one quarter of Mexico's population (of 84.5 million people in 1990) live in forest areas, often subsisting in extreme poverty. Several states with more than 50% forest cover (Chiapas, Oaxaca, Guerrero, Yucatan) have 40% of their population below the poverty line of USD 350 yr^{-1}. The most common form of land tenancy in forest areas is communal, either in ejidos accounting for 70% of the total forest area or indigenous communities which account for 5% of forest area (14), the remainder being small-holder private property.

Significant changes in ownership and control are likely over the next few years. The recently approved North American Free Trade Agreement (NAFTA) will put increasing pressure on traditionally protected and less competitive sectors of the Mexican economy. Reforms of ownership of land precipitated by NAFTA have the objective of allowing the allocation of land among competing uses to become more responsive to interna-

Wet tropical forest cleared for agriculture, Chiapas, Mexico. Photo: J. Barkham.

tional market forces. In the absence of mitigation policies, this is likely to set in motion a process of major structural change in the traditional agriculture and communal forest management, possibly with high social costs.

The process of reform will be facilitated by recent legislative changes, including the reform of Article 27 of the constitution and new agrarian and forestry laws. These changes redefine ownership rights of ejidos and communities, allow the disposal of private property and enable outside investors to enter into joint venture with communities. The ultimate outcome of this transformation is still uncertain. Consideration of the full range of forest values should therefore be an important input in the forthcoming debate on the further evolution of the legislative, institutional and policy framework.

DIRECT-USE VALUES

Tourism and Recreation

Conventional (hotel-beach based) tourism is an important, well-established economic activity in Mexico. In the period 1990–1992 total foreign tourism visits averaged 6.4 mill. per year, with average annual expenditure of USD 3.68 billion (Secretaria de Turismo, pers. comm. 1993). Average growth of 4.8% per year over the period 1970–1992 places the sector in the top three foreign currency earners, contributing 1.5% to GNP (average 1987–1989) and employing 7.3% of the country's active population (15, 16).

Despite the sector's importance, quantitative information is quite limited in detail and disaggregation. Official statistics do not divide visitation or expenditure figures by type of recreational experience, so that there are no specific data on nature-related tourism as such. Furthermore, data on tourism are only systematically collected in the major tourist areas or resorts and rarely at the state level.

Ecotourism and Forest Visitation Rates: Tourism to nature resorts in Mexico can be categorized in two ways: multipurpose tourism, which comprises, as part of a larger trip, visits to national parks or other protected areas, lasting one or two days; and specialized nature tourism (or ecotourism), organized in tours of longer length (between one week and 10–12 days). The two categories can be characterized by opposite patterns of quantity and price. A large volume of short-term visitors to national parks pay significantly less than the fewer numbers of ecotourists. This difference can partly be explained by the relative incomes of the two groups. Participation by Mexican nationals tends to be higher in multi-purpose trips relative to foreign visitors.

For a group of six forest resorts (El Triunfo, Sian Ka'an, Izta-Popo, Arrareko, Mariposa Monarca and Barranca del Cobre), it was possible to gather data on area size, annual average numbers of visitors, price paid or estimated willingness to pay for the recreational experience, revenues per year, revenues per hectare per annum. For those areas, plots of the number of visits against estimated willingness to pay or actual price paid, illustrates the distinction between "ecotourists" and multi-purpose tourists explained above (Fig. 1).

For many protected forest areas in Mexico the distinction between ecotourist and multi-purpose visits is often unclear. The distinction can be made according to the proximity to established resorts or the existence of established infrastructure and communications; and the stated rationale for visits ascertained through written records or established by sample survey. Thus, for example, remote sites such as El Triunfo, Chiapas, would be placed in the ecotourist category; similarly Sian Ka'an, Quintana Roo, would also be an ecotourist destination because of the nature tourist elements in tours to the Biosphere Reserve.

Since these data refer to a limited number of areas, some inference is required to provide a national picture. In the case of ecotourism, we use an estimate of 20 000 yearly visits, an average stay of 10 days, and a price of USD 70 per day (17–20). From this, we obtain revenues of USD 14 mill. yr^{-1}. Compared with total expenditure of foreign visitors for the last three years, USD 3.6 billion, this amounts to a modest 0.4% of the total. It should be noted that some part of the total will include forest-based activities.

In the case of multipurpose tourism, it is estimated that total number of visits to national parks and biosphere reserves ranges between 5.19 and 6.13 mill. visitors per year (21), with individual consumer surplus of around USD 3.20, as ascertained by a survey conducted for this study in Barranca del Cobre (22). This gives an overall consumer surplus ranging between USD 16.6 and 19.6 mill. Summing the potential priced benefits of ecotourism (USD 14 mill. yr^{-1}) and the unpriced benefits of multipurpose tourism, from the above travel cost estimate (USD 16.6–USD 19.6 mill. yr^{-1}), produces a total benefit range of USD 30.6–USD 33.6 mill. yr^{-1}.

Use Value of Nontimber Forest Products

Ethnobotanical studies indicate that over 2000 plant species are utilized from Mexico's forests. These range from products such as resins, terpentine and pitch from Pine forests, and chicle from the chicle forests in the Yucatan, which are internationally traded,

Figure 1. Visitation and potential revenue for nature tourism and multipurpose tourism at six Mexican forest sites.

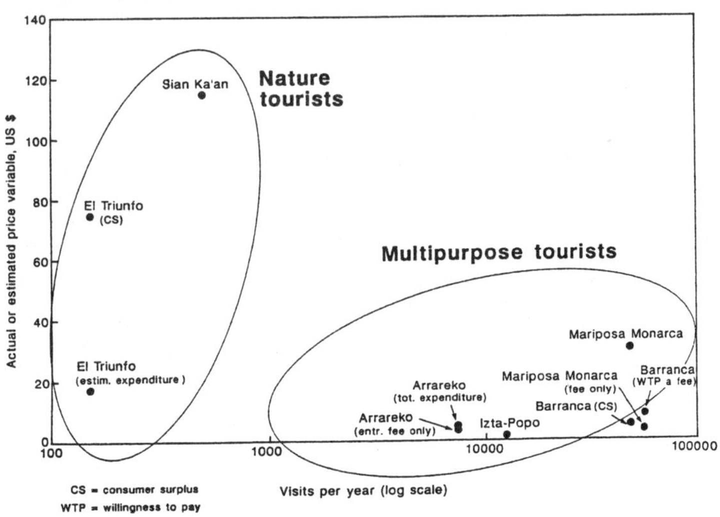

Table 1. Use values of selected nontimber forest products from Mexico's forests.

Product	Region/area	Use value ha^{-1} yr^{-1} (USD 1989)	
Te'lom grove + coffee	San Luis Potosi	1537	Net of costs
Te'lom grove - coffee	San Luis Potosi	330	Net of costs
Chicle	chicle forests Yucatan	4.9	Average for 3 ejidos in Quintana Roo (net of costs)
Allspice	Rain forest, Peten, Guatemala	3.3	Estimate for Peten
Building materials	San Luis Potosi	20.3	Net of costs
Medicines1	San Luis Potosi	51.7	Cost of avoided doctor visits
Fuelwood	San Luis Potosi	14.5	Shadow price of labour method
Fruits	San Luis Potosi	401	Net of costs, half the product marketed
Mushrooms	Pine forest	3.1–281	Costs not subtracted, 1 sp only
	Fir-Pine forest	20.1–97	Costs not subtracted, 1 sp only
Game meat	Quintana Roo	9.0	One ejido
Birds	all Mexico	0.01	Indicative only
Game trophies	Chiapas	0.02	Indicative only

[1] Falls within the range of USD 19–61 calculated by Balick and Mendelsohn (26) in Belize (27).
Sources: Own calculations and (25, 27).

to a multiplicity of medicinal plants and wildfoods utilized at subsistence level.

The task of estimating the use values of these resources is a complex one. Each species has a range of products and different uses, most of which do not have a market price, being used as substitutes for other products, for example. The most in-depth economic analysis of use values of these products in Mexico is Alcorn's work in the southeastern part of San Luis Potosi (23–25). This examines the management and utilization of areas of forest groves, known as te'lom, in cloud forest and tropical rainforest ecosystems, by indigenous Huastec people. Alcorn estimates the net return to te'lom management as being in the region of USD 1540 ha^{-1} yr^{-1} including revenue from coffee, and USD 330 ha^{-1} yr^{-1} excluding coffee (1989 prices). This value has been calculated using shadow prices to assign values to products consumed by the household or used on the farm. This estimate provides an indicative figure of the use value of cloud forest and tropical rainforest environments within a traditional, yet fairly intensive management regime.

Table 1 gives estimates of the use value of selected nontimber forest products (NTFPs) in Mexico. It shows the figures calculated for the Huastec te'lom, some of the component values within that use value, and selected figures from other forest areas for comparison. Some of these figures give indications only of likely magnitudes, as the estimates are highly site specific. Bases for comparability between different forest types and different figures are limited. The values are not necessarily additive. However, the figures accord well with the few other such studies which exist to date, such as those by Peters et al. in Amazonia (7), and Balick and Mendelsohn (26) in Belize. The additive value of the various NTFPs extracted from certain forests may exceed the value of commercial timber.

The use value of NTFPs is likely to be relatively high compared to other values, and possibly very high in certain regions, given particular modes of management. However, many of the figures presented are indicative of particular sites. There are problems in extrapolating from small case study or sample areas to regional or national scale since expanded supply will lower unit values, and data on the amount of time spent collecting and processing NTFPs and other costs, especially opportunity costs, are required. The use of NTFPs may make substantial contributions to livelihood strategies of rural people in certain areas, particularly providing building materials, fodder and wildfoods. This contribution is likely to be especially important for poorer sections of the community, which might be able to benefit from policies designed to maximize the values of NTFPs.

Table 2. Once for all carbon fluxes from changing land uses in Mexico.

Forest Type	Converted land use: Permanent Pasture (tC ha^{-1})	Agriculture (tC ha^{-1})
Temperate Coniferous	168.4	167.1
Temperate Deciduous	34.0	31.5
Tropical Evergreen	164.8	163.5
Tropical Deciduous	92.5	91.25

Source: Own calculations and (31–33).

FUNCTIONAL VALUES

Two functional values of forests are estimated here: the value of forests in their role in the global carbon cycle and the value of the functions in the hydrological cycle. The appropriate economic technique on estimating TEV is that of the loss in economic activity avoided by conserving the resource. In the case of the global carbon cycle it is the avoidance or postponement of the impact of future climate change through the build up of atmospheric greenhouse gases; and for watershed protection it is the offsite and onsite costs of soil erosion, which inevitably occurs when forests are lost in critical upland areas.

Carbon Storage

The carbon sequestration benefits of forest conservation and of plantation forestry is estimated in a two stage process. First, the carbon sequestration and storage is estimated through physical models of forest type and land-use change. The carbon sequestration depends on the species mix; the organic matter content of the species; the age distribution of the stand, and soil and climate factors. Below ground biomass also is a critical part of the net flux. The second stage is to place a monetary value on this forest function in terms of global-warming damage avoided.

To estimate the carbon storage in forests, and hence the fluxes involved in the loss of forests, several methods are available, such as extrapolation from experimental plots and modelling from inventory data. Forest inventory data can be used to estimate the above and below ground biomass of regional areas (28, 29). A difficulty with any accounting approach, however, is in the assumption of the simple dynamics of land-use change. There

Dying epiphytes on edge of cleared cloud forest, Yerbabuena Nature Reserve, Chiapas, Mexico. Photo: J. Barkham.

tends to be a sequence of conversion activity; although the forest area may be converted to agriculture, this may after a few years be converted into permanent pasture for cattle. Even this use may be transient, if the land cannot sustain ranching and is eventually eroded. It is assumed in estimating carbon fluxes that the next land use continues for an indefinite period. Although this may not reflect the reality of the conversion process, the major carbon fluxes occur at deforestation and the subsequent changes are therefore assumed to be insignificant (unless reforestation or regeneration occurs) (30).

The conversion process of forests to other activities generally involves a series of intermediate stages as discussed above. The predominant subsequent uses are agricultural - ranching or cropping. The once for all carbon fluxes associated with those land-use changes are shown in Table 2. Other reasons for forest loss include road building and erosion, though this is relatively small in areal terms (< 10% of the forest area lost). The results in Table 2 show a range from > 30 tC ha^{-1} lost with forest loss to > 160 tC ha^{-1}, for the predominant subsequent uses.

The economic value of carbon sequestration: As discussed above, the forests have a role in the global carbon cycle, but net atmospheric accumulation of greenhouse gases due to growing emissions will cause global warming in the future. Global warming and climatic change will have a number of serious impacts on ecosystems, economies and regions, such as accelerated sea-level rise; unknown impacts on agriculture due to temperature and precipitation changes; impacts on demand for energy and other economic activities; health-related impacts and others (34–36). Of greatest concern is the uncertainty surrounding the magnitude and distribution of any changes, and several catastrophic scenarios have been suggested due to changes in prevailing winds and ocean currents. Activities which reduce emissions or enhance sinks of the trace gases, primarily CO_2, have then a value to society equal to the expected damage avoided.

Several studies exist that attempt to put a money value on global-warming damages (37–39). The procedure is generally to relate temperature changes to physical impacts, then apply economic values to these impacts (40). These studies are associated with various emissions scenarios and damage and abatement functions but rely critically on extrapolation from one economy (the US) across all regions of the world. The earlier studies by Nordhaus (38) and Cline (39) have been refined by Fankhauser (37) who sub-divides climatic-change impacts on the world economy by region. Fankhauser's estimates are based on a study which explicitly allows for uncertainty by incorporating random variables, such as the climate sensitivity to the doubling of atmospheric concentrations of CO_2 and the discount rate and damage estimates. The randomization is generally a closer representation of current scientific understanding of the greenhouse effect than the use of best guess values. We take Fankhauser's central estimate for current emissions of USD 20 per tC and apply them to the once for all carbon-flux changes to give a capital value to the world of the global-warming damage avoided in maintaining the four generalized Mexican categories of forest.

The results show that the value of avoiding the carbon fluxes associated with changing land use range from USD 650 to USD 3400 (equivalent to USD 20 to USD 100 ha^{-1} yr^{-1}), depending on the forest type and the subsequent land use. This is a social value in that this is the benefit of avoiding global warming impacts in the global sense. It is not therefore either the benefit captured by Mexico for carbon sequestration, or the financial or opportunity cost of forest conservation from the perspective of the forest owners.

These estimates are conservative for a number of reasons. First, with a time-related profile of carbon-flux changes, the value of fluxes occurring in future time periods increases in the models of global-warming damage. Second, the carbon-flux estimates are mid-range estimates of the differences between biomass and soil accumulation by different forest types. Other studies have estimated higher carbon fluxes from land-use change, though ongoing experimental research will shed light

Table 3. Value of carbon sequestration from forest conservation (Present value USD per ha).

Forest Type	Converted land use: Permanent Pasture (USD ha^{-1})	Agriculture (USD ha^{-1})
Temperate Coniferous	3436	3410
Temperate Deciduous	693	643
Tropical Evergreen	3633	3337
Tropical Deciduous	1887	1863

Notes: Based on social value of USD 20 per tC and once for all fluxes as estimated in Table 2.

on these results. Third, non-CO_2 trace gases are emitted in biomass burning, the process by which forest is lost in many cases (41, 42). These gases such as CH_4 are more potent greenhouse gases, and add to the climate change-impact. Again this and the other economic factors mentioned above give a conservative estimate of the economic cost of global warming damage, and hence of the value of the carbon storage function of Mexico's forests.

Watershed Protection

Deforestation, hydrology and soil erosion: Although the damaging effects of deforestation have been confirmed in experimental trials, site-specific valuation of forest hydrological functions is complex due to the heterogeneity of forest types, geological and climatic factors, and to difficulties in valuing impacts (43). Deforestation exposes forest soils, and erosion effects are worsened by clearance and extraction, using logging tracks and new roads. Quantifiable damages include increased soil erosion entailing on-site and off-site sedimentation costs; accelerated water runoff leading to localized flooding; and reduced hydrological cycling and recharge of groundwater and watercourses. The costs of mitigation or reparation of damages provides a surrogate measure of the value of these protective forest functions (44). Data limitations confine estimates for Mexico to those from sedimentation damages on infrastructure; damages from increased flooding and reduced water remain unquantified, giving lower bound estimates of the true value of this forest function.

Sedimentation costs: There is considerable uncertainty in linking regional water-quality degradation to deforestation. This uncertainty is primarily due to the nonpoint nature of the sources of sedimentation. In Mexico, the historical and current loss of forest cover in watersheds is undoubtedly a principal factor in accelerating soil loss. Maas et al. (45) report up to a 100-fold increase in soil loss through forest conversion for experimental trials in Jalisco (0.20 t ha^{-1} yr^{-1} compared to a highest value of 130 t ha^{-1} yr^{-1} under maize, see Table 4). Extensive measurement in watersheds across hydrological sub-regions indicates a typical average range of 0.169 to 7.43 t ha^{-1} yr^{-1} although averaging across catchments disguises extreme cases of up to 50 t ha^{-1} yr^{-1}. Quantification of external costs of soil erosion have received far less attention than more easily observed on-farm productivity benefits of erosion control (46). In Mexico, little is known of the relative magnitudes of on-site and off-site costs although limited US estimates (47, 48) indicate that off-site costs are significant.

For water quality some degree of the benefit of forest cover can be estimated by considering the likely extent of defensive expenditures required nationally for water purification of suspended sediment. Martínez-Menez and Fernández (49) estimate an average annual sediment production figure of 365 mill. tonnes, although without specifying where this originates. Of this an estimated 69% is carried to the sea and 31% is discharged in waterways and other hydrological infrastructure. Assuming that all 113 mill. tonnes of deposition induces some form of purification and using an average figure calculated for the US for induced treatment costs of USD 20 per 1000 tonnes of discharge (48), a rough lower bound cost is estimated as USD 2.3 mill. yr^{-1} (1989 prices) for all sedimentation in Mexico.

Existing research suggests that likely savings to the water-treatment industry from reduced erosion are a relatively minor benefit (48). This is confirmed by the estimates derived here. It is noted however, that more vulnerable watersheds with high sediment loading rates may have significant problems. Specific examples of the benefit of forest cover due to avoided replacement purification costs may be in the range of USD 50-USD150 ha^{-1}. Since these represent one-off or periodic costs avoided, they do not approximate annual losses. In the absence of more detailed cost information this is assumed as the typical range for vulnerable watersheds.

Infrastructural damage: Despite the consensus on the severity of infrastructural damage from sedimentation, there is little organized information on which to base damage estimates. Casual evidence from national electricity generators responsible for major dam projects in Mexico suggests that sedimentation rates have been (and possibly still are) underestimated leading to curtailed life-spans and misguided investment signals. Such costs are often amplified by the social costs of habitat destruction and community relocation to make way for flood sites (for example, the case for the Aguamilpa hydropower project in Nayarit and Zimapán project in Queretaro and Hidalgo). Nevertheless, the existence of comprehensive state erosion figures (50) and information on vulnerable hydrological areas suggest that there is considerable information with which to direct watershed management. Some representative economic studies are required to demonstrate the extent of losses and mobilize resources to tackle the problem (51).

Watershed regularization: The hydrological and flood regulatory benefits of forest cover are well known although there appears to be little information on Mexican watersheds. Measurable benefits are likely to accrue in three areas: *i*) mitigation of flood damage; *ii*) groundwater recharge; *iii*) assurance of urban water supplies. Flood damage to crops is a periodic occurrence in several low-lying areas and there is evidence that watershed deforestation has aggravated the volume of runoff (most notably, the watershed of the Rio Paloapan, Veracruz). Estimated damage mitigation benefits are site specific and are difficult to attribute to discrete areas of forest cover. Existing protected area studies

Table 4. Selected soil erosion estimates for Mexico.

Authors	Measurement
Estrada and Ortíz-Solorio (54)	Average losses in excess of 50 t ha^{-1} yr^{-1} using FAO methodology; location unclear.
Martínez-Menez and Fernández (55)	Average losses of only 5 t ha^{-1} yr^{-1} using suspended sediment measurement in main watersheds.
Maas and García-Oliva (56)	From a review of studies state an average range of 0–15 t ha^{-1} yr^{-1} in cultivated areas with modest slopes.
Raa (57)	Study in mountain area; cuenca alta del rio La Antigua (Veracruz) range 13.5–1296 t ha^{-1} yr^{-1}.
Torres (58)	Range 30.4-151.7 t ha^{-1} yr^{-1} over 4 months; cuenca alta Rio San Marcos (Tamaulipas).
Maas et al. (45)	Range 30 - 130 t ha^{-1} yr^{-1}, Chamela research station, coastal Jalisco.

Sources: (45, 54–58)

Table 5. Option value of pharmaceuticals from Mexico's forests.

	Option value moist tropical forests USD ha^{-1} yr^{-1}	Option value of moist tropical forests USD mill. yr^{-1}	Option value, all forests USD mill. yr^{-1}
Low estimate	1	5	26
Central estimate	6	66	332
High estimate	90	875	4646

Assumptions:
5000 species is the lower estimate of the number of species in Moist Tropical Forests in Mexico, and therefore represents the lowest estimate for all forests. Forest areas: 9.7 mill. ha MTF; 51.5 mill. ha all forests. Probability of a hit is 0.0005, royalty rate 0.05 (see text)
Low estimate: appropriation rate = 0.1, drug value = USD 0.39 mill. yr^{-1}
Medium estimate: appropriation rate = 0.5, drug value = USD 1 bill. yr^{-1}
High estimate: appropriation rate = 1, drug value = USD 7 bill. yr^{-1}

in other countries attempt to make broad with/without estimates of likely flood damages and water supply benefits (52, 53).

The balance of existing information suggests that the external costs of soil loss and implicitly some proportion of the benefits of forest preservation impact in various ways. Measurement of the economic costs may be by several methods, some of which are more relevant in the Mexican context. The most significant off-site damages from forest conversion are sedimentation costs and alterations to hydrological balances.

OPTION VALUE OF PHARMACEUTICALS

A major argument for the preservation of biological diversity, when many species and their properties remain unknown, is the *potential* uses of these resources. This diversity is therefore valued for the options that it represents. Option value can be conceptualized as being similar to an insurance premium, paid to ensure the supply of an asset, the availability of which would otherwise be uncertain (1). Option value reflects the willingness of a risk-averse society to pay a premium, on top of the use value itself, for guaranteeing access to a resource of uncertain future supply.

An indication of the potential information value of genetic diversity in Mexico is illustrated by the example of *Zea diploperennis*. This wild species of maize (known as a teosinte) was discovered in the Mexican state of Jalisco in the 1970s, and its most valuable characteristics are that it is perennial and shows resistance to certain viruses. Fisher and Hanemann (59) have estimated that *Zea diploperennis* may have a value of USD 6.83 billion per year for its use in the creation of perennial maize only. The discovery of the teosinte has resulted in the establishment of a 135 000 ha Reserva Biosfera de la Sierra de Manantlan in 1987 (60), which should protect not only this teosinte, but also the other species of the mountain habitat.

The option value of forests as habitats for species likely to yield pharmacologically active substances is often cited as a reason for promoting forest conservation. Although the medicinal value of plants and their derivatives has been recognized for millennia, the estimation of the importance and economic value of the biodiversity which gives rise to the possibility of more discoveries is a very recent field of interest (61–63).

The option value, or more strictly quasi-option value (3, 4), of pharmaceuticals derived from genetic material in Mexico's forests is estimated using a model which takes into account the number of species present in the forests; the probability of a species yielding a useful product; the royalty rate on sales of such a product which would be payable to Mexico; a coefficient of rent capture; the likely value of an internationally traded pharmaceutical product; and the area of forest. The model, developed by Pearce and Puroshothaman (64), estimates option values as a function of the number of species at risk, the number of drugs based on plant species, and the number of ha likely to support medicinal plants (which they take to be total area of tropical forest). They suggest commercial values in the range from very low to around USD 20 ha^{-1}. These values can be construed as lower bound estimates, given that these relate only to species at risk, and the loss of large tracts of tropical forest would place many other plant species at risk.

The methodology used here to estimate the option value of pharmaceuticals from Mexico's forests is adapted from Pearce and Puroshothaman and calibrated for Mexico. The function for the value on a per ha basis is:

$$Vp(L) = \{ N \cdot p \cdot r \cdot a \cdot V/n \}/H \ yr^{-1}$$

Where:
Vp (L) = The pharmaceutical value of 1 ha of forest (USD ha^{-1})

N = The number of plant species in forests
p = The probability of a "hit"
r = The royalty rate

Table 6. Evidence of existence value for Mexico's forests.

Type of Evidence	Organization Involved
Debt for Nature Swap	Conservation International
Donations for Nature Conservation Organizations	Amigos de Sian Ka'an (Cancun, Quintana Roo)
	Pronatura (Mexico City)
On-site evidence, visitors'	
US Contributions to biodiversity conservation projects in Mexico	World Resources Institute Washington DC

Sources: Tajbakhsh, M. Bezauri, J. Herman, H. pers. comm. 1993 and (69)

continued overleaf

a = The appropriation rate, or rent capture
V/n = The average value of drugs developed (USD yr^{-1})
H = The area of forest (ha)

These components and their empirical magnitudes for Mexico are dependent on the diversity of species, including vascular plants and micro-organisms and animal species (65)—extremely high in a megadiverse country like Mexico (66–68)—institutional factors affecting the return to Mexico from a discovery with global implications; the area of Mexico's forests; and factors exogenous to Mexico such as the technology employed by drug prospecting companies. Based on the best guess factors for the Mexico case, a range of potential values given in Table 5 can be identified.

EXISTENCE VALUES

Existence value derives from the knowledge of a resource's continued existence, independent of any use. There are several plausible underlying motivations for existence value which have been widely covered in social science literature (2). From the methodological standpoint, measurement of existence value is perhaps the least tractable and most contentious aspect of natural resource valuation. The available developed country evidence, suggests that nonusers have indicated values in the range USD 1.2–64 yr^{-1} per person for wild species, while WTP for scenic and wilderness areas range between USD 9 and USD 107 yr^{-1}. To the extent that the latter estimates could be extended to large groups of the populations involved, the resulting aggregate figures are likely to be quite significant, and increased effort to measure and appropriate existence value seems to be warranted. For the time being existing studies provide indicative values where full-blown studies are not possible at different sites. The wisdom of transferring benefit estimates from other similar sites continues to be the subject of scrutiny in valuation literature (10, 11). An important caveat for this study is that most valuation studies have been carried out in the US or in other developed countries, and may therefore have a limited applicability to Latin American countries like Mexico.

Table 6 presents elements of revealed value for a sample of transactions related to natural area conservation in Mexico. Reported elements include contributions to conservation organizations and programs, a tourism survey carried out by the authors, and debt for nature swaps. On the basis of this evidence, indicative per ha values would appear to be in the range (0.03–10.4 USD ha^{-1}). Multiplying the upper bound of that range over the extent of Mexico's protected areas (approx. 5.8 mill. ha) provides a conservative estimate of USD 60.2 mill. Given the diversity of protected areas it is fair to suggest that only a proportion of this value is attributable to forest areas.

Some comparison with other Latin American country data can be made using estimates of existence value implicit in debt-for-nature (DFN) transactions. As reported by Pearce and Moran (70), the range of implicit values is from around USD 0.01 ha^{-1} to just over USD 4 ha^{-1}. The idea of valuing the biodiversity so conserved through DFNs is that the payment made reflects some kind of willingness to pay on the part of the conservation body purchasing the debt. Different DFNs can be expected to derive different implicit valuations since the nature of the 'good' being bought will vary. For example, the quality of the area protected will vary, and different packages of measures will be involved.

AGGREGATE VALUES

Table 7 provides aggregate estimates for the elements of TEV discussed above. In general, conservative estimates have been presented. For example, for option value the upper bound figure is many times greater than the mid-range estimate used. Similarly, the true population over which existence value should be estimated is the world, and higher estimates would result. No aggregate estimate has been extrapolated for NTFPs as explained above. In summary, the results show that the benefits of nontimber forest products are significant in forest-management decisions.Te'lom intensive management, for example can produce USD 330 ha^{-1} yr^{-1} value added. Important forest products include chicle, medicines, mushrooms, and building materials. Nonmarket functional values have the greatest global value, higher than option, existence or indirect-use values.

Table 6. continued

Area Involved	Size(ha)	Amount of transaction (USD)	WTP per hectare (USD)	Comments	Source
Selva Lacandona; Montes Azules (Chiapas)	385 000	4 000 000	10.38		Tajbakhsh (1993) pers. comm.
Sian Ka'an biosphere reserve (Quintana Roo)	528 147	34 000	0.06 0.05	The first figure is the total amount of donations for 1992; Sian Ka'an report that only 10% of donors have actually visited the reserve.	Bezauri (1993) pers. comm.
	Not Applicable	809 622	Not Applicable	The figure refers to total 1992 receipts of Pronatura, including donations and other sources of income	Herman (1993) pers. comm.
Barranca del Cobre (Chihuahua)	450 000	100 100	4.4	A survey carried out in Barranca del Cobre suggests a willingness to pay per $person^{-1}$ yr^{-1} related to existence value of USD 1.82. Multiplying this by the number of visitor per annum (55 000 in 1992) gives a total of USD 100 100; and at a 5% discount rate, a present value of USD 4.4	Authors' Survey
Several areas	190 869 000	5 528 809 (1989 USD)	0.029	Areas involved in the calculation are those contained in a WRI biodiversity projects database.	Abramovitz (69)

The carbon sequestration function of forests provides large global benefits. A range of capital carbon value from USD 650–USD 3400 ha^{-1} is estimated (shown as annual value in Table 7 of USD 20–USD 100 ha^{-1} yr^{-1}). This functional value alone is greater than typical land values in tropical forest areas in many parts of the world. Watershed protection is estimated at USD 2.3 mill. yr^{-1} from sedimentation costs alone. Other functions such as watershed regulation would expand this lower bound figure. Although this is lower than other values, it is a highly localised problem.

The option value of pharmaceuticals is estimated in the range from USD 26-USD 4600 mill. with a central estimate of USD 330 mill. in Table 7. This value stems from the probability of discovering genetic material in the average ha of forest which may yield billion dollar benefits. Existence value can be imputed from the willingness to pay for conservation of international agencies, or from direct survey responses. A value of USD 60 mill. is estimated for Mexico. The nonpriced benefits of ecotourism and the priced benefits of multipurpose tourism, the total benefits range obtained is USD 30.6-USD 33.6 mill. yr^{-1}. The size of these values leads us to conclude that there are considerable returns (both to private decision-makers and to society) if these can be captured. The following section outlines some mechanisms available to maximize and capture some of the component values of TEV.

CAPTURING THE VALUES

Nontimber Forest Products: Enhancing the value of NTFPs and increasing the capture of their value could bring significant benefits to poorer sections of rural communities, both in terms of providing subsistence goods, such as building materials, where substitutes might be expensive and produced outside the area, and in terms of direct income generation. Many NTFPs, for example chicle, provide the bulk of cash income for some households. However, producers of such products are vulnerable to shifts in demand and price fluctuations. However, there are opportunities for win-win policies which provide income generation for local people, and encourage sustainable utilisation of forests, through developing NTFPs (71, 72).

The market potential for many valuable NTFPs is little known. For example, ornamental palms (*xate*), allspice and mushrooms have been identified in previous studies (73) as marketable species. The development of markets is required, which ensures that local gatherers and managers are able to capture more of value of NTFPs, but which recognizes that the expansion of these markets increases vulnerability to international price fluctuations.

Carbon Values: Offset deals are one possible mechanism for the appropriation of global carbon values. These agreements involve a commitment of forest users to a forestry management plan which maximizes carbon sequestration in exchange for an annual payment which represents a carbon credit for the financier. In the past, success of schemes has crucially depended upon coordination of the individuals involved in forest management. Existing community structures of ejidos, association of ejidos and indigenous communities are institutional structures already in place and which can guarantee the coordination of individual actions necessary for the operation of offset deals (74, 75). Indeed, there are examples of such schemes proposed in Mexico. The Union de Communidades y Ejidos Forestales de Oaxaca (UCEFO) proposes to promote growth of biomass and of forest area to sequester 2.7 mill. tC by 2030. The cost of undertaking this management strategy translates to approximately USD 3 per tC sequestered (74).

Although these offset deals will become more frequent in the context of the UN Framework Convention on Climate Change (76), and do provide one source of finance for implementing management plans, their impact in terms of the large global value of the carbon cycling function of forests is small. This is due to the institutional problems associated with such offset deals and also to the fundamental property rights issues involved in trading this carbon sequestration function, which are seen as inequitable by many parties.

Tourism: Ecotourism in Mexico is currently characterized by low number of visits and high willingness to pay. Operating just on the quantity side, i.e. by encouraging a higher number of visits, could lead to a significant increase in revenue. For instance, just diverting 1% of foreign tourists from conventional tourism, with an average of expenditure USD 40 $person^{-1}$ day^{-1}, to ecotourism, with expenditure of USD 70 $person^{-1}$ day^{-1}, would generate an increase in revenues of about USD 8.8 mill. yr^{-1}.

In the case of multipurpose tourism, we observe the opposite pattern: higher visitor numbers but lower price paid per visit. Here a strategy could be followed, which would increase, or in some cases establish, charges associated with the recreational experience. Entrance fees to protected areas should be higher for international visitors. This could be justified on efficiency grounds (higher foreign recreational values) and on equity grounds (per capita income levels of US and European visitors are higher than domestic incomes). By using the lower bound non-Mexican willingness to pay obtained from the mentioned survey in Barranca del Cobre (USD 8) as a benchmark for an average entrance fee, and assuming that about 25% of visitors to protected areas are foreign, it is estimated that the resulting additional revenues would lie somewhere between USD 10.4 mill. and USD 12.6 mill. yr^{-1}.

Option Values: The scope for biodiversity prospecting contracts with conservation of genetic resources have been highlighted by recent developments in other countries (77), and given the high option value of Mexico's biodiverse forests, Mexico is

Table 7. Estimated values of elements of total economic value of Mexican forests (USD values).

	Area (m ha)	Tourism	NTFPs	Carbon	Watershed protection	Option value	Existence
Tropical Evergreen	9.7		330 ha^{-1} yr^{-1}	100 ha^{-1} yr^{-1}		6.4 ha^{-1} yr^{-1}	
Tropical Deciduous	16.1			56 ha^{-1} yr^{-1}			
Temperate Coniferous	16.9			103 ha^{-1} yr^{-1}			
Temperate Deciduous	8.8		330 ha^{-1} yr^{-1}	20 ha^{-1} yr^{-1}			
Total (Sm yr^{-1})	51.5	32.1 mill.	n/a	3788.3 mill.	2.3 mill.	331.7 mill.	60.2 mill.

Sources and Notes: Forest areas from (33). Excludes open forest and scrubland. Option value is a conservative estimate based on the quasi-option value of genetic resources in tropical forests. No estimate of aggregate value of non-timber forest products is available. The USD 330 ha^{-1} yr^{-1} (1989 prices) in the table represents returns from intensively managed te'lom systems only as an indication of the multiple use of forests. Watershed protection (1989 prices) includes water quality only (not infrastructure costs). The carbon figures are annuitized sums (assuming a 3% discount rate) of the capital values of carbon sequestration (USD 650-USD 3400), reported in Table 3.

well placed to capture this value. It has large amounts of expertise compared with other developing countries, but requires bolstering of research and development facilities, and clarification of property rights. Agreements at the moment would seem likely to be developed through private organizations, perhaps in collaboration with educational and research institutions. If linked in with conservation projects at a local level, then there are opportunities for innovative involvement perhaps of communities or groups of ejidos.

Existence Value: The international community may reveal its willingness to pay for conservation by engaging in some kind of trade of land-use rights. Landowners would give up rights on some uses of land, e.g. burning the forest, or developing land beyond a given level of intensity, *in exchange* for a money payment. Land users would retain rights to conservation compatible activities such extraction of nontimber forest products. Those carrying out these activities would then have two sources of income: the proceeds of the activity itself, plus the fee paid by the conservation agency for giving up the right to other uses.

Several alternative avenues offer potential for viable trades in conservation between Mexico and the international community. Options differ according to the choice of the *intermediary structure* linking the buyer (the international community) and seller (the resource users). One possibility would be to have federal or local government as intermediary between the international community and resource users, as in the proposed Franchise Agreements (78) scheme. An alternative would be to let private entities play the role of intermediary. This appears to be the basis of the program of the newly constituted Foundation for the Conservation of Biotic Resources (FUNDAREB) (D. Laird, pers. comm.). Although still in the program design stage, FUNDAREB's intention is to pool in a company the financial resources required to initiate a program of land-use rights purchase. In a second stage, its share in the company could be partly or totally sold in the market, and should the initial experiences be successful, new acquisitions could be started. To make the program possible, FUNDAREB has proposed the introduction of *conservation easements* or of other forms of property rights attenuation in the Mexican legislation (79).

CONCLUSIONS

Forests are multi-functional ecosystems with numerous uses by local people and more distant users. Failure to recognize and account for these multiple uses have led to patterns of global forest use with many detrimental environmental consequences. In economic terms, it is the public good nature of the environmental services which leads to this detrimental resource use. This paper has demonstrated the economic techniques for valuing these nonmarketed services and has applied these techniques to the Mexican forest case. The results show an annual lower bound value of the services of Mexico's forests to be in the order of USD 4 billion. These values stem from the nonmarketed services provided by nonconsumptive use; from future potential uses of the genetic resources and from pure existence values; and the largest proportion of economic value coming from the functional values of hydrological and carbon cycling.

The implications for Mexico's forest policy stem from the observation that a large proportion of these values accrue globally: that global human society benefits as much from the existence of Mexico's forests as Mexicans do. Two conclusions follow from this observation. First, to capture at least part of this global value, international mechanisms would be required. These may be government levies on international tourists to Mexico, or international transfers for carbon offsets or biodiversity conservation, though it should be recognized that the globalization of property rights tends to undermine local management and control. Second, a clearer understanding is required of the economic and other 'values' of forests which are inherent in forest users' decisions. This paper has demonstrated that a strong case can be made for forest conservation in the Mexican case, based on local, regional and global values of forests, and that these values should be incorporated into decisions on the future management of this important resource.

References and Notes

1. Following Pearce, D.W. 1993. *Economic Values and the Natural World.* Earthscan, London.
2. Total economic value was first defined in Randall, A. and Stoll, J. 1983. Existence values in a total valuation framework. In: *Managing Air Quality and Scenic Resources at National Parks and Wilderness Areas.* Row, R. D. and Chestnut, L. G. (eds). Westview Press, Boulder Co.
3. Arrow, K.J. and Fisher, A.C. 1974. Environmental preservation, uncertainty and irreversibility. *Quart. J. Econ. 88,* 312–319.
4. Henry, C. 1974. Investment decisions under uncertainty: the irreversibility effect. *Am. Econ. Rev. 66,* 1006–1012.
5. Tobias, D. and Mendelsohn, R. 1991. Valuing ecotourism in a tropical rainforest reserve. *Ambio 20,* 91–93.
6. Maille, P. and Mendelsohn, R. 1993. Valuing ecotourism in Madagascar. *J. Environ. Mgmt 38,* 213–218.
7. Peters, C.M., Gentry, A.H. and Mendelsohn, R.O. 1989. Valuation of an Amazonian rainforest. *Nature 339,* 655–656.
8. Brown, K. and Pearce, D.W. 1994. The economic value of non-marketed benefits of tropical forests: carbon storage. In: *The Economics of Project Appraisal and the Environment.* Weiss, J. (ed.). Edward Elgar, London, pp. 102–123.
9. Hanley, N.D. and Ruffell, R.J. 1993. The contingent valuation of forest characteristics: two experiments. *J. Agric. Econ. 44,* 218–229.
10. Brookshire, D. and Neill, H. 1992. Benefit transfers: conceptual and empirical issues. *Water Resour. Res. 28,* 651–655.
11. Krupnick, A. J. 1993. Benefit transfers and valuation of environmental improvements. *Resources 110,* Winter 1993, 1–7.
12. Synnott, T.J. 1993. *Forestry Subsector Study in Mexico: Diagnostic Study.* Report to World Bank, Washington DC.
13. Barbier, E., Burgess, J., Collins, M. and Clancy, C. 1993. *Mexico Forestry Sector and Conservation Review: Substudy of Economic Issues for Implementation of the New Forestry Policy.* Report to the World Bank, London Environmental Economics Centre, London.
14. The Ejido system is a product of post-revolutionary land reforms. Land under the law of ejido tenure, was until recently a non-negotiable resource. *Ejidatarios* were granted a plot which was neither sold, transferred nor mortgaged and subject to confiscation if left for more than 2 years.
15. World Resources Institute. 1992. *World Resources 1992–1993.* Oxford University Press, New York.
16. Boo, E. 1990. *Ecotourism: The Potentials and Pitfalls. 2 Volumes.* Vol. 2. WWF International, Washington DC.
17. According to Ceballos Lascurain, H. 1988. *El Turismo Ecologico en Mexico como Instrumento de Desarrollo Socioeconomico y de Conservacion,* mimeo, a 12 day ornithological tour costs about USD 700 per person (not including airfare), i.e. USD 58 per day.
18. Another source reports a cost of USD 1000 for a 10 days ecological tour in the Selva Lacandona (Chiapas), i.e USD 100 per person per day. A reasonable intermediate value could then be USD 70 per person per day. Vásquez-Sánchez M.A., March, I. and Lazcano-Barrero, M. 1992. Caracteristicas socioeconomicas de la Selva Lacandona. In: *Reserva de la Biosfera Montes Azules, Selva Lacandona: Investigacion para su Conservacion.* M.A. Vásquez-Sánchez and M. Ramos (eds). Ocas. Ecosfera. (In Spanish).
19. Regarding the number of visits, again Ceballos Lascurain suggests a tentative estimate of about 20 000 people per year participating to ecotourism tours. This 1988 figure refers to all sort of ecotourism activities in Mexico, not only to forest related ones.
20. However, the number of ecotourists in Mexico has been in all likelihood growing in parallel to the growth of the sector worldwide, reported as growing at a rate of about 30% per year. Whelan, T. (ed.) 1991. *Nature Tourism: Managing for the Environment.* Island Press, Washington DC. The number of ecotourists participating to forest related recreation has been plausibly growing as well. In the absence of any better information, the figure of 20,000 visits per year is then used.
21. The only available data from the ministry for social development (Secretaria de Desarrollo Social, SEDESOL) refer to estimated visits per year in each of 18 protected areas comprised in SINAP, the National System of protected areas (Constitucion 1857, Lagunas de Montebello, Cañon del Sumidero, Palenque, Cascadas de Bassaseachic, El Chico, Tula, Miguel Hidalgo y Costilla, Izta Popo, Zoquiapan, Lagunas Zempoala, El Tepozteco, Lagunas Chacahua, Pico de Orizaba, El Vizcaino, Islas Golfo California, Mariposa Monarca, Isla Contoy). By calculating simple and area-size weighted average of this data, and multiplying the result by the total of national parks and biosphere reserves, it was calculated that the range of yearly visitors to protected areas is 5.19–6.13 million people.
22. The Copper Canyon (Barranca del Cobre) is a pine-oak forested area of 450 000 ha in the Sierra Madre mountain region in the south-western part of the state of Chihuahua. In order to assess present and potential economic benefits from recreation in the Barranca, a tourism survey was carried out by the authors in May of 1993. A sample of 60 visitors to the area were interviewed in the train travelling between Chihuahua and El Devisadero, and in hotels in both Creel and El Devisadero. Respondents were asked to fill in a questionnaire consisting of 13 questions, available both in English and in Spanish. The first group of questions were designed to directly elicit the visitor's travel cost pattern, as the sum of individual transportation and time cost components. The purpose of these questions was to construct a demand function for visits to Barranca del Cobre, enabling evaluation of an average yearly consumer surplus. This was estimated at USD 3.27. The second set of Contingent Valuation questions were intended to elicit respondents' willingness to pay (WTP) for conserving the forest area. Respondents to the WTP question were also invited to allocate their stated WTP by four distinct value categories, roughly corresponding to: i) use value related to forest

recreation, ii) use value related to enjoyment of cultural diversity, iii) option value and iv) existence value. The average overall WTP was estimated at USD 8.20.
23. Alcorn, J.B. 1984. *Huastec Mayan Ethnobotany.* University of Texas Press, Austin, Texas.
24. Alcorn, J.B. 1989. Process as resource: The traditional agricultural ideology of Bora and Huastec resource management and its implications for research. *Adv. Econ. Bot. 7,* 63–77.
25. Alcorn, J.B. 1989. An economic analysis of Huastec Mayan forest management. In: *Fragile Lands of Latin America.* Browder, J.O. (ed.). Westview Press, Boulder, Co. pp. 182–206.
26. Balick, M.J. and Mendelsohn, R. 1992. Assessing the economic value of traditional medicines from tropical rainforests. *Conserv. Biol. 6,* 128–130.
27. Bye, R., Caballero, J., Linares, E., Mapes, C., Martinez, M. A. and Mendoza, M. 1993. *Non-timber Forest Resources in Mexico: An Overview.* Mimeo, Jardin Botanico, Universidad Nacional Autonoma de Mexico, Mexico.
28. On alternative methods of estimating carbon contents of forests see for example Brown, S. and Lugo, A.E. 1992. Above ground biomass estimates for tropical moist forests of the Brazilian Amazon. *Interciencia 17,* 8–18.
29. Fearnside, P.M., Leal, N. and Moreira Fernandes, F. 1993. Rainforest burning and the global carbon budget: biomass, combustion efficiency and charcoal formation in the Brazilian Amazon. *J. Geophys. Res. 98,* 16733–16743.
30. Adger, W.N. and Brown, K. 1994. *Land Use and the Causes of Global Warming.* John Wiley, Chichester.
31. Brown, K. 1992. *Carbon Sequestration and Storage in Tropical Forests.* Global Environmental Change Working Paper 92–24, Centre for Social and Economic Research on the Global Environment, University of East Anglia, University College London.
32. Houghton, R.A., Skole, D.L. and Lefkowitz, D.S. 1991. Changes in the landscape of Latin America between 1850 and 1985: II Net release of CO_2 to the atmosphere. *For. Ecol. Mgmt 38,* 173–199.
33. Masera Cerutti, O., Ordonez, M. d. J. and Minjarez, R.D. 1992. *Carbon Emissions and Sequestration in Forests: Case Studies from Seven Developing Countries. Volume 4: Mexico.* USEPA and Energy and Environment Division, Lawrence Berkeley Laboratory, Berkeley, Ca.
34. Haines, A. 1993. The possible effects of climate change on health. In: *Critical Condition: Human Health and the Environment.* Chivian, E., McCally, M., Hu, H. and Haines, A. (eds). MIT Press, Cambridge Mass, pp. 151–170.
35. Rosenzweig, C. and Parry, M.L. 1994. Potential impact of climate change on world food supply. *Nature 367,* 133–138.
36. Schneider, S.H. 1992. Global climate change: ecosystem effects. *Interdiscip. Sci. Rev. 17,* 142–148.
37. Fankhauser, S. 1995. *Valuing Climate Change: the Economics of the Greenhouse Effect.* Earthscan, London.
38. Nordhaus, W.D. 1992. An optimal transition path for controlling greenhouse gases. *Science 258,* 1315–1319.
39. Cline, W.R. 1992. *The Economics of Global Warming.* Institute for International Economics, Washington DC.
40. These values may be based on market prices or estimates of non-market values such as the "statistical value of life" for mortality and morbidity effects.
41. Crutzen, P.J. and Andreae, M.O. 1990. Biomass burning in the tropics: impact on atmospheric chemistry and biogeochemical cycles. *Science 250,* 1669–1677.
42. Cofer, W.R., Levine, J.S., Winstead, E.L., Stocks, B.J., Cahoon, D.R. and Pinto, J.P. 1993. Trace gas emissions from tropical biomass fires: Yucatan Peninsula, Mexico. *Atmos. Environ. 27A,* 1903–1907.
43. Bruijnzeel, L. A. 1990. *Hydrology of Moist Tropical Forests and Effects of Conversion: A State of Knowledge Review.* International Hydrological Programme, UNESCO, Paris.
44. Strictly, reparation costs are a minimum estimate of damage only if it can be assumed that it is worth repairing the damages.
45. Maas, J., Jordan, C. and Sarukhan, J. 1988. Soil erosion and nutrient losses in seasonal tropical agroecosystems under various management techniques. *J. Appl. Ecol. 25,* 595–607.
46. Schramm, G. 1982. *The Economics of Soil Conservation in a Semi-Arid Country: Mexico.* School of Natural Resources, University of Michigan, Ann Arbor MI.
47. Clark, E.H., Haverkamp, J.A. and Chapman, W. 1985. *Eroding Soils: The Off-farm Impacts.* The Conservation Foundation, Washington DC provide an estimate of about USD 2.2 billion per year as a cropland share of total offsite soil erosion damage.
48. A more modest figure of USD 353 million per year based only on water treatment costs, can be found in Holmes, T.P. 1988. The offsite impact of soil erosion on the water treatment industry. *Land Econ. 64,* 356–366.
49. Martinez-Menez, M.R. and Fernandez, J. 1984. Evaluacion de la Degradacion Especifica a Nivel Nacional a Partir de Cuencas Hidrologicas. In: *XVII Congreso Nacional de la Ciencia del Suelo in Mexico.* Sociedad Mexicana de la Ciencia del Suelo. (In Spanish).
50. SARH (Secretaria de Agricultura y Recursos Hidraulicos) 1987. *La Erosion y Conservacion del Suelo en Mexico: Realidades y Perspectivas.* SARH, Mexico DF. (In Spanish).
51. An application dealing with these economic costs in the Philippines is Cruz, W.D., Francisco, H.A. and Conway, Z.T. 1988. *The On-site and Downstream Costs of Soil Erosion in the Magat and Pantabangen Watersheds.* College of Economics, University of Philippines, Los Baños, Philippines.
52. Gutman, P. 1992. *Venezuela Natural Resources Management Project: Costs and Benefits of the Project.* World Bank, Washington DC.
53. Ruitenbeek, J. 1989. *Republic of Cameroon: the Korup Project.* Ministry of Planning and Regional Development, Cameroon.
54. Estrada, J.W. and Ortiz-Solorio, C.A. 1982. Plano de erosion hidrica del suelo en Mexico. *Geografia Agricola 3,* 23–37. (In Spanish).
55. Martinez-Menez, M.R. and Fernandez, V.J. 1983. *Jerarquizacion de Acciones de Conservacion de Suelos a Partir de Cuencas Hidrologicas.* Mimeo, Mexico DF. (In Spanish).
56. Maas, J.M. and Garcia-Oliva, F. 1990. La conservacion de suelos en zonas tropicales: el caso de Mexico. *Ciencia y Dessarrollo 15,* 21–36. (In Spanish).
57. Raa, T. 1983. Estudio de la erosion en la cuenca alta del Rio La Antigua, Veracruz, Mexico. *Biotica 8,* 361–380. (In Spanish).
58. Torres, J. 1987. *Ecologia y Uso de los Recusos Renovables de la Cuenca del Rio San Marcos, Tamaulipas, Mexico.* Thesis Facultad de Ciencias, Universidad Nacional Autonoma de Mexico, Mexico DF. (In Spanish).
59. Fisher, A.C. and Hanemann, W.M. 1985. *Option Value and the Extinction of Species.* Mimeo, California Agricultural Experiment Station, Berkeley, Ca.
60. Iltis, H.H. 1988. Serendipity in the exploration of biodiversity: What good are weedy tomatoes? In: *Biodiversity.* Wilson, E.O. (ed.). National Academy Press, Washington DC, pp. 98–105.
61. For estimates of the number of plant species used as sources of drugs in western medicine, see: Bell, E.A. 1993. Mankind and plants: The need to conserve biodiversity. *Parasitology 106,* 47–53.
62. Principe, P. P. 1991. Valuing the biodiversity of medicinal plants. In: *The Conservation of Medicinal Plants.* Akerele, O., Heywood, V. and Synge, H. (eds). Cambridge University Press, Cambridge, pp. 79–124.
63. Other studies evaluate the commercial value of drugs derived from higher plants; see Farnsworth, N.R. and Soejarto, D.D. 1985. Potential consequences of plant extinction in the United States on the current and future availability of prescription drugs. *Econ. Bot. 39,* 231–240.
64. Pearce, D.W. and Puroshothaman, S. 1992. *Protecting Biological Diversity: The Economic Value of Pharmaceutical Plants.* Global Environmental Change Working Paper 92–27, CSERGE, University of East Anglia and University College London.
65. The biggest group of therapeutics drugs, antibiotics, are derived from *Penicillium* molds for example. Oldfield, M. L. 1984. *The Value of Conserving Genetic Resources.* Sinauer Associates Inc, Sunderland, Mass.
66. World Conservation Monitoring Centre 1992. *Global Biodiversity: Status of the Earth's Living Resources.* Chapman and Hall, London.
67. Bye, R. 1993. The role of humans in the diversification of plants in Mexico. In: *Biological Diversity in Mexico: Origins and Distribution.* Ramamoorthy, T.P., Bye, R., Lot, A. and Fa, J. (eds). Oxford University Press, New York, pp. 707–731.
68. Toledo, V.M., Batis, A.I., Becerra, R., Martinez, E. and Ramos, C.H. 1992. Products from the tropical rainforests of Mexico: An ethnoecological approach. In: *Sustainable Harvest and Marketing of Rainforest Products.* Plotkin, M. and Famolare, L. (eds). Conservation International, Washington DC, pp. 99–109.
69. Abramovitz, J.N. 1991. *Investing in Biological Diversity: US Research and Conservation Efforts in Developing Countries.* World Resource Institute, Washington DC.
70. Pearce, D.W. and Moran, D. 1994. *The Economic Value of Biological Diversity.* Earthscan, London.
71. Anderson, A.B. (ed.). 1990. *Alternatives to Deforestation.* Columbia University Press, New York.
72. Redford, K.H., and Padoch, C. (eds). 1992. *Conservation of Neotropical Forests.* Columbia University Press, New York.
73. Reining, C. and Heinzman, R. 1992. Nontimber forest products in Peten, Guatemala: Why extractive reserves are critical for both conservation and development. In: *Sustainable Harvesting and Marketing of Rainforest Products.* Plotkin, M. and Famolare, L. (eds). Conservation International, Washington DC, pp. 110–117.
74. Faeth, P., Cort, C. and Livernash, R. 1994. *Evaluating the Carbon Sequestration Benefits of Forestry Projects in Developing Countries.* World Resources Institute, Washington DC.
75. Dixon, R.K., Andrasko, K.J., Sussman, F.A., Lavinson, M.A., Trexler, M.C. and Vinson, T.S. 1993. Forest sector carbon offset projects: near-term opportunities to mitigate greenhouse gas emissions. *Water, Air Soil Pollut. 70,* 561–577.
76. Brown, K. and Adger, W. N. 1994. Economic and political feasibility of international carbon offsets. *For. Ecol. Mgmt 68,* 217–229.
77. Reid, W.V., Laird, S.A., Meyer, C.A., Gamez, R., Sittenfeld, A., Janzen, D., Gollin, M. and Juma, C. 1993. *Biodiversity Prospecting: Using Genetic Resources for Sustainable Development.* World Resources Institute, Washington DC.
78. An IFA is a concession, by the state, of exclusive rights on land to a franchisee, with limitation on allowable uses in the interest of a third party. In particular, the franchise would be a limited term of use of a specified property within a given region, or state, subject to specific restrictions. The grantor would be the owner-state (O-S), e.g. Mexico. The franchisee would be the entity allocated the franchise. The third party is the global community, represented by bilateral agreement between an international organization such as the World Bank, and Mexico. The IFA operates by the O-S dividing the total development rights for the particular parcel of land between the GC and the Franchisee in a way that maximizes the O-S's return from that land. The O-S then collects a rental payment from the franchisee for its use of the franchise, and a rental payment from GC for the restrictions placed upon that franchise. For details see Cervigni, R. 1993. *Biodiversity, Incentives to Deforest and Tradeable Development Rights.* GEC Working Paper 93-07, CSERGE, University College London and University of East Anglia.
79. The use of land easements (US) or covenants (UK) are well established use right attenuation schemes for areas of environmental, cultural or historical interest.
80. Acknowledgement: The Centre for Social and Economic Research on the Global Environment is a designated research centre of the UK Economic and Social Research Council. Funding for the research for this paper came from the World Bank, Latin America and the Caribbean Division. The authors would like to thank staff at SARH and SEDESOL for assistance and David Pearce and Kerry Turner for comments on earlier drafts. The authors retain sole responsibility for the remaining errors and the opinions expressed in the paper.
81. First submitted 18 July 1994. Accepted for publication after revision 1 February 1995.

Neil Adger is a senior research associate at the Centre for Social and Economic Research on the Global Environment (CSERGE). His address: University of East Anglia, Norwich, NR4 7TJ, UK.

Katrina Brown is a lecturer in natural resource management in the School of Development Studies, University of East Anglia and Senior Research Fellow in CSERGE. Her research focuses on development and biodiversity conservation in South Asia, Africa, Central America and the Caribbean. Her address: School of Development Studies, University of East Anglia, Norwich NR4 7TJ, UK.

Raffaello Cervigni is an economist specializing on the costs, benefits and incentive mechanisms of biodiversity conservation. He is a research associate at CSERGE, and a consultant of the Global Environment Facility. His address: Global Environment Facility Secretariat, Room G6-143, The World Bank, 1818 H Street NW, Washington DC 20433 USA.

Dominic Moran is a research associate at CSERGE with research interests in biodiversity, agriculture and economic valuation. His address: CSERGE, University College London, Gower Street, London WC1E 6BT, UK.

[13]

On the Value of the Condition of a Forest Stock

Thomas D. Crocker

I. INTRODUCTION

A request commonly made of practitioners of benefit-cost analysis and other types of applied welfare analysis is to evaluate the economic consequences of alternative use patterns of natural ecosystems. With respect to forested systems, economists have responded with a number of analytical and empirical studies on the demand for legally designated wilderness areas, especially the outdoor recreational manifestations of this demand.[1] Because it has focused on estimating the values associated with the entire wilderness experience package, this work has implicitly presumed that recreational behavior is influenced largely by economic and demographic factors as opposed to variations in the landscape where the recreation occurs [Clawson and Knetsch (1966)]. While this information aids in determining the optimal stock of wilderness areas, it is of considerably less help in managing existing publicly owned forested areas, or in defining which areas should receive or retain wilderness designation.

Most public forest lands must now be operated under multiple–use management schemes.[2] Land managers must respond to changing patterns of demands for the various products and services of the forest which, in addition to harvested timber, include water storage, wildlife, grazing, and recreational and amenity services that depend on the condition of the standing vegetation. The forest manager can manipulate vegetation condition and its development by the spraying of insecticides, timing and type of thinning, harvesting, and regeneration practices, access restrictions, and trail and facility construction. If the manager lacks information on how nontimber values vary with the condition of the standing stock of trees, he may choose to respond only to the marketlike and readily available price data reflecting the derived demand for harvested timber. Forest managers relying only on this market-related information will allocate insufficient resources to manipulating the condition of the standing stock.

Any workable method to assess the economic value of the condition of a standing forest stock can help managers formulate policies and regulations for allocating scarce budgets between furthering the flow of harvested timber and the condition of the stock.[3]

Land Economics, Vol. 61, No. 3, August 1985
0023-7639/85/001-0244 $1.50/0

Department of Economics, University of Wyoming, Laramie. The advice and cooperation of Henry J. Vaux, Jr., is gratefully acknowledged. Scott Atkinson, Don Coursey, Bruce Forster, Edna Loehman, and an anonymous referee have provided helpful comments and suggestions. This study was partly funded by a grant from the U.S. Environmental Protection Agency.

[1]Krutilla (1967) is the analytical wellspring. Empirical studies, such as Cicchetti and Smith (1973), Smith (1975), Shecter and Lucas (1978), Smith and Kopp (1980), and Walsh, et al. (1981), report consumer surpluses of $8 to $18 per visitor day to wilderness areas in 1980 dollars. Other than Davis (1963), no similar estimates of the consumer surpluses from recreational uses of forested areas not legally designated as wilderness appear to exist.

[2]For example, under the Multiple Use-Sustained Yield Act of 1960, the Forest and Rangeland Renewable Resources Planning Act of 1974, and the National Forest Management Act of 1976, the U.S. Forest Service must try to maximize benefits from sustained yield, multiple use production, with due consideration to the relative values of all forest resources and uses [U.S. Forest Service (1978)].

[3]By no means are the two objectives always competitive. See Hartman (1976) and Berck (1981) for recent treatments of some efficiency conditions that describe solutions to this allocation problem.

In this paper, a contingent valuation survey is employed to derive the economic values a sample of self-selected outdoor recreationists attach to alternative air pollution-induced health states of a southern California national forest. Results therefore cannot be extrapolated to nonparticipants. The setting of the study and the model on which the measurement of value is based are described in the next section. A third section is a discussion of the structure of the questionnaire and the procedures used to gather data. Empirical results are reported in a fourth section. The final section is devoted to the implications the results might have for studies of the nonharvest values of standing forest stocks and similar environmental amenities.

II. THE SETTING

The Background. Even a cursory review of the literature on the biological impacts of air pollution turns up expressions of concern about the effects upon forests of oxidants and sulfur compounds. In spite of these concerns, no published study of the possible economic losses appears to exist. Nearly always mentioned as the leading exhibit of what has been happening and what could happen on a far larger scale is the San Bernardino National Forest (SBNF) of Southern California.

Some 30 years ago, observers first noticed a yellow mottling of the needles and extensive needle drops on the ponderosa and Jeffrey pines that dominate much of the SBNF. It has since been conclusively demonstrated that the malady is caused by ambient oxidants, the precursors for which are emitted in the vast upwind Los Angeles urban complex lying beneath and as much as 80 miles to the west [Miller and McBride (1975)]. Among the ecosystem stresses that the needle mottlings and drops represent are declines in conifer reproduction [Kichert and Gemmill (1980)], reduced resistance to fungal [James, et al. (1980)] and beetle [Taylor, et al. (1980)] attacks, mortality rates that are three times greater than normal [Miller (1973)], and reductions of up to 84% in standing average wood volume [Miller, et al. (1977)]. In addition, greater fire frequencies and intensities, reduced water storage capacities, and substitutions of deciduous and shrub species for pines and firs have been occurring [Miller et al. (1982)]. Attention here is limited to an economic assessment for current outdoor recreators of the oxidant-induced damages to the *existing* stock of ponderosa and Jeffrey pine trees that comprise the dominant vegetation types in the most popular areas of the forest.[4]

A Model. The model which drives this application presupposes that the individual annually engages in a two-stage budget process, where, in accordance with Becker (1965), the "full" income that he allocates is a combination of his money income and his leisure time. In the first stage, the individual allocates this full income between recreational activities and other commodities such that the necessary and sufficient conditions for weak separability are fulfilled [Deaton and Muellbauer (1980, chapter 5)]. Given this initial allocation, he then decides how the full income to be devoted to recreation will be allocated across specific recreational activities at specific sites, including activities in the SBNF.

These private, site-specific recreation activities are represented by a vector, $x \equiv (x_1, \ldots, x_n)$, where each x is the annual frequency with which the individual makes visits of fixed length to a site suitable for the activity. Central to the enjoyment of any one incident of one of these activities, SBNF recreation, is the exogenously determined biological health status of the forest. The many dimensions of this health status are represented by a consistent scalar index, Q.[5] This index varies inversely with actual air

[4]According to Bolsinger (1980), these two species also comprise 59% of the forest's sawtimber volume. Though little actual logging takes place, the forest contains nearly 1.2 billion board feet of sawtimber. Originally established in 1925, the SBNF has since become the most heavily used unit for outdoor recreation in the National Forest System [Hilb (1976)].

[5]See Deaton and Muellbauer (1980, chaps. 7 and 8) and Forster (1981) for discussion of the conditions under which a vector of quantities can be treated as a scalar quantity. The extent to which these conditions are met in this study is unknown.

pollution-induced damages, A; that is, $Q = Q(A)$, and $Q'<0$.

The individual obtains utility U, from more x and more Q. His utility function is continuous and quasi-concave in x. He must select a particular bundle from his budget set, $B = [x|px < M]$, where $p \equiv (p_1, \ldots, p_n)$ is the vector of exogenous prices such as travel costs and access fees that he confronts, and M is the individual's full income, which includes the predetermined opportunity cost of his leisure time. As in Cornes (1980), the individual's second-stage decision problem can then be stated as:

$$V(p, Q(A), M) = \max_x [U(x, Q(A)) \mid x \in B, \; Q(A) \text{ preassigned}] \quad [1]$$

$V(\cdot)$ is a "restricted" or "rationed" indirect utility function which shows the maximum utility available, given a parametric full income, price vector, and environmental amenity, and hence, air pollution. For a particular visit to the SBNF, increases in air pollution-induced damages will reduce the maximum utility the individual can achieve since:

$$\frac{\partial V}{\partial A} = \frac{\partial V}{\partial Q} \cdot Q' < 0. \quad [2]$$

Upon differentiating [1] with respect to Q and M, setting the change in utility equal to zero, and using Roy's Identity, one obtains the individual's marginal valuation, v_A, of air pollution damage for the visit in question.

$$v_A(p, Q(A), M) = \frac{-\dfrac{\partial V}{\partial Q} \cdot Q'}{\partial V / \partial M} \quad [3]$$

The basis of [3] is easily perceived in straightforward verbal terms. Suppose that the p vector is fixed. If Q declines, say, from Q_0 to Q_1, then the individual's maximum attainable utility will be $V_1(p, Q_1, M)$ rather than $V_0(p, Q_0, M)$, where $V_1(\cdot) < V_0(\cdot)$. With p fixed, the only way to restore the individual to his original utility level, V_0, is to increase his income from M_0 to M_1, given that $\partial V/\partial M > 0$. The amount $M_1 - M_0$ is then the income amount for the particular visit by which the individual must be compensated for the reduction in Q. Alternatively, it is the reduction in expenditures on the forest visit that must be allowed him if he is to recover his original utility level.[6]

Of interest is the way in which the individual's marginal valuation, v_A, in [3] changes when air pollution damages, A, change.

$$\frac{\partial v_A}{\partial A} = \frac{-1}{\partial V/\partial M} \left[\frac{\partial^2 V}{\partial Q^2} (Q')^2 + \frac{\partial V}{\partial Q} \cdot Q'' \right] \quad [4]$$

The conventional assumption is that $\partial^2 V/\partial Q^2 > 0$, that is, the marginal utility of higher environmental amenity levels is decreasing. If Q'' is also negative, then the entire bracket on the right-hand side of [4] is negative, and hence, $\partial v_A/\partial A > 0$, which is the orthodox result. Increases in air pollution damages are supposed to require progressively increasing levels of compensation if a constant utility level is to be maintained. However, if $Q(A)$ is linear, then the second set of terms within the brackets of [4] disappears, and $\partial^2 V/\partial Q^2$ is left to determine the sign of the entire expression. Contrary to the conventional assumption, it is by no means evident that a 100-foot tall scruffy tree will provide the pleasure to the eye of a 10-foot tree that has rich foliage. Similarly, that which an individual considers to be a spectacular forest scene may provide him more utility than what he regards as two moderately nice scenes.

III. THE DESIGN

The questionnaire was built around three 6 × 8-inch color photographs of near-

[6]This is because $\partial C/\partial Q = (-\partial V/\partial Q)/(\partial V/\partial M)$, where C is the expenditure or cost of utility function, the solution to the dual of the standard direct utility maximization problem.

identical large ponderosa and Jeffrey pine stands similar in size and configuration to those present in the most popular areas of the Forest. Each photograph was taken at ground level from a distance of 30 to 40 feet. The three photographs, labeled *A*, *B*, and *C*, were simultaneously presented to respondents in a left-to-right ordering. In terms of the linear injury scoring system developed by Miller (1973, 104), *A* represented "very slight injury" with a score of 1 to 8, *B* depicted "very severe injury" having a score of 29 to 35, and *C* showed "moderate injury" with a score of 15 to 21.[7] The photographs used to represent the forest environments were thus tied to a cardinal and linear index that biologists have frequently used to characterize oxidant damages, e.g., Taylor, et al. (1980).

The purpose of the questionnaire was to generate truthful respondent messages that could then be compared to the aforementioned environments when the institutional rules, times, and locations to which the messages refer were held constant. As Brookshire, et al. (1982) note, the air pollution problem in Southern California is well understood by local residents. So as to enhance the likelihood that this knowledge would be applied, and to provide among respondents the common frame and editing whose importance Tversky and Kahneman (1981) emphasize, each respondent was initially told that air pollution was the source of any differences he perceived in the health states of the depicted forest environments. In order to temper their subjective costs of participating, respondents were also told that the interview would take no more than five or ten minutes. The questionnaire had to be brief because all respondents were engaged in forest recreation. Moreover, it was expected that nearly all respondents would be daytrippers or overnight campers having severely limited opportunities to adapt their schedules to the time demands of an interview.

Interview brevity was possible because detailed explanations of hypothetical problems removed in time, space, and institutional detail from the respondent's immediate experience were unnecessary. He was asked to state his maximum willingness-to-pay in terms of a fee to be added to a daily $6.00 or $7.00 access fee he had, in fact, already paid on the interview day. Much of the environmental and preference information the respondent had to process in order to estimate this additional willingness-to-pay had therefore already been used by him in his decision to pay the original access fee. Basically, the close accord between the respondent's hypothetical and real situations reduced the likelihood of substantial discrepancies between the expected and the realized consequences of his choices [Brookshire and Crocker (1981)] and allowed a degree of control to be exercised over the respondent's total valuations.

A further attempt was made to ease the respondent's information-processing burden by asking him initially to rank-order the depicted environments from his most to his least preferred. This had the further advantage of allowing the respondent to frame as well as to solve his decision problem. Separation of the ordering and value estimation process conforms to the Tversky and Kahneman (1981) view of decision making in which the editing or framing phase precedes the evaluation phase. The sequence in which the respondent was asked to estimate his willingness-to-pay for the three environments corresponded to the rank-ordering he had already assigned.

Given that the respondent's estimate of his maximum willingness-to-pay declines as his perception of the quality of the depicted environment falls, his estimates are compensating surplus measures.[8] The iterative value esti-

[7]These color print enlargements were selected from among hundreds of 35 mm slides owned by Dr. Paul R. Miller of the Pacific Southwest Forest and Range Experiment Station. The injury scores were assigned by Dr. Miller on June 17, 1983.

[8]This is not accurate, since the fee change measures the price change required to maintain the ex ante utility level. Hicksian compensating measures refer to the income required for this maintenance. However, as opposed to some unidentified visit, the fees in terms of which the respondents made their estimates refer to specific visits to which the respondents were already committed. The difference between the income equivalent of the fee change and the fee change is therefore likely to be trivial.

mation procedure was applied only to the respondent's most preferred environment. In order to reduce its arduousness, it was presumed that once the respondent had settled upon this maximum payment, he had enough prior information about his preferences to allow him to state without further prodding his maximum willingness-to-pay for a visit to his next most preferred and his least preferred environments.

Following the respondent's estimation exercise, he was asked whether he would be willing to go to his least preferred environment if doing so were the only way to avoid crowds in his most preferred environment. A substantial number of positive (yes) answers to this question would suggest that individualized perceptions of crowding were more important to respondents than were air pollution damages.

There is an additional questionnaire feature worthy of note in that the respondent's estimated willingness-to-pay explicitly refers to an increment to an access fee for his current visit. The respondent did not confront the problem of deciding to which of his possibly several visits over a time period his value estimates would refer. Moreover, the emphasis on the current visit did not allow some respondents to adjust visit frequencies and value estimates simultaneously while others adjusted only their value estimates. The introduction of an element of noncomparability across the valuations of different respondents was thus avoided.

Nevertheless, the acquisition of information on visit frequency and its changes is useful. The adoption of substitute activities, with consequent alterations in SBNF visit frequency, is a major means by which respondents expand their opportunities to minimize their losses or maximize their gains from a changed forest environment.

IV. EMPIRICAL RESULTS

Data Properties. Some features of the data obtained from interviewing 36 respondents during a June 1983 weekday, and 64 respondents over one day of a July 1983 weekend, are reviewed in Table 1. The interviews for both samples were conducted at the same set of forest locations.[9] Weather conditions were similar on each interview day. The heavy local day-trip and overnight camping use of the forest that Rausser and Oliveira (1976) remark upon is evident. Fifty-eight of the respondents had a one-way trip shorter than the arithmetic mean trip of 76 miles. Given the heavy preponderance of one-day trips of local origin, it does not seem unreasonable to assume that the sample represents single-purpose recreational visits. Retirees, schoolteachers, students, and the unemployed dominated the weekday respondents. They appeared much less frequently with the weekend respondents.

Only 7 respondents reported a household annual income exceeding \$50,000. One can only conclude that the break-point for the binary income question was placed at too high a level.[10] This conclusion is reenforced by those respondents who said they could have worked on the day of the interview. Only 3 of these 23 respondents who could have worked reported an opportunity cost sufficient to exceed \$50,000 annually.

Forty-eight respondents stated they would have stayed home if they had not visited the forest on their respective interview days. The remaining 52 respondents would have chosen to engage in a variety of activities ranging from going to the beach [19], visiting regional fresh water bodies [11], going to another forested area [12], or to the desert [5], to playing golf and visiting friends. Even though indoor and outdoor commercial activities are plenti-

[9]The interviews were supervised by Professor Henry J. Vaux, Jr. No more than 10 individuals refused to participate. The two interviewers had conducted hundreds of interviews in another contingent valuation study involving forest fire damages.

[10]The binary form of the annual income question was prompted by two considerations: (1) the fact that, when trying to explain willingness-to-pay for aesthetic environmental commodities, most contingent valuation studies have obtained small and statistically insignificant coefficients for annual income; and (2) serious doubt that one can define adequately within the confines of a five- or ten-minute interview what is to constitute annual income. The binary measure enhances the likelihood that different respondents employed similar definitions.

TABLE 1
SAMPLE ATTRIBUTES

	Number	Mean	Standard Deviation
Sample size	100		
Daytrippers	5		
Overnight campers	95		
Southern California residents	98		
Would substitute to avoid crowding	77		
One-way distance (miles)		76.01	45.88
Usually visits "very slight injury" sites	47		
Usually visits "moderate injury" sites	27		
Usually visits "very severe injury" sites	7		
No resemblances to sites usually visited	9		
Annual visits now		5.14	6.85
Annual visits if only least preferred site available		4.50	6.78
Could have worked on interview day	23		
Earnings for those who could have worked		$144.38	$78.33
Expenditures on substitute activity		$ 28.85	$63.71
Annual earnings > $50,000	7		

ful in the Los Angeles area, only one lonely individual volunteered that he would substitute such an activity, an amusement park visit.

Preference Expressions. Eighteen respondents expressed indifference among the three environments, while 76 respondents most preferred the "very slight" injury environment to either the "moderate" or "very severe" injury environments. However, the latter 76 respondents were nearly equally divided in their rank orderings of the moderate and severe injury environments. Two of the remaining 6 respondents placed the moderate injury environment at the top of their rank ordering, while the other 4 most preferred the severe injury environment. Clearly, in spite of substantial differences in the associated values of the biological damage index, the collection of respondents was unable to express an unambiguous preference ordering between the moderate and severe injury environments.

The above preference pattern also emerges with respect to the respondents' cardinal evaluations.[11] Table 2 includes the means and the standard deviations of the willingnesses-to-pay additional access fees. The mean willingness-to-pay for the slight injury environment is about three times greater than the willingness-to-pay for either the moderate or severe injury environments. The hypothesis of the equality of willingnesses-to-pay for the elements in the three possible pairs of the depicted environments was tested statistically. At the 95% confidence level, the hypothesis could not be rejected for the moderate and severe injury environments.

[11]All those who were not indifferent among the environments, yet who refused to state a positive willingness-to-pay, replied with "We pay too much already," "Present fees are high enough," and similar sentiments. One individual estimated he would pay an additional $20, and five individuals were willing to pay an additional $10 for the slight injury environment. One $10 estimate of willingness-to-pay was for the moderate injury environment. Removal of these "outliers" does not alter the monotonically increasing form of the marginal benefits function.

TABLE 2
WILLINGNESS-TO-PAY

	Injury Level		
	Slight	Moderate	Severe
Injury score midpoint	4.5	18	32
Arithmetic mean payment	$2.09	$0.66	$0.74
Standard deviation of payment	$2.80	$0.78	$1.00

No respondents exhibited differences between the place they assigned slight injury in their rank ordering and the ordering of their cardinal estimates of value; however, 12 respondents who preferred moderate to severe injury, or vice versa, made identical positive value estimates for the moderate and the severe injury environments.

Bradford (1970) has demonstrated that marginal versions of the values found in Table 2 can be interpreted as points on indifference curves with slopes representing marginal rates of substitution between an income equivalent and the good of interest. His demonstration does not require that the indifference curves exhibit a diminishing marginal rate of substitution, a requirement that indifference curves underlying Table 2 would clearly fail. Moreover, the convex form of the total willingness-to-pay function in this table cannot have originated in the biological damage index since Taylor, et al. (1980) state that it is linear. Any monotonic transformation of this index will continue to yield a convex total willingness-to-pay function for improvements in environmental quality. Thus, *if* respondent perceptions of the qualities of the depicted environments are some monotonic transformation of the biological damage index, the source of the willingness-to-pay pattern displayed in Table 2 must reside in a positive sign being attached to the $\partial V^2/\partial Q^2$ term in expression [4]; that is, it must reside in the underlying preferences for the depicted environments.

Explaining Willingness-to-Pay. The willingness-to-pay expression in [3] has the prices of recreation activities, full income, and the environmental amenity as its arguments. For the expression to be estimated, the simple sum of respondent travel costs and on-site costs for the current forest visit, and respondent-estimated expenditures for a single incident of his chosen substitute activity, are employed as price measures. Full income is represented by a variable for the respondent's foregone income on the day of the interview and a variable for the annual income of his household. Any influences of differences in states of mind and tastes among respondents that are not captured in the random error are presumed to be mirrored in variables depicting an aversion to congestion, the number of annual forest visits, and the day of the interview. As earlier noted, the demographic characteristics of the weekday and weekend respondents differed substantially.

Each of the preceding right-hand-side terms is presumed to be exogenous and to be linear in the variables with respect to willingness-to-pay. However, the natural logarithm of Miller's (1973) air pollution damage index is used as the indicator of the magnitude of the environmental amenity variable, Q. This semi-logarithmic form accords with the relation between willingness-to-pay and air pollution damages that is reported in Table 2. Since respondents were being asked to react to a visual representation of an aesthetic phenomenon, the form also receives prior support from numerous psychophysical experiments involving human perceptions of sensory events such as light and color changes. Specifically, the results of these experiments have been codified in the form of Fechner's law, which states that the strength of a just noticeable increment in a sensation is proportional to the logarithm of its stimulus [Baird and Noma (1978)]. These psychophysical results are consistent with that class of arguments in the economics literature which posits a "threshold of sensitivity" [Georgescu–Roegen (1936), (1958)] or a "zone of indifference" [Luce (1956); March (1978)] within which choices are randomly made or an "inertia of choice" prevails [Devletoglou (1971)].

The interviewing process resulted in 153 statements of a zero willingness-to-pay out of a total 300 statements, where each respondent value estimate-air pollution damage combination is treated as a distinct observation. There was no upper bound to the willingness-to-pay that a respondent could state. These facts justify the application of the Tobit transformation to the version of [3] that is to be estimated. Each parameter estimate thus accounts for changes in the probability of making a positive willingness-to-pay statement as well as for changes in the magnitude of a statement of positive willingness-to-pay. The parameter estimates are set forth in Table 3. Travel costs were assumed to be 20¢ a mile, and on-site costs were the $6.00 or $7.00 access fees.

Given that the marginal utility of money was everywhere positive, and that respondents had already solved for the optimal number of annual visits, two of the three price variables have the expected signs attached to their coefficients; that is, as travel costs and foregone incomes are increased, more compensating income is required to restore the individual's ex ante utility level. For the same reason, a positive sign was expected for substitute expenditures, the proxy for the unit price of the most preferred substitute. Since 48 respondents stated that they would have spent nothing in their most preferred substitute activity, the measurement error that this plausibly represents would attenuate the coefficient. The a priori incorrect sign in Table 3 could, as Garber and Klepper (1980) show, be due to measurement error in another covariate such as annual income that is collinear with substitute expenditures.

Of the remaining right-hand-side variables in Table 3, aversion to congestion, annual visits, and air pollution damages have coefficient magnitudes exceeding their standard errors. The first two of these variables are consistent with a heightened sensitivity to differences among the "qualities" of forest environments, whatever the sources of these differences. Frequent visitors and those who desire to avoid crowds are more finely calibrated. This finer calibration presumedly increases the compensation its possessors require for a quality reduction.

Reductions in environmental quality, as represented by more air pollution damages, increase required compensation. However,

TABLE 3

TOBIT ESTIMATES OF WILLINGNESS-TO-PAY[a] FOR CURRENT VISIT

Variable		Normalized Coefficient	Asympotic Standard Error
Constant		−.0807	.2703
Travel costs plus on-site costs[a]		.0065*	.0033
Substitute expenditures[a]		−.0012	.0011
Foregone income[a]		.0021*	.0009
Annual income[a]		.0416	.0881
Aversion to congestion[b]		1.0271*	.1938
Annual visits		.0116	.0077
Weekend interview[c]		−.1180	.1394
ln (Air pollution damage)[d]		−.4322*	.0792
Log—likelihood function	−451.6491		
Observations at limit	153		
Observations not at limit	147		
Standard error of estimate	2.9536		
$\bar{R}^2$	.1852		

*Significant at least at the .05 level of the one-tailed *t*-test for the null hypothesis of no association.
[a]Dollars.
[b]1 = Would go to least preferred environment if it were the only way to avoid crowds = 1; otherwise = 0.
[c]Weekend interview = 1; otherwise = 0.
[d]Midpoints of Miller (1973) index: Slight injury = 4.5; moderate injury = 18; severe injury = 32.

since 77% (77 respondents) of the sample (see Table 1) would have been willing to go from their most preferred to their least preferred environment if the former were perceived to be crowded, it seems that the compensation respondents would demand for crowding exceeds that which they would demand for air pollution damages. The "very slight" damage category was the most preferred environment for 70 of the aforementioned individuals.

Numbers of Visits. Because [1] is an indirect utility function, the estimates of the preceding two subsections presume that the respondent has optimally adjusted x, the vector representing his annual number of visits of fixed length to the SBNF and other sites, to the air pollution damages in the SBNF; that is, given the values of the other parameters, there is only one value of each element of x that will allow the respondent to attain the given utility level. These comparative static responses in terms of optimal adjustments in visit frequency would presumedly also be of interest to the forest manager. Although the previous two subsections demonstrated otherwise for this study, it is perfectly conceivable that air pollution would have no welfare effect whatsoever because of the substitution possibilities available across sites [(Courant and Porter (1981)].

An expression for the individual's optimal number of visits of fixed length to the SBNF is easily derived from [1]. In particular, let x_i $(i = 1, \ldots, n)$ be the number of annual visits to a particular site, and alter the meaning of Q so that it is now a vector $Q \equiv (Q_1, \ldots, Q_n)$ of indices of environmental quality levels at all sites.[12] Differentiating [1] with respect to P_i and M, setting the change in utility equal to zero, and again exploiting Roy's Identity, generates an ordinary demand function for x_i.

$$x_i(p, Q, M) = -\frac{\partial V}{\partial p_i} / (\partial V / \partial M). \qquad [5]$$

In principle, this demand function for visits could be used, in the manner outlined by Freeman (1979) and many others, to estimate the ordinary surpluses associated with various levels of Q_i. Unfortunately, the earlier-mentioned desire for a brief and relatively uncomplicated questionnaire did not allow the acquisition of information leading to measures of the environmental quality levels of substitute sites. The ordinary-least-squares regression in Table 4 of annual visit frequency upon measures of price (travel costs plus on-site costs and substitute expenditures), full income (annual income and unrationed work time), and SBNF air pollution damages therefore cannot be interpreted as a demand function. It nevertheless includes variables that would enter any ordinary demand function for SBNF visits.

TABLE 4

OLS Estimates of Numbers of Annual Visits

Variable	Coefficient Estimates	Standard Error
Constant	1.7899	2.6099
Travel costs plus on-site costs[a]	−0.0671	0.0377
Substitute expenditures[a]	0.0246*	0.0112
Annual income[a]	−4.9433	3.0585
Unrationed work time[b]	3.3514	1.7366
Air pollution damages[a]	−0.1486*	0.0704

n = 180; $\bar{R}^2 = 0.15$; F = 13.82

*Significant at least at the .05 level of the one-tailed t-test for the null hypothesis of no association.

[a]See notes to Table 3 for units of measure.

[b]Work time was unrationed on the day of the interview = 1; otherwise = 0.

Since each respondent was asked only how frequently he now visited the SBNF and how frequently he would visit it if only his least preferred environment were available, 200 observations were available on the number of SBNF annual visits, the left-hand-side variable. However, 20 observations were deleted because 10 respondents professed never to visit an environment similar to those depicted. Six of the remaining 180 observations had visit frequencies of zero. In Table 4, al-

[12]Recall that the earlier discussion referred only to the respondent's estimated values conditional upon a particular visit to the SBNF alone.

though the coefficient signs of all but one of the right-hand-side variables (annual income) conform to standard expectations for a Marshallian demand function, only substitute expenditures and air pollution damages achieve statistical significance at traditional levels. When evaluated at the relevant means, the coefficient attached to the air pollution damage variable implies an elasticity of approximately -0.52, meaning that annual visit frequency is not highly sensitive to air pollution damages. Even this is probably an exaggeration since it is based on an empirical expression that did not assign any responsibility for stated variations in SBNF visitations to environmental quality levels at substitute sites.

V. CONCLUSIONS

With the single exception of Calish, et al. (1978), the literature is devoid of empirical work relating the condition of forest stocks to measures of nonharvest economic value. Forest managers will nonetheless make decisions affecting the conditions of forest stocks. This and similar studies can assist in giving empirical form to the economic content of these decision problems. If one keeps in mind that the values obtained here represent surpluses over and above the $6.00 or $7.00 access fees the respondents had already been paid, the valuations are very similar to the $5.00–$20.00 values obtained in the numerous travel cost and contingent valuation studies of wilderness and fishing activity days. Nevertheless, this comparability does not imply that values derived for activity days at one site are readily extrapolated to another site. This study has shown that such extrapolations must be adjusted for variations in the landscape as well as for the traditional pecuniary and demographic factors.

References

Baird, James C., and Noma, Edward. 1978. *Fundamentals of Scaling and Psychophysics*. New York: John Wiley and Sons.

Becker, Gary. 1965. "A Theory of the Allocation of Time." *The Economic Journal* 75 (Sept.):493–517.

Berck, Peter. 1981. "Optimal Management of Renewable Resources with Growing Demand and Stock Externalities." *Journal of Environmental Economics and Management* 8 (June):105–17.

Bolsinger, Charles L. 1980. *California Forests: Trends, Problems and Opportunities*. Resource Bulletin PNW–89. Berkeley, Calif.: Pacific Southwest Forest and Range Experiment Station (Aug.).

Bradford, David. 1970. "Benefit Cost Analysis and Demand Curves for Public Goods." *Kyklos* 23 (4):775–91.

Brookshire, David S., and Crocker, Thomas D. 1981. "The Advantages of Contingent Valuation Methods for Benefit-Cost Analysis." *Public Choice* 36 (2):235–52.

———; Thayer, Mark A.; Schulze, William M.; and d'Arge, Ralph C. 1982. "Valuing Public Goods: A Comparison of Survey and Hedonic Approaches." *The American Economic Review* 72 (Mar.):165–77.

Calish, S.; Fight, R. D.; and Teegarden E. 1978. "How Do Nontimber Values Affect Douglas-Fir Rotations." *Journal of Forestry* 76: 217–21.

Cicchetti, Charles J., and Smith, V. Kerry. 1973. "Congestion, Quality Deterioration and Optimal Use: Wilderness Recreation in the Spanish Peaks Primitive Area." *Social Science Research* 2 (Mar.):15–31.

Clawson, Marion, and Knetsch, Jack L. 1966. *Economics of Outdoor Recreation*. Baltimore: The Johns Hopkins Univ. Press.

Cornes, Richard. 1980. "External Effects: An Alternative Formulation." *European Economic Review* 13:25–42.

Courant, Paul N., and Porter, Richard C. 1981. "Averting Expenditure and the Cost of Pollution." *Journal of Environmental Economics and Management* 8 (Dec.):321–30.

Davis, Robert K. 1963. "Recreation Planning as an Economic Problem." *Natural Resources Journal* 3 (Oct.):239–49.

Deaton, Angus, and Muellbauer, John. 1980. *Economics and Consumer Behavior*. New York: Cambridge Univ. Press.

Devletoglou, Nicos E. 1971. "Thresholds and Transactions Costs." *The Quarterly Journal of Economics* 85 (Feb.):163–70.

Forster, Bruce A. 1981. "Separability, Functional Structure and Aggregation for a Class of Models in Environmental Economics." *Journal of Environmental Economics and Management* 8 (June):118–33.

Freeman, A. Myrick III. 1979. *The Benefits of Environmental Improvement*. Baltimore: The Johns Hopkins Univ. Press.

Garber, Steven, and Klepper, Steven. 1980. "Extending the Classical Normal Errors in Variables Model." *Econometrica* 48 (Sept.):1541–6.

Georgescu-Roegan, Nicholas. 1936. "The Pure Theory of Consumer's Behavior." *The Quarterly Journal of Economics* 50 (Feb.):18–33.

———. 1958. "Threshold in Choice and the Theory of Demand." *Econometrica* 26 (Feb.):156–68.

Hartman, R. 1976. "The Harvesting Decision When a Standing Forest Has Value." *Economic Inquiry* 4 (Mar.):52–8.

Hilb, Lionel. 1976. *Natural Forest Guide*. Chicago: Rand McNally and Co.

James, Robert L.; Colb, Franklin W.; Miller, Paul R.; and Parameter, James R., Jr. 1980. "Effects of Oxidant Air Pollution on Susceptibility of Pine Roots to 'Fomes assosus'." *Phytopathology* 70:560–63.

Kickert, R. N., and Gemmill, B. 1980. "Data-Based Ecological Modelling of Ozone Air Pollution Effects in a Southern California Mixed Conifer Ecosystem." In *Effects of Air Pollutants on Mediterranean and Temperate Forest Ecosystems*, ed. Paul R. Miller. General Technical Report PSW–43. Berkeley, Calif.: Pacific Southwest Forest and Range Experiment Station (Oct.).

Krutilla, John V. 1967. "Conservation Reconsidered." *The American Economic Review* 57 (Sept.):777–86.

Luce, R. Duncan. 1956. "Semiorders and a Theory of Utility Discrimination." *Econometrica* 24:178–91.

March, James G. 1978. "Bounded Rationality, Ambiguity, and the Engineering of Choice." *The Bell Journal of Economics* 9 (Autumn):587–608.

Miller, Paul R. 1973. "Oxidant-Induced Community Change in a Mixed Conifer Forest." In *Air Pollution Damage to Vegetation*, ed. James A. Maegela. Advances in Chemistry Series 122. Wash., D.C.:American Chemical Society.

———, and McBride, James R. 1975. "Effects of Air Pollutants on Forests." In *Responses of Plants to Air Pollution*, eds. J. Brian Mudd, and Theodore T. Kozlowski. New York: Academic Press.

———; Kickert, Richard N.; and Taylor, O. Clifton, et. al. 1977. *Photochemical Oxidant Air Pollutant Effects on a Mixed Conifer Forest Ecosystem*. EPA–600/3–77–104. Corvallis, Ore.: U.S. Environmental Protection Agency (Sept.).

———; Taylor, O. Clifton; and Wilhour, Ray G. 1982. *Oxidant Air Pollution Effects on a Western Coniferous Forest Ecosystem*. EPA–600/D–82–276. Corvallis, Ore.: U.S. Environmental Protection Agency (Aug.).

Rausser, Gordon C., and Oliveira, Ronald A. 1976. "An Econometric Analysis of Wilderness Area Use." *Journal of the American Statistical Association* 71 (June):276–284.

Schecter, Mark, and Lucas, Robert C. 1978. *Simulation of Recreation Use for Park and Wilderness Management*. Baltimore: The John Hopkins Univ. Press.

Smith, V. Kerry. 1975. "Travel Cost Demand Models for Wilderness Recreation: A Problem of Non-Nested Hypotheses." *Land Economics* 51 (May):103–111.

———, and Kopp, Raymond J. 1980. "The Spatial Limits of the Travel Cost Recreation Demand Model." *Land Economics* 56 (Feb.):64–72.

Taylor, O. Clifton; Miller, Paul R.; et. al. 1980. *Photochemical Oxidant Air Pollution Effects on a Mixed Conifer Forest Ecosystem*. EPA–600/3–80–002. Corvallis, Ore.: U.S. Environmental Protection Agency (Sept.).

Tversky, Amos, and Kahneman, Daniel D. 1981. "The Framing of Decisions and the Psychology of Choice." *Science* 211 (Jan. 30):453–58.

U.S. Forest Service. 1978. *The Principal Laws Relating to Forest Service Activities*. Agricultural Handbook No. 453. Wash., D.C.: U.S. Department of Agriculture.

Walsh, Richard G.; Gillman, Richard A.; and Loomis, John B. 1981. *Wilderness Resource Economics: Recreation Use and Preservation Values*. Fort Collins: Department of Economics, Colorado State Univ. (May).

C Temperate and Tropical Wetlands

[14]

Valuing Environmental Functions: Tropical Wetlands

Edward B. Barbier

ABSTRACT. *Recent studies have suggested that tropical wetland systems may have a crucial economic role to play in development. The following paper provides an overview of these benefits, using the general framework of cost-benefit analysis as the methodological approach to assessing wetland values. An analysis of trade-offs between conserving or converting tropical wetlands demonstrates that taking into account the opportunity cost of wetland loss leads to a lower level of conversion than would otherwise be the case. Finally, the paper discusses the extensions and limitations of the production function approach as applied to valuing nonmarketed wetland benefits.* (JEL Q20)

I. INTRODUCTION

Since 1900, over half of the world's wetlands may have disappeared. The United States alone has lost an estimated 54 percent (87 million hectares) of its original wetlands, of which 87 percent has been lost to agricultural development, 8 percent to urban development and 5 percent to other conversions (Maltby 1986). The total area and status of tropical wetlands are still unknown, but the available evidence suggests that the pattern of wetland conversion in developing countries may be similar to that of the United States—and perhaps proceeding at even a faster rate in some regions.

Natural wetlands perform many important functions for humankind—prevention of storm damage, flood and water flow control, support of fisheries, nutrient and waste absorption and so forth. Wetlands can also be used for recreation and water transport, and their diverse resources can be directly exploited for fishing, agriculture, wildlife products, wood products and water supply. When properly measured, the total economic value of a wetland's ecological functions, its services and its resources may exceed the economic gains of converting the area to an alternative use. Some economic studies have valued the benefits of *temperate* wetlands (see Turner 1991; Turner and Jones 1991; Farber and Costanza 1987). But to date, little analysis of *tropical* wetland benefits has been undertaken. Yet recent studies across the developing regions of the world have suggested that tropical wetland systems—whether inland freshwater systems or coastal, mangrove systems—may have a crucial role to play in economic development (Barbier 1993).

In particular, economic analysis of the environmental functions of tropical wetlands—the support and protection they provide for economic activity and property—is underdeveloped. A major problem is the lack of scientific data on ecological relationships and functions in developing countries. However, recent advances in the methodology for valuing nonmarketed environmental goods and services suggest that in many instances the data requirements for valuing environmental functions may not be too overwhelming. The production function approach to valuation may be especially promising as an approach to valuing certain environmental functions of tropical wetlands.

The following paper discusses the valuation of environmental functions of tropical wetlands, focusing in particular on their *regulatory ecological functions* in support or protection of economic activities. The valuation problem is illustrated through the use of a basic model indicating the costs and benefits of converting or diverting wet-

Professor, Department of Environmental Economics and Environmental Management, University of York, United Kingdom.

This article was taken from a paper presented to the Biodiversity Programme Workshop, The Beijer Institute for Ecological Economics of the Swedish Royal Academy of Sciences, Stockholm, 29–31 July 1992. The author is grateful to Carl Folke, Karl-Göran Mäler, and Charles Perrings and to the comments provided by two anonymous referees. All errors and omissions are, of course, solely those of the author.

Land Economics • May 1994 • 70(2): 155–73

land resources to an alternative use. The model indicates that failure to consider the foregone net benefits of *in situ* use of wetland resources can lead to an underestimation of the costs of alternative uses of these resources and their excessive appropriation from the wetlands. Given that many of the direct and indirect uses of wetland resources are nonmarketed, estimation of their value may be best approximated through the production function approach. Potential applications of this approach in the simple case of a single-use system are examined, along with the potential modifications and problems encountered. The full valuation problem of a multiple-use system presents further difficulties, which are also discussed. The paper concludes by highlighting the additional steps required to develop analytical techniques and applications which may have wider relevance to many valuation problems in developing countries.

II. VALUATION APPROACHES AND METHODOLOGY—AN OVERVIEW

"Valuing" a wetland essentially means valuing the characteristics of a system. Any system, whether natural or human-made, can be characterized by three concepts: stocks, flows, and the organization of these stocks and flows. These three system characteristics have parallel concepts in both ecology—structural components, environmental functions and diversity; and economics—assets, services and attributes. Table 1 summarizes the linkages between these basic system characteristics and their ecological and economic counterparts.

In ecology, a distinction is usually made between the regulatory environmental *functions* of an ecosystem (e.g., nutrient cycles, microclimatic functions, energy flows, etc.) and its *structural components* (e.g., biomass, abiotic matter, species of flora and fauna, etc.). This distinction is useful from an economic perspective, as it corresponds to the standard categories of resource *stocks* or *assets* (i.e., the *structural components*) vs. environmental *flows* or *services* (i.e., the ecological functions). In addition, ecosystems as a whole often have certain *attributes* (biological diversity, cultural uniqueness/heritage) that have economic value either because they induce certain economic *uses* or because they are valued in themselves.

It is therefore helpful to distinguish between:

(a) *direct use* values (e.g., the values derived from direct use or interaction with a wetland's resources and services);
(b) *indirect use* values (the indirect support and protection provided to economic activity and property by the wetland's natural functions, or regulatory "environmental" services); and
(c) *nonuse* values (the values derived neither from current direct or indirect use of the wetland).[1]

[1] A special category of value is *option value*, which arises because an individual may be uncertain about his or her future demand for a resource and/or its availability as a wetland in the future. There is a general consensus in the economics literature that option values are not a separate form of value but represent a difference between *ex ante* and *ex post* valuation (Smith 1983; Freeman 1984). If an individual is uncertain about the future value of a wetland but believes it may be high or that current exploitation and conversion may be irreversible, then there may be *quasi-option value* derived from delaying the development activities. Quasi-option value is simply the expected value of the information derived from delaying exploitation and conversion of the wetland today. Again, there is a consensus that quasi-option value is not a separate component of benefit but involves the analyst properly accounting for the implications of gaining additional information (Fisher and Hanemann 1987). In contrast, however, there are individuals who do not currently make use of tropical wetlands but nevertheless wish to see them preserved "in their own right." Such an "intrinsic" value is often referred to as *existence value*. It is a form of nonuse value that is extremely difficult to measure, as existence values involve subjective valuations by individuals unrelated to either their own or others' use, whether current or future. An important subset of nonuse or preservation values is *bequest value*, which results from individuals placing a high value on the conservation of tropical

TABLE 1

General, Ecological and Economic System Characteristics

General System Characteristics	Ecological System Characteristics	Economic System Characteristics
Stocks	Structural components	Assets
Flows	Environmental functions	Services
Organization	Biological and cultural diversity	Attributes

Source: Adapted from Aylward and Barbier (1992).

Thus the *total economic value* of a tropical wetland system may comprise all three types of values—use (direct and indirect), option, and existence values. The standard techniques available for measuring the various economic values of wetlands are summarized in Figure 1.[2]

The use and nonuse values of temperate wetlands, which exist largely in developed countries, may differ significantly from those of tropical wetlands, which occur mainly in the developing world. For example, many tropical wetlands are being directly exploited, often through nonmarket, "informal" economic activity, to support human livelihoods, e.g., through fishing, hunting, fuelwood extraction, and so on, whereas recreation/tourist use may often be limited. In contrast, direct exploitation to support livelihoods—except perhaps commercial fishing or forestry in some areas—may be small for most temperate wetlands, but their recreational value is often significant. Valuation of the noncommercial direct use of wetlands by local populations is often critical in determining the economic value of tropical wetlands to developing countries. The failure to take this value into account is often a major factor behind policy decisions that lead to overexploitation or excessive degradation of tropical wetland systems.

Direct uses of the wetlands would therefore include both *consumptive* uses of its resources (e.g., livestock grazing, fuelwood collection, forestry activities, agriculture, water use, hunting and fishing) and *nonconsumptive* uses of wetland "services" (e.g., recreation, tourism, *in situ* research and education, navigation along water courses). Direct uses of wetlands could involve both commercial and noncommercial activities, with some of the latter activities often being important for the subsistence needs of local populations. Commercial uses may be important for both domestic and international markets. In general, the value of marketed products (and services) of wetlands is eas-

wetlands for future generations to use. Bequest values may be particularly high among the local populations currently using a wetland, in that they would like to see the wetland and their way of life that has evolved in conjunction with it passed on to their heirs and future generations in general.

[2] Note that in Figure 1, option and quasi-option values are indicated with a dotted line, since these values are not strictly a separate component of total economic value in the sense that direct and indirect use values can be separated from existence value. Moreover, in most cases, the preferred approach for incorporating option values in the cost-benefit analysis would be to develop well-specified models of individual choice, through reasoning about how marginal utilities of income differ in the various contingency states (Freeman 1984). Similarly, quasi-option value can be calculated by an analysis of the conditional value of information in the decision problem (Fisher and Hanemann 1987). Such approaches are particularly valid in the case of uncertainty surrounding "decrements" in natural environments, e.g., the losses in value that might occur if a tropical wetland is converted to an alternative use. In contrast, Brookshire, Eubanks and Randall (1983) have argued that in cases of "supply-side uncertainty" surrounding "increments" in natural environments, e.g., a project to protect wetlands might provide additional future values such as recreation, contingent valuation methods (CVM) might be employed. However, Freeman (1985) has argued that this scenario is only one of four possible patterns of supply uncertainty, and has cautioned against the use of CVM in cases where either the project can only reduce but not eliminate supply uncertainty or where there is a positive probability of supply even without the project.

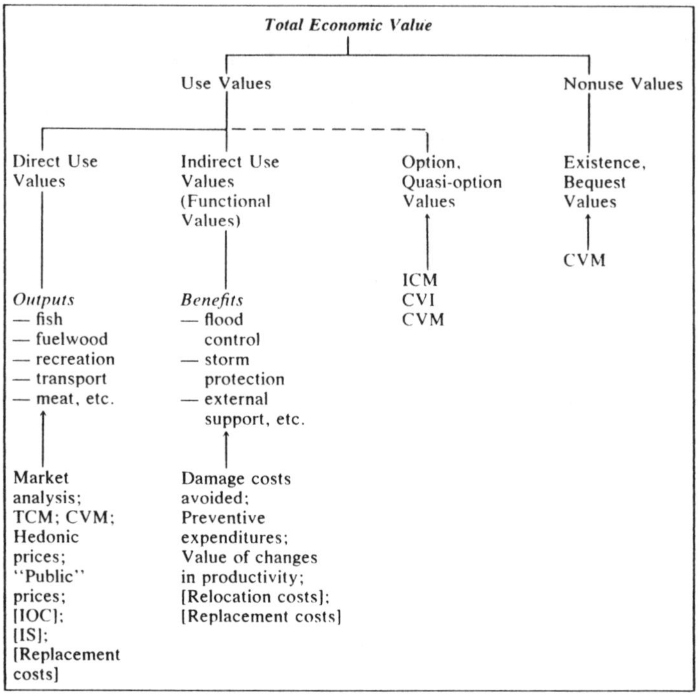

FIGURE 1
VALUING WETLAND BENEFITS

Notes: ICM = individual choice models
CVI = conditional value of information
CVM = contingent valuation method
TCM = travel cost method
IOC = indirect opportunity cost approach
IS = indirect substitute approach
[ ] = valuation methodology to be used with care

Source: Adapted from Barbier (1989a).

ier to measure than the value of non-commercial and subsistence direct uses. As noted above, this is one reason why policymakers often fail to consider subsistence and informal uses of tropical wetlands in many development decisions.

Various *regulatory ecological functions* of tropical wetlands may also have important indirect use values. Their values derive from supporting or protecting economic activities that have directly measurable values. For example, the groundwater recharge function of floodplain wetlands may have indirect use value through its replenishment of aquifer systems that supply water for domestic use and agriculture. The storm prevention function of mangrove swamps may also have indirect use value through the protection afforded coastal property and economic activity. The biological diversity of a wetland ecosystem may also have an important role in maintaining regulatory functions, e.g., changes in species diversity may affect how well support and protective services function, and in some cases, even their availability. The indirect use value of an environmental function is related to the change in the value of production or consumption of the activity or property that it is protecting or supporting. However, as this contribution is nonmarketed, financially unrewarded and only indirectly connected to economic activities, these indirect use values may be difficult to value (Aylward and Barbier 1992). Although few valuation studies of key environmental functions in tropical wetlands have been conducted, the available evidence suggests that the economic value of regulatory environmental functions may be highly significant (Adams and Hollis 1988; Barbier 1989a and 1993; Barbier, Adams, and Kimmage 1991; Lal and Dixon 1990; Ruitenbeek 1992; Twilley 1991; Yañez-Arancibia and Day 1988).

Despite the range of benefits provided by tropical wetlands, they are currently under threat from a number of sources. In many cases, the loss of wetland resources, functions and even entire systems may be justified. To make consistent choices between wetland conservation, preservation and development options—or between a decision to halt, modify or continue with an activity that is inflicting damage on a wetland—requires the application of a consistent economic appraisal methodology for evaluating the alternative options. At the heart of the appraisal methodology is the determination of the various costs and benefits associated with each option. Critical to this assessment of costs and benefits is the choice of appropriate valuation techniques, as indicated in Figure 1 and discussed above.

Barbier (1993) suggests essentially three broad categories of assessment, with each category corresponding to each major type of policy decision concerning wetland use that generally needs evaluating:[3]

(a) *impact analysis*—an assessment of the damages inflicted on the wetland from a specific environmental impact (e.g., oil spills);
(b) *partial valuation*—an assessment of alternative resource allocations or project options involving wetland systems or resources (e.g., whether to divert water from the wetlands for other uses, or to convert/develop part of the wetlands at the expense of other uses); and,
(c) *total valuation*—an assessment of the total economic value of the wetland system (e.g., for national income accounting or to determine its worth as a protected area).

Under the first approach, assessing a specific environmental impact involves valuing the changes in the wetland resulting from that impact. For example, assume that discharges of oil are regularly polluting an estuarine wetland, affecting both fish production and water quality in the wetlands. The *costs* of this activity are the *losses in wetland values* arising from damage to the ecosystem and its resources. These damages would amount to the losses in *net production benefits* (i.e., the economic benefits of production less the costs) from the impacts of the oil spills on the fishery plus the losses in *net environmental benefits* in terms of poorer quality water supplies for wetland and neighboring settlements, as well as for general ecosystem functioning. Thus, by assessing and valuing these losses, we would arrive at an estimate of the net production and environmental benefits of the wetlands, NB^W, that are affected by the oil spills. The total cost of this impact, C^I, in terms of damage to the wetland are these foregone net benefits:

$$C^I = NB^W. \quad [1]$$

Dixon and Hufschmidt (1986) and Dixon et al. (1988) provide case studies of applying this particular approach in the overall context of economic appraisal of environmental impacts. For example, in the analysis of the cost-effectiveness of various options for disposing of wastewater from a geothermal power plant on the island of

[3]In what follows, it is assumed that all costs and benefits are discounted at some positive rate into present value terms.

Leyte in the Philippines, it was necessary to decide which means of wastewater disposal from the plant would protect the environment in the most cost-effective manner. For some of the options, the costs of the environmental impacts in terms of lost marine fishery and rice production were quantified. Other environmental costs, such as energy loss, lost riverine fishery production, human health effects and amenity impacts, were not possible to quantify. For example, the analysis showed that the quantifiable environmental costs of releasing untreated waste into the Bao River or into the Mahiao River were quite high, accounting for 41 percent and 35 percent of total measurable costs of these options, respectively. Both options may also seriously contaminate the marine ecosystem with unknown and unquantifiable effects.

An impact analysis was also conducted to examine the downstream effects of logging-induced sedimentation on the marine environment of Bacuit Bay, Palawan, the Philippines (Hodgson and Dixon 1988).[4] The analysis indicated that continued logging of the Bacuit Bay watershed would result in a reduction in gross revenues of more than US$ 40 million from tourism and marine fisheries over a ten-year period. The present value of these lost earnings exceeded US$ 11 million. The major impact was on coral cover and diversity, which have an important indirect use value in supporting marine fisheries. Increased sedimentation of coastal waters reduces both coral cover and diversity, eventually causing a marked fall in fish biomass. Loss of pristine coral reefs and clear water also affects tourism, which is largely from an international clientele.

The second assessment approach, i.e., partial valuation of wetland benefits, may be required when one or more development options may lead to alteration or conversion of wetland systems. That is, choices involving diversion, allocation or conversion of wetland resources should compare the opportunity costs of the proposed options in terms of the subsequent loss in wetland benefits. For example, assume that there is an upstream irrigation project on a river that is providing water for agriculture. If this project diverts water from a wetland downstream, then any resulting loss in wetland benefits must be included as part of the overall costs of the project. Given *direct benefits* (e.g., irrigation water for farming), B^D, and *direct costs* (e.g., costs of constructing the dam, irrigation channels, etc.), C^D, then the *direct net benefits* of the project are:

$$NB^D = B^D - C^D. \qquad [2]$$

However, by diverting water that would otherwise flow into the downstream wetlands, the development project may result in losses to floodplain agriculture and other primary production activities, less groundwater recharge and other external impacts. Given these reductions in the *net production and environmental benefits*, NB^W, of the wetlands, then the true net benefits of the development project (NB^P) are $NB^D - NB^W$. The development project can therefore only be acceptable if:

$$NB^P = NB^D - NB^W > 0. \qquad [3]$$

If the foregone wetland benefits are significant, then the failure to assess the loss of wetland benefits will clearly lead to an overestimation of NB^P. This is tantamount to assuming that there is no opportunity cost of diverting floodwater from the wetlands, which is rarely the case. Moreover, it may not be necessary to measure all affected wetland benefits—for example, if one or two impacts prove to be sufficiently large to render the development project uneconomical. In any case, it is not necessary to measure all wetland benefits but only those benefits which are affected by the development project—which is why this approach is called a "partial valuation."

A partial valuation was conducted to assess the economic importance of the Hadejia-Jama'are wetlands, and thus the opportunity cost to Nigeria of its loss, by

[4] See Aylward and Barbier (1992) for a critical review.

estimating some of the key direct use values the floodplain provides to local populations through crop production, fuelwood and fishing (Barbier, Adams, and Kimmage 1991).[5] The economic analysis indicates that these benefits are substantial on both a *per hectare* basis and a *water input* basis—i.e., the minimum and maximum amount of floodwater required to sustain them. This proves to be the case even when the agricultural benefits were adjusted to take into account the unsustainability of much pump-irrigated wheat production within the wetlands. As indicated in Table 2, the present value of the aggregate stream of agricultural, fishing and fuelwood benefits were estimated to be around N850 to N1280 per ha, or around N240 to N370 per $10^3 m^3$ (with "maximum" flood inputs).[6]

The economic importance of the wetlands suggests that the benefits it provides cannot be excluded as an opportunity cost of any scheme that diverts water away from the floodplain system. When compared to the net economic benefits of the Kano River Project, the economic returns to the floodplain appear much more favorable (see Table 2). This is particularly the case when the relative returns to the Project in terms of water input use is compared to that of the floodplain system. The result should cause some concern, given that the existing and planned water developments along the Hadejia-Jama'are river system, such as the Kano River Project, will continue diverting water from the floodplain.

As the name implies, the final assessment approach involving total valuation of a wetland system requires an appraisal of all the net benefits of a wetland system. If the objective of the total valuation is to measure, say, the economic contribution of the wetlands to the welfare of society as part of a resource accounting exercise, then the objective should be to value as many of the net production and environmental benefits, NB^W, of the wetlands as possible. Another objective requiring total valuation would be the need to determine whether or not the wetlands should become a protected area with restricted or controlled use. The total net wetland benefits would therefore have to exceed the direct costs, C^P, of setting up the protected area (including any costs of relocating or compensating existing users) plus the net benefits foregone, NB^A, of *alternative uses of the wetlands:*

$$NB^W > C^P + NB^A. \qquad [4]$$

Ruitenbeek (1989) has followed this approach in determining the economic value of Korup National Park in Cameroon, where the alternative use would be logging of the forest area.[7] Ruitenbeek (1992) uses a similar approach in evaluating the tradeoffs between different forestry options in a mangrove system in Bintuni Bay, Irian Jaya, Indonesia—although in this example the comparison is between the total economic value of a wetlands preserved through a cutting ban and the total economic value generated by various forestry development options ranging from partial, selective cutting to clear-cutting.[8]

An important feature of the analysis is that it explicitly incorporates the linkages between mangrove conversion, offshore fishery productivity, traditional uses and the imputed benefits of erosion control and biodiversity maintenance functions. To the extent that these linkages exist, some of these direct and indirect uses become mutually exclusive with more intensive mangrove exploitation through forestry options. The "optimal" forest management option will therefore depend on the strength of the environmental linkages. The results indicate that the clear-cut option is optimal only if no environmental linkages exist—a highly unrealistic assumption. At the other

[5]See Barbier, Adams and Kimmage (1991) for further details on the analytical approach of the study, including the difficulties encountered, and Barbier (1993) for a retrospective review. Both papers also discuss possible alternative approaches to valuing the groundwater recharge function of the floodplain.

[6]In 1989/90 prices, 7.5 Nigerian Naira (N) = US$ 1.

[7]See Aylward and Barbier (1992) for a critical review.

[8]The "production function" methodology of the Bintuni Bay case study is discussed in more detail later in the paper. For a critical review, see Barbier (1993).

TABLE 2
COMPARISON OF PRESENT VALUE NET ECONOMIC BENEFITS
KANO RIVER PROJECT PHASE I AND HADEJIA-JAMA'ARE FLOODPLAIN, NIGERIA

(N7.5 = US$ 1, 1989/90)

Per Hectare[a]	*(8%, 50 yrs)*	*(8%, 30 yrs)*	*(12%, 50 yrs)*	*(12%, 30 yrs)*
HJF (N/ha)	1,276	1,176	872	846
KRP (N/ha)	233	214	158	153
Per Water Use[b]				
HJF ($N/10^3m^3$)	366	337	250	242
KRP ($N/10^3m^3$)	0.3	0.3	0.2	0.2

Notes: [a]Based on a total production area of 730,000 ha for Hadejia-Jama'are floodplain (HJF) and a total crop cultivated area of 19,107 ha in 1985/86 for the Kano River Project Phase I (KRP).
[b]Assumes an annual average river flow into Hadejia-Jama'are floodplain (HJF) of 2,549 Mm^3 and an annual water use of 15,000 m^3 per ha for the Kano River Project Phase I (KRP).
Source: Barbier, Adams and Kimmage (1991).

extreme, a cutting ban is only optimal if the linkages are very strong, i.e., mangrove alteration and conversion would lead to immediate and linear impacts throughout the ecosystem. Even if weak interactions exist, an 80 percent selective cutting policy with replanting is preferable to clear-cutting. However, given the considerable uncertainty over the dynamics of the mangrove ecosystem, and the fact that alteration and conversion may be irreversible and exhibit high economic costs, the analysis concludes that there is little economic advantage to cutting significant amounts (e.g., more than 25 percent) of the mangrove area.

III. THE VALUE OF ALTERNATIVE TROPICAL WETLAND USES—A MORE FORMAL ANALYSIS[9]

A simple model can be invoked to indicate more formally the need for consistent choice in decisions to convert tropical wetlands, or to allocate their resources (e.g., water) to other uses, when such decisions involve the loss of wetland benefits. As the previous sections indicate that many conflicts in tropical countries are over wetland *resource use values*—i.e., whether to convert or exploit wetland resources to alternative productive uses or to maintain current direct and indirect uses of tropical wetlands—the following model will focus on this choice of alternative wetland use.

At any time, t, scarce tropical wetland resources can be represented as a stock, $S(t)$. Which resources are to be represented by S will of course depend on the allocation problem. For example, if a coastal mangrove is being converted to shrimp ponds or irrigated rice production, then S could represent the total stock of land area within the mangrove system. Alternatively, if the mangrove forest itself is valued, e.g., for woodchip production, then S could be either the stock of mangrove biomass or forest area. Finally, in cases where the water flowing into the wetland is being reallocated, e.g., floodplain water diverted by upstream developments for irrigation and industrial use, then S might represent the total volume of water stored in the wetlands.

Equally, the amount of wetland resources converted (or diverted) from the wetlands in any time period, $D(t)$, can also be defined depending on the nature of the conversion (or diversion) activity.[10] To sim-

[9]The analysis of the following section benefitted greatly from discussions held with Karl-Göran Mäler and Carl Folke at the Beijer Institute of Ecological Economics, Royal Swedish Academy of Sciences, in preparation of a project proposal on mangrove valuation. Based on these discussions, Mäler (1992) has also extended a production function model to incorporate mangrove valuation.

[10]In what follows, notation is simplified by omitting the argument of time-dependent variables, by representing a derivative of a function by a prime, by em-

plify matters, it is assumed that the conversion/diversion activity leads to an *irreversible loss* in the wetland resource, S, which is essentially nonrenewable:

$$\dot{S} = -D. \qquad [5]$$

Thus, by virtue of [5], the analysis can focus on the costs and benefits of productive activities that lead to wetland loss.[11]

Two production activities are assumed to be competing for wetland resources. One activity combines resources extracted from the wetlands, D, with other inputs, Z, to produce a commodity, Y. The other activity either *directly* uses the remaining stock of resources, S, or is *indirectly* supported or protected by the wetland resources. The various direct and indirect use values of tropical wetlands were discussed in detail in previous sections and are outlined in Figure 1. The stock of wetland resources, S, can therefore be considered one input in the production function of the activity, which along with other inputs, X, produce a commodity Q.[12]

If the production functions for Q and Y are assumed to be increasing, strictly concave, twice differentiable and linearly homogeneous, and if their units are normalized so that X and Z are equal to 1, then:

$$Q = F(S, 1) = f(S), \quad f' > 0, \quad f'' < 0, \qquad [6]$$

$$Y = G(D, 1) = g(D), \quad g' > 0, \quad g'' < 0. \qquad [7]$$

Thus, the tropical wetlands country is assumed to maximize the present value of future welfare, W:

$$\underset{D}{\text{Max}}\ W = \int_0^\infty \{U(Q, Y) - C\}e^{-\delta t}dt, \qquad [8]$$

subject to [5], [6], and [7], and

$$C = c(D, S), \quad c_D > 0, c_S > 0, \quad c_{DD} \geq 0, c_{SS} \geq 0, \qquad [9]$$

$$S(0) = S_0 \quad \text{and} \quad \lim_{t\to\infty} S(t) \geq 0. \qquad [10]$$

The control variable of the problem is D, and δ is the social rate of discount. The "utility" function, U, essentially represents aggregate *gross consumer surplus*, and is assumed to have the standard properties with respect to its partial derivatives, $U'(\cdot) > 0$, $U''(\cdot) \leq 0$ and $U'(\cdot) \to \infty$ as $D, S \to 0$.[13] Equation [9] represents the aggregate *direct* costs to the country of produc-

ploying numbered subscripts to indicate partial derivatives of a function, and by denoting the time derivative of a variable by a dot.

[11] This is an obvious simplification in some problems of wetland resource conversion or diversion. For example, mangroves and other wetland forests do regenerate, and even the total land area of some wetland systems has been known to grow in size as more sediment (and water) is "trapped" by the system. Thus for some wetland resource allocation problems, a more realistic formulation of Equation [1] would be:

$$\dot{S} = g(S) - D, \qquad [1]'$$

where g represents the rate of natural growth, or accumulation, of the wetland resource, with $g_S > 0$ and $g_{SS} < 0$. Alternatively, as noted in the previous section, wetland systems and resources may be indirectly affected by the environmental impacts of neighboring activities, such as oil spills and other forms of pollution. Equation [1] could therefore be written in the form of a damage function:

$$\dot{S} = -h(D), \qquad [1]''$$

where $h_D > 0$ and $h_{DD} < 0$.

[12] See, for example, the following section and Mäler (1991) for a discussion as to why this formulation may be appropriate for capturing the major use values of wetland resources. As the next section will discuss, including the stock of resources, S, explicitly as an argument in the production function for Q is particularly appropriate for the indirect use values of wetland environmental regulatory functions.

[13] For example, it is conceivable the U takes the form:

$$U(Q, Y) = \int_0^Q B_1(Q)dQ + \int_0^Y B_2(Y)dYdY$$

where $B_1(Q) = p_1 = d_1^{-1}(Q)$ and $B_2(Y) = p_2 = d_2^{-1}(Y)$ are the *inverse demand functions* for Q and Y determining their prices p_1 and p_2, respectively, if $d_1'(Q)$ and $d_2'(Q) < 0$. It follows that $U'(Q) = p_1 > 0$ and $U'(Y) = p_2 > 0$. Note that the objective function [8] would also result if U is a *quasi-linear utility function*, i.e., if it is linear in some good m (e.g., "money") which serves as the numeraire and is assigned the price 1, such that $U = U(Q, Y) + m$. For proof, see Varian (1984).

ing Q and Y, using the (normalized) inputs of S and D.[14] Equation [10] contains the initial and terminal boundary conditions.

The current value Hamiltonian of the above optimal control problem is:

$$H = U(f(S), g(D)) - c(S, D) - \lambda D, \quad [11]$$

where λ is the costate variable of the shadow price of the "unextracted" wetland resource.

Assuming an interior solution, the maximum principle yields the following conditions:

$$\lambda \geq U_Y g' - c_D, \quad [12]$$

$$\dot{\lambda} = \delta\lambda - (U_Q f' - c_S), \quad [13]$$

where $\partial U/\partial Q = U_Q$ and $\partial U/\partial Y = U_Y$. Equation [12] reflects the nonnegativity constraint $D \geq 0$. If $D > 0$, then the shadow price of the "unextracted" wetland resource must equal the difference between the value of the marginal product of the resource appropriated to produce Y and the marginal costs of this production. However, if λ exceeds the net value marginal product of the appropriated resource, then conversion of wetland resources is not worth it and $D = 0$.[15] Equation [13] indicates the optimal rate of increase in the value of the (unextracted) wetland resource. Assuming $D > 0$, [13] can be rewritten as:

$$\dot{\lambda}/\lambda = \delta - (U_Q f' - c_S)/(U_Y g' - c_D). \quad [14]$$

Thus, the rate of change in the shadow price of the tropical wetland, $\dot{\lambda}$, is determined not only by the social rate of discount, δ, as is the case in standard nonrenewable resource problems, but also by an additional factor indicating the relative social value of *in situ* wetland resources $(U_Q f' - c_S)/(U_Y g' - c_D)$. In other words, δ represents the opportunity cost of "holding onto" wetland resources today—as opposed to appropriating them for current production of Y—whereas $(U_Q f' - c_S)/(U_Y g' - c_D)$ represents the social gains from "holding onto" wetland resources in terms of their relative direct and/or indirect use values. Note that the right-hand side of [14] can be positive, negative or equal to zero. If $\dot{\lambda}/\lambda < 0$, then the optimal rate of extraction should increase over time; however, $\dot{\lambda}/\lambda > 0$ implies that $\dot{D} < 0$.[16]

Essentially, condition [14] confirms the arguments in the previous section: if policymakers fail to take into account the "opportunity" costs of wetland loss in terms of foregone direct and indirect use values, then they are misrepresenting the true social value of tropical wetlands. The failure to consider the foregone net benefits of *in situ* use of wetland resources can lead to an underestimation of the costs of alternative uses of these resources and their excessive extraction from the wetlands. The latter point will be illustrated explicitly in a moment. First consider the case where the net

[14] For example, C may take the form $c(D, S) = c(D) + c(S)$. It is conceivable that $c(D) = w_1 D$, where w_1 is the average cost of "extracting" or "converting" a unit of wetland resource D to produce commodity Y. It is also possible that $c(S) = w_2 S$ is the costs of "maintaining" or "utilizing" the remaining stock of wetland resource, S for activity $f(S)$. However, if this latter activity only *indirectly* uses the resource S, then there may be no direct costs of utilization or maintenance and $c(S)$ is effectively zero. In either instance, as the next section discusses, the correct approach to valuing the *net* welfare contribution of S is through the production function approach applied to the (nonnormalized) production of $Q = F(X, S)$.

[15] However, it follows from the standard economic theory of exhaustible resources that Y can still be produced if D is not essential to its production (Dasgupta and Heal 1974, 1979). Note also that conditions [12] and [13] depend on the assumption that $S(t) > 0$ in finite time. This assumption can be revoked in cases where D is not essential to producing Y, and either S is not essential to producing Q or Q can be perfectly substituted by another commodity (e.g., Y) in social utility, W.

[16] If it is assumed as before that $U(Q, Y)$ is additively separable, and additionally that $Y = g(D)$ is constant over all periods, then totally differentiating [12] with respect to time yields $\dot{\lambda} = (U_{YY}(g')^2 + U_Y g'' - c_{DD})\dot{D}$. Thus when $\dot{\lambda}$ is negative in expressions [13] and [14] then $\dot{D} > 0$. If positive, then $\dot{D} < 0$. However, note that $\dot{D} > 0$ is infeasible over an infinite time horizon as wetland resources are fixed and must eventually be exhausted. Nevertheless, $D(t)$ is not necessarily monotonic over the planning horizon; e.g., depletion could conceivably increase in earlier and decrease in later periods.

benefits of *in situ* wetland resource use are ignored.

If tropical wetlands are viewed only as a source of resources that essentially have no economic value unless they are converted or diverted to a "productive" activity, then the planning problem as represented by [5]–[10] effectively reduces to the even simpler problem of nonrenewable resource extraction. As the previous sections of the paper have indicated, this is precisely the view of tropical wetlands that is often taken in policy and investment decisions that determine the allocation of wetland resources. The Hamiltonian [11] therefore becomes:

$$H = U(g(D)) - c(D) - \lambda D. \quad [15]$$

Assuming $D > 0$ and $S > 0$ in finite time and an interior solution, the standard first-order conditions for optimal nonrenewable resource extraction result:

$$\lambda = U_Y g' - c_D, \quad [16]$$

$$\dot{\lambda} = \delta\lambda \quad \text{or} \quad \lambda(t) = \lambda_0 e^{\delta t}, \quad [17]$$

where $\lambda_0 = \lambda(t)$. Condition [16] has the same interpretation as [12] for $D > 0$. Comparison of [17] and [14] confirm the difference between the standard nonrenewable extraction problem and the problem where the foregone net benefits of *in situ* wetland resources are an opportunity cost to appropriating these resources for an alternative productive use.

Following Dasgupta and Heal (1979), the extraction path $D(t)$ can be characterized as a demand for wetland resources that is inversely related to its price, $\lambda(t)$. Assuming this demand is isoelastic, then:

$$D(t) = \lambda(t)^{-1/\alpha}, \quad [18]$$

where $\alpha > 0$. Substituting [17] into [18] yields:

$$D(t) = \lambda_0^{-1/\alpha} e^{-\delta t/\alpha}. \quad [19]$$

However, the condition that D is essential to production of Y such that $D > 0$ over finite time also implies that:

$$\int_0^\infty D(t)dt = S_0 \quad \text{or} \quad (\alpha/\delta)\lambda_0^{-1/\alpha} = S_0. \quad [20]$$

Utilizing [17]–[20], optimal paths for D and λ can be determined:

$$D^*(t) = [\delta S_0/\alpha]e^{-\delta t/\alpha}, \quad [21]$$

$$\lambda^*(t) = [\delta S_0/\alpha]^{-\alpha} e^{\delta t}. \quad [22]$$

Thus, if the initial level of wetland resource appropriation is set at the optimal level $\delta S_0/\alpha$, $D(t)$ will decline thereafter at the rate δ/α. The value of "unextracted" resources will begin at $[\delta S_0/\alpha]^{-\alpha}$ and rise at the rate δ.

Similar optimal paths can also be constructed for the original wetlands problem. To facilitate comparison with the paths just derived for the "myopic" wetlands problem, assume that $\dot{\lambda} > 0$ in the first-order conditions of the original model. Returning to [13] and integrating yields:

$$\lambda(t) = \lambda_0 e^{\delta t} - \int(U_Q f_S - c_S)dt. \quad [23]$$

The second term on the right-hand side of [23] represents the cumulative stream of net benefits from *in situ* wetland resources over time. Effectively, this term is the undiscounted *asset value* of "conserved," or "unappropriated," wetland resources at time t. Defining $\varphi(t) = \int(U_Q f_S - c_S)dt$ and substituting [23] into [18] yields:

$$D(t) = \lambda_0^{-1/\alpha} e^{-\delta t/\alpha} - \varphi^{-1/\alpha}. \quad [24]$$

Letting $\varphi(t) = \varphi_0 e^{\beta t}$, where $\beta > 0$,[17] then

$$\int_0^\infty D(t)dt = (\alpha/\delta)\lambda_0^{-1/\alpha} + (\alpha/\beta)\varphi_0^{-1/\alpha}. \quad [25]$$

[17]This expression assumes that the (undiscounted) asset value of wetland resources through production activity $Q = f(S)$ will increase with time over its initial period value. The presumption is that $U_Q f_S - c_S > 0$ over time, which is a reasonable assumption provided that the activity, Q, is "sustainable"—i.e., it does not

It is possible for expression [25] to be greater than or equal to S_0. Assuming first that the entire stock of wetland resources is appropriated in the long run, then [25] suggests that:

$$\lambda_0^{**} = [\delta S_0/\alpha]^{-\alpha} - (\alpha/\beta)\varphi_0 < \lambda_0^{*}. \qquad [26]$$

From [23] and [24], the optimal paths for D and λ can therefore be determined as:

$$D(t)^{**} = [(\delta S_0/\alpha) - ((\alpha/\beta)\varphi_0)^{-1/\alpha}]e^{-\beta t/\alpha} - \varphi^{-1/\alpha} < D(t)^{*}, \qquad [27]$$

$$\lambda(t)^{**} = [(\delta S_0/\alpha)^{-\alpha} - (\alpha/\beta)\varphi_0]e^{\delta t} - \varphi < \lambda(t)^{*}. \qquad [28]$$

Taking into account the asset value of "unappropriated" wetland resources has the effect of shifting down both the optimal path of resource extraction, $D(t)$, and that of the shadow price, $\lambda(t)$. The difference in the two extraction paths [24] and [27] is:

$$\Delta(t) = D(t)^{*} - D(t)^{**} = ((\alpha/\beta)\varphi_0)^{-1/\alpha}e^{-\beta t/\alpha} + \varphi^{-1/\alpha}, \qquad [29]$$

where $d\Delta/dt < 0$. Thus as shown in Figure 2, the two paths eventually will converge in the long run.

In the second case, where some of the stock of wetland resources remain even in the long run, then:

$$\int_0^{\infty} D(t)dt + R = S_0, \qquad [30]$$

where R is the remaining long-run stock of wetland resources. Defining $\hat{D}(t)$ as the new extraction path for wetland resources, it follows that:

$$\hat{D} = [(\delta/\alpha(S_0 - R)) - ((\alpha/\beta)\varphi_0)^{-1/\alpha}]e^{-\beta t/\alpha} - \varphi^{-1/\alpha} < D(t)^{**} < D(t)^{*}. \qquad [31]$$

In sum, the above analysis has demonstrated formally that, when the loss of wetland benefits is taken into account in decisions to convert tropical wetlands, a lower level of wetland conversion will result than would be the case if these wetland benefits are ignored. In other words, policymakers should be absolutely certain that the opportunity cost of wetland resource conversion and appropriation is zero before embarking on investment and policy decisions that are based on this assumption and lead to irreversible wetland loss.

If the opportunity cost of wetland loss is not zero, then the wetland benefits foregone must be explicitly accounted for in development decisions. This again raises the issue as to what is the best method for valuing these benefits, given that the economic contribution of wetland resources—particularly their environmental functions—is often nonmarketed.

IV. VALUATION OF WETLAND ENVIRONMENTAL FUNCTIONS: THE PRODUCTION FUNCTION APPROACH

The preceding analysis noted the importance of wetland resources for production by including them explicitly in the production function for Q. The assumption was that wetland resources were being either *directly used* in production as inputs or *indirectly used* in the sense that the environmental functions of wetland resources were supporting or protecting economic activity. Earlier sections have noted the important economic contribution of many wetland indirect use values, such as support of fisheries, groundwater recharge, flood and water flow control, prevention of storm damage, and so forth. The results of the previous analysis would suggest that decisions con-

itself lead to degradation of the wetland resource base, S. For instance, some direct uses of wetland resources, such as harvesting for fuelwood and timber or certain farming systems, may prove to be "unsustainable" with time. As discussed by Barbier (1989a, 1993), it is important that any calculation of the future stream of net benefits from these activities takes into account their "unsustainability"; otherwise, the asset value of wetland resources as reflected in the economic value of these activities will be overestimated. For example, the analysis of agricultural benefits in the Hadejia-Jama'are floodplain of Nigeria took into account the impact on land degradation of pump-irrigated wheat production within the wetlands that could arise within three to four years (Barbier, Adams and Kimmage 1991).

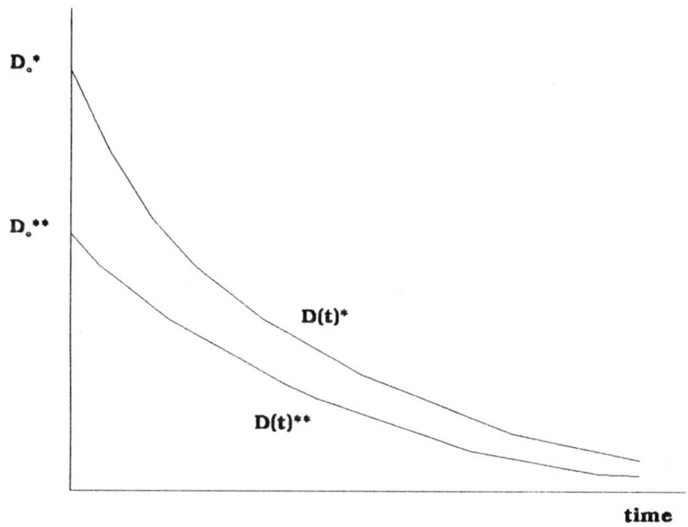

FIGURE 2
OPTIMAL EXTRACTION PATHS FOR WETLAND RESOURCES

cerning wetland conversion and resource use do require proper valuation of the environmental functions of tropical wetlands where these functions are instrumental in supporting or protecting economic activity.

One promising method is to employ the *production function approach*,[18] as implied by equation [6] above, to capture the indirect use value of regulatory ecological functions:

$$Q = F(X_i \ldots X_k, S), \quad [32]$$

where $F_S > 0$, $F_{SS} < 0$. For example, a common ecological function of mangroves is support of offshore fisheries by serving as both a spawning ground and a nursery for fry (see, for example, Yañez-Arancibia and Day 1988). The area of mangroves in a coastal region, S, may therefore have a direct influence on the catch of mangrove-dependent species, Q, which is independent from the standard inputs of a commercial fishery, $X_i \ldots X_k$. Including mangrove area as a determinant of fish catch may therefore "capture" some element of the economic contribution of this important ecological support function.

[18]Another term for this method is the *household production function* approach, which is a more appropriate term for applications based on the derived demand by households for environmental quality (Smith 1991). In such applications, F can be thought of as a "household" production function that appears directly in the household utility function:

$$U = U(F(X_i \ldots X_k, S), Y),$$

where Y is a vector of other goods and services for which both prices and quantities are known. Thus by explicitly incorporating nonmarketed environmental functions in the modelling of individuals' preferences, household expenditures on private goods can be related to the derived demand for environmental functions. Some well-known techniques in applied environmental economics—such as travel cost, recreation demand and averting behavior models—are based on this approach. However, applications at the household level in developing countries may be limited given the detailed data requirements for household patterns of expenditures, time allocations, commodity prices and wage rates, along with measures of levels of environmental quality experienced by the same households (Smith 1991).

Applying the production function approach to the various indirect use values of tropical wetlands may prove a useful method of estimating these nonmarketed—but often significant—economic values. However, it is extremely important that the relationship between the environmental regulatory function of the tropical wetlands and the economic activity it protects or supports is well understood.

Mäler (1991) distinguishes between applications of the production function approach. When production, Q, is measurable, and either there is a market price for this output or one can be imputed, then determining the marginal value of the resource is relatively straightforward. If Q cannot be measured directly, then either a marketed substitute has to be found, or possible complementarity or substitutability between S and one or more of the other (marketed) inputs, $X_i \ldots X_k$, has to be specified explicitly. Although all these applications require detailed knowledge of the physical effects on production of changes in the wetland resource, S, and its environmental functions, applications that assume complementarity or substitutability between the resource and other inputs are particularly stringent on the information required on physical relationships in production. Clearly, cooperation is required between economists, ecologists and other researchers to determine the precise nature of these relationships.

Applications of the production function approach may be most straightforward in the case of *single use systems*—i.e., wetland systems in which the predominant economic value is a single regulatory function or a group of ecological functions providing support or protection for an economic activity in concert. For example, Ellis and Fisher (1987) use this approach to model explicitly the environmental function of Gulf Coast wetlands in support of the commercial blue crab fishery. Taking the sum of consumers' and producers' surpluses as the measure of economic value, they hypothesize that an increase in wetland area increases the abundance of crabs and thus lowers the cost of catch. The value of the wetlands' support for the fishery—which in this case is equivalent to the value of increments to wetland area—can then be imputed from the resulting changes in consumers' and producers' surplus.

In the Ellis and Fisher model, equation [32] above assumes the specific Cobb-Douglas form:

$$Q = AX^a S^b, \qquad [33]$$

where Q is the quantity of crab catch in pounds, X is catch effort measured by traps set and S is area of wetlands. The corresponding cost function is:

$$C = WA^{-1/a} S^{-b/a} Q^{1/a}, \qquad [34]$$

where W is the unit cost of effort and S is determined exogenously. Assuming an isoelastic demand for crabs and either private ownership or optimal public management (i.e., price equals marginal cost in both cases), Ellis and Fisher solve the two-equation model for the incremental value of the wetlands' support function.

However, using the Ellis and Fisher model for the blue crab fishery, Freeman (1991) has made the important additional point that the values imputed to the wetlands are influenced by the market conditions and regulatory policies that determine the conditions of access and rate of utilization of the fishery. For example, under open access, rents in the fishery would be dissipated, and price would be equated to average and not marginal costs. As a consequence, producer surplus is zero and only consumer surplus determines the value of increased wetland area. When the demand for crabs is inelastic, the social value of an increase in area is higher under open access than under optimal regulation, whereas the wetlands are more valuable under optimal regulation when demand is elastic. This result stems from the role of price changes in allocating welfare gains between producers and consumers: in the case of optimal regulations, part of the consumers' gain is a transfer from producer surplus, whereas under open access and zero producer surplus, any reduction in the price of fish associated with the average cost curve shifting down (in response to an increase in wetland area) results in a gain in consumer surplus and increased wetland value.[19] These different impacts of market conditions and regulatory policies for the production function approach to valuing wetlands are an important consideration in the application of this approach to tropical wetlands in many developing regions, where open access and imperfect markets for resources are common.

In the case of *multiple use systems*—i.e., wetland systems in which a regulatory function may support or protect many different economic activities, or which may have more than one regulatory ecological function of important economic value—applications of the production function approach may be slightly more problematic. In particular, assumptions concerning the ecological relationships among these various multiple uses must be carefully constructed.

For example, an important feature of the analysis of the mangrove wetlands of Bintuni Bay, Irian Jaya, Indonesia was that it attempted to incorporate explicitly the possible ecological linkages between indirect and direct use values (Ruitenbeek 1992). Specifically, the mangroves may support many economic activities within the wetlands, such as commercial shrimp fishing, commercial sago production and traditional household production from hunting, fishing, gathering and cottage industry; they may also have indirect use value through controlling erosion and sedimentation, which protects agricultural production in the region; and they have an indirect role in supporting biodiversity. To the extent that the ecological linkages in terms of support or protection of these activities are strong, then the opportunity cost of forestry options that lead to the depletion or degradation of the mangroves will be high. Thus, as discussed above, the "optimal" forest management option—whether clear-cutting, selective cutting or complete preservation—depends critically on the strength of the ecological linkages.

In the absence of any ecological data on these linkages, Ruitenbeek developed several different scenarios based on different linkage assumptions. This essentially amounted to specifying more precisely the relationship between Q and S in the production function [32] for each productive activity at time t, Q_{it}:

$$Q_{it}/Q_{i0} = (S_{t-\tau}/S_0)^{\alpha}, \qquad [35]$$

where S_t is the area of remaining undisturbed mangroves at time t, α and τ are impact intensity and delay parameters respectively, $Q_{i0} = Q_{it}(t = 0)$ and $S_0 = S_t(t = 0)$. For example, for fishery-mangrove linkages, a moderate linkage of $\alpha = 0.5$ and $\tau = 5$ would imply that shrimp output varies with the square root of mangrove area (e.g., a 50 percent reduction in mangrove area would result in a 30 percent fall in shrimp production), and there would be a delay of five years before the impact takes effect. If no ecological linkages are present, i.e., there is no indirect use value of mangroves in terms of supporting shrimp fishing, then $\alpha = 0$. At the other extreme, very strong linkages imply that the impacts of mangrove removal are linear and immediate, i.e., $\alpha = 1$ and $\tau = 0$. As discussed above, the analysis concluded that the assumption of no environmental linkages is unrealistic for most economic activities in the wetlands. Moreover, given the uncertainty over these linkages and the high costs associated with irreversible loss if environmental linkages prove to be significant, then only modest selective cutting (e.g., 25 percent or less) of the mangrove area was recommended.

Two major difficulties in specifying ecological-economic relationships for the application of the production function ap-

[19] Freeman (1991) also calculates the social value of the marginal product of wetland area, which is given by:

$$VMP_S = bPQ/S,$$

where P is the price of crabs. As optimal regulation should lead to a higher price than open access, an inelastic demand means that VMP_S is higher under optimal regulation.

proach to estimating indirect use values in multiple use tropical wetland systems are the problems of "double-counting" and "trade-offs" between various direct and indirect use values (Aylward and Barbier 1992). The problem arises when analysts attempt to "aggregate" the total economic value (TEV) of the wetlands from the different use value subcomponents.

Aylward and Barbier provide an example of both on-site and off-site double-counting in terms of the nutrient retention function of a coastal wetland. Coastal wetlands often absorb organic nutrients from sewage and other waste emitted into waterways further upstream. Suppose that the nutrients held by the wetland are indirectly supporting both shrimp production within the wetland area and the growth of fish fry that supply an offshore fishery. If the full value of the shrimp production is already accounted for as a direct use value of the wetland's resources, adding in the share of the nutrient retention service as an indirect value and aggregating to obtain TEV would double-count this indirect use. In other words, the value of shrimp production already "captures" the value-added contribution of nutrient retention. If instead one wanted to explicitly account for the value-added contribution to shrimp production of the nutrient retention function, then the direct value of the shrimp must be decreased to account for the return in value now attached to the nutrient retention service.

Similarly, if the fish fry supported through nutrients retained in the wetland eventually migrate to an offshore fishery, then the indirect contribution during the fry's stay in the wetland is included as an off-site component of the service's value. That is, the nutrient retention function of the wetland produces an "external" benefit in terms of supporting an offshore fishery. Again, care must be taken to adjust the value of harvested fish in any companion analysis of the adjoining fishery to avoid misrepresenting the total economic value of the wetland and the fishery taken together.

Trade-offs between two or more indirect use values of a given ecosystem may also occur. For example, as discussed above, the Hadejia-Jama'are floodplain in Northern Nigeria supports a number of important agricultural, forestry and fishing activities within the wetland area. The floodplain also contributes to the recharge of groundwater, which is in turn drawn off by numerous small village wells throughout the region for domestic use and agricultural activities. However, concerns have recently been expressed about the excessive water use of pump-irrigated wheat production within the floodplain (Barbier, Adams and Kimmage 1991). Increasing use of the floodplain water to support this activity may mean less water available for natural groundwater recharge—and thus for village wells outside the floodplain. If there are trade-offs between the two environmental support functions, then adding the full value of the wetland's contribution to pump-irrigated wheat production within the floodplain to the full value of groundwater recharge of wells in neighboring regions would overestimate the total benefit of these two environmental functions.

In sum, the problems posed by double-counting and possible trade-offs need to be sorted out in any measurement of the indirect use values of environmental regulatory functions in multiple use tropical wetland systems. While it is necessary to disaggregate the goods, services and attributes of an ecosystem for valuation purposes, any complementarity and substitutability of these services must be accounted for in arriving at either total indirect use or total use values. Otherwise these values may be grossly overstated.

Finally, a major difficulty that any interdisciplinary effort to improve economic valuation of ecological functions must overcome is reconciling the different "worldviews" of economics and ecology. Despite the apparent unifying methodology over system concepts implied by Table 1, fundamental disagreements do arise between economists and ecologists over what these concepts actually entail and how they relate to the ecosystem as a whole, especially in a dynamic context of ecological change. For example, Common and Perrings (1992) have argued that disagreements over the interpretation of economic and ecological "sustainability" have hindered cooperation between the two disciplines on the decision rules concerning economic use of ecosystem goods and services. Bridging these conceptual differences is fundamental to progress in economic valuation of ecological functions, which is best accomplished through more applied interdisciplinary research on such problems.

V. CONCLUSIONS

Increasingly, tropical wetlands are seen to perform many important ecological regulatory functions that protect or support economic activity. Valuing this indirect use value may be important to decisions concerning wetland conversion and the diversion of wetland resources to other uses. When the economic value of environmental functions are significant, less "extraction" of wetland resources appears to be more optimal than in the case where the opportunity cost of extraction is zero.

Consistent choice in the allocation of wetland resources therefore requires careful analysis of the economic values provided by tropical wetlands, in particular the indirect use values of environmental regulatory functions. Because the economic services of these functions are generally nonmarketed, special valuation techniques may need to be employed to assess their economic contribution. Although the production function approach is promising, care must be taken to specify correctly the ecological-economic relationships determining the role of wetland environmental functions in production, the regulatory policies and market conditions that influence the value of this production and the problems of double-counting and possible trade-offs encountered when the indirect use values of multiple use systems are analyzed.

However, there are certain ecosystem valuation issues that are problematic for a cost-benefit framework to address without adequate guidance from policymakers.[20] One is the possibility that human activities can lead to changes in ecosystems with uncertain and irreversible consequences. As discussed previously, economic concepts such as option and quasi-option value can be incorporated into the cost-benefit analysis to allow for certain contingencies arising from environmental losses. However, economists have increasingly recognized that valuation of whole-scale ecosystem changes must explicitly or implicitly make judgements involving intergenerational equity, with implications for the choice of discount rate and/or the valuation methodologies chosen.[21] When these changes may potentially endanger key species, or even whole ecosystems, then some recent economic thinking has advocated the more precautionary approach of respect for "safe minimum standards" and "ecological threshold effects," particularly where there is reason to believe that the social costs of species loss or system collapse may be high.[22] It is conceivable that a cost-benefit framework involving marginal valuations is only relevant when development decisions influencing environmental losses operate within the bounds established by "safe minimum standard" and "ecological threshold" criteria. Nevertheless, any approach that puts the requirements of an ecological-economic system above those of the individual again requires an ethical judgement about the role and rights of present and future generations—a judgement that is beyond the scope of the analyst alone to make.

Addressing such major issues requires not only guidance from policymakers, but

[20] There is always the danger, however, that this guidance may be dominated by special-interest groups. For an interesting review of how the methodology of appraisal of natural resource projects can become a public "bargaining" process, and not always for the worse, see Henry (1989).

[21] For the general issues involved in economic appraisal, see Krutilla and Fisher (1985) and Porter (1982). For issues related to the specific concerns of developing countries, see Markandya and Pearce (1988).

[22] On "safe minimum standards" and endangered species, see Bishop (1978), Ready and Bishop (1991), and Tisdell (1990). On the implications of ecological threshold effects for economic development, see Barbier (1989b) and Perrings and Pearce (1992).

also close cooperation between economists and ecologists to determine the policy-relevant analytical framework and the decision choices open to society. There is clearly a strong case to be made for more interdisciplinary research to bridge the gap between ecological and economic views of the world. As this paper has sought to demonstrate, such cooperation has already begun to make a worthwhile contribution to developing a more complete cost-benefit analysis, particularly with regard to valuation of important ecological regulatory functions that protect or support economic activity.

References

Adams, William M., and G. E. Hollis. 1989. "Hydrology and Sustainable Resource Development of a Sahelian Floodplain Wetland." Report prepared for the Hadejia-Jama'are Wetlands Conservation Project, Nguru, Nigeria.

Aylward, Bruce A., and Edward B. Barbier. 1992. "Valuing Environmental Functions in Developing Countries." *Biodiversity and Conservation* 1:34–50.

Barbier, Edward B. 1989a. *The Economic Value of Ecosystems: 1—Tropical Wetlands*. LEEC Gatekeeper 89-02. London: London Environmental Economics Centre.

Barbier, Edward B. 1989b. *Economics, Natural Resource Scarcity and Development: Conventional and Alternative Views*. London: Earthscan.

Barbier, Edward B. 1993. "Valuing Tropical Wetland Benefits: Economic Methodologies and Applications." *Geographical Journal*, Part 1, 59 (Mar.):22–32.

Barbier, Edward B., William M. Adams, and Kevin Kimmage. 1991. *Economic Valuation of Wetland Benefits: The Hadejia-Jama'are Floodplain, Nigeria*. LEEC Discussion Paper DP 91-02. London: London Environmental Economics Centre.

Bishop, Richard C. 1978. "Endangered Species and Uncertainty: The Economics of a Safe Minimum Standard." *American Journal of Agricultural Economics* 57:10–18.

Brookshire, David S., Larry S. Eubanks, and Alan Randall. 1983. "Estimating Option Prices and Existence Values for Wildlife Resources." *Land Economics* 59 (Feb.):1–15.

Common, Michael S., and Charles Perrings. 1992. "Towards an Ecological Economics of Sustainability." *Ecological Economics* 6: 7–34.

Dasgupta, Partha S., and Geoffrey R. Heal. 1974. "The Optimal Depletion of Exhaustible Resources." *Review of Economic Studies, Symposium on the Economics of Exhaustible Resources* 41:3–28.

———. 1979. *Economic Theory and Exhaustible Resources*. Cambridge: Cambridge University Press.

Dixon, John A., Richard A. Carpenter, Louise A. Fallon, Paul B. Sherman, and Supachit Manopimoke. 1988. *Economic Analysis of the Environmental Impacts of Development Projects*. London: Earthscan Publications in association with the Asian Development Bank.

Dixon, John A., and Maynard M. Hufschmidt, eds. 1986. *Economic Valuation Techniques for the Environment: A Case Study Workbook*. Baltimore: Johns Hopkins University Press.

Ellis, Gregory M., and Anthony C. Fisher. 1987. "Valuing the Environment as Input." *Journal of Environmental Management* 25:149–56.

Farber, Stephen J., and Robert Costanza. 1987. "The Economic Value of Wetland Systems." *Journal of Environmental Management* 24: 41–51.

Fisher, Anthony C., and Michael W. Hanemann. 1987. "Quasi-Option Value: Some Misconceptions Dispelled." *Journal of Environmental Economics and Management* 14:183–90.

Freeman, A. Myrick III. 1984. "The Sign and Size of Option Value." *Land Economics* 60 (Feb.):1–13.

———. 1985. "Supply Uncertainty, Option Price, and Option Value." *Land Economics* 61 (May):176–81.

———. 1991. "Valuing Environmental Resources under Alternative Management Regimes." *Ecological Economics* 3:247–56.

Henry, Claude. 1989. "Investment Projects and Natural Resources: Economic Rationality in Janus' Role." *Ecological Economics* 1: 117–35.

Hodgson, Gregor, and John A. Dixon. 1988. "Logging Versus Fisheries and Tourism in Palawan." East-West Environmental Policy Institute Occasional Paper No. 7. Honolulu: East-West Center.

Krutilla, John V., and Anthony C. Fisher. 1985. *The Economics of Natural Environments:*

Studies in the Valuation of Commodity and Amenity Resources. Washington, DC: Resources for the Future.

Lal, Padma N., and John A. Dixon. 1990. "The Management of Coastal Wetlands: Economic Analysis of Combined Ecologic-Economic Systems." Australian Bureau of Agricultural and Resource Economics, Canberra. Mimeo.

Mäler, Karl-Göran. 1991. "Measuring Environmental Damage—The Production Function Approach." In *Valuing Environmental Benefits in Developing Countries*, eds. J. R. Vincent, E. W. Crawford, and J. P. Hoehn. Special Report 29. East Lansing: Michigan State University.

———. 1992. "Multiple Use of Environmental Resources: A Household Production Function Approach to Valuing Resources." Paper presented at the International Society of Ecological Economics Meeting on Investing in Natural Capital: A Prerequisite for Sustainability, 3–6 August, Stockholm.

Maltby, Edward. 1986. *Waterlogged Wealth: Why Waste the World's Wet Places?* London: Earthscan Publications.

Markandya, Anil, and David W. Pearce. 1988. "Environmental Considerations and the Choice of Discount Rate." Environment Department Working Paper No. 3. Washington, DC: The World Bank.

Perrings, Charles, and David W. Pearce. 1992. "Biodiversity Conservation and Economic Development." Paper presented at the Economics and Ecology of Biodiversity Loss, 29–31 July, Beijer Institute for Ecological Economics, Stockholm.

Porter, Richard C. 1982. "The New Approach to Wilderness Preservation through Benefit-Cost Analysis." *Journal of Environmental Economics and Management* 9:59–80.

Ready, Richard C., and Richard C. Bishop. 1991. "Endangered Species and the Safe Minimum Standard." *American Journal of Agricultural Economics* 70:309–12.

Ruitenbeek, H. J. 1989. "Social Cost-Benefit Analysis of the Korup Project, Cameroon." Report prepared for the World Wide Fund for Nature and the Republic of Cameroon, London.

———. 1992. *Mangrove Management: An Economic Analysis of Management of Options with a Focus on Bintuni Bay, Irian Jaya*. Jakarta and Halifax: Environmental Management Development in Indonesia Project.

Smith, V. Kerry. 1983. "Option Value: A Conceptual Overview." *Southern Economic Journal* 50:654–68.

———. 1991. "Household Production Functions and Environmental Benefit Estimation." In *Measuring the Demand for Environmental Quality*, eds. J. B. Braden and C. D. Kolstad. Amsterdam: North-Holland.

Tisdell, Clem. 1990. "Economics and the Debate about Preservation of Species, Crop Varieties and Genetic Diversity." *Ecological Economics* 2:77–90.

Turner, R. Kerry. 1991. "Economics and Wetland Management." *AMBIO* 20 (2):59–63.

Turner, R. Kerry, and Tom Jones. 1991. *Wetlands: Market and Intervention Failures*. London: Earthscan Publications.

Twilley, Robert R. 1991. "Properties of Mangrove Ecosystems Related to the Energy Signature of Coastal Environments." Department of Biology, University of Southwestern Louisiana, Lafayette. Mimeo.

Varian, Hal R. 1984. *Microeconomic Analysis*, 2d ed. New York: W. W. Norton.

Yañez-Arancibia, Alejandro, and John W. Day, eds. 1988. *Ecology of Coastal Ecosystems in the Southern Gulf of Mexico: the Terminos Lagoon Region*. Mexico: UNAM Press.

[15]

Ecological Economics, 1 (1989) 335–361
Elsevier Science Publishers B.V., Amsterdam – Printed in The Netherlands

VALUATION AND MANAGEMENT OF WETLAND ECOSYSTEMS

ROBERT COSTANZA [1], STEPHEN C. FARBER [2] and JUDITH MAXWELL [3]

[1] *Coastal and Environmental Policy Program, Center for Environmental and Estuarine Studies, Chesapeake Biological Laboratory, University of Maryland, Solomons, MD 20688-0038 (U.S.A.)*
[2] *Economics Department, Louisiana State University, Baton Rouge, LA 70803 (U.S.A.)*
[3] *School of Natural Resources, 2021 Coffey Road, The Ohio State University, Columbus, OH 43210-1085 (U.S.A.)*

ABSTRACT

Costanza, R., Farber, S.C. and Maxwell, J., 1989. Valuation and management of wetland ecosystems. *Ecol. Econ.*, 1: 335–361.

We recently completed a study of wetland values in coastal Louisiana that employed both willingness-to-pay and energy analysis-based methodologies and were able to bracket a range of values within which we feel fairly confident the true value lies. However, a large amount of uncertainty remains. Our current estimates of the total present value of an *average* acre of natural wetlands in Louisiana are US $2429–6400 per acre (assuming an 8% discount rate) to $8977–17000 per acre (assuming a 3% discount rate). At the lowest value, the current annual rate of loss of Louisiana wetlands (50 sq miles per year) is worth about $77 million. At the largest value it is worth about $544 million.

In this paper we (a) discuss the fundamental theoretical and practical problems underlying natural resource valuation; (b) summarize our methods and findings for Louisiana wetlands; and (c) elaborate on some of the more recalcitrant problems attending applied natural resource valuation, including discounting and dealing with uncertainty and imprecision.

The discount rate makes more difference in the final result than any other one factor, and yet there is much disagreement about the appropriate approach to discounting natural resources. We discuss the discounting problem as applied to natural resources and argue for lower discount rates for valuing renewable natural resources than apply for other aspects of the economy.

It now seems clear that no reasonable amount of effort will produce very precise estimates of wetland values, and we suspect this is also the case for several other classes of natural resources. We elaborate a Wetlands Assurance Bonding system to address these problems.

BACKGROUND

Imperfect markets and prices

In market economies, prices are relied on to provide signals regarding resource scarcity. These prices are considered to provide a true measure of

0921-8009/89/$03.50

economic value only if the market is characterized by a large number of buyers and sellers and by private property rights which are enforceable and transferable. The consumption choice must be made from a basket of 'rival' goods and services, that is, the choice to consume one good comes at the expense of some other good. If these conditions hold, prices can be relied on to direct the allocation of scarce resources to their highest valued use. Unfortunately, for the many ecosystem goods and services that humans *do* care about, such markets are of limited usefulness and/or fail to perform efficiently. The absence of efficient markets for these outputs is the major cause of what can be called 'inefficient' habitat modification. This is the conversion from those land use activities that have large, real, but uncapturable benefits (i.e., natural ecosystems) to land use activities that have smaller but capturable income streams. In this most prevalent case, individual preferences diverge from those of society as a whole and collective action is warranted.

Even when the preferences for the resources provided by existing natural ecosystems are strong, their common property nature makes it impossible to exclude those who do not pay for enjoying or using them. As a result there is no incentive to conserve such resources and overuse and even exhaustion can occur when utilization or harvest rates exceed the population growth rates of species, as in a fishery, or rates of forage production in common grazing areas. For some resources, such as camping sites and scenic views, utilization levels result in congestion and subsequent declines in resource quality (Cicchetti and Smith, 1973). For congestible goods, exclusion results in efficient use only if consumers pay for their use in accordance with their own valuation for that use plus congestion costs imposed on others. Because of obstacles to both collecting payment and in eliciting true evaluations, public or collective management is usually required.

Lack of provision for equity and sustainability

Another deficiency of markets is that the choices and prices observed are a function of types and levels of personal endowments. The distribution of endowments and the relative importance imparted to these endowments in the marketplace can cause significant bias in environmental valuations from the individual, societal and regional viewpoint. For example, threatened species will be valued very differently by the local population in poor countries characterized by dependence on natural resource trade to secure foreign exchange and in rich countries characterized by use of imported natural resources. Unfortunately, when consumers and producers are geographically separated, consumers are able to internalize the benefits of natural resource trade while externalizing the cost of habitat destruction.

Markets may also fail to encourage ecologically sound management when very long time horizons are involved, resulting in intergenerational inequity and irreversible damage to ecosystems. The choice of interest rate is critical in determining the optimal harvest rate for renewable resources and in turn determines their renewability. Clark (1973) has shown that the high discount (or interest) rates used by individuals to assure short-term profit maximization may cause overexploitation and exhaustion of species characterized by slow growth rates. However, even if the discount rate were zero, i.e. future values equal current values, the short lifespan of humans will result in management practices which favor current values and incomes, even if such practices have a potential to cause high social costs to future generations. For example, fluorocarbons permit packaging under pressure and a finer spray than that provided by the alternative pump spray. These benefits are directly reflected in market prices and are quite modest when compared to the catastrophic cost which would result from ozone depletion. Page (1978) contends that the inability to estimate the probability of ecocatastrophy creates indifference to the huge asymmetry between current benefits and the potential for future catastrophic costs. This makes even political solutions difficult.

Where the allocation of scarce ecological resources is concerned there will invariably be a clash between individual preferences and social benefits. Markets are just one of many arenas for resolution of these disputes. Public resource management, legislation, and regulatory agencies offer alternative methods or forums for directing management decisions. In spite of the absence of market prices, environmental decision making increasingly emphasizes the quantification of benefits and costs. This has resulted in a proliferation of new resource value concepts and valuation methods.

Willingness to pay and extending existing markets

For the individual, one estimate of the economic value of an increment in any good or service is the maximum amount that he or she is willing to pay (WTP) for it. Conversely, the value of a decrement is the minimum amount that the individual is willing to accept (WTA) for it. The prices formed in well functioning markets are one source of WTP and WTA estimates of marginal increments or decrements of goods and services. Where markets fail to provide appropriate measures of environmental values, the WTP and WTA concepts of economic value are not invalidated, but alternative 'pseudomarkets' must be used to elicit these values from individuals.

The notion that an alternative chosen will be at the expense of some opportunity foregone is central to economic decision making. For example, the cost of providing a scenic view can be directly derived from the net value of outputs and uses foregone such as timber harvest and dispersed grazing. This is referred to as an *opportunity cost*. For scenic view preservation to be economically efficient, the scenic view must be preferred over other uses. In other words, its value must exceed the cost of providing it, including opportunity costs.

Ecological goods and amenities are valued by individuals for a variety of reasons: *utilitarian (or use) value* refers to the value of using an ecosystem's products and amenities to derive both current and future benefits; these benefits include commercial outputs such as timber, outdoor activities and experiences, wildlife, and aesthetics (for examples of raw material evaluation see Bartlett (1984), who discusses valuation assumptions and methods for range forage). Individuals may also be willing to pay now for the option of using a resource in the future. Such an option price includes an amount equivalent to the expected use value plus a premium, similar to a risk premium, which a person would pay over and above the expected use value. This premium is referred to as *option value*, and is due either to uncertainty surrounding the individual's preferences or to uncertainty regarding the price or availability of the resource. This premium may be positive, negative, or zero (as in the case of preference uncertainty) but it will always be positive in the case of supply availability for a risk averse person – see Greenley et al. (1981), Bishop (1982) and Brookshire et al. (1983) for the theory and empirical studies of option value. The passage of time will likely reduce the uncertainty surrounding resource usefulness. When resource use is irreversible, individuals would be willing to sacrifice current irreversible use until uncertainty about its cost has been reduced. They would be willing to pay for increased information. This payment is termed *quasi-option value* (Arrow and Fisher, 1974; Conrad, 1980). It is not attributable to risk aversion, like option value, but is due to the value of information; this value arises in the case of resource use decisions that create irreversible damages, such as species extinction or large-scale deforestation. A final, pure non-use value is what a person may be willing to pay simply to know that a resource exists even when there is no intention of use. This *existence value* has nothing to do with preserving options for future use or paying to delay use until more information is available (see Randall and Stoll, 1980, and Brookshire et al., 1983).

In practice, the measurement of these value concepts has remained difficult and largely limited to the valuation of environmental commodities and amenities which produce direct benefits to humans. An alternative approach is Norton's (1986) concept of *contributory value*, which assigns value to environmental resources not due to their *direct* value to humans, but according to their *indirect* role in maintaining and accentuating the ecosystem processes which support these direct benefits. These include the maintenance of atmospheric and aquatic quality, the amelioration and control of climate, flood control, the maintenance of a genetic library, and the supportive role of food webs and nutrient cycling. Contributory value recognizes both the long time horizons involved in many ecosystem processes and the synergism which can result from the interaction of two or more species creating benefits of which neither is individually capable.

Though empirically elusive, contributory value does provide a useful framework for conceptualizing how natural ecosystems might be evaluated. However, as Randall (1986) contends, human preferences are focussed more on life forms than on life processes. This bias is further distorted by the fact that humans, in general, will assign higher preferences to species with commercial value, to wild relatives of domesticated species and to those which are most familiar and/or easy to empathize with, such as large mammals. Lovejoy (1986) refers to this bias against invertebrates as vertebrate chauvinism, while others point to interspecies inequity (Costanza and Daly, 1987). If it is accepted that each species, no matter how uninteresting or lacking in direct usefulness, has a role in natural ecosystems (which do provide many direct benefits to humans), it is possible to shift the focus from the imperfect perceptions of individuals to the contributory value of ecosystems as expressed through their ecological relationships. One might argue that this contributory value is an estimate of the value individuals would place on environmental services if they were fully informed about the functioning of the environment in their behalf.

Ecosystem function and economic value

Assessing the contributory value of ecosystems involves the ability to understand and model the ecosystem's role in an integrated ecologic–economic system and its response to perturbations. The models must be at a level of detail and resolution that allows the assessment of impacts (marginal products) on economically important ecosystem commodities and amenities. Several types of ecological modeling can be used for this purpose, which we define under the general heading of 'ecological–economic' models. They range from relatively simple, static, linear input–output models (Isard, 1972; Hannon, 1973, 1979; Costanza and Neill, 1984; Costanza and Hannon, 1990) to multiple regression models (Farber and Costanza, 1987) to more sophisticated nonlinear, dynamic spatial simulation models (Costanza et al., 1986). Braat and Van Lierop (1986) provide a summary of ecological–economic models currently in use.

The point that must be stressed is that the economic value of ecosystems is connected to their physical, chemical, and biological role in the overall system, *whether the public fully recognizes that role or not*. Standard economics has too often operated on the assumption that the only appropriate measures of value are the current public's subjective preferences. This yields appropriate values only if the current public is *fully informed* (among a host of other provisios). The public is most likely far from being fully informed about the ecosystem's true contribution to their own well being, and they may therefore be unable to directly value the ecosystem's services (Constanza, 1984). However, scientists may be able to derive estimates of the values that a fully informed public would produce by analyzing the structure and function of ecosystems. Economists also tend to assume that local optimizing of independent individuals will produce optimal results for the society. We have noted that this only works when there are perfect markets and that perfect markets are the exception rather than the rule in the natural resource area. Therefore, once appropriate values for ecosystem services are derived, that information must be inserted into the decision making process in order to correct the market signals.

WETLAND VALUATION

This section provides an example of the methods and problems in deriving empirical estimates of natural wetland values (details are available in Costanza and Farber, 1985a,b; and Farber and Costanza, 1987). Two different wetlands valuation techniques were used: willingness-to-pay (WTP) valuation and energy analysis (EA) valuation. These methodologies are thoroughly explained elsewhere (Costanza and Farber, 1985a,b; Farber and Costanza, 1987). We provide a brief summary here.

WTP valuation refers to valuing the particular dimensions of benefits or projects by determining society's willingness-to-pay (WTP) for those particular benefits. It requires a listing of the types of benefits and an estimate of the WTP for each one. Our analysis of the WTP for wetlands concentrated on four major categories of benefits of wetlands: commercial fishing and trapping, recreation, and storm protection. Waste treatment benefits are partially included in the other benefit estimates to the extent that water quality affects recreation, fishing, and trapping values. We were not able, as part of this analysis, to place values on the existence and option value of wetlands.

The methodology for estimating commercial productivity consisted of estimating the marginal productivity of an acre of wetlands. Our estimates concentrated on the following commercial products: shrimp, menhaden, oysters, blue crab, and furs. The critical problem in estimating the marginal productivity of wetlands was to separate the effect of human effort from the effect of the intrinsic wetland productivity, when we can only observe the total effect as reflected in harvests. Failure to make this separation results in a potentially very large overestimate of the contribution of the wetlands to commercial production. A second problem in this estimation procedure is to distinguish between the average and marginal productivity of the wetlands. This is important if there exist decreasing returns to wetlands productivity (i.e. if the productivity of a unit area of wetlands depends on the amount of remaining wetlands). Perhaps average productivity is appropriate for valuing very large wetlands projects, but valuation of small projects should use marginal productivity.

The estimation of WTP for recreational value is also complex. Two techniques can be used to make this valuation. First, one can simply ask recreational users what they would be willing to pay to use the wetlands in the project area. The problem with this technique is that respondents may engage in strategic responses. For example, if they think they may have to actually pay what they say they are willing to pay, they may state a value lower than their true value. On the other hand, if they think their response may positively impact the probability of implementing a project, they may state a value higher than their true value. A second technique is to estimate recreational users' WTP based on observations of what it actually costs them to use the project area. This technique is called the travel cost method and was our primary means of assessing the recreational value of wetlands.

The value of the wetlands for hurricane protection was obtained from a methodology which determined the reduction in expected property damages in populated areas along a gradient relative to distance from the coast. In principle, people would be willing to pay for a wetlands project according to the reduction in expected property damages attributable to the project.

The energy analysis valuation technique looks at the total biological productivity of wetland vs. adjacent open water ecosystems as a measure of their total contributory value. Primary plant production is the basis for the food chain which supports the production of economically valuable products such as fish and wildlife. It is converted to an equivalent economic value based on the cost to society to replace this energy source with fossil fuel as measured by the overall energy efficiency of economic production. This technique is comprehensive and does not require a detailed listing of all the specific benefits of wetlands, but it may overestimate their value if some of the wetland products and services are not useful (directly or indirectly) to society.

We acknowledge that the overall level of precision of our current estimates of total wetland values is not as high as we would like (due mainly to cost and data constraints), especially for important categories such as option and existence values. We suggest methods for dealing with this uncertainty in a later section.

342

Willingness to pay-based wetland value estimates

Commercial fishery production value of wetlands. The commercial value of wetlands for fish and wildlife harvesting is attributable to the fact that wetlands provide food and habitat necessary for the production and survival of various species. This section summarizes estimates of the value of wetlands for the harvesting of shrimp, blue crab, oysters, menhaden, and furs.

Attempts to estimate the productivity of wetlands must address the problem of separating the productivity of the wetlands themselves from the productivity of human inputs which are used in harvesting those wetlands. The problem can be outlined as follows. Suppose the harvest from a given ecological system, Q, is a function of the level of some environmental variable, W (for example, wetland acreage), and some level of human effort, M:

$$Q = f(M, W) \tag{1}$$

Also, the level of human effort will be a function of the price per unit of biomass harvested, P, the cost per unit of human effort, C, and the resource abundance, which, in turn, is dependent on the level of the environmental variable:

$$M = g(P, C, W) \tag{2}$$

Any change in W will result in a change in both M and Q. The productivity of the environment alone is the change in Q, holding M constant. This is referred to as the Marginal Productivity (MP) of the environmental variable. Likewise, the MP of human effort is the change in Q, holding W constant.

Simply estimating (1) will not control for the effect of changes in W on the level of human effort across observations. In addition, simply observing Q/W, the Average Product (AP) of the environmental variable will more than likely overestimate the marginal productivity of the environmental variable since there are probably decreasing returns to the environmental variable when human effort is held constant; i.e., AP > MP. More importantly, the average product implicitly includes the productivity of the human effort embodied in the dependent variable Q.

A conceptual problem also arises in the estimation of the system composed of (1) and (2). It is difficult to specify the proper environmental level (W) for migratory species because it is not clear which environment or set of environments they are dependent on and for how long. It is easiest to estimate this system for more stationary species because: (a) the environmental linkage is clearer; and (b) there will be more observations over which to estimate the system of equations.

The economic value of a unit of the environment, V, for example, an acre of wetland, is the product of MP and the dockside price, P:

$$V = \text{MP} * P \tag{3}$$

Estimates of this value (V) based on the average value of harvest per unit of environment (AP) are inaccurate since they do not account for the difference between AP and MP. Human inputs have costs and productivity, so AP will most likely overestimate the social value of a unit of the environment. Another measure has been $[(Q * B) - (M * C)]/W$, the net rent to harvesters. In a perfectly competitive harvesting industry, which is probably close to what prevails, this value is approximately zero; therefore, it will most likely underestimate the social value of a unit of the environment.

Lynne et al. (1981) developed a model of catch in which catch in year t, Q_t, depends on marsh acreage in year $t-1$, W_{t-1}, catch in year $t-1$, Q_{t-1}, and effort in year t, M_t:

$$Q_t = \beta_0 + \beta_1(\ln W_{t-1})M_t + \beta_2(\ln W_{t-1})M_t^2 + \beta_3 Q_{t-1} + \epsilon_t \tag{4}$$

The marginal productivity of marshland is then the derivative of (4) with respect to W. The economic value (V) of that marginal acre of marsh is then the marginal productivity times the dockside market value of catch.

In what follows we present summaries of our estimates of marginal productivities of marshlands with respect to commercial fishery production. We made estimates only for shrimp, menhaden, oysters, and blue crab. These species account for approximately 95% of commercial fishery harvest in Louisiana (USDC, 1983).

In our study, brown and white shrimp catch in offshore and inshore waters in Louisiana was measured using National Marine Fisheries Service (NMFS) data for subareas of Louisiana from 1963 to 1976. These data are by area of catch, not landing. There were 24 inshore subareas and four offshore areas designated by NMFS prior to 1976. Catch is measured in heads-off pounds. Effort is measured in 24-hour days. Marsh area was measured for years 1955 and 1978 by the U.S. Fish and Wildlife Service for 7.5′ quad sheets (see Costanza et al., 1983). Although data were collected on both saline and fresh marsh, this study uses the sum of the two. Area data for years between these two years were estimated by simple linear interpolation.

---

* acre ≈ 0.4047 ha.
* lb, pound (avdp) ≈ 0.454 kg.

344

TABLE 1

Estimated marginal productivities per acre of marsh for brown and white, inshore and offshore shrimp (heads-off pounds/acre)

Type of model	Inshore		Offshore	
	brown	white	brown	white
Non-lagged	1.60	1.44	0.90	1.23
Lagged	1.25	0.86	–	–

lb/acre ≈ 1.12 kg/ha.

Our simplest estimates of the model excluded the lagged catch value. Separate regressions were performed for brown and white, and for inshore and offshore shrimp. There was no consideration of possible changes in the structure of the equations over time. We fit equation (4) for both the lagged and non-lagged case to estimate the values of the parameters β_1 and β_2. All the parameters and equations were significant with R^2 ranging from 0.67 to 0.90.

The marginal productivity of marsh based on equation (4) is then given by:

$$\mathrm{MP_m} = (\beta_1 + \beta_2 M)M/W \tag{5}$$

For inshore areas, the sample means of M and W for brown shrimp were 1426 days and 18 249 ha, respectively. The marginal productivity estimates based on these data are summarized in Table 1.

Unfortunately, there are not sufficient data to estimate the model employed above for menhaden, oysters, or crabs. For menhaden, we assumed that the marginal and average products of estuarine area (marsh plus open water) are equal for menhaden harvest. The entire 1983 Gulf Coast catch was 2036 million lb. Using total Gulf Coast estuarine area of 14 million acres, this implies an average and marginal productivity of approximately 145 lb/acre. This estimate could be biased upward because marginal product is generally lower than average product. It could also be biased downward since we divide through by the sum of marsh plus open water area. If marsh is more productive of menhaden per acre than open water (as we suspect it is) then this procedure underestimates the per-acre production of marsh. In lieu of better data, however, we feel our current estimate is the best that can be done.

We used the results from a study of the blue crab productivity of western Florida salt marshes for estimating the marginal product of Louisiana wetlands for blue crab. Lynne et al. (1981) estimated a bioeconomic model for blue crab harvesting that is similar to the system outlined above. They estimated that the marginal productivity of the Florida salt marsh is 2.3 lb

TABLE 2

Acres of marsh open water and total recorded pounds of oyster meat harvested from 1948 to 1978

State	(a) Area of marsh and open water (million)	(b) Recorded total harvest (million lb)
La.	7.3	283.5
Fla.	3.0	78.1
Texas	2.7	66.1
Miss.	0.6	47.7
Ala.	0.4	33.2
Total	14.0	508.6

Source: Allen et al. (1984, p. 120).

of blue crab per acre per year. Although they do not specifically derive the average salt marsh productivity, one can use their figures to derive an average productivity of approximately 28 lb of blue crab per acre of salt marsh per year. In contrast, they estimate that the marginal productivity of effort is 214 lb per trap per year, while the average productivity of traps is 435 lb per year. This illustrates the inaccuracy of using average productivities to measure environmental values.

We would have liked to estimate a model for oysters similar to the one for shrimp above. Data were by area of oyster landing rather than area of catch. Without data on the origins of oyster harvest it was not possible to relate harvest to the associated level of the environmental variables (marsh area).

The results of several other studies were therefore used to estimate the marginal product of wetlands for oysters. Batie and Wilson (1978) estimated the oyster productivity of Virginia wetlands using an equation similar to (1) and found a marginal product of 1.66 lb/acre in the county with the greatest wetlands area in their sample. Aggregate data presented in Allen et al. (1984) and shown in Table 2 can be used to estimate average oyster productivity of estuarine areas. This table shows estuarine area (marsh plus open water) in acres and pounds of oyster meat harvested as reported between 1948 and 1978 in five Gulf Coast states. Using the total values, the annual recorded harvest per acre of marsh plus open water over the 30-year period was 1.21 lb. A more sophisticated estimate of marginal product is obtained by regressing harvest in Table 2 (H) on estuarine acreage (A) shown in that table. The resulting least-squares equation is:

$$H = \underset{(2.3)}{2.03} + \underset{(3.4)}{35.60}A \qquad F = 35.44 \tag{6}$$

346

TABLE 3

Summary of estimated economic value of wetland productivity for commercial fishery harvest.

(a) Species	(b) Basis	(c) Annual MP estimate (lb/acre)	(d) 1983 ex-vessel price [a] ($/acre)	(e) Value of annual MP ($/acre)
Shrimp	Marsh Area			
brown inshore		1.60	$2.10	$3.36
white		1.44	2.10	3.02
brown offshore		0.90	2.10	1.89
white		1.23	2.10	2.58
Menhaden	Marsh and open water area	145.00 [b]	0.04	5.80
Oyster	Marsh and open water area	6.00	1.34	8.04
Blue crab	Salt marsh area	2.30	0.29	0.67
Total				$25.36

[a] USDC (1983).
[b] Assuming MP-average product.

Since the harvest data are for a 30-year period, the annual marginal product of estuarine area implied by equation (6) is 1.19 lb of oyster meat per acre. This estimate is roughly equal to the cruder average product estimate above.

The data in Table 2 are only for recorded harvests collected by NMFS. This includes oysters used only for canning. Sack and counter oysters are unreported. Lindall et al. (1972, p. 139) found that the total oyster production is five times the reported harvest. We assume a canning oyster marginal product of 1.20 lb, so the total marginal product is estimated to be 6.00 lb per acre of estuarine area per year.

Table 3 summarizes the commercial fishery productivities of marsh and estuarine areas by the four primary species. Column (b) shows the environmental variable being evaluated. For example, shrimp productivity was estimated by using the total of fresh and salt marsh acreage. Column (c) shows our annual marginal productivity estimates. Column (d) shows the 1983 ex-vessel price per pound. We assumed all shrimp were valued at the same price per pound. Column (e) is derived from columns (c) and (d). It shows the economic value of the marginal acre, and is the product of columns (c) and (d). For example, the estimated economic value of the annual marginal productivity for brown inshore shrimp was $3.36 per acre of marsh. The sum of all these economic values is $25.36 per acre of marsh per year. Discounting at 8% gives a present value per acre of $317. Discounting at 3% gives a present value per acre of $845.

Fur trapping value of wetlands. Muskrat and nutria accounted for 78% of the value of Louisiana trapping in 1977 (Davis, 1978, p. 87). The muskrat is primarily a brackish-to-intermediate marsh species. In this environment, the muskrat yield is approximately 0.98 pelts per acre. Freshwater marshes, the primary habitat of nutria, yield 0.88 pelts per acre (Davis, 1978, p. 85).

There is some difficulty placing a price on these pelts. The recent increase in the value of the dollar has reduced prices to less than one-half their 1979–81 value. We used the 1980–81 price of $6 for eastern muskrat pelts and $7 for eastern nutria pelts as expected average prices and assumed these prices to be constant over time. Using the pelt productivity values above, the annual value of pelt productivity is $12.04 per acre. At a 8% discount, this generates a present value of $151 per acre; at 3%, a present value of $401 per acre.

Recreation value of wetlands. Wetlands have values as recreational areas. They are used for fishing, hunting, photographing, boating, etc. These are non-commercial uses of wetlands. A survey of recreational users was undertaken on various days over a 1-year period to determine willingness-to-pay to preserve wetlands for recreational purposes in Terrebonne Parish, Louisiana. The survey was designed to utilize the travel cost method of evaluating consumer surplus from use of a site, and the contingent valuation method. The details of the survey and data are contained in Costanza and Farber (1985b) and Farber and Costanza (1987). Here we summarize the method and results.

The sampling procedure consisted of placing self-addressed, stamped questionnaires on windshields of all vehicles parked in the morning at 27 boat launch facilities in Terrebonne Parish on various dates throughout the period July 1984 to June 1985. The sum of the average number of vehicles per day across all sites was 563.29 on weekends and 132.1 on weekdays.

A total of 7837 questionnaires were distributed, and 1126 were returned for a response rate of 14.4%. There were 6248 questionnaires distributed on weekends, with a 15.0% response rate; and 1589 on weekdays with a 11.7% response rate.

There was no explicit attempt to determine non-response bias. However, we thought that persons placing a higher value on these particular wetlands would be more likely to return questionnaires. This could mean that response rates may decline with distance from the site. In order to test this, we instituted a lottery with several prizes ($50, $30, and $20) on weekends in

February and March, 1985. The response rate for weekend samples increased from 18.5% in the pre-lottery period to 24.3% in the lottery period.

In order to implement the travel cost methodology, seven rings of 35-mile * increments in radii were constructed centered at Dulac, Louisiana. Each parish or county of Louisiana, Texas, Mississippi and Alabama was placed in one of the rings or in a rest-of-world (ROW) category. The localized use of these wetlands is apparent from the fact that 78% of the respondents came from ring 1, and 98% from rings 1 through 3. This localized use may make the travel cost methodology inadequate for determining willingness-to-pay.

We estimated the value of travel time by determining the total cost of travel time to the typical user group in the sample. The typical user group consisted of 2.72 persons, with one male head, 0.66 female spouses, 0.51 children, and 0.55 non-family members. The average wage rate of male respondents in the sample was $13.37. Using the fact that a previous study found southern female wages to be 70% those of males, the average spouse wage was assumed to be $9.36. The non-family member was assumed to be male. Children were assumed to have zero wages. Weighting these wages by the number of members yielded average total hourly wages of the typical user group of $26.90. The estimated annual WTP using this travel cost method was $3.9 million/year.

A second method was to use respondents' answers to the direct question regarding their willingness-to-pay for the preservation of the Terrebonne Parish wetlands (contingent valuation). In particular, persons were asked what they would be willing to pay annually for the sake of their entire household. The mean response in the sample was $103.48 per household annually.

The number of different households using these wetlands annually must be estimated indirectly. An independent survey of recreational saltwater fishermen (Bertrand, 1980) found that 110 373 licensed fishermen used the LaFourche/Terrebonne area at least once in the 1982–83 season. Since Terrebonne constitutes approximately one-half of the total LaFourche/Terrebonne wetlands (Costanza and Farber, 1985b), we assumed only 55 186 of these persons used the Terrebonne area. Multiplying this by the annual WTP per household ($103.48) gives an annual total WTP of $5.7 million. This is slightly above the travel cost annual estimates. However, since the direct question referred to the head of household's WTP and several family members may have licenses, this direct estimate is not too far from the travel cost estimate. For example, if we assume the 2.17 members of a typical

mile ≈ 1.609 km.

family user unit are licensed, the $103.48 amounts to $47.69 per person. The resulting annual WTP is then $2.6 million, only slightly less than the travel cost method result.

Depending on assumptions made about population growth, the present value estimates range from $34.2 million in the no growth situation to $40.8 million if population grows at a 1.3% annual rate when the discount rate is 8%. Assuming the population growth case, this amounts to $46 per acre of estuarine area (890 000 acres) using a 8% discount rate, and $181 per acre using a 3% discount rate.

Storm protection value of wetlands. Wetlands provide storm protection for urbanized areas in several ways. The marshes create surface friction for both tidal surges and winds. They also aid in reducing the heat source of energy for the storms. The result is that marshes help reduce tidal surge levels and wind velocities of storms. A very important value of coastal marshes is therefore their value in protection against personal and property damage due to storms

Costanza and Farber (1985b) contains our analysis of the hurricane protection value of wetlands for protection from storm surge and wind damage. Farber (1987) contains a more detailed analysis of the value of coastal wetlands for protection from wind damage only. We used U.S. Army Corps of Engineering estimates of property damages resulting from four recent Gulf Coast hurricanes, and adjusted these to 1983 dollars using a construction price index. These damage estimates included both wind and flooding damages. Damages were given by county. By measuring the distance of the center of a county from the coast, and controlling for population size, hurricane strength, distance from the path of the hurricane, and probability of hurricane incidence, we estimated the increase in expected property damages resulting from being closer to the landfall of a hurricane.

If Terrebonne wetlands receded by one mile, expected damages in a four-parish area would increase by $5 752 816 annually. The loss of a 207-ft wide strip running the length of Terrebonne parish coastline would increase expected damages by $128.30 per acre of the coastal strip annually. Discounting these values by 8% yields a present value of the strip of $71.9 million; and a present value of 1 acre in the coastal strip of $1604. Assuming population grows at 1.3% annually, the present value of such a strip for storm protection is $1915 per acre, or a total of $85.9 million for the entire strip. This is a relevant value for consideration of projects designed to stop rapid, wide-area recession and coastal erosion. It is not really appropriate for evaluating projects designed to affect erosion in areas that are not direct coastal areas; unless, of course, these internal areas eventually have an impact on coastal recession.

Another way of looking at the protection value of Terrebonne wetlands is to determine how much higher expected property damages would be if they were permitted to continue receding at their present rates. Using recent recession rates (approximately 12 ft/year), the present discounted value (at 8%) of increased expected property damages lies between $2.1 and $3.1 million. This is relevant in determining whether to initiate projects directed toward stopping long-term recession.

Energy analysis of wetland values

The theoretical basis for energy analysis as an economic value estimation tool is discussed in Costanza and Farber (1985b) and Turner et al. (1988). The method looks at the total amount of energy captured by natural ecosystems as an estimate of their potential to do useful work for the economy. It yields an estimate of a comprehensive (in that it should include all possible useful outputs) upper bound on the economic value of the products of natural systems. It is an upper bound because not all of the work done by the system is necessarily useful to the economy.

Gross primary production.

The energy analysis methodology can be quite complicated (cf. Costanza, 1980, Costanza and Herendeen,1984; Costanza and Neill, 1984). For the purposes of this study we employ a simplified technique that is readily calculable (we discuss its shortcomings later). This technique uses the Gross Primary Production (GPP) of the whole ecosystem as an index of the solar energy captured by the system, and converts this energy value into dollars using a single dollar–energy conversion factor (described below). GPP is a measure of the solar energy that is used by the plants in the system to fix carbon into organic molecules. This 'primary' production is then used to power all the plants and animals in the system. The plants and animals in the system also moderate water flow, sedimentation and other variables. GPP for an ecosystem can be thought of as analogous to GNP for an economy: both are crude (but essential) measures of overall system performance that say nothing about the internal distribution of production and must therefore be used with caution. GPP and GNP measure the value of inputs (or outputs) of ecological and economic systems, respectively.

The energy analysis procedure can be summarized as follows:

(1) Determine by field measurements and laboratory experiments the GPP of the natural area in question, under the with and without project conditions.

(2) Convert this estimate (usually measured in grams of carbon fixed per time unit or the heat equivalent energy content of the carbon) to fossil fuel equivalents (FFE) by considering the fuel efficiency of each source.

(3) Convert the FFE value into dollars using an economy-wide ratio of economic value per unit of energy, usually the ratio of GNP to total-economy energy use (measured in FFE).

All three steps involve uncertainty. Below we discuss the steps in more detail, pointing out the potential sources of error.

GPP measurement methods. Gross Primary Production (GPP) in plants is usually measured by analysis of gas exchange. During photosynthesis plants take up carbon dioxide from the air and fix it into higher energy organic molecules. They also give off oxygen as a by-product. GPP is a measure of how much carbon plants take up (or conversely, how much oxygen they give off) and, therefore, how much high-energy organic matter they create. The measurement is usually done with an infrared gas analyzer that detects carbon dioxide concentration differences for terrestrial plants, or with oxygen meters for submerged aquatic plants. Like most field measurements, GPP measurement is much more complicated than the simple description given above. Hopkinson (1978) gives a discussion of the technique and some of the problems and inaccuracies. Despite problems, GPP remains a popular and useful index of overall ecosystem energy capture. Table 4 gives some examples of GPP estimates for Louisiana wetlands and aquatic habitats.

Conversion to fossil fuel equivalents. GPP estimates are frequently stated in grams of carbon or calories of plant biomass per unit area per unit time. The first step in converting this to a measure of equivalent economic value is to convert it to energy units more directly relevant as input to the economy, i.e.

TABLE 4

Gross primary production (GPP) estimates for Louisiana wetland and open water habitats.

Habitat type	Gross primary production (cal m^{-2} $year^{-1}$)
Salt marsh	48 000
Brackish marsh	70 300
Fresh marsh	48 500
Average marsh	55 600
Salt aquatic	6 600
Brackish aquatic	5 100
Fresh aquatic	9 300
Average aquatic	7 000
Coastal plankton	3 600

All values are directly from Hopkinson (1978). cal, calorie (International Table) = 4.1868 J (def).

TABLE 5

Energy quality conversion factors (from Odum and Odum, 1976)

Type of energy	(a) Calorie cost (calories of heat to make 1 FFE calorie)	(b) Fossil-fuel equivalents (FFE calories/per heat calorie) (1/column a)
Heat from sun's rays, uncollected	10000	0.0001
Sunlight	2000	0.0005
Gross plant production	20	0.05
Wood, collected	2	0.05
Coal and oil, delivered for use	1	1.0
Energy in elevated water	0.33	3.0
Electricity	0.25	4.0

fossil fuels. Fossil fuels are a much more concentrated, higher quality form of energy than plant biomass. One way of seeing this is to consider how much extra energy is required to upgrade biomass to fossil fuel; for example, in a biogas process. Another way is to consider the relative number of calories of biomass that would have to be burned in a power plant to produce the same amount of electricity as a given quantity of oil. Both of these methods have been used to estimate the 'energy quality factor' of biomass relative to fossil fuel. An approximate average is 0.05 cal biomass/cal fossil fuel (Odum and Odum, 1976), indicating that unprocessed biomass, such as that measured by GPP, is about 20 times less concentrated than fossil fuel. Table 5 lists energy quality conversion factors for various energy forms.

Conversion to economic value. One can look at the overall ratio of the value of economic output to energy input in the economy as a crude way to convert plant production to an equivalent economic value. This step is certainly the most controversial, with critics arguing that energy consumption and economic value are not necessarily related (Huettner, 1982). But recent studies provide supportive evidence that total direct and indirect energy consumption (embodied energy) and dollar values are indeed highly correlated in the U.S. economy (Costanza, 1980, 1984; Cleveland et al., 1984; Costanza and Herendeen, 1984). We therefore use a conversion factor based on these studies to give a crude estimate of the economic value of ecosystem production from their GPP estimates converted to FFE.

Habitat interdependence. The GPP technique outlined above does not account for the interdependence between habitats or differences in productivity within the same habitat type. For example, all salt marsh is assumed to have the same GPP, regardless of what other habitats it is adjacent to or to the special conditions of the site. This is an approximation, but the level of current knowledge is such that it is the best available approximation.

Louisiana wetland value estimates via energy analysis. Table 6 lists the gross primary production estimates for relevant Louisiana wetlands and aquatic systems and their estimated economic value using the simplified energy analysis methods outlined above. The values range from \$47 acre^{-1} year^{-1} for open ocean (coastal plankton) to \$914 acre^{-1} year^{-1} for brackish marsh. The relevant numbers are the change in value from wetland systems to open water systems, since this is the major impact of coastal erosion. These are listed in column (c). They range from \$509 acre^{-1} year^{-1} (for fresh marsh to fresh aquatic conversion) to \$847 acre^{-1} year^{-1} for brackish marsh to brackish aquatic conversion, with an average of \$631 acre^{-1} year^{-1}. The

TABLE 6

Gross primary production and EA-based economic value estimates for relevant Louisiana wetland and marine habitats

Habitat type	Total energy captured measured by GPP [a] (kcal m^{-2} year^{-1})	Annual equivalent dollar value [b] (\$ acre^{-1} year^{-1})	Net marsh-aquatic change in annual value (\$ acre^{-1} year^{-1})	Present value (\$ acre^{-1}) assuming specified discount rate [d]	
				8%	3%
Salt marsh	48 000	624			
Salt aquatic	6 600	86	538	6 700	18 000
Brackish marsh	70 300	914			
Brackish aquatic	5 130	67	847	10 602	28 200
Fresh marsh	48 500	630			
Fresh aquatic	9 300	121	509	6 400	17 000
Coastal plankton	3 600	47	Average 631	7 900	21 000
Spoil banks [c]	13 000	169			

[a] GPP is gross primary production. Values are from Hopkinson 1979.

[b] Based on conversion factors of 0.05 coal equivalent (CE) kcal/GPP kcal (Table 5), and 15 000 CE kcal/1983 dollar (from Costanza, 1980) and 4047 m^2 acre^{-1}. The overall conversion factor from GPP (in kcal m^{-2}) to estimated economic value (in \$ acre^{-1} year^{-1}) is therefore: $(0.05 \times 4047)/15\,000 = 0.013$.

[c] Estimated from values for upland systems.

[d] Rounded to nearest \$100.

present value of an infinite stream of these values using 8% and 3% discount rates are given in columns (d) and (e), respectively. These yield values for average marsh/aquatic conversions of \$7900/acre and \$21 000/acre, respectively.

Summary wetland value estimates

The estimation of the economic value of natural wetlands is a difficult and complex task, but it is essential to rational management. We have employed both willingness-to-pay (WTP) and energy analysis (EA) techniques to improve this estimate. Table 7 summarizes our findings.

One problem in communicating our findings is that individual elements of the overall estimates are of varying precision and yet they must be added to produce a total average value. In Table 7 we list totals for the WTP-based analysis. We have not included any estimate for option and existence values, which studies in other areas have shown to be potentially large components of the total. This indicates that our total WTP estimate is probably still very conservative.

Our 'best estimate' values are given as a range in Table 7. These are \$2429–6400 per acre (assuming an 8% discount rate) and \$8977–17 000 per acre (assuming a 3% discount rate). Which discount rate may be most appropriate under given conditions is discussed further on, as are methods for dealing effectively with the current range of uncertainty in wetland valuation by using a assurance bonding fund system.

TABLE 7

Summary of wetland value estimates (1983 dollars)

Method	Per-acre present value (\$) at specified discount rate	
	8%	3%
WTP based		
commercial fishery	317	846
trapping	151	401
recreation	46	181
storm protection	1915	7549
Total	2429	8977
Option and existence values	?	?
EA-based GPP conversion	6400–10 600	17 000–28 200
'Best estimate'	2429– 6400	8977–17 000

DISCUSSION OF CURRENT VALUATION RESULTS

This section discusses what we perceive as the potential estimation problems that we encountered in valuing wetlands and outlines future studies that would be most effective in improving the estimates.

WTP-based estimates

First, the estimates of commercial values of wetlands may be overstated for some species since we used the average, rather than marginal, product of the marsh. Second, recreational value may be underestimated. The travel cost method works well if a site attracts persons from a wide range of distances. However, our evidence is that there is very intensive localized use of the Terrebonne wetlands. Since travel costs to local users are small, our estimate of their willingness-to-pay for access is also small. However, there may be no better techniques at the moment since directly asking persons their willingness-to-pay is plagued with many problems. Third, the value of wetlands for hurricane protection is important but our estimation methodology is crude. We feel it is very important to spend more effort in improving the accuracy of the flood damage benefits of wetlands. The estimation methodology we used probably overstates hurricane protection value since some of the hurricanes studied passed over coastal areas with significantly more upland area than those passing over Louisiana wetlands. Fourth, loss of wetland reduces habitats and food sources and reduces natural waste treatment capacity. We would need to know the WTP for loss of treatment capacity. Finally, we were unable to measure the value placed on wetlands by persons who do not directly use the wetland and/or who do not consume products produced by wetlands. For example, some people may be willing to pay to preserve wetlands in order to preserve the option of using them in the future. Also, people may obtain pleasure from knowing the wetlands exist; perhaps because of indirect pleasure they receive from such things as photographs, literature, etc. pertaining to those wetlands. These values may be considerable. These option and existence values would be good topics for future research. We attempted to estimate these through a statewide survey; however, the response rate was so low that we realized we would not have enough funds to complete the estimation.

EA-based estimates

The EA estimate may be an upper bound on the total value of wetlands. In practice there is enough imprecision in the data and uncertainty in the methods to make it difficult to tell whether the actual numbers are over or underestimates of the true value. It was encouraging that the EA based estimate was higher than the total WTP based estimate by an amount that seemed reasonable, given the known omissions from the WTP estimate. But more research on the general relationship of energy use to economic activity, and the specific relationship of GPP to economic value are needed to improve the quality of this estimate, and to determine its role in the overall valuation procedure.

DISCOUNTING

Often the present vs. future issue is thought to be objectively decided by discounting. But discounting at best only reflects the subjective valuation of the future to presently existing individual members of human society. Discounting is simply a numerical way to operationalize the value judgment that: (a) the near future is worth more than the distant future to the present generation of humans, and (b) beyond some point the worth of the future to the present generation of humans is negligible. Economists tend to treat discounting as rational, optimizing behavior based on people's inherent preferences for current over future consumption.

There is evidence, however, that discounting behavior may be symptomatic of a kind of semi-rational, sub-optimizing behavior known as a 'social trap'. A social trap is any situation in which the short-run, local reinforcements guiding individual behavior are inconsistent with the long-run, global best interest of the individual or society (Platt, 1973; Cross and Guyer, 1980; Costanza, 1987). We go through life making decisions about which path to take based largely on the 'road signs', the short-run, local, reinforcements that we perceive most directly. These short-run reinforcements can include monetary incentives, social acceptance or admonishment, and physical pleasure or pain. Problems arise, however, when the road signs are inaccurate or misleading. In these cases we can be trapped into following a path that is ultimately detrimental because of our reliance on the road signs. Discounting may allow individuals to give too little weight to the future (or other species, other groups or classes of humans, etc.) and thus helps to set the trap. Economists, while recognizing that individual behavior may not always lead to optimal social behavior, generally assume that discounting the future is an appropriate thing to do. The psychological evidence indicates, however, that humans have problems responding to reinforcements that are not immediate (in time and space), and can be led into disastrous situations *because* they discount too much.

It can therefore be argued that the discount rate used by the government for public policy decisions on common property resources (like wetlands) should be significantly lower than the rate used by individuals for private investment decisions. The government should have greater interest in the future than individuals currently in the market because continued social existence, stability, and harmony are public goods for which the government is responsible, and for which current individuals may not be willing to fully pay (Arrow, 1976).

Discounting future value by the rate of interest also provides a tight link between ecological destruction and macroeconomic policy. Any exploited species whose natural rate of population growth is less than the real rate of interest is under threat of extinction, even in the absence of common property problems. While Paul Voelker and the Federal Reserve probably do not worry about the effect of U.S. interest rate policy on deforestation in the Amazon or destruction of Louisiana wetlands, such links really do exist, and they probably should not.

In terms of our wetland valuation problem all this merely increases the uncertainty concerning the total present value of wetlands, because the appropriate discount rate is uncertain and it makes a big difference in the results. We have stated estimates for a range of discount rates (3%–8%) in order to demonstrate how much uncertainty is introduced by uncertainty in the discount rate, and have given arguments for why a lower discount rate may be more appropriate for wetland valuation decisions. Below we consider a method to deal with this uncertainty.

DEALING WITH IMPRECISE VALUES USING AN ASSURANCE BONDING SYSTEM

One fundamental problem with our current methods of wetlands management is that those parties that damage and destroy wetlands are not charged for the true social cost of that damage or destruction. Two reasons often given for this are: (a) we don't know what to charge them and/or (b) charging them anything implies that wetlands can be sold and they will just pay the money and continue on their merry destructive ways. Therefore we cannot accept any charging scheme and must simply prevent all destruction of wetlands.

These ideas have guided wetlands management over the last several decades and the results have been abysmal. Wetland destruction has accelerated, not abated. Canal dredging and other hydrologically disruptive activities have resulted in a current land loss rate of over 100 km^2/year in coastal Louisiana (Craig et al., 1979; Gagliano et al., 1981; Scaife et al., 1983). To understand why it has not worked we must understand the structure of the short term reinforcements facing the various players in the game of wetlands management. The situation is a 'social trap' because the narrow, short term incentives of those damaging the wetlands are inconsistent with the long term good of the system (Costanza, 1987).

To turn this trap into a trade-off one should charge the parties responsible the full cost of the ultimate environmental damage, at the time the damage is done. To do this one needs to know the economic value to society of coastal marshes and the amount of marsh destroyed by each activity. Since there is much uncertainty involved in these estimates the 'worst case' costs should be assumed and the burden of proof that the damages are in fact less than the worst case shifted to the parties who benefited or stand to benefit (see Costanza and Perrings, 1990, for a more complete discussion of this approach).

For example, our study concluded that each acre of coastal wetlands in Louisiana has a present value to society of roughly \$2500–\$17000 per acre. This range of values was due to uncertainties in the valuation procedures and the discount rate. Increasing the accuracy of the valuation estimates is an expensive proposition, and one that would stress the research budget of the state.

To effectively eliminate this trap one could charge the parties responsible for marsh destruction (ie. oil companies for dredging access canals through wetlands or developers wishing to convert wetlands to other uses) the \$17 000 per acre worst-case cost. These fees would go into an assurance bond to be returned to the developers in the event that damages are less than the worst case, or to be used for mitigating environmental damages by purchasing marshland elsewhere, backfilling canals, freshwater diversion, etc. (Costanza and Perrings, 1990). It would allow managers much more flexibility than the current 'mitigation banking' schemes because it is not limited to simple acre for acre tradeoffs. The responsible parties could lower the fee or secure a return of the bond by proving that the damages are actually less than the worst case assumption (by funding independent studies) or by minimizing the amount of wetlands they damage in the process of accomplishing their goal (by directional drilling, immediate backfilling, etc.). In either case the cause of wetland conservation would be served without unduly hindering the search for oil and gas, by converting the trap to a trade-off. Wetland developers would have strong economic incentives to minimize their damage to wetlands, and to fund additional needed research into the functions of wetlands (in order to narrow the range of uncertainty in wetland valuation estimates). Money would be held in the bond to be paid back to developers if future research aimed at reducing the uncertainty of the valuation estimates proves that they were overcharged.

SUMMARY AND CONCLUSIONS

Markets for ecological goods and services are far from perfect and we cannot rely on the free market to efficiently allocate these resources. The current system, which misallocates these resources, is better described as a social trap. Escaping from the trap involves turning it into an economic trade-off by making long run social costs, risks, and uncertainty incumbent on individuals and firms in the short run using fees and subsidies. Quantifying the proper level of fees and subsidies should be tied to models of the ecological impact of activities, but the burden of proof as to the magnitude of these damages should fall on the parties that stand to profit from them, not the general public. This addresses the distribution and incentive issues inherent in the problem in an equitable way. A flexible ecological cost charging and assurance bonding scheme can be designed that induces ecotechnological innovation by making it the most economically attractive option in the short run (as well as the long run). Implementation of such a scheme could go a long way toward allowing the development of a more ecological economics.

ACKNOWLEDGEMENTS

This research was funded in part by the Coastal Protection Section of the Louisiana Department of Natural Resources and the Terrebonne Parish Commission. Dr. Donald Boesch of the Louisiana Universities Marine Consortium and Dr. Donald Davis of Nichols State University provided much useful input data, encouragement, and advice. We are grateful to D. Pearce, K. Roberts, D. Chambers, K.J. Meyer-Arendt, D. Boesch, D. Davis, and J. Edmonson and J. Bartholomew for useful comments on earlier drafts.

REFERENCES

Allen, R.L., Turner, R.E. and Day, J.W., 1984. Oyster management and environmental influences in the Gulf of Mexico, 1880–1978. Center for Wetland Resources, Louisiana State University, Baton Rouge, La.

Arrow, K.J., 1976. The rate of discount for long-term public investment. In: H. Ashley, R.L. Rudman and C. Shipple (Editors), Energy and the Environment: A Risk Benefit Approach. Pergamon, New York, pp. 113–141.

Arrow, K.J. and Fisher, A.C., 1974. Environmental preservation, uncertainty, and irreversibility. Q. J. Econ., 55: 313–319.

Bartlett, E.T., 1984. Estimating benefits of range for wildland management and planning. In: G.L. Peterson and A. Randall (Editors), Valuation of Wildland Benefits. Westview, Boulder, Colo., pp. 143–157.

Batie, S.S. and Wilson, J.R., 1978. Economic values attributable to Virginia's coastal wetlands as inputs in oyster production. South. J. Agric. Econ., 10: 111–117.

Bertrand, A.L., 1980. Marine recreational finfisherman in Louisiana. Coastal Ecology and Fisheries Institute, Louisiana State University, Baton Rouge, La., 15 pp.

Bishop, R., 1982. Option value: an exposition and extension. Land Econ., 58: 1–15.

Braat, L.C. and Van Lierop, W.F.J., 1985. 1986. Economic–ecological modeling: an introduction to methods and applications. Ecol. Modelling, 31: 33–44.

Brookshire, D.S., Eubanks, L.S. and Randall, A., 1983. Estimating option prices and existence values for wildlife resources. Land Econ., 59: 1–15.

Cicchetti, C.J. and Smith, V.K., 1973. Congestion, quality, deterioration, and optimal use: Wilderness recreation in the Spanish Peaks primitive area. Soc. Sci. Res., 2: 15–30.

Clark, C.W., 1973. The economics of overexploitation. Science, 181: 630–634.

Cleveland, C.J., Costanza, R., Hall, C.A.S. and Kaufmann, R., 1984. Energy and the United States economy: a biophysical perspective. Science, 255: 890–897.

Conrad, J.M., 1980. Quasi option value and the expected value of information. Q. J. Econ., 94: 813–820.

Costanza, R., 1980. Embodied energy and economic valuation. Science, 210: 1219–1224.

Costanza, R., 1984. Natural resource valuation and management: Toward an ecological economics. In: A.-M. Jansson (Editor), Integration of Economy and Ecology: An Outlook for the Eighties. University of Stockholm Press, Stockholm, pp. 7–18.

Costanza, R., 1987. Social traps and environmental policy. BioScience, 37: 407–412.

Costanza, R. and Daly, H.E., 1987. Toward an ecological economics. Ecol. Modelling, 38: 1–7.

Costanza, R. and Farber, S.C., 1985a. Theories and methods of valuation of natural systems: a comparison of willingness-to-pay and energy analysis based approaches. Man Environ. Space Time, 4: 1–38.

Costanza, R. and Farber, S.C., 1985b. The economic value of coastal wetlands in Louisiana. Final Report to the Louisiana Department of Natural Resources. Center for Wetland Resources, Louisiana State University, Baton Rouge, La.

Costanza, R. and Hannon, B., 1990. Multicommodity flows in ecosystem analysis: Dealing with the 'mixed units' problem In: B.C. Patten and S.E. Jørgensen (Editors), Complex Ecology. Prentice Hall, Englewood Cliffs, NJ (in press).

Costanza, R. and Herendeen, R.A., 1984. Embodied energy and economic value in the United States economy: 1963, 1967, and 1972. Resour. Energy, 6: 129–163.

Costanza, R. and Neill, C., 1984. Energy intensities, Interdependence, and value in ecological systems: a linear programming approach. J. Theor. Biol., 106: 41–57.

Costanza, R. and Perrings, C.H., 1990. A flexible assurance bonding system for improved environmental management. Ecol. Econ. (in press).

Costanza, R, Neill, C., Leibowitz, S.G., Fruci, J.R., Bahr, L.M. and Day, J.W., 1983. Ecological models of the Mississippi Deltaic Plain Region: Data collection and presentation. FWS/OBS-82/68, Division of Biological Services, Fish and Wildlife Services, Washington, D.C., 342 pp.

Costanza, R., Sklar, F.H. and Day, J.W., Jr., 1986. Modeling spatial and temporal succession in the Atchafalaya/Terrebonne marsh/estuarine complex in South Louisiana. In: D.A. Wolfe (Editor), Estuarine Variability. Academic Press, New York, pp. 387–404.

Craig, N.J., Turner, R.E. and Day, J.W., Jr., 1979. Land loss in coastal Louisiana (U.S.A.). Environ. Manag., 3: 133–144.

Cross, J.G. and Guyer, M.J., 1980. Social Traps. University of Michigan Press, Ann Arbor, Mich.

Davis, W., 1978. Wetlands trapping in Louisiana. Geosci. Man, 19: 81–92.

Farber, S.C., 1987. The value of coastal wetlands for protection of property against hurricane wind damage. J. Environ. Econ. Manage., 14: 143–151.

Farber, S. and Costanza, R., 1987. The economic value of wetlands systems. J. Environ. Manage., 24: 41–51.

Gagliano, S.M., Meyer-Arendt, K.J. and Wicker, K.M., 1981. Land loss in the Mississippi River Deltaic Plain. Trans. Gulf Coast Assoc. Geol. Soc., 31: 295–299.

Greenley, D.A., Walsh, R.G. and Young, R.A., 1981. Option value: empirical evidence from a case study of recreation and water quality. Q. J. Econ., 95: 657–673.
Hannon, B., 1973. The structure of ecosystems, J. Theor. Biol., 41: 535–546.
Hannon, B., 1979. Total energy costs in ecosystems. J. Theor. Biol., 80: 271–293.
Hopkinson, C.S., Jr., 1978. The relation of man and nature in Barataria Basin, Louisiana, Ph.D. dissertation, Louisiana State University, Baton Rouge, La., 236 pp.
Huetner, D.A., 1982. Economic values and embodied energy (with reply by R. Costanza). Science, 216: 1141–1143.
Isard, W., 1972. Ecologic–economic analysis for regional development. Free Press, New York, 270 pp.
Lindall, N., Hall, J.R., Sykes, J.E. and Arnold, E.L., 1972. Louisiana coastal zone: analyses of resources and resource development needs in connection with estuarine ecology. National Marine Fisheries Service, St. Petersburg, Fla., 315 pp.
Lovejoy, T.E., 1986. Species leave the ark one by one. In: B.G. Norton (Editor), The Preservation of Species. Princeton University Press, Princeton, NJ., pp. 13–27.
Lynne, G.D., Conroy, P.D. and Prochaska, F.J., 1981. Economic valuation of marsh areas for marine production processes (Florida). J. Environ. Econ. Manage., 8: 175–186.
Norton, B.G., 1986. On the inherent danger of undervaluing species. In: B.G. Norton (Editor), The Preservation of Species. Princeton University Press, Princeton, N.J., pp. 110–137.
Odum, H.T. and Odum, E.P., 1976. Energy Basis for Man and Nature. McGraw-Hill, New York, 297 pp.
Page, T., 1978. A generic view of toxic chemicals and similar risks. Ecol. Law Q., 7: 207–244.
Platt, J., 1973. Social traps. Am. Psychol., 28: 642–651.
Randall, A., 1986. Human preferences, economics, and the preservation of species. In: B.G. Norton (Editor), The Preservation of Species. Princeton University Press, Princeton, N.J., pp. 79–109.
Randall, A. and Stoll, J., 1980. Consumer's surplus in commodity space. Am. Econ. Rev., 70: 449–555.
Scaife, W.W., Turner, R.E. and Costanza, R., 1983. Coastal Louisiana recent land loss and canal impacts. Environ. Manage., 7; 433–442.
Turner, M.G., Costanza, R., Springer, T.M. and Odum, E.P., 1988. Market and nonmarket values of the Georgia landscape. Environ. Manage., 12: 209–217.
USDC 1983. Fisheries of the United States. U.S. Department of Commerce, 117 pp.

[16]

Marine Resource Economics, Volume 9, pp. 291–310
Printed in the USA.
0738-1360/94 $3.00 + .00

Renewable and Nonrenewable Resource Theory Applied to Coastal Agriculture, Forest, Wetland, and Fishery Linkages

STEPHEN K. SWALLOW

Department of Resource Economics,
University of Rhode Island,
Kingston, RI 02881

Abstract *This paper addresses tradeoffs in wetland development using a framework that integrates economic theory of renewable and nonrenewable resources. The theory treats wetland development as use of a nonrenewable resource, while wetland preservation protects critical fishery habitat. The framework recognizes that wetland quality may vary for either development or fisheries. An illustrative application assesses tradeoffs in converting pocosin wetlands to agriculture rather than maintaining wetlands to protect salinity in estuarine nursery areas. Results reveal the marginal value of salinity protection may be substantial, while location may affect a wetland's value to an estuarine shrimp fishery. Comparisons between agricultural and forestry land-uses show that ecological links may cause wetland values to depend upon the land-use chosen for the developed state. Future assessments of other development may reveal additional impacts through impacts on salinity.*

Keywords nonrenewable, renewable, fishery, wetland value, pocosin, Pamlico Sound

Economic development, such as drainage of coastal wetlands, may impose external losses on renewable resource production, such as commercial or recreational fisheries. Several studies address the opportunity costs of preservation (*e.g.*, Brown 1976, Batie and Mabbs-Zeno 1985, Danielson and Leitch 1986, Shabman and Bertelson 1979) and several others address the external costs of development (*e.g.*, Batie and Wilson 1978; Farber 1987; Kahn and Kemp 1985; Lynne *et al.* 1981). Relatively few studies address both preservation and development, perhaps due to the natural division between proponents of each or perhaps due to the difficulty of simultaneously estimating benefits for both. For example, although Stavins (1990) provides a partial exception, most studies (*e.g.*, Gupta and Foster 1975) emphasize the taxonomy of costs and benefits rather than ecological inter-

Robert F. Conrad, Diane P. Dupont, James E. Easley, Jr., William F. Hyde, Thomas A. Grigalunas, David H. Newman, Peter J. Parks, Mark G. Smith, James R. Waters, and the referees motivated improvements to the paper. The National Wildlife Federation; the U.S. National Oceanic and Atmospheric Administration grant #NA86AA-D-CZ015; the Duke University Research Council through Duke's School of the Environment; and the Rhode Island Agricultural Experiment Station (RI AES) funded this research. North Carolina statistical data were provided at the author's request by the N.C. Division of Marine Fisheries (DMF). Analyses of these data and conclusions drawn therefrom are those of the author and do not necessarily represent the views of the N.C. DMF or any financial sponsor. RI AES contribution no. 3024.

dependencies implicit in the wetland allocation issue (*cf.* Crocker and Tschirhart 1992).

This paper addresses tradeoffs implicit in wetland development using a framework suggested by integrating the basic theories of renewable and nonrenewable resource economics (Hotelling 1931; Clark and Munro 1975; Swallow 1990). The model is tailored to assess issues raised by the historical controversy surrounding the drainage of coastal wetlands for agriculture on the Pamlico-Albemarle Peninsula of North Carolina (Heath 1975, NRCD 1987, Street and McClees 1981). Ecologists and fishermen believe that these *freshwater* wetlands, known as pocosin wetlands, may be critical to sustaining commercial fisheries in the Pamlico Sound estuary, particularly the Penaeid shrimp fishery (Street and McClees 1981). This debate arises, in part, because human-induced ecosystem responses may limit economic activity and sustainability (see Crocker and Tschirhart 1992) in the Pamlico fisheries.

Three objectives motivate this paper: 1) to develop a resource-theoretic approach to marginal preservation-development choices; 2) to apply this framework in an assessment of the public issues raised by historical concern regarding pocosin development; 3) to suggest that cognizance of resource interdependencies can clarify policy debates. After providing background on the study area, the paper addresses these objectives in order.

The application relies heavily on the Penaeid shrimp fishery of Pamlico Sound because both biological and economic data are available and because ecological or "natural history" theories are sufficient to suggest a plausible approach to assess biophysical relationships. Furthermore, public concerns often focus on the shrimp fishery because it is by far the largest of Pamlico's fisheries. The assessment-level application offers substantial applied insights as well as illustration, including empirical results showing that wetland values may depend on the alternative land-use under consideration. Such key results are revealed by intermediate steps taken in the empirical assessment of the pocosin development issue.

Study Area: Background and Context

This paper applies a model of interdependent renewable and nonrenewable resources to coastal zone development and the Pamlico Sound, NC, Penaeid shrimp fishery. The background for this application will motivate the theoretical structure for the empirical analysis in subsequent sections.

In the U.S., coastal zone development causes up to 90% of losses in estuarine acreage and indirectly diminishes estuarine productivity through off-site impacts (Tiner 1984). The Pamlico-Albemarle Estuarine Complex is one of the largest and most productive estuarine systems in North America (Epperly and Ross 1986, NRCD 1987). Throughout the U.S., estuarine-dependent fish species comprise 50–90% of commercial landings, with North Carolina landings nearly 90% estuarine-dependent (Epperly and Ross 1986, Street and McClees 1981, Tiner 1984). In Pamlico Sound, Penaeid shrimp support the most prized fishery, producing about 25% of gross dock-side revenues (Purvis and McCoy 1974, Street and McClees 1981). Regionally, however, North Carolina produces a small share (2–4%) of U.S. landings and ex-vessel prices follow the larger Gulf of Mexico or U.S. markets (Waters *et al.* 1980).

Recent public debates concern the conflicts between coastal development and estuarine-dependent vocations (see NRCD 1987). On the preservation side, discussions focus on commercial marine fisheries, especially the highly-valued shrimp fishery (Street and McClees 1981). On the development side, forestry and agriculture dominate the converted freshwater pocosins (peat-bog wetlands) near the estuary (Heath 1975). However, substantial site preparation costs make forestry and agricultural uses profitable marginally, but these uses remain the dominate threat to pocosin wetlands (Heimlich and Langner 1986). No consensus exists concerning what future developers will propose, but predictions range from urbanization to hog and poultry production to peat mining for electricity generation, with public officials anxious to diversify the area's impoverished economy (Richardson 1981, Tiner 1984, NRCD 1987). Agriculture still motivates policy analyses (*i.e.*, Palmquist and Danielson 1989) and therefore appears to provide the highest return.

In this case study, coastal development requires up to 20 miles of drainage canals per square mile (Heath 1975). Drainage of pocosin wetlands irreversibly alters the local hydrologic system by eliminating the vegetative and peat-bog structure that inhibits water flow, causing a decline in the salinity level of estuarine nursery areas (Heath 1975, Skaggs *et al.* 1980, Jones and Sholar 1981, Epperly and Ross 1986). Rehabilitation of the hydrologic function is viewed as impractical since the peat-bog structure may take millennia to regenerate.[1] Thus, wetland development may be assumed irreversible, which is broadly consistent with the ecology of pocosin wetlands and the engineering of their drainage (Heath 1975, NRCD 1987). Irreversibility places wetland development as a nonrenewable resource sector (Krutilla 1967; Fisher and Krutilla 1974).

On the fishery side, Pamlico's juvenile shrimp stock annually arrives via onshore currents; these currents carry juveniles from an open ocean breeding ground to estuarine nurseries along the South Atlantic and Gulf coasts (Epperly and Ross 1986, Williams 1955). Furthermore, if shrimp escape from the Pamlico Sound fishery,[2] they generally will not be harvested elsewhere (Williams 1955, McCoy 1972, Purvis and McCoy 1974, Hettler and Chester 1982, Babcock and Mundy 1985, Matylewich and Mundy 1985). Finally, an altered hydrologic system causes the salinity level to decline in estuarine nurseries, thereby diminishing the survival rate of juvenile shrimp (Jones and Sholar 1981, Street and McClees 1981; Williams 1955). Thus, while a fishery is generally viewed as a renewable resource sector, Pamlico's annual shrimp crop fits the independent-generations fishery model (Wilen 1985). This fishery depends upon environmental factors rather than on a breeding stock.

Economists recognize this process in other fisheries by linking shrimp production to environmental variables (Blomo *et al.* 1982, Griffin *et al.* 1976), but their analyses have not linked estuarine environmental variables directly to development. Irreversible hydrologic impacts suggest wetland development is a type of nonrenewable resource extraction, the key analogy here.

[1] Native Americans used "pocosin" to indicate that these wetlands are perched atop a hill; damming of drainage canals is not expected to completely simulate the sponge-like storage capacity of the original peat soils. Heath (1975) focused on agriculture, while Campbell and Hughes (1991) note that modern forestry practices may still preserve hydrologic functions.
[2] Including Pamlico Sound proper and the Pamlico and Neuse Rivers (U.S. National Marine Fisheries Service areas 6354, 6355, and 7011).

Theory

The theoretical foundation for this analysis derives from well-known theories of exploitation for renewable and nonrenewable resources (Hotelling 1931; Clark and Munro 1975), particularly when these resource sectors are interdependent, such as when the nonrenewable resource provides wetland habitat for the renewable resource (Swallow 1990). With independent generations in the fishery, the shrimp population X is modeled to depend only upon the habitat stocks E:

$$X_t = X(E^H(t), E^N(t)), \tag{1}$$

where E^H and E^N quantify stocks of "high quality" habitat and "normal quality" habitat, respectively; development determines the habitat available at time t. In the renewable sector, fishermen maximize returns or profits R from their labor L and the stock of shrimp, so that renewable resource benefits are:

$$R_t = R(L^*, X(E^H, E^N)) = R^*(X), \tag{2}$$

where L^* is the benefit-maximizing quantity of labor conditional on X, so R^* represents fishery benefits conditional on the shrimp stock X in a given year. In the wetland development sector, nonrenewable resource benefits B^i ($i = H,N$) depend upon the development rate d^i and the available stock of wetlands of a particular type:

$$B^i(d^i, E^i) = C^i(E^i)\, d^i,\ i = H,N, \tag{3}$$

where $C^i(\bullet)$ denotes the net marginal benefit of an additional acre of development for wetland type i, which depends on the acres remaining, E^i. E^i indexes the quality of remaining acres from the perspective of development.

Development of pocosin wetlands requires permits through environmental managers, state officials and the U.S. Corps of Engineers under Section 404 of the U.S. Clean Water Act. Economically, environmental managers must consider the balance between the present value of wetlands preserved for fisheries and the return to wetlands development, at the margin. Then, the economic choice depends on the net opportunity cost (NOC) of wetland development:[3]

$$NOC^i(X, E^H, E^N) = (\partial R^*/\partial X) \cdot (\partial X/\partial E^i) - r \cdot C^i(E^i) \tag{4}$$

where NOC^i is the net opportunity cost of developing wetlands of type i ($i = H,N$). NOC^i measures the loss of fishery profits if a marginal unit of E^i is developed, net of the annualized return to development. If NOC^i is positive, denial of development permits maximizes social returns from wetlands; if NOC^i is negative, then approval of development permits returns marginal development benefits that more than offset losses in the renewable sector. Since wetland quality may be heterogeneous for either preservation or development, NOC may differ for different wetland qualities. The next sections assess the potential for managers to

[3] This marginal condition assumes appropriate concavity of (1)–(3).

find NOC is negative, favoring development, for pocosin wetlands near Pamlico Sound.

Application

The empirical analysis assumes the Pamlico Sound shrimp fishery: 1) is economically and ecologically independent of neighboring shrimp populations; 2) is affected by development; 3) and is a price taker. The link between fishery benefits and coastal development is generated in a stepwise process which 1) estimates the potential fishing rents for a given stock of shrimp, 2) links the shrimp stock to estuarine salinity, and 3) links estuarine salinity to development (Fig. 1). Loomis (1988) used a stepwise model in a forestry-fishery context. Our stepwise approach combines two approaches suggested by Kahn (1987), by including environmental variables in the harvest function while also using an (simple) ecosystem model. This approach permits improvements at any step as new scientific or policy information warrants.

Figure 1 illustrates the basic empirical model. Development initiates changes in potential fishery benefits: drainage of pocosin wetlands alters estuarine salinity which then lowers the expected annual, estuarine production of shrimp; lower estuarine productivity alters the conditions of profit maximization; finally, the shrimp fishery realizes a change in expected annual rents. Figure 1 summarizes the consensus (see above or NRDC 1987) among managers in the Pamlico region.

This consensus (Figure 1) implies three basic relationships:

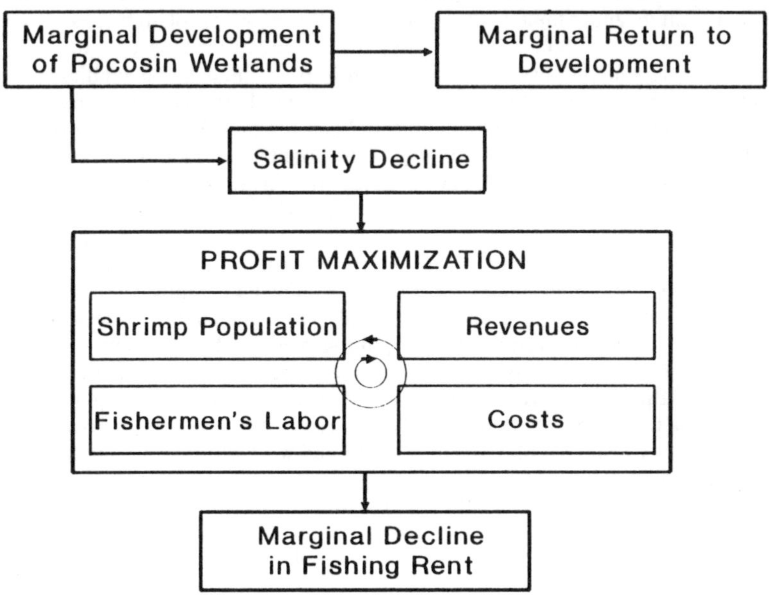

Figure 1. Outline of Empirical Model: Pocosin Wetland Development Impacts Estuarine Shrimp Fishery

$$\pi = f(XFB, L) = \text{total revenue} - \text{total costs}; \quad (5)$$

$$XFB = g(SAL); \quad (6)$$

$$SAL = h(POC); \quad (7)$$

where π is annual "profit" or quasi-rent, XFB indexes the shrimp population X, L represents craft-days of fishermen's labor, SAL is salinity in estuarine nursery areas, and POC is the acreage of pocosin wetlands on the Pamlico-Albemarle Peninsula. Rent (5) estimates renewable resource benefits, R, as a function of X and L; here π represents R and XFB proxies for X in (2). The combination of (6) and (7) empirically represent (1), where POC denotes the stock of undeveloped pocosin wetlands ($E = (E^H, E^N)$ above).

Consistent with (2), the results below derive from a simulation of Pamlico's shrimp fishery under efficiently restricted access. Benefits based on restricted access tend to estimate an upper limit to benefits under open access. The current study assumes labor may easily switch to competing fisheries, so the marginal unit of labor earns the opportunity cost available elsewhere. Available data do not identify specific vessels, so that the empirical harvest equation (below) is for a "representative vessel."[4]

The Data and Key Variables

The NC Division of Marine Fisheries and the NC Land Resources Information System (Lukin and Mauger 1983) provided the majority of data (see Swallow 1988). These weekly data included: price, catch, and effort measures for 1978–86; a fishery-independent survey of juvenile shrimp abundance and salinity levels in Pamlico's nursery areas, 1979–86; and the acreage of land in various uses, based on analysis of 1980–81 air photos and tabulated for each USGS topographic map of the Pamlico-Albemarle Peninsula. Additional data included daily rainfall records (Wiser 1983) and USDA price indices for shrimp and farmed meats (1978–84). Dollar values were adjusted to December 1986 using the producer price index (monthly series).

Pocosin wetlands may contribute differently due to their geographic location relative to the estuarine salinity regime. The Peninsula extends eastward, serving as the northern boundary of Pamlico Sound. Thus, wetlands on the south shore may buffer freshwater inflows to shrimp nursery areas, but the southeast shore's wetlands are close to inlets for undiluted seawater and may, therefore, be less critical (see Epperly and Ross 1986; Giese et al. 1985). These geographic concerns defined, for this study, wetland quality relative to shrimp, with a stock of "normal" wetlands near the southeastern shore and a "high quality" stock near the southwestern shore.

The shrimp fishery includes three Penaeid species, but brown shrimp (*Peneaus aztecus*) are the focus of this study. Brown shrimp comprise the bulk of total catch (always >65%; ≥80% in 7 of 9 years of available data) and are particularly sen-

[4] This limitation introduces a downward bias in the estimated fishery rents, since rents accruing to more skilled fishermen cannot be estimated separately (see Copes 1970).

sitive to salinity changes (see Street and McClees 1981). Brown shrimp inhabit nursery areas exclusively during the spring wet-season when estuarine salinity is sensitive to previous wetland drainage and juveniles of other species are largely absent (Heath 1975, Williams 1955). Fishermen and consumers do *not* differentiate among species, so that species-specific harvest functions are inappropriate. The empirical model estimates expected total harvest of all Penaeids.

The data permitted calculation of a population index for juvenile brown shrimp. The annual index, XFB, is simply the average of all trawl samples (no. brown shrimp caught per min.)[5] taken by NC Division of Marine Fisheries (NC DMF) in juvenile nursery areas during weeks 18 to 25 of each year. This average included samples which contained zero brown shrimp, but only considered sampling locations in known shrimp habitat (locations producing at least one shrimp over the eight years (1979–86) of available data).

Parameter Estimates and Simulation Results

A chain of marginal effects across (5)–(7) empirically links irreversible wetland development and renewable resource benefits (Fig. 1):

$$(\partial R^*/\partial X)(\partial X/\partial E) = (\partial \pi/\partial XFB)(\partial XFB/\partial SAL)(\partial SAL/\partial POC). \qquad (8)$$

This section estimates these linkages.

Rent Function

The empirical model estimates and sums weekly quasi-rents to obtain annual rents π for the brown shrimp index existing in each year. This calculation quantifies the main cell in Figure 1. Weekly rents are:

$$\pi_i = p_i C_i - w_s L_i = p_i k_i (XFB)^{\varphi} (L_i)^{\beta} - w_s L_i \qquad (9)$$

where i indexes weekly observations; p_i is the price per pound (heads-off) for shrimp of average size; C_i is the catch of all shrimp; w_s is the marginal opportunity cost of a craft-day during sub-season s; L_i corresponds to fishermen's labor in 24-hour craft-days; k_i (defined in Table 1) measures exogenous fluctuations in shrimp abundance as shrimp migrate from nurseries through Pamlico Sound to the Atlantic (see Appendix); XFB is the brown shrimp index for the current year.

Table 1 summarizes parameter estimates for the catch equation (implicit in (9)). The opportunity cost of a craft-day was estimated, following Bell (1986), by recognizing that the data were generated under open access, so that total revenues would approximate total costs. Using Bell's (1986) approach, Swallow (1988, p. 137) estimated w_s at \$714, \$1131, \$959 per 24-hour craft-day for early, middle, and late seasons (before July, July to September, after September, respectively).[6] The

[5] The sample only includes sites where NC DMF uses a 3.2 m head-rope for towing the trawl. DMF uses this head-rope in shallower, upstream nurseries where land-use changes have the most direct effect.

[6] The physical productivity of a craft-day is assumed constant, but its opportunity cost varies, e.g. due to seasonality in competing fisheries. Preliminary regressions examined

Table 1
Estimated Weekly Harvest Equation (N = 222, 1979–1986).[a]

Variable	Model 1 OLS	Model 1 GLS[c]	Model 2[b] OLS	Model 2[b] GLS[c]
Components of k_i:				
Intercept	3.22	3.28	3.20	3.26
(ln[A])	(0.190)	(0.154)	(0.188)	(0.144)
Juv. white	0.186	0.198	0.187	0.199
shrimp index	(0.0464)	(0.0329)	(0.0462)	(0.0322)
(XFW)[d]				
No. of	0.341	0.399	0.381	0.419
landings (N_i)[e]	(0.0888)	(0.0724)	(0.0595)	(0.0417)
Lag catch	0.144	0.130	0.152	0.133
(C_{i-1})	(0.0279)	(0.0222)	(0.0247)	(0.0180)
Juv. brown	0.315	0.312	0.300	0.310
shrimp index	(0.0529)	(0.0334)	(0.0460)	(0.0295)
(XFB)				
Fishermen's	0.744	0.709	0.700	0.690
labor (L_i)	(0.0850)	(0.0712)	(0.0460)	(0.0295)
F	870.	1162.	1091.	1460.
R-square	0.953	0.964	0.953	0.964
SSE/216	0.18517	0.18585	0.18462	0.18610

[a] Catch in week i is $C_i = A\ (XFW)^a\ (N_i)^b\ (C_{i-1})^c\ (XFB)^{\varphi}\ (L_i)^{\beta}$. Index i denotes weekly observations, XFW and XFB are annual observations. N and L were standardized by Swallow (1988, pp. 112–21). S.e. given in parentheses; all t-tests and F-tests were significant ≤0.001. The equation was estimated in log form so the estimate of ln(A) produces a biased estimate of A (Goldberger 1968) (see Appendix).

[b] Restricts $\varphi + \beta = 1$. Lagrangian multiplier test of this restriction was not significant, even at 0.50 (1 df t-statistic = 0.604 for OLS and 0.202 for GLS).

[c] OLS regressions were heteroscedastic, assuming the multiplicative model of Harvey described by Judge et al. (1985, pp. 439–441). This model assumed the logged-variance for each observation was a function of an intercept and $\ln(L_i)$. The test statistic was significant at 0.01 (1 df Chi-square = 20.83 and 24.34 for models 1 and 2, respectively).

[d] Calculated by methods used for XFB (see text), but with juvenile samples from weeks 28 to 35 for white shrimp (*P. setiferus*). A similar variable for pink shrimp (*P. duorarum*) was not significant at 0.05.

[e] Number of times craft docked and unloaded shrimp.

Appendix provides more details. The remaining discussion uses the GLS Model 2 (Table 1) to estimate π_i in (9), because that model is jointly concave in XFB and L. In Model 2, concavity is assured by the statistically insignificant restriction that $\varphi + \beta = 1$ ($P > 0.50$; Table 1).

Maximum annual rents for π in (5) are the sum of the maximized weekly rents

seasonal effects in the catch equation, but results were inconsistent with ecology and produced trivial differences in the model fit ($R^2 > 0.95$ in Table 1) (see Swallow 1988, pp. 122–127).

for π_i in (9);[7] this value was calculated for a typical year, using means for exogenous variables. Weekly rents are maximized by L_i^* such that:

$$\beta\ p_i\ k_i\ (XFB)^{\varphi}\ (L_i^*)^{\beta-1} = w_s. \tag{10}$$

Substituting L_i^* into (9) and summing over i, one obtains the annual value of marginal product of the brown shrimp population index (VMP_B):

$$VMP_B = \partial\pi^*/\partial XFB = \varphi\ \Sigma_i\ p_i\ k_i\ (XFB)^{\varphi-1}\ (L_i^*)^{\beta} = [\varphi/XFB]\ (TR), \tag{11}$$

where π^* without a subscript denotes annual maximized rents as in (5) and the last equality derives from the definition of annual total revenues (TR) (with (9)). VMP_B in (11) is a simple function of revenues, the shrimp index, and the elasticity of rents (φ) with respect to the index. For the mean observed shrimp index XFB (12.98), parameters in Model 2 (Table 1), and mean exogenous conditions (Appendix), VMP_B equals \$131,952/index point/year.[8] With $\varphi + \beta = 1$, simple analytical or numerical analyses confirm that VMP_B remains constant.

Shrimp versus Salinity

A two-step process estimated (6), the effect of salinity on the shrimp index. The statistical step estimated brown shrimp abundance in estuarine nurseries as a function of salinity. The second step used statistical results to estimate the change in the brown shrimp index XFB and fishery benefits that might result from marginal reductions in salinity. The results quantify the effect of salinity on rents (Figure 1). This section summarizes the key results for the present discussion (details are in citations below).

Using data from monitoring sites in shrimp nursery areas, the statistical step regressed the number of juvenile brown shrimp caught per minute (NBRW) against salinity, water temperature, and week (weeks 18–25; 1979–86). For samples with at least one brown shrimp, final results yielded:

$$NBRW = k_B + 1.476\ SAL \tag{12}$$

where the intercept, k_B, represents independent variables that proxy for salinity-independent effects on juvenile shrimp production.[9]

This regression (12) allows an estimate of the relationship between the shrimp index and salinity because the typical (mean) shrimp index XFB is the mean of NBRW across years:

[7] Kellogg et al. (1986) analyze the effects of discounting and shrimp growth. These effects are beyond the scope of this paper.

[8] Based on GLS Model 1 (Table 1), estimated VMP_B is only \$99,893.

[9] Since 449 of 1074 observations contained zero brown shrimp, the regression followed a censored-data approach (Lee et al. 1980). The final regression is significant ($P < 0.01$; with $R^2 = 0.12$, with a significant salinity coefficient (s.e. = 0.4862; $P < 0.01$). k_B depends on dummy variables for each year, water temperature, and week number; salinity is in parts per thousand. A quadratic term for SAL was rejected ($P > 0.05$) by an MSE test (Toro-Vizcarrondo and Wallace 1968). Swallow (1988, pp. 192–99) gives details.

$$XFB = [\Sigma_y\ FB_y \cdot (\Sigma_j\ NBRW_{jy})/N_y]/8 \tag{13}$$

where N_y is the number of NBRW samples from year y, summation j is over those samples, summation y is over the eight years of data, and FB_y is the fraction of samples in year y with at least one brown shrimp. Regression (12) estimates the effect of a marginal change in salinity on NBRW, such that

$$\partial NBRW/\partial SAL = 1.476. \tag{14}$$

Using (13) and (14), the link between an across-the-board change in salinity (at all sites) and the brown shrimp index for an average year is estimated as

$$\partial XFB/\partial SAL = [\Sigma_y\ FB_y \cdot (N_y \cdot 1.476)/N_y]/8 = 1.476\ (\Sigma\ FB_y)/8 = 0.8635, \tag{15}$$

where the last equality uses the observed mean of the annual fraction of samples containing at least one brown shrimp (mean of $FB_y = 0.5850$).[10]

Slope (15) proxies for the marginal physical product of salinity in the production of juvenile shrimp (MP_{SAL}). Equation (15) with (11) permits estimation of the value of the marginal product of the salinity level in the shrimp fishery, VMP_{SAL}:

$$VMP_{SAL} = VMP_B\ (\partial XFB/\partial SAL) = VMP_B \cdot 0.8635. \tag{16}$$

For a typical year's conditions, VMP_{SAL} equals \$113,941/salinity point/year.

Salinity versus Wetland Development

A similar, two-step process estimates relationship (7), linking salinity in nursery areas to wetland acreage stocks. The result links the cells, in Figure 1, for development and salinity.

The regression step related estuarine salinity in nursery areas along the Pamlico-Albemarle peninsula to adjacent land-use (Hyde and Dare Counties, NC), including the proportion of adjacent land in general agriculture (AGRIC), forestry-forest cover (FOREST), and pocosin wetlands (WETLAND). Lukin and Mauger (1983) define these land-uses. Results showed a statistically significant ($P < 0.05$) relationship between salinity in nursery areas and the proportion of adjacent land in three land-use categories:

$$SAL = k_{SAL} - 13.86\ AGRIC - 10.08\ FOREST - 5.444\ WETLAND - 5.104\ SESHORE \cdot WETLAND \tag{17}$$

where the intercept, $k_{SAL} > 0$, incorporates variables that describe the hydrologic

[10] See Swallow (1988, p. 205). This calculation assumes $\partial FB_y/\partial SAL = 0$, which is consistent with the interpretation that the presence of brown shrimp in a sample proves habitat suitability at a site and that a marginal salinity change will not eliminate the whole site as habitat. Alternative assumptions would not leave $\partial XFB/\partial SAL$ constant, but a numerical procedure using predicted values from the statistical model can handle alternative assumptions.

Table 2
Parameters for Estimating the Impact of Wetland Conversion on Estuarine Salinity and Value of Marginal Product (VMP) of Preserving Pocosin Wetlands Rather than Converting Acres to Agriculture or Forestry-Forest Land-Uses.[a]

	Conversion Impact on Salinity			
Conversion of Wetlands to:	Symbol	Estimate	S.e.	VMP
Agriculture				
Normal quality wetlands	$\beta^{N}_{AGR\text{-}WET}$	3.312*	0.936	0.282
High quality wetlands	$\beta^{H}_{AGR\text{-}WET}$	8.416*	1.01	3.37
Forestry-Forest				
Normal quality wetlands	$\beta^{N}_{FOR\text{-}WET}$	−0.468‡	0.907	—
High quality wetlands	$\beta^{H}_{FOR\text{-}WET}$	4.636*	0.476	1.85
Constant of proportionality (K^{i}_{WET}) to calculate $\partial XFB/\partial WETLAND^{i}$:				
Normal quality wetlands (i = N)	0.4376×10^{-6}			
High quality wetlands (i = H)	2.054×10^{-6}			

[a] Based on equation (17) and estimated covariances.
* Significantly different from zero at $P < 0.01$.
‡ Not significantly different from zero, even at $P < 0.25$.

conditions for the year, and dummy variable SESHORE[11] equals 1 for sites near the peninsula's southeastern shore.[12]

The coefficients on the land-use variables in (17) show that converting pocosin wetlands (lowering WETLAND) to agriculture or forestry (raising AGRIC and/or FOREST) causes a *net decline* in the salinity of nearby shrimp nursery areas. Table 2 gives estimated parameters for the net effects on salinity. For example, converting wetlands on the southwestern shore (SESHORE = 0) to agriculture decreases mean salinity in adjacent estuarine nurseries by 8.4 units (13.86 − 5.44; see (17) and Table 2). These results are consistent with the expectation that south*western* wetlands are of "high quality" (*i.e.*, E^{H}) while south*eastern* wetlands are of "normal quality" (i.e., E^{N}).[13]

One can show that developing peninsular wetlands has an estimated impact on

[11] Southeastern shore is from longitude 75° 53′ 25″ W to 76° 07′ 30″ W.

[12] See Swallow (1988, pp. 179–87, 213–19) for variable definitions and discussion of preliminary regressions. The standard errors for (17) are, respectively, 1.912, 1.057, 1.159, and 0.7768. These estimates pertain to an equivalent GLS model which corrects for heteroscedasticity across years of data. All variables were significant ($P < 0.01$) as is the model ($F_{13,301} = 439.014$; $P < 0.001$; $R^2 = 0.9499$; $N = 315$). The intercept term is given by

$$K_{SAL} = \underset{(0.7918)}{24.81} - \underset{(0.3650)}{14.47}\,D79 - \underset{(0.2481)}{13.79}\,D80 - \underset{(0.7815)}{3.363}\,D81 - \underset{(0.2713)}{5.118}\,D82$$
$$- \underset{(0.2984)}{12.40}\,D83 - \underset{(0.2804)}{11.79}\,D84 - \underset{(0.3899)}{1.504}\,D85 - \underset{(0.08329)}{0.2170}\,RAIN\,10\ \ \underset{(0.4365)}{4.702}\,SESHORE$$

where standard errors are given in parentheses, dummy variables for each year are Dnn (nn = 1979 to 1985), and RAIN10 is the rainfall in the ten days preceding the salinity sample.

[13] Land-use variables are proportions (acreage of land-type divided by total acreage sampled, TOTACRE). E.g., converting 1 acre of wetlands to agriculture decreases the numerator of WETLAND by 1 and increases the numerator of AGRIC by 1.

the shrimp index that is proportional to the coefficients in Table 2. Doing so uses (13) and splits the summation over sampling sites j into sites near one shore of the peninsula (j:SESHORE $\rightarrow$ i; i = H,N)[14] and sites away from that shore,

$$\partial XFB/\partial WETLAND^i = (1/8)\Sigma_y\{(FB_y/N_y) \cdot [\Sigma_{j:SESHORE\rightarrow i}(\partial NBRW_{jy}/\partial SAL)(\partial SAL/\partial WETLAND^i)_j + 0]\} \quad (18)$$

where the zero relates to sampling sites away from that shore; and $\partial SAL/\partial WETLAND^i$ (i = H,N) is estimated from (17) (*i.e.*, the βs in Table 2 with SESHORE = 1 for i = N). For agricultural development, (18) simply becomes

$$\partial XFB/\partial WETLAND^i = K^i_{WET}\ \beta_{SAL}\ \beta^i_{AGR\text{-}WET},\ i = H,N \quad (19)$$

where β_{SAL} is the coefficient on salinity in (12), $\beta^i_{AGR\text{-}WET}$ comes from (17) with Table 2, and K^i_{WET} captures the remaining terms in (18)[15] (see Table 2).

Valuation of Pocosin Wetlands for Shrimp Fishery

By (11) and (19), the marginal value of peninsular wetlands for brown shrimp production is estimated by:

$$\begin{aligned} VMP_i &= (\partial\pi^*/\partial XFB)\,(\partial XFB/\partial WETLAND^i) \quad (20) \\ &= VMP_B \cdot (\partial XFB/\partial WETLAND^i),\ i = H,N; \end{aligned}$$

where this annual value of marginal product is given in Table 2 for each wetland quality and each potential land-use that wetlands displace (*cf.* Fig. 1). For example, the highest losses to the shrimp fishery are estimated as \$3.37/acre/year for developing agriculture on wetlands near the sou'hwestern shore (Table 2). Results also show that conversion to forestry-forest cover may only be a concern for high quality wetlands, from the perspective of protecting shrimp nurseries. Results for normal quality (southeastern) pocosins remain consistent with forestry research suggesting that modern management may maintain the hydrologic role of these wetlands (Campbell and Hughes 1991), but results for the high quality (southwestern) wetlands suggest that forestry in some locations alters wetland functions and values.

The estimates in Table 2 represent values for an average year based on the statistical parameters. Table 3 provides a sensitivity analysis in the form of upper and lower bound estimates derived by using the 95% confidence bounds for each of the three key parameters. Value estimates are most sensitive to potential estimation error in the linkage between the shrimp index and salinity, equation (12). These bounds also show that conversion of normal wetlands to forestry-forest land-uses may cause fishery losses up to \$0.22/acre-year, despite statistical insignificance of the impact on salinity.

[14] That is, j identifies sites with one value of SESHORE and that value determines quality index i.

[15] For any sampling site and for the agriculture example, $(\partial SAL/\partial WETLAND^i)_j = \beta^i_{AGR\text{-}WET}/TOTACRE_j$; $TOTACRE_j$ is the total acreage of land in the land-use sample near site j. K^i_{WET} includes $\Sigma_{j:SESHORE\rightarrow i}(1/TOTACRE_j)$.

Table 3

Sensitivity Analysis of Parameters on the Value of Marginal Product of Normal (VMP_N) and high (VMP_H) Quality Wetlands for Protection of Salinity in Brown Shrimp Nursery Areas[a]

Conversion of wetlands:	Using S.E. of (equation no.)[b]	VMP_N Upper	VMP_N Lower	VMP_H Upper	VMP_H Lower
To agriculture	φ (9)	0.335	0.230	4.00	2.74
	β_{SAL} (12)	0.464	0.100	5.54	1.19
	$\beta_{AGR\text{-}WET}$ (17)	0.439	0.126	4.16	2.58
	All three	0.856	0.036	8.12	0.743
To forestry-forest[c]	φ (9)	—	—	2.20	1.51
	β_{SAL} (12)	—	—	3.05	0.657
	$\beta_{FOR\text{-}WET}$ (17)	0.112	−0.191	2.23	1.48
	All three	0.218	—	4.35	0.427

[a] Upper and lower bounds calculated using, respectively, +1.96 or −1.96 times the standard error of estimated parameters, as indicated.

[b] β_{SAL} is the coefficient on SAL in (12); $\beta_{AGR\text{-}WET}$ and $\beta_{FOR\text{-}WET}$ are defined in Table 2 using coefficients in (17). "All three" bounds are calculated using the standard errors of φ, β_{SAL}, and either $\beta_{AGR\text{-}WET}$ or $\beta_{FOR\text{-}WET}$.

[c] Some estimates are omitted since results in (17) cause the base value of $\beta_{FOR\text{-}WET} < 0$ and $\beta_{FOR\text{-}WET}$ is statistically insignificant, so some bounds provide no meaningful information.

Finally, these results refute the common notion that wetlands may be valued as a homogeneous group. The results show variation in value, despite the absence, in the analysis, of a detailed accounting of the various ecological types of pocosin wetlands (Lukin and Mauger 1983; Richardson and Gibbons 1993). The variation here depends not only upon geographic location, which may correlate with ecological types, but also upon the type of developed land-use contemplated. The results illustrate that wetlands values depend not only on their role in the ecosystem but also on the role that the alternative land-use would play in the ecosystem.

Tradeoff Assessment and Implications

This section assesses the preservation and development tradeoffs in pocosin wetland development by combining the empirical results with Heimlich and Langner's (1986) analysis of returns to agriculture. The objective is illustrative, rather than prescriptive, highlighting advantages of a resource-theoretic framework and a stepwise empirical approach.

Some assumptions are necessary to adjust the available results for a resource-theoretic framework. First, on the preservation side, we assume the results for shrimp are indicative for other fisheries and that economic impacts on other Pamlico fisheries are proportional to their gross dock-side value;[16] then impacts of salinity or wetland changes on shrimp represent 25% of total impacts of agricul-

[16] Some precedent for this type of assumption exists in Kahn and Kemp (1985); see also arguments in Gupta and Foster (1975).

tural development. Second, on the development side, we build on Heimlich and Langner (1986) and assume returns to wetland conversion decline as development proceeds eastward, moving further from mainland transportation networks. Finally, an 8% discount rate is assumed.

Empirical results show that, in Pamlico's estuarine nursery areas, the marginal loss due to an across-the-board reduction in average salinity is about \$114,000 annually or \$1.4 million in present value for shrimp, or about \$5.7 million in present value for all fisheries. Since policy makers (NRCD 1987) link coastal zone development to salinity reductions, these estimates justify debate over *net* benefits of development.

This debate has emphasized wetland development for agriculture. Such development annually impacts shrimp fisheries by \$3.37 1986-dollars per acre of high quality wetlands developed and \$0.28 per acre of normal quality wetlands, with respective present values of \$42.13 and \$3.50 per acre. These estimates initially may appear low, but they apply to diffuse and indirect impacts. The annual marginal loss is comparable to previous studies: losses from direct destruction of habitat for Florida blue crabs are about 0.30/acre-yr 1975-dollars (Lynne *et al.* 1981); wetlands that mitigate hurricane damage provide benefits of 0.40/acre-yr 1980-dollars (*cf.* Farber 1987). In 1986-dollars, these comparison values are \$0.44–\$0.48/acre-yr. Finally, Heimlich and Langner (1986) identify pocosin development as a marginal investment, so that our estimated fishery losses may be large relative to development values.

The stepwise empirical model offers one significant lesson. Pocosin wetlands comprise one resource that benefits the shrimp fishery. Yet the intermediate results indicate a real potential for *general* coastal zone development to cause substantial *aggregate* losses via salinity. A localized focus on specific types of development (*e.g.*, pocosin development) may miss significant impacts from development in the full watershed of an estuary.

Following the resource-theoretic framework, we now assess the net opportunity cost (NOC in (4)) of agricultural development. Adjusting the per-acre wetland values (Table 2) to acknowledge all Pamlico fisheries, the preservation value becomes \$13.48 and \$1.12 per acre annually for high and normal quality pocosins. Heimlich and Langner (1986) suggest that converted wetlands sell for \$1350 per acre, while Barnes (1981) estimates acquisition and development costs around \$1190. However, Heimlich and Langner (1986) estimate "typical" development costs above \$1500, suggesting that more than half of remaining wetlands are not economic for agriculture. Our illustration assumes 20% of wetlands remain economic for development, with $C^i(E^i) > 0$ in (4) ($i = H,N$); this acreage includes 2809 acres of high quality, southwestern wetlands and 11,009 acres of normal quality, southeastern wetlands (13,818 acres total). The illustration also assumes the first acre earns an annualized return of \$12.8 (*i.e.*, $rC^H(2809) = 0.08 \cdot [1350-1190]$) and the return declines at a constant rate per acre so that $rC^N(E^N) = 0$ for acre number 13,818. These assumptions yield example equations for the returns to development:[17]

$$C^H(E^H) = 160 - 0.01158 \cdot (2809 - E^H) \tag{21a}$$

[17] Available data on returns to wetland conversion omit geographic location. Readers may evaluate other assumptions at their discretion.

$$C^N(E^N) = 127.47 - 0.01158 \cdot (11009 - E^N), \qquad (21b)$$

where it is assumed that the return to the last marginal unit of high quality wetlands equals the return to the first marginal unit of normal quality wetlands; that return is $10.20 annualized (0.08 times $127 present value).

In this example, the "efficient" policy would be to preserve all high quality (southwestern) wetlands because, despite their higher value for agriculture, their value for fisheries is higher still and the net opportunity cost of development is negative ($N^H < 0$). In contrast, the policy would allow development of normal quality (southeastern) wetlands, but would halt development when the marginal net value of development fell to an annualized $1.12 per acre (or $14 in present value), where the net opportunity cost of development just equals zero ($N^N = 0$). In this example, developers use 9800 of 11,009 acres of southeastern wetlands, leaving 1209 acres preserved.

The example compromises development and preservation interests. Developers forego their most profitable wetlands because these same wetlands are most valuable to fisheries. However, preservationists lose 89% of the normal quality wetlands that development threatens,[18] preserving 11%.[19]

Concluding Summary

A resource-theoretic approach to development and preservation tradeoffs merges renewable and nonrenewable resource theories, highlighting that both preservation and development contribute positively to social welfare. The framework is applied to preservation and development tradeoffs between agricultural development of pocosin wetlands and its impact on estuarine fisheries. An illustrative assessment supports preservation of wetlands that are most attractive to both development and fishery sectors, while development of some less highly valued wetlands may be efficient.

Furthermore, a stepwise approach to link freshwater wetland development and estuarine shrimp production reveals a potential for substantial welfare losses if estuarine salinity declines across-the-board. This result encourages research to identify impacts of development throughout the estuarine watershed. This stepwise approach also offers a number of stages where future biophysical-economic models may enter the evaluation. Such flexibility may be important as methods for estimation of total values improve to account for non-consumptive uses (*e.g.* Costanza *et al.* 1989; Whitehead 1993). Finally, supporting Crocker and Tschirhart (1992), empirical results reveal that tracing human impacts through ecosystem linkages affects resource valuation: wetland values depend not only on their ecological type, but also on factors such as their geographic location and the land-use (agriculture versus forestry) in their developed state.

[18] Recall the example assumes 80% of southeastern wetlands are not threatened and are preserved by default.

[19] Given the linearity in (21) and the $160 present development value of the first acre, both the preservation of all high quality wetlands and the proportion of normal wetlands preserved are *invariant* to alternatives (e.g. 50%) to the assumption that 20% of all remaining wetlands offer positive marginal returns to development.

References

Babcock, A. M. and P. R. Mundy. 1985. A quantitative measure of migratory timing applied to the commercial brown shrimp fishery in North Carolina. *North American Journal of Fisheries Management* 5(2A):181–196.

Barnes, J. S. 1981. Agricultural adaptability of wet soils of the North Carolina Coastal Plain. *Pocosin Wetlands: An Integrated Analysis of Coastal Plain Freshwater Bogs in North Carolina*, ed. C. J. Richardson. Stroudsburg, PA: Hutchinson Ross Publishing Company.

Batie, S. S. and C. C. Mabbs-Zeno. 1985. Opportunity costs of preserving coastal wetlands: a case study of a recreational housing development. *Land Economics* 61(1):1–9.

Batie, S. S. and J. R. Wilson. 1978. Economic values attributable to Virginia's coastal wetlands as inputs in oyster production. *Southern Journal of Agricultural Economics* 10(1):111–118.

Bell, F. W. 1986. Competition from fish farming in influencing rent dissipation: the crawfish fishery. *American Journal of Agricultural Economics* 68(1):95–101.

Blomo, V. J., J. P. Nichols, W. L. Griffin, and W. E. Grant. 1982. Dynamic modeling of the eastern Gulf of Mexico shrimp fishery. *American Journal of Agricultural Economics* 64(3):475–482.

Brown, R. J. 1976. A study of the impact of the wetlands easement program on agricultural land values. *Land Economics* 52(4):509–517.

Campbell, R. G. and J. H. Hughes. 1991. Impact of forestry operations on pocosins and associated wetlands. *Wetlands* 11(Special Issue):467–479.

Clark, C. W. and G. R. Munro. 1975. The economics of fishing and modern capital theory: a simplified approach. *Journal of Environmental Economics and Management* 2(2): 92–106.

Constanza, R., S. C. Farber, and J. Maxwell. 1989. Valuation and management of wetland ecosystems. *Ecological Economics* 1:335–361.

Copes, P. 1970. The backward-bending supply curve of the fishing industry. *Scottish Journal of Political Economy* 17(1):69–77.

Crocker, T. D. and J. Tschirhart. 1992. Ecosystems, externalities, and economics. *Environmental and Resource Economics* 2:551–567.

Danielson, L. E. and J. A. Leitch. 1986. Private vs public economics of prairie wetland allocation. *Journal of Environmental Economics and Management* 13(1):81–92.

Epperly, S. P. and S. W. Ross. 1986. *Characterization of the North Carolina Pamlico-Albemarle estuarine complex*. National Oceanic and Atmospheric Administration NOAA Technical Memorandum NMFS-SEFC-175, U.S. Department of Commerce.

Farber, S. 1987. The value of coastal wetlands for protection of property against hurricane wind damage. *Journal of Environmental Economics and Management* 14(2):143–151.

Fisher, A. C. and J. V. Krutilla. 1974. Valuing long run ecological consequences and irreversibilities. *Journal of Environmental Economics and Management* 1(2):96–108.

Giese, G. L., H. B. Wilder, and G. G. Parker, Jr. 1985. Hydrology of major estuaries and sounds of North Carolina. U.S. Geological Survey Water-Supply Paper 2221.

Goldberger, A. S. 1968. The interpretation and estimation of Cobb-Douglas functions. *Econometrica* 36(3–4):464–472.

Griffin, W. L., R. D. Lacewell, and J. P. Nichols. 1976. Optimum effort and rent distribution in the Gulf of Mexico shrimp fishery. *American Journal of Agricultural Economics* 58(4 Part I):644–652.

Gupta, T. R. and J. H. Foster. 1975. Economic criteria for freshwater wetland policy in Massachusetts. *American Journal of Agricultural Economics* 57(1):40–45.

Heath, R. C. 1975. *Hydrology of the Albemarle-Pamlico Region, North Carolina—A Preliminary Report on the Impact of Agricultural Development*. U.S. Geological Survey, Water-Resources Investigations 9–75.

Heimlich, R. E. and L. L. Langner. 1986. *Swampbusting: Wetland Conversion and Farm Programs*. Economic Research Service, Natural Resource Economics Division, Agricultural Economic Report No. 551, U.S. Department of Agriculture.

Hettler, W. F. and A. J. Chester. 1982. The relationship of winter temperature and spring landings of pink shrimp, *Penaeus duorarum*, in North Carolina. *Fishery Bulletin* 80(4): 761–768.

Hotelling, H. 1931. The economics of exhaustible resources. *Journal of Political Economy* 39(2):137–75.

Jones, R. A. and T. M. Sholar. 1981. *The Effects of Freshwater Discharge on Estuarine Nursery Areas of Pamlico Sound*. Division of Marine Fisheries, Completion Report for Project CEIP 790-11, North Carolina Department of Natural Resources and Community Development.

Judge, G. G., W. E. Griffiths, R. C. Hill, H. Lütkepohl, and T.-C. Lee. 1985. *The Theory and Practice of Econometrics*, 2nd ed. New York: John Wiley & Sons.

Kahn, J. R. 1987. Measuring the economic damages associated with terrestrial pollution of marine ecosystems. *Marine Resource Economics* 4(3):193–209.

Kahn, J. R. and W. M. Kemp. 1985. Economic losses associated with the degradation of an ecosystem: the case of submerged aquatic vegetation in Chesapeake Bay. *Journal of Environmental Economics and Management* 12(3):246–263.

Kellogg, R. L., J. E. Easley, Jr., and T. Johnson. 1986. *Application of a Seasonal Harvesting Model to Two North Carolina Shrimp Fisheries*. Sea Grant College Program Pub. UNC-SG-86-03, North Carolina State University.

Krutilla, J. V. 1967. Conservation reconsidered. *American Economic Review* 57(4):777–786.

Lee, L.-F., G. S. Maddala, and R. P. Trost. 1980. Asymptotic covariance matrices of two-stage probit and two-stage tobit methods for simultaneous equations models with selectivity. *Econometrica* 48(2):491–503.

Loomis, J. B. 1988. The bioeconomic effects of timber harvesting on recreational and commercial salmon and steelhead fishing: a case study of the Siuslaw National Forest. *Marine Resource Economics* 5(1):43–60.

Lukin, C. G. and L. L. Mauger. 1983. *Environmental Geologic Atlas of the Coastal Zone of North Carolina: Dare, Hyde, Tyrrell, and Washington counties*. Office of Coastal Management, CEIP Report No. 32, North Carolina Department of Natural Resources and Community Development.

Lynne, G. D., P. Conroy, and F. J. Prochaska. 1981. Economic valuation of marsh areas for marine production processes. *Journal of Environmental Economics and Management* 8(2):175–186.

Matylewich, M. A. and P. R. Mundy. 1985. Evaluation of the relevance of some environmental factors to the estimation of migratory timing and yield for the brown shrimp of Pamlico Sound, North Carolina. *North American Journal of Fisheries Management* 5(2A):197–209.

McCoy, E. G. 1972. *Dynamics of North Carolina Commercial Shrimp Populations*. Division of Commercial and Sports Fisheries, Special Scientific Report No. 21, North Carolina Department of Natural and Economic Resources.

NRCD 1987. *Draft Source Document for the Albemarle-Pamlico Estuarine Study, 5-year Plan*. North Carolina Department of Natural Resources and Community Development, 5 Feb. 1987 (unpublished manuscript).

Palmquist, R. B. and L. E. Danielson. "A hedonic study of the effects of erosion control and drainage on farmland values." *American Journal of Agricultural Economics* 71(1): 55–62.

Purvis, C. E. and E. G. McCoy. 1974. *Population Dynamics of Brown Shrimp in Pamlico Sound*. Division of Commercial and Sports Fisheries, Special Scientific Report no. 25, North Carolina Department of Natural and Economic Resources.

Richardson, C. J., ed. 1981. *Pocosin Wetlands: An Integrated Analysis of Coastal Plain Freshwater Bogs in North Carolina*, Stroudsburg, PA: Hutchinson Ross Publishing Company.

Richardson, C. J. and J. W. Gibbons. 1993. Pocosins, Carolina Bays, and Mountain Bogs. *Biodiversity of the Southeastern United States: Lowland Terrestrial Communities*, W. H. Martin, S. G. Boyce, and A. C. Echternacht, eds. New York: John Wiley & Sons, pp. 257–310.

Shabman, L. A. and M. K. Bertelson. 1979. The use of development value estimates for coastal wetland permit decisions. *Land Economics* 55(2):213–222.

Skaggs, R. W., J. W. Gilliam, T. J. Sheets, and J. S. Barnes. 1980. *Effect of Agricultural Land Development on Drainage Waters in the North Carolina Tidewater Region*. Raleigh, NC: Water Resources Research Institute UNC-WRRI-80-159, University of North Carolina.

Stavins, R. N. 1990. Alternative renewable resource strategies: A simulation of optimal use. *Journal of Environmental Economics and Management* 19(2):143–159.

Street, M. W. and J. D. McClees. 1981. North Carolina's coastal fishing industry and the influence of coastal alterations. *Pocosin Wetlands: an Integrated Analysis of Coastal Plain Freshwater Bogs in North Carolina*, ed. C. J. Richardson. Stroudsburg, PA: Hutchinson Ross Publishing Company.

Swallow, S. K. 1988. Economically interdependent stocks and joint optimal allocation of renewable and exhaustible natural resources, with an application to coastal zone development. Ph.D. dissertation, Duke University.

Swallow, S. K. 1990. Depletion of the environmental basis for renewable resources: The economics of interdependent renewable and nonrenewable resources. *Journal of Environmental Economics and Management* 19(3):281–296.

Tiner, R. W., Jr. 1984. *Wetlands of the United States: Current Status and Recent Trends*. U.S. Department of the Interior, Fish and Wildlife Service, National Wetlands Inventory.

Toro-Vizcarrondo, C. and T. D. Wallace. 1968. A test of the mean square error criterion for restrictions in linear regression. *Journal of the American Statistical Association* 63:558–572.

Waters, J. R., J. E. Easley, Jr., and L. E. Danielson. 1980. Economic trade-offs and the North Carolina shrimp fishery. *American Journal of Agricultural Economics* 62(1): 124–129.

Whitehead, J. C. 1993. Total economic values for coastal and marine wildlife: Specification, validity, and valuation issues. *Marine Resource Economics* 8(2):119–132.

Wilen, J. E. 1985. Bioeconomics of renewable resource use. *Handbook of Natural Resource and Energy Economics*, eds. A. V. Kneese and J. L. Sweeney, chapter 2. Amsterdam: North-Holland.

Williams, A. B. 1955. A contribution to the life histories of commercial shrimps (*Penaeidae*) in North Carolina. *Bulletin of Marine Science of the Gulf and Caribbean* 5(2): 116–146.

Wiser, E. H. 1983. *HISARS Hydrologic Information Storage and Retrieval System User's Guide*. Raleigh, NC: Water Resources Research Institute UNC-WRRI-83-201, University of North Carolina.

Appendix

This appendix provides details on the data and the parameters used to simulate potential rents from the Pamlico Sound, NC shrimp fishery (Table A.1). This material will facilitate replication of the empirical results.

The empirical model treats the number of times craft landed shrimp (docked and unloaded), N_t, as one index for shrimp migratory timing based on a bioeconomic argument detailed in Swallow (1988, pp. 121–40). In the context of Babcock and Mundy (1985), bioeconomic support exists for using observed catch (C_{t-1}) as a simple proxy for migratory timing since the data derive from time-invariant harvest regulations (open access).

Finally, based on Williams' (1955) life history research, the empirical analysis uses a large white shrimp index (XFW) as an indicator of prolonged high salinity conditions in nursery areas; the observed mean XFW was 0.629. If salinity is high in nurseries during summer and early fall, the survival rate of white shrimp population will be higher. This event indicates a better year for brown shrimp production since spring salinity levels are correlated with late-season salinity levels (Heath 1975).

Catch effort data covered 1978–86, with annual series beginning in weeks 17–25 and ending in weeks 48–52 (see Swallow 1988, pp. 105–6). Shrimp indices only covered 1979–86. Estimates in Table 1 derive from 1979–86 data, while means in Table A.1 include available 1978 data. Since 1978 produced a below average shrimp harvest, including 1978 in the estimate of mean conditions lowers the estimate of rents and marginal values.

The expected price in week i, p_i (Table A.1), was estimated as a function of a national price index (available from U.S. National Marine Fisheries Service for 1978–84) and shrimp size. In turn, average size was estimated as a function of week number and seasonal dummy variables. See Swallow (1988, pp. 140–48). For

Table A.1

Weekly Mean Values Used to Simulate Exogenous Migratory Timing of Shrimp (N_t, C_{t-1}) and Ex-Vessel Price (p_t) (December 1986-dollars, heads-off) in Quasi-rent Equation (9) with Table 1

Week (t)	No. of landings (N_t)	Catch (lbs) (C_t)	Price ($/lbs) ($p_t$)	Week (t)	No. of landings (N_t)	Catch (lbs) (C_t)	Price ($/lbs) ($p_t$)
17	0.4	42.7	3.145	35	160.6	84325.4	3.870
18	0.9	72.7	3.217	36	121.8	66180.7	3.932
19	1.3	54.6	3.078	37	103.4	52424.3	3.941
20	1.9	165.0	3.168	38	99.4	46276.8	3.940
21	3.3	320.1	3.301	39	82.9	38719.1	3.929
22	5.9	1218.6	3.271	40	82.4	40984.8	3.918
23	10.9	2459.9	3.057	41	88.3	44101.0	3.449
24	14.0	3292.4	3.141	42	72.7	33787.6	3.414
25	28.6	9860.2	3.226	43	63.0	33343.3	3.371
26	73.0	33773.3	3.309	44	63.2	28316.6	3.275
27	138.3	95070.4	3.409	45	61.8	23384.8	3.213
28	190.6	118663.1	3.490	46	42.7	12278.0	3.150
29	229.6	168633.6	3.565	47	25.8	8410.0	3.083
30	223.0	160208.0	3.635	48	17.7	6121.2	3.001
31	198.5	139427.2	3.718	49	10.2	4174.7	2.930
32	230.1	162642.7	3.773	50	6.4	2266.4	2.854
33	205.0	120820.1	3.820	51	2.3	742.9	2.778
34	164.9	93128.4	3.857	52	1.1	72.1	2.703

simplicity, this study used the weekly average size of shrimp rather than disaggregated size classes; however, statistical results were consistent with those of Kellogg et al. (1986). Furthermore, USDA indices for prices received by farmers for hogs and for all meat animals made no statistical improvements.

For the simulation results, the intercept in Table 1 was adjusted for bias by modifying Goldberger's (1968) procedure. This modification approximates the regression variance used by Goldberger (1968) with the sum of squares of the regression errors from the GLS model, divided by error df. The correction multiplies A (Table 1) by 1.0974 and 1.0975 for models 1 and 2, respectively. While the modification is *ad hoc* in the presence of heteroscedasticity, omitting the correction biased rents downward by >200%. (See Swallow 1988, pp. 154–56).

D Marine Resources

[17]

Pressures, Trends, and Impacts in Coastal Zones: Interactions Between Socioeconomic and Natural Systems

R. K. TURNER*
S. SUBAK
W. N. ADGER
Centre for Social and Economic Research on the Global Environment (CSERGE)
University of East Anglia
Norwich, UK and
University College London, UK

ABSTRACT / This paper assesses the status of coastal zones in the context of expected climate change and its related impacts, as well as current and future socioeconomic pressures and impacts. It is argued that external stresses and shocks relating to sea-level rise and other changes will tend to exacerbate existing environmental pressures and damage in coastal zones. Coastal zones are under increasing stress because of an interrelated set of planning failures including information, economic market, and policy intervention failures. Moves towards integrated coastal zone management are urgently required to guide the coevolution of natural and human systems. Overtly technocentric claims that assessments of vulnerability undertaken to date are overestimates of likely future damages from global warming are premature. While it is the case that forecasts of sea-level rise have been scaled down, much uncertainty remains over, for example, combined storm, sea surge, and other events. In any case, within the socioeconomic analyses of the problem, resource valuations have been at best only partial and have failed to incorporate sensitivity analysis in terms of the discount rates utilized. This would indicate an underestimation of potential damage costs. Overall, a precautionary approach is justified based on the need to act ahead of adequate information acquisition, economically efficient resource pricing and proactive coastal planning.

People have often favored coastal locations for settlement because, among other benefits, these areas tend to contain the greatest biological productivity. Over half of the world's population lives within 60 km of the shoreline, and coastal populations in many countries are growing at double the national rate. Historic human intervention within the coastal zone and upstream in catchment areas has often led to nonsustainable levels of resource exploitation. Sewage and siltation linked to human development, for example, are among the most significant causes of coral reef and other ecosystem degradation in much of tropical Asia, Polynesia, and the Caribbean (Lundin and Linden 1993).

The sea-level rise that has been observed over the last century poses a significant threat to such coastal zones (Wigley and Raper 1992). Human activities can also influence sea-level directly and may have led to a net change in sea-levels over the past century, although even the direction of this change is disputed. It has recently been estimated that a combination of groundwater withdrawal, surface water diversion, and land-use change (wetland loss and deforestation) may have contributed at least 0.54 mm each year since 1960 to the observed sea-level rise (Sahagian and others 1994). Other estimates suggest that irrigation and water impoundment in reservoirs and dams may have led to a net reduction in sea-level rise of −1.63 mm each year (Gornitz and Rosenzweig 1994). The latter supports the view that climate-related effects may be larger than has previously been supposed. In short, many of the world's coastal ecosystems are already under intense and growing pressure, regardless of any climate-related changes. When climate-induced change is also present, complex interrelationships and feedbacks are generated between human and environmental driving forces and impacts and climate changes and effects.

Changes expected to accompany global warming, such as accelerated sea-level rise and more frequent and severe storm events represent potential additional stresses on coastal areas. Destructive climate-related events accompanying sea-level rise include increased beach erosion, storm surges, saltwater intrusion, and

KEY WORDS: Coastal zone management; Environmental damage valuation; Climate change; Sea-level rise

*Author to whom correspondence should be addressed.

Table 1. Climate change events and impact categories

	Climate-related events[a]					
	Sea-level rise					
Impact categories	Beach erosion	Storm surges	Salt-H_2O infusion	Siltation	Eutrophication	Cyclones
Tourism	$				$	$
Freshwater supplies			$	$	$	
Fishing	'$'	'$'		$	$	
Coastal residences	$	$				$
Commercial buildings	$					$
Wetlands	$, nm	$, nm	$, nm	$, nm	$, nm	
Agricultural and drylands		$	$	$		$
Life expectancy		$, nm			$, nm	$, nm
Culture and heritage sites	nm	nm			nm	

[a]nm = non-market impacts; $ = market priced major impacts; '$' = minor impacts.

siltation. In addition, warming of the atmosphere and oceans can lead to coral bleaching, increased eutrophication in wetlands and freshwater supplies, and damage to coastal structures and agricultural lands due to more frequent and powerful cyclones. These climate-related impacts will reduce human access to fresh water and land area for agriculture, commerce, and residences in the zone and increase the risk and scale of destruction from tropical storms. These problems in turn place in jeopardy tourism revenues, fisheries, commercial and agricultural enterprises, and human life. Both the market and nonmarket impact linkages with the different climate change effects are summarized in Table 1.

Feedbacks between human and environmental driving forces and climate-induced changes will continue to influence environmental impacts in the coastal zone, and many of these feedbacks are positive, leading to increased vulnerability. Human actions, for example, have led to the degradation of natural buffers from storms and waves, such as salt marshes, mangroves, coral reefs and dunes, the world over. To date much less human effort has been expended trying to reduce vulnerability through strategies such as beach and dune nourishment, and the adoption of insurance policies that discourage construction in low-lying areas.

Increasingly, planned and unplanned development in the coastal zone will respond both to growing human-induced pressures and to the anticipation of heightened climate change-related stress. Given access to information on potential and actual climate change, individuals and communities can launch a complex feedback process of adaptation. For this reason, assessments of potential impacts in the coastal zone will not be able to distinguish precisely between changes wrought by humans and those caused by long-term climate change. Nor can we neatly model consequences "with" and "without" climate change. Conventional analyses that have tried to make this distinction have provided a misleading view of environmental damages in coastal areas and of the costs and benefits of prevention.

In the following sections, the potential impacts of climate-related events in the coastal zone are evaluated, drawing upon recent case studies. Pressures in the coastal zone are discussed in terms of three interrelated planning failures—information deficiencies, economic market pricing, and policy intervention failures—and the corrections necessary for a more integrated approach to coastal planning. Because of the high degree of uncertainty that surrounds climate-related impacts, research efforts that increase information should be a high priority. Studies estimating the scale of assets at risk in coastal zones, a form of economic benefit analysis, have failed to encompass a range of impacts, such as wetland loss, reduced life expectancy, disruption of traditional communities, and ecosystem deterioration, because of assessment and pricing difficulties. The overall primary value of healthy evolving ecosystems, and not just individual component use values, also needs to be recognized (Turner and others 1995b). Furthermore, institutional arrangements should be aimed at increasing diversity and the potential for flexible response to changes in the coastal zone. In some cases these arrangements should help to preserve traditional communities that have developed distinct and effective strategies for responding for extreme events. Unless these issues are properly addressed, nonsustainable resource usage in coastal zones will not be reversed and damage will only increase.

Populations at Risk

During the twentieth century urbanized coastal populations have increased because of the many economic opportunities and environmental amenities that coastal zones can provide, as well as because of the general urbanization process. Thus, in the US, for example, the Mississippi river delta was roughly in a state of dynamic balance before the twentieth century, but since then human intervention in the form of large-scale engineering works, levees, dams, canals, and water diversions has effectively starved the wetlands of needed freshwater and sediments. Relative sea level is rising at a rate of 1 m/100 y in this region and up to 100 km² of wetlands are lost each year (Warrick and Rahman 1992, Titus 1991).

Other human interventions have contributed to degradation of coastal natural resources. In Bangladesh, run-off, sediment flow, and deposition rates have been attributed to changes in the flood defense systems as well as deforestation in the headwaters of the Ganges–Bramaputra–Meghna river system (see Warwick and Rahman 1992, Ives and Messerli 1989). These changes have then contributed to detrimental effects on coastlines and fisheries and have played a part in the increased frequency and severity of inland flooding. Small island states (many of which are low-lying) face particularly severe threats. Pollution and mining of coral in the Maldives, for example, will further serve to inhibit that country's capability to respond to sea-level rise.

The trend of increasing coastal urban population is expected to continue globally, with the most dramatic increases forecasted to be in Asia, Africa, and South America (World Resources Institute 1994). Perhaps as much as 66% of the present human population lives within 60 km of the shore and this figure could be 70% by the year 2000 (Pernetta and Elder 1992). Coastal populations will continue to increase at a higher relative rate (double in some developing countries) than the population in general. Two-thirds of the population of developing countries (3.7 billion) is expected to be living along the coast by the year 2000 (IPCC 1994). Sixty-five percent of cities with populations above 2.5 million inhabitants are located along the world coasts, and several of these are already at or below present sea level.

To take some examples of this population pressure, in the United States over the period 1960–1990, the population in the coastal counties grew by 41 million (43%). A similar picture emerges from Australia. Between 1983 and 1991, 90% of all building activity in Australia took place within the coastal zone (Resource Assessment Commission 1993). The developing countries of the Eastern African Region (Somalia, Kenya, Tanzania, and Mozambique plus the islands of Comoros, Seychelles, Madagascar, Reunion, and Mauritius) provide another stark example. Some 55 million people live in the region with most of the major towns and cities located in the coastal strip. In Mozambique, for example, 75% of the total population is concentrated in the coastal strip 40 km wide (Ngoile and Horrill 1993). Small island states typically have experienced population growth rates of around 3% per annum, although the rate of emigration is also high in some cases. Many islands are also densely populated, with the capital island of the Maldives, Malé, providing an extreme example. It is home to 56,000 people despite being only 1700 m long and 700 m wide (Pernetta 1992).

On a global basis, the IPCC (1994) has estimated that between 100 million and 250 million people live below the maximum storm surge level and some 60 million may be subject to significantly increased flooding risks from relatively small increases in mean sea level (and excluding the potential effects of increased storm frequency and severity). Because of regional differences in storm surge events, the increased risk of flooding due to sea-level rise is greater than average for Asia (especially the Indian Ocean coast), the south Mediterranean coast, the African coast, and the coasts of the Caribbean and other small island states.

Human interactions with extreme events can determine whether these phenomena merely pose hazards or lead to major economic losses. Thus, a natural hazard is an extreme event that threatens human well-being, while a natural disaster is an event that overwhelms societal coping capabilities and therefore imposes a heavy cost burden on society (Mitchell and Ericksen 1992). On average about 45 fully developed tropical cyclones occur each year somewhere on Earth. Approximately 15 of these storms approach or cross the coastal zone and represent potential hazards, but only two or three of these inflict substantial losses and are recorded as natural disasters. Annual damages have been estimated at around US$1.5 billion, with a death toll of 15,000–23,000 (Smith 1992, Bryant 1991).

Resources at Risk

The coastal zone (8% of the global surface area) may account for as much as 25% of global primary production. Ninety percent of the current world fisheries harvest (5%–10% of world food production) comes from within national exclusive economic zones (i.e., from within 200 miles of the coast), and most of what is caught comes from within 9 km of the shore. In Australia, for example, the gross value of production in the fishing industry was $1.4 billion in 1992–1993; and recreational fishing was worth $2.2 billion in the 1980s. Fish and seafood are important sources of protein. In some countries around the tropical oceans, people rely on seafood for 40%–95% of their protein intake. In the Southeast Asian region, seafood averages 60% and in the Pacific and Caribbean region the average is 90% of the diet (Lunden and Linden 1993). Fisheries are also important economic assets in the East African region, with Mozambique and Tanzania, for example, relying on annual harvests of more than 50 kT/yr (Ngoile and Horrill 1993).

The fishing industry will be affected both by sea-level rise, storms, and other climate-related changes. Fishing activities will be affected by potential beach erosion, and facilities will be threatened by sea-level rise and more intense storms. The loss of coastal wetlands may be of particular importance to the fishing industry. Such wetlands are important breeding grounds for coastal fish species but are currently being lost at rates of more than 1% per annum in some areas (IPCC 1994). Bigford (1991) estimated that a 50% reduction in wetland productivity would lead to a 15%–20% loss in estuarine dependent fish harvests.

Demographic and other socioeconomic trends have combined to produce a growing urban and industrial concentration in the coastal zones worldwide. The use of the coast and small island states for tourism has also expanded at a very rapid rate. International tourist demand has risen from 285 million arrivals in 1980 to 443 million in 1990, and coastal zones have attracted a significant proportion of this tourist pressure (WTO 1991). While precise patterns of settlement and mixes of economic activities vary from place to place, the general urbanization and industrialization trend is common to most coastal zones.

Since the late 1970s tourism has become the second most important item in world trade, surpassed only by oil. By the late 1980s global tourism was generating expenditures and receipts of US$ billion, or around 5% of global GNP. Gross tourism financial benefits will continue to be very important for small island economies. The loss of beach tourism and recreation areas because of sea-level rise, and more storms leading to increased beach and reef erosion, would represent a significant economic effect for many developing countries and especially for small island states.

Coastal wetlands are likely to be particularly severely affected by rises in sea level and have already experienced significant cumulative losses (Turner and Jones 1991). Potential wetlands loss will also be affected by the form of response option chosen by policy makers. Thus, a complete protection policy in the United States may mean a 50% loss of wetlands (Titus 1991).

In developing countries, mangroves, which extend to around 174,000 km^2 world wide (World Resources Institute 1994), are important wetland ecosystems both in terms of direct use values (fishing and forestry) and local/international recreation value. On this basis, the estimated annual revenue from one mangrove swamp in Trinidad has been put at US$681/ha (1993) (Brown and Adger 1994 based on Rambial 1980). Ignoring the storm buffering capacity of mangroves, their loss would represent globally some US$118 billion (undiscounted) over the next century.

An OECD study has quantified the more significant economic resources at risk from sea-level rise in OECD countries (OECD 1991). Total dryland loss in OECD countries following a 1-m rise in sea level might be around 76,000 km^2. The value of this land depends on its current and future use. Land values differ widely across different use categories. Marginal agricultural land may have a value of no more than US$1 million/km^2, whereas prime residential sites can fetch up to US$500 million/km^2. A 5-km stretch of beach close to Sydney is estimated to have a real estate value of US$35 million/km (O'Neill 1990), while the 500-km shoreline in the San Francisco Bay area has a real estate value of US$50 billion (US$500 million/km) (Gleick and Maurer 1990).

If extreme weather events become more frequent and intense, or their distribution changes, then urban areas and related capital assets will be at risk in countries such as Japan, Australia, the United States, and in some countries bordering the North Sea in Europe. In the Netherlands, for example, the financial impacts of a 10% change in the intensity of storms may be worse than those of a 60-cm rise in sea level (Nijkamp 1991).

Rising sea levels can lead to saltwater infusion into freshwater sources and aquifers. Urban areas in low-lying coastal areas are especially vulnerable, but areas substantially inland along tidal rivers may also be affected. For example, a study of the potential effects of sea-level rise on drinking water supplies in Philadelphia, which is situated some 30 miles from the Atlantic ocean along the Delaware River, concluded that a 0.73-m rise in sea-level would increase chloride concentrations in the city above drinking water standards (Hall and others 1986, Thatcher 1986).

The magnitude of the economic assets at risk in coastal zones depends not only on the expected severity of the external stress (i.e., sea-level rise and extreme storm events) but also on a number of other factors including the level of protection already in place in the zone. An OECD study (OECD 1991) indicates that most of the urbanized and industrialized sections of coastline in the high-income economies are already heavily protected and so capital asset losses in these areas are unlikely to be significant unless the climate-induced threats turn out to be very severe or protection costs become prohibitively expensive (because of upgrading and maintenance and operational costs) relative to protected assets values.

All coastal zones of the world are vulnerable to the range of possible impacts from sea-level rise and storm

Table 2. Indicators of vulnerability to climate-change-induced 1-m rise in mean sea level for selected low-lying states (adapted from Nicholls 1995)

Country	People affected		Capital value at loss		Land at loss			Wetland at loss (km^2)	Adaptation costs (million US$)	Costs cost/yr (% GNP)
	# People (×1000)	% Total	Million US$	% GNP	km^2 1 m	km^2 0.5 m	% Total			
Antigua	38	50	—	—	2		0.4	3	76	0.32
Argentina	—	—	5,600[a]	6[a]	3,400[b]	1,600	0.1	1,100	>1,800	>0.02
Bangladesh	71,000	60	—	5[c]	25,000		17.5	5,800	>1,700[d]	>0.06
Belize	70	35	—	—	1,939		8.4	—	—	—
Benin	1,350	25	126	12	230		0.2	85	>430	>0.41
China	72,000	7	—	—	35,000		—	—	—	—
Egypt[e]	4,700	9	59,272	204	5,800		1.0	—	13,133	0.45
Guyana	600	80	4,000	1,115	2,400		1.1	500	200	0.26
Japan	15,400	15	807,000	72	2,000		0.5	—	>159,000	>0.12
Kiribati	9	100	2	8	4		12.5	—	3	0.10
Malaysia	—	—	—	—	7,000		2.1	6,000	—	—
Marshall I.	40[f]	100	175	324	9		80	—	>380	>7.04
Mauritius	6	1	—	—	10		0.5	—	—	—
Netherlands	10,000	67	186,000	69	2,165		5.9	642	12,286	0.05
Nigeria	3,200[g]	4	18,000[a]	52	18,600[h]	(9,000)	2.0	16,000	>1,400	>0.04
Poland	235	1	24,000	24	1,700		0.5	36	1,500	0.02
Senegal	>110[g]	>1	700	14	6,100[h]	(1,700)	3.1	6,000	>1,000	>0.21
St. Kitts-Nevis	—	—	—	—	1		1.4	1	53	2.65
Tonga	30	47	—	—	7		2.9	—	—	—
Uruguay	13[g]	<1	1,800[a]	26	96[b]	(44)	0.1	23	>1,000	>0.12
USA	—	—	—	—	31,600[h]		0.3	17,000	>143,000	>0.03
Venezuela	56[g]	<1	350[a]	1[a]	5,700[b]	(2855)	0.6	5,600	>1,700	>0.03
Subtotal	177,892		1,107,025		148,765			58,790	>337,961	
Total (local studies)	935		18,000		16,894			3,250	951	
Total	178,827		1,119,525		165,659			63,040	>338,912	

[a]Minimum estimates—capital value at loss does not include ports.

[b]Land loss from 0.5 m mean sea-level rise (Nicholls and Leatherman 1995).

[c]Source: Broadus (1993).

[d]Adaptation for Bangladesh only provides protection against a 1 in 20 year event.

[e]Adaptation costs for Egypt include development scenarios.

[f]National estimate.

[g]Minimum estimates—number reflects estimated people displaced.

[h]Best estimate is that 20,000 km^2 of dry land is lost, but about 5400 km^2 is converted to coastal wetlands.

surges, although to different degrees. As a rule, small island developing states are judged to be the most vulnerable, followed by low-lying delta areas in developing countries. For island states the major economic threats to small economies are dominated by tourism. Table 2 shows estimates of land, wetland, and capital loss as indicators of vulnerability for selected low-lying countries, as estimated by Nicholls (1995) for the IPCC. The results show that for small island states, 100% of the population is at risk from future climate change impacts.

Many small island countries would lose a significant part of their land area with a sea-level rise of 0.5–1.0 m. The Maldives, for example, have average elevations of 1.0–1.5 m above existing sea level (Pernetta 1992). Submergence and erosion can convert many small islands to sandbars and significantly reduce the usable dry land on the larger, more populated islands. Saltwater intrusion and loss of the freshwater lens may be an equally binding constraint on human habitation in some islands, particularly smaller atolls (Leatherman 1994).

About a dozen small island states in the Pacific Ocean region alone have been classified as profoundly or severely vulnerable to accelerated sea-level rise of 1 m by 2100 (Pernetta 1992). Impacts would be on the scale of major social and economic disruption, with a subset of these countries possibly ceasing to exist.

Even the less vulnerable small islands would suffer significant economic effects from the loss of beach tourism and recreation areas because of sea-level rise and more storms leading to increased beach and reef erosion. Among the Caribbean islands, for example, in 1988 income from tourism as a percentage of GDP was highest for Antigua and Barbuda (69%) and for the Bahamas (53%), but a dozen Caribbean islands reported that tourism revenues made up more than 10% of the GDP (ECLAC 1991, quoted in Hameed 1993). The Indian Ocean islands of the Seychelles and the Maldives have also seen a steady growth in tourism. Total receipts from tourism generated foreign-exchange earnings of US$94 million in the Maldives (US$15 million in 1981). This represented some 60% of the country's total foreign-exchange earnings. Since 1985 tourism has been the single biggest contributor to the GDP of the Maldives. Tourism to developing countries has increased significantly in recent years, and small island developing states have experienced a particularly rapid increase. Tourist numbers to Mauritius, for example, have increased from 1800 visitors in 1968 to 180,000 in 1988 (UNEP 1991).

Significant land loss to the sea through inundation of deltas is one of the main results from a study of selected middle- and lower-income countries that utilized a new rapid and low-cost reconnaissance technique—aerial videotape-assisted vulnerability analysis (Nicholls and Leatherman 1995). Five countries were included in the analysis—Senegal, Argentina, Uruguay, Nigeria, and Venezuela—and a sensitivity analysis was also undertaken encompassing current sea-level rise (0.18 m per century), 0.5 m and 2.0 m by 2100. Even with current rates of sea-level rise (0.2-m rise), there is significant land loss. Senegal, Venezuela, and most particularly Nigeria have the largest amount of land loss amounting to 350 km², 1140 km², and 2,800 km², respectively. More than 95% of their total loss involves inundation of land in deltas or around estuaries.

Some of the world's most productive agricultural land is found in low-lying coastal areas and in the flood plains of major rivers. Thus about 85% of the world's rice is produced in the South, Southeast and East Asia, and around 10% of this output is derived from areas that may be vulnerable to accelerated sea-level rise. Less favorable hydraulic conditions caused by a 1-m rise in sea level would affect the food supply of more than 200 million people, if no adaptive measures are taken. The large deltas of Vietnam, Bangladesh, and Myanmar (Burma) seem to be especially vulnerable (IPCC 1994).

In Bangladesh some 110 million people densely populate an area of 144,000 km², 80% of which is made up for the complex Bengal delta system created by the Ganges, Brahmaputra, and Meghna rivers. It has been calculated that a 1-m sea-level rise (13 cm global sea level plus local subsidence of 70 cm by 2050) would threaten about 7% of the country's habitable land (5% of the total population) and a 3-m rise (2 m global sea level plus 1 m local subsidence) would threaten 26% of the land and over 27% of the total population (Broadus 1993).

In Bangladesh, 16% of the rice production area would be inundated and 13 million people would be displaced. It would also be difficult to replace croplands lost to the sea because the countryside is already extensively cultivated. Of the 24.5 million acres in the nation thought to be cultivable, more than 90% is already in cultivation. The Sunderbans, the second largest mangrove swamp in the world, could also be lost, In the Nile delta, sea-level rise combined with subsidence would destroy 12%–15% of the agricultural land and displace over six million people (Nicholls and Leatherman 1995).

When developing economies are hit by weather-related disasters such as cyclones (with coastal zones bearing the brunt of such storms), they face a disproportionate impact on human populations. An estimated 85% of the people killed, injured, or made homeless as a result of storms, cyclones, and floods between 1960 and 1990 were from developing countries (Berz 1991). In the four coastal plains of China, millions of people, including inhabitants of cities such as Shanghai, and extensive areas of agricultural land would be at risk. National economic losses as a percentage of total GNP (but not as absolute costs) follow a similar pattern. In 1989, for example, US$7.6 billion of losses from natural disasters were recorded in the United States, with more than 50% due to one event, Hurricane Hugo, but these losses represent less than 0.1% of the gross national production (GNP) of the United States. Three countries in Latin America—Bolivia, Ecuador, and Peru—had to face costs of US$4 billion due to the El Niño weather of 1982–1983. This cost burden, however, represented 10% of their combined GNP (Mitchell and Ericksen 1992). In other respects, developing countries face lower costs because many communities have infrastructure that is more adaptable and mobile than those of developed nations, e.g., dirt tracks are more easily rerouted than eight-lane highways.

Information, Economic Market, and Policy Failures

Vulnerability is a multidimensional concept encompassing biophysical, socioeconomic, political, and ethical factors. It includes some notion of the susceptibility of a coastal area or small island state to the risk of external shock and damage posed by climate-induced

Table 3. World coast conference vulnerability classes[a]

Impact categories	Vulnerability classes			
	Low	Medium	High	Critical
Population living in risk zone (no/total) × 100%	<1%	1–10%	11–50%	>50%
Population at risk from 1 metre rise in sea-level (no. × probability)/1000	<10	10–100	100–500	>500
Capital assets value at risk (total loss/GNP) × 100%	<1%	1–3%	3–10%	>10%
Land loss (area/total area) × 100%	<3%	3–10%	10–30%	>30%
Protection/adaptation costs (total cost/GNP) × 100%	<0.05%	0.05–0.25%	0.25–1%	>1%
Wetland loss (area/total area) × 100%	<3%	3–10%	10–30%	>30%

[a]Source: IPCC (1994)

sea-level rise and increased storm frequency and severity. The susceptibility status of an area is conditioned by past, current, and future population and settlement patterns and rates of change, together with its aggregate and per capita level of economic wealth (i.e., the scale and distribution of economic entitlements possessed by the local population). In general, the higher the level of economic development and per capita entitlements provided, and the greater the economic and resource diversity within the system, the less susceptible any given coastal zone should be. Paradoxically, however, areas with higher levels of economic development, but a low coping potential, may be more vulnerable, because they have more to lose.

Table 3 relates proposed vulnerability classes of low to critical vulnerability, as set out in the World Coast Conference 1993 Report, to economic and physical indicators discussed above. This table therefore defines high vulnerability as 10%–30% loss of existing wetlands, over 10% of the population living in the hazard zone and other economic indicators. The classes are necessarily arbitrary and do not indicate interaction between the components. For example, a high proportion of the population being forced to migrate owing to stress in the coastal zone would have implications for the other economic and social spending categories: central government expenditure on coastal protection or on wetland conservation may have to be augmented by redirected funds in the face of severe adjustments. The physical and social systems therefore coevolve in the face of external shocks, and classification of vulnerability in separate parts of the system is inappropriate.

Vulnerability also includes the institutional capability or capacity of a region or a country to cope with or manage (prevention and alleviation measures) the impacts as well as the relevant physical and socioeconomic dimensions. This capacity itself is determined by a range of factors including national wealth and indebtedness and cultural and other factors.

Interrelated failures—information deficiencies, policy intervention failures and economic market failures—have led to resource damage and losses and have constrained the effectiveness of management policies aimed at mitigating this damage. Unless these failures are properly addressed, the current nonsustainable trends in coastal zones will not be reversed and losses will increase. At the core of the failures dilemma is a problem of information, either the complete absence of data or inadequate data. Fundamentally, the interface between natural science research and models and social science research and models needs to be better defined and elaborated (Turner 1991). The great variability in the quantity and quality of environmental and other data available to assess the possible impacts of accelerated sea-level rise and more frequent extreme events at a regional level underscores the urgent need for greater data compatibility. The compatibility should apply to different scientific disciplines and to different national agencies charged with data collection and interpretation.

Scientific uncertainties continue to shroud the scale and significance of potential combined sea-level rise and storm-event risks at the regional level. Existing knowledge about severe weather climatology and about the nexus between weather and its impacts are inadequate. Emanuel (1987), for example, has suggested increased hurricane intensities as a result of higher sea-surface temperatures, but in some regions interannual variations in tropical cyclone activity are dominated by Southern Oscillation (SO) effects (Holland and others 1988). Uncertainty over SO predictions under climate-change scenarios, however, means that cyclone behavior prediction in the region remains an uncertain art. Nevertheless, the combined accelerated sea-level rise and storm-surge event is a potentially very significant threat.

The legislation that regulates construction and development in coastal zones is often contradictory and insufficiently comprehensive to respond adequately to the complex relationships that exist (OECD 1993a,b). The lack of coordinated policies and ineffective coastal management policies has merely served to compound value and land-use conflicts among the various interested parties. The result has been increasing levels of resource degradation and loss, which an OECD study has cataloged in case studies covering southeast Tasmania, Australia, Izmir Bay, Turkey, and the Aveiro estuary in Portugal (OECD 1993b).

Policy conflict is a common outcome of multisector, multiagency, and multilevel government involvement in coastal zone management (OECD 1993a). In the United States, for example, agricultural policy has contributed to the coastal eutrophication process; subsidized flood insurance has encouraged inefficient and potentially hazardous coastal development and redevelopment; and federal tax and resource development policies conflict with wetlands and other environmental resource conservation requirements. Coastal urban communities have displaced wetlands and added to the pollution loading of coastal waters (see Turner and Jones 1991, for a survey of the relevant literature).

On the other hand, legislation in the United States is succeeding in discouraging uncontrolled development in some coastal areas. In 1980, 50% of the 280 coastal barrier islands along the US Atlantic and Gulf coasts were at least partially or heavily developed. A law passed in 1982 effectively removed all federal subsidies for construction on many undeveloped barrier areas. Consequently, a study of 157 units in the Atlantic and Gulf coasts found that over the period 1982–1989 major development had occurred on only 10 of the units (Weber 1990).

A lack of coordination of the different resource uses and constraining policy regimes, together with inadequate knowledge of the dynamics of the coastal processes and systems, has resulted in inadequate overall management of highly pressured zones. The pressures have been added to and the negative resource impacts compounded by a further set of market failures.

Whenever resource allocation decisions are made without the guidance of information on the externalities involved (i.e., due to the complete absence of prices or heavily distorted prices), they are prone to result in economic and environmental loss or damage. Many of the resources located in coastal zones provide important functions and services that are not fully accounted for in coastal planning. The rapid loss of coastal wetlands is a classic example of this failure situation. The result of the underpricing of the waste assimilative functions of rivers, estuaries, and coastal waters, and therefore the inappropriate disposal of a variety of household and industrial wastes into the water medium, causes the assimilative capacity to be breached. Groundwater loss, in turn, is more the result of compensatory behavior exhibited by individuals suffering from an underprovision of water services. The proliferation of private wells in northern Jakarta, for example, has lowered the water table to such an extent that saline intrusion now affects a 10-km-wide zone along the coastal plain. Overpumping of groundwater has also led to land subsidence and asset damage in Jakarta, Bangkok, and Los Angeles County.

Existing regulatory instruments in coastal zone management systems therefore need to be augmented by economic incentive instruments (with their pricing effects). Some OECD countries are moving in this policy direction with a tax *de séjour* in some tourist areas, transferable quotas for fish, fines for noncompliance with legislation, and the establishment of resource rentals for the occupation of water space.

In summary, vulnerability to present and future threats is manifest in increasing frequency of events with negative impacts on human welfare. Susceptibility to these events is caused by failure phenomena: failure to implement appropriate planning and management, enhanced by the market's failure to provide environmental protection for essentially public goods, but measures can be taken to support and augment the resilience of coastal systems (natural and human) to better cope with external stresses and shocks. Economic approaches to finding appropriate strategies and values for coastal resources are now discussed.

Economic Valuation

In economic terms natural assets in the coastal zone can yield direct use values and indirect use values, as well as nonuse values (Turner 1988, 1991, Barbier 1994), many of which are generally ignored in coastal planning. A range of methods is available to evaluate such environmental benefits. At a generic level, a spectrum of evaluation methods and techniques can be distinguished (Figure 1) and used depending on the objective of study. The more comprehensive the technique, the more information that will be made available to assist in the appraisal of policy options from a societal perspective.

The so-called total economic value of an ecosystem can be estimated via a number of monetary valuation methods (see Figure 2 for a wetland example). A full range of valuation techniques is required to quantify the economic value of wetlands. Even if all the elements are appraised, another evaluation problem arises when the individual private values are aggregated to obtain a total social/economic value. This is the argument that

1. Financial Analysis:	2. Economic Cost-Benefit Analysis:	3. Extended Cost-Benefit Analysis	4. Environmental Impact Assessment	5. Multi-Criteria Decision Methods
financial profitability criterion; private costs and revenues; monetary valuation	economic efficiency criterion; social costs and benefits; monetary valuation	sustainable development principles; economic efficiency and equity trade-off environmental standards as constraints; partial monetary valuation	quantification of a diverse set of effects on a common scale, but no evaluation	multiple decision criteria, monetary and non-monetary evaluation

Increasing Complexity: multiple decision criteria; multiple effects; increasing scale of analysis

Adapted from Pearce and Turner (1992)

Figure 1. Evaluation methods.

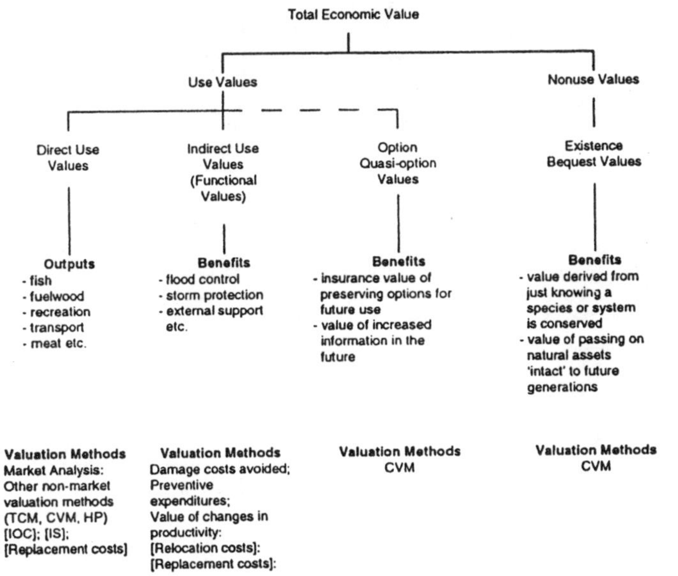

Notes:
Market Analysis: based on market prices
HP = hedonic pricing, based on land/proprty value data
CVM = contingent valuation method based on social surveys designed to elicit willingness to pay values
TCM = travel cost method, based on recreationalist expenditure data
IOC = indirect opportunity cost approach, based on options foregone
IS = indirect substitute approach
[ ] =valuation methodology to be used with care
Source: Adapted from Barbier (1989)

Figure 2. Valuing wetland benefits.

ecosystems have primary value that is conditional on the existence and maintenance of the healthy ecosystem rather than on any individual human use (Turner and others 1995b). Preserving some overall ecosystem size and structure at the landscape level is probably required if resource conservation and utilization is to be sustainable. The scientific details on habitat size, diversity, and threshold limits are not currently available, and therefore a precautionary approach opting for conservation rather than development has much to recommend it.

Empirical studies have confirmed that conserved ecosystems such as mangroves and other coastal wetlands possess significant economic value, both in terms of use and nonuse outputs identified in Figure 2. Mangrove forests have been shown to sustain more than 70 direct human activities ranging from fuelwood collection to artisinal fisheries (Dixon 1989), with the value of these functions estimated to be up to $680 ha/yr (Brown and Adger 1994).

Many of the functions and services that wetlands provide are nonmarket goods and therefore do not carry appropriate market prices and value. Although wetlands are capable of yielding a range of goods and services, some uses preclude others and so some caution is necessary when utilizing total economic value estimates. Published estimates vary from US$1.5 million/km² to US$13 million/km², but the average is US$2–5 million/km² for OECD countries and US$1.25 million/km² for developing countries (Fankhauser 1995). One recent study took a median value of US$8 million/km² as an approximate total wetland value (OECD 1991). This resulted in an estimated potential damage due to loss of coastal wetlands in OECD countries (excluding Canada, Australia, and New Zealand) of US$400–500 billion (undiscounted) by the end of the next century. Taking recreational value and storm buffering function value only, estimated at US$216/ha/yr in 1989, the wetland loss in OECD countries would represent some US$2 million/km², or US$100–125 billion in total (Costanza and others 1989). Economic losses resulting from the threat of climate change and other stresses are therefore significant for natural areas as well as for urbanized or developed coastal land.

In principle, a global picture of the potential damage costs in coastal zones and small island states can be built up via the aggregation of the growing number of existing local assessments (see IPCC 1992, Turner and others 1995a, Gleick and Maurer 1990, Milliman and others 1989, Den Elzen and Rotmans 1992, Rijkswaterstaat 1991, Nicholls and Leatherman 1995, Nijkamp 1991). Table 4 summarizes the results of some studies that have assessed protection and damage costs for a range of sea-level rise scenarios in the Netherlands, United States, United Kingdom, and Uruguay. The damage cost estimates in particular are only partial and therefore are almost certainly underestimates. The general picture that emerges for the industrialized countries is that for sea-level rise up to and approaching 1 m by 2100, current defense systems with some modifications are sufficient to protect the bulk of dryland assets. Only wetland losses are significant.

Fankhauser (1995) has calculated protection costs for OECD countries facing a sea-level rise of 50 cm by 2100. The revised protection cost is US$521 million/yr. The overall conclusion is that for OECD countries it is cost-effective to protect most of their coasts (Fankhauser 1995). The situation facing developing countries is, however, substantially different. In general they face greater risks of experiencing adverse impacts due to sea-level rise because their existing sea and coastal defense systems are relatively less well established. On the other hand, the lack of engineered defense systems can be counterbalanced in some countries by the natural adaptability of traditional means and ways of life.

Substantial protection costs would be required in some developing countries in order to protect urban areas and related activities such as beach tourism. Some nine small island states appear in the IPCC list of countries facing the highest coastal protection costs as a percentage of their total economic wealth. The global average percentage of GNP required annually for coastal protection is 0.037% (OECD 1991), but the percentage for the Maldives is 34% and for Kiribati, Tuvalu, Tokelau, Anguilla, Guinea Bissau, Turks and Caicos, Marshall Islands, Cocos Islands, and the Seychelles it is between 5% and 35% (OECD 1991).

Several caveats are in order to put the protection and damage cost estimates presented in Table 4 in perspective. The baseline scenarios on which all the studies depend assume that existing settlement and related economic assets patterns and development trends continue largely unchanged into the future, but the mere threat of extensive loss of land and other assets may stimulate macroeconomic effects within the national economy. Some assets may be relocated, others may be adapted so as to reduce the damage implications. On the other hand, the damage cost estimates will represent underestimates because they neglect many nonmarket environmental assets values and costs such as the cost of resettling of refugees. Finally, besides the scientific uncertainties surrounding sea-level rise, the impact studies do not adequately capture the risks posed by combined sea-level rise and possibly more frequent or intense extreme weather events. It is also assumed in most models of sea-level rise impacts that the rise in sea level will be a slow and gradual process, although this may not be the case for all regions. Scientific uncertainties are compounded by the socioeconomic adaptation uncertainties referred to above and by the fact that the economic cost estimate results are very sensitive to changes in the discount rate that is applied.

Table 4. Sea-level rise damage and protection costs: Netherlands, UK, USA, and Uruguay

Response option costs: damage costs (DC) and protection costs (PC) (US$ billion, 1991)	Sea-level rise scenarios (m)									
	0.2 by 2050	0.2 by 2100	0.4 by 2100	0.5 by 2100	0.6 by 2050	0.8 by 2050	0.9 by 2100	1.0 by 2100	1.0 m (by 2100) 5.0 m (by 2300)	2.0 by 2100
Netherlands[a]										
Maintain current defenses (DC)		$19 bill						$126 bill	>$2324 bill	
Improve current defenses (PC + (DC)		$19 bill						$335 bill	$115–148 bill	
New improved defenses (PC + DC)		$71–120 bill						$71–120 bill	$76–117 bill	
Retreat (DC)		$33–67 bill						$40–79 bill	$91–148 bill	
Netherlands[b]										
Improve defenses (PC)			$17 bill				$21 bill			
United Kingdom–East Anglia region[c]										
Maintain current defenses (PC)	$247 mill		$257 mill		$283 mill	$294 mill				
Improve current defenses (PC)	$350 mill		$435 mill		$547 mill	$908 mill				
Retreat (DC)	$2489 mill		$2530 mill		$2624 mill	$2680 mill				
Maintain defenses option: net benefits	$1888 mill		$1809 mill		$1762 mill	$1671 mill				
Improve defenses option: net benefits	$2006 mill		$1969 mill		$1935 mill	$1624 mill				
USA[d]										
Retreat (DC)				$88–213 bill				$187–511 bill		$466–159 bill
Maintain current defense (PC)				$70–100 bill				$156–1543 bill		$418–668 bill
Uruguay[e]										
Improve defences (PC)				$2 bill				$3 bill		$5 bill

[a]Nijkamp (1991); estimates include both protection and damage (human health, capital assets and environmental impacts) costs depending on the given response option. No discount rate cited.

[b]Den Elzen and Rotmans (1992); dike, dune, and water management defensive expenditure; 2% rate of discount.

[c]Turner and others (1995a); agricultural land, property and selected environmental impact damage costs; mainly seawall defense system costs; 6% rate of discount.

[d]Titus (1991); costs refer to property loss or protection only (75% confidence interval).

[e]Nicholls and Leatherman (1995); protection costs for urban areas only, 80% of which is for beach nourishment.

Discounting and Equity

The discounting procedure, in the context of long-run costs and benefits, has been a controversial issue in economics and public policy. Coastal zone management raises a series of equity concerns both within contemporary social systems and across intergenerational time. Climate-change policy in particular highlights intergenerational equity because future generations are affected by current decisions made in response to climate and other environmental change effects but cannot influence such decisions. Thus, for example, if the current generation decides to adopt a wait-and-see policy in the context of climate change and other change phenomena, a potentially significant cost burden is transferred to the future. On the other hand, the adoption of a precautionary approach now (greenhouse gas emissions reduction measures and proactive protection investments in coastal zones and elsewhere) would mean certain near-terms costs and only potentially significant large long-term benefits.

It is not within the scope of this paper to summarize all the salient points of debate in the discounting and discount rate controversy. Rather we would argue that in the context of coastal zone management, cost–benefit analysis of policy options should be undertaken via a range of discount rates. Further, in this resource management context, which is characterized by equity concerns, long-run cost and benefit effects and complex environmental considerations, the weight of argument is in favor of low discount rates (i.e., relative to market rates of interest). The actual range of discount rate that should be adopted in the economic assessment of projects and policy options should be closer to the social rate of time preference (i.e., pure time preference, the myopian effect, plus a measure of diminishing marginal utility of consumption) than to productivity of capital (market) rates favored by finance ministries (e.g., 8%–10%.

The social rate of time preference discount rate depends on whether one assumes that the pure rate of time preference is zero or positive and on what one estimates to be the rate at which utility diminishes as the level of consumption rises, as well as the rate of increase in per capita incomes. According to Cline

(1992) the effective discount rate in the global warming policy responses context is 2% per annum. We would argue that a range from 0.5% to 3% seems reasonable and is compatible with a sustainable development goal (Broome 1992).

Institutional Analysis

It is the case that conventional impact evaluation techniques and indicators for assessing potential damages from climate-related events, such as GDP and population at loss, protection costs, and cost–benefit analysis, reflect a largely Western perspective based on economics. Alternative approaches from anthropology and sociology seek to assess changes in culture, community, and habitat and their interactions. For example, a study of coastal vulnerability and resilience to sea-level rise and climate change in Fiji and Western Samoa computes a sustainable capacity index (Nunn and others 1994). This is based on the sum of ratings of vulnerability and resilience for many categories of cultural, social, agricultural, and industrial impacts, at the local, regional, and national level. Areas with higher concentrations of assets are judged to be more vulnerable, while areas with diversity and flexibility in the system, whether natural or managerial, tend to be viewed as more resilient. The study evaluated potential impacts to subsistence economies according to the view that communities in which people feed and clothe themselves with little cash exchange are more vulnerable, but that subsistence economies in which staples can be replaced with other crops tend to be more resilient. In addition, cultural sites are ranked according to the level of national interest in their preservation. The study concludes that subsistence economies and cultural assets are more vulnerable in Fiji and that conventional and quick analyses of relatively high-lying islands such as Fiji would tend to underestimate the potential vulnerability of these areas, given that most people live in the low-lying coastal plain and the majority of cash and subsistence economic activities take place in the low-lying areas (Nunn and others 1994).

The vulnerability of traditional societies will vary, however, according to their past experience with extreme events and flooding and the degree of social cohesion that is preserved in these communities. Some traditional societies have developed networks for support and reciprocity that are more effective than the natural disaster programs of even the wealthiest countries. For example, in 1971 the rivers of the Red River delta in Vietnam burst their banks, leaving a half million people homeless. The coastal population had centuries of experience in responding to droughts and floods, however, and had evolved time-tested communal management arrangements. As a result, few deaths resulted from an event that would have been catastrophic in some other regions (Wisner 1978). Paradoxically, changes to facilitate the spread of a market economy, such as decollectivization of agricultural holdings, may lessen Vietnam's ability to recover from some disaster events in the future (Pingali and Xuan 1992).

Conclusions

Climate change in the coastal zone can lead to impacts such as sea-level rise, beach erosion, saltwater intrusion, and storm surges. In addition, the coastal zone can be affected by eutrophication, enhanced by rising temperatures, and changing frequencies, intensities, and distributions of severe weather, such as cyclones. The physical impacts of sea-level rise enhance vulnerability of human populations and of physical environments to nonsustainable stresses. Further, the physical changes can exacerbate the existing social and economic stresses on many coastal populations. In general, the processes of climate change and associated sea-level rise will tend to have greatest impact on small island developing states, followed by low-lying deltas in low-income developing countries and coastal lowlands throughout the developing world. The socioeconomic circumstances that lead to these threats being ignored or underestimated are policy, information, and market failures. If these failures are reduced through coordinated coastal planning and the redefinition of property rights to avoid open-access utilization of coastal assets, then the effects of the physical environmental changes will be ameliorated, as natural systems and socioeconomic systems coevolve in the vulnerable coastal zones.

Such actions as the strengthening of transboundary forecasting and warning systems, hazard insurance schemes, resource pricing of groundwater, reduction of subsidies for agricultural drainage, and other policy failure corrections would reduce future vulnerability (Pearce and Warford 1993).

Because a disproportionate share of the global economic wealth is now located in the coastal zone, policy makers have been under pressure to both protect and enhance the wealth-creation potential of the zone. The result has been additional coastal construction and modification of physical shoreline processes, often with negative biophysical impacts. A set of interrelated failure phenomena (information, policy, and market failures) have both contributed to resource damage and loss, as well as constraining the effectiveness of management policies aimed at mitigating the problems.

Estimates suggest that most of the urbanized and

industrialized sections of coastline in the higher income countries are already heavily protected, and so capital asset losses in these areas are unlikely to be significant if best guess sea-level rise estimates prove accurate. However, these regions are still vulnerable to events such as hurricanes, storm surges, and droughts. The overall combined event risk would also pose a threat to some coastal areas in the developed world in Japan, Australia, the United States, and parts of Europe. About 50 million people experience flooding annually, based on estimates of existing protection levels. Increases in the frequency and severity of storms would further increase the population subject to annual flooding.

Extreme weather events inflict a disproportionate cost impact on developing countries, whose economies are already highly stressed and fragile and therefore less able to afford effective response mechanisms. Disasters related to climate change have a greater probability of more deaths in poorer than in richer countries. Additionally, proportionate economic losses follow a similar pattern.

The optimal response strategy will depend on the costs of the various options, the nature of the risks, the assets potentially at risk, and the consequent damage costs. It will therefore vary from region to region. Protection and potential damage cost studies suggest that for the OECD countries sea-level rise up to 1 m by 2100 is unlikely to necessitate radical upgrading of existing defenses protecting developed coastal areas (OECD 1993b). Wetland losses in this region may, however, be significant, with secondary implications for fisheries production. For those regions where industrialization has occurred in the coastal zone, it is cost-effective to protect most of their coasts.

Nonindustrialized regions face greater difficulties because their coastal zone defense systems are less well established, the costs of protection are relatively high (between 5% and 35% of gross national products in the poorest countries such as small islands in the Indian and Pacific oceans), and coastal assets such as agriculture and fishing represent a larger proportion of overall wealth. In terms of the number of people protected from flooding, and protection costs as a proportion of GNP, the coastal regions of continental South and Southeast Asia, the Pacific coast of South America and the southern Mediterranean will have greater opportunities for adaptation than the coastal regions of small island states.

While it is possible to claim that most vulnerability impact assessments undertaken to date are overestimates of the likely effects of climate change, such a claim is premature. This claim is based on the argument that previous studies take a 1-m rise in sea level as the basis for analysis rather than the recently revised best guess estimate of 48 cm with a range of 15–90 cm in the year 2100 (Wigley and Raper 1992). In fact some studies model a range of scenarios, including 25-cm and 50-cm sea level rises, and are not restricted to a 1-m assessment. Threshold effects are also not usually accounted for, i.e., the relationship between impacts and sea-level rise is taken to be linear, and the low-probability/high-impact and cost effects are effectively ignored.

All of these studies, however, fail to make a full (total economic value) assessment of possible impacts. The valuation analysis usually fails to encompass a range of nonmarket impacts because of the inherent pricing difficulties. The discount rates applied to the analysis will also greatly affect the magnitude of the damage estimate outcomes. Less than full valuation of inputs and high discount rates will both serve to underestimate true damage costs. The important point to note here is that changes in the socioeconomic parameters used in the assessment procedure are as significant as changes in the physical parameters such as rates of sea-level rise.

Any assessment therefore should be treated with caution because the scientific uncertainties relating to the climate-related threats are compounded by socioeconomic uncertainties. The major physical uncertainties relate to the overall impact of climate change on precipitation (and hence on the sedimentation budget of coastal deltas and estuaries), on ocean currents and the associated coastal impacts, on temperature impacts on coastal ecosystems such as coral, as well as the uncertainties over the direct impacts of sea-level rise and storm-surge frequency. The socioeconomic uncertainties center on the rate and extent of human adaptation. The time dimension and the rate of the physical changes and impacts are key factors. Precautionary action taken now therefore postpones impacts and allows greater opportunities for adaptation.

Acknowledgments

We are grateful to Robert Nicholls and to three referees for their comments on an earlier draft of this paper, remaining errors are our responsibility.

Literature Cited

Barbier, E. B. 1989. The economic value of ecosystems: 1—Tropical Wetlands, LEEC Gatekeeper 89-02, London Environmental Economics Centre, London.

Barbier, E. B. 1994. Valuing environmental functions: tropical wetlands. *Land Economics* 70:155–173.

Berz, G. 1991. Natural disasters in 1991: Claims burdens reaching new records. *Schadenspiegel* 1:14–17.

Bigford, T. E. 1991. Sea-level rise, nearshore fisheries and the fishing industry. *Coastal Zone Management* 19:417–437.

Broadus, J. M. 1993. Possible impacts of, and adjustments to, sea-level rise: The cases of Bangladesh and Egypt. Pages 263–275 *in* R. A. Warrick, E. M. Barrow and T. M. L. Wigely (eds.), Climate and sea-level change: Observations, projections and implications. Cambridge University Press, Cambridge.

Broome, J. 1992. Counting the cost of global warming, The White Horse Press, Cambridge.

Brown, K., and N. Adger. 1994. Incremental costs and biodiversity conservation in Trinidad and Tobago. Centre for Social and Economic Research on the Global Environment, University of East Anglia, Norwich.

Bryant, E. A. 1991. Natural hazards. Cambridge University Press, Cambridge.

Cline, W. 1992. The economics of global warming. Institute for International Economics, Washington DC.

Costanza, R., S. Farber, and J. Maxwell. 1989. Valuation and management of wetland ecosystems. *Ecological Economics* 1:335–362.

Den Elzen, M., and J. Rotmans. 1992. The socioeconomic impact of sea-level rise on the Netherlands: A study of possible scenarios, *Climatic Change* 20:169–195.

Dixon, J. A. 1989. The value of mangrove ecosystems. *Tropical Coastal Area Management, ICLARM* 4:5–8.

Emanuel, K. A. 1987. The dependence of hurricane intensity on climate. *Nature* 326:483–485.

Fankhauser, S. 1995. Protection vs. retreat: Estimating the costs of sea-level rise. *Environment and Planning A* 27:299–319.

Gleick, P. H., and E. P. Maurer. 1990. Assessing the costs of adapting to sea-level rise: A case study of San Francisco Bay. Pacific Institute for Studies in Development, Environment and Security, and Stockholm Environment Institute, Berkeley, CA.

Gornitz, V., and C. Rosenswieg. 1994. Is sea-level rising or falling? *Nature* 371:481.

Hall, C., M. Thatcher, and R. Tortoriello. 1986. Salinity in the Delaware Estuary. *In* C. Hall and J. Titus (eds.), Greenhouse effect, sea-level rise and salinity in the Delaware Estuary. US Environmental Protection Agency and Delaware River Basin Commission, Washington DC.

Hameed, H. 1993. Sustainable tourism in the Maldives. M Phil. thesis. University of East Anglia, Norwich.

Holland, G., J. McBridge, and N. Nicholls. 1988. Australian regional tropical cyclones and the greenhouse effect. Pages 438–455 *in* G. Pearman (ed.), Greenhouse: Planning for climate change. CSIRO, Canberra.

IPCC (Intergovernmental Panel on Climate Change). 1992. Global climate change and the rising challenge of the sea. Response Strategies Working Group, IPCC, Geneva.

IPCC (Intergovernmental Panel on Climate Change). 1994. Preparing to meet the coastal challenges of the 21st century. Conference report, World coast conference 1993. Ministry of Transport, Public Works and Water Management, The Hague.

Ives, J. D. and B. Messerli. 1989. The Himilayan dilemma: Reconciling development and conservation. Routledge, London.

Leatherman, S. 1994. Rising sea levels and small island states. *EcoDecision* 11:53–54.

Lundin, C. G., and O. Linden. 1993. Coastal ecosystems: Attempts to manage a threatened resource. *Ambio* 22:468–473.

Milliman, J. D., J. M. Broadus, and F. Gable. 1989. Environmental and economic implications of rising sea-level and subsiding deltas: The Nile and Bengal examples. *Ambio* 18:340–345.

Mitchell, J. K., and N. J. Ericksen. 1992. Effects of climate change on weather-related disasters. Pages 141–151 *in* I. M. Mintzer (ed.), Confronting climate change: Risks, implications and responses, Cambridge University Press, Cambridge.

Ngoile, M., and C. Horrill. 1993. Coastal ecosystems, productivity and ecosystem protection: Coastal ecosystem management. *Ambio* 22:461–467.

Nicholls, R. J. 1995. Synthesis of vulnerability analysis. Pages 181–216 *in* WCC '93, preparing to meet the coastal challenges of the 21st century. Proceedings of the World coast conference 1993, Noordwijk. Ministry of Transport, Public Works and Water Management, The Hague.

Nicholls, R. J., and S. P. Leatherman (eds.). 1995. Potential impacts of accelerated sea-level rise on developing countries. *Journal of Coastal Research*, Special Issue No. 14.

Nicholls, R. J., K. C. Dennis, C. R. Volonte, and S. P. Leatherman. 1993. Methods and problems in assessing the impacts of accelerated sea-level rise. Pages 193–205 *in* The world at risk: Natural hazards and climate change. American Institute for Physics conference proceedings no. 277, New York.

Nijkamp, P. 1991. Climate change, sea-level rise and Dutch defence strategies. *Project Appraisal* 16:143–148.

Nunn, P. D., A. D. Ravuvu, W. Aalbersberg, N. Mimura, and K. Yamada. 1994. Assessment of vulnerability and resilience to sea level rise and climatic change. Case study: Yasawa Islands, Fiji. South Pacific Regional Environment Programme, Environment Agency Japan, Overseas Environment Cooperation Centre, Japan.

OECD. 1991. Responding to climate change: Selected economic issues. OECD, Paris.

OECD. 1993a. Coastal zone management: Integrated policies. OECD, Paris.

OECD. 1993b. Coastal zone management: selected case studies. OECD, Paris.

O'Neill, B. 1990. Cities against the sea. *New Scientist* 3 February, pp. 46–49.

Pearce, D. W., and R. K. Turner. 1992. Benefits estimates and environmental decision-making. OECD, Paris.

Pearce, D. W., and J. Warford. 1993. World without end. Oxford University Press, Oxford.

Pernetta, J. C. 1992. Impacts of climate change and sea-level rise on small island states. *Global Environmental Change* 2:19–31.

Pernetta, J. C., and D. L. Elder. 1992. Climate, sea-level rise and the coastal zone: Management and planning for global changes. *Ocean and Coastal Management* 18:113–160.

Pingali, P. L., and V. T. Xuan. 1992. Vietnam: Decollectivisation and rice productivity growth. *Economic Development and Cultural Change* 40:697–718.

Rambial, B. S. 1980. The social and economic importance of the Caroni mangrove swamp forests. 11th commonwealth forestry conference, Trinidad.

Resource Assessment Commission. 1993. Coastal zone inquiry. Final report. Canberra, Australia.

Rijkswaterstaat. 1991. Rising waters, impacts of the greenhouse effect for the Netherlands. Dutch Ministry of Transport and Public Works, Tidal Waters Division, The Hague.

Sahagian, D. L., F. W. Schwartz, and D. K. Jacobs. 1994. Direct anthropogenic contributions to sea-level rise in the twentieth century. *Nature* 367:54–57.

Smith, K. 1992. Environmental hazards: Assessing risk and reducing disaster. Routledge, London.

Titus, J. G. 1991. Greenhouse effect and coastal wetland policy: How America could abandon an area the size of Massachusetts at minimum cost. *Environmental Management* 15:39–58.

Thatcher, M. L. 1986. Modifications of salinity model geometry for larger rises in sea-level. *In* C. Hall and J. G. Titus (eds.), Greenhouse effect, sea-level rise and salinity in the Delaware Estuary, US Environmental Protection Agency and Delaware River Basin Commission, Washington DC.

Turner, R. K. 1988. Wetland conservation: economics and ethics. Pages 121–159 *in* D. Collard, D. W. Pearce, and D. Ulph (eds.), Economics, growth and sustainable environments. Macmillan, London.

Turner, R. K. 1991. Economics and wetland management. *Ambio* 20:59–63.

Turner, R. K., and T. Jones (eds.). 1991. Wetlands, market and intervention failures: Four case studies. Earthscan, London.

Turner, R. K., P. Doktor, and W. N. Adger. 1995a. Assessing the costs of sea-level rise. *Environment and Planning A* 27 (in press).

Turner, R. K., C. Folke, I. M. Gren, and I. Bateman. 1995b. Wetland valuation: Three case studies. Pages 129–149 *in* C. A. Perrings, K. G. Mäler, C. Folke, C. S. Holling and B. O. Jansson (eds.), Biodiversity loss: Economic and ecological issues. Cambridge University Press, Cambridge.

UNEP. 1991. Environmental data report. Oxford University Press, New York.

Warrick, R. A., and A. A. Rahman. 1992. Future sea-level rise: Environmental and socio-political considerations. Pages 97–112 *in* I. M. Mintzer (ed.), Confronting climate change: Risks, implications and responses. Cambridge University Press, Cambridge.

Weber, M. 1990. Using common sense to protect the coasts: The need to expand the coastal barrier resources system. Coast Alliance, Washington DC.

Wigley, T., and S. Raper. 1992. Implications for climate and sea-level of revised IPCC emission scenarios. *Nature* 357: 293–300.

Wisner, B. G. 1978. An appeal for a significantly comparative method in disaster research. *Disasters* 2:80–82.

World Resources Institute. 1994. World resources 1994–95: A guide to the global environment. Oxford University Press, Oxford.

WTO (World Tourism Organisation). 1991 Yearbook of tourism statistics, 1991. World Tourism Organisation, Madrid.

[18]

ELSEVIER

Ecological Economics 10 (1994) 233–247

ECOLOGICAL ECONOMICS

Modelling economy–ecology linkages in mangroves: Economic evidence for promoting conservation in Bintuni Bay, Indonesia

H. Jack Ruitenbeek

R.R.#2 – Site 52, Gabriola, B.C., Canada V0R 1X0

(Received 5 September 1992; accepted 18 September 1993)

Abstract

Woodchip exports pose a potential threat to a 300 000-ha mangrove ecosystem in the Bintuni Bay area of Irian Jaya, Indonesia. The bay supports an important shrimp export industry, and coastal areas support 3000 households. Traditional non-commercial uses of mangroves have an estimated value of Rp20 billion/yr (US$10 million/yr); commercial fisheries are valued at Rp70 billion/yr (US$35 million/yr) and selective commercial mangrove cutting schemes have a maximum value of Rp40 billion/yr (US$20 million/yr).

Forest management options, ranging from clear cutting to a cutting ban, are evaluated in a cost–benefit analysis incorporating linkages among mangrove conversion, offshore fishery productivity, traditional uses, and benefits of erosion control and biodiversity maintenance functions. "Linkage scenarios" are developed that reflect potential ecosystem component interactions in Bintuni Bay. Clear cutting is optimal only if linkages are ignored. A cutting ban is optimal if linear and immediate linkages between ecosystem components exist. Under a scenario with linear but delayed linkages of 5 years, selective cutting of 25% of the harvestable mangrove is the optimal strategy; it has a present value of Rp70 billion (US$35 million) greater than the clear cutting option, and more extensive cutting would yield no additional net benefits.

Strong economic arguments exist for conservative mangrove clearing. Where strong ecological linkages occur, severe restrictions on clearing activities will be economically optimal. Where ecosystem dynamics are uncertain, programs reducing linkage effects – such as greenbelts, replanting, or selective cutting – will minimise potential economic losses.

Key words: Cost–benefit analysis; Mangrove; Modelling

1. Introduction

Mangroves in Indonesia are under intense pressure from competing resource uses. Their exploitation for charcoal, wood, fish ponds, or similar "productive uses" is often done on the grounds of narrow economic evaluations that focus on only a single use of the mangroves. Economic analyses that focus on the multiple-use aspects of mangroves will prescribe management practices which have lower rates of conversion and exploitation; some level of "conservation" makes economic sense. This study relies on such a "multiple-use" focus to develop – and apply –

SSDI 0921-8009(93)E0077-T

Table 1
Mangrove areas in Indonesia, 1990

Island	Area (ha)
Irian Jaya	2943000
Sumatra	391335
Kalimantan	383450
Riau	276000
Maluku	100000
Sulawesi	72800
Java	50000
Nusa Tenggara (east and west)	5500
Bali	2000
Total Indonesia	4225000

a framework which assists in selecting an economically optimal mangrove management strategy for any given area.

Indonesia's mangrove resources are significant both in their areal extent and in their importance to economic production. Of 144 million ha of forest in Indonesia, wetland forests comprise almost 30 million ha, and mangrove forests account for over 4 million ha of these forests. The most extensive mangrove forests are found on the islands of Irian Jaya, Sumatra, and Kalimantan (Table 1). Although Java and Sulawesi also at one time had extensive natural mangroves, they have been increasingly depleted by human population growth and other pressures.

The resource conflicts which occur in mangrove use are by no means peculiar to Indonesia: experience in other tropical countries, those in Southeast Asia in particular, has shown that proper management of the resource can avoid conflict and ensure long-term sustainability of the mangrove ecosystems (Rahman and van de Vusse, 1992). It is equally clear, however, that no *single* solution to proper mangrove management can be applied to all mangrove areas in Indonesia; both the problems, and solutions, are quite diverse. But failure to find and implement the appropriate management strategy can lead to substantial economic losses, ecological degradation, and – where mangroves support important traditional livelihoods – increased social and political instability.

Both the economic and ecological importance of mangroves are becoming well recognised within Indonesia. A National Mangrove Committee exists which has the function of identifying critical mangrove areas. Indonesian Forestry Departments, which in most cases have authority over mangrove exploitation, are seeking technical solutions for sustainable mangrove use. Where mangrove areas have been proximate to industrial sites, they have at times been the focus of environmental impact research. More recently, the Ministry of State for Environment (MENLH) has started to assess the role of mangrove management in a broader program of Marine and Coastal Environmental Management. This latter focus has been one component of the Environmental Management Development in Indonesia (EMDI) Project, which is funded by the Canadian International Development Agency (CIDA) and is jointly designed and implemented by MENLH and Dalhousie University, Halifax, Canada.

Interest in mangroves in Indonesia stems from a number of reasons, but two stand out as being particularly important in the broader context of developing a mangrove management strategy: (i) the *economic value* of some single component of the resource is often very significant, although not always obvious; (ii) the *ecological linkages* between different components of the mangrove are also very significant, although often fraught with uncertainty. These two factors alone imply that improper management of one component of the resource, such as forestry, can result in significant economic losses elsewhere, such as offshore fisheries. This suggests that the use of some form of economic analysis, one capable of incorporating the ecological linkages, can provide important information on the optimal use of the entire resource base.

Such analyses are not uncommon for establishing the conditions under which conservation of some parts of a critical area are economically justified (Hodgson and Dixon, 1988; Dixon, 1989; Ruitenbeek, 1992b; Turner, 1993). Key aspects of such analyses, however, include recognising that some components have economic value even if there are no "marketed" goods or services involved, and establishing the ecological linkages

H.J. Ruitenbeek / Ecological Economics 10 (1994) 233–247 235

between the various resource components. A review of traditional uses of non-timber forest products in Southeast Asia shows, for example, that traditional untraded uses of forest products can be a substantial component of local economies (de Beer and Mcdermott, 1989). Further, in the

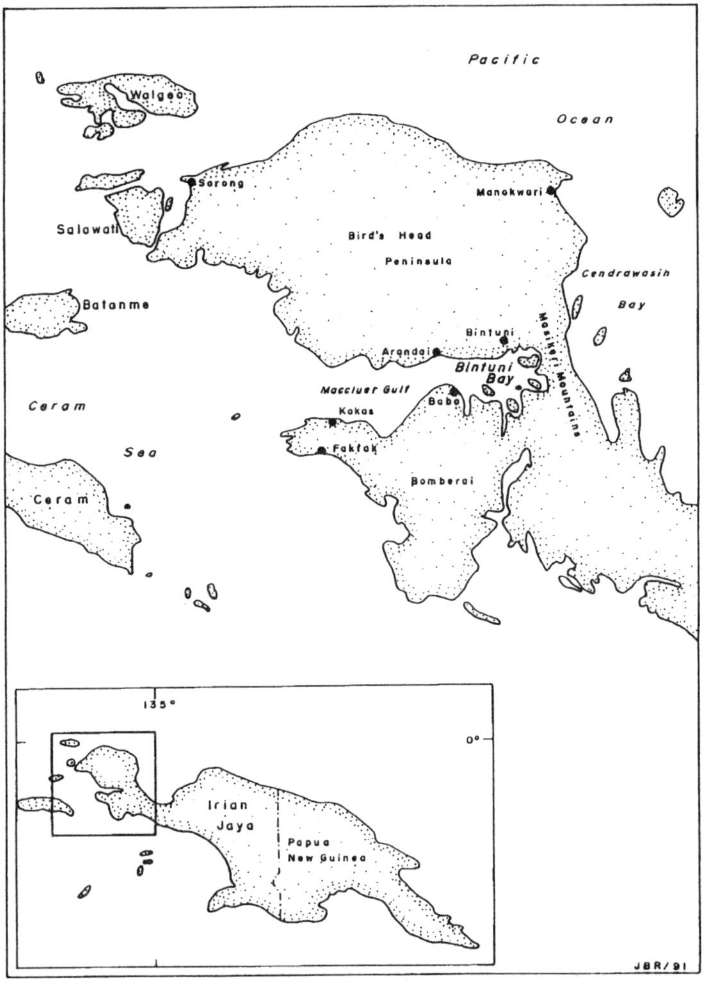

Fig. 1. Location map of Bintuni Bay, Irian Jaya, Indonesia.

case of mangroves, broad-based correlation studies such as those in West Java (Naamin, 1987; Hehanussa and Hehuwat, 1980) and the Philippines (Eusebio et al., 1987), and specific studies of linkage mechanisms (Macnae, 1986; Robertson et al., 1988a,b; Robertson and Daniel, 1989; Robertson and Duke, 1990; Williams and Cappo, 1990), demonstrate the dependencies which exist between different parts of the ecosystem. An important conclusion of this work is that – although the precise linkage mechanisms may differ from one site to the next – offshore fishery productivity is strongly correlated to the area of mangroves; declines in mangrove area – through reducing fish or shrimp nursery habitats or losses in other ecological functions – decrease near-shore and offshore fishery yields. In areas of Sumatra this linkage is so strong and so obvious that local fishermen are voluntarily replanting mangroves – in places where they have been depleted – in an attempt to re-establish fishery productivity which has been lost.

In summary, it is suggested that, where economic analyses recognise the untraded goods and services associated with mangrove ecosystems, and also recognise the linkages existing among their various components, they can provide important information to decision-makers in selecting optimal management strategies.

2. Bintuni Bay mangroves

Empirical work focuses on approximately 300 000 ha of mangroves in the Bintuni Bay area of Irian Jaya, in eastern Indonesia (Fig. 1). The bay supports an important shrimp export industry, and coastal areas support some 3000 house-

Table 2
Examples of uses and environmental functions of mangroves

Mangrove uses	Mangrove functions
Sustainable production uses	**Regulatory or carrier functions**
Timber	Erosion prevention (shoreline)
Firewood	Erosion prevention (riverbanks)
Woodchips	Storage and recycling of human waste and pollutants
Charcoal	Maintenance of biodiversity
Fish	Provision of migration habitat
Crustaceans	Provision of nursery grounds
Shellfish	Provision of breeding grounds
Tannins	Nutrient supply
Nipa	Nutrient regeneration
Medicine	Coral reef maintenance and protection
Honey	Habitat for indigenous people
Traditional hunting, fishing, gathering	Recreation sites
Genetic resources	
Conversion uses	**Information functions**
Industrial/urban land-use	Spiritual and religious information
Aquaculture	Cultural and artistic inspiration
Salt ponds	Educational, historical and scientific information
Rice fields	Potential information
Plantations	
Mining	
Dam sites	

The general categories are similar to those enumerated by de Groot (1987), in that they distinguish between productive uses of goods (some of which are sustainable and some of which involve conversion to a different land-use) and the production of some service in terms of regulatory, carrier, or information functions. Although in principle each of these uses or functions might be attributed to *any* mangrove area, in practice a first step in mangrove management will involve selecting those key uses and functions which appear a priori to be economically or ecologically most important for that area.

holds in a mixed economy of farming, wages, and traditional mangrove uses. Pressures from a woodchip export industry pose a direct threat to the mangrove ecosystem. Recent interest in the area has led to a proposed Bintuni Bay Nature Reserve which would protect approximately 267 000 ha of the ecosystem, 60 000 ha of which is in the bay itself. Also, in early 1991, MENLH initiated a pilot Integrated Regional Environmental Development Program (INREDEP) in the area which is intended to provide a basis for integrated management of all components of the resource base.

The specific objectives of the empirical work relating to Bintuni Bay were: (a) to conduct a household survey which would assist in providing a quantification of the value of traditional uses of mangroves in the Bintuni Bay area; (b) to conduct correlation studies to describe the "economic linkages" between the formal sector economy and the traditional economy; and (c) to conduct cost–benefit analyses which incorporate "ecological linkages" and appropriate constraints to evaluate different management options for the forestry component of the mangrove resource. Specific recommendations relating to Bintuni Bay and general recommendations relating to mangrove management elsewhere in Indonesia are based on these analyses (Ruitenbeek, 1992a).

3. A linkage framework

Experience with mangrove management in Indonesia and in other Southeast Asian countries has provided a rich literature which describes the potential uses for mangrove areas (Silvius et al., 1987; Soemodiharjo, 1987; Umali et al., 1987; Chua and Pauly, 1989). Work in the Bintuni Bay area builds on earlier conservation site identification work (Petocz and Raspada, 1984) and a preliminary resource inventory (Erftemeijer et al., 1989) that led to a proposal to establish the Bintuni Bay Nature Reserve.

Most conventional approaches to mangrove management focus entirely on the different mar-

Table 3
Examples of linkages among mangrove components

Type and nature of impact	Description	Example
Biogeophysical linkages		
Direct pre-emptive use	One mangrove use immediately pre-empts other use because they are incompatible uses which share same land area.	Conversion to fishpond pre-empts land for sustainable wood production.
Indirect partial or delayed impact	Activity in one component of mangrove partially affects productivity of some other system component.	Conversion to fishpond increases erosion which, over a number of years, increases siltation and destroys coral reef habitat offshore.
Indirect linear impact	Activity in one component of mangrove has immediate effect on productivity of some other system component.	Conversion to fishpond destroys nursery ground and reduces offshore fishery production in proportion to lost area of mangrove.
Indirect catastrophic impact	Activity in one component of mangrove irreversibly destroys critical ecosystem component.	Conversion to fishpond of a critical area of breeding ground causes collapse of offshore fishery.
Socio-economic linkages		
Activity substitution outside mangrove ecosystem	Availability of external income causes changed local use patterns of mangrove.	Expanded wage economy reduces traditional reliance on mangrove harvesting.
Activity substitution inside mangrove ecosystem (economic displacement)	Change in availability of one mangrove component causes local substitution for other mangrove component.	Loss of on-shore productivity for hunting and gathering due to mangrove conversion forces increased reliance on offshore fishing.

ketable uses of the mangrove wood: timber, poles, firewood, charcoal, or woodchips for pulp. We must also, however, recognise that mangroves support other productive uses – such as fisheries or traditional activities – or perform important functions such as preventing coastal erosion or maintaining biodiversity (Table 2). In particular, it must be recognised that not all of these uses or functions flow through formal markets and have some price attached to them. Even where some market transactions occur, the market prices do not necessarily reflect the value of the particular good or service being provided by the mangroves.

Once the primary uses and functions are identified, the next step involves identifying the linkages among the various components. If possible, this step should also describe the nature of the impact and how changes in one component might affect changes in other components. To provide some guidance in defining the types of linkages which might arise, we broadly consider: "biogeophysical" linkages; and "socio-economic" linkages (Table 3). The term "biogeophysical linkage" refers to a growing concern in Indonesia that – in evaluating environmental impacts of projects – attention must be paid to those impacts which are basically biological in nature (such as fishery productivity) as well as to those which are geophysical in nature (such as erosion). The second type of linkage – that involved with socio-economic adjustments – is potentially much more important than the biogeophysical linkages; response times for socio-economic adjustments can often be expected to be quite rapid. These linkages may be particularly pronounced in areas where there is high population density.

Pre-emptive biogeophysical linkages are relatively straightforward in that they reflect a linkage where one land-use directly conflicts with another land-use; conversion uses such as those described in Table 2 fall into this category. Direct immediate linkages are also relatively straightforward in that they reflect instances where changes in one "building block" of the mangrove (forest area, for example) immediately affects another "building block" (such as fishery productivity) because of the ecological ties which exist between these two components. Where these ties are relatively weak, or involve substantial response delays, a "partial" linkage may be specified. The last category – that involving a catastrophic impact – might arise because of the non-linear behaviour of ecosystems; it might be characterised, for example, by the complete collapse of one component as a result of only an apparently small change in some other component. If, for example, mangrove cutting eliminated a relatively small but critical fish breeding or nursery habitat, and the entire offshore fishery collapsed as a result, this could be described as a catastrophic impact.

Distinctions between "partial", "direct" and "catastrophic" impacts are – at best – a simplification of the complex ties which might exist between ecosystem components. The distinction is offered here primarily to assist policy-makers in understanding how different parts of ecosystems might interact. Much of the correlation work undertaken in Indonesia suggests that there is a direct and immediate relationship between mangrove area and offshore fishery productivity (Naamin, 1987).

A distinction is made between two types of impacts which might occur through socio-economic adjustments: one involving interactions between traditional uses of the mangrove and an "external" formal sector economy; and one involving a substitution of activity between different mangrove components. The first of these is often presumed to be important and is regularly put forward as a rationale for disregarding traditional uses of mangroves. It is argued, for example, that "economic development" through expansions in the wage sector economy will decrease dependence on traditional uses, and it is thus inconsequential if the mangroves are destroyed in the process. Rather than taking this as an assumption, however, evidence should be gathered to support the actual degree to which this linkage exists. Analysis suggests that this argument would not hold in Bintuni Bay (Ruitenbeek, 1992a).

The second socio-economic linkage relates to adjustments between different ecosystem components. Planners should pay close attention to this if traditional uses of mangroves are highly diversified. The most obvious case of such displacement is where a strong near-shore fishery exists in conjunction with a strong on-shore fishery in the mangrove areas. As mangrove areas are depleted and on-shore catch declines (because of biological linkages), a process of economic displacement may drive more fishermen into near-shore areas and increase the pressure on those areas. If this leads to subsequent collapse of the near-shore fishery because of over-fishing, it is clear that this "socio-economic" linkage can be quite strong even if there are weak or delayed biological ties between the mangrove forest and the near-shore fishery. Indeed, this underlines the fact that *people* are in such cases an integral part of the ecosystem, and that they can in fact be a strong "link" in any part of the stress-response chain.

A potentially important set of linkages also exists completely *outside* of this system, however, relating to the linkages between broad macro-economic policies and productive activities in the mangrove. Trade policy, fiscal policy, monetary policy, foreign exchange rate policy, and most other macro-economic policies can have an effect on mangrove systems. A shrimp export subsidy would, for example, provide an incentive for increased shrimp trawling in mangrove areas; this might in turn disrupt traditional fisheries. Promoting woodchip exports could, similarly, increase mangrove cutting and thereby disrupt other components of the ecosystem through ecological linkages. In the South Sulawesi mangrove areas of Indonesia, for example, low land taxes induce conversion to tambak which, through both ecological and socio-economic linkages, eventually causes degradation of near-shore fisheries.

4. Assumptions and methods

The key uses and functions ascribed to the Bintuni Bay mangrove resource, and the linkages among them, are shown in Fig. 2. As the evalua-

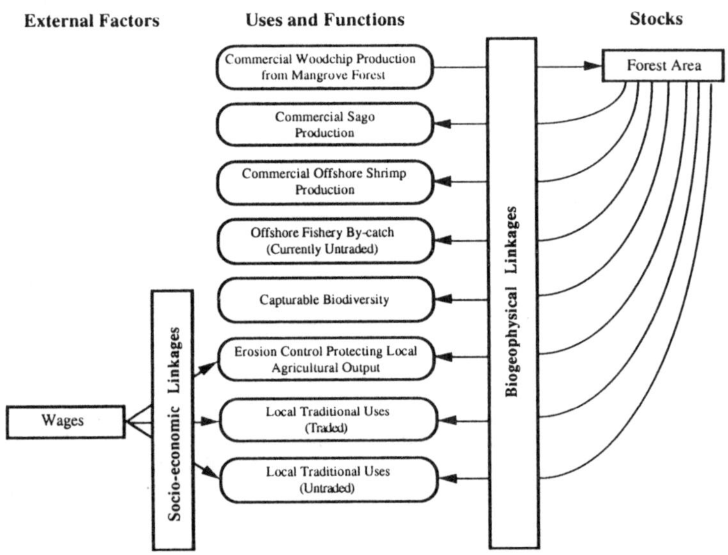

Fig. 2. Key uses, functions and linkages in Bintuni Bay mangrove.

Table 4
Mangrove area in Bintuni Bay

	Area outside proposed nature reserve (ha)	Area inside proposed nature reserve (ha)	Total area (ha)
Harvestable mangrove	97000	143000	240000
Unharvestable mangrove	0	64000	64000
Total mangrove	97000	207000	304000
Bintuni Bay to 10 m depth	0	60000	60000
Total system	97000	267000	364000

tion is intended to concentrate on selecting optimal forestry development strategies, and as most of the biogeophysical linkages are conjectured to relate to the mangrove area left intact, potential linkages between other parts of the resource base are not considered (such as between traditional and commercial fisheries, or between fishery bycatch and shrimp productivity). The primary socio-economic linkages investigated relate to the linkages between external sector wages (as a proxy for "external development" through, for example, commercial production of woodchips, sago or fisheries) and local production from farming and traditional mangrove uses.

To supplement published work, extensive discussions were held with government departments in Jakarta and the regions. In particular, this involved collecting up-to-date data from departments of forestry, fishery, trade, industry, statistics, as well as local government authorities. Although significant amounts of information were available for traded goods – such as shrimp and chipwood exports – few data were available on the value of the local uses. A household survey, consisting of 101 households in 6 villages, was therefore conducted in 1991 to provide primary data estimating the scale and value of traditional mangrove uses in the Bintuni Bay area.

The household data were analysed, along with other information on traded goods, to estimate both the traded and untraded value of goods and services associated with Bintuni Bay mangroves. In addition, correlation studies using qualitative dependent variable techniques of econometric analysis were conducted to identify some of the key economic linkages which exist. In particular, these techniques attempt to address whether individuals' reliance on traditional mangrove hunting, gathering and fishing will be affected by increased development in the area. These analyses were all undertaken with a view to providing input to a cost–benefit analysis (CBA) of different forestry management options. The CBA reflects both economic and ecological linkages, and assesses forestry options in the various mangrove areas (Table 4) ranging from clear cutting the mangroves to the imposition of a complete cutting ban (Table 5).

Table 5
Clearing options evaluated

Option	Description
20-year clear cut	Total harvestable area is cut over 20-year period, once only cut
30-year clear cut	Total harvestable area is cut over 30-year period, once only cut
30-year rotation 80% selective cut	80% of total harvestable area (= 192000 ha) is cut in perpetuity on 30-year rotation
30-year rotation 40% selective cut	40% of total harvestable area (= 96000 ha) is cut in perpetuity on 30-year rotation (equivalent to 100% of total harvestable area outside proposed nature reserve)
30-year rotation 25% selective cut	25% of total harvestable area (= 60000 ha) is cut in perpetuity on 30-year rotation (equivalent to 62% of total harvestable area outside proposed nature reserve) (equivalent to operating current chipwood plant at 80% capacity for 20 years)
Cutting ban	Entire mangrove area is maintained in natural state

The analyses generally assume a 30-year rotation for the selective cutting, as well as for one of the clear cutting scenarios. A 30- to 35-year rotation is typically regarded as technically feasible for mangrove resources if replanting and selective cutting is followed. In Bintuni Bay, as elsewhere in Indonesia, most concessions have been granted for a 20-year period; a 20-year clear cut is therefore also evaluated. In the absence of any regulations or controls, this "accelerated clear cut" would often be the strategy selected by a concessionaire who focuses purely on maximising profits.

In both of the clear cut scenarios, it is assumed that a once-only cut occurs and that no further stock is available after the clear cut is completed. This reflects the idea that such cutting is usually not sustainable in a mangrove ecosystem. The selective cutting scenarios are, however, assumed to be sustainable *from a forestry perspective* and thus allow cutting in perpetuity on a 30-year rotation. The "80% selective cut" approximates the maximum sustainable yield from the resource. The "40% selective cut", as indicated, is meant to approximate harvesting of the entire harvestable area outside of the proposed Bintuni Bay Nature Reserve.

The "25% selective cut" represents a more conservative strategy involving the clearance of only about 60 000 ha outside of the reserve area. This scenario was designed in a way that it ensures that the current concessionaires will have adequate mangrove stock to keep an existing woodchip plant operating at 80% of capacity for the 20-year life of the concession (without any extension to the life of that concession). If the plant operates at a higher capacity factor, it implies that the concessionaire would need to be limited to between 16 and 20 years of operation. Given that some shutdowns will be inevitable for even routine operations, it is expected that this scenario provides a realistic compromise between strict conservation and uncontrolled cutting.

In addition to the mangrove area, commercial sago production in the Bintuni Bay area commenced in late 1990. A total of 15 000 ha of concession has been allotted, and it is assumed that this amount will neither increase nor decrease. It is projected that production will reach a sustainable level of 225 000 tonnes per year by 2001.

The analysis assumes that real chipwood export prices stay constant over the period at a level of Rp80 000/m^3 (US\$40/m^3), and that sago prices also stay constant at a price of Rp300/kg (US\$0.15/kg). Production costs were based on investment costs provided by the companies, and on operating costs estimated from typical operations elsewhere. Royalties, taxes, and compensation payments were excluded from the costs in the cost–benefit analysis as they represent a direct transfer and are not regarded as a drain on society's resources.

Although the mangrove concessions in the Bintuni Bay area have been licensed for only 3 years, the commercial shrimp fishery has been operating in Bintuni Bay since approximately 1970. Production at historical levels has shown no evidence of falling off, and, although no estimates have been made of the size of the resource stock, popular opinion is that current levels of production are probably approaching the maximum sustainable yield. Output is currently frozen and packaged in one of four cold storage facilities in Sorong, or in a new facility recently opened near Babo. Most of the output is exported to Japan.

It is assumed that the fishery can still support a modest expansion to support one more cold storage facility. This expansion would involve an increase of output of about 15%, and allow for a sustainable shrimp harvest of approximately 5500 tonnes annually. Based on recent average prices, the CBA assumes that the shrimp value remains constant in real terms at Rp12 500/kg (US\$6.25/kg). As with forestry, fishery costs are based on investment and operating costs provided by the companies to government authorities. Royalties, taxes, and compensation payments are excluded from the costs in the cost–benefit analysis.

In addition to the commercial shrimp harvest, boats harvest a substantial by-catch which is currently thrown back into the bay or, on some occasions, eaten by the ships' crews or traded to local people. This by-catch represents more than 90% of the trawling catch by weight and, although it is currently not used, it does offer commercial opportunities as fish meal or fertiliser. Recent identification of the disposal practice has prompted local authorities to consider promoting the commercial use of this by-catch. In this cost–benefit analysis, it is assumed that some commercial development will eventually occur, although the by-catch is assigned an imputed value of only Rp300/kg (US$0.15/kg).

The valuation of local mangrove uses was based on traditional household production from hunting, fishing, gathering, and manufacturing, as estimated through the household survey. Wages, transfers, farming income, livestock sales, and other miscellaneous income was excluded from this category. One important component of the cost–benefit analysis is the imputed benefit of erosion control. In this study, the benefit is based on the value of agricultural output from local production. In 1991 this was estimated from the household survey to be Rp1.9 million (US$950) per household.

A value of Rp3 million (US$1500) per km^2 of mangrove area per year was imputed to capturable biodiversity benefits (Ruitenbeek, 1992b). Because some of the mangrove area is in very isolated areas and is not expected to be harvested even if there were no controls on cutting in the region, this benefit will never totally disappear even if the "accelerated clear cut" scenario is followed. One attribute of the modelling in the CBA, however, is that in some of the linkage scenarios which are modelled, it is assumed that this benefit is only capturable if there exists an expectation that the area will remain intact in the future. For example, if there are 300000 ha in 1991, and if there are expected to be only 150000 ha in the year 2000, the imputed value in 1991 is that which would accrue only to the 150000 ha. This assumption is made primarily in light of existing institutions which are likely to transfer funds to protect biodiversity. It is expected, for example, that international conservation groups or the GEF will provide funds to an area only if there exists some reasonable expectation that the area will be protected.

In addition to the above assumptions, the cost–benefit analysis generally assumes that there will be no real increase or decrease in costs or prices. Cost and benefit streams are extended over a 90-year time horizon to allow three full rotations in the forestry evaluations, and to accommodate potential delays in linkage effects among ecosystem components. All future costs and benefits are discounted to 1991 at a real discount rate of 7.5%. This discount rate was selected based on discussions with planning authorities in Indonesia, and it reasonably reflects the opportunity cost of risk-free investments. Sensitivity tests are undertaken at a higher (10%) and lower (5%) rate.

Because ecological linkages have often been ignored in past decisions, a key component of the CBA is to assess the "economically optimal strategy" under different assumptions of ecological linkages. Although estimating the value of each of the different components in the absence of any linkages is relatively straightforward, significant uncertainty exists relating to both the nature and degree of the linkages. Empirical work on the socio-economic linkages is conducted for this study, but no data specific to the Bintuni Bay area exist for providing estimates of the extent of the ecological linkages. To address this problem, "linkage scenarios" are developed which describe potential ecological linkages ranging from "no linkages" to "very strong" linkages similar to those which have been demonstrated on the north coast of Java (Table 6). Specification of such scenarios is also useful because they demonstrate what happens if development decisions are based on an assumption that no linkages exist when, in fact, very strong ecological ties might exist between components.

The basic procedure in specifying the linkages involves both an impact intensity parameter (α) and an impact delay parameter (τ). An impact parameter of $\alpha = 1$ would imply that a 50% reduction in mangrove area would yield a 50% reduction in fishery production. This is the result typically found in Java. From the formula in Table 6, it is clear that an impact parameter of $\alpha = 0.5$ would imply that fishery output varies as the square root of mangrove area; a 50% reduction in mangrove area would result in only a 30% reduction in fishery production. This would occur if only "partial" ecological linkages existed, or if

Table 6
Definition of ecological linkages

	Impact parameter α	Delay parameter τ (years)
No linkages	0.0	–
Weak non-linear linkages	0.5	10
Moderate non-linear linkages	0.5	5
Strong (delayed) linear linkages	1.0	5
Very strong (immediate) linear linkages	1.0	0

The parameters in this case show the defined linkage scenarios for (a) local uses; (b) erosion impacts; and (c) fisheries impacts of clearing mangroves. For productivity of any activity centre in year t, the linkage to mangrove cutting is defined as follows:

$$\left(\frac{\text{PRODUCTIVITY}_t}{\text{PRODUCTIVITY}_{t=0}}\right)=\left(\frac{\text{MANGROVE}_{t-\tau}}{\text{MANGROVE}_{t=0}}\right)^{\alpha}$$

where MANGROVE is the area of undisturbed mangrove, and α and τ are, respectively, intensity and delay parameters as specified above. Linkages for mangroves to offshore productivity generally fall into the "linear" category ($\alpha = 1$) with quite rapid response; the selected "most likely" scenario is, therefore, the "strong" or "very strong" linkage scenario case.

stresses on the fisheries were somehow buffered by other factors. The specification of thi. relationship does not, however, constrain α to being less than or equal to unity. Indeed, if a particularly critical habitat were being threatened, it is quite conceivable that α would take on a value greater than unity. An impact parameter of $\alpha = 2$, for example, implies that fishery output varies as the square of mangrove area; a 50% reduction in mangrove area in this case would result in a 75% drop in fishery output.

The time at which events occur is, because of the discounting procedure in CBA, of critical importance. An ecological linkage which takes 20 years to manifest itself is, other things equal, far less serious in economic terms than one which occurs immediately. The delay parameter τ is thus specified with a view to showing how delayed impacts might affect the optimal management choice. In the case of fisheries, most analyses suggest that the most serious consequences (for adjacent fisheries) of mangrove depletion would occur in under 5 years.

In principle, the "severity" of impacts leading from the linkages must also be considered. Cost–benefit analysis can only be applied if costs and benefits are finite. A linkage which leads to multiple "infinite" costs or benefits could not be evaluated using CBA; other decision criteria must be used in such instances.

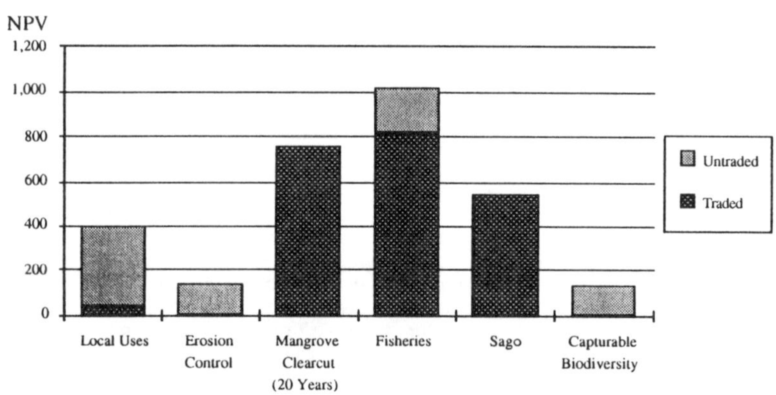

Fig. 3. Total economic value of reference case with no linkages. Note at bottom Fig. 3: NPV = Net present value in billions of 1991 Rp at 7.5% discount rate; 2000 Rp = US$1.

5. Results and discussion

A useful starting point, or reference case, involves the "no linkage" case where each resource component is assumed to function independently (Fig. 3). This reference scenario also shows the "accelerated clear cut" as the option that would be selected by forestry companies in the absence of any controls; other options have lower values from purely a forestry perspective. Taking into account all of these components, the total asset value of the resource approaches Rp3 trillion (US$1.5 billion). This is substantial and, given this high value, the resource represents a natural target for proper resource management. Two things stand out from this presentation: the commercial forestry and fishery resource have the same order-of-magnitude value of approximately Rp800 billion (US$400 million); the traditional local uses have a substantial, although untraded, value of approximately Rp400 billion (US$200 million).

A total of 26 cases can be analysed under different cutting and linkage assumptions and, from this, an optimal strategy can be determined (Table 7). First, the clear cut example provides the greatest net benefits if one ignores linkages. This result is quite common for biological resources when the growth rate of the resource stock is less than the discount rate. The optimal short-term economic strategy under such circumstances is to "mine" the resource. Forestry departments realise, however, that although this approach might maximise the net present value, it does not provide sustainable income from forestry. A preferred management option is therefore often to cut close to the forest's maximum sustainable yield, as shown in the "80% selective cut" case. Under this option, the NPV appears less than the clear cut although, strictly speaking, this implies that the resource managers are imputing some other value to resource sustainability which is not obviously reflected in the NPV calculations. In any event, if there are no linkages, or if linkages are ignored, the cutting ban option does not appear to be optimal.

If we take into account the linkages, however, the situation changes dramatically. Under a "very strong" linkage scenario, which involves linear and immediate impacts on other resource compo-

Table 7
Determination of optimal mangrove management strategy

	Linkage scenario				
	Very strong	Strong	Moderate	Weak	No linkages
Linkage parameters					
Impact parameter α	1	1	0.5	0.5	0
Delay parameter τ (years)	0	5	5	10	–
NPV of cutting strategy (billions of Rp) [a]					
Cutting ban	2230	2230	2230	2230	2230
25% selective cut	2190	2240	2320	2350	2400
40% selective cut	2160	2240	2370	2420	2500
80% selective cut	2070	2240	2480	2590	2770
Clear cut (30 years)	1960	2170	2440	2590	2820
Accelerated clear cut (20 years) [b]	1920	2180	2490	2690	2990
Optimal cutting strategy [c]	Ban	25% cut	80% cut	80% cut	Clear cut

[a] NPV is net present value at 7.5% discount rate; 2000 Rp = 1US$.
[b] The accelerated clear cut case is often regarded as unsustainable and is not usually allowed as a matter of government policy; results are presented for the sake of completeness.
[c] The optimal cutting strategy, under any given ecological assumptions of the delay (τ in years) and extent (α) of impacts, is determined by selecting the strategy with the highest NPV under those assumptions; the accelerated clear cut is excluded from this selection as politically unacceptable.

nents, the clear cut option is manifestly the worst of the alternatives, and the cutting ban option provides benefits which are Rp160 billion (US$80 million) greater than the 80% selective cut option, *even accounting for the lost forestry revenue*. Furthermore, it is clear that the 80% selective cut cannot be regarded as a sustainable option; the impacts which it has under this linkage scenario are substantial.

It is clear that cutting bans are optimal if very strong linkages occur, and clear cut options are optimal if one totally ignores linkages. If one rejects the "accelerated clear cut" option as unsustainable out of hand, and if one accepts that there are even weak linkages ($\alpha = 0.5$ with a 10-year delay) at play, then selective cutting options are preferred even over the 30-year clear cut. If moderate linkages exist ($\alpha = 0.5$ with a 5-year delay), then such selective cutting is little worse than even the accelerated clear cut case.

A striking result is that even in the "strong" linkage case ($\alpha = 1$ with a 5-year delay) there is no economic advantage to proceeding beyond a 25% cut of the harvestable area. If linkage effects occurred more rapidly than 5 years, or if the intensity parameter α exceeded unity, then it would be optimal to select an even more conservative cutting strategy than one involving a 25% harvest.

The procedure of selecting an optimal strategy depends, to some degree, on the discount rate used in the analysis (Table 8). In general, lower discount rates place a higher weight on future income streams, and in this case they result in more conservative cutting strategies. It is interesting, however, that at a 5% discount rate the clear cut option is *never* the option of choice. Under higher discount rates, less weight is placed on future benefit streams and, predictably, clear cut strategies dominate where substantial delays of the linkage effects occur. A notable result is that, at all discount rates investigated, the cutting ban is the optimal strategy if immediate linear linkages exist.

Table 8
Optimal strategy as a function of discount rate

	Discount rate		
	5%	7.5%	10%
No linkages	80% cut	Clear cut	Clear cut
Weak linkages	80% cut	80% cut	Clear cut
Moderate linkages	80% cut	80% cut	Clear cut
Strong linkages	Ban	25% cut	80% cut
Very strong linkages	Ban	Ban	Ban

6. Conclusion

When linkages between ecosystem components are not taken into account, or when mangrove cutting is left unregulated, mangrove resources will often be over-exploited. The existence of linkages between mangrove area and overall ecosystem productivity means that strong economic arguments can be made for conservative mangrove clearing. In instances where strong ecological ties occur, severe restrictions on clearing activities will be economically optimal. This generic result will apply to many situations where resource development needs place conflicting demands on mangrove resources.

In the Bintuni Bay area of Indonesia, the analysis provides additional rationale for setting aside some of the mangrove area in a conservation area. If strong ecological interactions exist, the analysis showed that the optimal amount of cutting was less than 25% of the harvestable area. Given the proposed area of the reserve and characteristics of the mangrove outside of the reserve, to achieve a "25% target" for the area as a whole would require that cutting outside of the reserve not exceed 60% of the harvestable area. Even so, if further research shows that ecological interactions are quite rapid, or if critical habitats are being disturbed, then it would be economically justified to reduce cutting below this 60% threshold.

It must be recognised that there is still considerable uncertainty in the dynamics of specific mangrove ecosystems. This study has demonstrated that *if we know* the nature of these interactions, an economically optimal strategy can be selected. The analysis also demonstrates that *if we do not know* the nature of the interactions, an incorrect guess can have substantial economic penalties. If, for example, we *assume* that there are weak delayed interactions and select an 80% selective cut on that basis, and if it turns out that the *actual* interactions are immediate and linear, then the economic value of such a decision in the Bintuni Bay case would be about Rp500 billion (US$250 million) *less* than what was expected, and Rp160 billion (US$80 million) *less* than the optimal strategy. Decision-makers must be aware of what the potential losses of incorrect decisions might be – even when information is incomplete or uncertain – and act accordingly.

Finally, the current situation of uncertainty raises two further issues: (a) that there is a need for information relating to the linkages between ecosystem components; and (b) that policies which mitigate the effects of the linkages will have economic as well as ecological merit. Where ecosystem dynamics are uncertain, programs reducing linkage effects – such as greenbelts, replanting, or selective cutting – will minimise potential economic losses.

One important aspect of this research relates to its overall role in the policy planning and decision-making process. The entire project – from inception, to field work, to analysis, to input into the decision-making process via a workshop of senior Indonesian policy-makers – took less than one-half of one year to complete. It demonstrates that meaningful modelling of economy–ecology linkages can be done within the time frame associated with typical political processes, and that such modelling can provide important information relating to a number of trade-offs.

Acknowledgements

This article is based on work conducted for the Environmental Management Development in Indonesia (EMDI) Project, which is jointly designed and implemented by the Ministry of State for Environment (MENLH) of the Government of Indonesia, and Dalhousie University, Halifax, Canada. The Project is funded by the Canadian International Development Agency (CIDA). The author acknowledges the kind permission of these parties for extracting this article from a more detailed study contained in the EMDI Environmental Report Series (Ruitenbeek, 1992a). A copy of the detailed study is available from the School of Resource and Environmental Studies, Dalhousie University, Halifax, Nova Scotia, Canada, B3H 3E2, or from the author.

The author is indebted in preparing this work to his Indonesian counterparts in MENLH for numerous discussions which assisted in its completion. In particular, the author thanks Ir. Aca S. Apandi, Ir. H. Uktolseya, Dr. H. Alikodra, Drs. Sudariyono, Ir. B. Widyantoro, and Ir. S. Wibowo. Assistance with household surveys used for this study was provided by Mr. C. Marlessy, Ms. C. Cartier, Ir. B. Widyantoro, Ny. W. Warwe, Ir. S. Djalimun and Ir. S. Loupatty.

Any opinions and conclusions expressed herein are those of the author, who takes full responsibility for any errors of commission or omission.

References

Chua, T.-E. and Pauly, D. (Editors), 1989. Coastal Area Management in Southeast Asia: Policies, Management Strategies and Case Studies. ICLARM Conference Proceedings 19, Ministry of Science, Technology and the Environment, Kuala Lumpur; Johore State Economic Planning Unit, Johore Bahru, Malaysia; and International Center for Living Aquatic Resources Management, Manila, Philippines.

de Beer, J.H. and Mcdermott, M.J., 1989. The Economic Value of Non-timber Forest Products in Southeast Asia: With Emphasis on Indonesia, Malaysia, and Thailand. Netherlands Committee for IUCN, Amsterdam.

de Groot, R.S., 1987. Environmental function as a unifying concept for ecology and economics. Environmentalist, 7(2): 105–109.

Dixon, J.A., 1989. Coastal resources: assessing alternatives. In: T.-E. Chua and D. Pauly (Editors), Coastal Area Management in Southeast Asia: Policies, Management Strategies and Case Studies. ICLARM Conference Proceedings 19, Ministry of Science, Technology and the Environment, Kuala Lumpur; Johore State Economic Planning Unit, Johore Bahru, Malaysia; and International Center for Living Aquatic Resources Management, Manila, pp 153–162.

Erftemeijer, P., Allen, G. and Zuwendra, 1989. Preliminary Resource Inventory of Bintuni Bay and Recommendations for Conservation and Management. PHPA/AWB, Bogor.

Eusebio, M.A., Tesoro, F.O. and Cabahug, Jr., D.M., 1987. Environmental impact of timber harvesting on mangrove ecosystem in the Philippines. In: R.M. Umali, P.M.

Zamora, R.R. Gotera, R.S. Jara, A.S. Camacho and M. Vannucci (Editors), Mangroves of Asia and the Pacific: Status and Management. Technical Report of the UNDP/UNESCO Research and Training Pilot Program on Mangrove Ecosystems in Asia and the Pacific. Natural Resources Management Center and National Mangrove Committee, Ministry of Natural Resources, Manila, pp. 337–354.

Hehanussa, P.E. and Hehuwat, F., 1980. Morphogenesis of the northern coastal plain of west Java between Cirebon and Jakarta: Its implications for coastal zone management. In: C.E.F. Bird and A. Soegiarto (Editors), Proceedings of the Jakarta Workshop on Coastal Zone Management. LIPI-UNU, Jakarta.

Hodgson, G. and Dixon, J.A., 1988. Logging versus Tourism in Palawan: An Environmental and Economic Analysis. East-West Environment and Policy Institute, Honolulu, 95 pp.

Macnae, W., 1986. A general account of the flora and fauna of the mangrove swamps and forests in the Indo-West-Pacific region. Adv. Mar. Biol., 6: 73–269.

Naamin, N., 1987. Impact of "tambak" aquaculture to the mangrove ecosystem and its adjacent areas with special reference to the north coast of West Java. In: R.M. Umali, P.M. Zamora, R.R. Gotera, R.S. Jara, A.S. Camacho and M. Vannucci (Editors), Mangroves of Asia and the Pacific: Status and Management. Technical Report of the UNDP/UNESCO Research and Training Pilot Program on Mangrove Ecosystems in Asia and the Pacific. Natural Resources Management Center and National Mangrove Committee, Ministry of Natural Resources, Manila, pp. 355–365.

Petocz, R.G. and Raspada, G.P., 1984. Conservation and Development in Irian Jaya: A Strategy for Rational Resource Utilization. PHPA, Bogor.

Rahman, M.J. and van de Vusse, F., 1992. Mangrove forests: a valuable but threatened Indo-Pacific resource. Agriculture Department Staff Paper 5, Asian Development Bank, Manila.

Robertson, A.I. and Daniel, P.A., 1989. The influence of crabs on litter processing in high intertidal mangrove forests in tropical Australia. Oecologia, 78: 191–198.

Robertson, A.I. and Duke, D.C., 1990. Mangrove fish-communities in tropical Queensland, Australia: Spatial and temporal patterns in densities, biomass and community structure. Mar. Ecol., 104: 369–379.

Robertson, A.I., Alongi, D.M., Daniel, P.A. and Boto, K.G., 1988a. How much mangrove detritus enters the Great Barrier Reef lagoon? In: J.H. Choat et al. (Editors), Proceedings of the 6th International Coral Reef Symposium. Australian Institute of Marine Science, Townsville, pp. 601–606.

Robertson, A.I., Dixon, P. and Daniel, P.A., 1988b. Zooplankton dynamics in mangrove and other nearshore habitats in tropical Australia. Mar. Ecol., 43: 139–150.

Ruitenbeek, H.J., 1992a. Mangrove Management: An Economic Analysis of Management Options with a Focus on Bintuni Bay, Irian Jaya. Dalhousie University, Halifax, 90 pp.

Ruitenbeek, H.J., 1992b. The rainforest supply price: A tool for evaluating rainforest conservation expenditures. Ecol. Econ., 6: 57–78.

Silvius, M.J., Steeman, A.P.J.M., Berczy, E.T., Djuharsa, E. and Taufik, A.W., 1987. The Indonesian Wetland Directory: A Preliminary Compilation of Existing Information on Wetlands of Indonesia. PHPA/AWB/INTERWADER/EDWIN, Bogor.

Soemodiharjo, S., 1987. Indonesia. In: R.M. Umali, P.M. Zamora, R.R. Gotera, R.S. Jara, A.S. Camacho and M. Vannucci (Editors), Mangroves of Asia and the Pacific: Status and Management. Technical Report of the UNDP/UNESCO Research and Training Pilot Program on Mangrove Ecosystems in Asia and the Pacific. Natural Resources Management Center and National Mangrove Committee, Ministry of Natural Resources, Manila, pp. 89–130.

Turner, R.K., 1993. Wetland valuation: three case studies. CSERGE GEC Working Paper Series. Centre for Social and Economic Research on the Global Environment, Norfolk.

Umali, R.M., Zamora, P.M., Gotera, R.R., Jara, R.S., Camacho, A.S. and Vannucci, M. (Editors), 1987. Mangroves of Asia and the Pacific: Status and Management. Technical Report of the UNDP/UNESCO Research and Training Pilot Program on Mangrove Ecosystems in Asia and the Pacific. Natural Resources Management Center and National Mangrove Committee, Ministry of Natural Resources, Manila, 538 pp.

Williams, D.McB. and Cappo, M., 1990. Life-histories of clupeids in north-eastern Australia: Preliminary data. In: S.J.M. Blaber and J.W. Copland (Editors), Tuna Baitfish in the Indo-Pacific Region. ACIAR, Honiara, Solomon Islands, pp. 70–74.

[19]

Marine Resource Economics, Volume 9, pp. 275–286
Printed in the USA. All rights reserved.

0738-1360/94 $3.00 + .00

Total Economic Values of Increasing Gray Whale Populations: Results from a Contingent Valuation Survey of Visitors and Households

JOHN B. LOOMIS

Department of Agricultural and Resource Economics
Colorado State University
Fort Collins, CO 80523

DOUGLAS M. LARSON

Department of Agricultural Economics
University of California-Davis
Davis, CA 95616

Abstract *The consistency of an individual's willingness to pay (WTP) responses for increases in the quantity of an environmental public good (whale populations) is tested along three lines. First, we test whether WTP for 50% and 100% increases in whale populations are statistically different from zero. Second, we ask whether the incremental WTP from a 50% increase to a 100% increase is statistically significant. Finally, we test whether there is diminishing marginal valuation of the second 50% increment in gray whale populations. The paired t-tests on open-ended WTP responses supported all three sets of hypotheses. Both visitors and households provided WTP responses that were statistically different from zero and increased (but in a diminishing fashion) for the second increment in WTP. In this survey both visitors and households provided estimates of total economic value (including non-use or existence values) for large changes in wildlife/fishery resources that were consistent with consumer theory.*

Keywords Existence value, contingent valuation, gray whale, California, willingness to pay.

Introduction

It is generally recognized that direct viewing of wildlife provides an economic value to the participant. However, preservation of the species for viewing is a joint product for the "users" and also may provide value to non-viewers, from the continued existence of the species for both themselves and (in many cases) for others in the future. That continued preservation of a species provides joint pro-

Funding for this research was provided by grants from Resources for the Future, the U.S. Environmental Protection Agency and Agricultural Experiment Station Regional Research Project W-133. We would like to thank Renatte Hagemann, San Diego State University, who supervised the collection of the San Diego portion of this data, and our visitor interceptors Paula Stevens and Dan Ritzman for their persistence in a variety of weather conditions. Ms. Stevens and Sunny Williams provided substantial effort in the data entry and follow up phone contacts of the household survey. Finally, the editor and two reviewers provided numerous suggestions for clarifying the paper. The usual disclaimer applies.

duction of viewing, existence and bequest values was first formalized by Randall and Stoll (1983) as "total economic value." Most economists recognize that existence of a species is a public good and, like any other public good, its total economic value should be included when calculating benefits and costs of proposed policies (Randall and Stoll, 1983; Kopp, 1992). Given this, two key questions arise. First, how is one to measure the total economic value? Second, is the marginal or incremental total economic value for changes in abundance of a particular species significantly different from zero? That is, once a minimum viable population is provided, does increased abundance beyond that add to society's well being?[1] In answering this latter question, we focus on whether the pattern of incremental valuation of gray whale populations is consistent with first principles of consumer demand theory. What we learn about the *patterns of an individual's valuation* of gray whales is useful for understanding how valuations of other marine mammals such as elephant seals, sea otters and dolphins would also change with their population levels. Such information may be helpful in dealing with a wide variety of policy decisions such as what magnitude of incidental take of marine mammals to allow in commercial fishing operations.

With respect to the first question, at present, the Contingent Valuation Method (CVM) appears to be the only way to measure total economic value, although Larson (1993) has proposed a revealed preference approach. While the reliability of CVM to measure existence values has been subject to much debate over the years, the journal literature provides several examples of the reliability (Loomis, 1989, 1990; Reiling, *et al.*, Kealy, *et al.*,) and validity (Brookshire, *et al.*, 1982; Welsh, 1986) of CVM derived benefit estimates. Nonetheless, the high stakes in natural resource damage assessment has brought renewed interest in CVM by a broad spectrum of economists. Concerns about inconsistency of CVM with economic theory appears to have been one of the reasons why the National Oceanic and Atmospheric Administration (NOAA) commissioned a "blue ribbon" panel to advise NOAA on the use of CVM for measuring non-use values (Augustyniak, Jones and Meade, 1992). After reviewing the available research, "The Panel concludes that CV studies can produce estimates reliable enough to be the starting point for a judicial or administrative determination of natural resource damages—including lost passive-use value." (Arrow, *et al.*, 1993), as long as certain sampling and survey design guidelines are adhered to. Since these guidelines were released about a year *after* our survey administration, we will only make reference to areas of consistency and inconsistency with these new guidelines.

This paper contributes to improving our understanding of the properties of individual's CVM responses as measures of total economic value, including non-use value. Specifically, we ask whether CVM-derived estimates of an individual's incremental total economic value for increases in wildlife populations are significantly different from zero, and whether they are consistent with expectations based on economic theory. These questions are also of practical importance to policy analyses aimed at increasing wildlife populations beyond their legally required minimum viable population. While numerous previous studies have found statistically significant marginal *use* values for consumptive users (Brookshire,

[1] This simplistic discussion assumes that once society is above the minimum viable population that risk of extinction is zero. In addition, a decidedly anthropocentric viewpoint is represented here.

Randall and Stoll 1980; Loomis and Cooper, 1990), and for bird viewing (Cooper and Loomis, 1991), few studies have attempted to measure incremental values for *total economic values* of wildlife.

Hypotheses About Marginal Values of Increases in Public Goods

Utility and consumer theory is primarily a theory about individual behavior. First principles of consumer theory require only that increases in quantities of economic goods provide non-negative values to individuals. It seems plausible that increases in populations of many native fish and wildlife species could have positive values given their currently low populations. For a given individual, the principle of diminishing marginal returns suggests that if the first increment has a positive value, the value of the second increment would normally have a lower marginal value than the first, *i.e.*, one would expect diminishing marginal value for greater and greater additions to the stock of the public good.

The first hypothesis to be tested is whether stated WTP for a given increase of size **n** in a population of a particular species is statistically different from zero. Specifically the null hypothesis is:

$$H_o\text{: } WTP_n \leqslant 0 \qquad (1)$$

against the alternative that $WTP_n > 0$. If the dollar amount of stated WTP is obtained from responses to open-ended CVM questions, then WTP is measured as a ratio level variable. Therefore (1) can be tested using parametric tests such as a one-tailed Student's t-test.

A second and related hypothesis test is whether WTP for a larger increase in the public good (WTP_{n+m}), where $m > 0$, is statistically different from WTP for a smaller increase (WPT_n). Formally,

$$H_o\text{: } WTP_{n+m} - WTP_n \leqslant 0 \qquad (2)$$

against the alternative that $WTP_{n+m} - WTP_n > 0$. This is a one-tailed *paired* t-test since it is a test of the consistency of an individual's behavior, a more direct test of consumer theory then comparing consistency of aggregate behavior.

If we obtain positive incremental values for $WTP_{n+m} - WTP_n$, and accept the alternative to the second hypothesis, then one would expect a diminishing incremental valuation. Specifically if both increments are of the same size $\mathbf{n} = \mathbf{m}$, then a diminishing marginal value implies that

$$H_o\text{: } WTP_{2n} - WTP_n < WTP_n \qquad (3)$$

This will also be tested as a paired t-test with the null hypothesis that

$$H_o\text{: } WTP_{2n} - WTP_n \geqslant WTP_n \qquad (4)$$

Case Study

The gray whales that migrate from their summer home in Alaska down the west coast past California to Baja were until recently a threatened species under the

Endangered Species Act. Whales are also a species with documented nonconsumptive use and strong evidence of existence/bequest values. Many conservation organizations such as Greenpeace and the Oceanic Society are maintained by contributions from people, many of whom rarely, if ever, see whales.

Our sample frame is comprised of two populations: (1) persons who went to the coast during the gray whale migration usually, although not exclusively, to view whales; and (2) California households. We include both current users and the general population of California households (some of whom may be past users) for two reasons. First, current viewers will have more first-hand experience "trading money for opportunities to view whales" and better information on whale abundance. These individuals generally meet the "Reference Operating Conditions" for CVM suggested by Cummings, *et al.* (1986). Hence CVM estimates of WTP should be more reliable for this group. In addition, we would expect that if any group would have significant total economic values, it should be this group. This user group contrasts with general households who may have never gone whale watching, possibly do not care about marine mammals, and have less experience with trading money for whales. However, it is this group for which even small estimates of total economic value per household often expand into large aggregate estimates, due to the sheer size of this group. Thus we are also interested to see whether this less-experienced and less-interested group also expresses statistically significant marginal values for successive increases in whale populations.

There have been two CVM studies to date attempting to measure the total economic value of whales. Hagemann (1985) used a payment card to elicit WTP for gray whales and blue whales as well as three other marine mammals. Hagemann's mail survey was of general households in California. Samples and Hollyer (1990) have applied CVM to value humpback whales in Hawaii. They asked individuals their WTP to provide one-time emergency assistance to protect whales in Hawaii. A dichotomous choice CVM format was used with in-person interviews of the general population.

While these studies both represent pioneering efforts in valuation of whales and presented interesting methodological comparisons, they generally involved either small sample sizes or low response rates. We informally compare our dollar values with these two studies in the results section. Formal comparisons are not possible since our survey evaluates large increases in number of whales while the two past studies involved WTP to avoid large reductions in whale populations.

Format of Whale Questionnaire

A major focus of the survey was questions regarding the total value for two increases in gray whale populations. First the respondents were told the current population of gray whales was 20,000. Then respondents were informed that the population could be increased by reducing coastal pollution and restricting commercial and industrial activities along the coast, particularly in calving habitat. The payment vehicle was payment into a Gray Whale Protection Fund. The survey stated that *"Legally the money could only be used to clean up coastal waters of pollution and drift nets, purchase additional calving habitat areas, etc."* Respondents were asked to state their WTP for a 50% and 100% increase in gray whale populations and corresponding increase in sightings of gray whales along the California coast. Specifically the question read *"What is the maximum you would pay each year into the Gray Whale Protection Fund to increase gray whale populations and your sightings by 50%?"* Respondents were asked to assume that a 50% increase in whale populations would translate into equal percentage increase in whale sightings. This WTP question format was open-ended and was followed by questions to help determine whether persons stating zero WTP were protesting some feature of the CVM market. On the adjacent page, the 100% increase in whale program was presented as follows: *"Program Two: This alternative program is a more comprehensive gray whale protection that would increase by 100% or double gray whale populations and your sightings along the California coast. This second program is illustrated in the graph below."* Respondents were then asked to state their WTP for Program #2 if it were the only program offered.

Thus, individuals were first asked to pay for a 50% increase, and then asked to consider an alternative second WTP question for a 100% increase. Reliance on a series of WTP questions has both methodological advantages and disadvantages. First, having the same individual give their WTP for both population levels provides a more efficient statistical control for estimating the value of the population change. In this setting, the difference in WTP by the same individual is related only to the change in whale population. The alternative questioning approach, of having different individuals answer one question for one of the two population levels would be statistically inefficient since the research would have to be able to completely control for differences among individuals before they could conclude if the differences in valuation related to differences in whale populations or differences in individual respondents. If separate samples were each asked one of the two questions, the required sample would need to be far greater than simply twice the current sample. Asking more than one quality level of the same respondent has been a standard CVM procedure for deriving Bradford bid curves (Bradford, 1970) for decades (Daubert and Young, 1981; Walsh, *et al.*, 1984) and continues to be used today (Hoehn, 1990; Hoehn and Loomis, 1993). Loosely speaking, this question format is like a panel design and has the same statistical advantages.[2] One methodological concern relates to possible order effects. Since everyone answered the 50% increase before the 100% increase this may create a conditional path in the WTP values, even though individuals were told that the 100% increase program was an alternative program. However, *if* there are sequencing effects such that the program valued second is worth less than the one valued first, this will tend to bias our hypothesis test #2 ($WTP_{50\%} < WTP_{100\%}$) against us because it would raise the WTP for the 50% increase and lower the WTP for 100% increase, thereby shrinking the differences.

A draft of the survey instrument was pretested using individuals who had gone whale watching in the previous year. A survey was given or mailed to them and they were asked to fill it out and make additional comments on it. The pretest indicated respondents had difficulty understanding what a 50% and 100% increase in whale populations represented. Therefore, two sets of bar charts were added to illustrate what a 50% and 100% increase in gray whale populations represented relative to the current situation (which was the no action or default level if they did not pay). The first bar chart showed the 50% increase next to the current popu-

[2] As a reviewer points out such a practice may not be an acceptable way to test for the presence of embedding. See Arrow, *et al.*, 1993 on this point.

lation, while the second chart showed the 100% increase next to the current population. We believe these bar charts significantly aided respondent understanding of the scope of the programs. However, we did not perform formal one-on-one debriefing or use verbal protocols on the revised survey instrument to verify this. Lastly, demographic information including attitudes toward the marine environment, work status, wage rates and household income was asked. The overall survey was a booklet.

Data Sources

Visitor Sample Frame

The visitor sample frame is defined by both time and space. We sampled over the months of the gray whale migration along the California coast. Whales start arriving in northern California in late December and migrate south to San Diego and Baja California during January. In February whales begin to migrate back north, passing through northern California by late March. Many people taking trips to the coast during these months are out to view whales rather than for traditional summer ocean activities. For cost effectiveness in sampling, we sampled weekends and holidays at the locations described below. The choice of whether to sample on Saturday or Sunday was random, but did allow for a balance of Saturdays and Sundays over the season. The interviewers were present from 10am until dark.

The visitor survey took place at four locations along the California coast: San Diego (Point Loma National Seashore), Monterey, Half Moon Bay (south of San Francisco) and Point Reyes National Seashore (north of San Francisco). At both Point Loma and Point Reyes people view whales from the shore. At both Monterey and Half Moon Bay, people took short boat trips being run at that time specifically for gray whale viewing by commercial operators in the area.

Visitor Data Collection Procedures

The survey was administered as follows. Every nth adult (age 16 or older) visitor (where n = 5 for boat trips and n = 10 for onshore viewers) was contacted by a trained interviewer as they returned from the viewing area or boat. The interviewer asked each sampled visitor to take a survey packet home and then recorded the visitor's name and address in case follow-up mailings were required. The interviewers explained that the survey was being done by the University of California, which is viewed as a non-advocacy, research-oriented organization by most Californians. This message was reinforced by official jackets and name tags worn by the interviewers. The survey packet include a cover letter, the questionnaire and a postage-paid return envelope.

We chose this survey approach because it combined the best of personal interview and mail questionnaires. The personal contact allowed the interviewer to stress the importance of the respondent to the study and obtain a "good faith" commitment to return the survey (Mitchell and Carson, 1989:110). The interviewer was able to answer any concerns the visitor had about the survey or how they were selected. However, we did not want to conduct the interview at the site itself, for several reasons. First, weather at the coast in the winter is not always conducive to lengthy outdoor interviews and weatherproof shelter was very limited at most of these interview sites. Second, some passengers coming off the boats (particularly at Half Moon Bay) were not in physical or mental condition to answer detailed questions about their trip and thought-provoking CVM questions. By giving the visitor the survey to take home, they could devote adequate thought to answering the CVM questions at their own pace (Bailey, 1987:148).

In total, 1,402 surveys were handed out at the four intercept locations. We obtained 1,003 back for an overall response rate of 71.3%. The response rate was reasonably similar across the four locations, varying from a low of 65.2% at Point Loma in San Diego to a high of 80.3% at Half Moon Bay. On-site refusals were not a problem. For example, at Point Reyes only 10 people out of roughly 600 refused to take a survey (about 1.5%).

Household Survey Design and Data Collection Procedures

The CVM questions in the household survey were identical to those asked of visitors. However, less information was requested on whale viewing trips.

A stratified sample of California households was purchased from Survey Sampling Inc. The sample was stratified between persons living in counties adjacent to the California coast and those living in inland counties. Given that the California population distribution is heavily centered along the coast, a simple random sample would not have given us an adequate representation of non-coastal residents who are more likely to be nonusers. The foundation for the sample drawn by Survey Sampling Inc. is telephone listings. To partially overcome the omission of unlisted numbers, Survey Sampling Inc. supplements the initial list with drivers license and utility records. Since our purpose here is to compare households to whale viewers rather than generalize our WTP results to give aggregate values, we feel this is an adequate sampling scheme to obtain non-users.

A total of 2,000 names and addresses were obtained. Following Dillman's repeat contact approach, a personally-addressed cover letter (with original signature) and postage-paid return envelope accompanied the questionnaire. We also attached a dollar bill to the survey as a token of appreciation. This first mailing was followed up with a reminder postcard one week later. Then a second mailing was carried out (without the dollar bill). Finally, a random sample of nonrespondents was phoned. The purpose of the phone call was to encourage nonrespondents to complete the survey or to find out the reasons why they would not complete a survey. We also noted any English language difficulty (which to our surprise was not a problem). We asked persons refusing to accept another survey whether they had ever been whale viewing and their education level for comparison purposes. As expected, non-responding households were less likely to have been whale watching and had slightly lower education levels.

Of the 2,000 household surveys, 301 were undeliverables, 41 were deceased and 16 turned out to be businesses or government agencies. Of the eligible sample of 1642, 890 questionnaires were returned after all three contacts. The overall response rate was 54% of deliverable questionnaires, slightly above the average for other mail CVM surveys in California (Loomis, 1987). In the household sample, 56% of the respondents had *never* gone whale watching, 35% had gone at some time in the past and only 9% had gone whale watching this year. This compares to the visitor survey where 100% had gone whale watching this year.

Table 1
Visitor and Household WTP for Increases in Whale Populations

	Visitor Sample		Household Sample	
	Mean	Std Error	Mean	Std Error
50% Increase in Population	$25.00	1.16	$16.18	1.07
100% Increase in Population	$29.73	1.39	$18.14	1.16
n=	672		519	

Thus the household sample is dominated by pure non-users or those with only past whale viewing activity.

Survey Results

WTP Estimates

Before calculation of WTP two steps are typically performed in most CVM studies. First, individuals stating a zero WTP and indicating it was a "protest" response (*e.g.*, it is unfair to ask people to pay for greater protection of gray whales, *etc.*) are dropped. This was only about 10% of the visitor sample and about 20% of the household sample. We offer two reasons why visitors were more willing to accept the trust fund scenario. First, visitors had experience trading money for whale viewing opportunities, so thinking about whales in dollars was not alien to them. Second, nearly half of the visitors were members of environmental organizations (more than double the rate in the household sample). Hence visitors were most used to seeing real appeals for money to protect particular species or natural areas.

Once protests are accounted for, the next step, which is especially important with an open-ended response format, is to check for outliers. There was only one bid in the visitor sample that would be considered an outlier: an annual WTP of $500 per year, where the next highest amount was $300.[3] In the household survey the highest WTP was $180. Both of the remaining upper dollar amounts are well within the household's payment capacity, as the sample average *household* income was $50,000.[4]

Table 1 presents the mean WTP and standard error of the mean for the visitor and household samples. Each sample was asked their WTP for both a 50% and a 100% increase in whale populations. As would be expected, the average WTP values for general households are lower than for current visitors. In both cases the WTP estimates are measured fairly precisely, with the coefficients of variation for mean WTP (standard error/mean) being 6 or 7% of the means in the household survey and about 4% in the visitor survey. These standard errors are generally

[3] The $500 WTP response was dropped from the empirical analysis that follows. However, none of the results of the hypothesis tests reported below for visitors were changed by inclusion of this $500 WTP amount.

[4] While by U.S. standards this average household income may seem high, it is fairly close to the average California household income of $48,100 in 1991 (based on updating the 1990 Census estimate for California).

smaller than what others (Walsh *et al.*, 1984; Desvousges, *et al.*, 1992; Loomis, *et al.*, 1993) have found in other open-ended CVM surveys. Table 2 presents the WTP distribution of visitors and households for the two increases in whale populations.

Table 3 presents the results of the hypothesis tests concerning mean WTP for increments n = 50% and 2n = 100% above current gray whale populations. With respect to whether the incremental valuations are statistically different from zero, we can reject the hypothesis of equality with zero for both the 50% and 100% changes in gray whale populations for both visitors and general households. The t statistics are all significant at well below the 1% level.

Since each individual surveyed provided WTP for the 50% and 100% increment, the second hypothesis test involves a paired t-test. Thus we test whether their *difference* in expressed WTP for the 50% and 100% increments is different from zero. That is, by forming the WTP difference $D = WTP_{100} - WTP_{50}$ we test whether $D \leq 0$. The t statistics reported in Table 3 are significant at the 1% level, indicating the incremental valuation from a 50% to a 100% increase is statistically different from zero.

While the incremental valuations are statistically different from zero, the third consistency test requires that the second increment (D) to be of less value than the first increment (WTP_{50}). Rearranging and collecting like terms allows us to use a paired t-test of $2(WTP_{50}) - WTP_{100} > 0$. The test statistics reported in Table 3 are statistically significant, indicating the marginal value of the public good in our study is decreasing. That is, we reject the null hypothesis that the valuation for the second 50% increase is greater than or equal to the value of the first 50% increase for both the visitor and household samples.

Thus results from our case study suggest that for substantial increases in wildlife populations that are above current (and minimum viable) levels, there appears to be a statistically significant positive but diminishing marginal valuation for individuals. This pattern of results was found for both current gray whale

Table 2
Distribution of WTP of Visitors and Households

	Number of Visitors		Number of Households	
WTP	50% change	100% change	50% change	100% change
\$0	107	121	195	216
\$10	201	145	145	115
\$20	127	121	79	63
\$30	119	117	60	65
\$40	15	38	15	24
\$50	71	74	29	34
\$60	6	10	2	2
\$80	10	10	3	5
\$100	39	48	18	25
\$150	7	15	3	4
\$200	1	2	1	1
\$250	0	1	0	0
\$300	1	2	0	0

Table 3
Hypothesis Tests Concerning WTP for Increases in Whale Populations

Null Hypothesis	Visitor Sample		Household Sample	
50% Increase				
$WTP_{50} = 0$	Reject	t = 21.45	Reject	t = 15.15
100% Increase				
$WTP_{100} = 0$	Reject	t = 21.31	Reject	t = 15.65
$WTP_{50} = WTP_{100}$	Reject	t = 9.14	Reject	t = 4.60
WTP 1st 50% gain = WTP 2nd 50% gain	Reject	t = 17.78	Reject	t = 12.47

viewers and for a sample of households, the majority of which were non-users and is consistent with expectations from basic principles of consumer demand theory.

There are of course several qualifiers to our results. First, we tested our hypotheses by comparing responses by the same individuals to the two increases in whale populations. Had we asked one sample their WTP for 50% increase and another sample their WTP for a 100% increase, the results of the hypothesis tests might have been different. In addition, our WTP values for the 100% increase in population may be conditioned on the responses given to the 50% increase. Since only one survey treatment for each population was performed, we do not have the ability to test for the significance of order effects. These are important methodological issues that should be evaluated in future research.

Comparison of Results to Previous Studies

Samples and Hoyller (1990) used in-person interviews of Hawaii residents and a dichotomous choice format. A one-time payment to prevent a loss of Humpback whales in Hawaii was elicited from two relatively small (n = 65 to 72) samples. They obtained values for general households ranging from $125 to $142, depending on the survey treatment. Annualized at 10% and updated for inflation, this value is somewhat higher than our annual estimates. In part this may be due to being asked study about their WTP to avoid a complete loss of humpback whales (Samples and Hoyller, 1990:183). Hagemann's mail survey of California households has some similarity to ours in terms of sample frame and evaluation of large changes in whale populations. Using a payment card, Hagemann (1985:72) obtained an annual value of $27 for gray whales in 1984 dollars or $36 in 1992 dollars to avoid a reduction in whales from their 1984 population of 16,000 to 1,300. However, there are some differences that may partially explain why she obtained higher values for the general population that we did. First, she asked WTP to avoid a 90% reduction in gray whale populations, rather than an increase in populations as we did. She also had a relatively low response rate (21%) which may imply only very interested respondents with correspondingly higher values. Compared to both studies, our value estimates for general households are relatively low, but similar in magnitude, and the differences may well be ascribed to the relative magnitudes and direction of changes in whale populations between the three studies.

Our annual values for doubling whale populations are lower than what Olsen, *et al.* found for doubling anadromous fish runs in the Columbia River basin. His non-users had values averaging $27 while users had values of $74. Of course his users, were consumptive users (*e.g.*, anglers) while our users where viewers. This comparison suggests that CVM studies of different marine resources do obtain values that seem to vary with the intensity of interest the public and users have in the resource.

Conclusions

At the level of the individual respondent, this research found that non-marginal increases in whale populations above current and minimum viable levels resulted in a statistically significant positive and diminishing marginal valuation. This pattern was found in samples of both current users (whale viewers during winter 1991-92) and of general households in California, which consists mainly of nonusers. The results suggest that carefully-performed CVM studies can obtain results consistent with principles of consumer demand theory for reasonably large changes in the quantity of a public good. Thus this study provides evidence that CVM can provide useful information on societal benefits from maintaining and increasing wildlife populations. The robustness of our conclusions should be tested in future research for order effects and use of independent samples for each increment, however.

In addition, the results suggest that citizens do derive a benefit from having a larger stock of whales, but from the standpoint of marine policy the costs (especially opportunity costs) of increasing whale populations would need to be considered before concluding that expanding populations beyond current levels is warranted on the grounds of economic efficiency.

References

Arrow, Kenneth, Robert Solow, Paul Portney, Edward Leamer, Roy Radner, and Howard Schuman. 1993. Report of the NOAA Panel on Contingent Valuation. Federal Register 58(10):4601–4614.

Augustyniak, Christine, Carol Jones, and Normal Meade. 1992. Nobel Laureat Panel on Contingent Valuation. Damage Assessment Center, National Oceanic Atmospheric Administration, Rockville, MD.

Bailey, Kenneth. 1987. Methods of Social Research. Free Press, New York, NY.

Bradford, David. 1970. Benefit-Cost Analysis and Demand Curves for Public Goods. Kyklos 23:775–791.

Brookshire, David, William Schulze, Mark Thayer, and Ralph d'Arge. 1982. Valuing Public Goods: A Comparison of Survey and Hedonic Approaches. American Economic Review 72:165–177.

Brookshire, David, Alan Randall, and John Stoll. 1980. Valuing Increments and Decrements in Natural Resource Service Flows. American Journal of Agricultural Economics 62:478–488.

Cooper, Joseph, and John Loomis. 1991. Economic Value of Wildlife Resources in the San Joaquin Valley: Hunting and Viewing Values. in Economics and Management of Water and Drainage, edited by A. Dinar and D. Zilberman. Kluwer Academic Publishers, Boston, MA.

Cummings, Ronald, David Brookshire, and William Schulze. 1986. Valuing Environmental

Goods: An Assessment of the Contingent Valuation Method. Rowman & Allenheld. Totowa, NJ.

Daubert, John, and Robert Young. 1981. Recreational Demands for Maintaining Instream Flows: A Contingent Valuation Approach, American Journal of Agricultural Economics 63(4):666–676.

Desvousges, William, F. Reed Johnson, Richard Dunford, Kevin Boyle, Sara Hudson, and K. Nicole Wilson. 1992. Measuring Nonuse Damages Using Contingent Valuation. Research Triangle Monograph 92-1. Research Triangle Park, NC.

Hagemann, Rhonda. 1985. Valuing Marine Mammal Populations: Benefit Valuations in a Multi-Species Ecosystem. Administrative Report LJ85-22. Southwest Fisheries Center, National Marine Fisheries Service, La Jolla, CA.

Hoehn, John. 1990. Valuing Multidimensional Impacts of Environmental Policy: Theory and Methods. American Journal of Agricultural Economics 73(2):544–551.

Hoehn, John, and John Loomis. 1993. Substitution Effects in the Valuation of Multiple Environmental Programs, Journal of Environmental Economics and Management 25(1):56–75.

Kealy, Mary Jo, John Dovidio, and Mark Rockel. 1988. Accuracy in Valuation is a Matter of Degree. Land Economics 64:158–170.

Kopp, Ray. 1992. Why Existence Values Should Be Used in Benefit-Cost Analysis. Journal of Policy Analysis and Management 11(1):123–130.

Larson, Douglas. 1993. On Measuring Existence Value. Land Economics 69(4):377–388.

Loomis, John. 1987. Expanding Contingent Value Sample Estimates to Aggregate Benefit Estimates. Land Economics 63(4):396–402.

Loomis, John. 1989. Test-Retest Reliability of Contingent Valuation: A Comparison of General Population and Visitor Responses. American Journal of Agricultural Economics 71(1):76–84.

Loomis, John. 1990. Comparative Reliability of the Dichotomous Choice and Open-Ended Contingent Valuation Techniques. Journal of Environmental Economics and Management 18:78–85.

Loomis, John, and Joseph Cooper. 1990. Comparison of Environmental Quality-Induced Demand Shifts Using Time-Series and Cross-Section Data. Western Journal of Agricultural Economics 15(1):83–90.

Loomis, John, Michael Lockwood, and Terry DeLacy. 1993. Some Empirical Evidence on Embedding Effects in Contingent Valuation of Forest Protection. Journal of Environmental Economics and Management 24:45–55.

Olsen, Darryll, Jack Richards, and R. Douglas Scott. 1991. Existence and Sport Values for Doubling the Size of Columbia River Basin Salmon and Steelhead Runs. Rivers 2(1): 44–56.

Randall, Alan, and John Stoll. 1983. Existence Value in a Total Valuation Framework. in R. Rowe and L. Chestnut, eds. Managing Air quality and Science Resources at National Parks and Wilderness Areas. Westview Press, Boulder, CO.

Reiling, Stephen, Kevin Boyle, Marcia Phillips, and Mark Anderson. 1990. Temporal Reliability of Contingent Valuation. Land Economics 66:128–134.

Samples, Karl, and James Hollyer. 1990. Contingent Valuation of Wildlife Resources in the Presence of Substitutes and Complements. in R. Johnson and G. Johnson, editors. Economic Valuation of Natural Resources. Westview Press, Boulder, CO.

Walsh, Richard, John Loomis, and Richard Gillman. 1984. Valuing Option, Existence and Bequest Demands for Wilderness. Land Economics 60(1):14–29.

Welsh, Michael. 1986. Exploring the Accuracy of the Contingent Valuation Method: Comparisons with Simulated Markets. Department of Agricultural Economics, University of Wisconsin, Madison.

E Grasslands and Agriculture

[20]

Biodiversity and Conservation 6, 59–74 (1997)

Plain tales from the grasslands: extraction, value and utilization of biomass in Royal Bardia National Park, Nepal

KATRINA BROWN
School of Development Studies and CSERGE, University of East Anglia, Norwich NR4 7TJ, UK

Received 11 July 1995; revised and accepted 20 November 1995

Royal Bardia National Park in Southern Nepal contains a number of internationally endangered wildlife species, and a mix of habitats, including areas of grassland known as phantas. The conservation and current use of these areas is briefly described, including the extraction of grass from inside the National Park by local people living in adjacent villages. The products collected fulfil a range of human needs, both practical and ceremonial. The study attempts to quantify the annual offtake in both material and monetary terms, and comments on the sustainability of the practice. Alternative sources of roofing materials are apparently outside the scope of poorer households. The use of these materials has particular significance given the initiation of a Parks and People Project to implement a Buffer Zone around the National Park and effectively to link conservation and development needs.

Keywords: conservation policy; economic value; grasslands; protected areas; sustainable utilization.

Introduction

This paper examines the use and management of natural resources in Royal Bardia National Park (RBNP) which lies in the southern lowlands of Nepal, the Terai. This protected area, which consists of *sal* (*Shorea robusta*) forest and areas of grassland, harbours a number of internationally important endangered wildlife species, including the Bengal tiger (*Panthera tigris*), Asian one-horned rhinoceros (*Rhinoceros unicornis*), Hispid hare (*Caprolagus hispidus*), and Bengal florican (*Houbaropis bengalensis*). Under the current management regime local people are permitted to enter the park for a period of 10 days during the dry season to harvest a number of grass species from which they make different products, chiefly to meet subsistence needs. However, concern has been expressed over the extent of this utilization and its impacts on the biological diversity of the grassland habitats. These concerns include the possible degradation of the grassland and changes in the flora; disturbance of wildlife; increasing numbers of people entering the park for the annual cutting; and the illegal poaching of other products, particularly wood. These put the sustainability of this practice into question. However, a contrary view is that habitat diversity is maintained, in part at least, by precisely this human intervention and that critical habitats such as the grasslands have persisted only as a result of the annual cutting. Furthermore, the annual cutting contributes significantly to the livelihoods of local people.

The paper describes and quantifies the extraction of the various grasses and comments on its desirability within the context of Nepal's evolving Protected Areas policy. The next section describes that context, and then historical and ecological perspectives and current management policies of the study site are outlined. Following sections discuss the extraction and use of the grasses, and the implications for conservation of the grassland habitats within the RBNP. Alternative sources of roofing materials, their availability and cost, are discussed, and development opportunities for implementation of income generating projects in the Buffer Zone surrounding the National Park are suggested.

Conservation Policy in Nepal

The organization of the state, and the ownership and control of land determined conservation policy in Nepal in the past. Economic and cultural factors were as important in this respect as biogeography. Prior to the 1950s national security played a prominent role in Nepal's conservation strategy according to Basnet (1992). Forests, for example, were used as a means of political protection – in the 18th century the Terai forests formed an effective barrier to Indian invasion, so the state had a vested interest in their preservation. These areas were also used for hunting by the Royal family, and indeed, the designation of the protected areas in the Terai was originally to provide hunting sanctuaries for use by the Nepalese Royal family and visiting dignitaries. Essentially, Nepal was a subsistence agricultural economy, with little contact with the outside world before 1950. Population growth rates were low, and relatively little pressure was put on forests for timber production, neither was there large-scale conversion of natural habitats for agricultural production. This situation has changed in recent decades, particularly in terms of demand for land, internal migration to lowlands, and technological changes in the use of resources.

Legislation governing National Parks and other protected areas was introduced in 1973 by the National Parks and Wildlife Conservation Act. This Act created the Department of National Parks and Wildlife Conservation (DNPWC), which was formerly an office of the Department of Forestry. Under the Act, the DNPWC has power to create and manage four types of protected areas: National Parks, Wildlife Reserves, Strict Nature Reserves, and Hunting Reserves. A later amendment gave the national government the power to recognise a fifth type of protected area in response to a newer proposal, called Conservation Areas. At present, approximately 10% of Nepal's land area is under some form of protection, and protected areas remain a key mechanism for conservation policy. The five Terai protected areas were among the earliest to be designated soon after the enactment of the 1973 legislation. However, even the Royal Chitwan National Park, the first in Nepal, has been designated for little over two decades. Resettlement is therefore still within living memory for the local population, and indeed, one village remains within the area designated and negotiations continue to relocate inhabitants. In Royal Bardia National Park, extensions to the park in the last decade involved resettlement of numerous villages. In many of the protected areas the sites of old villages and cultivated areas are recognisable within the park by their vegetation (having a higher concentration of *Imperata*). In fact, this characteristic is part of the management conundrum, whereby human intervention may be necessary to maintain some types of grassland. The dilemma appears to be how that intervention is managed, by whom, and for whose benefit.

Nepal is now considered a leader among developing nations with regard to conservation programmes and legislation (Heinen and Kattel, 1992); it was among the first Asian nations to develop national conservation legislation, sign CITES, and develop a national conservation strategy. The conservation policy has also followed the international tide and shifted away from a single species focus (identifying and legislating protection of endangered species, and re-location programmes such as the tiger re-location programme of 1970s), towards a more comprehensive approach to biodiversity protection, for example, in the designation of the Annapurna Conservation Area in 1988 (see Heinen and Yonzon, 1994). In a recent statement of policy, the Nepal Environmental Policy and Action Plan (HMG, 1993), HMG clearly links conservation and development, and the integration of environment and development through appropriate institutions, adequate legislation and economic incentives, and the balancing of 'development efforts and environmental conservation for the sustainable fulfillment of basic needs of the people' (p. xi) are among its central aims. This commitment to policy integration is manifest in the development of a 'Parks and People' initiative to implement Buffer Zones around the five protected areas in the Terai. The main objective of this project is to enhance the capacity of local communities and the DNPWC to jointly manage the protected areas and their buffer zones, and to improve socio-economic conditions of people living in Village Development Committees adjoining the protected areas (HMG/UNDP, 1994). [Village Development Committees (VDCs) are the form of local administration which consist of nine wards each, a ward being a village. There are 16 VDCs in the Buffer Zone, ranging in size from under 3000 to nearly 15 000 people. Implications of this project are discussed later in this paper.]

The Royal Bardia National Park: historical, ecological and human perspectives

The RBNP covers an area of 968 km^2 in Bardia and Benke Districts in the Southern Terai in the mid-Western region of Nepal (see Fig. 1). The Park extends East of the Geruwa branch of the Karnali River to the West; its Northern boundary runs along the crest of the Churiya Ridge, and its Southern boundary is demarcated by a fringe of cultivated land and government forest (Upreti, 1994). The park can be roughly divided into the Churiya Ridge which dominates the Northern portion, the Bhabar area which lies at the foothills of the Churiyas, the Terai flatlands and the alluvial floodplain to the South and West, and the Babai Valley to the East. The South-Western floodplains and flatlands are most easily accessible during the grass-cutting period, although the construction of a tarmac road through the middle of the RBNP has enabled access to the interior, as shown in Fig. 1.

Seven major vegetation types have been identified inside the RBNP (Pokharel, 1993): four types of forest and three different grasslands habitats. The *sal* (*Shorea robusta*) forest covers approximately 70% of the park area, dominating the alluvial floodplain, and is also found on parts of the South-facing slopes of the Churiya. Khair-sissoo forests, composed of *Dalbergia sissoo* and *Acacia catechu* is restricted to major water courses and flood plain islands. Moist riverine forest is characterized by evergreen species such as *Ficus racemosa*, *Syzigium cuminii* and *Mallotus phillipinensis*.

Three types of grassland have been identified: wooded grassland, open areas of grassland known as *phanta*, and floodplain grassland. The first two types consist of tussock-forming perennials such as *Imperata cylindrica*, *Saccharum spontaneum*, *Erianthus ravennae* and *Vetivera zizaniodes*. Wooded grassland is savannah type, often dotted with *simal (Bombax ceiba)*, or silk cotton trees. The *phantas* consist of open areas; the most significant of these are shown in Fig. 1. These habitats are thought to represent previously disturbed and cultivated sites. The third grassland habitat is found in the floodplain and along the banks of the Geruwa, Babai and Aurai rivers in areas commonly inundated during the monsoon. These areas are characterized by tall grasses, including *Saccharum*

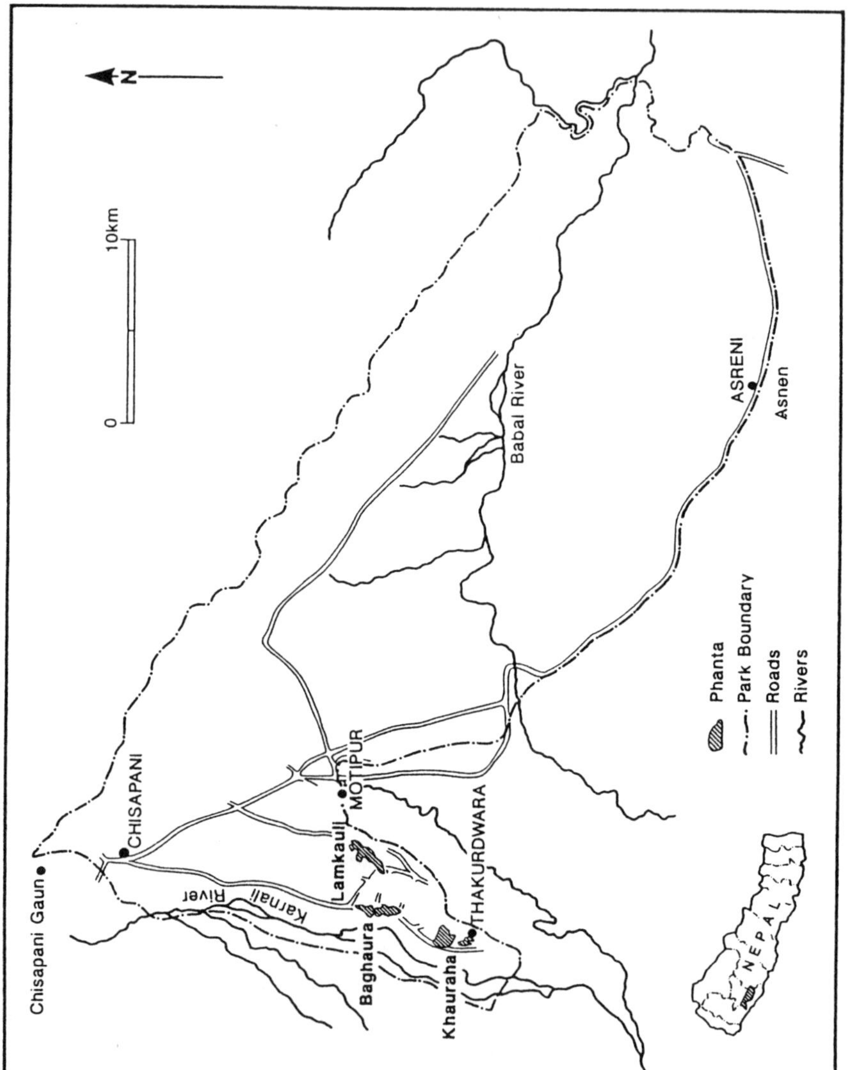

Figure 1. Royal Bardia National Park.

spontaneum, Saccharum bengalense and *Phragmites karka*. More detailed descriptions of the vegetation of the park can be found in Dinerstein (1979) and Pokharel (1993).

In terms of the human population, there are a range of different ethnic and caste groups inhabiting the Terai, although this has been changing over time. The Tharu are perhaps the oldest and original inhabitants of the Terai; according to Cox (1990), Tharu people have lived in the region for more than 600 years. Ghimire (1992) states that the Tharu migrated from India during the 15th and 16th centuries; there is speculation that the Tharu originally came from Rajisthan; indeed, their dress and customs are similar to tribal people in that region. There are several clans of Tharu in the Terai, including Dangora, Kathariya, Dhakar and Rana Tharu. Much of the available literature alludes to these groups as living in harmony with nature, consistent with an idealized view of indigenous people as 'ecosystem people' (Dasmann, 1991; see also Gadgil, 1993). For example, Upreti (1994) claims that Tharu people are part of the jungle ecosystem and that their needs and habits do not have significant negative impacts on the natural environment. Tharu are traditionally dependent on the forest for timber, thatch grass and fodder, and also collect fibres, leaves, canes and reeds, mushrooms, honey, vegetables, medicinal herbs and fruits. They snare deer, pigs, birds, and fish and aquatic snails are important additions to their diets. While the Tharu utilize a range of wildlife and other natural resources in the region, the long-term impact of these practices is not known.

It seems likely that scattered groups of Tharu were the most numerous inhabitants of the area up to the 1960s. Although there had been migration and land colonization during the Rana regime, according to Ghimire (1992:50) 'due to the existence of malaria, endemic disease and other dangers (wild animals and bandits) in the plains, the response from both landlords and peasants from the Hills to land colonisation measures remained fairly limited. The peasants showed most interest in clearing and cultivating lands only in the close adjacent plains so they could get to higher ground to sleep, and the landlords came only to the plains during the winter when malaria was less prevalent and when harvests and rents were collected.' Agricultural labourers were recruited from India in the second half of the 19th century, and there was increased immigration from India during the 1930s and 1940s, from the densely populated plains across the border. A malaria-eradication programme, and land shortages in the uplands of Nepal, led to rapid expansion of the population of the Terai from the 1960s onwards, principally as a result of migration from the Hills.

Human pressures on the natural resources of the Terai have thus changed considerably over time, primarily as a result of the migration of the last two decades. This movement continues, and the areas bordering the RBNP have experienced high rates of population increase in the past 20 years. The population in Bardiya District rose from 199 000 in 1981, to 290 000 in 1991, an increase of nearly 46% (Central Bureau of Statistics, 1993). Over the same period the population of Nepal rose by 23%. The Village Development Committees adjoining the Park, i.e. the area defined by the HMG/UNDP Parks and People Project as the Buffer Zone, consist of a total of 137 600 people in 18 600 households.

The region that is now RBNP has undergone a number of shifts in property rights and changes in conservation status. Prior to the nationalization of the forests in 1956, most of the forests in Bardia were *birta* (private) forests (Upreti, 1994). As personal property of the Rana rulers, these areas were under strict protection. In 1969 the area was declared a Royal Hunting Reserve, although local people had free access to the forest and to graze their livestock. An area of 386 km^2 was officially gazetted as the Royal Karnali Wildlife

Reserve in 1976 and renamed the Royal Bardia Wildlife Reserve in 1982. In 1984 it was extended to the east to include the Babai Valley. It was upgraded to national park status in 1988.

Little documented evidence exists describing the human use of these areas, although we know that the area which is now the RBNP was inhabited until 20 years ago. Before the area was declared a reserve in 1976, two villages were inside the park area, one at Chisipani and another at Auraini. Before the establishment of the hunting reserve, the villages of Bagaura Phanta and Lamkauli Phanta were also relocated outside the reserve boundary (Upreti, 1994). Recent expansion of the Park resulted in more widespread resettlement, and Resources Nepal (1994) reports that: 'Between 1979–1984 during the extension of the park a total of 1572 families, comprising over 9500 people were moved out of the Babai Valley. Some 220 families did not want to leave their homesteads, however, they were forcibly removed. The total population from the 20 villages had left 1200 ha of farmland, 1096 permanent houses and 1793 semi-permanent structures behind'.

In terms of management of the grassland habitats within the area, relatively little is known of the human ecology. Pokharel (1993) describes the history of the *phantas* as being quite obscure. Dinerstein (1979) reported that Baghaura and Lamkauli *phantas* were under cultivation prior to 1975. Pokharel states that a number of people used to cultivate the *phantas* and are still residing outside the park, though very close to it. One of these informants is recorded by Pokharel (1993: 13):

> 'According to Gopi, in 1965 when he was one year old, his father and friends came down from Surkhet in search of cultivable land. Obtaining land at that time was no problem. They found Baghaura a very suitable open place to live and cultivate. Later they settled there with 15 different families from Surkhet. They were the first inhabitants of Baghaura. Cultivation of Lamkauli could have taken place much later than Baghaura, according to Gopi. He did not know if Khauraha was cultivated before. Baghaura could have been a small open and flat grassland in 1965, which was expanded according to the need of new inhabitants by peripheral deforestation in the jungle.'

Grass cutting was a regular activity in Bardia, but it is difficult to trace when this began. In RBNP the *phantas* were cultivated and grazed between 1965 and 1975. Records from the RBNP show authorised grass cutting since 1983, 6 years before the National Park was created. Livestock grazing in the phantas and other areas prior to 1975 was apparently very common. After 1976 the area was fenced and livestock grazing prohibited. Fire has also been used as a management tool, and in the past uncontrolled burning in the park was described. Fire is generally set at the end of the dry season when grass cutting is completed; in RBNP this is mainly carried out by park authorities. To what extent maintenance of the sward requires cutting and burning – and therefore the necessity of human intervention to maintain the patchwork of habitats – is relatively unknown. It can be hypothesized that grazing by domestic and wild animals, but especially the destructive browsing of elephants, was instrumental in maintaining grassland in the past when large herds were prevalent. The extent to which annual cutting and human-set fires recreate or mimic these natural processes can only be postulated, and whether the resulting grassland communities resemble those present in the past is uncertain (see Peet, 1994).

The grasses and their uses

A number of different grass species are utilized by people from the RBNP. The three major products extracted are thatch grass, building materials and rope. Thatch grass, known as *khar*, consists of a mixture of the shorter grass species, predominantly *Imperata cylindrica* and, to a lesser extent, the shortest of the tall grasses, *Saccharum spontaneum*. Canes and reeds are extracted for house building and a number of other uses. They are known as *kharai* and include taller grasses, *Saccharum bengalense, Narenga* and *Phragmites* spp. The third product is rope, made from *Eulaliopsis binata*, known as *buncas* or *babiyo*. Table 1 shows the most important grass species harvested from RBNP, and their main uses. *Khar* is collected primarily from the wooded savanna areas and the *phantas, kharai* from the floodplains, and *buncas* from the foothills. Much of the analysis presented in this paper concerns the collection and use of thatch or *khar*.

The list of grasses and products in Table 1 is by no means exhaustive and does not detail all of the different species of grass and reeds used by local people, or all of the uses. There are certain species which have highly specialized and localized uses. For example, one type of grass, known locally as *bus heri* is used for constructing fish traps because of its properties of resisting rot even when repeatedly saturated in water. These different resources are not examined here, but rather this analysis is focused on the products which are used in large amounts by a majority of people.

A number of complications arise in obtaining local names for the grasses. First, a particular species may be given a different name, according to its use or which part of the plant is used, or its method of harvest or treatment. In the table above, for example, *Narenga porphorycoma* may be known by a generic term for cane or reed, *kharai* or *khadai*. The same species has a number of alternative names, according to its different uses and whether it is burned before it is cut, and whether the whole, or part of the plant is used. Second, complications arise because of different local languages and dialects. There are a number of different Tharu clans in the area, who give different names to the grasses. There are marked differences between the terms used in Chitwan and those used around Bardia.

Table 1. Grasses collected from Royal Bardia National Park

Grass species	Local name	Uses and names
Narenga porphorycoma	*khadai* or *kharai*	Cane–if burned, used for building etc. *sakhata* Cane, not burned, whole used for grain silo *phank* Cane, top part only *silicili*
Saccharum bengalense	*narkat*	Cane for walls and ceiling *chatati* Cane for baskets *kenari*
Phragmites kharka		Cane for lamp stand for diwali festival
Tifa augustifolia/Tifa elephantina	*pat* or *pater*	Mats, fans, mattresses, howdahs
Imperata cylindrica	khar	Thatch, brooms; flower head used for ceremonial lamp *kuwar bati*
Eulaliopsis binata	*buncas, sabai, babiyo*	Rope used for beds, chairs, bullock carts, tying thatch; paper
Saccharum spontaneum	*khans*	Reeds and canes for walls, bed, thatch, fodder

Table 1 gives the terms most commonly reported and used in Bardia; some are Tharu names, others Nepali. A number of different terms are used locally for the same grass, as in the case of *Eulaliopsis binata*. This may reflect the uses and cultural importance of the species. Martin (1995) observes that species of great cultural significance are *over-differentiated*, or split into distinct categories by local people, whereas species that are less important are usually *under-differentiated*, or lumped together in a single generic. Use values will also be reflected by the number of different uses repeatedly cited by respondents (see Phillips *et al.*, 1994, for a discussion of quantitative ethnobotanical techniques) and this would confirm the importance of the three products discussed above.

These findings give some indication of the number of different uses for the grasses, and how important these resources are for local households. This is especially true in the case of house building, where the grassland provides raw material for walling (both small and large cane) and roofing (cane and thatch). However, the grass products also have important cultural functions. For example, they are used in ceremonial lamps, and the flower head of *Imperata cylindrica* is used as a wick in lamps for temples or shrines. The grass cutting is an extremely important event in the annual calendar; preparations are made by the community in advance, and Tharu people celebrate the Maghi festival at the end of the grass-cutting. *Saccharum spontaneum* has an important role in Hindu worship and plays a role in a number of festivals (Majupuria and Joshi, 1989). Most of the discussion in this paper is confined to the collection and use of thatch grasses from the *phantas*.

The extent of harvesting in RBNP

Since 1983 permits have been sold which enable local villagers to enter the protected areas in the Terai and collect grass and cane. This policy was initiated as a means of granting people partial recompense for the loss of access to these and other products as a result of protected areas implementation, and to improve relations between the local community and the Parks Authorities. In the past a nominal fee of 1 NR (75 NR = £1, in 1994) was imposed, which acted as a means of controlling the total numbers of grass collectors, and also enabled the DNPWC to keep a tally of the numbers entering the protected areas over the 15-day period, and to cover administration costs. However, as the numbers of people entering the protected areas increased over the years, the fee has been increased, and the cutting period has been shortened to 10 days (in 1994/5 season these spanned 2–11 January 1995). In RBNP in 1994/5 people reported paying 12 NR for a permit (equivalent to over 25% of 1 day's wage of 40 NR), which would enable one person to enter the National Park each day for the 10 days. Each person can carry as much grass or cane as possible, but no carts or other means of transportation are allowed inside the protected areas. Around the entrance points, bullock carts are loaded to transport the biomass back to the village, but only people are allowed in the Park.

Figure 2 shows the number of permits sold annually for grass cutting in RBNP since 1983. Permits are sold in advance of the cutting days, and are checked at each of 44 different entrance points around the Park perimeter by DNPWC and army officials during the grass cutting. Figure 2 shows a steady increase in the number of permits sold, representing more than a doubling in the years 1983–1993 from 21 000 to 45 500 permits (i.e., greater than the increase in population), with 1994 showing the first drop in numbers, a decrease of some 4% of the 1993 figure. Over 43 000 permits were sold in 1994.

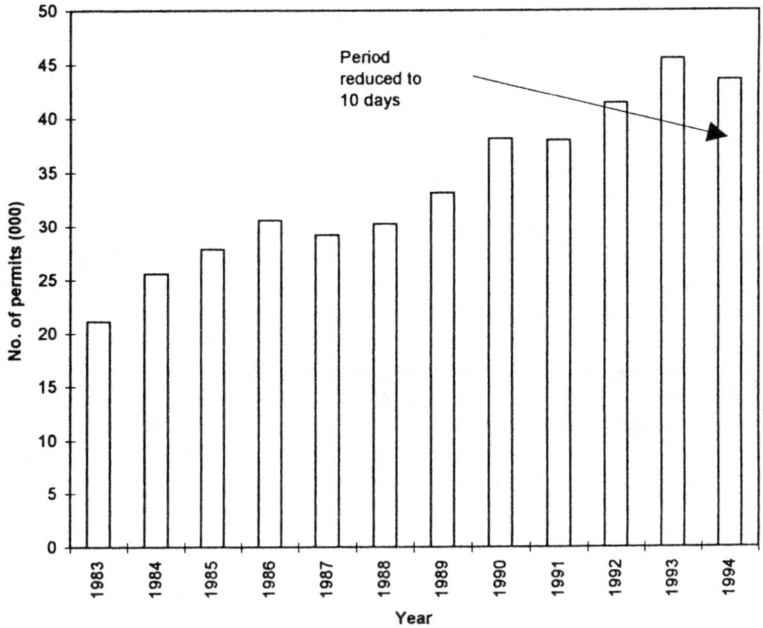

Figure 2. Grass-cutting permits sold in Royal Bardia National Park. Source: Royal Bardia National Park.

A series of household interviews were conducted in sample villages in VDCs adjacent to the RBNP and inside the proposed Buffer Zone. (Structured interviews were undertaken in 26 households in six villages by the researcher, an assistant and two Nepali collaborators. In addition, interviews and discussions were held with a range of key informants, including VDC chairmen, local teachers, DNPWC personnel, and information obtained from group interviews carried out in the sampled villages. Data on numbers of permits and bundles of grass collected were obtained by direct observation, supplemented by records of the DNPWC, during the 10-day cutting period.) Interviews were undertaken in the weeks prior to the cutting period, and direct observations and enumeration of collection of grasses during the 10-day cutting period carried out. All of the households interviewed, with one exception, were sending adult members into the RBNP to collect grass. Many households committed all their adult labour to this task, and although both women and men participated, some households send just adult males. From the survey, the average number of permits per household was 3.4, and the household size ranged from 5 to 40 (n + 26). The total numbers of permits sold indicate that the average for households adjacent to the Park (in the Buffer Zone) is 2.34 per household. Total numbers indicate that 32% of the total population of the defined Buffer Zone (i.e. the 16 VDCs adjacent to the RBNP) participate in the grass cutting. This is confirmed by the findings of the household survey: much of the available adult labour is deployed to the task of grass cutting and collection.

How does the demand for the grass by local communities compare with the supply of biomass inside the RBNP? There is an assumption by conservationists that over-exploitation, caused by increased numbers of cutters entering the park, is leading to changes in the composition of the grassland, and there is some anecdotal evidence of a decline in the quality of the thatch grass, perhaps caused by a change in the species mix. (The change in quality may be a result of changes in the species mix, in the growth habit or quality of the plants themselves, or because cutters are utilizing more marginal areas of grassland.) However, cutting and/or burning is required to discourage the colonization by taller grass species, shrubs and small trees, and possibly therefore slow the succession to *sal* forest. On-going ecological experiments aim to examine the impact of different cutting and burning regimes on the grassland diversity. There are likely to be other ecological impacts of the regular removal of such large quantities of biomass, which will depend in part on local environmental conditions, including, for example, the removal of nutrients.

Using the observations and measurements from the survey, and data from Pokharel's study (1993) on the area of the *phantas*, the total amount of thatch grass available from the southern *phantas* (Upper and Lower Baghaura, Upper and Lower Khauraha, and Lamkauli) is calculated, and this is shown in Table 2. This gives the estimated production of biomass from the five southern *phantas*, using measurements from the 1994 survey. (Bundles of grass harvested from Lower Baghaura *phanta* were sampled and weighed just prior to the cutting period. The mean weight of a bundle was 2.52 kg. The same proportion of yield/productivity, i.e. the same rate of offtake as Pokerel's study (1993) were assumed, and confirmed by field observations.) These calculations indicate that a total of 841 000 bundles of *khar* were available from the five southern *phantas* in RBNP in January 1995. Observations of the grass collection, based on a sample of cutters entering and leaving the Park at Thakadwara Gate in the south west, indicate that the possible total harvest of *khar* may be in the region of 1 050 000 bundles, or approximately 2650 tonnes. This figure is larger than the estimates for the *phantas* alone, as small patches of grassland occur throughout the RBNP and will be utilized by local people.

This estimate is likely to be an overestimate for a number of reasons. First, the offtake rate may well be lower than estimated, as the observations may have been biased to the higher quality parts of the *phanta*. Second, the observations from Thakadwara gate were used, and then extrapolated to give a figure for the whole Park according to the number of

Table 2. Estimates of biomass production in five southern phantas in RBNP

	Area (ha)	Production (t/ha)[a]	Yield (t/ha)	Number of bundles available
U Baghaura	41.6	13.98	8.5	140 673
L Baghaura	58.7	8.06	4.9	114 442
U Khauraha	94.7	11	6.7	251 973
L Khauraha	34.5	8.57	5.2	71 517
Lamkauli	111.3	9.74	5.9	262 220
				840 825 Total bundles

[a]Production is net primary production.

permits sold each gate. However, Thakadwara is situated close to the *phantas* so a higher proportion of *khar* may be removed through this gate. Third, assumptions have been made about the amount of time spent collecting *khar* and other products, and (as indicated by observations) that *khar* collection was concentrated in the first 3 days of harvest. It has therefore been assumed that each of the 43 580 people with permits will have collected *khar* for 3 days out 10 days. Such an assumption is confirmed by household interviews, although again this may be biased due to the geographical proximity of the surveyed villages to the southern *phantas*.

How do these estimates of harvest relate to the predicted demand? According to the village surveys, many households have at least one thatched roof, and, in response to household interviews, a typical roof might require 250 bundles of *khar* for thatching every 2–3 years. Given the estimates of the yield, this implies that only 56 bundles are available per household in the Buffer Zone each year, or enough to thatch one roof per household every 4–5 years. These figures indicate a short fall in thatch available, which is confirmed to a certain extent by the interviews and information obtained from key informants. It may also go some way to explain the drop in the number of permits sold. A perception of scarcity of the grass means that people are not willing to invest the permit fee when they are not sure of collecting enough grass for their needs.

Assessing the economic importance and value of grass collection

The grass and cane collected from the protected areas of the Terai undoubtedly have importance, especially, as the foregoing discussion has shown, in terms of their uses in house building. Information on the relative importance of these products, and how households would fare if access were stopped, is essential for future management strategies. A number of different approaches can be made to assess the economic value of extracted biomass (see for example, Godoy *et al.*, 1993; Martin, 1995). Three approaches are most commonly used; based on the marketed value of products, the cost of time or labour involved in their extraction, or the cost of substitutes. For the purposes of this paper some indication of the relative value of these products and their importance in the livelihood strategies of local people is required. Limited data on the quantities of materials extracted, their local prices, and the costs of substitutes, primarily for *khar*, were collected and are now briefly discussed. This represents only a cursory survey of the economic values rather than a full analysis, and concerns only the direct use value of these resources. For a discussion of the range of economic values of non-timber forest products see Chopra (1995).

In an earlier study by Lehmkuhl *et al.* (1988) the gross value of cane and grass collected from Chitwan National Park (RCNP) in January, 1986, was calculated to be in the region of US$451 836. The net value, with labour and permit costs subtracted, amounted to US$252 103, as shown in Table 3. The study also examined the costs of alternative building materials, and concluded that most villagers close to the park do not have enough capital to invest in other supplies, but they do have the time to cut their own grasses. The standard of living of these people would therefore be lower, without access to park resources.

Similar calculations for the amount of *khar* harvested from RBNP in January, 1995, can be made using the estimates outlined in the preceding section. If it is assumed that 1 050 000 bundles were harvested at a cost of 1.5 NR per bundle (prices indicated in our

Table 3. The value of grasses and cane collected from Royal Chitwan National Park, 1986

Product	Number of loads	Weight ('000 kg)	Value (million Nrupees)[a]	Value ('000 US$)[b]
Khar	162 592	6406	4.55	206.9
Kharai	134 265	4726	5.39	244.9
Total gross value			9.94	451.8
Net value[c]			5.46	252.1

Source: data from survey reported by Lehmkuhl *et al.* (1988).
[a]Local prices for *khar* 0.71 NR/kg, *kharai* 1.14 NR/kg.
[b]Exchange rate of US$1 = 22 NR (1986).
[c]Permit (1 NR pp/day) and labour costs (20 NR pp/day) subtracted.

household interview) then the gross value of production is 1 575 000 NR. If a labour cost of 40 NR per day (the agricultural wage rate) is assumed, and a permit cost of 12 NR per person, but that grass cutters spent only 3 days collecting *khar*, then the net value of *khar* from RBNP is in the region of 819 000 NR, or £10 900 or US$16 400 (75 NR = £1, 50 NR = US$1 in 1994). This figure is considerably lower than the estimate of Lehmkuhl *et al.* for RCNP in 1986. This is for a number of reasons, not only the vagaries of international exchange rates. The biomass production and harvest, and the actual area of grassland, are considerably lower in RBNP than RCNP. Fewer people enter RBNP to cut grass than they do in RCNP, where some 80 000 permits were sold in 1994. The harvest period is now restricted to 10 days, whereas it was 15 days at the time of the study of Lehmkuhl *et al.* The cost of permits and labour are now higher than in 1986. The prices of the grasses differ, and around Bardia a relatively smaller proportion of the grasses are sold. In addition there are a number of problems associated with using the price of labour to impute the value of extracted materials. Although the agricultural wage rate of 40 NR per day is used, this does not accurately reflect the opportunity cost of time, as the 2 weeks when grass collection takes place are the slack season for farm labour, and it is unlikely that there are many employment opportunities at that time.

These are just some of the factors which make attaching a monetary value to the collection of these materials problematic. In addition, the economic value may not reflect or be indicative of the cultural significance of these resources (see Brown, 1994 for a discussion). Despite the apparently low economic value of the grass using this type of analysis, these findings indicate that the grass is a very important resource, especially for the Tharu people. This is confirmed by the wide range of uses – most important in housing – grain silos, beds, mats, bullock carts and other uses. There are few – or no – alternative sources of grass outside the protected areas. Perhaps surprisingly, the survey found only limited local markets for thatch and cane, although there are many reciprocal arrangements between friends and neighbours. The grass cutting has cultural significance – in terms of the uses (for example, in ceremonial lamps in religious activities) – and as a social event, and is linked to the Maghi festival. So the direct use value as estimated here indicates only part of the social and livelihood significance of these resources. Cultural values of the grasses and products, and the full range of ecological and wildlife benfits of the grasslands should also play important roles in the consideration of management of these resources. Continuing to confine the analysis to direct use values, the costs and use of tiles as an alternative to thatch is now briefly discussed.

Availability of alternative roofing materials

A sizable number of households in the vicinity of RBNP use clay and, to a lesser extent, cement tiles as an alternative to thatch grass as roofing for houses and other buildings. The household survey found almost exactly half of the buildings of the households sampled were tiled, and that nearly 75% of the households had at least one tiled building (most households had between two and five buildings which consisted of a dwelling and cattle shed). These figures were confirmed by direct observation in the villages visited.

Tiles provide a number of advantages over thatch; they last longer, needing to be replaced only every 20–25 years compared with 5 years or less for thatch; they reduce fire hazards, a characteristic noted by several informants; and, if of reasonable quality, they will keep a house drier than thatch during the monsoon. In general, tiles are generally perceived to be desirable, as 'modern', and reflect the high status of a household. The disadvantages of tiles primarily concern their cost, the lack of skills available locally to produce them, and access to materials such as clay and sand for their construction, and fuel for firing.

The reported costs of tiles varied widely. This is for a number of reasons. First, households had installed tiles at different times through the years, the earliest of those reported being manufactured almost 20 years ago. Second, there were different sizes and types of tiles, some made from different materials. Lastly, the prices varied, as some people bought ready-made tiles, whereas others paid for tiles to be made and installed. The different costs reported are not directly comparable, and are only indicative of the magnitude of costs. Households reported that the cost of tiling a house ranged from 4000 to 20 000 NR. The cost of tiles ranged from less than 0.5 NR per tile when hiring labourers to make tiles using clay from a houshold's own land, to 5 NR per tile to buy the finished product. One household reported the cost of tiles as 500 NR per 1000 to buy, or 350 NR per 1000 to make. Most households who had tiled buildings had hired labourers to make tiles using clay collected from their own land. Sand was reported to be freely available locally, and tiles were dried in the sun for a few days and then fired using cattle dung. Four or five labourers, working for 4 or 5 days, could produce enough tiles (10 –20 thousand) to roof an average size building. Lehmkuhl *et al.* (1988) report a cost of between 2500–3700 NR for ceramic tiles to roof a typical 30 m^2 house in 1986.

Most of the households which had tiled buildings had hired labourers to make the tiles. The 'tile season' was defined as December–February, as would be expected, the dry season, but also the slack time of the year in the agricultural calendar. Many of the households hired itinerant labourers from India, not local casual labour, which implies a lack of skills in tile manufacture, and this is confirmed by the observations of Lehmkuhl *et al.* (1988) around Chitwan. This is despite the fact that tile use is not a particularly recent phenomenon, that it is relatively widespread, and that materials appear to be available locally. Use of tiles is undoubtedly increasing, and whether this is spurred by a perception of increasing scarcity and/or higher costs of thatch is uncertain. However, the loss of access to grasslands has in the past provided an incentive to change from thatch to tiling. A few households reported that they had first thatched their buildings about 18 years ago, and when asked about the reasons for their changes, said that when the Wildlife Reserve was first designated they had lost access to the grasslands and had not expected that it would be regained. One man told how his father, the leader of the Panchayat, had been the first person to use tiles in his village, and had actively encouraged his neighbours to use tiles for fear of losing access to source of thatch grass.

The most important feature of tiles as an alternative to thatching, and one which has particular implications for poorer households in the community, is that installation of tiles involves, in every case reported, a shift from a mainly subsistence activity to a cash transaction. In most cases this cash transaction involved a large lump sum investment, so even though the net costs of tiles annually may be competitive with thatch, they are unlikely to be readily available or attractive to poorer households.

More investigation into the viability of providing training in tile-making, and in setting up tile works as income generating projects as part of the Buffer Zone Management Plan is required. In addition, the wider environmental implications of more widespread tile making need to be considered; the large-scale removal of clay and sand, and the availability of fuel associated with tile manufacture could result in a new set of ecological, environmental and socio-economic problems. Encouraging tile production may appear to be a logical step to decreasing pressure on grassland resources, but it requires careful examination.

Conclusions

The grassland habitats inside the protected areas of the Terai are a most important resource for both wildlife and local villagers. Grasses are utilized by local people, particularly Tharu groups, for a wide range of uses. The case study of RBNP has shown that although the direct use value of the main grass products, thatch and cane, may not be especially high in financial terms, they are important subsistence products which make significant contributions to the welfare of local villagers. Without access to the grassland products, traditional methods of house building could no longer be maintained and there are virtually no sources of thatch and cane outside the protected areas. The grassland products play an important role in livelihood strategies and have cultural value in terms of ceremonial uses.

The ecological impacts and sustainability of biomass extraction are currently uncertain, although on-going ecological experiments intend to shed light on any changes occurring in the grasslands. On the one hand, biologists argue that the disturbance associated with cutting and burning will have detrimental impacts on small animals, insects and birds, while, on the other, the flush of short grass resulting from burning may encourage ungulate species. These in turn attract large predators such a tigers, which are most highly prized by tourists. There is therefore much uncertainty, further clouded by the vested interests of the parties concerned, of the relative benefits and costs of current management regimes. It may well be that certain types of cutting and burning, at particular times of the year, produce the mosaic of habitats which optimise multiple uses by both animals and humans. The diversity of the grasslands, where a patchwork of diverse habitats is provided by micro-scale differences of site, means that management of the protected area needs to take localized features into account, rather than attempting to implement one regime across all the Terai protected areas.

In summary, the rich biological diversity of the Terai protected areas is maintained in part by human intervention, and current cutting practices bring benefits to both local communities and to wildlife. Whether the practices are sustainable, and in what way they can be made so, is open to question, but the current focus of conservation and development policy is to relieve perceived pressure on resources inside the protected areas by encouraging the more intensive use of those in the designated Buffer Zone surrounding the National Park (HMG 1993; HMG/UNDP 1994). The findings of this study have highlighted the importance of grass collection for local people, and also indicate that alternatives for one of the main uses, thatch, are available in the form of ceramic or concrete tiles. However, the cost of tiles may well be prohibitive for poorer households, and as yet, the skills for their production are not prevalent locally, and in addition the environmental and economic impacts of more widespread production are unknown. This is one area which could usefully be explored as part of the community development activities of the Parks and People Project. A critical issue for implementation of this project in the Buffer Zone is how effective participation of local people can be facilitated and how benefits, especially for poorer households, can be maintained.

Progress in understanding the ecological and socio-economic dimensions of the conservation–development conundrum, their dynamics and potential trade-offs, demands extensive research which transcends traditional disciplinary boundaries. This research takes a step towards this aim. Such work is necessary to inform policy if sustainable utilization and long-term management is to succeed in conserving areas such as the RBNP.

Acknowledgements

Fieldwork was conducted under the auspices of the UEA Tall Grasslands Research Project which is funded by the UK Department of Environment Darwin Initiative for Survival of Species, in collaboration with the Department of National Parks and Wildlife Conservation in Nepal. The author would like to thank the following people for their assistance and advice: Dr Uday Sharma, Dr Bijaya Kattel, Mr Karki, and Mr B. Pathak at DNPWC; Mr T. Khatri at the King Mahendra Trust for Nature Conservation; Mr Gopal Sharma, Ram Din Mahato, Nic Peet and Neil Adger. Views expressed here are the authors alone, and any omissions or errors remain sole responsibility of the author. CSERGE is a designated research centre of the Economic and Social Research Council. An earlier version of this paper was presented at the International Association for the Study of Common Property Conference, Norway, May 1995.

References

Basnet, K. (1992), Conservation practice in Nepal: past and present. *Ambio*, **21.6**, 390–3.

Brown, K. (1994) Approaches to valuing plant medicines: the economics of culture or the culture of economics? *Biodiv. Conserv.* **3**, 734–50.

Central Bureau of Statistics (1993). *Statistical Yearbook of Nepal.* Kathmandu: CBS, HMG.

Chopra, K. (1995) *The Valuation and Pricing of Non-timber Forest Products: a study in Raipur District of Central India.* Delhi: Mimeo, Institute of Economic Growth.

Cox, T. (1990) Land rights and ethnic conflict in Nepal. *Econ. Polit. Weekly* June 16–23, 1318–20.

Dasmann, R.F. (1991) The importance of cultural and biological diversity. In *Biodiversity: Culture, Conservation and Ecodevelopment* (M. Oldfield and J.B. Alcorn, eds) pp. 7–16. Boulder: Westview Press.

Dinerstein, E. (1979) An ecological survey of the Royal Karnali-Bardia Wildlife Reserve, Nepal. Part 1: vegetation. modifying factors and successional relationships *Biol. Conserv.* **15**, 127–50.

Gadgil, M. (1993) Biodiversity and India's degraded lands. *Ambio* **22**, 167–72.

Ghimire, K. (1992) *Forest or Farm? The Politics of Poverty and Land Hunger in Nepal.* Delhi: Oxford University Press.

Godoy, R., Lubowski, R. and Markandya A. (1993) A method for the economic valuation of non-timber tropical forest products. *Econ. Bot.* **47**, 220–33.

Heinen, J.T. and Kattel, B. (1992) A review of conservation legislation in Nepal: past progress and future needs. *Environ. Manage.* **16**, 723–33.

Heinen, J.T. and Yonzon, P.B. (1994) A review of conservation issues and programs in Nepal: from a single species focus toward biodiversity protection. *Mountain Res. Dev.* **14**, 61–76.

HMG (1993) *Nepal Environmental Policy and Action Plan: Integration Environment and Development.* Kathmandu: Environment Protection Council.

HMG/UNDP (1994) *Parks and People Project Document.* Kathamandu: DNPWC.

Lehmkuhl, J.F., Upretti, R.K. and Sharma, U.R. (1988) National parks and local development: grasses and people in Royal Chitwan National Park, Nepal. *Environ. Conserv.* **15**, 143–8.

Majupuria, T.C. and Joshi, D.P. (1989) *Religious and Useful Plants of Nepal and India.* Lashkar (India): M. Gupta.

Martin, G.J. (1995) *Ethnobotany.* London: Chapman and Hall.

Peet, N. (1994) *Fire and Biodiversity Conservation in the Tall Grasslands of Nepal and Northern India.* Norwich: Mimeo, School of Biological Sciences, University of East Anglia.

Phillips, O., Gentry, A.H., Reynel, C., Wilkin, P. and Galvez-Durand, C.B. (1994) Quantitative ethnobotany and Amazonian conservation. *Conserv. Biol.* **8**, 225–49.

Pokharel, S.K. (1993) Floristic Composition, Biomass Production and Biomass Harvest in the Grassland of the Royal Bardia National Park, Nepal. MSc thesis, Agricultural University of Norway, As, Norway.

Resources Nepal (1994) *Habitat Himalaya Resource Nepal Newsfile*, January–February, p. 2. Kathmandu: Resources Nepal.

Upreti, B.N. (1994) *Royal Bardia National Park.* National Planning Commission, HMG, Nepal, in Collaboration with IUCN.

[21]

The non-market value of the Swedish agricultural landscape*

LARS DRAKE

The Swedish University of Agricultural Sciences

(received August 1989, final version received August 1991)

Summary

The contingent valuation method has been used to estimate the Swedes' willingness to pay to preserve the agricultural landscape. They seem willing to pay about 541 SEK/person year (78 ECU) or 975 SEK/hectare* year (140 ECU) (May 1986). The results are judged to be reliable but with a low degree of precision. It is shown that willingness to pay is significantly correlated to income, age (negatively), level of education and to positive attitudes towards preservation of the agricultural landscape. Willingness to pay per hectare differs due to land use and regional location. Subsidies based on acreage instead of price support may be an adequate political solution.*

Keywords: Welfare economics, contingent valuation method, landscape value, agricultural policy.

1. Introduction

1.1. The Swedish agricultural landscape

The Swedish landscape is dominated by forest (54%), whereas only 8% is used in agriculture, in the northern parts less than 2%. Privately owned family farms dominate very heavily. The agricultural landscape has a high degree of variety, even if the trend is towards more grain production on bigger fields with less interruptions. This trend affects both scenic and biological variety. Two-thirds of the 419 endangered, vulnerable, rare or

* The author is indebted to Alan Randall at Ohio State University (earlier draft), the reviewers and the editor for constructive comments on this paper. The author bears the full responsibility for remaining imperfections.

Euro. R. Agr. Eco. 19 (1992) 351–364 0165–1587/92/0019–0351 $2.00

care-demanding vascular plant species in Sweden exist on and depend on agricultural land (Databanken, 1991).

Between 1951 and 1981, 11% of the agricultural land was abandoned and/or converted into forest. The most frequent use is dense spruce forest, which is the alternative considered in this study. Due to regulation and the presently slow expansion of our cities, development on the urban fringe is of minor importance. The highest proportional decline was for a special type of wooded pasture (*hagmark*, Swed.), which in several cases has been used for grazing since long before the Vikings. On this type of low productivity land there are normally a few big trees, some bushes, some big rocks and it is often hilly. This land use, or type of nature, is one of the richest in species in Sweden. Some plant species have even developed on this type of land.

In Sweden anybody has the right to walk around in the countryside, except in gardens and on growing crops. Wooded pastures and the borders between forests and agriculture are popular places for picnics. There is also some concern in Sweden about the preservation of our cultural inheritance.

1.2. The problem

The primary product of agriculture is, of course, food. With or without intention farmers also produce or maintain a specific landscape, with its scenery, species, recreational facilities and historical continuity. The quality of this byproduct is dependent on what is produced on the land and by what means. Farmers are not paid specifically for this product which makes it an externality. The landscape as such is undoubtedly a public good. Once it has been produced the social cost of adding additional consumers of the landscape is low, limited to the costs of congestion is some areas and stress in some sensitive biotopes, due to recreational activities.

If this public good is of significant value, agricultural and/or environmental policy should consider this market imperfection, if social efficiency is to be attained. The agricultural landscape would not be produced in sufficient quantity or of desirable quality, if the market were the sole allocative instrument in use. The aim of this study is to estimate the value of preserving the agricultural landscape in Sweden.

1.3. Previous studies

In a study by Halstead (1984) the WTP to avoid residential development on agricultural land was measured in three counties in Massachusetts, USA. The iterative bidding game technique was used in a household survey. The mean annual bid to avoid low-density development ranged from US $28 to $60 and from $70 to $176 to avoid high-density development. Urban development is of course not the same as converting the land into dense spruce forest and there are also other differences between Massachusetts and Sweden that make comparisons hard.

A study by Bergstrom et al., (1985) measured the WTP to protect 1/4 (or 1/2, or 3/4, or all) of the agricultural land in a county in South Carolina, USA. The payment card procedure was used in a mail survey. The main bid to prevent development on all land was below $10. It is hard to judge if this difference between these two American studies is caused solely by differences between the two states, i.e., in the landscape, levels of income or in attitudes.

2. Method

2.1. *Choice of method*

Two methods often used for estimating environmental and recreational values, the travel cost method and the hedonic price method, have the advantage of being based on actual behaviour. The travel cost method (Clawson, 1959), however, cannot be used to estimate the value of preserving the agricultural landscape. The objects whose values are to be estimated are far too scattered, and a considerable part of the value is due to the existence of an everyday surrounding that does not require any extra travel costs to be consumed. Another approach is called the hedonic price method (Rosen, 1974). This method would fail to estimate the full value of preserved agricultural landscape; only values that are enjoyed exclusively by the landowner would be estimated. These methods can in principle estimate use values but not option and existence values which could be of significant size for agricultural landscape.

The contingent valuation method (CVM) (Randall et al., 1974; Cummings et al., 1986), which uses citizen's self-reports of willingness to pay (WTP) in hypothetical situations, seems to be the only appropriate method. The method has been used to estimate the value of recreation sites, local air and water quality and hunting licenses, local agricultural landscape, etc. Only a few studies have, like this study, estimated the values of more general characteristics of the environment, e.g., Randall et al., (1985) for national air and water quality programmes. Use values as well as option and existence values are included in estimates from this method. A serious problem with CVM is the difficulty with value formulation and value statement and the possibility of a number of biases or unsystematic distortions, such as starting point, interview and vehicle bias.

Several studies have been made using the contingent valuation method together with one of the other methods mentioned on the same object. The estimated WTP was in most cases on the same level (Cummings et al., 1986). These results indicate that the CVM method can provide reliable results, given that it is rigorously applied.

2.2. Research design

This study is made up of three separate surveys. The first one, with 1,089 persons interviewed in all parts of Sweden, was carried out by Sifo, The Swedish Institute of Opinion Research, in May 1986. The other surveys, with 152 and 49 participants, respectively, were carried out in Uppsala county by the author and two assistants. The main motive for the second study was to enable testing of potential biases and to obtain relative values of different land use. The third one was carried out five months later to check the cause of difference between the other two.

The Swedish Institute of Opinion Research has selected a large number of representative local administrative areas. Randomly chosen birth dates are used in selecting participants. In the Uppsala survey participants were selected directly from the population register. The register is organized in a day-by-day age order and every (pop./300) was selected. Every third person on the list was left out and the remaining 200 used in the first Uppsala survey. The sample for the other Uppsala survey was selected in a similar way from those (101) omitted from the list. The populations in all surveys included all persons between 16 and 74 years of age.

2.3. Value formulation and value statement problems

The main question, the one concerning WTP, was preceded by some information about the problem, the questions concerning attitudes, and a picture of a typical part of the agricultural landscape. The WTP question was formulated as follows:

> In this question you should imagine that the government pays the farmers to preserve agricultural land that otherwise would be brought out of production. Imagine that the alternative is that half of all agricultural land would be cultivated with spruce forest.
>
> Such a change may look like this. (Two drawings were shown, 'before' and 'after').
>
> How much would you at the most be willing to pay per year in income tax to prevent half of all agricultural land from being cultivated with spruce?

The respondent has to decide what his/her WTP is and what WTP to state. The first is the value formulation problem and the second the value statement problem. Hoehn and Randall (1987) argue that unfamiliarity with the good to be evaluated and limitations in time and resources devoted to the value formulation problem may make the contingent choices difficult. Concerning the equivalent measure, which is relevant in this study, they conclude from utility theory that 'either understatement or overstatement is possible' (Hoehn and Randall, 1987: 233). Cummings et al., (1986) assert that familiarity with the good is a condition for a reasonable valuation. The good to be valued, i.e., the agricultural landscape, is quite well known to the

Swedes. On the other hand, Swedes have no experience in valuing the good in question in monetary terms. Given that the time limit for valuation is not too restrictive, it should be possible for the respondents to reveal their WTP reasonably well. The fact that 88% of the respondents in the main survey answered the WTP question indicates their acceptance of the question and their confidence in giving a reasonable answer, but it does not prove that they have been able to reveal their 'true' preferences.

Strategic behaviour may appear when stating the value. Some previous studies show that strategic behaviour is not as great a problem as economists earlier believed (Bohm, 1972; Marwell and Ames, 1981; Brookshire et al., 1982; Rowe et al., 1980; Mitchell and Carson, 1981; Hoehn and Randall, 1987). In this study real payments are expected to be nearly proportional to individual income, which makes incentives for strategic behaviour very small. Rational behaviour would, in this setting, be to overstate if you think you are above average and understate if below. This should result in a tendency towards high and low bids and few near the average. Such a tendency has not been found in the data. The payment for an individual would be very close to $(C/N)^*(y_i/y)$, where C is the total cost of a programme to preserve the agricultural landscape, N the size of the affected population, y_i individual i:s income and y average income. As there were more than 1,000 participants, the respondent with average WTP and income would only affect the average bid by less than 0.1% if he doubled his bid.

In the first Uppsala county surveys the respondents were asked, after the interview and in an informal way, if they had argued in a strategic way before they gave their bid ($N = 152$). All answers were negative. This result does not, of course, prove that everyone has revealed his or her true preferences, but it gives some evidence that strategic bias is of minor importance.

3. Results

3.1. *Attitudes*

The respondents were asked about their general attitudes towards the agricultural landscape. One question was as follows: 'How important do you think it is to preserve the agricultural landscape?' The alternatives 'very important' (56.4%) or 'fairly important' (38.4%) were chosen by 95% of the respondents. This indicates a clearly positive attitude towards preservation.

In order to obtain the relative importance of different motives for the estimated WTP the following question was asked: 'Those who wish to preserve the agricultural landscape emphasize different motives. What do you think is most important? ... second? ... third?" The alternatives were: 'The landscape is beautiful'; 'Many animals and plants depend on the agricultural landscape'; 'More open areas to be in'; 'Preserve the memories of the ancestors' work'; and 'Another alternative'. These motives will be referred to as, respectively, the aesthetic, the nature conservation, the recreation and the cultural-historic motives. Sixty-nine percent gave the highest priority to the nature conservation motive, 13% to the aesthetic motive and 5% to each of the recreation and the cultural-historic motives. For each motive the following sum was calculated: frequency for highest priority plus frequency for second priority multiplied by 0.5 plus frequency for third priority multiplied by 0.25. These weighted sums were transformed into shares as follows:

the nature conservation motive 47%
the aesthetic motive 22%
the recreation motive 13%
the cultural-historic motive 12%
other motives 7%

A sensitivity analysis has shown that the proportions are reasonably stable with some variations in the weights. The list of shares is a very rough estimate of the relative importance of different motives for the WTP to preserve the agricultural landscape.

3.2. *Willingness to pay*

Fifteen alternative bids were given, ranging from 0 to 3,000 SEK/year, i.e., a payment card procedure was used. The average bid in the main survey was 468.5 SEK/person*year (68 ECU/pers*year); see Table 1. The respondents were then asked if they were willing to increase their bid by 20% if their 'and the others' bids were not sufficient to preserve the agricultural landscape'. Eighty-two percent indicated a positive response in terms of willingness to pay the proposed increase, and the average bid thereby rose to 546.9 SEK/person*year (79 ECU) (*Statistical Abstracts of Sweden*, 1991). The precision in these estimates are probably not better than ±50% (Cummings et al., 1986).

Table 1. *Primary bids, medians and standard deviation in the surveys (SEK/year* person)*[a]

	Sifo	B1	B2	C
bid	468.5	729.4	712.9	686.5
std. dev.	(670.9)	(932.5)	(708.1)	(762.6)
median	200	375	500	500

[a]Sifo denotes the main survey; B and C were only carried out in Uppsala county. In B1 the alternative bids were 50% higher than in the Sifo survey. In B2 the respondents were informed that the agricultural policy would change in such a way that his or her budget would increase by 700 SEK/year. The interviews in Sifo and B took place in April and May, 1986. The C survey with questions identical to those in Sifo was carried out in the autumn of 1986.

In order to check the reliability in the estimated WTP the respondents were asked to list, according to their preferences, six different ways for the government to use tax revenue. The alternatives concerned sport buildings, nurseries, preservation of the agricultural landscape, road maintenance, relief work and parks in the cities. The alternatives were formulated in such a way that they would each cost an additional 5 to 10 million SEK/year in a normal county. The idea was that a high WTP may be questioned if preserved agricultural landscape received a low priority. The result was, however, that the landscape alternative got the highest ranking in all surveys.

3.3. *Variables explaining willingness to pay*

Individual WTP ought to be systematically related to income, age, level of education, place of residence and attitudes. This has been tested using multiple regression analysis. The hypotheses were that household income or personal income, high level of education, being female, place of residence in rural areas or areas dominated by agriculture and stated positive attitudes towards preserving the agricultural landscape should increase the individual WTP. No hypotheses were made concerning age and place of residence in the north–south dimension. Results from two regression equations, where only significant variables are included, are presented in Table 2. The equations were:

$$WTP = f(INCOME(B)(or\ HOUSEHINC(Sifo)),\ AGE,\ EDUCATION,$$
$$AB\text{-}ALT(Sifo),\ ATTITUDE) + e$$

The variables have all been commented on above except AB-ALT in which the respondents were to choose between two different statements. In the first, profitability was the sole criterion for continued agricultural production (1), in the second other motives were emphasized (2). ATTITUDE refers to attitudes towards preserving the agricultural landscape, where (1) indicates 'very important', (2) 'fairly important' and (3) two lower priorities. EDUCATION has three levels; only primary school = 1, secondary school = 2, and university = 3. INCOME is individual pre-tax income and HOUSEHINC is pre-tax income of the household.

At least five relevant variables are significantly correlated to WTP, which indicates that the respondents do not answer arbitrarily. All relations had the assumed sign. The marginal willingness to pay is 0.44% of individual pre-tax income. A rough point estimate for income elasticity for demand for agricultural landscape can be calculated to 0.5 (average income in B was 87,791 SEK and the primary bid 721.15 SEK/pers.).

The participation rate in Sifo surveys is as high 78%; about 7% are impossible to reach and 15% refuse to participate. It is important to point out that they did not know that landscape values were part of the survey.

Table 2. *Result of regressions*

	Sifo	B
R^2	.112	.135
n	922	143
Model F-value	16.46	3.54
Intercept	170	134
HOUSEHINC	.00227(.00052)**	
INCOME		.00438(.00212)**
AGE	−4.70(1.62)**	−12.4(5.3)**
EDUCATION		
1	−242**	−208
2	−136*	−505**
3	0	0
AB-ALT		
1	−139**	
2	0	
ATTITUDE		
1	591**	1569**
2	283	1289*
3	0	0

Estimates and standard error of estimates for continuous variables. Estimates of all class variables are biased and only indicate relative values, standard errors of estimates are not relevant. Significance, Student t-test. on the .10 level is marked with * and on the .05 level with **. Sifo indicates the main survey and B the first Uppsala county survey.

The OLS equation (Sifo) in Table 2 was used to calculate the probable WTP among the 12% of the participants who refused to answer the WTP-question. They were assumed to have the same WTP as persons with the same income, age, education and attitudes. Their WTP was 49 SEK lower than the average bid, implying that average bid should be reduced by 6 SEK or 1% to 541 SEK/person*year (78 ECU/pers*year). This is relevant for the scaling up to aggregate WTP in Section 3.5 below.

3.4. Analysis of biases

Starting point bias implies that estimated WTP may be affected by the original bid in an iteration process (Boyle et al., 1985; Cummings et al., 1986). In this study 15 possible answers ranging from 0 to 3,000 SEK/year were given, which means that starting point bias should be smaller than in bidding games with a single starting point. Starting point bias was estimated in this study by comparing means from two surveys in Uppsala county, one with questions identical with the main survey (mean = 686.5, std dev = 762.6, n = 49, C in Table 1), carried out in the autumn, and one with the 15 original bids increased by 50% (mean = 729.4, std dev = 932.5, n = 72, B1 in Table 1), carried out in the spring. The difference between average estimated WTP was not significant on the .10 level (Student t-test). There is no evidence that the respondents have been affected by the choice of original bids.

On the other hand a significant difference in WTP does exist between the main survey and those carried out in Uppsala county. The comparison was made between respondents in regions dominated by agriculture in the main survey (mean = 492.0, std dev = 673.2, n = 678) and the Uppsala county survey with identical questions (see C above). The difference between mean bids was significant on the .10% level (Student t-test). This could to some extent be explained by the fact that the main survey was carried out by The Swedish Institute for Opinion Research (Sifo), and the Uppsala county surveys were carried out in the name of The Swedish University of Agricultural Sciences. The respondents may have interpreted the first one as more neutral than the second. The difference could alternatively be explained by the fact that respondents in the Uppsala county surveys who spontaneously said they would prefer not to pay were asked the same questions again. The result was fewer than zero WTP answers but, assuming that those respondents stated an average WTP the second time the question was asked, this only explains about one-third of the difference. There is yet another possible explanation which cannot be completely ruled out: the time between the surveys, which was five months.

The measurements of WTP reported below are exclusively from the main survey, which in this sense implies an underestimation. Analyses of the impact of missing observations in the main survey show that this factor has increased estimated WTP by approximately 1%, which has been corrected for in calculations of aggregate WTP.

3.5. *Value of agricultural land use*

The average individual WTP, 541 SEK/person and year, could be summarized in a simple way for the whole country through multiplying by the number of inhabitants aged between 16 and 74 years (6,219,690, *Statistical Abstracts of Sweden*, 1989). The total WTP is approximately 3.4 billion SEK/year which is approximately equal to 485 million ECU. If this sum is divided by the total acreage of agricultural land, the resulting quota will be the per hectare value of agricultural land use. The result is 975 SEK/hectare*year or approximately 140 ECU/hectare*year. This is more than the net value of agricultural production in most parts of Sweden. This scaling up can be done in more than one way and it may increase the already low precision somewhat (Loomis, 1987). This is also true for the calculations for regional and land use valuations. In spite of this problem it ought to be meaningful to derive the order of magnitude of these values.

It is also possible to calculate the regional distribution of the value of agricultural land use. When the respondents of the main survey were subdivided into three regions, it was found that average WTP differs, though not significantly, among regions. WTP in each region has been multiplied by the number of citizens aged between 16 and 74 years in the region and divided by the number of hectares in the region. The major influence on regional differences in the value of agricultural land use are caused by the uneven distribution of agricultural land and population (Table 3).

These calculations assume that people are solely interested in preserving the agricultural landscape in their own region. This is probably not the full truth, but the pilot study, a questionnaire, gives some indication that it is a reasonable assumption. In the pilot study the respondents where asked about their WTP to preserve agricultural land in their county. The resulting average WTP was almost the same as in the Uppsala county surveys (Drake, 1985b).

In the Uppsala county surveys, the respondents were also asked about their preferences concerning different forms of land use. Pictures of grain production on arable land, of grazing on arable land and of a type of wooded pasture land were shown together with information about the normal occurrence of wild plants and birds, grazing animals and about the differences in the right to walk on the land. The respondents were asked how they value the latter two types of land use, given that the value of grain production was fixed to 25 points. The results in relative values were; grain production index = 100, grazing land 191 (140) and wooded pastures 241 (200) — medians in parentheses. These figures and the acreage of the different types of land use could be used to transform the general WTP in SEK/person and year into a value of each of the different types of land use in SEK/hectare and year. The following equation was used:

$$WTP^* \; Pop = X(igr * Gr + igz * Gz + iwp * Wp)$$

where *WTP* is the average willingness to pay from the main study (SEK/year*hect.), *Pop* is the Swedish population between 16 and 74 years, *X* is a factor for transformation of the relative values to absolute, *igr*, *igz* and *iwp* are the index values as above for, respectively, grain production, grazing on

Table 3. *Willingness to pay for preservation of the agricultural landscape; regional distribution*

	SEK/pers.*yr.	SEK/hect.*yr.
Northern Sweden	524	1287
Central Sweden	557	1114
Southern Sweden	532	703
Regions dominated by agriculture	571	1046
Other regions	485	862
Sweden	541	975

Source: *Statistical Abstracts of Sweden*, 1986; *Yearbook of Agricultural Statistics*, 1986; own calculations.

arable land and wooded pasture; *Gr*, *Gz* and *Wp* are acreage of the respective type of land use. The equation can be solved for X, and then $X * igr$ (or *igz* or *iwp*) will give the willingness to pay to preserve an average hectare of each type of land use. The results are listed in Table 4.

4. Conclusions and policy implications

4.1. Results

This study shows quite clearly that Swedes have a positive attitude towards the existing agricultural landscape. The most important motive to preserve it is that 'many animals and plants depend on the agricultural landscape'. Another important motive is that 'the landscape is beautiful'.

As an average, Swedish citizens aged between 16 and 74 years seem willing to pay about 541 SEK/year (78 ECU/year). For the whole country the sum would be 3.36 billion SEK/year (485 million ECU). The results can be judged to be reliable but with a low degree of precision probably not better than $\pm 50\%$ (Cummings et al., 1986). The reason for this first judgement is the fact that the method has normally given results comparable to those derived from travel cost and hedonic price methods that are based on actual behaviour, the Swedes familiarity with the good in question and the fact that strategic behaviour is not expected or experienced in this study. The hypothetical character of the questions or decision problem and the respondents' lack of prior experience in valuing the good in monetary terms is the main reason for low precision. It can be shown that willingness to pay is significantly correlated to income (or income of household), age (negatively), level of education and to positive attitudes towards preservation of agricultural landscape.

Willingness to pay per hectare differs significantly between different forms of land use, for example grain production approximately 860 SEK (123 ECU), grazing land approximately 1,600 SEK (237 ECU) and a special type of wooded pasture land almost 2,100 SEK (299 ECU). Willingness to

Table 4. *Willingness to pay for preservation of the agricultural landscape 1986; different land use. SEK/hectare and year.*

	Area (hectares)	Mean
Grain production and comparable	2,741,000	860
Grazing on arable land and cult. pastures	370,000	1,643
Wooded pasture	338,000	2,076

Source: *Statistika meddelanden* J10 SM 8601; *Statistical Abstracts of Sweden*, 1989; own calculations.

pay per hectare also varies due to location, from 700 SEK (101 ECU) in the southern parts to 1,300 SEK (187 ECU) in the northern parts of Sweden.

This study was carried out in Sweden and the results are of course not directly applicable to other countries. The special type of landscape, attitudes and traditions, the relative scarcity of agricultural land and the high standard of living are among the factors that give rise to the high WTP in this study.

4.2. Policy implications

The present agricultural policy is based on import levies and heavy domestic price regulations similar to CAP, which mainly increase the price of grain. The effect is among other things a higher fertilizer and pesticide intensity (partly offset by fertilizer and pesticide taxes), a larger share used for grain production and also larger total area used in agriculture (Drake, 1989; Jonasson, 1990), all compared with a free market situation. The first three effects are mostly negative for the environment, i.e., through leaching and reduced natural diversity. The effect of a larger area in agriculture is, however, positive. Present agricultural policy has been heavily criticized in Sweden for being inefficient in attaining goals such as national food security, farm income parity and environmental quality (Bolin et al., 1986; Drake, 1989).

The decision on a new Swedish farm policy in 1990 was based on domestic deregulation and slightly lower border protection and includes a small landscape protection subsidy. In this process of deregulation one should consider the effects on the landscape, especially in areas where profitability will be close to or below zero at the new expected prices. According to a recent Swedish study (Jonasson, 1990) 25% of all agricultural land will be below zero profitability if this policy is implemented. Another study concludes that about 30% of the arable land will not be used for agriculture in the year 2000 (Johansson, 1991). Much of the threatened land is arable land in areas dominated by forest and grazing lands in general. It has high values according to this study and bears a high biological diversity. One could argue that production of the collective good 'open and varied landscape' should be financed by the state which could pay the farmers who provide the good. This is just the opposite of taxing a negative externality like pollution.

Subsidies, or compensation for a service, based on acreage to all grazed land or grasslands, and to all agricultural land in weaker areas, as an alternative to price support, would change the relative prices in such a way that extensive meat production would be more profitable than it is today and intensive grain production less profitable. The decline in profitability caused by the domestic deregulation and lower border protection would be capitalized in lower land values. On the average, these income effects could be neutralized by capital transactions. The allocative effects of the suggested policy change would probably be more agricultural land still in use, less nitrogen loss due to the decrease in grain production and lower intensity, a more diversified landscape, preservation of endangered species, and healthier cattle mainly due to the increase in grazed acreage or grasslands. The suggested corrective policy instruments will no doubt push the system towards a socially more efficient resource use compared to the present situation. There is however not enough knowledge as to how far or how firmly this policy ought to be enforced in order to reach an optimal solution. It must of course be a political decision, but the values derived in this study can be used as guidelines.

References

Bergstrom, J., Dillman, B.L., and Stoll, J. (1985). Public environmental amenity benefits of private land: The case of prime agricultural land. *Southern Journal of Agricultural Economics* 17(1): 139–149.

Bohm, P. (1972). Estimating demand for public goods: An experiment. *European Economic Review* 3: 111–130.

Bolin, O., Meyerson, P.-M. and Ståhl, I. (1986). *The Political Economy of the Food Sector.* Stockholm: SNS.

Boyle, K., Bishop, R. and Welsh, M. (1985). Starting point bias in contingent valuation bidding games. *Land Economics* 61(2): 188–194.

Brookshire, D., Thayer, M., Schulze, W. and d'Arge, R. (1982). Valuing public goods: A comparison of surveys and hedonic approaches. *American Economic Review* 72(1): 165–176.

Clawson, M. (1959). Methods of measuring the demand for and value of outdoor recreation. Reprint No. 10, Resources for the Future. Washington, DC.

Cummings, R., Brookshire, D. and Schulze, W. (1986). *Valuing Environmental Goods: An Assessment of the Contingent Valuation Method.* Totowa, NJ: Rowman & Allanheld.

Databanken för hotade arter (1991). Hotade växter i Sverige 1990 [Threatened plants in Sweden 1990]. Stockholm.

Drake, L. (1985a). Jordbrukets naturresursutnyttjande — En ekonomisk analys [The use of natural resources in agriculture: An economic analysis]. Report no. 261, Department of Economics and Statistics, The Swedish University of Agricultural Sciences, Uppsala.

— (1985b). Scenic Values of Agriculture. Småskriftsserien no. 4, Department of Economics and Statistics, The Swedish University of Agricultural Sciences, Uppsala.

— (1987). Värdet av bevarat jordbrukslandskap — resultat av intervjuundersökningar. [The Value of preserving the Agricultural Landscape). Report no. 289, Department of Economics and Statistics, The Swedish University of Agricultural Sciences, Uppsala.

— (1989). Swedish agriculture at a turning point, *Agriculture and Human Values*, VI(1&2): 117–126.

Halstead, J. (1984). Measuring the nonmarket value of Massachusetts agricultural land: A case study. *Journal of Northeastern Agricultural Economic Council* 13(1): 12–19.

Hoehn, J. and Randall, A. (1987). A satisfactory benefit cost indicator from contingent valuation. *Journal of Environmental Economics and Management* 14: 226–247.

Johansson, V. (1991). Vad kosar det att behålla åkerareal? [What is the cost of preserving arable land?]. Report (forthcoming), Department of Economics and Statistics, The Swedish University of Agricultural Sciences, Uppsala.

Jonasson, L. (1990). Livsmedelspolitikens effekter på framtidens jordbruk och jordbrukslandskap [The effects of the farm policy on agriculture and farm landscape] forthcoming in

Det svenska jordbrukslandskapet inför 2000 talet — bevaras eller försvinna? The Royal Academy of Agriculture and Forestry, Stockholm.

Loomis, J. (1987). Expanding contingent value sample estimates: Current practices and proposed solutions. *Land Economics* 63(4): 396–402.

Marwell, G. and Ames, R.E. (1981). Economists free ride. Does anyone else? Experiments on the provision of public goods. *Journal of Public Economics* 15(3): 295–310.

Mitchell, R. and Carson, R. (1981). An Experiment in Determining Willingness to Pay for National Water Quality Improvements. Draft report to the US Environmental Protection Agency, Washington, DC.

Randall, A., Ives, B. and Eastman, C. (1974). Bidding games for valuation of aestetic environmental improvements. *Journal of Environmental Economics and Management* 1: 132–149.

— Blomquist, G., Hohen, J. and Stoll, J. (1985). National Aggregate Benefits of Air and Water Pollution Control. Interim report to the US EPA, Washington, DC.

Rosen, S. (1974). Hedonic prices and implicit market: Product differentiation in pure competition. *Journal of Political Economy* 82: 34–55.

Rowe, R., d'Arge, R. and Brookshire, D. (1980). An experiment on the economic value of visibility. *Journal of Environmental Economics and Management* 7: 1–19.

Statistical Abstracts of Sweden (1989). Statistiska Centralbyrån, Stockholm.

— (1991). Statistika Centralbyrån, Stockholm.

Yearbook of Agricultural Statistics (1986). Statistika Centralbyrån, Stockholm.

Lars Drake
The Swedish University
of Agricultural Sciences
Department of Economics
Box 7013
S-75007 Uppsala
Sweden

Part III
Capturing the Value of Ecosystem Functions and Services

A Property Rights, Institutions and Policy Instruments

[22]

ELSEVIER

Ecological Economics 12 (1995) 107–123

ECOLOGICAL ECONOMICS

Third-world debt and tropical deforestation

James R. Kahn [a,b,*], Judith A. McDonald [c]

[a] Department of Economics, University of Tennessee, Knoxville, TN 37996-0550, USA
[b] Energy Division, Oak Ridge National Laboratory, PO Box 2008, Bldg 4500N, MS 6205, Oak Ridge, TN 37831-6205, USA
[c] Department of Economics, Rauch Business Center, 621 Taylor Street, Lehigh University, Bethlehem, PA 18015-3117, USA

Received 7 December 1993; accepted 9 April 1994

Abstract

The deforestation and degradation of tropical forests are taking place at an extremely rapid pace. According to the Food and Agriculture Organization (FAO), the estimated annual rate of tropical deforestation during the 1981–1985 period was 113 846 square kilometers or 0.6% of the 1981 total forested area. The implications of the loss of these forests are staggering (Myers, N., 1989. Deforestation Rates in Tropical Forests and Their Climatic Implications. Friends of the Earth, London). Tropical forests are extremely rich ecosystems which support a disproportionately large share of the world's plant and animal species. Forests play a crucial role in both nutrient and hydrological cycling and may provide sustainable economic benefits through managed harvesting of timber and the collection of non-timber products such as fruits, nuts, and rubber. Also, deforestation is a significant source of global warming through its effects on the global carbon cycle. This paper focuses on the relationship between debt and deforestation, examining conceptual and empirical arguments that debt is a source of deforestation pressure. Our study develops a behavioral model which suggests that debt can lead to myopic behavior, leading to deforestation rates that may not be optimal in the long run, but are necessary in the short run to meet current constraints. Then, country-by-country data on debt, deforestation, and other variables are analyzed with regression analysis. It is shown that debt is significantly correlated with deforestation under a wide variety of assumptions and specifications. Our results indicate that debt is an important factor in the deforestation of tropical countries. There are certainly other sources of deforestation, both micro- and macroeconomic, which may vary significantly from country to country. However, we focus on debt because of its dominant role in the economies of developing countries, and because of the increased use of debt-for-nature swaps. The link between debt and deforestation that is suggested in this paper implies that debt-for-nature swaps may have a dual effect on deforestation. First, the contractual agreement is designed to preserve forests as part of the swap. Second, the reduction in debt may itself reduce the pressure to deforest, although this indirect effect is small. Our research provides evidence that reducing debt reduces deforestation, which may be an argument to offer deforesting third-world countries some form of debt relief, and to utilize more fully debt-for-nature swaps as a tool for preserving environmental quality.

Keywords: Debt; Deforestation; Developing countries

* Corresponding author.

SSDI 0921-8009(94)00024-P

108 J.R. Kahn, J.A. McDonald / Ecological Economics 12 (1995) 107–123

1. Introduction

In recent years, there has been considerable discussion of the role of third-world debt in the rapid deforestation which is currently taking place in tropical countries (see Table 1). One aspect of this discussion is the effectiveness of using "debt-for-nature swaps" in preventing future deforestation. [1] Another aspect is whether the existence of large levels of debt leads to more rapid deforestation. Although both areas are important areas of investigation, we are focusing on the latter, the role of debt as a source of deforestation pressure. Von Moltke (1990) concurs in the need for this type of analysis, stating that (p. 5) "... the precise dynamics of the debt servicing and debt reduction, and their potential impact on tropical forest conversion, have not yet been assessed. In other words there is widespread sensibilization [of the existence, direction and magnitude of the relationship] without corresponding amounts of analysis."

In the 1970s and early 1980s, many developing countries overborrowed relative to their ability to repay the debt. [2] This accumulation of debt interfered with the development process in many countries. For example, the World Resources Institute (1992, p. 12) argues that "Debt payments often constitute a large share of the national budget, squeezing out needed investments in environmental protection or economic and human development. In some countries, debt service obligations are so large that they exceed new loans and private external investment, meaning that financial resources are flowing out of, rather than into, the country."

There are many different options available to a country to deal with its debt problem, such as debt rescheduling, debt repudiation, increased borrowing, devaluation of the domestic currency, restricting imports, and increasing exports. Both restricting imports and increasing exports could lead to increased deforestation. The actual number of options that are available to a country depends on its creditworthiness, export base, and level of development. In this paper we are focusing on explaining the economic pressures that are generated by debt which may lead to increased deforestation. We do not attempt to explain the conditions under which a country would choose alternative policies to deal with its debt. Bird (1989, p. 29), among others, feels that "there are compelling reasons to believe that an initiative based on the provision of some relief to debtors is necessary if the debt problem is to be managed successfully." [3] If debt does indeed increase deforestation, then our research will provide additional support to the proposal that many heavily-indebted developing countries should be extended some form of debt relief.

It is not immediately apparent that debt would lead to increased deforestation. Not only are there non-forest related options to deal with debt, but one could construct an argument that suggested that if a country could maximize its income or net benefits (including intra-country public goods and externalities) through a particular rate of deforestation, then it would be optimal for it to pursue this deforestation rate regardless of the debt position. In the next section, we present an economic model that suggests that under certain conditions, it is possible for debt to drive a wedge between strategies that must be pursued in the short run and strategies that are optimal in the

[1] See Chambers et al. (1992) for a discussion of the incentives of all participants and the long-term stability of debt-for-nature swaps.

[2] See Corbridge (1993, pp. 11–86) or Loxley (1986, pp. 1–87) for a comprehensive discussion of the history of the debt crisis and its effects on the economies of developing countries. As noted by Horowitz (1987, p. 103), the debt crisis began with developing countries' inability to repay foreign loans. When caught in a pattern of borrowing to meet interest payments (rather than to repay the principal loans), third-world borrowers find themselves in a downward, demoralizing spiral. As their creditworthiness plummets, and therefore new money is not forthcoming, developing debtor countries must institute ever tighter austerity or adjustment programs.

[3] Czerkawski (1991, pp. 132–135) also advocates a gradual, partial reduction in the total stock of debt, along the lines proposed by Brady (see p. 132). Past debt-reduction packages (for example, those instituted in Costa Rica, Mexico, the Philippines, and in 12 sub-Saharan African countries in 1988 and 1989) have created a favorable environment for foreign investment and stimulated new lending (p. 133).

long run. Of course, one can argue that these apparently non-optimal deforestation rates are merely an attempt by a country to redistribute income from one segment of society to another. The latter hypothesis is certainly a possibility, but it is very difficult to test. We focus on the hypothesis that debt leads countries to pursue myopic policies with respect to their tropical forests.

There are several alternative uses of a tropical forest, some of which are sustainable and some of which are nonsustainable. Sustainable activities include selective harvesting of timber, collection of fruit, nuts, honey, and rubber, environmentally benign agro-forestry, and conversion to sustainable agricultural land. Nonsustainable uses include mining the forest for timber and fuelwood or converting the forest to nonsustainable agriculture. Many social, political, and economic factors may affect the choice among these activities and influence the subsequent deforestation patterns. These factors include population growth, development patterns (roadbuilding, etc.), export prices (timber and other raw materials), import prices (energy and manufactured goods), foreign-exchange needs, and cultural factors.

The above factors will affect a country's socially optimal level of deforestation, but in many cases will not cause the actual level of deforestation to deviate from the socially optimal level. It is important to distinguish between a level of deforestation that is optimal from a forested country's point of view and a level that is optimal from a global point of view, as the global perspective incorporates the global public good aspects of tropical forests. When we use the adjective "non-optimal," it will mean non-optimal from the perspective of the forested country.

In the past several years, researchers have made several important attempts to understand the market failures leading to non-optimal levels of deforestation. These studies can be classified into three broad categories. First, there have been the country case studies, such as Repetto (1988) and Repetto and Gillis (1989). Repetto and Gillis show that a variety of inefficient government policies (investment incentives, credit concessions, tax provisions, agricultural pricing policies, and the nature of lease or sale of forest exploitation rights) creates incentives to engage in faster deforestation. Second, there have been studies that attempt to explain microeconomic behavior and how that can lead to inefficiently high levels of deforestation. For example, Mendelsohn (1990) examines the role of property rights and settler and squatter behavior.

The third category, which focuses on macroeconomic causes, is not as well developed as the first two. Some researchers, such as Von Moltke, discuss the importance of macroeconomic conditions as a constraint on the intertemporal allocation of resources, which may affect deforestation behavior. Although Von Moltke (1990) does not derive a formal behavioral model, he advances a hypothesis of myopic behavior (p. 6): "The imbalance created by overwhelming levels of debt causes countries to adopt short term policies, designed to meet the next payment of interest or principal, or to sustain positions in rescheduling negotiations. Such policies are frequently contrary to the [long term] interests of conservation."

Our contribution to this third strand of investigation is to model formally some of the macroeconomic effects which before have only been advanced intuitively. This is done by developing and estimating a constrained macroeconomic optimization model for forested tropical countries.

This study should be regarded as a complement to both the country case studies and the conceptual microeconomic studies, as all three approaches further the understanding of tropical deforestation. In fact, our study helps to explain the existence of the phenomena discussed in these other approaches. For example, Repetto and Gillis discuss case after case of inefficient government policies. However, they do not address why these seemingly irrational policies exist. Our model develops a general framework of individual country maximizing behavior under which governments might institute such short-sighted policies. This behavior is one possible explanation for the existence of these policies.

The research emphasis in this paper on the relationship between debt and deforestation should not be construed to mean that we feel that debt is the only or the most important economic factor that causes tropical deforestation. Rather, we are examining the degree to which debt may be a contributing factor to a country's level of deforestation.

2. The behavioral model

The model is based on the assumption that countries maximize the discounted stream of consumption over their planning horizon. Eq. 1 represents an identity that shows that consumption (C) is equal to GNP less investment (I), government expenditures (G), and interest payments on the stock of debt (iD). [4] The change in debt (δ) is added to this equation since changes in consumption (or competing uses of GNP) can be funded through borrowing.

$$C(t) = \mathrm{GNP}(t) - I(t) - G(t) + \delta(t) - iD(t-1) \quad (1)$$

Eq. 2 contains the production function for GNP, with time subscripts eliminated for simplicity:

$$\mathrm{GNP} = \alpha_0 R^{\alpha_1} F^{\alpha_2} (T-F)^{\alpha_3} K^{\alpha_4} L^{\alpha_5},\ \alpha_0 > 0 \quad (2)$$

The production function is a multiplicative formulation, where the greater the amount of any input (level of deforestation (R), forested land (F), non-forested land ($T-F$), capital (K), or labor (L)), the greater the resulting output. Although the forest sector is not modeled explicitly in this production function, it is included in both the R and F terms. Forestry activities that do not result in the reduction of forested area (such as selective harvesting of certain exotic hardwoods, collecting rubber from wild trees, non-destructive subsistence agriculture, and harvesting fruit, meat, fiber, nuts, and honey) would be included in F. Forestry activities that result in deforestation (clearcutting and subsequent conversion of the land to non-forested scrub or agricultural land) would be represented by the contribution of deforestation (R) to GNP. [5]

Eq. 3 contains the maximization problem and the constraints on the maximization process:

$$\max \int_0^T C(t)\, \mathrm{e}^{-rt}\, \mathrm{d}t \quad (3)$$

subject to:

$$\frac{\mathrm{d}F}{\mathrm{d}t} = -R,\quad \frac{\mathrm{d}D}{\mathrm{d}t} = \delta,\quad \frac{\mathrm{d}K}{\mathrm{d}t} = I,$$
$$C(t) \geqslant C^*,\quad C(t) \geq 0,\quad F(0) = F_0,$$
$$D(0) = D_0,\quad K(0) = K_0$$

The control variables in this model are the level of deforestation (R), the change in debt (δ), and the level of investment (I). As mentioned previously, we are focusing on debt and deforestation, and not examining all possible policies that countries may use to deal with their debt problem. For example, the amount of debt rescheduling does not appear as a control variable in our model, although it will influence interest payments on the stock of debt (iD).

The constraints on the maximization problem are relatively straightforward, except for the constraint on consumption, $C(t) > C^*$. This constraint stipulates that consumption in each period must remain above some minimum level (C^*). In

[4] It may be surprising to find a net export variable absent from this equation. The reason for this is that net exports is a component of the change in debt variable (δ). We let net exports be subsumed in the change in debt variable, which facilitates the construction of a state equation governing D. As a check, a net export variable was included in some empirical specifications, but its inclusion did not quantitatively or qualitatively alter the results. In addition to a conventional net export variable, a net export variable equal to total exports minus total imports minus net exports of forest products was tested with similar results.

[5] It should be noted that the conversion of forest land to non-forest land is much more complicated than our simple model suggests. One factor that generates this complexity is that the soil (and economic productivity) of converted tropical forest land degrades rapidly. Differential productivities (as vintages or functions of time) of converted land could thus appear in the production function to model real world phenomena more truly. However, in our model, this level of complexity has been suppressed in order to concentrate on the economic relationships outlined above. As can be seen as one works through the model to Eq. 12, this omission will not qualitatively affect the relationship between debt and deforestation. For those interested in seeing how the existence of soil degradation affects the deforestation decision, see Ehui et al. (1990).

other words, countries cannot reduce consumption to very low levels in some periods in exchange for higher levels of consumption in future periods, even though such a path may represent an unconstrained maximum. This gives an incentive for short-sighted behavior beyond that specified by Von Moltke (1990, p. 6), who blames much of the short-sighted behavior on the discounting process.

One could derive a model in which debt may influence deforestation through debt's impact on the social discount rate. However, in order to depict accurately the economic behavior in many developing countries in the 1980s (where debt and debt service dominated the economy [6]), a minimum consumption constraint is needed. There are several reasons for this. A heavily-indebted country's ability to invest in its future (either conventional investment or leaving resources intact) is severely limited by the need to pay debt service *and* meet current consumption needs. Also, as Shilling (1992) notes, many developing countries have attempted to increase current consumption at the expense of long-run sustainable consumption. Shilling suggests that the demand for these unsustainable consumption levels was due, in some cases, to "subsistence demands of the poorest segments of the population." In other cases, Shilling (p. 28) claims that it was "the relatively well off trying to increase their living standards quickly." Both of these demands would represent the type of social or political pressure that would generate a minimum acceptable consumption level in the short run. Note that it is not important to model the social, economic, and political processes that define C^* for each country. It is only necessary to hypothesize that a minimum consumption level exists.

[6] Our sample includes many countries with high debt burdens. In Bolivia and Costa Rica, debt is approximately equal to GNP (when both are measured in current U.S. dollars in 1985). There are many countries in which debt is a significant fraction of GNP (for example, Malaysia, 49%, Mexico, 42%, Brazil, 32% and Indonesia, 32%). The sample also includes some countries where debt is less important (for example, Thailand, 24%, Cameroon, 20%, and India, 12%) (see World Bank, 1989a).

The Hamiltonian equation associated with this maximization problem is represented by:

$$\begin{aligned} H = &\left[\alpha_0 R^{\alpha_1} F^{\alpha_2} (T-F)^{\alpha_3} K^{\alpha_4} L^{\alpha_5} - G - I \right.\\ &\left. -iD + \delta\right] e^{-rt} + \lambda_1(-R) + \lambda_2(I) + \lambda_3(\delta) \\ &+ \lambda_4\left[\alpha_0 R^{\alpha_1} F^{\alpha_2} (T-F)^{\alpha_3} K^{\alpha_4} L^{\alpha_5} - G - I \right.\\ &\left. -iD + \delta - C^*\right] \end{aligned} \tag{4}$$

with time subscripts eliminated for simplicity. Note that consumption, C, has been eliminated by substituting the right-hand side of Eq. 2 into the right-hand side of Eq. 1 and substituting the right-hand side of the resulting equation into Eq. 3 for $C(t)$. The first, second, and third constraints (λ_1 through λ_3) in Eq. 4 represent the state equations for the stock of forests, capital stock, and stock of debt, respectively. The fourth constraint (λ_4) embodies the minimum consumption constraint.

The necessary conditions for a constrained optimization are contained in Eqs. 5–11. Eqs. 5–7 are the optimality conditions, Eqs. 8–10 are the co-state equations, and Eq. 11 represents the Kuhn–Tucker condition. The optimality conditions are solved twice: once when the minimum consumption constraint is not binding ($\lambda_4 = 0$, $C \neq C^*$), and then when the constraint is binding ($\lambda_4 > 0$, $C = C^*$). In the binding case, Eq. 6 is solved for λ_4 and then substituted into Eq. 5. The minimum consumption constraint ($C = C^*$ from Eq. 1) is then substituted into the resulting equation. This equation is then solved for R to yield the constrained solution in Eq. 12 for the optimal level of deforestation as a function of the state variables (F, K, D), the remaining control variables (I, δ), and the minimum consumption level (C^*). In the non-binding constraint case, Eq. 6 is solved for e^{-rt} and then substituted into Eq. 5, which is then rearranged to yield the unconstrained solution for optimal R in Eq. 12.

$$\begin{aligned} \frac{\partial H}{\partial R} = &\left[\alpha_1 \alpha_0 R^{\alpha_1 - 1} F^{\alpha_2} (T-F)^{\alpha_3} K^{\alpha_4} L^{\alpha_5}\right] \\ &\times\left(e^{-rt} + \lambda_4\right) - \lambda_1 = 0 \end{aligned} \tag{5}$$

$$\frac{\partial H}{\partial I} = -e^{-rt} + \lambda_2 - \lambda_4 = 0 \quad (6)$$

$$\frac{\partial H}{\partial \delta} = e^{-rt} + \lambda_3 + \lambda_4 = 0 \quad (7)$$

$$\frac{d\lambda_1}{dt} = -\frac{\partial H}{\partial F} = -\left[e^{-rt} + \lambda_4\right]\left(\alpha_0 R^{\alpha_1} K^{\alpha_4} L^{\alpha_5}\right) \times\left[\alpha_2 F^{\alpha_2 - 1}(T-F)^{\alpha_3} - \alpha_3 F^{\alpha_2}(T-F)^{\alpha_3 - 1}\right] \quad (8)$$

$$\frac{d\lambda_2}{dt} = -\frac{\partial H}{\partial K} = -\left[e^{-rt} + \lambda_4\right] \times\left(\alpha_4 \alpha_0 R^{\alpha_1} F^{\alpha_2}(T-F)^{\alpha_3} K^{\alpha_4 - 1} L^{\alpha_5}\right) \quad (9)$$

$$\frac{d\lambda_3}{dt} = -\frac{\partial H}{\partial D} = i\, e^{-rt} + i\lambda_4 \quad (10)$$

$$\lambda_4[C(t) - C^*] = 0, \quad \lambda_4 \geqslant 0 \quad (11)$$

Eq. 12 describes the decision rule that defines the current level of deforestation.

unconstrained solution:

$$R = \left(\left[\frac{\lambda_2}{\lambda_1}\right]\alpha_0 \alpha_1 F^{\alpha_2}(T-F)^{\alpha_3} K^{\alpha_4} L^{\alpha_5}\right)^{-1/(\alpha_1 - 1)}$$

constrained solution:

$$R = \alpha_1\left(\frac{\lambda_2}{\lambda_1}\right)[C^* + G + I + iD - \delta]$$

$$\lambda_1 > 0, \lambda_2 > 0 \quad (12)$$

Although optimal control models of this nature are often employed to examine comparative steady states, we focus on the current optimization problem (which takes into account future costs and benefits). In the context of the rapid deforestation of the 1980s, the current deforestation decision is more policy-relevant than steady-state behavior (which many experts predict to be near complete deforestation sometime in the not so distant future). In other words, we are looking for information as to how current rates of deforestation are related to the level of other economic variables.

The first expression of the unconstrained solution in Eq. 12, $(\alpha_0\alpha_1\lambda_2/\lambda_1)$,is positive since the exponent of deforestation (α_1), the technology coefficient (α_0), and the shadow prices of forested land (λ_1) and of the capital stock (λ_2) must all be positive. Given that $(\alpha_0\alpha_1\lambda_2/\lambda_1)$ must be positive, it is possible to identify the direction of the direct effects that changes in the inputs (F, $T-F$, K, L) will generate in the level of deforestation. If $\alpha_1 < 1$, then increases in any input will increase the optimal level of deforestation, and if $\alpha_1 > 1$, then the same change will reduce the optimal level of deforestation.

When the maximization problem is solved with the minimum consumption constraint serving as a binding constraint, the current optimal level of deforestation is no longer a function of inputs, but it is a function of competing uses of GNP, as shown above in the constrained solution of Eq. 12. Given that λ_2, the shadow price of capital, is positive, the greater the competing uses of GNP, and the greater the minimum acceptable level of consumption, the greater the level of deforestation.

To guard against the possibility that these results might be peculiar to the choice of functional form, we solved the maximization problem for two other types of production functions. These include a quadratic function (where interaction terms are limited to interactions with R) and a hybrid function which combines some of the desirable properties of both the quadratic and log-linear production functions. The quadratic production function and the solutions to the maximization problem are contained in:

quadratic production function:

$$\text{GNP} = \gamma_0 + \gamma_1 R + \gamma_2 F + \gamma_3(T-F) + \gamma_4 K + \gamma_5 L + \gamma_6 R^2 + \gamma_7 F^2 + \gamma_8 (T-F)^2 + \gamma_9 K^2 + \gamma_{10} L^2 + \gamma_{11} RF + \gamma_{12} R(T-F) + \gamma_{13} RK + \gamma_{14} RL$$

unconstrained solution:

$$R = \frac{1}{2\gamma_6}\left[\frac{\lambda_1}{\lambda_2} - \gamma_1 - \gamma_{11} F - \gamma_{12}(T-F) - \gamma_{13} K - \gamma_{14} L\right]$$

J.R. Kahn, J.A. McDonald / Ecological Economics 12 (1995) 107–123 113

Table 1
Sample countries, average annual deforestation, and public debt levels, 1981–1985

Country (Code)	Deforestation (1000s of hectares)	Public debt (billions of U.S. dollars)
Bangladesh (13) [a]	8.0	5.264
Belize (16)	8.5	0.077
Benin (17) [a]	67.2	0.610
Bolivia (20) [a]	117.2	3.306
Botswana (21) [a]	20.0	0.266
Brazil (22) [a]	2530.0	63.856
Burkina Faso (172) [a]	80.0	0.430
Burundi (27) [a]	1.1	0.321
Cameroon (29) [a]	110.0	1.998
Central African Rep. (32) [a]	55.0	0.255
Chad (34)	80.0	0.165
Colombia (38) [a]	890.0	7.905
Comoros Islands (39)	0.5	0.099
Congo (40) [a]	22.0	2.062
Costa Rica (41) [a]	65.0	2.991
Dominican Rep. (47) [a]	4.0	2.165
Ecuador (48) [a]	340.0	5.965
El Salvador (50) [a]	4.5	1.253
Ethiopia (52) [a]	88.0	1.403
Fiji (53)	1.7	0.281
Gabon (58) [a]	15.0	0.860
Gambia (59)	5.2	0.169
Ghana (62) [a]	72.0	1.274
Guatemala (69) [a]	90.0	1.614
Guinea-Bissau (71)	57.0	0.194
Guyana (72)	2.5	0.706
Haiti (73) [a]	1.8	0.473
Honduras (74) [a]	90.0	1.782
India (78) [a]	147.0	23.647
Indonesia (79) [a]	620.0	23.017
Ivory Coast (85) [a]	510.0	5.353
Jamaica (86) [a]	2.0	2.459
Kenya (89) [a]	39.0	2.730
Liberia (96) [a]	46.0	0.793
Madagascar (100) [a]	156.0	1.938
Malawi (101) [a]	150.0	0.759
Malaysia (102) [a]	255.0	11.863
Mali (104) [a]	36.0	1.096
Mauritania (107) [a]	13.3	1.201
Mauritius (108) [a]	−0.1	0.367
Mexico (109) [a]	615.0	63.354
Nepal (114)	84.0	0.436
Nicaragua (120) [a]	121.0	3.624
Niger (121) [a]	67.1	0.729
Nigeria (122) [a]	400.0	12.135
Pakistan (125) [a]	9.0	10.154
Panama (126) [a]	36.0	3.098
Papua-New Guinea (127)	23.0	0.952
Paraguay (128) [a]	212.0	1.236
Peru (129) [a]	270.0	8.706
Philippines (130) [a]	92.0	12.092
Rwanda (138) [a]	5.20	0.266
Samoa (516)	1.6	0.061
Senegal (141) [a]	50.0	1.616
Sierra Leone (143) [a]	6.0	0.388
Solomon Islands (23)	0.8	0.038
Somalia (146)	13.5	1.455
Sri Lanka (149) [a]	58.2	2.441
Sudan (153) [a]	504.0	5.842
Tanzania (159) [a]	130.0	2.796
Thailand (160) [a]	379.0	7.847
Togo (162) [a]	12.1	0.813
Trinidad-Tobago (164) [a]	0.8	1.112
Uganda (168)	50.0	0.706
Venezuela (175) [a]	245.0	16.417
Zaire (182) [a]	370.0	4.610
Zambia (183) [a]	70.0	2.811
Zimbabwe (136) [a]	80.0	1.369

[a] Indicates that the country is included in the 1981–1985 sample.

constrained solution:

$$R = \frac{\dfrac{\lambda_1}{\lambda_2} \pm \sqrt{\left(\dfrac{\lambda_1}{\lambda_2}\right)^2 - 4\gamma_6(Z)}}{2\gamma_6}$$

$$\text{where} \quad Z = \gamma_0 + \gamma_2 F + \gamma_3(T-F) + \gamma_4 K + \gamma_5 L + \gamma_7 F^2 + \gamma_8 (T-F)^2 + \gamma_9 K^2 + \gamma_{10} L^2 - G - I - iD + \delta - C^*$$

$$\lambda_1 > 0, \quad \lambda_2 > 0 \tag{13}$$

The hybrid production function and corresponding solutions are contained in:

hybrid production function:

$$\text{GNP} = \text{B}_0 F^{\beta_0} K^{\beta_4} L^{\beta_5} + \text{B}_1 (T-F)^{\beta_1} K^{\beta_6} L^{\beta_7} + \text{B}_2 R^{\beta_2} K^{\beta_8} L^{\beta_9}$$

unconstrained solution:

$$R = \left(\frac{\lambda_2}{\lambda_1} \beta_2 \text{B}_2 K^{\beta_8} L^{\beta_9}\right)^{-1/(\beta_2 - 1)}$$

constrained solution:

$$R = \left[\beta_2 \frac{\lambda_2}{\lambda_1} \right] \times \left(-B_0 F^{\beta_0} K^{\beta_4} L^{\beta_5} - B_1 (T-F)^{\beta_1} K^{\beta_6} L^{\beta_7} + G + I + iD - \delta + C^* \right)$$

$$\lambda_1 > 0, \quad \lambda_2 > 0 \tag{14}$$

The solutions to the maximization problem when the quadratic production function of Eq. 13 is used provide interesting results, especially when compared to the solutions presented in Eq. 12. The unconstrained optimal level of deforestation is still a function of the inputs, but it is unambiguously an inverse relationship. The greater the level of any input, the lower the optimal level of deforestation. The constrained solution of Eq. 13 represents a bigger departure from the corresponding solution of Eq. 12 because in the quadratic version of the problem, the constrained optimal level of deforestation is a function of both the inputs (F, $T-F$, K, L) and the competing uses of GNP (G, I, iD, C^*, $-\delta$). Of course, since the radical in the solution has both positive and negative roots, the level of deforestation could be negatively or positively related to any of these variables.

The hybrid production function is different from those normally seen in GNP models due to the nature of the inputs. In a conventional GNP model (when the inputs are labor, capital, energy, etc.), it is acceptable to have a multiplicative formulation, where a zero level of any input leads to a zero level of GNP. However, the same thing cannot be said of forested or deforested land. If non-forested land ($T-F$), forested land (F), or the level of deforestation (R) is zero, there is no reason for GNP to be zero. Consequently, the GNP production function requires additive terms to capture these properties.

Capital (K) and labor (L) are included multiplicatively in each term so that the marginal products of F, ($T-F$), and R are functions of the levels of K and L. However, just as in conventional GNP models, zero levels of K or L will cause GNP to vanish. It should be noted that the functional form of these three additive terms in Eq. 14 is not designed to represent three distinct outputs from three separate sectors of the economy. It is simply a mathematical formulation which allows the marginal product of each input to be a function of all other inputs, and at the same time, zero levels of F, R, or $T-F$ do not cause GNP to vanish.

The unconstrained solution to the optimal control problem with the hybrid production function is analogous to the unconstrained solution for the log-linear production function (Eq. 2). If the exponent on R (β_2 in this case) is greater than one, increases in the levels of the inputs (K, L) will reduce the optimal level of deforestation. If $\beta_2 < 1$, then increases in the levels of the inputs will increase the optimal level of deforestation. The constrained solution in the hybrid case contains unambiguous comparative statics. If inputs (F, $T-F$, K, L) increase, optimal deforestation falls, and if competing uses of GNP increase, optimal deforestation increases.

The comparative statics are summarized in Table 2. Although the three sets of solutions contain different comparative statics for the effect of other variables on deforestation, they contain significant information that can be used to structure a regression equation which is designed to explain the current levels of deforestation. Input variables (F, K, L, $T-F$) have an effect on deforestation in all three unconstrained solutions and in two of the three constrained solutions. Competing uses of GNP (I, G, iD, C^* and $-\delta$) are only important when a country's optimization strategies are constrained by the existence of a minimum acceptable consumption constraint.

Since input variables are relevant in all but one of the six cases, it is not difficult to make a case to include these variables in an estimating equation designed to explain deforestation. However, the variables which constitute competing uses of GNP (I, G, iD, C^* and $-\delta$) only affect deforestation when a country's optimization possibilities are constrained by a minimum acceptable consumption constraint. The ideal way to deal with this dichotomy would be to divide the sample of countries into those which are constrained and those which are unconstrained, and specify different sets of variables for each sample.

The estimating equation appropriate for the sample of unconstrained countries would contain only inputs as explanatory variables, while the corresponding equation for the constrained countries would contain both inputs and competing uses of GNP as explanatory variables.

Unfortunately, it is not possible to observe which countries are constrained or unconstrained, although it is possible to observe variables that may influence the existence of a minimum acceptable consumption constraint. Consequently, all countries will be included in one sample, with both inputs and competing uses of GNP as explanatory variables. In addition, other variables will be included in the estimation in an attempt to control for the likelihood of the existence of this constraint.

There is one other complication associated with the interpretation of the comparative statics. Ideally, one would like to substitute the other choice variables (I, δ) out of Eqs. 12–14, but Eqs. 6 and 7 are too sparse to allow for this. It is conceivable that the multiplier equations (Eqs. 8–10) could be used to eliminate I and δ, but this process would replace these variables with initial values of the state and exogenous variables, rather than current values of these variables. This would not be satisfactory since the question of interest is the relationship between current levels of deforestation and the current economic situation. Also, it would be difficult to implement empirically a model based on initial values, since there is no guide as to when the initial period was. Additionally, even if one could determine the initial period, it is unlikely that sufficient data would exist for most developing countries if one moves very far into the past.

The difficulty in interpreting comparative statics arises because if one is interested in finding how a change in an explanatory variable, such as capital, affects deforestation, this cannot be done simply by differentiating a solution (either the constrained or unconstrained solutions to Eq. 14, for example) with respect to K, since this would only give the direct effect of K on R. Since a change in K would affect the other choice variables (I, δ), there would also be an indirect effect of K on R through I and δ. A priori predictions can therefore only be made on the direct impacts of changes in explanatory variables. This yields some important information, but it is conceivable that the direct effects of a variable on deforestation could be overwhelmed by an indirect effect of the opposite direction. In the empirical section of the paper, regression equations based on Eqs. 12–14 will be estimated both with and without I and δ as explanatory variables.

Eqs. 12–14 describe the decision rules that define the current level of deforestation. Although optimal control models of this nature are often employed to examine comparative steady states, we focus on the current optimization problem (which takes into account future costs and benefits). In the context of the rapid deforestation of the 1980s, the current deforestation decision is more policy-relevant than steady-state behavior (which many experts predict to be near complete deforestation sometime in the not-so-distant future).

Although Eqs. 12–14 are derived in an effort to examine the relationship between debt and deforestation, the specific policies by which governments might service debt through increased deforestation are not incorporated into the model. This is in keeping with our effort to look at this issue in a cross-sectional context, rather than a case-study approach. Additionally, insufficient data exist to incorporate these policies in a cross-sectional framework.

There are many different forest-related policy mechanisms through which a government could encourage the generation (or conservation) of foreign exchange in order to service debt. These policies (which would increase deforestation) include: relying on fuelwood to reduce the import of energy, encouraging the export of timber products, addressing population growth and urban problems by subsidizing urban–rural migration, and promoting the conversion of forest land to agricultural land to increase exports and reduce reliance on food imports.

Similarly, there are many different non-forest policy options to deal with debt. These policy options include: devaluing domestic currency in an attempt to encourage exports, obtaining International Monetary Fund or other international assistance, imposing tariffs to discourage imports, subsidizing exports, and rescheduling or repudiating external debt obligations.

The extent to which any particular country relies on any specific forest or non-forest policy depends on that country's willingness and ability to pursue that policy. Consequently, a regression based on Eqs. 12–14 will not depict exactly any particular country's deforestation response to debt. Rather, in its empirical implementation, it will represent an average response across countries.

Ideally, all the variables in our Eqs. 12–14 would be included in our empirical analysis. Unfortunately, data on capital (K) are not available for many tropical countries, so including this variable would drastically reduce the sample size. Also, the minimum acceptable consumption level (C^*) is excluded from our empirical work as it is an unobserved variable. Regional dummy variables, income level dummy variables, an index of development, population, and urban population are used as controls and proxies for C^* in our estimations. [7]

Other unobserved variables include the production function coefficients. If production functions were identical across countries, the omission of these coefficients would not create any problems. However, this assumption is not likely to hold since countries with different physical environments, cultures, and stages of development are likely to employ different technologies. It is possible to develop reasonable proxy variables to control for this variation. Many of the proxies used to control for the variation in C^* will also control for variation in these production function coefficients, as well as control for the likelihood that the minimum acceptable consumption constraint is binding. It should be noted that the regional dummy variables may capture some of the specific motivations for deforestation which may vary across regions. For example, the primary activities that remove the forest tend to be fuelwood and overgrazing in Africa, timbering in South Asia, and conversion to agricultural land in Latin America. Since these activities are likely to be differentially driven by debt, the regional dummy variables will help to account for these and other sources of variation in deforestation levels.

Table 2
Comparative statics: Sign of direct effect on optimal level of deforestation due to changes in inputs and competing uses of GNP

	Inputs ($F,K,L,$ $T-F$)	Competing uses of GNP ($I,G,iD,C^*,-\delta$)
Unconstrained solution		
Production function:		
(i) log-linear (see Eq. 12)	$-$ if $\alpha_1 > 1$	zero
	$+$ if $\alpha_1 < 1$	zero
(ii) quadratic (see Eq. 13)	$-$	zero
(iii) hybrid (see Eq. 14)	$-$ if $\beta_2 > 1$	zero
	$+$ if $\beta_2 < 1$	zero
Constrained solution		
Production function:		
(i) log-linear (see Eq. 12)	zero	+
(ii) quadratic (see Eq. 13)	+ or −	+ or −
(iii) hybrid (see Eq. 14)	−	+

3. Data

The data used in our empirical analysis come primarily from Food and Agriculture Organization (FAO) and World Bank sources. Our sample consists only of those tropical countries for which both deforestation and debt data are available. Tropical countries are defined as those countries whose land masses lie entirely or primarily between the Tropics of Cancer and Capricorn. For example, Argentina and South Africa are not included in our sample. The time dimension of

[7] The formula for the index of development is presented below, where "min" refers to the minimum value for all countries in our sample and "max" refers to the corresponding maximum value. Data is obtained from World Bank sources (1989a,b,c). Index = min(crude death rate)/crude death rate + min(crude birth rate)/crude birth rate + per capita caloric intake/max(per capita caloric intake) + min(infant mortality rate)/infant mortality rate + primary school enrollment rate/max(primary school enrollment rate) + secondary school enrollment rate/max(secondary school enrollment rate) + life expectancy/max(life expectancy) + energy consumption per capita/max(energy consumption per capita) + GNP per capita/max(GNP per capita) + personal consumption per capita/max(personal consumption per capita).

the sample is limited by the availability of deforestation data. Deforestation is available as an average of the 1975–1980 period and an average of the 1981–1985 period. However, the need for lagged variables as instruments to explain right-hand side endogenous variables prevents regression analysis of the 1975–1980 data. The countries that are included in our sample are listed in Table 1.

Data on deforestation and total forested area are taken from FAO sources. [8] The FAO defines deforestation as the permanent conversion of forest land to other uses. Although this is the most comprehensive data set pertaining to forests and deforestation, it has several shortcomings. First, it is difficult to determine whether clearcutting for logging, clearcutting for conversion to agriculture, or even selective logging results in permanent conversion to non-forest uses, as this may depend on future actions. Second, the data is largely self-reported by the tropical countries, and therefore may be biased. Both the deforestation and total forested area variables are measured in thousands of hectares (ha). Tropical open and tropical closed forests are aggregated in our deforestation and forested area variables.

The primary debt variables used in our empirical analysis are total debt service (which represents actual payments of interest and principal on debt), relative debt service (debt service/exports), and the percent change in public external debt. [9] The total debt service variable corresponds to the iD term in Eqs. 12–14. In the late 1970s and 1980s many countries rescheduled their debt obligations. [10] This debt rescheduling reduced countries' debt-servicing burdens, and is reflected in our iD variable. The percent change in public debt corresponds to the δ term in these equations. Our debt data are taken from the World Bank's *World Tables* (1989a) and *World Debt Tables* (1989b). We convert these data from millions of current U.S. dollars to their real (1980) equivalent. Table 1 contains a listing of countries, average deforestation levels, and average public debt levels for the 1981–1985 period.

For our labor force variable we use the 1988 percent of population which comprises the work force (as reported by the United Nations' *Human Development Report 1990*) multiplied by total population. Alternative measures exist, but are not employed in the analysis. Other measures such as working age population are not as complete and would generate reductions in the number of countries included in our sample.

The government expenditure and investment variables we use are from the World Bank's *World Tables* (1989a). [11] Although these variables are in current units of domestic currency, we calculate the real U.S. dollar (1980) equivalent. The investment variable which is employed is gross domestic investment, which the World Bank defines as the sum of gross domestic fixed investment and the change in stocks.

Two compromises are required to implement our model. Both compromises are necessitated by the high degree of collinearity between pairs of variables. The simple correlation coefficient between total forested area (F) and non-forested

[8] Data for the period 1975–1980 are obtained from the FAO Forest Resources reports (1981a,b,c). Data for the period 1981–1985 are obtained from Lanly (1988).

[9] Public external debt is the sum of all external obligations of public debtors (including the national government, its agencies and autonomous public bodies) that may or may not be guaranteed by a public entity. Although total external debt (the sum of public, publicly guaranteed and private external debt) is a better measure than public and publicly guaranteed debt, there are too many gaps in the private external debt measure. The use of the public debt series is not a critical shortcoming however, since the correlation between public/publicly guaranteed external debt and private external debt is 0.899 for the countries in our sample for which both are available. Also, for most countries for which data is available, public external debt is approximately 70–90% of total external debt. It also should be noted that there is substantial variation in public external debt in our sample of countries.

[10] See Korner et al. (1986, pp. 65–71, pp. 188–191), Loxley (1986, pp. 77–79) or Corbridge (1993, pp. 50–53) for an analysis of the economic effects of debt restructuring, as well as a listing of the countries that rescheduled.

[11] It should be noted that (World Bank, 1989a, p. 5) "inadequate statistical coverage of state, provincial and local governments has dictated the use of central government data only" in the compilation of the government expenditure data. However, the World Bank indicates where a distortion may be caused by this omission of these lower levels of government, and this occurs for only four countries in our sample of 55 (Brazil, Congo, India, and Pakistan).

area $(T-F)$ is -0.9996. This high correlation coefficient means that these two variables contain the same information. Consequently, we dropped non-forested area $(T-F)$ from our empirical analysis. Similarly, the simple correlation coefficient between public debt (D) and the change in public debt (δ) is 0.956. The change in public debt variable (δ) is redefined as the relative or percent change in public debt to reduce this collinearity problem. However, since δ is a choice variable, it is excluded from our primary estimations. This modification only matters when looking at the estimations aimed at determining the direct effects and does not matter when the sum of direct and indirect effects is examined.

4. Empirical results

The discussion of our empirical results will focus on the 1981–1985 period. This is when both deforestation rates and external debt accumulation increased dramatically, leading to the speculation that they were linked. It should be cautioned that since our paper is based on cross-sectional data from 1981–1985, our results may not necessarily be applicable in the 1990s.

The regressions to be estimated are based on the solutions to the optimization models contained in Eqs. 12–14. As discussed above, it was not possible to derive solutions in which the other choice variables (I and δ) were algebraically eliminated. Consequently, the estimations have been conducted in two fashions. First, the regressions are estimated with I and δ as explanatory variables. [12] These results (columns 1 and 2 of Tables 3–6) show the direct effects on deforestation of a change in an explanatory variable. Second, the regressions are estimated with I and δ eliminated from the regressions. These results (columns 3 and 4 of Tables 3–6) show the total (direct and indirect) effects on deforestation of a change in an explanatory variable. Since columns 3 and 4 of these tables contain the total effects, they represent our most important regression results.

Additionally, since government spending may be an endogenous variable (it is treated as endogenous in some studies and exogenous in others), it is treated as an endogenous variable in columns 1 and 3 and an exogenous variable in columns 2 and 4 of Tables 3 through 6. It should be highlighted that since column 3 in each table measures both direct and indirect effects and treats government spending as endogenous, it represents the most meaningful regression in each table.

All the regressions in Tables 3 through 6 are scaled to account for size. The reason for doing this is to eliminate the generation of a relationship between deforestation and debt that is merely based on size. In other words, large countries could tend to have large debt, large forests, and large deforestation while small countries could tend to have small debt, small forests, and small deforestation. An unscaled regression could establish a relationship between debt and deforestation that is completely driven by country size and has nothing to do with economic relationships. Two procedures were used to adjust for size. In the first, all relevant variables (that is, all variables except dummy and proportional variables) were divided by population. These regressions are contained in Tables 3 and 5. The second procedure scales by real U.S. dollar GNP, with the regression results presented in Tables 4 and 6.

Our least impressive regression results are contained in Table 3, where the debt variable is measured as total debt service per million population. The t-statistic for the coefficient of the debt variable is only 1.28 in the primary regression of column 3. In this regression, the coefficients on regional dummy variables suggest that there are significant differences between South America and the other regions, but no other variable is significant with the possible exception of the labor force variable, which is borderline significant with a t-statistic of approximately -1.5. It is interesting to note that when scaled by

[12] These regressions are estimated using the two-stage least-squares procedure of the personal computer version of LIMDEP, with investment and percent change in debt treated as right-hand side endogenous variables.

Table 3
Two-stage least-squares regressions explaining deforestation (1981–1985) using total debt service (t-statistics in parentheses). Dependent variable: deforestation (1000s ha) per million people

Independent variables	Direct effects (investment and change in debt variables included)		Direct & indirect effects (investment and change in debt variables excluded)	
Intercept (South America)	37.15 (3.25)	36.35 (3.13)	40.30 (4.26)	38.38 (4.13)
Africa dummy	−13.39 (−2.08)	−13.82 (−2.11)	−13.87 (−2.62)	−12.91 (−2.48)
Central America & Caribbean dummy	−14.35 (−2.52)	−15.89 (−2.77)	−13.55 (−2.45)	−14.43 (−2.65)
Asia dummy	−19.76 (−3.20)	−18.74 (−3.00)	−21.19 (−3.54)	−20.41 (−3.46)
Forested land area (10^3 ha/10^6 people)	0.00036 (1.15)	0.00043 (1.34)	0.00028 (0.94)	0.00027 (0.94)
Labor force (% of population)	−0.34 (−1.11)	−0.24 (−0.80)	−0.39 (−1.53)	−0.37 (−1.51)
Government spending (real US$/$10^6$ people)	$-0.55*10^{-8}$ [a] (−0.20)	$0.20*10^{-7}$ [b] (0.83)	$-0.23*10^{-7}$ [a] (−1.57)	$-0.85*10^{-8}$ [b] (−0.66)
Investment (real US$/$10^6$ people)	$-0.16*10^{-7}$ (−0.98)	$-0.26*10^{-7}$ (−1.61)	-	-
Total debt service (real US$/$10^6$ people)	$0.94*10^{-7}$ (1.41)	$0.93*10^{-7}$ (1.37)	$0.51*10^{-7}$ (1.28)	$0.41*10^{-7}$ (1.03)
Annual % change in public external debt	1.56 (0.02)	−28.21 (−0.36)	-	-
R^2	0.292	0.268	0.286	0.305
Number of observations	55	55	55	55

[a] Government spending is treated as an endogenous variable.
[b] Government spending is treated as an exogenous variable.

Table 4
Two-stage least-squares regressions explaining deforestation (1981–1985) using total debt service (t-statistics in parentheses). Dependent variable: deforestation (1000s ha) per million of real U.S. dollar GNP

Independent variables	Direct effects (investment and change in debt variables included)		Direct & indirect effects (investment and change in debt variables excluded)	
Intercept (South America)	0.025 (1.21)	0.026 (1.30)	0.011 (0.93)	0.0068 (0.60)
Africa dummy	−0.0095 (−0.62)	−0.014 (−0.95)	−0.0039 (−0.30)	−0.0070 (−0.56)
Central America & Caribbean dummy	−0.015 (−1.14)	−0.019 (−1.47)	−0.013 (−0.99)	−0.017 (−1.32)
Asia dummy	−0.0225 (−1.35)	−0.025 (−1.49)	−0.021 (−1.51)	−0.023 (−1.71)
Forested land area (10^3 ha/10^6 of US$ real GNP)	0.00089 (2.43)	0.00079 (2.23)	0.00093 (2.58)	0.00082 (2.39)
Labor force (per 10^6 of US$ real GNP)	0.000014 (2.94)	0.000013 (2.80)	0.000014 (3.03)	0.000013 (2.90)
Government spending (real US$/$10^6$ of US$ real GNP)	$-0.17*10^{-7}$ [a] (−0.13)	$0.85*10^{-7}$ [b] (0.84)	$-0.86*10^{-7}$ [a] (−0.81)	$0.12*10^{-7}$ [b] (0.16)
Investment (real US$/$10^6$ of US$ real GNP)	$0.58*10^{-9}$ (0.007)	$-0.11*10^{-8}$ (−0.01)	-	-
Total debt service (real US$/$10^6$ of US$ real GNP)	$0.21*10^{-6}$ (0.72)	$0.94*10^{-7}$ (0.35)	$0.33*10^{-6}$ (1.86)	$0.24*10^{-6}$ (1.49)
Annual % change in public external debt	−0.11 (−0.80)	−0.15 (−1.14)	-	-
R^2:	0.321	0.333	0.310	0.349
Number of observations	55	55	55	55

[a] Government spending is treated as an endogenous variable.[b] Government spending is treated as an exogenous variable.

population, the size of the country's forests has little effect on deforestation.

The government spending variable has a t-statistic of -1.57 in this regression, but it does not have the predicted sign (see Table 2). In all the other regressions of all four tables, the government spending variable is insignificant. This lack of significance and the negative tendencies of the government spending variable may be due to several factors. First, the government spending variable may have a significant consumption component, which would contribute to the meeting of consumption needs. Second, current government spending may be related to past government spending and actually proxying for an infrastructure variable. Since both of these effects could be negative, they may counteract the hypothesized positive effect.

Table 4 contains regressions that are based on the same variables as Table 3, except that the regression is scaled by real U.S. dollar GNP, rather than population. This scaling procedure leads to better regression results with higher R^2 values and more of the important variables (particularly total debt service) becoming statistically significant in the important regression of column 3. In addition to the improved performance of total debt service, one of the interesting differences between Tables 3 and 4 is that the scaling by real U.S. dollar GNP eliminates much of the difference among regions, as evidenced by the insignificance of all but one of the regional dummy variables in Table 4. Another interesting difference is that the scaling by real U.S. dollar GNP leads to the forested land area variable becoming statistically significant, with more forests leading to more deforestation, ceteris paribus.

The labor force variable (labor force relative to GNP) is statistically significant, with a greater labor force (relative to GNP) leading to more deforestation. An examination of Table 2 suggests that the direct effects of expanding the labor force (unscaled) are unclear or ambiguous. A negative impact was observed (although insignificant) in Tables 3 and 5. However, the labor force variable in Tables 4 and 6, which is the labor force divided by real U.S. dollar GNP (or the multiplicative inverse of GNP per unit labor)

Table 5

Two-stage least-squares regressions explaining deforestation (1981–1985) using total debt service relative to exports (t-statistics in parentheses). Dependent variable: deforestation (1000s ha) per million people

Independent variables	Direct effects (investment and change in debt variables included)		Direct & indirect effects (investment and change in debt variables excluded)	
Intercept (South America)	24.85 (1.84)	26.04 (1.89)	26.26 (2.37)	25.55 (2.34)
Africa dummy	−13.78 (−2.28)	−14.59 (−2.38)	−13.82 (−2.66)	−12.67 (−2.49)
Central America & Caribbean dummy	−13.76 (−2.42)	−15.52 (−2.71)	−13.71 (−2.49)	−14.66 (−2.72)
Asia dummy	−19.41 (−3.21)	−18.56 (−3.02)	−19.63 (−3.26)	−18.83 (−3.19)
Forested land area (10^3 ha/10^6 people)	0.00058 (1.83)	0.00063 (1.97)	0.00054 (1.92)	0.00049 (1.80)
Labor force (% of population)	−0.25 (−0.86)	−0.15 (−0.50)	−0.27 (−1.07)	−0.27 (−1.12)
Government spending (real US$/$10^6$ people)	$-0.74*10^{-8}$ [a] (−0.27)	$0.21*10^{-7}$ [b] (0.88)	$-0.12*10^{-7}$ [a] (−0.88)	$0.10*10^{-8}$ [b] (0.09)
Investment (real US$/$10^6$ people)	$-0.30*10^{-8}$ (−0.20)	$-0.14*10^{-7}$ (−0.98)	–	–
Relative debt service (total debt service/exports)	70.26 (2.17)	62.00 (1.90)	67.80 (2.31)	61.96 (2.15)
Annual % change in public external debt	3.46 (0.05)	−35.44 (−0.49)	–	–
R^2	0.296	0.271	0.292	0.316
Number of observations	55	55	55	55

[a] Government spending is treated as an endogenous variable.

[b] Government spending is treated as an exogenous variable.

has a positive effect on deforestation. Although there is no a priori prediction of the sign of the coefficient of this variable, one would expect the sign to remain constant across specification. This may be due to the scaling by GNP changing the nature of the variable, as labor per unit output may be serving as a proxy for development or technology, with lower values of the variable implying higher states of development or more sophisticated technologies. This interpretation then means that countries that have more developed economies (holding debt and other variables constant) tend to deforest less.

Another way of measuring the burden of the debt is to measure debt relative to the ability to repay it, which would be related to foreign-exchange earnings. The debt variable used in the regressions of Tables 5 and 6 is designed to capture this, with the debt variable constructed as total debt service divided by total exports.

In Table 5, the regressions are scaled by population, with more significant results than the regressions in Table 3 (which were also scaled by population). In particular, the relative debt service variable becomes highly significant, with the t-statistic equal to 2.31 in column 3. This debt variable (total debt service/total exports) also worked well when the regressions were scaled by real U.S. dollar GNP. These results are contained in Table 6.

In order to test the robustness of our results, additional regressions were estimated with different functional forms, different measures of debt, and additional control variables. These variables included an index of development (see footnote 7), total population as well as population density, GNP, and dummy variables pertaining to the income level of each country. These regressions had comparable results and are not reported to conserve space.

As noted in the discussion of data, the regional dummy variables are included to control for differences in production technologies and differences in non-debt related motivations for deforestation (and other factors that may vary by region). After controlling for these region-specific

Table 6

Two-stage least-squares regressions explaining deforestation (1981–1985) using total debt service relative to exports (t-statistics in parentheses). Dependent variable: deforestation (1000s ha) per million of real U.S. dollar GNP

Independent variables	Direct effects (investment and change in debt variables included)		Direct & indirect effects (investment and change in debt variables excluded)	
Intercept (South America)	−0.013 (−0.51)	−0.013 (−0.49)	−0.022 (−1.28)	−0.025 (−1.50)
Africa dummy	−0.0022 (−0.14)	−0.0054 (−0.37)	0.0024 (0.18)	−0.0014 (−0.11)
Central America & Caribbean dummy	−0.013 (−0.92)	−0.016 (−1.26)	−0.011 (−0.82)	−0.015 (−1.16)
Asia dummy	−0.017 (−1.04)	−0.016 (−1.06)	−0.013 (−0.96)	−0.016 (−1.14)
Forested land area (10^3 ha/10^6 of US$ real GNP)	0.0011 (3.01)	0.0011 (2.91)	0.0012 (3.10)	0.0011 (2.98)
Labor force (per 10^6 of US$ real GNP)	0.000013 (2.80)	0.000012 (2.74)	0.000012 (2.88)	0.000012 (2.91)
Government spending (real US$/$10^6$ of US$ real GNP)	$-0.23*10^{-7}$ [a] (−0.21)	$0.59*10^{-7}$ [b] (0.70)	$-0.34*10^{-7}$ [a] (−0.37)	$0.33*10^{-7}$ [b] (0.46)
Investment (real US$/$10^6$ of US$ real GNP)	$0.16*10^{-7}$ (0.23)	$-0.79*10^{-8}$ (−0.12)	–	–
Relative debt service (total debt service/exports)	0.23 (2.44)	0.22 (2.38)	0.26 (3.03)	0.24 (2.94)
Annual % change in public external debt	−0.062 (−0.52)	−0.074 (−0.64)	–	–
R^2	0.305	0.341	0.304	0.332
Number of observations	55	55	55	55

[a] Government spending is treated as an endogenous variable.

[b] Government spending is treated as an exogenous variable.

reasons for deforestation, there exists an additional effect due to debt which is strongly significant (see column 3 of Tables 4 through 6).

In summary, our empirical results suggest that there is a strong statistical relationship between debt and deforestation. An examination of the third column of Tables 3 through 6, which represents the most meaningful regression models, shows that a strong positive statistical relationship exists between deforestation and debt, which is robust across scaling procedures and measures of debt burden. The only exception to this is in Table 3, where the positive coefficient on debt has a t-statistic of 1.28, which is associated with an alpha level of 0.2, which is relatively low, but higher than normally associated with the rejection of a null hypothesis. However, since this is the worst regression of the four primary regressions contained in the third column of each table, and since the other regressions have positive coefficients on the debt variables with t-statistics of 1.86, 2.31, and 3.03, it may be reasonable to reject the null hypothesis that debt and deforestation are unrelated and accept the alternative hypothesis that there is a positive relationship between debt and deforestation.

5. Conclusions

There are many factors that affect the level of deforestation and generate a disparity between the actual and optimal level of deforestation. Our study develops a behavioral model which suggests that debt can lead to myopic behavior, leading to deforestation rates that may not be optimal in the long run, but are necessary in the short run to meet current constraints. The constrained optimizing equations for deforestation which were derived from the models were estimated using two-stage least-squares regressions, which showed that the relationship between deforestation and debt was statistically strong.

Our results indicate that debt is an important factor in the deforestation of tropical countries. It should be emphasized that this paper is not suggesting that there are not important microeconomic causes of deforestation, which may vary significantly from country to country.

The link between debt and deforestation which is suggested in this paper implies that debt-for-nature swaps have a dual effect on deforestation. First, the contractual agreement is designed to preserve forests as part of the swap. Second, the reduction in debt may itself reduce the pressure to deforest, although this indirect effect is small.

Although our empirical work establishes a correlative (rather than a causative) link between debt and deforestation, the results can be used to examine the effect of debt reduction. Based on the public debt coefficient in the third column of Tables 3 through 6, a one billion dollar reduction in a country's debt level may lead to reduced annual deforestation of 51 km^2 (Table 3), 330 km^2 (Table 4), 115 km^2 (Table 5) or 930 km^2 (Table 6). [13] These numbers can be put in perspective in two ways. First, they can be compared to existing deforestation levels, such as 25 300 km^2 per year in Brazil, 6200 km^2 per year in Indonesia, and 1100 km^2 per year in Cameroon (see Table 1 for a complete list of deforestation levels). Second, debt elasticities for deforestation can be calculated. For example, the elasticity based on column 3 of Tables 3 through 6 varies between 0.17 to 0.31, implying that a 10% reduction in total or relative debt service may reduce deforestation by 1.7 to 3.1%.

Our research provides evidence that reducing debt reduces deforestation. This finding may be an argument to offer deforesting developing countries some form of debt relief. Many deforesting debtor countries have unmanageable debt burdens and are caught in a debt trap. Debt reduction would not only improve the welfare of tropical countries (and countries that value tropical forests) by reducing deforestation, but it would

[13] In these calculations, an interest rate of 10% is used to convert a one billion dollar change in debt to a one hundred million dollar change in total debt service. When the debt variable is defined as total debt service relative to exports, exports are defined as the mean level for our sample. Changing the interest rate would generate a proportionate change in these results.

also help the tropical countries escape from the "vicious circle of indebtedness" [14] by encouraging investment and perhaps new lending. Conservation agencies and developing governments should also be encouraged to utilize more fully debt-for-nature swaps as a tool for preserving environmental quality. However, the global benefits to be realized through debt-for-nature swaps can be better understood with the conduct of more research in this area. In particular, more research needs to be done on intercountry differences in the relationship between debt and deforestation, and on the intercountry differences in the costs and benefits of these swaps. This information can allow the debt reduction process to be conducted in a fashion which maximizes its net social benefits.

Acknowledgments

This research was primarily supported by the Climate Change Division, Office of Policy Analysis, U.S. Environmental Protection Agency, which is not responsible for the positions taken in the paper. We would like to express our appreciation to D. Pearce for his helpful insights, and to thank S. Ferry, W. Branson, W. O'Neill, M.J. Kealy, and K. Andrasko for their comments and suggestions, and T.H. Ha for his comments, suggestions and research assistance. Anonymous reviewers also made important contributions to the paper. The authors are solely responsible for any remaining errors.

References

Bird, G., 1989. Loan–loss provisions and third-world debt. Essays in International Finance, No. 176, Princeton University, Economics Department, International Finance Section, Princeton, NJ.

Chambers, P.E., Jensen, R. and Whitehead, J.C., 1992. Debt-for-nature swaps, Nash equilibrium and the bargaining outcome. Unpublished.

Corbridge, S., 1993. Debt and Development. Blackwell, Oxford.

Czerkawski, C., 1991. Theoretical and Policy-Oriented Aspects of the External Debt Economics. Springer-Verlag, Berlin.

Ehui, S.K., Hertel, T.W. and Preckel, P.V., 1990. Forest resource depletion, soil dynamics and agricultural productivity in the tropics. J. Environ. Econ. Manage., 18: 136–150.

Food and Agriculture Organization of the United Nations, 1981a. Forest Resources of Tropical Africa. UN32/6.1201-78-04 technical report no. 2. FAO, Rome.

Food and Agriculture Organization of the United Nations, 1981b. Forest resources of Tropical Asia. UN32/6.1201-78-04 technical report no. 3. FAO, Rome.

Food and Agriculture Organization of the United Nations, 1981c. Los recurses forestales de la America tropical. UN32/6.1201-78-04 technical report no. 1. FAO, Rome.

Horowitz, I.L., 1987. The "Rashomon effect:" Ideological proclivities and political dilemmas of the IMF. In: R.J. Myers (Editor), The Political Morality of the International Monetary Fund. Transaction Books, Oxford, pp. 93–109.

Korner, P., Mass, G., Siebold, T. and Tetzlaff, R., 1986. The IMF and the Debt Crisis: A Guide to the Third World's Dilemma. Zed Books, London.

Lanly, J.P., 1988. An Interim Report on the State of Tropical Forest Resources in the Developing Countries. FAO, Rome.

Loxley, J., 1986. Debt and Disorder: External Financing for Development. Westview Press, Boulder, CO.

Mendelsohn, R., 1990. Market failure and tropical forest management. Unpublished manuscript presented at the Allied Social Science Association Meetings, Washington, DC.

Repetto, R., 1988. The Forest for the Trees? Government Policies and the Misuse of Forest Resources. World Resources Institute, Washington, DC.

Repetto, R. and Gillis, M., 1989. Public Policies and the Misuse of Forest Resources. World Resources Institute, Washington, DC.

Shilling, J.D., 1992. Reflections on debt and the environment. Finance Dev., 29: 28–30.

United Nations Development Programme, 1990. Human Development Report 1990. Oxford University Press, Oxford.

Von Moltke, K., 1990. International economic issues in tropical deforestation. Unpublished paper presented at the Workshop on Climate Change and Tropical Forests, Sao Paulo.

World Bank, 1989a. World Tables. The Johns Hopkins University Press, Baltimore, MD.

World Bank, 1989b. World Debt Tables. The Johns Hopkins University Press, Baltimore, MD.

World Bank, 1989c. World Development Report. Oxford University Press, Oxford.

World Resources Institute, 1992. World Resources 1992–93. Oxford University Press, Oxford.

[14] Korner et al. (1986), p. 5.

[23]

Ecological Economics, 6 (1992) 57–78
Elsevier Science Publishers B.V., Amsterdam

The rainforest supply price: a tool for evaluating rainforest conservation expenditures

H. Jack Ruitenbeek

Suite 303, 116 Albert Street, Ottawa, Ont. K1P 5G3, Canada

(Accepted 10 December 1991)

ABSTRACT

Ruitenbeek, H.J., 1992. The rainforest supply price: a tool for evaluating rainforest conservation expenditures. *Ecol. Econ.*, 6: 57–78.

Economic policies are often suggested as mechanisms for promoting rainforest conservation in developing countries. To help decide whether international resources should be used to protect specific rainforests, the calculation of a 'rainforest supply price' (RSP) is proposed. Korup National Park in Cameroon contains the oldest rainforest in Africa and — as a haven for important endangered species — it is the subject of active international conservation efforts. A cost–benefit analysis of a conservation project to protect Korup from increased land-use pressures suggests that it is not in Cameroon's interest unless a 5.4 million ECU inducement is transferred to Cameroon. Given the protection afforded, the transfer is equivalent to a RSP of 1060 ECU per km^2 per year. Evaluations of six other tropical rainforest projects suggest that international donors made transfers having values ranging from 15 to 1575 ECU per km^2 per year. It is thus concluded that the inducements required are within a range that conservation interests are apparently willing to mobilize.

INTRODUCTION

In addition to the impacts on local ecology and economic well-being, tropical deforestation gives rise to significant effects outside of any particular developing country. Biodiversity loss leads to potentially substantial losses in internationally important food, medicinal and material products

---

Correspondence to: H.J. Ruitenbeek, Suite 303, 116 Albert Street, Ottawa, Ont. K1P 5G3, Canada.

Financial assistance for portions of this work was provided by the Social Sciences and Humanities Research Council of Canada and the Government of Alberta Heritage Fund. The author also thanks the World Wide Fund for Nature — UK for providing access to the Korup National Park Project in Cameroon.

0921-8009/92/$05.00

(Prescott-Allen and Prescott-Allen, 1983; Farnworth et al., 1983; McNeely, 1988). The deforestation process itself leads to regional climate disruptions and contributes to the emission of greenhouse gases thought to cause global warming (Prance, 1986). Moreover, the loss of forest stocks represents a loss in amenity or existence value to all those individuals in the world who ascribe some importance to the presence of wild places and habitats. All of these factors have contributed to an increase in international efforts to promote conservation of rainforests in developing countries.

One manifestation of this effort is the growing realisation that there may be a need for significant transfers of resources to developing countries to ensure that the activities that cause these effects are reduced. Foreign aid programs are being directed towards sustainable forestry practices, non-governmental organizations (NGOs) are mobilizing substantial transfers to establish protective parks, and — in late 1990 — the Global Environment Facility was established by the World Bank to provide one channel for transferring funds to developing countries (World Bank, 1990). This paper, therefore, is broadly concerned with providing some means by which the rest of the world can evaluate in a monetary fashion whether it will be beneficial to make such transfers to a particular developing country.

Within the economics discipline, there has been considerable effort directed to evaluating the global benefits of conservation (Pearce, 1990). Although progress is being made in measuring the value of certain local environmental functions (Dixon and Sherman, 1990; Barbier, 1991), the global benefits of 'biodiversity value' or 'climate control value' are far from being quantified. It is often asserted that the genetic and climate control value attributable to rainforests in developing countries is immense. It is not hard to imagine, therefore, that sufficiently large numbers might be applied to evaluate genetic resources or the integrity of the global climate such that, if one applies these values in a global cost–benefit analysis, one can readily justify saving every rainforest in the world. While this result is of interest, it is not particularly helpful to planners faced with limited conservation budgets; some allocation criteria are still needed to distribute limited conservation funds.

To provide a meaningful decision-making tool, this paper develops the concept of a rainforest supply price (RSP). Quite simply, the RSP is the amount that a developing country must capture through direct transfers from the rest of the world in order for it to save a particular rainforest. Through selecting a common denominator (one hectare of rainforest), a tool is provided for evaluating rainforest conservation expenditures. While this does not allow us to select the optimum amount of rainforest to be conserved (which would require an estimate of the global benefits), it does allow us to make cross-country comparisons, thereby giving donor agencies and international funding bodies an objective assessment of how much must be deployed in any given area to save a rainforest.

The RSP is essentially a monetary measure of the amount that a developing country must be given by the rest of the world to induce it to institute the reforms required to conserve a particular rainforest. In principle it is based on a one-time lump sum transfer. But to permit comparative analyses between projects, the RSP is expressed as what might be regarded as an annual rental payment — such as ECU per hectare per year [1] — which reflects the amount that would be transferred to the developing country for the particular level of conservation that occurs, since the amount required depends entirely on the net costs of the necessary reforms within the developing country; we can address separately the issue of whether it is worth this amount to the rest of the world to save a particular rainforest.

In the following sections the paper will first discuss the general basis for the calculation of the RSP. The paper then illustrates how the RSP can be calculated for a specific rainforest conservation project: Korup National Park in Cameroon. Starting from a cost–benefit analysis (CBA) of this project, a numerical estimate of the RSP is generated. To help determine whether it would be economically reasonable for the rest of the world to compensate Cameroon this amount, a comparative analysis of several projects then shows what has actually been paid in direct compensation to ensure rainforest conservation. Finally, the paper provides some general discussion regarding the versatility and usefulness of this tool.

DEFINITION OF THE RSP

As a basis for an operational definition of the rainforest supply price, we start with the idea that we are attempting to compensate a host developing country for conserving some amount of rainforest. This can be represented in fairly conventional terms of CBA: the net present value (NPV) of the compensation must be adequate to offset the NPV of any net losses associated with the conservation reforms. Formally, we must ensure that the compensation is such that,

$$\text{NPV}\{\text{compensatory transfers}\} \geq -\text{NPV}\{\text{project to host excluding transfers}\}$$

[1] Foreign currency transactions are converted to the European Currency Unit (ECU) to allow comparisons. The conversion factor used for mid-1989 transactions is 1 ECU = US$1.20 = £0.68 = 340 CFAF.

The starting point for any estimate of the required transfer must therefore be an evaluation of the project (or reform) from the host country's perspective.

In some cases a project or reform will be beneficial to the host country even in the absence of transfers — the NPV of the project is then positive — and the above condition will be fulfilled even if no transfers take place. The host country might still, however, oppose the reform if it is unaware of some of the local benefits arising from it. This is often seen as the primary justification for 'educating' developing countries about the local importance of the rainforests. Indeed, in many analyses the evaluation work concentrates on demonstrating the important economic benefits attributable to local or regional ecological functions of the forest, such as erosion control, soil fertility maintenance, or protecting offshore fisheries (Hodgson and Dixon, 1988; Ruitenbeek, 1991). Throughout the remainder of this discussion, however, it shall be assumed that the NPV of the project in the absence of transfers is negative.

Formally, we define the RSP as the minimum real amount of compensation that would need to be paid to a developing country for every year that a given hectare of rainforest is conserved. A discrete annual structure is used to allow the calculations to be easily accommodated within the cost–benefit accounting of rainforest conservation projects. Conducting a CBA requires that we define two explicit scenarios: a status quo case of what happens if we *do not* undertake the project, and a case of what happens if we *do* undertake the project. For the 'without project' case we define rainforest area as a vector $\mathbf{E} = \{E_0, E_1, E_2, \ldots, E_t, \ldots\}$, where E_t is the area (in hectares or km^2) of undisturbed rainforest, in any given year t, if the project does not go ahead. Similarly, for the 'with project' case we define $\mathbf{E}' = \{E_0' \; E_1', E_2', \ldots, E_t', \ldots\}$ as undisturbed rainforest area if the project does go ahead. A conservation project then generates some change in the amount of undisturbed rainforest, where,

$$\Delta\mathbf{E} = \mathbf{E}' - \mathbf{E} = \{E_0' - E_0, E_1' - E_1, E_2' - E_2, \ldots, E_t' - E_t, \ldots\}$$
$$= \{\Delta E_0, \Delta E_1, \Delta E_2, \ldots, \Delta E_t, \ldots\}$$

We refer to $\Delta\mathbf{E}$ as the 'protection scenario', as it represents the amount of rainforest conserved compared to some base scenario without a conservation project; ΔE_t is thus the area, in year t, that remains undisturbed as a direct consequence of the project. It is important to note that $\Delta\mathbf{E}$ does not require net additions to forested land; the prevention of deforestation by a conservation project will be sufficient to generate a protection scenario with a positive ΔE_t in every year.

The RSP can be regarded conceptually as a year's rental for a hectare of conserved rainforest. In calculating it, we must select a rental value such that, when paid to the host country, it will be exactly enough to bring the total NPV of the project plus the transfers to zero. Introducing simple cost–benefit accounting conventions, this results in the condition that,

$$0 = \sum_{t=0}^{\infty} (1+r)^{-t} \cdot (B_t - C_t) + \sum_{t=0}^{\infty} (1+r)^{-t} \cdot \mathrm{RSP}\Delta E_t$$

where B_t is the stream of local benefits at shadow prices (excluding transfers) to the host country as a result of the rainforest conservation project, C_t is the stream of local costs at shadow prices, and r is the real discount rate. The first term on the right-hand side is the NPV of the project in the absence of transfers; it is important that all of the *local* conservation costs and benefits (for example, soil fertility maintenance and flood control) be included in this formulation. The second term is simply the NPV of the transfers.

Assuming that the RSP is a constant payment for each hectare saved in any given year, this can be rearranged to provide an operational definition for the rainforest supply price as,

$$RSP = -\frac{\sum_{t=0}^{\infty} (1+r)^{-t} \cdot (B_t - C_t)}{\sum_{t=0}^{\infty} (1+r)^{-t} \cdot \Delta E_t}$$

This formulation allows one to calculate the minimum amount that we have to pay to a developing country to conserve a unit of intact rainforest. The reader will note that the denominator involves calculating 'discounted hectares'; while this may seem worrisome to some, it is simply the result of solving for the particular level of RSP required to make the present value of external transfers adequate to cover the economic losses otherwise generated by the project. The formulation is similar to supply cost concepts used in defining the long-term costs of resource production (Bradley, 1979) and has a number of advantages in applied work.

First, it is readily added to cost–benefit analyses of conservation projects as the calculations required for such analyses can be extended to computing the supply price. The costs and benefits must, however, be identified as those that are clearly accruing to the host country. From a donor's perspective, this may actually simplify the CBA as it does not require identification of conjectural benefits outside of the country.

Second, the explicit exclusion of the external benefit stream from the supply price calculation allows this estimate to be relegated to a secondary activity. Depending on the circumstances, it may never be necessary actually to estimate this external benefit stream (Table 1).

TABLE 1

Comparison of two hypothetical conservation projects

Conservation area	Project A 100000 ha	Project B 200000 ha
Present value to host country of:		
Local benefits of conservation	10 million ECU	20 million ECU
Local costs of conservation	30 million ECU	40 million ECU
Net present value (NPV) to host	−20 million ECU	−20 million ECU
Foreign transfers required to make NPV = 0	20 million ECU	20 million ECU

Note: a donor has 20 million ECU available to give to developing countries for rainforest conservation. Two projects (A and B) are being considered with the above characteristics. As long as the foreign benefits are dependent only on conservation area, one would invest in Project B. This decision would be independent of the actual external benefits of the rainforest (assuming that a hectare of rainforest generates the same external benefits independent of where the rainforest is located).

Finally, the supply price is sensitive to the timing of conservation efforts in that it explicitly requires identification of a 'protection scenario'. This allows one to make comparisons between projects that have markedly different impacts on conservation. A national park investment program which has expected permanent conservation benefits can be compared to projects that, for example, require short-term bridge funding to maintain parts of the biome intact while other planning efforts proceed for the long-term initiatives.

CALCULATING THE RSP FOR KORUP NATIONAL PARK

Background, assumptions and methods

The purpose of this section is to illustrate how the rainforest supply price can be calculated for a specific conservation initiative. The Korup Project centers on a program initiated by the World Wildlife Fund (WWF) to promote the conservation of the rainforest in Korup National Park in Southwest Province, Cameroon. The rainforest region in this part of West Africa, which includes Korup National Park in Cameroon and Oban Park in Nigeria, faces the risk of succumbing to persistent land-use pressures from within both Cameroon and Nigeria. These pressures include timber production, fuelwood gathering, hunting, trapping, and forest farming.

Interest from the international aid and conservation community has centered on Korup because of its importance as an international reserve for biodiversity. It is the oldest remaining rainforest in Africa — having survived over some 60 million years — and is characterized by very high species richness believed to have arisen as a result of its isolation as a rainforest refuge during Pleistocene glaciation incidents. It is home to over 1000 species of plant and 1300 animal species, including 119 mammals and 15 primates. Of all species in the area, 60 occur nowhere else and 170 are currently listed as endangered or vulnerable. These include the drill, angwantibo, otter shrew, and Preuss' red colobus. The economic significance of this diversity is illustrated by the fact that at least one major international crop originated in the area (the oil palm, *Elaeis guineensis*) and endemic co-evolved insects that pollinate this species are currently being used in Southeast Asia to safeguard crop performance. In addition, the forested watersheds of Korup and Oban perform a vital role in sustaining mangrove fishery production downstream of the forests. The region's physiography is such that the two countries exchange major rivers, and the maintenance of forest cover on both sides of the border is believed to be essential to maintaining hydrological stability downstream (WWF–UK, 1989b).

The Korup Project is significant because, through the efforts of the Government of Cameroon and external funding agencies, an economic development plan is being pursued for the area that is meant to complement the conservation goals of the park. In line with the World Conservation Strategy of 1980, the attitude has been that conservation of the biome and its products can promote economic development in the region. The protection strategy that has been selected is to commence an active program of economic development in a buffer zone around the national park area. To this end, a management area has been defined consisting of 126 000 ha of core park area plus a buffer zone of 300 000 ha around this core (Fig. 1).

The starting point for the RSP calculation is a detailed CBA of the conservation project undertaken from the perspective of the Government of Cameroon (Ruitenbeek, 1990b). The CBA uses WWF plans for park development as a design basis, and it attempts to make a preliminary estimate of the value of a number of local environmental functions to the Cameroonian economy. The actual key analytical criteria and shadow prices were selected through discussions with Government of Cameroon planners, they are thus consistent with other project evaluation criteria used in the country. Because of the paucity of data available for the CBA, some of the costs and benefits are to a degree conjectural, representing best the estimates at the time that the park planning was being undertaken. Nonetheless, the exercise demonstrates how the technique can be used to calculate the rainforest supply price and, as with many situations involving project evaluation, refinements would normally be expected as planning progressed and projects became better defined.

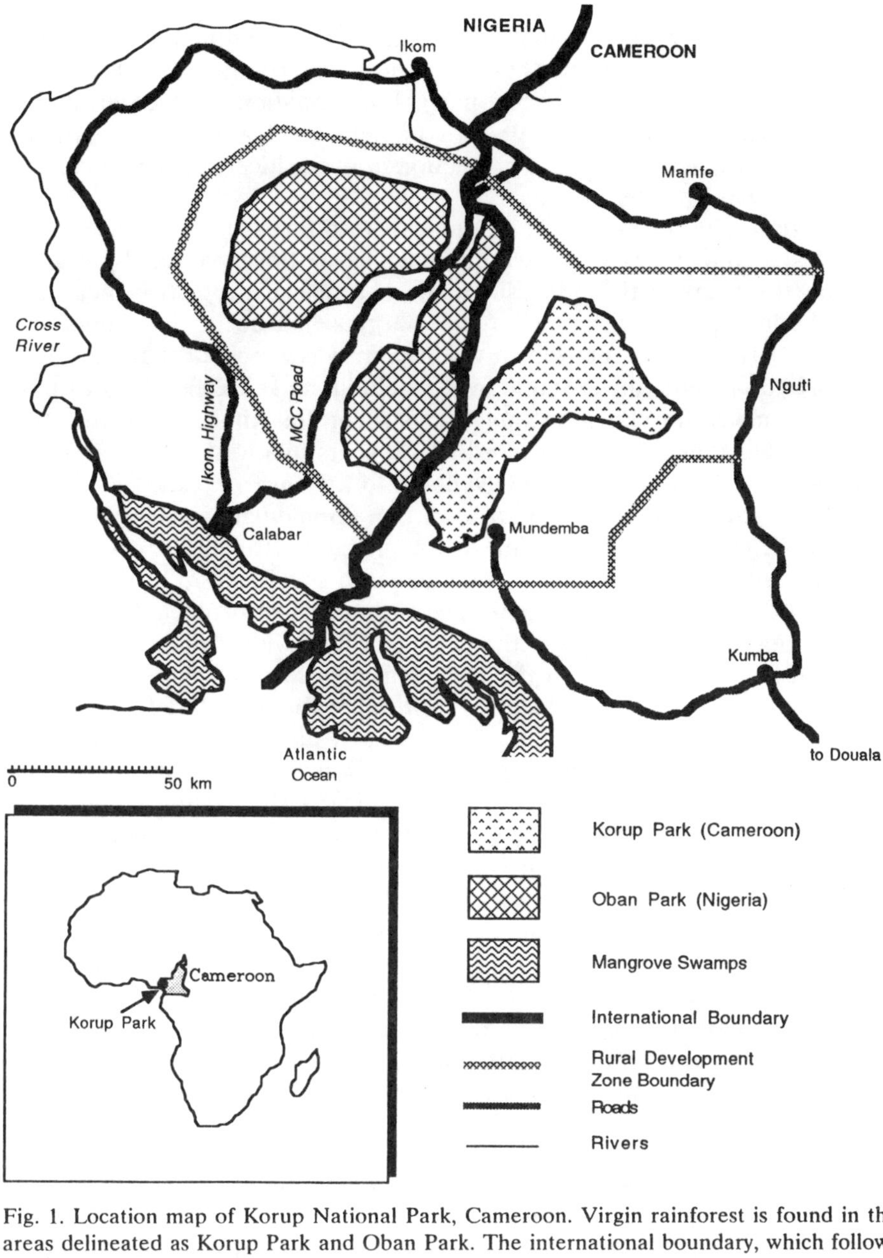

Fig. 1. Location map of Korup National Park, Cameroon. Virgin rainforest is found in the areas delineated as Korup Park and Oban Park. The international boundary, which follows a river course, divides the forest in two. Korup National Park was gazetted in October, 1986, giving it the highest protection possible under Cameroon law. In August 1987, an agreement between WWF and the Republic of Cameroon was executed which called for the development of a management plan for the park.

The first step in the evaluation of the Korup National Park Project involves defining the impacts that the project has compared with a status quo case. Without conservation, normal exploitation of biological and forest resources eventually results in complete elimination of the primary forest. The park actively targets protection at two critical ecosystem components: the trees and the animals.

The primary potential threat to the tree resource is far into the future; the forest's timber resources would most likely be exploited by logging between the years 2010 and 2040 in the absence of protective measures. This scheduling is consistent with the marginal nature of the timber and the current rate of exploitation in the *region*; it is expected that the forest on the Nigerian side, for example, will be eliminated by 2000–2010 without protective measures. By gazetting the park at this time, it ensures that timber concessions — often granted 10–20 years in advance of their actual exploitation — will not be made available to logging interests.

Protecting the animals in the area is a more immediate conservation goal and is, logistically, considerably more complex. Gazetting a park will not in itself prevent continued hunting of endangered species. Two broad options are being considered by park planning authorities to protect the animals: (i) relocating people away from the park; or (ii) providing direct compensation payments to those within the park in return for their compliance with hunting regulations. The relocation option, while mandated by current Cameroon legislation, is disruptive and costly; efforts are being made to introduce legislative reforms that will allow individuals to live in the park while pursuing some traditional activities. The CBA assumes that some compensation mechanism will be instituted to promote compliance with the regulations and that relocation will not be necessary. The compensation payments are not included as costs in the CBA because hunting losses *are* explicitly counted as lost production. Reducing hunting would still, however, involve some regulation of hunting activities. The enforcement programs designed for Korup — and the costs in the CBA — include: park boundary demarcation and maintenance; game guards with the mandate to patrol the park and enforce regulations; guard posts within the park; and adequate radio and other equipment to support the activities of the game guards.

A key component of the CBA addresses three local environmental functions performed by an intact rainforest: fisheries protection, flood control, and soil fertility maintenance.

Both the Korup Project rainforest area and its complement in Nigeria — Oban Park — protect a vast onshore and offshore fishery that has been estimated to exceed £100 million annually (Reid, 1988). The connection between the rainforest area and this fishery provides an interesting example of an environmental service provided by the rainforest, and it is not atypical of how coastal communities might depend on the integrity of inland environmental resources. The vast area of mangrove swamp (Fig. 1) along the shores of Nigeria and Cameroon supports a wide range of species — including whitefish, shrimps, and alligators — which are commercially harvested. The connection that the mangroves have to the rainforest 100 km away is that three major river courses collect water from the rainforest and flow into the ocean at this point. Large variations in water levels of these rivers result from fluctuations in the amount of rainfall occurring over the forest. At periods of high water the mangrove areas are inundated with fresh water, whereas during low periods the salt water from the ocean finds its way inland along these river courses. Because of the large variations in river volumes the mangrove area is larger than would typically arise from just tidal variations. But the entire mangrove ecosystem has evolved and flourished as a result of these phenomena: major changes in the freshwater environment would likely change the mangroves. For example, removing the rainforest canopy of Korup forest would increase the peak flows of the rivers which, in addition to decreasing the salinity of the mangrove area, would carry a greater volume of sediment into it. Major changes in the freshwater flows — such as that arising from deforestation — would alter the ecological balance in the mangroves to the extent that many of its commercial species could be lost. The CBA therefore estimates the loss in onshore fishery value that could arise if deforestation of Korup occurred. While there is some speculation that the offshore fishery resource depends to a degree on food fish that breed in the mangroves, impacts on this offshore fishery are excluded from the CBA since they fall primarily in Nigeria.

For a single rainforest such as Korup, it is possible to estimate the potential benefits from distinct functions such as flood control. As rainforests disappear the peak and trough levels of surface water flows become more pronounced. In a high rainfall area such as the Southwest Province of Cameroon, the potential flooding increases with deforestation. While it is impossible to predict exactly when a devastating flood might occur, the CBA here estimates the *expected* net present value of benefits. As a first approximation it was assumed that the expected value loss of flooding in any year t is:

$$\text{flood control benefit}_t = \frac{N_{fc} \cdot (A_d/A) \cdot Y_{fc}}{T}$$

where N_{fc} is the number of people expected to be affected by a flood event, and A_d/A is the proportion of the deforested area of forest to the total originally forested area. Y_{fc} is the per capita income in the region lost due to a flood event. The flood event is assumed to occur every $T = 5$ years. For calculating the NPV benefit, it was assumed that $N_{fc} = 20\,000$, which corresponds to the population in the Korup Park rural development zone as well as about 8000 people living along river courses between Korup and the coast (it excludes those in the mangrove estuaries). The income loss is estimated to be that corresponding to all livestock income and one season of cash crop income, as the flooding would likely occur in July or August and the major harvest occurs in October.

Although the soils in the area are not generally of very high quality, the rainforest does contribute to some maintenance of soil fertility through various mechanisms. Although these benefits are somewhat conjectural, an estimate is made based on the regional agricultural product. Specifically, it is assumed that without the Korup forest there would be a 10% loss in the value of the cash crop income in the area. This loss would commence in 2010 and manifest itself fully by the year 2040.

CBA requires that any losses or benefits reflect changes in income as a result of a project. Changes in certain input costs — or other opportunity costs — to derive this income must also be accommodated. For example, lost hunting revenues are partially offset by decreased hunting costs and, in the case of Korup, increased hunting levels at other locations outside of the protected areas. The methods used in this CBA reflect such adjustments; all estimates of changes in local incomes — for hunting, gathering, fishing, and farming — were based on a detailed survey of 357 households conducted in 1988 (Ruitenbeek, 1990a). Short-term impacts of the park restrictions have an effect primarily on hunting activities, whereas the long-term impacts of conservation ensure a sustainable source of gathered and farmed products in perpetuity.

Results

The infrastructure and programs mobilized by the Korup Project would not show immediate conservation benefits, and it is important in calculating the rainforest supply price to be explicit about how much forest is actually protected. Specifically, we must estimate the amount of deforestation that would occur in the absence of the park program (Table 2). In the very near term no protection is afforded because the population currently living in the park will continue their activities. After about 5 years, restrictions on these activities effectively protect 20 000 ha of forest area that otherwise would have been lost. It is not until after 2010, however, that the major protection benefits might be realised, corresponding to the restrictions on logging imposed by the park. Over the longer term the existence of the park effectively protects 351 000 ha in perpetuity.

TABLE 2

Korup National Park protection scenario

Year	Area of primary forest converted to secondary forest by local populations (ha)		Area of primary forest harvested for timber (ha)		Net protected area (ha)
	Without Park (1)	With Park (2)	Without Park (3)	With Park (4)	With Park [a] (5)
1989	20000	20000	0	0	0
1990	20000	20000	0	0	0
1991	20000	20000	0	0	0
1992	20000	20000	0	0	0
1993	20000	10000	0	0	10000
1994	20000	10000	0	0	10000
1995	20000	0	0	0	20000
2010	8300	0	11700	0	20000
2011	0	0	23400	0	23400
2020	0	0	128700	0	128700
2030	0	0	245700	0	245700
2040	0	0	351000	0	351000
2041–∞	0	0	351000	0	351000

[a] Column (5) = [(1) − (2)] + [(3) − (4)].

Notes: two separate sources of deforestation are considered: (i) a reduction in the area of primary forest through its non-destructive 'conversion' to secondary forest by populations living within the park; (ii) physical destruction of primary forest from harvesting timber.
The conversion component occurs because land is cleared for village sites, temporary shelters, farms, and general access trails. This conversion is estimated to be about 25 ha per person, although the forest recovers quickly once people leave an area provided that there is still a large area of primary forest adjacent to the disturbed area (Devitt, 1988). Without the park, the converted area associated with its 800 inhabitants is about 20000 ha. With the park, it is estimated that 50% of this would recover in 1993 and the remainder in 1995.
The second component corresponds to the protection of the intact forest which would otherwise be clear cut for timber over the period 2010–2040. Some 11700 ha of primary forest would be cleared annually over a 30-year period commencing in 2010. By the year 2040, in the absence of the park, some 351000 ha of primary forest would have been lost. The actual 'protection scenario' in column 5 is derived through inspection of these two components. The present value of the net protected area at an 8% discount rate is 513800 hectare-years.

The cost and benefit streams that form the basis for the CBA also reflect this protection scenario (Table 3). The value attributable to the protection of the key environmental functions associated with the rainforest (fisheries protection, flood control, and soil fertility maintenance) is not manifested until after 2010.

TABLE 3

Summary of Korup Project cost and benefit streams (millions of constant CFAF: 1989 values)

	1989 -1995	1996–2009	2010–2039	2040–∞
Direct costs (at market prices)				
Capital and running costs	2338			
Infrastructure costs	3685			
Operating costs		280/year	280/year	280/year
Opportunity costs (at shadow prices)				
Lost timber value			147/year	
Lost forest use	0–34/year	34/year		
Direct benefits (at shadow prices)				
Sustained forest use			0–132/year	132/year
Tourism	3–24/year	24–48/year	48/year	48/year
Fisheries protection			54–1628/year	1628/year
Flood control			8–244/year	244/year
Soil fertility maintenance			4–120/year	120/year

Note: Summaries of direct costs for 1989–1995 are period totals (WWF–UK 1989b). Direct costs include capital costs of park infrastructure (895), park vehicles and equipment (79), roads and airstrips (3350), labour (690), administration (205), compensation payments (256) and a contingency (548). The amount entering the CBA, however, is only 3272 million CFAF: costs related to compensation payments are purely pecuniary and are thus excluded; and a portion (67%) of the road costs would be spent in any event because of other regional imperatives.

Opportunity costs for lost timber value include that associated with a commercially marginal timber resource having an undiscounted net standing value of about 4400 million CFAF which would otherwise have been harvested over a 30-year period commencing in 2010. Lost forest use corresponds to the lost subsistence income to the region's residents, associated with restrictions on selected hunting, trapping, gathering and farming activities.

Direct benefits include, over the longer term, sustained gathering incomes to local residents, as well as the benefits arising from three environmental functions of the rainforest. Fisheries protection benefits are the net changes in income to Cameroonian fishermen in the downstream mangrove area. Flood control benefits are the estimated value of agricultural production which would otherwise be lost due to more frequent flooding. Soil fertility maintenance represents the net value of agricultural output which would otherwise be lost due to declines in soil fertility. Tourism benefits are a conservative estimate of net revenues from incremental wilderness-based tourism and scientific research.

The CBA shows that the net present value of the project to Cameroon, at an 8% rate of discount, is about *minus* 1852 million CFAF (Table 4). Sensitivity analyses undertaken at a 6% rate of discount result in a NPV of *plus* 319 million CFAF. The benefit and cost streams at shadow prices correspond to a real internal rate of return of 6.2% from Cameroon's perspective. Since the NPV using the Cameroonian base case discount rate of 8% is negative, the CBA indicates that the project would not be

TABLE 4

Summary of Korup National Park cost–benefit analysis

	Net present value (million CFAF)
Social costs	
Total capital costs, excl. roads	−1 988
Total capital costs, roads	−1 022
Total long-term operating costs	−2 381
Labour adjustment for direct costs	916
Lost timber value	−353
Lost forest use	−223
Total costs	−5 051
Social benefits	
Sustained forest use	354
Tourism	680
Fisheries protection to Cameroon	1 770
Control of flood risk	265
Soil productivity maintenance	130
Total benefits	3 199
Net benefit	−1 852
Present value of net protected area	513 800 hectare-years
Supply price	3 605 CFAF/hectare/year

Note: all values are expressed in 1989 constant terms, with discounting conducted at an 8% discount rate. Calculations are based on cost and benefit streams in Table 3. Capital costs of the roads included 33% of the budgeted costs, corresponding to that estimated to be incremental to those expenditures which would be incurred in any event. A labour adjustment was applied to correspond to Cameroonian planning practice to value labour in this region at 50% of the market wage.

economically acceptable to Cameroon if all of the costs were incurred internally. In short, one would not expect Cameroon to take steps to conserve Korup unless some transfers or other incentives were forthcoming from the rest of the world. The analysis indicates that the current value of such incentives must be at least 1852 million CFAF to provide an adequate inducement to Cameroon to pursue the park initiative.

A sensitivity analysis was undertaken to illustrate the effect of excluding the protective benefits of the forest to the onshore mangrove fisheries. Excluding these benefits essentially doubles the (negative) social cost of the project to −3622 million CFAF. This result underlines the importance of the value of this environmental function of the rainforest in the overall project evaluation.

We are now in a position to calculate the rainforest supply price. The ‘present value’ of the protection scenario (Table 2) is 513 800 ha-years.

Given that the required transfer was calculated to be 1852 million CFAF in the CBA, the RSP for Korup is about 3600 CFAF per hectare per year. This implies that if the rest of the world were to give Cameroon anything above 3600 CFAF for each hectare protected in any given year, Cameroon would have an incentive to protect the park.

To allow a more meaningful comparison with other projects, we can express this in some common international currency. The European Currency Unit (ECU) is used which, at the time of preparing the budgets for the Korup plan, stood at an exchange rate of 340 CFAF/ECU. The rainforest supply price for Korup is then 10.60 ECU per hectare per year (or 1060 per km^2 per year) of rainforest.

It should be stressed that, although the RSP is a fixed payment per hectare per year, the RSP need not be interpreted as a program that involves the transfer of the same amount of funds to a host country every year. In theory the exercise says simply that *any* program of transfers with a present value in excess of 1852 million CFAF will be adequate to conserve the forest. In practice this program would likely include a combination of capital transfers to fund the initial infrastructure as well as ongoing transfers to underwrite the operations of the park.

COMPARING THE RSP WITH INTERNATIONAL DIRECT TRANSFERS FOR RAINFOREST CONSERVATION

Background, assumptions and methods

The operative question is now whether it is worth just over 1000 ECU per km^2 per year to the rest of the world to induce Cameroon to conserve Korup. One means of answering this question is to investigate the global benefits of conserving rainforests. But the rationale for concentrating on the supply price is to allow comparisons to be made with other projects even when the global benefits are unknown or uncertain. Therefore, other conservation projects where actual transfers to developing countries *have* occurred are reviewed here. A specific objective of this section is to investigate whether Korup would then be a bargain or relatively dear from the point of view of the rest of the world.

In this section we shall not strictly be looking at the RSP since we are investigating actual transfers. One would normally presume that these transfers are more than the RSP for the project and, depending on the bargaining positions of the various parties, it is conceivable that the host country would have collected substantially more than they would have been willing to accept by way of compensation. As we are not in the position of evaluating the bargaining positions of the various parties in the negotiations, the evaluations that follow cannot necessarily be construed as RSP evaluations. They can, however, be taken as some measure of how much the rest of the world is willing to pay for a given conservation program. As such, they provide a useful comparison with projects for which the RSP is known (such as Korup).

Six specific transfers are investigated, all of which occurred over the period 1987 to 1989. Four of these transfers involve debt for nature (DFN) swaps, and two are targeted fund raising campaigns to conserve specific rainforest projects. They were selected because in each case there was a simple direct transfer of funds intended to conserve a particular rainforest area. In all cases, reasonably good relations existed between the initiating donor agency and the host government, and the expectations were that the transfers would indeed promote effective conservation of the rainforest, albeit in some cases for a limited time only.

The first project involves the Beni Nature Reserve (WWF–US, 1988a) in the Amazonian region of northern Bolivia; it supports 13 of Bolivia's 18 endangered animal species. The program involved a DFN swap in July 1987 in which debt paper having a face value of US$650 000 was purchased by Conservation International for US$100 000. The Bolivian government in return elevated the protected status of the reserve and turned the equivalent of US$100 000 in local currency over to a local NGO for establishing the reserve. In addition, a direct transfer of US$150 000 was provided by USAID to the NGO. The total funds are not anticipated to be adequate to protect the reserve in perpetuity as their primary intent was to initiate protection of the reserve and remove very short-term land-use pressures.

The second project involves about one million hectares of rainforest in two areas of the Amazonian region in Ecuador that are threatened by spontaneous colonization and industrial encroachment from oil exploration (WWF–US, 1987). The project was part of a DFN swap that included projects in the Andes and in the Galapagos Islands, although the initial injections of funds were meant to alleviate some of the immediate pressures on the Amazonian portions.

Key features of the third project — St. Paul Park in the Philippines — include mountain and coastal forests and a unique underground river (WWF–US, 1988c). The area is threatened by shifting cultivation and illegal logging, and a DFN swap augmented local park department resources to ensure its protection.

The last DFN swap investigated involves the Santa Rosa National Park project in Costa Rica (WWF–US, 1988b,d). This is one of many DFN swaps in which Costa Rica has been involved; most of the others involved multiple projects with unprogrammed budgets that precluded analysing specific components.

The fifth project involved a direct transfer to purchase critical lands from local farmers in the Monte Verde Cloud Forest area of northwest Costa Rica (WWF–US, 1988b,d). The forest contains rare and endangered species found in typical mid-elevation rainforests and it protects the watershed of Costa Rica's largest hydroelectric facility. Funds were raised by WWF–Canada through a direct mail campaign and individual donations.

The last project investigated involves the Oban National Park project in Nigeria (WWF–UK, 1989a). During the long-term planning of Oban Park, it became evident that immediate pressures from logging and bush fallow agriculture were threatening the integrity of the forest. A fund raising campaign based on the sale of an 'Oban Park Founder's Bond' was thus initiated in the UK; participants would receive a certificate acknowledging their assistance in establishing Oban Park.

Since the projects were undertaken at separate times and used different currencies, all calculations were converted to ECU at 1989 values using appropriate adjustments for both inflation and currency exchange rates. An 8% discount rate was used to allow comparisons to be made with the Korup evaluation. Finally, it is particularly important in these evaluations to specify the protection scenario carefully as all of the six programs have a rather short effective life. The estimates assume that effective protection occurs for a period consistent with the duration of the local currency funding mechanisms used in each case.

Results

The willingness-to-pay estimates for these transfers displayed a considerable range of values even within this small sample (Table 5). The Beni Reserve deal was the least expensive at about 15 ECU per km^2 per year, whereas the amount contributed for protecting the Monte Verde Cloud Forest approaches 1600 ECU per km^2 per year. Although the sample is not meant to provide a statistically significant cross-section of conservation, it does provide some indication of how much individuals or organizations have been implicitly willing to pay for conservation. These estimates should not be construed as the minimum amount that would be necessary to induce conservation nor should they necessarily be construed as the maximum that the world is willing to pay; one can say only that they fall between these extremes.

DISCUSSION

This paper has suggested an analytical mechanism — the rainforest supply price — for evaluating the level of transfers that are required to

TABLE 5

Estimates of willingness-to-pay and rainforest supply price

Project description	Protected area (km^2)	Protection scenario	NPV of actual transfers ($\times 10^3$ ECU)	PV area protected ($\times 10^3$ km^2-year)	Revealed willingness-to-pay (ECU/km^2/year)
Beni Reserve, Bolivia (1987) DFN Swap	15420	1 year	234	15.42	15.18
Amazonian Parks, Ecuador (1987) DFN Swap	10000	1 year	332	10.00	33.24
St. Paul Park, Philippines (1988) DFN Swap	57.5	9 years	353	0.39	910.34
Santa Rosa Park, Costa Rica (1988) DFN Swap	160	5 years	265	0.69	383.84
Monte Verde Forest, Costa Rica (1988) Direct Mail	36	5 years	245	0.16	1575.76
Oban Park, Nigeria (1989) Founder's Bond	2500	1 year	1029	2.50	411.76

Rainforest supply price estimate

Project description	Protected area (km^2)	Protection scenario	NPV of required transfers ($\times 10^3$ ECU)	PV area protected ($\times 10^3$ km^2-year)	Rainforest supply price (ECU/km^2/year)
Korup Park, Cameroon (1989)	3510	Protection of 200 km^2 from 1993 until 2011 phased in to full protection by 2040	5448	5.14	1060.15

Note: this table compares the RSP estimate for Korup with six other transfers which took place for promoting tropical rainforest conservation. The willingness-to-pay estimates show the implied value of these transfers to the recipient countries. They can be regarded as the equivalent amount paid for each unit of protected area for the period of years shown as the 'protection scenario'. All discounting is conducted at a real rate of 8%, and all transactions have been converted to a 1989 ECU equivalent. The following conversion factors were used: 1 ECU = US$1.20 = £0.68 = 340 CFAF.

induce a developing country to conserve a particular rainforest project. One particular advantage of this tool is that it allows one to address questions of allocative efficiency without necessarily requiring that broad issues relating to the valuation of *global* benefits of rainforest conservation be discussed. Also, it allows one to make comparisons between many types of rainforest conservation projects on a consistent basis.

One issue that must invariably be addressed relates to the measurement of the actual rainforest 'commodity'. We have spoken relatively loosely of rainforest area, tacitly assuming that rainforests can be treated equally, e.g. conserving a hectare of rainforest in Brazil is of no greater or lesser value than conserving a hectare of rainforest in West Africa. In fact, however, this is an oversimplification and conservationists often assert that some rainforests are more important than others: those with high numbers of endangered endemic species, for example, are generally of greatest interest. While this point is well taken, this paper relies on the idea that some area unit of rainforest can be considered as a common numéraire. The ecological literature is actually quite divided on what an appropriate numeraire indicator is of any ecosystem's health. While population and species counts have traditionally provided a convenient and tractable measure, a survey of recent work (Kay, 1991) suggests that such measures are not necessarily of much relevance to an ecosystem's overall health or integrity. It is, in any event, conceptually not difficult to adjust for rainforest quality by introducing some form of hedonic pricing. If we let ϑ be a dimensionless index for rainforest quality, normalized such that $\vartheta = 1$ for a standard hectare-year for some standard rainforest, then we can define a hedonic rainforest supply price RSP^{h} as RSP/ϑ, and higher quality rainforests would then appear as having lower hedonic supply prices. As with all RSP criteria, donors would select the 'least-cost' candidates first: those with the lower RSP^{h} and concomitant higher quality. One would presume that ϑ could itself be a function of many criteria, and there is thus some risk of arbitrariness in specifying such quality. It would, nonetheless, give individual donors a means of ranking conservation alternatives.

The numerical estimates of RSP and willingness-to-pay actually cover a wide range of values. Abstracting from potential differences in bargaining position, a number of potential reasons for these differences can be offered. The first is that the amounts paid do actually reflect the supply prices of these projects fairly closely, in which case we may conclude that some of the readily identifiable 'bargains' had been spotted quite early. The projects in Bolivia and Ecuador were indeed the first to be initiated and subsequent projects are going to cost one to two orders of magnitude more. A second reason for the differences may be in the quality of the area being protected. If we assume that all six of these transfers have the same hedonic price, it implies that we regard the quality of a hectare in Monte Verde to be about one hundred times greater than a hectare in Amazonia. This may not be an unreasonable result given that cloud forest such as that found in Costa Rica is rarer than Amazonian rainforest. A third potential reason for the differences is that they reflect expectations of success in the particular protection scenario. In this case one might argue along the lines that the chances of success in Monte Verde were expected to be one hundred times greater than the chances in Amazonia. This means that the protection scenario has been mis-specified (recall that it is the *expected* protection), but it suggests that there may be some merit in explicitly identifying the probability of success when evaluating such projects.

While there is little, if any, basis for testing these conjectures at this stage, the framework that has been used does allow us to start making comparisons and asking relevant questions. For example, in the case of Korup National Park we see that, at 1000 ECU per km^2 per year, it is not necessarily one of the great bargains in the world, yet the level of transfers that is required is not out of line with that raised or paid elsewhere. As Korup is regarded by many to be one of the most important rainforest areas on the African continent, 'quality' considerations may suggest that the necessary funds to pursue the project could be raised.

Finally, one other result arose from the CBA which relates to the *types* of transfers that one might wish to consider for a project such as Korup. The long-term cost and benefit projections for Korup indicated that annual financial costs of the project would be about 280 million CFAF, and that the level of benefits at shadow prices approached 2200 million CFAF annually. On the surface it might appear that once the capital outlays were over and the project was well established there would be a significant incentive for Cameroon to continue funding the park program. A common problem with projects of a 'social' nature in developing countries, however, is that local governments do not meet ongoing current budgetary needs for programs that generate few revenues. Korup Park would fall into this category, as most of the environmental benefits are to maintain income or production levels elsewhere in the economy and they do not directly increase the amounts in government coffers. This suggests that one potential area for transfers is to pay for the ongoing expenses of the national park. While this approach deviates from both the short-term nature of DFN swaps and traditional aid programs, which involve injections only at the start of projects, it may provide a greater chance of successfully promoting rainforest conservation. In addition, continuous transfers are more in line with the ongoing benefit that the rest of the world purportedly receives from the environmental functions of rainforests.

CONCLUSION

The RSP effectively provides a preliminary measure to which standard cost-efficiency principles can be applied: if funding to save the rainforests is limited by a fixed budget or by some decreasing willingness-to-pay as more forests are saved, then efficient allocation of this funding quite simply involves applying it to those forests that are the cheapest to save.

Specific applied work in a cost–benefit analysis framework relating to Korup National Park illustrated that the project was not in Cameroon's interest if Cameroon were required to pay for the entire project. Transfers in the amount of just over 1000 ECU per km^2 per year would be required to compensate Cameroon for the area actually protected by the park program. Transfers of twice this level would be necessary were it not for selected local environmental benefits afforded by rainforest conservation — in particular the protection of Cameroon's onshore mangrove fisheries.

An analysis of six other rainforest projects pursued between 1987 and 1989 indicated that funds had been raised for these projects that fell into a range of 15 to 1575 ECU per km^2 per year. It is concluded that the cost of supporting transfers for the Korup project is not a misuse of resources given that similar projects have attracted transfers of this order. Indeed, as Korup is regarded by many conservation interests as being of high quality, supporting transfers for Korup would be an efficient use of scarce conservation resources.

REFERENCES

Barbier, E., 1991. Environmental degradation in the third world. In: D. Pearce (Editor), Blueprint 2: Greening the World Economy. Earthscan, London, pp. 75–108.

Bradley, P., 1979. Production of Depleting Resources: A Cost Curve Approach. University of British Columbia, Department of Economics, Vancouver.

Devitt, P., 1988. The people of the Korup Project area: report on phase 1 of the socio-economic survey. WWF–UK, Godalming (unpubl.).

Dixon, J.A. and Sherman, P., 1990. Economics of Protected Areas: Approaches and Applications. Island Press, Washington/Covelo, 234 pp.

Farnworth, E.G., Tidrick, T.H., Smathers, Jr. W.M. and Jordan, C.F., 1983. A synthesis of economic and ecological theory toward more complete valuation of tropical moist forests. Int. J. Environ. Stud., 21: 11–28.

Hodgson, G. and Dixon, J.A., 1988. Logging versus Tourism in Palawan: An Environmental and Economic Analysis. East–West Environment and Policy Institute, Honolulu, 95 pp.

Kay, J.J., 1991. The concept of ecological integrity, alternative theories of ecology, and implications for decision support indicators. In: Canadian Environmental Advisory Council (CEAC), Economic, Ecological, and Decision Theories: Indicators of Ecologically Sustainable Development. CEAC, Ottawa, pp. 23–58.

McNeely, J.A., 1988. Economics and Biological Diversity: Developing and Using Economic Incentives to Conserve Biological Resources. International Union for the Conservation of Nature, Gland, 236 pp.

Pearce, D., 1990. An Economic Approach to Saving the Tropical Forests. London Environmental Economics Centre, London, LEEC Paper 90-05.

Prance, G.T. (Editor), 1986. Tropical Rainforests and the World Atmosphere. American Association for the Advancement of Science, New York, 105 pp.

Prescott-Allen, R. and Prescott-Allen, C., 1983. Genes from the Wild: Using Wild Genetic Resources for Food and Raw Materials. Earthscan, Nottingham, 101 pp.

Reid, G.M., 1988. The fishes and the rainforest. WWF–UK, Godalming (unpubl.), 81 pp.

Ruitenbeek, H.J., 1990a. Evaluating Economic Policies for Promoting Rainforest Conservation in Developing Countries. Ph.D. dissertation, London School of Economics and Political Science, London.

Ruitenbeek, H.J., 1990b. The Rainforest Supply Price: A Step Towards Estimating a Cost Curve for Rainforest Conservation. Suntory-Toyota International Centre for Economics and Related Disciplines, Development Economics Research Programme, London, Discussion Paper 29, 39 pp.

Ruitenbeek, H.J., 1991. Mangrove Management: An Economic Analysis of Management Options with a Focus on Bintuni Bay, Irian Jaya. Environmental Management Development in Indonesia, Halifax, Environmental Report Series, 90 pp.

World Bank, 1990. Discussion notes: funding for the global environment. World Bank, Washington (unpubl.).

World Wildlife Fund–UK, 1989a. Cross River National Park (Oban Division): Plan for Developing the Park and its Support Zone. WWF, London, 98 pp.

World Wildlife Fund–UK, 1989b. Korup National Park Master Plan. WWF, Yaoundé.

World Wildlife Fund–US, 1987. Ecuador Fundación Natura: debt-for-nature agreement. WWF, Washington (December 14), 7 pp.

World Wildlife Fund–US, 1988a. Debt-for-nature swaps: a new conservation tool. World Wildlife Fund Letter, 1: 1–9.

World Wildlife Fund–US, 1988b. Government of Costa Rica: debt-for-nature agreement. WWF, Washington (March 4), 7 pp.

World Wildlife Fund–US, 1988c. Republic of the Philippines: debt-for-nature agreement. WWF, Washington (June 24), 8 pp.

World Wildlife Fund–US, 1988d. World Wildlife Fund and the Conservation Foundation: a quarter century in Costa Rica. WWF, Washington (unpubl.), 13 pp.

[24]

Institutions, Governance and Incentives in Common Property Regimes for African Rangelands

BRENT M. SWALLOW[1] and DANIEL W. BROMLEY[2]
[1] *International Livestock Centre for Africa, P.O. Box 46847, Nairobi, Kenya;* [2] *University of Wisconsin-Madison, 427 Lorch Street, Madison, Wisconsin, U.S.A.*

Abstract. The general distinctions between open access, state property, common property and private property are now well established in the academic literature. When applied to African rangelands, however, common property admits a wide variety of resource management regimes. To formulate effective policies it is necessary to understand the structure and operations of particular regimes. In this paper we discuss three examples of common property regimes, two from the southern African nation of Lesotho and one from the west African nation of Senegal, to illustrate some of the key characteristics of common property regimes. In particular, it is important to understand the structure of governance, the types of institutions that govern behavior, and the compatibility between governance, institutions and individual incentives. A common property regime can only be effective if its institutions are compatible with the structure of governance. The extent of its effectiveness also depends upon the incentives and expectations of individuals expected to enforce the rules of the institutions or comply with their terms. At present, most African governments lack tne organizational capacity and political will necessary to implement state property regimes, official regulations on resource use, or individual property rights for rangelands resources. In many cases it is more appropriate for governments to define and enforce group rights to particular resources, then help to establish conditions in which internal group dynamics yield efficient resource management outcomes.

Key words: Common property, property rights, rangelands, Africa, Lesotho, Senegal

1. Introduction

Africans keep about 14 percent of the world's cattle and 21 percent of the world's sheep and goats on a land base that comprises 25 percent of the world's total area of rangelands. The number of people engaged in extensive livestock production is higher in Africa than in any other region of the world. Of the 30 to 40 million pastoralists in the world – people who rely on extensive livestock production for at least half of their income and subsistence needs – over half reside in Africa (Galaty and Johnson, 1990; Sandford, 1983). Livestock contribute about 35 percent of Africa's agricultural domestic product and 17 percent of the dietary protein in human diets (Winrock International, 1992: pp. 12–16).

The forage and water resources emanating from most African rangelands are shared by several individuals, households or village groups. This shared use of natural resources received its early indictment in the fisheries literature when writers began, unfortunately, to refer to the open access fishery as

Environmental and Resource Economics **6**: 99–118, 1995.

a 'common property resource' (Gordon, 1954; Scott, 1955). Garrett Hardin, seeking to make a point about uncontrolled population growth, coined the allegorical phrase, 'the tragedy of the commons' (Hardin, 1968). To make his point, he invoked the proverbial grazing pasture in which each separate owner of livestock has little incentive to control livestock numbers. For some reason, few questioned the Hardin link between parents deciding to have more children, and a community of grazers deciding how many animals to place on the range. Indeed, it is precisely this difference which is central to natural resource policy – and which Hardin should have used in his appeal for population control. By mixing metaphors, Hardin missed a wonderful opportunity, and he managed to induce an entire generation of economists and development assistance experts to believe that livestock management was no different from couples having babies. While it may be difficult to control the breeding behavior of humans, social groups certainly have many ways to control greedy and obstreperous livestock keepers. Those intent on individualizing (i.e., 'privatizing') African land tenure so as to make it conform with ideological precepts found much comfort in the Hardin allegory. A paper about population policy became, instead, the 'theoretical' foundation for grazing policy across an entire continent.

Repeated challenges to the applicability of the 'tragedy of the commons' model has prompted a surge of interest in common property (Berkes, 1989; Bromley and Cernea, 1989; Ciriacy-Wantrup and Bishop, 1975; McCay and Acheson, 1987; National Academy of Sciences, 1986). Analysts familiar with particular situations often point out that collectively-used natural resources are governed by resource management regimes that approach the conceptual essence of common property (*res communes*). For example, analysts have described common property regimes governing rangeland use in Lesotho (Shoup, 1987), Ethiopia (Helland, 1982), Tanzania (Lane, 1991), Morocco (Gilles *et al.*, 1992) and Kenya (McCabe, 1990; Stiles, 1992; Kituyi, 1990). Here we consider a *common property regime* to be a set of institutional arrangements that define the conditions of access to, and control over, a range of benefits arising from collectively-used natural resources (Bromley, 1989, 1991; Bromley and Cernea, 1989; Ostrom, 1986; Young, 1989).

In this paper we argue that there are resource management regimes with a wide range of institutional arrangements and governance structures that are consistent with this definition of common property. Different types of economic and environmental problems may arise depending upon the structure of governance and the types of institutions. Different policies and interventions may thus be appropriate for solving those problems. Herein we offer a political-economy framework for analysis of resource management regimes that fall within the broad class of common property. The framework recognizes the importance of the socio-political structures that set the institutional context for individual behavior and collective action. We apply the framework to common property regimes for African rangelands. We begin with brief reviews of rangeland regimes from Lesotho and Senegal to motivate the framework.

2. Examples of Common Property Regimes for African Rangelands

In this section we review three regimes, two from Lesotho and one from Senegal, that represent some of the variety of rangeland regimes that prevail across Africa. In these descriptions we draw heavily upon our previous analyses of those regimes (see Swallow *et al.*, 1987; Swallow and Brokken, 1990; Swallow, 1990, 1991; Bromley and Cernea, 1989). While all of the regimes meet the definition of common property regime as presented above, all have different political and institutional structures. At the outset it is important to note that all three regimes operate in geographical areas that receive high average annual rainfall compared to rangelands in many other areas of Africa.

2.1. LESOTHO'S MABOELLA REGIME

Lesotho's extensive grasslands are utilized under two resource management regimes: the *maboella* regime governs the use of village common lands in the lowlands and foothills and the cattlepost regime governs the use of mountain grazing areas. The lowlands and foothills comprise a narrow belt of land around the western borders of the country with about 25 percent of the country's land base. Elevation ranges from 1,520–1,830 metres above sea level and average annual rainfall is 750 mm. Over half of the country is classified as mountainous. Elevation in the mountains varies from 1,830 to over 2,700 metres above sea level and the average annual rainfall is 1,000 mm. The foothills are a transition zone between the lowlands and mountains.

Maboella was established in the mid-1800s by Lesotho's first Paramount Chief, Moshoeshoe, at the height of his political and economic power. *Maboella* was instituted in order to protect areas of common lands for special uses during particular times of the year. Some forms of *maboella* governed the utilization of natural stands of edible plants, medicinal herbs, woodlots, grasses, and reeds, while other forms reserved pasture areas for utilization during certain seasons or prohibited grazing on certain areas to allow their regeneration (Eldredge, 1986; Sheddick, 1954).

The rules for *maboella* were promulgated at the national level by the Paramount Chief and enacted at the village level by the Paramount Chief's designates. At the local level, the *molisa oa lethobo*, or warden of enclosures, was appointed by the village chief to be responsible for the enforcement of *maboella* rules. Wardens were allowed disproportionate allocations of thatching grass as compensation for enforcing *maboella* rules. Village chiefs had incentives to impose fines on those who violated the rules since they received income from fines levied in their courts.

In most villages there is now relatively little enforcement of *maboella* rules on village residents. The external and internal bases for *maboella* have gradually weakened during the last hundred years. Externally, the advent of the colonial and post-colonial governments and courts reduced the political and legal support for village chiefs and wardens to enforce *maboella* rules. A number of alternative policy organizations – political parties, village development committees, donor agencies and government agencies – have come to compete with chiefs for external authority and legitimacy. Many of these organizations have attempted to 'modernize' or otherwise 'reform' the customary institutions.

Internally, chiefs and wardens no longer receive financial rewards for *maboella* enforcing rules. Legislation passed in 1946 resulted in a 90 percent reduction in the number of recognized indigenous courts and a prohibition on chiefs receiving income from fines levied in their courts (Machobane, 1990: pp. 226–277). The primary motivations for chiefs to enforce *maboella* regulations have become personal and political. Chiefs who support the government of the day, or own large numbers of livestock, have greater interests in collective grazing lands. Contemporaneous changes in the distribution of wealth and economic opportunities in South Africa have made rural residents less dependent upon the economic power of chiefs. Households with migrant workers have become more dependent upon remittances and less dependent upon natural resources for their subsistence or economic livelihoods (Lawry, 1988).

Shoup (1987) documents the case of one lowland village, Ha Nchele, that has initiated a successful village grazing regime with very little external assistance. In about 1976 the village leaders initiated a rotational grazing scheme on village-grazing lands as an alternative to taking their animals to cattleposts in the mountains. The regime has been successful in enforcing rules on its members regarding access and allocation of the village-grazing areas. Shoup (1987) suggests several reasons for this success: (1) The village chief had a personal interest in the success of the regime because he held the greatest number of livestock, but did not abuse his power as chief. (2) The chief and the Village Development Committee shared leadership in defining the operations and evolution of the regime. (3) The regime built upon the customary institution of *maboella*.

2.2. LESOTHO'S CATTLEPOST REGIME

Increased exploitation of the country's expansive mountainous regime enabled Lesotho's human and livestock populations to expand at the end of the nineteenth century and beginning of the twentieth century (Swallow *et al.*, 1987). By the beginning of the twentieth century, a system of seasonal transhumance between the lowlands and mountains was established. Animals were kept on village grazing areas and harvested fields during the winter months (governed by *maboella*) and moved to mountain rangelands areas – cattlepost areas – for the summer months.

The cattlepost regime was established in the 1920s to coordinate the activities of the increasing number of people grazing their animals on mountain pastures and to protect the mountain rangelands from over-exploitation. The regime was enforced by the Paramount Chief and the Principal Chiefs. Any livestock owner who sought to use a specific cattlepost area had to request permission from the relevant chief. Chiefs were not allowed to withhold permission to any Masotho on the basis of the person's residential location or social status. With permission, the person could build a cattlepost hut and paddock and graze their animals in specific areas during designated times of the year. Exclusive rights were not granted to individuals; individuals were prohibited from blocking access to others who had also been granted permission (Sheddick, 1954).

It appears that the cattlepost system performed the coordination function more effectively than the regulation function. Lesotho's populations of cattle, sheep and goats grew rapidly during the first three decades of the twentieth century so that cattle and smallstock populations reached their historic peaks in 1921 and 1931 respectively (Swallow and Brokken, 1987). Concern about the ecological impacts of these large numbers of livestock (e.g., Pim, 1935) prompted the colonial government to commission an extensive survey of the mountainous region. Staples and Hudson (1938) estimated that the overstocking of Lesotho's mountain pastures had resulted in the encroachment of inedible bushes, noxious weeds and less palatable grasses into areas formerly dominated by palatable perennial grasses. The 'carrying capacity' of the pastures was dramatically reduced as a result.

In the early 1970s a grazing-permit system was introduced to help monitor seasonal movements of livestock and to restrict the numbers of livestock kept on particular areas of rangeland. Permits issued by, or on behalf of, the principal chiefs specify the location where an individual is permitted to graze his animals and also the number of animals allowed. Under the Range Management and Grazing Control Regulations of 1980 and 1986, Principal Chiefs operated under the direction of officers of the Ministry of Agriculture to enforce a strict set of controls including: grazing permits, limitation of livestock numbers to the 'carrying capacity' of the local area, specifications on the location of cattlepost kraals, dates at which cattlepost areas were opened and closed for grazing, and prohibitions on grass burning.

The stated objectives of these interventions were to arrest the deterioration of both lowland and mountain rangelands, to bring livestock numbers in line with the 'carrying capacity' of rangelands, and to increase the value of livestock products produced in Lesotho. The evidence – continued rangelands deterioration (Swallow, 1991: pp. 238–241), a decline in the quantity and quality of livestock products (Hunter, 1987), ignorance of the rules (Shoup,

1987) – indicated that these interventions had, in general, not achieved the (perhaps unrealistic) objectives set for them.

An institutional innovation that has achieved some success in governing resource management in the mountain grazing areas is the grazing association. The grazing association that is now being used as a model for Lesotho and other areas in southern Africa is the Sehlabathebe Grazing Association.

The Sehlabathebe Grazing Association was established in 1982 with support from the U.S.A.I.D.-funded (United States Agency for International Development)-Land Conservation and Range Development Project as a mechanism for the residents of mountain areas to take greater control over the grazing resources in their areas. The area allocated to the grazing association, 30,720 hectares, included village grazing areas, cattle post grazing areas, and 11 villages. Residents of the 11 villages were granted exclusive grazing rights to the area, a concept project which personnel labelled 'controlled' communal grazing (Weaver, 1986). The enforcement of exclusive use by local residents resulted in an immediate reduction in the number of livestock grazed in the area. The consequent decrease in competition for forage has resulted in a recuperation of the rangelands and improvement in the condition and productivity of local animals.

The chief, the project manager, and the grazing association have been quite successful in enforcing the regime's boundary. They have been far less effective, however, in enforcing rules within the regime. Frustration over the project with the grazing association resulted in the project playing an increasing role in the active management of the association. As of August 1986, it was the project manager, not the executive committee, who directed the range riders, set the dates for grazing rotation, and administered the finances of the association (Lawry, 1988).

2.3. EASTERN SENEGAL'S PASTORAL UNITS

Eastern Senegal is the site of one of the World Bank's most successful experiences with collective management of African rangelands. The project covers an area of approximately 14,000 square kilometres that is populated by about 70,000 people of the Peul, Wolof and Mandingo ethnic groups. Average annual rainfall is 700 mm in the north of the area and 1,400 mm in the south. Most of the permanent residents are agro-pastoralists who live in permanent villages. Approximately four percent of the area is planted to crops (Shanmugaratnam *et al.*, 1992: pp. 34–35).

The traditional authority system that governed access and use of the pasture lands was based on lineage and caste. A project pre-appraisal undertaken by the World Bank during the mid-1970s noted that the traditional authority system was no longer effective. The nationalization of land ownership in 1964 undermined the basis of the customary system. Drought conditions in the Fero area of northern Senegal and Mauritania resulted in an influx of transhumant pastoralists. Problems of local over-grazing and land-use conflicts between livestock and cultivation were on the rise. The former common property regime had essentially become and open access situation (Bromley and Cernea, 1989: p. 34).

During the first phase of the project (1976–1983), 53 pastoral units were established in the area. Each pastoral unit was comprised of eight to ten small villages and 1–2,000 agro-pastoralists. The pastoral units collaborated with the project in the delivery of supplemental feed and veterinary supplies and served as collective credit guarantors. Loan repayment rates were excellent.

During the project's second phase (1984–88), the pastoral units began to take on common-property rangeland management as a secondary activity. In 1984, a Protocol Agreement was passed that permitted the pastoral units to register legally-defensible communal rights to grazing lands and watering points (Vedeld, 1992). The pastoral units began to administer grazing rotations to coordinate the grazing and watering activities of their members as well as agricultural and livestock activities in the area (Associates in Rural Development, 1989; Bromley and Cernea, 1989).

2.4. SUMMARY AND COMPARISON OF THE CASE STUDIES

The common property regimes from Lesotho and Senegal have general similarities and specific differences. First, in each case the effectiveness of the customary regime depended upon the strength of the *authority structure* on which it was based. The *maboella* and cattle post regimes of Lesotho were based on the authority of the chieftaincy, while the customary regime of eastern Senegal was based on lineage and the caste system of nobles and slaves. In both countries the customary authority structures weakened with the advent of colonialism and the increased commercialization of the rural economies. Weaker authority structures resulted in less effective implementation of the *institutions* that governed the behavior of individuals and groups.

Second, each of the current cases of effective common property management was based on a combination of customary authority, new organizations and the de-centralized actions of individual livestock owners. The success of the Ha Nchele regime in the lowlands of Lesotho depended upon the actions of the chief as customary leader and wealthy livestock owner, the leadership of the Village Development Committee, and the willingness of all livestock owners to comply with the grazing regulations and enforce compliance by their neighbours (Shoup, 1987). The Sehlabathebe regime in the mountains of Lesotho depended upon the leadership and strength of the Principal Chief, cooperation between several village chiefs, technical advice and assistance provided by representatives of the Ministry of Agriculture, and the actions of the elected Range Management Association (Lawry, 1988).

The pastoral units of eastern Senegal are new organizations (Groupements d'Interêt Economique) that are based on the customary caste system (i.e., all board members of the new organizations are former nobles) and are empowered by new national laws that provide them with recourse to the national legal system. About half of the potential members of a GIE in a given locality are actual members (Shanmugaratnam *et al.*, 1992).

Third, each regime consists of a constellation of institutional arrangements. In the Ha Nchele area every permanent resident of the village has the *right* to graze animals on village pastures – but they must follow certain regulations restricting where they can graze their animals at different times. Those regulations are a combination of *rules* enforced by village authorities and *contracts* enforced by all livestock owners. In the Sehlabathebe area, every permanent resident of the eleven villages in the area has the right to graze animals in the village areas and mountain pastures covered by the Range Management Association (RMA). Residents of other areas are not permitted to bring animals into the RMA. Rules covering the use of the areas by members of the RMA are specified but are seldom enforced. In Senegal the members of the pastoral units have recognized rights to use land and water resources in their local areas. The pastoral units also provide mechanisms to resolve conflicts between crop and livestock uses of land. It is not clear, however, whether the pastoral units have any right to exclude outsiders.

3. Alternative Institutional Forms

The case studies illustrate that African rangeland regimes consist of diverse constellations of institutional arrangements that are supported by various external and internal enforcement mechanisms. The sub-section 2.4 that summarizes and compares the case studies introduces several concepts that are useful for analyzing the operation of common property regimes. Rights, rules and contracts were mentioned as types of institutions. In this section we define those and other types of institutions and discuss examples. Wherever appropriate, we draw upon the case studies presented above.

A *rule* is a 'standard setting forth actions that agents are expected to perform (or refrain from performing) under appropriate circumstances' (Young, 1989: p. 16). African governments are often tempted to pass legislation to enact strict rules designed to govern livestock owners' behavior. For example, Lesotho's Range Management and Grazing Control Regulations of 1980 and 1986 decreed that principal chiefs should operate under the direction of officers of the Ministry of Agriculture to enforce a strict set of controls including: grazing permits, limitation of livestock numbers to the 'carrying capacity' of the local area, specifications on the location of cattleposts, dates at which cattlepost areas are opened and closed for grazing, and prohibitions on grass burning (Government of Lesotho, 1980). In practice these rules have had little impact on rangeland utilization in the country. For example,

Shoup (1987) found that the ward chief in the area he studied made no attempt to limit either the number of animals grazed by any individual livestock keeper or the total number of livestock in any cattlepost area. In fact, he considered it his duty to facilitate his subject's access to mountain pastures.

To understand African rangeland regimes sufficiently to design appropriate institutional innovations, it is essential that we move beyond the formalism of official rules and into another domain.

3.1. RIGHTS

Much of the current literature on common property speaks in terms of the concept of 'property rights.' To clarify, we consider a *right* to be a guarantee given by a collective authority system to those who comprise – are part of – the entity. An individual or group has a right to opportunities or benefits from a policy that advances the enjoyment of that opportunity is preferred over one that does not, regardless of the implications for the collective goals of the society (Dworkin, 1977). *Property rights* – rights to potential future benefits – are rarely unconditional guarantees. Usually they are qualified by: (1) the types of agents holding rights and bearing duties; (2) the conditions necessary for their realization; and, (3) their level of abstractness. Rights and duties vis-à-vis rangeland resources are held by states, directly by individuals, or indirectly by individuals through their membership in groups. Property rights can be held by all individuals in a society or only by the occupants of certain social roles. Rights can be superior to collective goals, conditional on the advancement of those goals, or derived from those goals (Becker, 1977; Dworkin, 1977).

There have been many situations, particularly in French West Africa, in which colonial governments, and later post-colonial governments, have declared customary right structures to be void and declared state 'ownership' of collectively-used rangelands. The results vary, but few post-colonial governments have been effective in governing local-level resource use and management. In eastern Senegal, state declaration of ownership may have had some positive consequences by reducing the power of the established elites and providing legal machinery for a more equitable distribution of rangeland rights (Bromley and Cernea, 1989). More common, however, have been the instances in which only negative social impacts were apparent. In the Niger River delta of Mali the *dina* common property regime has been undermined by French colonial rule and by the 1960 declaration of state ownership of all natural resources. In some areas of the delta, powerful groups and individuals – founding lineages, merchants, retired soldiers – have been able to establish exclusive access to the most productive resources. Other areas of the delta now have characteristics of open access. These areas have come under increased stress from herders, fishers and rice producers. The level of conflict among users has been high and the quality of the resource base has deteriorated (Lawry, 1989; Moorehead, 1989).

Rights can also be general, specific, or particular. While a general property right may entitle an individual to some unspecified ownership right to rangeland benefits; a specific property right would identify the specific benefit – forage and water generated by the rangelands – to which the individual is entitled. A particular property right would further indicate the particular area of rangelands over which the right extended.

From the case studies presented above, it appears that customary rangeland rights are often expressed in specific but not particular terms. Under Lesotho's cattlepost regime, Basotho livestock keepers have specific rights to products emanating from Lesotho's mountain rangelands. To be effective, however, those rights had to be enforced by someone from a national-level institution such as the Paramount Chief. With the deterioration in the authority of the chieftaincy, however, it has become important that rangeland regimes can be enforced at the local level. For locally-enforced institutions to be effective, rangeland rights must be specified in particular terms. The grazing rights of individual livestock keepers must be associated with particular areas of the rangeland. This has been a primary outcome of the Sehlabathebe grazing association.

3.2. CONVENTIONS

Runge (1981, 1986) hypothesizes that the internal order of common property regimes can often be self-enforced by the actions and expectations of the individual resource users themselves. In general, a self-enforcing social institution that provides agents with assurance regarding others' behavior is called a *convention*. Runge used the customary rangelands tenure regime in Botswana to support this hypothesis:

> . . . the traditional institutional structure provided a relatively stable and cooperative basis for resource use and the exchange and trade of basic commodities. These 'rules of the game' were sufficient to assure each individual that his rights, duties, liberties and exposures would be respected by others (1981: p. 301).

3.3. CONTRACTS

An institutional arrangement that has received little explicit attention by analysts of common property regimes is the *contract*. A contract is an agreement among agents that is supported by the actions of a third party, in the case of externally-enforced contracts, or by the actions of the agents themselves, in the case of internally-enforced contracts. The need for a contract can arise in any situation in which there is 'transactional insecurity,' that is, any situation in which agents' actions of compliance or deviation from the terms of an agreement do not occur simultaneously (Kronman, 1985).

Alternatively, agents' actions may occur simultaneously but information about those actions may be expensive to collect. This inter-temporal dimension makes contracts inherently a dynamic phenomena.

Because common property has generally been neglected in the current literature, we focus primarily on internally-enforced contracts in this paper. Internally-enforced contracts can either be explicit or implicit: the difference is the type of enforcement mechanism supporting the contract. Explicit contracts are enforced by the deliberate actions of the contracting agents, while implicit contracts are enforced by implicit threats of future retaliation for current deviations. Deliberate actions taken to secure explicit contracts might include hostage taking, collateral requirements, hand-tying, gain sharing, or partial surrender of autonomy.

Dynamic contracts are enforced by strategies containing credible threats of future punishments for current deviations from the terms of the (explicit or implicit) agreement. Punishment strategies considered in the theoretical literature include trigger strategies, stick-and-carrot strategies, exclusion, and tit-for-tat. Briefly, an agent following a trigger strategy will comply as long as everyone else complies, but will revert to Nash equilibrium behavior if anyone ever deviates (Friedman, 1986). An agent following a stick-and-carrot penal code will respond to others' deviations by imposing a harsh but short-lived punishment, followed by compliance (Abreu, 1986). An agent following an exclusion penal code will take steps to exclude deviants from future access to the resource (Hirshleifer and Rasmusen, 1989). And an agent following a tit-for-tat penal code will mimic others' actions: cooperation will be rewarded with cooperation, deviation will be punished by deviation (Axelrod, 1984).

The customary rangeland tenure system of the pastoral Fulani of Senegal and elsewhere in West Africa has been described by a number of analysts in terms that are very similar to the conceptual notion of the dynamic contract. From a review of empirical studies of African pastoralists, Niamir (1989) distinguishes between what she calls active coordination and passive coordination among herding units. Active coordination refers to situations where herding units follow formal and informal rules that are created and recognized by a group of herding units. Passive coordination refers to situations where 'no formal agreements are made between tribes but where coordinated movements result from the wish to avoid other tribes' (p. 40). Passive coordination is maintained through 'informal rules,' or, in the terminology suggested here, dynamic contracts (Swallow and Bromley, 1994).

4. Governance and Institutions

Fundamental to the operations of common property regimes are the authority systems that sanction rights, enforce rules, and define the contexts in which conventions and contracts are negotiated. Authority systems are concerned with *governance*, and governance is the process of deciding what a collective will do and how it will do it. *Governments* exist for the process of governance; that is, governments are created to carry out governance. *Institutions* are defined by the process of governance to provide order to the relations among the members of a collective and to constrain what governments may do to the members in the name of governance (Bromley, 1989). To understand institutions, therefore, it is necessary to consider the nature of the authority system, process of governance, and the implications of those processes for the implementation of institutions.

4.1. GOVERNMENTAL DIVERSITY

Besides institutional diversity, rangelands regimes also exhibit considerable governmental diversity. Mair (1977) identified three types of customary government that prevail (or have prevailed) among the livestock-keeping peoples of Africa: (1) centralized governments; (2) diffused governments; and (3) minimal (or non-existent) governments. Societies with centralized governments have hierarchical political structures in which '. . . a single supreme authority is recognized, and public affairs are regulated, decisions taken, and obedience claimed by persons acting in his name' (Mair, 1977: p. 138). Many agro-pastoral groups in Eastern and Southern Africa have or had centralized governments in which chiefs perform executive, legislative and judicial functions. During the days of Moshoeshoe I (circa 1796–1886) the people of Lesotho had one of the most hierarchical governments of all the livestock-keeping peoples of Africa. The Paramount Chief had authority to perform a variety of functions:

> He everywhere represents his people in their dealings with outsiders, and organizes such communal activities as war, collective labour, and certain types of ritual. . . . He is both legislator and judge, with power to inflict capital punishment; he claims many forms of tribute, in both labour and kind; he controls the distribution and use of land, coordinates agricultural activities, provides for the poor and needy, and rewards those who serve him well (Shapera, 1956: pp. 40–41).

In contrast to this centralized structure, most pastoralist groups in Eastern and Western Africa have customary governments that are characterized as being 'diffused' or 'minimal.' Those with diffused customary governments include the Maasai, Kipsigi, Pokot, Nandi and Samburu of Kenya, and the Karimojong of Uganda (Mair, 1977; Quam, 1976). In those societies legislative and judicial authority is held by relatively egalitarian elders' councils. The enforcement of elders' decisions depends upon the type of case: the coercion of younger age-sets is used to enforce judgements in criminal cases, while social sanctions and coercion of individual litigants combine to enforce judgements in most civil cases (Mair, 1977). For example, among the Maasai:

> . . . the primary institution for the resolution of conflict and social accountability is a council of elders. . . . The range of affairs dealt with include conflicts between local residents, access

to communal resources, and interpretation of proper conduct as deemed consistent with being Maasai. The council derives its powers from and is a practical expression of the age-based principle of organization. The council's authority is underpinned by its collective curse on any member who persistently fails to heed decisions and the cohesion of age-set leaders who enforce subordination to the collective wisdom manifest in council decisions (Kituyi, 1990: p. 114).

Societies identified as having minimal governments include the Turkana of Kenya, the Fulani of West Africa, and the Western Dinka, Nuer, and Mandari of Sudan. These groups have neither chiefs nor elders' councils with the authority or power to enforce rules. Designated individuals or councils may arbitrate disputes but have no power to enact punishments. Individuals and coalitions have responsibilities to assess the fairness of the judgements reached by their chiefs or elders' councils and undertake actions necessary to impose sanctions on deviants. Each individual or coalition will therefore consider their own welfare (short-term or long-term) when they make such decisions.

The nomadic Fulani, as well as most of the semi-nomadic ones, seem to be anarchic in their political organization. Their chiefs act as guides, arbitrators, and spokesmen for extended family groups that tend to move about more or less together. These chiefs have no real power of coercion over their followers, nor can they make binding agreements on their behalf (Riesman, 1978: p. 27).

4.2. IMPLEMENTATION OF INSTITUTIONS

The structure of government under which a common property regime operates – whether it be centralized, diffused, or minimal – determines the type of institutions that can be implemented to govern relations among members and between members and non-members. The most demanding type to govern is that concerning rights. To be implemented rights must be: (1) formulated and promulgated, (2) communicated to those to whom they apply (members and non-members alike), (3) interpreted, (4) enforced, (5) legitimized with the persons or groups to which they apply, and (6) adapted to changing needs and circumstances (Bull, 1977: pp. 56–57).

To implement the rights of groups – that is, those at the boundaries of regimes – a body performing these functions must also be able to interpret the aims of the larger society in which the groups are established, to judge between the rights and duties of competing groups, and enforce sanctions on individuals, groups and collectives of groups. Such a body must be capable of enforcing sanctions on the individuals and groups who hold rights and duties and on collectives of groups. To implement the rights of individuals as members of groups – that is, rights within regimes – the body must interpret the aims of the group, judge between competing rights, and enforce sanctions on individuals and collectives of individuals.

Evidence from Lesotho, Senegal and Zimbabwe (Cousins, 1992) suggests that national governments are better suited to the enforcement of group rights than individual rights. In those countries, the national governments have achieved some successes in defining boundaries between group resources and in supporting groups to identify and punish those who violate those boundaries. Governments have been less successful in enforcing order on the internal operations of rangeland regimes. The Sehlabathebe Grazing Association has been keen to use external authority to exclude outsiders but reluctant to apply that same authority to enforce internal grazing rules.

Conventions and contracts, because they are not 'rights,' can be implemented without central government units. They can be effective, therefore, in social groups with central, diffused, minimal or non-existent governments. Conventions and internally-enforced contracts can also govern relations between groups of resource users (at the boundaries of regimes) and within groups (within regimes). It is unlikely, however, that internally-enforced contracts will be simultaneously effective on both levels in the absence of some other institutional arrangements. That is, for a group of resource users to be successful in maintaining an implicit internal agreement they must have confidence that future entry into the group will be restricted, or at least limited to a known number of potential entrants. A minimum condition for the effective operation of dynamic implicit contracts may therefore be that there is some social authority that enforces rights or rules regulating the entry of new individuals or groups, the mobility of individuals between groups, and the mobility of groups between sovereign polities.

An example that supports this hypothesis is provided by Ha Nchele in Lesotho. Because the boundaries between most Lesotho villages are relatively clear and well-known, the residents of Ha Nchele have encountered few difficulties in excluding outsiders from their village common areas. Membership of the grazing association was naturally restricted to the resident households who kept cattle. At Ha Nchele this was a relatively small number, 29 out of a total of 128 households. Even with that small number of households, grazing rules were not self-enforcing. People violated rules and were punished. Anyone who violated the clearly-articulated rules had their animals impounded and was charged a fine. Group members enforced rules themselves by identifying deviants and helping to collect fines. No external enforcement was necessary.

5. Expectations, Incentives and Institutions

The structure of governance determines the necessary, but not sufficient, conditions for the successful implementation of institutions regarding common property regimes. Implementation of these institutions requires action on the part of particular individuals. Individuals respond to their own incentives and expectations.

Consider first the incentives and expectations of the individuals who are responsible for defining and enforcing the rights and rules that comprise common property regimes. Local leaders – customary, appointed or elected

– often perform important roles in common property regimes. Their incentives depend upon their relationships with the external political and legal system and their relationships with the group of resource users and other local constituents. The incentives they have to enforce restrictions on resource use also depend directly on their use of the resource and the benefits and costs they derive from enforcing the rules.

Government employees are often responsible for advising local groups in group organization and technical aspects of resource management. They may also be responsible for mediating relationships between groups of resource users. The incentives and expectations of government employees depend upon the government system of incentives, the tolerance of nepotism and corruption, the resources that are available to them to fulfil their assigned roles, their personal affiliations with one group or another, and possibly their own goals regarding resource use.

Now consider the incentives and expectations of resource users themselves: Swallow (1991: pp. 156–156) shows that the user of a collectively-used rangeland has an incentive to demand a resource use rule; at the same time he or she also has an incentive to deviate from such a rule. The incentive to deviate depends upon the external gains from deviation, the probability that other resource users will comply or deviate, the probability of being caught and punished, and the likely severity of the punishment.

Runge (1986) hypothesizes that the incentive of an individual to deviate from a stocking-rate rule generally decreases with the number of other agents with which he expects to comply. Results from a multiple-period simulation model, presented in Swallow and Bromley (1994), challenge the validity of that hypothesis. Those results indicate that, *ceteris paribus*, the incentive for an individual to deviate from a stocking-rate rule actually increases with the number of agents with which he expects to comply. Further, individuals have incentives to deviate from stocking-rate rules in almost all cases, and time periods. Generally, therefore, internal stocking-rate rules will not be self-enforcing.

The literature on internally-enforced contracts suggests other mechanisms by which internal agreements (explicit or tacit) on resource allocation can be maintained without external enforcement. Internally-enforced contracts can operate within a set of articulated rights, rules, and codified contracts (Bowles and Gintis, 1988; Williamson, 1985) or within a 'state of nature' in which 'individuals and groups must arrange their transactions without the aid of an independent enforcement mechanism whose powers are significantly greater than their own' (Kronman, 1985: p. 9).[1]

Swallow and Bromley (1994) show that the strength of a dynamic contract – the difference between the payoff to compliance and the payoff to deviation – is positively related to: (1) the marginal cost of production and (2) the duration and severity of the punishment, and negatively related to: (1) the discount rate, (2) the size of the group, (3) the price of the output, and

(4) the number of other agents expected to comply. The strength of contract is very sensitive to changes in group size. Dynamic contracts are unlikely to be effective in groups with more than 30–40 members.

6. Policy Implications

To address the problems of African livestock-rangelands development, feasible policy objectives must be established and feasible policy instruments designed to advance those objectives. A nation-wide system of individual property rights and duties is only a feasible objective if legitimate central authorities have the power, will (incentive) and managerial capacity – external legitimacy – necessary to protect individuals against collectives and adapt institutions to changing circumstances. To be enforceable under current conditions, property rights must be defined in specific terms. For example, the common property innovations that have been successful in Lesotho and Senegal have defined specific rangelands and water resources to which people have rights. The more variable the environment, the more difficult the tradeoff between the need for specificity and the need to allow flexibility in grazing patterns (van den Brink *et al.*, 1995).

Group property rights are feasible under a broader range of circumstances. Implementation of group property rights requires a central authority that can arbitrate between the interests of broad groups of the population and enforce boundaries and agreements between groups. Unambiguous membership criteria are a pre-requisite for effective group rights.

Effective group rights will only lead to efficient resource allocation if there are effective mechanisms for internal governance. The effectiveness of a formal structure of internal governance will depend upon the incentives of the individuals whose roles are to enforce rules and other individuals who are expected to comply with the rules. Internally-enforced contracts are mechanisms for internal governance that can be effective with little or no internal authority system. Internal contracts can be enforced through such explicit mechanisms as collateral, hostages, or gains sharing arrangements, or through implicit threats of future retaliation for current deviations (dynamic contracts). External agencies can facilitate internal contracts by: (1) providing assurance that agents will have long-term interactions with the same group of resource users; (2) increasing information about agents' preferences and their compliance or deviation from contracts; (3) increasing the points of contact between resource users; or (4) absorbing some of the transaction costs associated with the maintenance of contracts.

Political structures and the institutions they support are continually changing: what might have been feasible with customary government structures may not be feasible with current structures. Common property regimes, polities, and ecologies are systems that are evolving, co-evolving through their repeated interactions, and responding to exogenous political, economic and ecological shocks. At the present time, the factors that are having the greatest impact on the common property regimes for African rangelands include: (1) greater competition for livestock, land and water between pastoralists, cultivators, and conservationists; (2) reductions in the economic and political power of customary authorities; (3) inter-ethnic and international conflicts between pastoralist groups; (4) declarations of 'ownership' of rangelands by African governments (Lawry, 1989); (5) adoption of new water provision techniques and tsetse-control techniques (Behnke, 1988); (6) increasing importance of non-livestock and non-agricultural opportunities for employment and income generation; and (7) increased concentration concerning the ownership of livestock among pastoralists and increased ownership of livestock by external investors (Little, 1987).

These processes are serving to undermine the efficacy of status quo institutional arrangements. In this context policy makers need to search for ways to facilitate institutional change that will make regimes more resilient and better able to allocate resources to the mutual benefit of those who share their access. Priority should be given to group rights and the internal institution-building capacity of local groups and communities.

Acknowledgements

Earlier revisions of this paper were presented at the 'International Conference on African Economic Issues', Abidjan, Côte d'Ivoire, in August 1992 and the 'Workshop on New Directions in African Range Management Policy', Matopos, Zimbabwe in January 1992. Useful comments have been provided by David Pratt, Peter de Leeuw, Roy Behnke, Maryam Niamer-Fuller and Ian Scoones. We also appreciate the constructive comments provided by the anonymous reviewers.

Note

[1] Kronman (1985) points out that his state of nature is somewhat different from the Hobbesian state of nature as a 'war of every man against every man.' A Hobbesian state of nature involves 'transactional insecurity' – in a non-simultaneous exchange, at least one party is exposed to the danger that another party will not complete the exchange – and 'vulnerability of possession' – the objects possessed by an individual are subject to attack at any time. Kronman's state of nature, on the other hand, involves transactional insecurity and possessive security.

References

Abreu, D. (1986), 'Extremal Equilibria of Oligopolistic Supergames', *Journal of Economic Theory* **39**, 191–225.

Associates in Rural Development (1989), 'Options for Promoting User-Based Governance of Sahelian Renewable Natural Resources', paper prepared for presentation at the CILSS-sponsored conference, 'Regional Encounter for a Better Socio-Ecological Balance in the Rural Sahel', Bamako.

Axelrod, R. (1984), *The Evolution of Cooperation*, Basic Books, New York.

Becker, L. C. (1977), *Property Rights: Philosophical Foundations*, Routledge & Kegan Paul, Boston.

Behnke, R. H. (1988), 'Range Enclosure in Central Somalia', Pastoral Development Network Paper No. 25b, Overseas Development Institute, London.

Berkes, F. (ed.) (1989), *Common Property Resources: Ecology and Community-based Sustainable Development*, Belhaven, London.

Bowles, S. and H. Gintis (1988), 'Contested Exchange: The Political Structure of Competitive Markets', Department of Economics, University of Massachusetts, mimeo, Amherst, MA.

Bromley, D. W. (1991), *Environment and Economy: Property Rights and Public Policy*, Basil Blackwell, Oxford.

Bromley, D. W. (1989), *Economic Interests and Institutions: The Conceptual Foundations of Public Policy*, Basil Blackwell, Oxford.

Bromley, D. W. and M. M. Cernea (1989), *The Management of Common Property Natural Resources: Some Conceptual and Operational Fallacies*, World Bank Discussion Paper 57, The World Bank, Washington, D.C.

Bull, H. (1977), *The Anarchical Society: A Study of Order in World Politics*, Columbia University Press, New York.

Ciriacy-Wantrup, S. V. and R. C. Bishop (1975), ' "Common Property" as a Concept in Natural Resources Policy', *Natural Resources Journal* **15**, 713–727.

Cousins, B. (1992), *Managing Communal Rangelands in Zimbabwe: Experiences and Lessons*, Commonwealth Secretariat, London.

Dworkin, R. M. (1977), *Taking Rights Seriously*, Harvard University Press, Cambridge.

Eldredge, E. A. (1986), *An Economic History of Lesotho in the Nineteenth Century*, Ph.D. dissertation, University of Wisconsin-Madison.

Friedman, J. W. (1986), *Game Theory with Applications to Economics*, Oxford University Press, Oxford.

Galaty, J. G. and D. L. Johnson (1990), 'Introduction: Pastoral Systems in Global Perspective', in J. G. Galaty and D. L. Johnson, eds., *The World of Pastoralism: Herding Systems in Comparative Perspective*, New York: The Guilford Press, pp. 1–15.

Gilles, J. L., B. E. Norton and J. T. O'Rourke (1992), 'Oukaimadene, Morocco: A High Mountain Agdal', in D. W. Bromley, ed., *Making the Commons Work: Theory, Practice and Policy*, San Francisco: ICS Press, pp. 229–258.

Gordon, H. S. (1954), 'The Economic Theory of a Common Property Resource: The Fishery', *Journal of Political Economy* **62**, 124–142.

Government of Lesotho (1980), *Range Management and Grazing Control Regulations 1980*, Legal Notice No. 39, Government Printer, Maseru.

Hardin, G. (1968), 'The Tragedy of the Commons', *Science* **162**, 1234–1248.

Helland, J. (1982), 'Social Organization and Water Control Among the Borana', *Development and Change* **13**, 239–258.

Hirshleifer, D. and E. Rasmusen (1989), 'Cooperation in a Repeated Prisoners' Dilemma with Ostracism', *Journal of Economic Behavior and Organization* **12**, 87–106.

Hunter, J. P. (1987), *The Economics of Wool and Mohair Production and Utilization in Lesotho*, Institute of Southern African Studies Research Report No. 16, National University of Lesotho and Research Division Report RD-R-80, Ministry of Agriculture and Marketing, Roma and Maseru.

Kituyi, M. (1990), *Becoming Kenyans: Socio-economic Transformation of the Pastoral Maasai*, ACTS Press, Nairobi, Kenya.

Kronman, A. T. (1985), 'Contract Law and the State of Nature', *Journal of Law, Economics and Organization* **1**, 5–32.

Lane, C. R. (1991), *Alienation of Barabaig Pasture Land: Policy Implications for Pastoral Development in Tanzania*, Ph.D. dissertation, University of Sussex.

Lawry, S. W. (1989), *Tenure Policy and Natural Resource Management in Sahelian West Africa*, LTC Paper 130, University of Wisconsin, Land Tenure Center, Madison.

Lawry, S. W. (1988), *Private Herds and Common Land: Issues in the Management of Communal Grazing Land in Lesotho, Southern Africa*, Ph.D. dissertation, University of Wisconsin-Madison.

Little, P. D. (1987), 'Land Use Conflicts in the Agricultural/Pastoral Borderlands: The Case of Kenya', in P. D. Little and M. M. Horowitz, eds., *Lands at Risk in the Third World*, Colorado: Westview Press, Boulder.

Machobane, L. B. B. J. (1990), *Government and Change in Lesotho, 1800–1966*, St. Martin's Press, New York.

Mair, L. (1977), *Primitive Government*, Indiana University Press, Bloomington.

McCabe, J. T. (1990), 'Turkana Pastoralism: A Case Against the Tragedy of the Commons', *Human Ecology* **18**, 81–103.

McCay, B. and J. M. Acheson (1987), *The Question of the Commons: The Culture and Ecology of Communal Resources*, University of Arizona Press, Tuscon.

Moorehead, R. (1989), 'Changes Taking Place in Common Property Resource Management in the Inland Niger Delta of Mali', in F. Berkes, ed., *Common Property Resources: Ecology and Community-Based Sustainable Development*, London: Belhaven Press.

National Academy of Sciences (1986), *Common Property Resource Management*, National Academy Press, Washington, D.C.

Niamir, M. (1989), *Herders' Decision-making in Natural Resources Management in Arid and Semi-Arid Africa*, Community Forestry Note 4, Food and Agriculture Organization of the United Nations, Rome.

Ostrom, E. (1986), 'Issues of Definition and Theory: Some Conclusions and Hypotheses', in National Academy of Sciences, *Proceedings of the Conference on Common Property Resource Management*, Washington, D.C.: National Academy Press, pp. 597–614.

Pim, A. W. (1935), *Financial and Economic Position of Basutoland*, His Majesty's Stationary Office, London.

Quam, M. D. (1976), *Pastoral Economy and Cattle Marketing in Karimoja, Uganda*, Ph.D. dissertation, Indiana University.

Riddell, J. C. (1982), *Land Tenure Issues in West African Livestock and Range Development Projects*, LTC Research Paper No. 77, University of Wisconsin-Madison, Land Tenure Center, Madison.

Riesman, P. (1978), 'The Fulani in a Development Context: The Relevance of Cultural Traditions for Coping with Change and Crisis', Report written under USAID contract, no. REDSO/WA-78-138, U.S. Agency for International Development, Washington, D.C.

Runge, C. F. (1981), 'Common-Property Externalities: Isolation, Assurance and Resource Depletion in a Traditional Grazing Context', *American Journal of Agricultural Economics* **63**, 595–607.

Runge, C. F. (1986), 'Common Property and Collective Action in Economic Development', *World Development* **14**, 623–635.

Sandford, S. (1983), *Management of Pastoral Development in the Third World*, John Wiley and Sons in association with the Overseas Development Institute, Chichester.

Scott, A. (1955), 'The Fishery: The Objectives of Sole Ownership', *Journal of Political Economy* **63**, 116–124.

Shoup, J. (1987), *Transhumant Pastoralism in Lesotho: Case Study of the Mapoteng Ward*, Land Conservation and Range Development Project, Maseru.

Shanmugaratnam, N., T. Vedeld, A. Mossige and M. Bovin (1992), *Resource Management and Pastoral Institution Building in the West African Sahel*, World Bank Discussion Paper 175, The World Bank, Washington, D.C.

Shapera, I. (1956), *Government and Politics in Tribal Societies*, Schocken Books, New York.

Sheddick, V. (1954), *Land Tenure in Basutoland*, Her Majesty's Stationary Office, London.

Staples, R. R. and W. K. Hudson (1938), *An Ecological Survey of the Mountain Area of Basutoland*, His Majesty's Stationary Office, London.

Stiles, D. (1992), 'The Gabbra: Traditional Social Factors in Aspects of Land-use Management', *Nomadic Peoples* **30**, 41–52.

Swallow, B. M. (1991), *Common Property Regimes for African Rangelands Resources*, Ph.D. dissertation, University of Wisconsin-Madison.

Swallow, B. M. (1990), *Strategies and Tenure in African Livestock Development*, LTC Paper 140, University of Wisconsin Land Tenure Centre, Madison, Wisconsin.

Swallow, B. M. and R. F. Brokken (1987), 'Cattle Marketing Policy in Lesotho', *African Livestock Policy Analysis Network*, Network Paper No. 14. Addis Ababa, International Livestock Centre for Africa, Ethiopia.

Swallow, B. M. and R. F. Brokken (1990), 'Information and the "Optimal Control" of an African Rangelands Resource', paper presented at the American Agricultural Economics Association Annual Meeting, Vancouver, Canada.

Swallow, B. M., R. F. Brokken, M. Motsamai, L. Sopeng and G. G. Storey (1987), *Livestock Development and Range Utilization in Lesotho*, Institute of Southern African Studies Research Report No. 18, National University of Lesotho and Research Division Report RD-R-82, Ministry of Agriculture and Marketing, Roma and Maseru.

Swallow, B. M. and D. W. Bromley (1994), 'Co-Management or No Management: The Prospects for Internal Governance of Common Property Regimes through Dynamic Contracts', *Oxford Agrarian Studies* **22**(2), 3–16.

van den Brink, R., D. W. Bromley and J.-P. Chavas (1995), 'The Economics of Cain and Abel: Agro-Pastoral Property Rights in the Sahel', *Journal of Development Studies* **31**(3), 373–399.

Vedeld, T. (1992), 'Local Institution-Building and Resource Management in the West African Sahel', *Forum for Development Studies* **1**, 23–50.

Weaver, L. C. (1986), 'Planning for Grazing Associations in Lesotho', in *Proceedings from the Workshop on Grazing Associations in Lesotho*, Maseru, Ministry of Agriculture and USAID Land Conservation and Range Development Project.

Winrock International (1992), *Assessment of Animal Agriculture in Sub-Saharan Africa*, Winrock International, Morrilton, Arkansas, U.S.A.

Williamson, O. (1985), *The Economic Institutions of Capitalism*, Free Press, New York.

Young, O. R. (1989), *International Cooperation: Building Regimes for Natural Resources and the Environment*, Cornell University Press, Ithaca, New York.

[25]

Article

Madhav Gadgil

Conserving Biodiversity as if People Matter: A Case Study from India

India has rich traditions of nature conservation as well as a vigorous official program of protection of nature reserves developed over the last 40 years. However, the official program suffers from total reliance on authoritarian management arrangements in which decisions are made centrally and coercion is used to implement them. At the same time, the state apparatus organises subsidized resource flows to the urban-industrial-intensive agricultural complex which promote inefficient, non-sustainable resource-use patterns that are inimical to conservation of biodiversity. These processes are illustrated within the concrete setting of the district of Uttara Kannada in southern India. It is suggested that the interests of conservation would be served far better by an approach that withdraws the subsidies to the elite so that a much more efficient, sustainable and equitable pattern of resource use, compatible with conservation of biodiversity, can be instituted. In conjunction with this, the larger society should involve local people in working out detailed plans for conservation of biodiversity and offer them adequate authority as well as appropriate financial incentives to implement these plans. The paper goes on to illustrate how such an approach may be implemented in the case of Uttara Kannada.

INTRODUCTION

With extensive tracts of humid tropical forests and great heterogeneity of environmental regimes, India is one of the top twelve megadiversity countries of the world. Given the large biomass needs of its tribal and rural populations and the exploding resource demands of its growing urban-industrial-intensive agriculture complex, conserving this heritage of biodiversity is a formidable challenge. India has rich traditions of nature conservation; following independence it has also developed an extensive network of nature reserves.

The Project Tiger with 17 reserves spanning the country has saved this magnificent animal from the brink of extinction. There are however signs that all is not well with the country's programs of conservation. India's cultural traditions have preserved an enormous network of trees of the genus *Ficus*, an important keystone resource- throughout the countryside. These trees are now being increasingly felled to bake bricks and to make crates. At the same time, key reserves in the network of Project Tiger such as Kanha and Manas are threatened by discontented local tribal people. The important questions that must now be tackled, relate not only to identifying the deficiencies in the coverage of the national network of nature reserves, but to how decisions are made on what elements of biodiversity are to be conserved and how this is to be accomplished (1, 2).

It is important to look critically at the processes affecting overall patterns of natural-resource utilization and not just at those impinging on nature reserves. Very broadly the ongoing processes may be summarized as follows:

a. Intensification of resource fluxes in favor of industry-organized services-intensive agricultural complex involving large-scale state subsidies.

b. Increasing biomass demands of a growing rural population forced to meet its requirements from open-access public lands.

Together these two kinds of pressures promote nonsustainable, inefficient use of natural-resources, resulting in decimation of biodiversity. This is coupled to

c. Attempts at conservation of biodiversity in a network of officially constituted nature reserves, relying on policing by the state apparatus, while the traditional practices of conservation are given short shrift.

How do these processes affect biodiversity? What specific elements are responsible for negative impacts? How might these problems be overcome and what would come in the way of overcoming these problems?

UTTARA KANNADA

The two richest humid tropical forest tracts of India fall in the biogeographic provinces of Eastern Himalayas and Malabar, the latter comprising the west coast and the hill range of Western Ghats (3, 4). The district of Uttara Kannada (13° 52' to 15° 30' N and 74° 05' to 75° 5' E) with an area of 10 200 km² lies at the center of Malabar (Fig. 1). It is a region of gentle undulating hills, rising rather steeply from a narrow coastal strip bordering the Arabian Sea to a plateau at an altitude of 500 m with occasional hills rising above 600 to 860 m. The annual precipitation largely confined to the monsoon months of June to September ranges between 3500 mm on the coast, rising to 5000 mm on the crestline and declining to 1000 mm on the eastern plateau. An interpretation of the Landsat imagery suggests that around 6900 km² of the district is under forest cover, around 2000 km² is under paddy and millet cultivation, 130 km² under coconut and betelnut orchards, 200 km² under rocky outcrops and the balance under habitation and reservoirs (5). This tract of 6900 km² of forest land is the largest single contiguous tract of humid tropical forest in Peninsular India. Today, it harbors 1741 species of flowering plants and 403 species of birds. Notable wildlife includes the tiger, elephant, gaur or Indian bison, liontailed macaque, Wynaad laughing thrush, Travancore tortoise, several species of legless amphibians and dipterocarp trees (6).

Colonial Period

Uttara Kannada has been well known historically for its forests and wildlife. In the 17th century both the British and Dutch had established trade stations on its coast. These dealt extensively in wild pepper and cardamom, sandal and teak wood and poon (*Calophyllum elatum*) for ship masts. Accounts of Europeans who worked at these stations mention the rich wildlife with an abundance of tiger, panther, elephant, bison and several species of deer. The district came into British hands in 1799. Buchanan, a naturalist in the employ of the East India Company travelled extensively through the district in 1800–1802. His very detailed accounts confirm that apart from the southern parts of the coastal tract, the district was thickly forested and abounded in wildlife. He also mentions, the cultural traditions of local people that focused on nature conservation such as sacred groves. However, he interprets the almost-total protection offered to these groves as a "contrivance" to prevent British rulers from laying a claim to what was now its rightful property (7).

The history of the British period can be summarized as a series of attempts by the colonial power to appropriate the rich forest resources of the district as cheaply as possible. To this end, the community-based systems of restrained use and conservation had to be scuttled. This was accomplished through refusal to recognize the legitimacy of all customary and community ownership rights. The British recognized only two forms of ownership, state ownership of all non-cultivated lands and private ownership of all cultivated lands. The bulk of state-owned lands were converted into reserve-forest lands. A fraction, about 25% was set aside as minor or leaf-manure forest lands for meeting the subsistence biomass needs of local people. These are substantial since agriculture in this hilly district with its laterized nutrient deficient soils depends heavily on organic manure inputs. But the minor/leaf-manure forest lands came to be treated as open-access resources and have consequently been subject to escalating degradation as envisaged in the tragedy of the commons scenario. The reserve forests were dedicated to supply cheap raw material, primarily, teak to serve colonial interests of shipbuilding, railways and other constructions. As a result, they were almost totally depleted of natural teak between the years 1800–1850; followed by depletion of other hardwoods, especially *Terminalia* and *Lagerstroemia* species, and conversion to single species plantations of teak. The evergreen tree species were of little commercial value until the 1940s, and up to that time forest working focussed on their replacement by the more valued timber species. Although a succession of management plans initiated in early 1900s professed sustainable harvests as their aim, in fact there was only further depletion. All management plans were set aside during the two world wars, permitting totally unregulated harvesting from reserve forests (8–10).

Since Independence

The British forest-management regime in Uttara Kannada was dedicated to export of teak and other timber as cheaply as possible. World War II brought about an important change, however, when the British decided to encourage plywood manufacture in India. With its rich evergreen forests providing abundant raw material, one of India's first plywood factories was set up in the Uttara Kannada district. This was followed by a paper factory and a polyfiber industry, both of which were established soon after independence.

The policy of dedicating state-owned forest lands to furnish a cheap supply of industrial raw materials was carried to further extremes after independence. Thus, in 1958, bamboo, earlier prescribed to be eradicated as it constituted a weed in teak plantations, in spite of its manifold rural uses, was sold to the paper industry. The price was as low as Rs. 1.50 (USD 0.30 at the then prevalent exchange rates) per tonne, i.e. over a thousand times less than the market value. Giant wild mango trees that regularly yielded much valued fruit worth more than Rs. 100 per year, were also made over to plywood industry for as little as Rs. 150 for a whole tree. The result has been rapid decimation of a whole range of species in the more humid tracts, especially on the steeper western hill slopes. Other natural-resources have also been made available at highly subsidized rates to the urban, industrial, intensive agriculture complex. For example, there have been a series of state sponsored hydroelectric projects within the district. The electricity so generated has been supplied to industrial consumers and urban households as well as used in water lifting for irrigated agriculture at greatly subsidized rates. At the same time the cultivators whose lands were submerged under the reservoirs have been poorly compensated and often forced to encroach on forest land to eke out a living. This whole system of subsidized resource use has ensured that neither resource managers nor resource users are concerned with resource-use efficiency.

While the commercial pressures have mounted, so have the subsistence demands of the rural population that has grown rapidly as a result of the eradication of malaria after World War II. These subsistence demands continue to be met from open-access public lands, with further erosion of traditional practices of disciplined harvests from common lands. Consequently, highly inefficient and wasteful patterns of natural-resource use have prevailed (11).

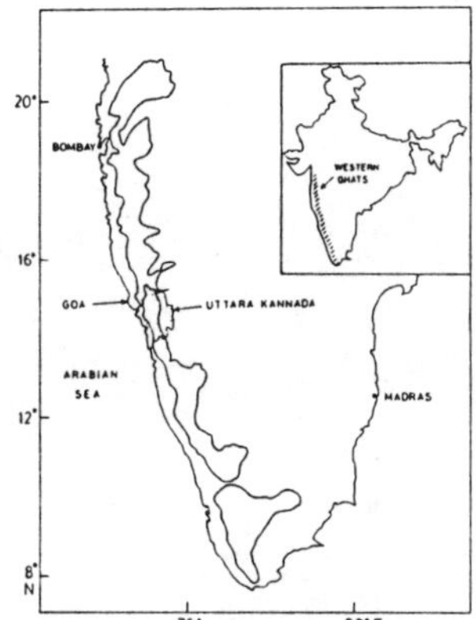

Figure 1. A map of peninsular India indicating the location of the hill range of Western Ghats. The district of Uttara Kannada stretches from the coast, across the hills to the peninsular Indian plateau.

Traditions of Conservation

In common with the rest of the Indian society, people of Uttara Kannada have inherited a rich variety of traditions of nature conservation (12, 13). These include:

- Protection of individual plants and animals considered to be sacred, e.g., trees of the genus *Ficus* or monkeys such as the hanuman langur and bonnet macaque.
- Protection of specific life history stages, such as birds breeding at heronaries.
- Protection of entire biological communities in the sacred groves and sacred ponds.
- Protection of animals from overhunting through devices, such as a ban on hunting of fruit bats at the roosting sites but not outside the sites.

The British colonialists and, following independence, Indian resource managers have both consistently attacked these traditions in many ways. For example, the plywood industry has harvested *Ficus nervosa* as a preferred species and many sacred groves have been clearcut to supply timber. Fisheries managers have poisoned sacred ponds removing indigenous fish and restocked them with exotic carps for sale to outside agencies.

State Initiated Conservation Efforts

The main focus of regulation of hunting during the British period was on the traditional subsistence/ritual hunting by local people, which aimed to ensure supply of game to the European hunters. These hunters began to hunt many species, such as elephant, that had previously been immune from hunting by local people. They also vigorously pursued extermination of species perceived as vermin such as the Indian wild dog. Indeed, there was a rapid decline in wild-animal populations during the British regime. The pace of this decline increased further after independence with the introduction of the jeep and widespread availability of firearms (14).

This ever accelerating depletion of wildlife led to the beginning of conservation measures in the 1950s with the establishment of an Indian Board for Wildlife and of Wildlife Wings in Forest Departments. The main tool of conservation that was adopted was a ban on hunting of endangered species everywhere together with the establishment of Wildlife Sanctuaries and National Parks.

A major Wildlife Sanctuary, Dandeli W. L. S., of over 5730 km^2, was established in 1953 in the northern parts of Uttara Kannada. However, little protection to natural vegetation and wildlife was provided since a large paper mill was established in the very heart of this sanctuary in 1958 and a series of dams of the giant Kali hydel project covered the rest of the sanctuary in the 1970s (15). The ban on hunting has not been very effectively implemented since it is operated in a centralized, rigid fashion without allowance for local conditions. For example, farmers claim that there are now too many wild pigs in the district with the newly arriving weeds, *Lantana* and *Eupatorium* providing excellent shelter, and depletion of panther and tiger populations, cutting down predation. These pigs cause severe crop damage and farmers hunt them relentlessly. Nevertheless, hunting wild pig remains an offence in law. So does the annual ritual hunting by local communities such as the *Halakki Vakkals* who engage in such a hunt as a religious observance. At the same time official managers continue to

The coastal landscape of Uttara Kannada. There is little left of natural vegetation. However the coastal stretch is covered by coconut plantations mixed with a tremendous diversity of cultivars of mango (*Mangifera indica*) and jackfruit (*Artocarpus integrifolia*) along with a number of fruit yielding tree species belonging to genera *Myristica, Spondias* and *Garcinia*. All of these also occur in the wild. Photo: M.D. Subash Chandran.

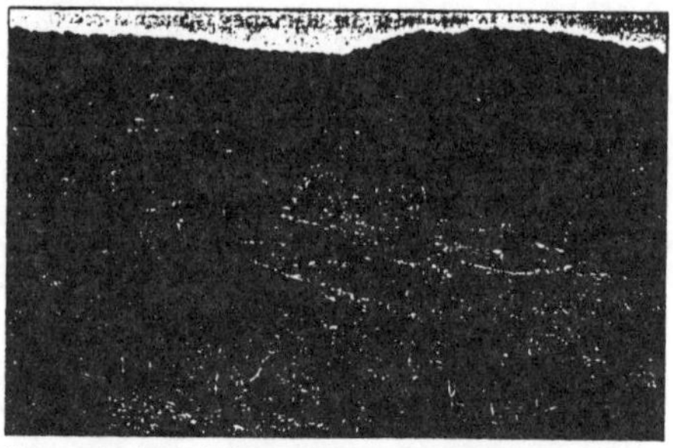

The mosaic of disturbed natural forest and anthropogenic grasslands near the crestline of Western Ghats in Uttara Kannada district. Such vegetation supports the greatest diversity of bird species within this district. Photo: R.J. Ranjit Daniels.

An irrigation pond in the eastern part of Uttara Kannada district. These ponds support a wide diversity of resident and migratory waterfowl. Photo: R.J. Ranjit Daniels.

violate traditions of conservation of the local people.

Social Conflicts

This whole system of resource management initiated under the British rule and further elaborated after independence is based on alienating local people from control of and access to resources. Its primary objective has been to make natural resources available as cheaply as possible to the elite, be it the British ruling classes or the industry-organized services-intensive agriculture complex of the Indian society. The elite that benefits from resource mobilization is shielded from the ill-effects of the degradation of the resource base, since it can shift to the use of other resources or resources from other regions as the occasion demands.

The pattern of natural-resource use promoted by centralized authority has been non-sustainable when viewed from local areas such as the Uttara Kannada district. It has also led to severe conflicts with the local populations attempting to maintain their customary rights over and access to resources (5, 11). In the process, the local traditions of resource conservation have been increasingly disrupted or have broken down altogether.

Nevertheless, the effectiveness of traditional conservation can be clearly seen in the landscape of Uttara Kannada, where a large number of monkeys still survive and thousands of *Ficus* trees dot the countryside due to their religious significance. The only remaining natural stand of the genus *Dipterocarpus* persists in a sacred grove, as does the last large patch of a *Myristica* swamp.

RE-ORIENTING CONSERVATION

Evidently, there is a case for working out a new approach to conservation of biodiversity in Uttara Kannada. It is clear that conservation cannot be considered in isolation from local people, and broader patterns of natural-resource use and development, but must be complemented by policies promoting sustainable and equitable development of the natural-resource base as a whole. To be effective, any approach must give a larger role to the local people. This would entail restoring to them much of the authority they have lost over the past two centuries along with appropriate financial incentives. Local people must be involved because their well being is still intimately linked to the health of the natural-resource base of their own localities, hence they have a real stake in its sustainable use. Thus, in Uttara Kannada people depend on natural vegetation to meet almost all the needs of domestic cooking and water heating, 90% of fodder needs for livestock, and 80–90% of nutrient-supply needs for cultivated lands. Local people are also dependent on natural vegetation for making ropes, baskets, agricultural and fishing implements and for thatching their huts and cattle sheds (5). They still collect many wild fruits, hunt birds and wild pigs and fish the rivers to provide a significant fraction of their nutrition (16). For them, degradation of natural-resources is a genuine hardship, and of all the people and groups who compose the Indian society they are the most likely to be motivated to take good care of the landscape and ecosystems on which they depend. The many traditions of nature conservation that are still practiced could form a basis for a viable strategy of biodiversity conservation (2).

Involving local people in conservation, however, does not imply as some Gandhian environmentalists have stated that India must abandon all efforts at industrialization and return all authority to village communities in an effort to recreate an agricultural society in balance with nature (17). Such a scenario is simply not feasible for a variety of reasons, including the pressures of the international economic and political system. The practical option is to continue the processes of industrialization and intensification of agriculture, but with an emphasis on efficient, sustainable and equitable resource use.

The central argument so far has been that inefficient, non-sustainable and inequitable resource use is promoted by the large-scale state intervention through subsidies to the industrial-organized services-intensive agriculture complex that is shielded from the costs of environmental degradation. These costs are passed on, for instance to the peasants and artisans of Uttara Kannada district who have been deprived of all authority over the local natural-resource base. The proper response would then be to withdraw the subsidies that presently flow to the elite. Instead, the elite should be made to

BOX

Existing Apparatus for Planning and Implementing Biodiversity Conservation in India.

CENTRAL GOVERNMENT

Ministry of Environment and Forests

Primarily advisory, and as disburser of funds for special projects such as Project Tiger *since Land Use, Forestry and Law and Order are responsibilities of State Governments. Advisory Boards provide minimal inputs to the Central Ministry.*

STATE GOVERNMENTS

Forest Departments: Territorial and Wildlife Wings

These two wings have joint jurisdiction throughout a state. Wildlife Wings *are manned by Forestry Personnel on temporary secondment. All the decision-making power of these as well as all other State Government agencies is concentrated in the hands of Ministers, Secretaries and and Departmental Heads operating from the state capitals. At the state level a Wildlife Advisory Board provides minimal inputs, there are no inputs from a more local level.*

Assessing conservation priorities for Uttara Kannada (6).

Sl No.*	Attribute	Criterion	Prescriptions for UK	The only or best representative localities of conservation interest
1.	Species richness of constituent biological communities	Greater value attached to communities with larger number of species	Evergreen forests harbor largest number of species of flowering plants and amphibians, and moist deciduous forest zone that of birds	Bharatnalli Madurahalli, Gunjavati Range, Bargadda-Patoli, Hulidevargodlu-Unchalli, Aganashini estuary, Kulgi-Virnoli
2.	Geographical range of constituent biological communities	Greater value attached to communities with more restricted range	*Myristica* swamps harbor communities with a restricted range	Kathlekan
3.	Spatial occurrence of constituent biological communities within their range	Greater value attached to communities with more restricted spatial occurrence	Communities of spray zones of waterfalls are of very restricted occurrence; Limestone outcrops	Unchalli waterfalls, Yan
4.	Identity of species making up consitutent biological communities	The more distinctive the component of species, the greater the value	Mangrove vegetation and beaches harbor a highly distinct set of sepcies	Sunkeri, Thenginagundi Aganashini estuaries Karwar-Kamat Bay
5.	Endangerment due to human pressures	The more endangered communities are of greater value	Riverine forests are greatly endangered by human pressures	Hulidevaragodlu-Unchalli waterfalls
6.	Attributes of component species			
(a)	Restricted geographical range	Greater value attached to species with more restricted range	Evergreen forests harbor birds with more restricted range	Suremane, Hulidevara-godlu
(b)	Narrow habitat preference	Greater value attached to species with narrow habitat preferences	Freshwater ponds and estuaries harbor birds with narrow habitat preferences	Madurahalli, Nyasergi, Salgaum, Sanikatta, Masur
(c)	Taxonomic uniqueness	Greater value attached to species with fewer related species	Gnetum, a gymnospermous climber of evergreen forests is the only member of its order in UK	Hulidevaragodlu
(d)	Endangerment due to human pressures	Greater value attached to species subject to greater pressures	Wintering waterfowl in marshes are hunted extensively	Aganashini estuary and Nyasergi
(e)	Ecological role	Species serving as keystone resources would be attached greater value	Ficus species are keystone resources of tropical forests throughout the district	Ficus trees protected on religious grounds are scattered
7.	Interaction with adjacent communities	Greater value attached to communities serving as links in maintaining higher diversity	Betelnut plantations serve as links between patches of evergreen forests	

* Sl = Specific locality

compensate the rural poor for the decline in their quality of life, consequent on degradation of the natural-resource base. At the same time the local people should be increasingly involved in controlling and managing the natural-resource base of their localities. This ought to promote a far more efficient, sustainable and equitable process of intensification of natural-resource use, a process that would be compatible with conservation of biodiversity.

Outside Control

The natural-resource development approach advocated here, would of course have to be complemented by an effort more specifically focused on biodiversity. The current approach treats conservation as a matter of keeping local people out of a few large nature reserves and preventing them from killing most larger species of wild reptiles, birds and mammals anywhere within the district. All the decisions pertaining to such regulations are made centrally, partly in the national capital of Delhi and largely in the state capital of Bangalore by the Forestry Service personnel (Box). Substantial amounts of funds are then placed at the disposal of the Forest Departments to discharge their regulatory function. The state apparatus has at its disposal little detailed locality-specific knowledge of either the distribution of biodiversity or the various pressures impinging on it. Its functionaries have no real personal stake in conservation of biodiversity, nor are they in a position to effectively discharge their regulatory function. It is then no wonder that these efforts have not been very fruitful.

Involving the Local People

A far more effective conservation effort would focus on the whole landscape instead of a few reserves, on the whole diversity of species and ecosystems instead of only on larger vertebrates and emphasize positive rewards for promoting conservation in place of regulation by the state apparatus. Above all, it would place the major responsibility for the task squarely in the hands of the local people, rather than with an impersonal centralized bureaucracy or technocracy.

The local people would not however be

in a position to operate entirely on their own and the conservation effort would have to be one of co-planning and co-management. Local people do possess very detailed information on local biodiversity, its history and the forces impinging on it. But they lack a broader, global perspective. Outside experts collaborating with local scientists, when these are available, could provide such a perspective (6).

The starting point of this exercise was to note that we might wish to conserve biodiversity, defined as the entire spectrum of variety and variability among living organisms and the ecological complexes in which they occur, for a variety of reasons. These reasons may include subsistence use value, e.g. herbal medicines used locally; commodity use value, e.g. cane or wild honey; non-consumptive use value, e.g. watershed services; option value, e.g. wild relatives of cultivated plants that may provide useful genetic material for future breeding programs; transformative value, provision of experience of unspoilt nature; and existence value, for its own sake. This rationale suggests certain conservation priorities. These priorities must be related to attributes of any locality to assess the locality's significance for conservation. In general, a locality will be valued more if it contains elements that are distinctive, threatened, rare or of restricted occurrence; and if it harbors many such elements (see Table).

The local people must be intimately involved to take this broad prioratization further and to decide on specific action points. For example, the evergreen forests of Uttara Kannada district are rich in wild relatives of cultivated fruit trees, as well as many cultivars of fruit trees. Quite a few farmers have developed large collections of such species, especially of genera *Mangifera, Artocarpus, Myristica, Garcinia* and *Spondias* on their own lands, out of personal interest. Many local people also know of individual wild mango trees that bear fruit of special flavor or have exceptionally high yields. Their inputs would obviously be of value in deciding on the focus of a program to conserve wild relatives of cultivated plants. Similarly some habitat types are now conserved only in sacred groves, e.g. *Myristica* swamps in Kathlekan. Local people know of such patches in forest interior; they also appreciate their religious significance.

Conservation priorities must also take account of the whole set of human demands on a given locality. For example, articificial freshwater irrigation ponds constructed in preBritish times in the eastern parts of the district are important habitats of migratory waterfowl. These ponds are now being encroached by paddy cultivators. The local people are fully aware of who is encroaching and what their economic motivations are. Today, the state apparatus tries to regulate such encroachment through coercion. This often fails. Instead, the resources devoted to enforcement could be offered to local communities so as to motivate them to continue to conserve the freshwater ponds in their own villages. The different villages concerned could submit bids based on the annual level of financial incentive they would need to accept the conservation option. If the state that speaks for the broader public interest finds these bids acceptable, it could decide to implement conservation programs in partnership with some or all of the villages involved. Conversely, it could decide to write off some of the freshwater ponds of the Uttara Kannada district to extend paddy cultivation. Local communities should of course receive payment only if they succeed in conserving the freshwater ponds involved.

Undoubtedly, more research and practical experience is required to design and implement an alternative to the centralized and authoritarian approach to conservation. An important obstacle is likely to be the resistance of the central authority because a large bureaucratic apparatus to implement programs locally would simply be unnecessary. Rather a much smaller, technically more sophisticated apparatus would be required, one which would have a capability to assess broader priorities, to help local authorities work out detailed plans and monitor to determine whether the local authorities are in fact implementing the conservation measures as agreed. The local authorities would have to be strengthened and better organized and capable of taking further programs of decentralization of political and administrative responsibilities. This would entail rendering local governments much more accountable to people than is presently the case. The central authority would also have to transfer a great deal of its jurisdicial and policing powers to the more locally-oriented institutions.

Such a local involvement management strategy is very much in the spirit of the proposal to involve local communities that is an important element of the *Global Biodiversity Strategy* being developed by WRI, IUCN and UNEP (18). It is also consistent with the microlevel planning initiative of the National Wastelands Development Board of the Government of India (19). Such international and national support is critical to the success of an alternative approach, for it is apt to run into serious opposition on the part of the political-economic-bureaucratic vested interests that benefit from the present-day pattern of non-sustainable resource use and of a rigid policing approach to conservation. Fortunately, the global trends are today exceptionally favorable for a decentralized, peopleoriented approach. As a result, more and more attempts like that exemplified by the Marine Conservation and Development Program on the Visayas islands, Philippines, are likely to be initiated in the coming years (18). Their success would be greatly enhanced if they are backed by a locality specific analysis of the situation in its concrete historical setting. Involvement of local people is absolutely essential for implementing the conservation measures on the ground.

References and Notes

1. Rodgers, W.A. and Panwar, H.S. 1988. *Planning a Wildlife Protected Area Network in India*, Vol.1. The Report. Wildlife Institute of India, Dehra Dun, India.
2. Gadgil, M. 1991. Conserving India's biodiversity : the societal context. *Evolutionary Trends in Plants 5*, 3–8.
3. Mani, M.S. (ed.). 1974. *Ecology and Biogeography in India*. W. Junk, The Hague.
4. Nair, N.C. and Daniel, P. 1986. The floristic diversity of the Western Ghats and its conservation : A review. *Proc. Ind. Acad. Scie. (Anim.Sci/Plan.Sci), Supplement*, November 1986, p.103–125.
5. Gadgil, M., Hegde, K.M. and Shetty, K.A. 1986. Uttara Kannada: A case study in hill area development. In: *Karnataka State of Environment Report-1985–1986*. Saldanha, C.J. (ed.). Centre for Taxonomic Studies, Bangalore, p. 155–170.
6. Daniels, R.J.R., Subash Chandran, M.D. and Gadgil, M. A strategy for conserving the biodiversity of Uttara Kannada : A district in South India. *Environ. Conserv.* (In Press).
7. Buchanan, F. 1870. *Journey Through the Northern Parts of Kanara (1800–1802)* 2. Higginbothams, Madras, India.
8. Gadgil, M. and Subash Chandran, M.D. 1989. On the history of Uttara Kannada forests. In: *Changing Tropical Forests*. Dargavel, J., Dixon, K. and Semple, N. (eds). Australian National University, Canberra, p. 47–58.
9. Gadgil, M. and Iyer, P. 1989. On the diversification of common property resource use by the Indian society. In: *Common Property Resources: Ecology and Community Based Sustainable Development*. Berkes, F. (ed.). Belhaven Press, London, p. 240–255.
10. Gadgil, M. 1989. Deforestation: problems and prospects. *Energy Environment Monitor 5*, 3–47.
11. Gadgil, M. 1991. Restoring India's forest wealth. *Nat. Res. 27*, 12–20.
12. Gadgil, M. 1991. Diversity: cultural and biological. *Trends Ecol. Evolut. 2*, 369–373.
13. Gadgil, M. and Berkes, F. 1991. Traditional resource management systems. *Resource Management and Optimization 18*, 127–141.
14. Nair, P.V.K. and Gadgil, M. 1980. The status and distribution of elephant populations in Karnataka. *Journal of Bombay Natural History Society 75 (Suppl.)*, 1000–1016.
15. Kamath, S.V. (ed.). 1985. *Karnataka State Gazetteer, Uttara Kannada District*. Government of Karnataka, India.
16. Subash Chandran, M.D. and Gadgil, M. State forestry and decline of food resources in tropical forests in Uttara Kannada in Southern India. *Proceedings of the International Symposium on Food and Nutrition in the Tropical Forest: Biocultural Interactions and Applications to Development*. Paris, 10–13, September 1991. (In press).
17. Shiva, V. 1991. *Ecology and the Politics of Survival*. Sage Publications, New Delhi, p. 365.
18. World Resources Institute et al. 1992. *Global Biodiversity Strategy*.
19. National Wastelands Development Board, Govt. of India. 1991. *Guidelines for Micro Planning*. National Wastelands Development Board, Ministry of Environment and Forests, Govt. of India, New Delhi 110003, p. 51.

20. I gratefully acknowledge support for this work by the Department of Environment, Government of India through a series of long-term research grants. I am also grateful to my colleagues R.J. Ranjit Daniels and M.D. Subash Chandran for many years of collaborative work that has contributed to the ideas embodied in this paper.

Madhav Gadgil holds a Ph.D. in biology from Harvard University and has served as a lecturer at Harvard University and a visiting professor at Stanford University. For the past 18 years he has been on the faculty of the Indian Institute of Science where he currently holds the Astra Professorship in Biological Sciences. His research interests encompass mathematical modelling as well as field studies in the areas of population biology, conservation biology and human ecology. He is also active in policy studies having served for 4 years on the Scientific Advisory Council to the Prime Minister of India. His address: Centre for Ecological Sciences, Indian Institute of Science, Bangalore 560012, India.

[26]

Estimating the Supply of Conservation Goods in Britain: A Comparison of the Financial Efficiency of Two Policy Instruments

Martin Whitby and Caroline Saunders

ABSTRACT. *Rapid uptake of agreements in Environmentally Sensitive Areas (ESAs), compared with negotiated management agreements on Sites of Special Scientific Interest (SSSI), has lead to renewed interest in obtaining public conservation goods through agreements in the U.K. Economic comparisons between these two instruments are made using a simple supply model. The paper reviews the legal/administrative basis for management agreements, states relevant theory, and compares these two means of securing conservation goods. Conclusions regarding the difference in public expenditure on these mechanisms are established and information on transactions costs presented. The conclusion is that SSSI are more cost-effective than ESAs.* (JEL Q28)

I. INTRODUCTION

The management of land for primary production in the U.K. is driven mainly by the objectives of farmers and foresters with regard to the level and stability of their incomes. Management reflects expected prices and costs during the production period, which are strongly influenced by government support policies. In recent decades conservation of semi-natural and historic phenomena have been seen as threatened by rising intensity of land use. Although the provision of such public goods was seen, rightly or wrongly, as an automatic outcome of the production process in the past, policies have increasingly implied that producers are no longer producing these goods jointly with primary products in sufficient supply.

Policies to ameliorate the unwanted impacts of agricultural externalities on public goods have now been in use in the U.K. for some decades. This paper compared two policies using management agreements, in terms of cost-effectiveness, using supply theory. The remaining sections of the article discuss the policy instruments, the relevant economic theory, and the empirical indications of producer surpluses.

II. MECHANISMS FOR SECURING CONSERVATION

Instruments have been developed over a long period to conserve identified areas, including designation as Sites of Special Scientific Interest (SSSI), Environmentally Sensitive Areas (ESAs), and other mechanisms. Typically these work through management agreements between farmers and other parties and involve compensation for following prescribed practices. These instruments vary widely in costs, benefits, and impact. This paper focuses more narrowly on the two most widespread forms of management agreement: SSSI and ESAs (Whitby, Coggins, and Saunders 1990; Whitby 1994; NAO 1994).

Management Agreements

The essence of the management agreement is that producers agree to modify their production activities to benefit the environment and are compensated for the costs they so incur. Most modifications imply a reduction in the intensity of production and the loss is usually conceived as income foregone by profit-seeking producers. The level of compensation offered must be sufficient to persuade producers to forgo production options and to replace the income they therefore lose.

The authors are, respectively, professor and lecturer in Countryside Management at the University of Newcastle upon Tyne.

This paper has benefited from comments from a number of colleagues, in particular Paul Allanson, Vic Adamowicz, Ken Thomson, Colin Price, and various members of staff in English Nature. Errors remain the authors' own responsibility.

The use of management agreements goes back at least to the 1949 National Parks and Access to the Countryside Act, which introduced them. The legal basis for these arrangements has recently been reviewed by Withrington and Jones (1992). The last major addition to the list of possible management agreements was the Environmentally Sensitive Areas introduced under Article 19 of EC Regulation 797/85 as modified regarding EC financing in Regulation 1760/87 and further extended under the EU Regulation of Measures Accompanying CAP Reform (Regulation 2078/92) (Whitby 1996). Socioeconomic studies have also been undertaken for 7 of the first 19 ESAs and they are summarized in Whitby (1994).

The two arrangements here compared differ both in their impact on the property rights of owners and in the method of determining compensation. Crucially, compensation for proprietors in SSSI is individually negotiated on each farm whilst in ESAs it is predetermined at flat rates for each area.[1] Whilst some SSSI agreements will receive less compensation per hectare than ESAs, others will receive more. It therefore remains an open question whether the cost of compensation per hectare of SSSI agreement would be higher, reflecting individual negotiation and enforcement costs, or lower, reflecting the success of negotiators in tracking the farmer's supply price of conservation goods. The remainder of the paper examines these hypotheses first in terms of national aggregate expenditures and then at a more local scale based on individual ESAs and SSSI.

Sites of Special Scientific Interest (SSSI). The present system of management agreements on SSSI involves compensating farmers when they are prevented from carrying out operations which would damage the scientific value of a site. The process now operating (see NAO 1994) is initiated either by a farmer notifying the conservation authority of intent to carry out a Potentially Damaging Operation (PDO) on a site or, in the past, when an Agriculture Minister refused an application for farm capital grant,[2] following objection by a conservation body. If the conservation agencies then wish to prevent the operation, they must, within four months, offer to enter into a management agreement providing compensation calculated in accordance with the published Guidelines (DoE 1983).

If the farmer persists in his intention to carry out the PDO, the conservation agency may serve a Conservation Order prohibiting this. The conservation authorities may take further steps, possibly ending in public purchase. These powers are rarely used, although they are an important reserve in negotiating agreements. Farmers may appeal against compulsion and compensation is available (on the same basis as for deteried expenditure).

The financial cost elements in this system are those of compensation, negotiation, and monitoring, and indirect cost savings to the Exchequer, following adjustments to farm output. Economic costs would include losses in resource productivity measured at social cost, including transactions costs, and there would be benefits to society from the public good of improved conservation. There may also be private benefits, for example to landowners who subscribe to a stewardship ethic (Colman 1994). The annual payments under the 1981 Act run only for the length of the management agreement (up to 21 years, though no time limit is prescribed). Annual rates of compensation are renegotiated every few years. Because new agreements are negotiated and some are terminated each year, future expenditure data do not exist for the program as a whole. The data presented here (Table 1) are therefore those released by the Nature Conservancy Council (NCC).

The reported expenditures on compensation for negotiation and monitoring of SSSI agreements are displayed in Table 1, showing the importance of compensation (76 percent of total expenditure). Also notable is the cost of professional advisers, which arises because the NCC pays the agreed

[1] The current average area of ESAs is 80,000 hectares.

[2] Capital Grants were used for some decades, up to 1994, to promote farm investment. They provided up to 50 percent of the cost of specified farm investments.

TABLE 1
ESTIMATED AGGREGATE ANNUAL EXPENDITURE OF SECTION 15 MANAGEMENT AGREEMENTS ON AGRICULTURAL LAND, 1988/89

	Aggregate Cost £m (1988/89 prices) or Area (thousand hectares)	Distribution of Cost or Area (percent of total)
Estimated Cost of Compensation	5.58	75.7
NCC Professional Advisors	0.24	3.3
Landowners' and Occupiers' Costs	0.25	3.4
NCC Staff and Overhead Costs	1.30	17.6
Total Expenditure	7.37	100.0
Total Area of SSSI (thousand ha of agricultural land)	915	100.0
Area under Agreement (thousand ha of agricultural land)	77	8.4
Cost per Hectare of SSSI (£)	8.05	—
Cost per hectare of Agreement (£)	95.7	—

Source: Whitby, Coggins, and Saunders 1990.

negotiating costs of owners and/or occupiers as well as those it incurs itself. By carrying negotiating costs the NCC risks encouraging protracted negotiations and, to that extent, transactions costs may be somewhat overstated by these data. Some negotiations are indeed protracted but this is generally believed to reflect reluctance on the part of the landowners. An estimate of the share of NCC staff "overhead" costs has also been made which brings the total of these administrative costs of setting up and running the system of SSSI to £1.79m. They are therefore equivalent to virtually one-third of the value of compensation but they do not include all relevant policy costs. Landowners and occupiers probably incur more costs than the amount (£0.25m) awarded to them by the NCC and there may well be administrative costs relating to this operation which have been omitted from the calculations because they could not be separated from those of the NCC's other functions. By reducing the amount of farm output these agreements cut the policy costs of supporting it and an element of Exchequer cost savings is achieved through them. An approximation of these savings, at £1.6m, was published by Whitby, Coggins, and Saunders (1990).

However, the costs reported in Table 1 are seriously incomplete because they relate only to the cost of operating management agreements and exclude expenditures on the establishment and designation of SSSI. They thus understate the administrative costs of establishing SSSI incurred over long periods prior to 1988/89. The capital cost of setting them up is an important element in the transactions costs of SSSI. Further, the major effort towards the renotification of SSSI[3] during the decade following the 1981 Wildlife and Countryside Act required further public expenditure. Although the relevant sums are not published, their magnitude may be inferred from NCC Annual Reports for 1986 and 1990 which mention extra costs amounting to £4m, in each of those years, associated mainly with SSSI renotification. Further insight is gained from staff salary expenditure by NCC during the 1980s. Converting the amounts to real terms (using the Retail Price Index) shows that whilst real staff costs fell during the period 1980/81 to 1983/84, thereafter they increased each year until 1988/89. The cumulative cost of that increase (over the 1983/84 staff cost) was some £13m. Assuming that

[3] The Act required that all SSSI then existing should be redesignated, which meant that all landowners within them had to be contacted and told explicitly which potentially damaging operations were proscribed on them.

sum was attributable to renotification and amortising £13m over five years at a real 5 percent interest gives an annual cost of some £3m during the 1980s. Perhaps the period over which the sum is spread should be longer but, against that, there would also be a case for including some of the earlier initial costs of designating SSSI as well as the other costs incurred by NCC. These arguments gain support from Whittaker, O'Sullivan, and McInerney (1991), who report that individually negotiated agreements account for a major share of the total costs of such arrangements compared with those based on standard assessments. Adding £3m to the administrative costs reported in Table 1 would increase them from 24 percent to 39 percent of total, though still excluding uncompensated private transactions costs.

Environmentally Sensitive Areas. The U.K. endorsed Structural Regulation (797/85) of 1985 by passing the Agriculture Act of 1986, providing for the designation of Environmentally Sensitive Areas (s 17). A later Regulation (1760/87) provided for a measure of reimbursement from FEOGA.[4] ESAs are areas of landscape, conservation, and/or archaeological interest within which farmers are compensated for complying with a package of prescribed management practices designed to secure conservation goals. Each ESA has a separate management package; they range from the simple, specifying only one level of required practices, to the complex, allowing for two or more "tiers" of incentives for particular practices (see Whitby 1994 for a more detailed review, and Whitby 1996 for a European perspective).

The "first round" of ESAs began in 1987 and, by 1992, they covered some 0.8 m ha containing agreements with 6,000 farmers which related to 285,000 hectares in the U.K. Agreements were taken up rapidly by farmers, which may partly be explained by the comparatively slight constraint they impose on the practices of many participants (Colman et al. 1992; Hodge, Adams, and Bourn 1992), although such comments should strictly be based on a comparison of farms *with* agreements and those on similar land but who were outside ESAs and *without* agreements.

The rate of growth of ESA agreements and the area they protect is displayed, for England, in Table 2, showing the high initial rate of uptake which leveled off after two years. The average cost per hectare of agreement fell steeply after the first year, but showed no trend for the rest of the period. The only obvious explanation for this outcome must be that the ESAs with the greatest initial response were those with higher per hectare payments, particularly those operating in arable areas.

A flat rate of payment is set for each tier in ESAs by the Ministry of Agriculture Fisheries and Food (MAFF) and, once the rates of payment have been determined, little scope for negotiation remains. The same rate of payments is available to all participants adopting the relevant practices. The total cost of payments for first-round ESAs in England was reported as £45.3 m (Table 2) for the five-year period: of this, some 24 percent consisted of administrative costs. How closely this compares with the administrative expenditures in Table 1 cannot be deduced from the original sources. Up to 25 percent of the compensation paid will have been reimbursed to the U.K. from the CAP budget.

In summary, provision of conservation through these two mechanisms is different in important respects. Whilst the individually negotiated arrangements on SSSI are tailor-made for small areas of land, compliance with the broader management prescriptions of ESAs is gained more simply by defining a generally acceptable package of management activities for each ESA and compensating those who agree to follow it. Whilst SSSI have been in operation for some time, the first ESAs were designated from 1987 and the newest "round" was designated in 1994. The question now considered

[4] The Common Agricultural Policy of the European Union is funded through a budget, consisting of two funds known by its French acronym FEOGA (Fonds Européen d'Orientation et Garantie Agricole): in English the European Agricutural Guidance and Guarantee Funds or EAGGF. The original ESAs were funded through the smaller Guidance section of FEOGA but, since 1992, have been paid for through the Guarantee Section.

TABLE 2
NUMBER AND AREA OF AGREEMENTS: ESAS IN ENGLAND, 1987–1991

	1987	1988	1989	1990	1991	Total
Number of New Agreements	1,330	1,090	350	186	105	3,660
New Area under Agreement (hectares)	31,230	68,700	8,970	2,650	1,560	113,110
Cumulative Area (Hectares)	31,230	99,930	108,900	111,550	113,110	113,110
Average Area of New Agreements (hectares)	22.7	55.4	25.6	14.2	14.9	
Total ESA Cost (£m)	4.9	9.3	9.5	10.5	11.1	
Mean Cost (£ per hectare)	156.9	93.1	87.2	94.1	98.1	

Source: MAFF 1990, 1991.

is to what extent the costs associated with these different arranagements may be meaningfully compared. This is addressed first through an examination of the relevant theory and second by relating this to the data on these arrangements.

III. INDICATIONS FROM THEORY

The *definition and enforcement of property rights* is usually a matter for governments to resolve. The resulting transactions cost enters into the identification of an optimum allocation, although owners have an important role in defending their rights. The essence of management agreements is that they recognize the rights of owners and a requirement for compensation arises if owners are forbidden to exercise some of these rights. Thus farmers have the right to cultivate their land and to develop it for agricultural use. But where their land lies within an SSSI, limits are imposed on their freedom to undertake various operations—PDOs—without permission from conservation authorities. Where permission is withheld, owners may be offered a management agreement as compensation. If landowners persist in their intention to carry out a PDO, they may be restrained by a Conservation Order and public purchase may follow under defined circumstances. Designation of an SSSI therefore constitutes a clear case of a reassignment of property rights because after designation and prescription of PDOs the rights of landowners and occupiers are less than before it.

By contrast, ESA designation does not change the individual's property rights, rather it provides those within it a further option of adopting prescribed management practices in return for payments. The farmers may discontinue such agreements and there is no back-up to persuade reluctant farmers to enter into agreements. Designation of an ESA does not therefore change the rights of landowners.

Management agreements bring inescapable transactions costs, to both the conservation agency and the landowner, associated with surveying the relevant areas of land, negotiating agreements, and monitoring compliance with them. In SSSI, to avoid such costs becoming a deterrent, the conservation authorities may pay the negotiating costs of those entering into agreements. There will be other substantive costs incurred both by the agencies themselves in connection with monitoring compliance and by the land managers working within agreed constraints. These *transactions costs* are an important argument in deciding to adopt particular policies. As the technology of monitoring and enforcement improves, however, some policies will become more attractive due to falling transactions costs. The role of aerial survey in monitoring ESA compliance is a case in point (Hooper 1992). Thus, as several authors have shown, where different assignments of property rights and levels of transactions cost are found, the "optimal" allocation of resources will also be different. Bromley (1991) demonstrates the point by reference to two radically different assignments of rights, then pointing out that although each allocation resulting from different right structures might be efficient, it remains a political decision as to which is the "correct" configuration of rights for a country. This conclusion suggests likely differences in transactions costs between ESAs and SSSI management agreements, which contrasts with the finding reported in Tables 1 and 2 that the administrative expenditures of ESAs and SSSI are similar as a proportion of total costs. Omissions noted (above) from the administrative expenditures reported would account for some of the relevant transactions costs, which are not officially recognized.

The *homogeneity of goods* subject to trade may at first sight seem insurmountable in the case of environmental and wildlife goods. Clearly each SSSI may be seen as unique in its physical and biotic characteristics; ESAs have been chosen to exemplify a variety of different situations (Colman et al. 1992). However, at higher levels of aggregation, we argue that groups of SSSI have enough common characteristics to allow them to be assigned to broad ecological categories and hence to be compared with specific ESAs. SSSI agreements, being negotiated individually, would be expected to have more stringent management prescriptions than ESAs. Nevertheless the degree of homogeneity of conservation goods is sufficient to allow the analysis of data from SSSI as if they were effectively the same. There is, of course, heterogeneity in the ecotypes: we merely *assume* that a sufficient degree of homogeneity exists to allow their treatment as one group. In the absence of that assumption, comparison of the two systems would not be meaningful. There is also diversity in the quality of management of SSSI, even within management agreement areas, and this has been a matter of concern to the nature conservation authorities who are seeking to develop alternative means of positively enhancing wildlife benefits (Felton 1993). However, there is less diversity of agricultural potential and therefore less problem in comparing profits foregone under the two arrangements.

Where land users are compensated for abstaining from PDOs it can be argued that the supply price of that abstention reflects the minimum amount of compensation they are prepared to accept. This notion is embedded in the negotiation of compensation which seeks to establish the extent of income forgone as a result of the agreement. It is acknowledged that substantial public and private transactions costs are present in management agreements and the available measure of income forgone reflects only the *private* supply price of conservation.

Compensation also raises a problem of moral hazard due to asymmetry of information between the participants. By basing compensation on income forgone in a hypothetical situation, those seeking compensation are encouraged to overstate potential losses in the expectation of gaining more compensation whilst those determining compensation outlays have only imprecise means of verifying claims. A means of checking claims would be to use standard budget data which are at best highly approximate because of the well-known variability in economic performance between farms. In the early 1980s, concerns about overstatement of costs lead to predictions of overspending on the budget for management agreements. The fact that such expenditure forecasts were not reached might suggest that farmers are resistant to such pressures, perhaps reflecting notions of "stewardship." There are, nevertheless, a few cases of particular management agreements which have seemed particularly costly: although, when standardized to a per-ha per-annum basis, they often appear comparable with modal rates of compensation.

IV. AN EMPIRICAL INTERPRETATION

Viewing management agreements as a form of market, where owners meet would-be buyers, a supply schedule of conservation goods is identified. The supply-price negotiated for the good translates to a supply model using data on the levels of compensation of an ESA with that of an SSSI on the same area of land, as shown in Figure 1.

The whole area of an ESA is represented here by *XY* on the horizontal axis. The area *XO* is land which does not need or receive protection because it would not be damaged by cultivation (bare rock for example) or is not capable of being farmed. By contrast an SSSI would have been designated as far as

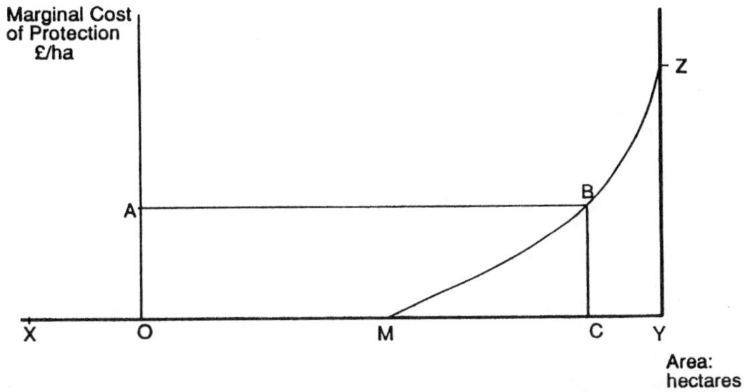

FIGURE 1
A THEORETICAL COMPARISON OF ESAS AND SSSI AGREEMENTS

possible to minimize *XO* as being of no interest from the point of view of conservation. Those managing the area from *O* to *M* have not signaled any intention to carry out PDOs and hence management agreements are not required to protect it. From *M* to *Y* management agreements have been negotiated and are arranged in order of increasing cost per hectare up to the highest level *YZ*. The curve *OMZ* then follows the supply price of the conservation goods generated within the SSSI, reflecting the increasing willingness of farmers and landowners to enter into such agreements as the compensation increases.

Because management agreements are at the farm level, the per-hectare rate of payment constitutes an average per farm. To use these averages as points on a supply curve is to imply that average and marginal costs are equal for each point on the curve. This assumption is least tenable on the larger areas of land under agreement, where the farmer might enter fewer hectares into agreement if the compensation rate was lower. Furthermore, the competitive model may be inappropriate if the conservation agency has monopsony powers, enhanced by the possibility of public purchase, which allow it to dictate the "price" to be paid for goods it "buys."

The area (*MYZ*) beneath this curve then represents the total compensation outlay for the SSSI excluding transactions costs. But the SSSI mechanism protects the whole area at risk from PDOs, namely *OY*; which is therefore the appropriate denominator to calculate the average cost per ha protected by the SSSI, rather than the smaller *MY*, which represents the area under agreement.

Where *OMZ* intersects the level of ESA payments *AB*—at *B*—shows how much of the land within an ESA would not be entered into agreements (*CY*) because the income foregone would exceed the payments offered. Comparing the cost of protection in ESA and SSSI then requires dividing the whole area protected into the total cost as reported in Tables 1 and 2. The appropriate area for dividing into ESA costs is the area subject to agreements, which means that the per-hectare rate of compensation payments, represented by *OA* or *BC*, indicates cost per hectare. The average of £93.1 per hectare in 1988 is reported in Table 2. But for SSSI a somewhat lower cost than the average per-hectare level of payments is appropriate, reflecting the size of the area protected, not just that under agreement.

For SSSI as a whole, the agricultural area under agreement was 77,000 hectares while the area protected (the agricultural area) amounted to 915,000 hectares. This latter area would imply an average cost of £8.05 per hectare which contrasts with the £95.7 per hectare for land under agreements; reflecting the small share of the agricultural area of SSSI which is under agreement. To the extent that quality of management for conservation in SSSI as a whole is inferior to that practiced on SSSI agreement land, the £8.05 per hectare should be seen as a lower estimate of value. Felton (1993) stressed the poor standard of conservation management on SSSI, from an analysis of the effectiveness of the recently introduced Wildlife Enhancement Schemes. Conversely, £95.7 per hectare is an upper limit of annual compensation costs. But a comprehensive measure of transactions costs, added to compensation costs, would bring the cost of SSSI agreements back towards those of ESAs.

The use of the inappropriate denominator (MY) in calculating average costs reflects the conventional approach of analyzing these arrangements at the level of the agreement, as displayed in Tables 1 and 2. Shifting the focus to the policy-relevant level of the area protected (OY), a different configuration of costs emerges and the conventional view becomes markedly less convincing. In terms of Figure 1 the difference between ESAs and SSSI is that the former allocates a producer surplus of $ABMO$ to farmers and landowners, whilst negotiating individual agreements secures (saves) that sum for society. Nevertheless, the problem remains that in comparing the average cost per hectare in ESAs and SSSI we are dealing with areas that vary, perhaps systematically, in the intensity of agriculture they support. The next step must be to apply this model to SSSI selected for their comparability with individual ESAs.

Costs by Type of Public Good Provided

To compare the compensation cost of ESAs with SSSI in more detail, various alternatives were considered. To measure the quantity of public good provided an obvious method would be an ecological or scientific index of site quality (Oglethorpe et al. 1995). The only one of these available for this study was habitat type. However, the NCC definition of habitat types is not refined enough to distinguish between quite different types of public good provision: for example, neutral grassland is the most common habitat type under agreement and this covers areas varying from marshes and wetlands in the South West to hay meadows in the North East of England. Whilst they may have some similar ecological characteristics there will also be important variability. Without ecological or scientific valuation, another means of comparing ESAs and SSSI would be to pick areas closely matched in terms of location and public good provision, with similar current uses and threats to land use. Ideally the SSSI would be located within an ESA and despite differences between the constraints of the two arrangements, comparisons would be more meaningful. A further consideration in choosing sites to compare was that the SSSI must have sufficient management agreements to allow construction of a supply curve.

However, the method of selecting areas with SSSI for comparison may have introduced bias, as these are only chosen for analysis if they have a large number of agreements and are close to an ESA in location and type. It also excludes the majority of SSSI, which have only a few agreements: this particularly reduces the number of SSSI to compare with ESAs in the uplands.[5] Thus the sites suitable for analysis will tend to be in the lowland where the cost of SSSI protection is substantially above average.

Data for two sites are presented here firstly in the Somerset Levels and Moors ESA (in the West of England), which allow a direct comparison between an ESA and the SSSI within it; and SSSI in the South East of England close to the South Downs ESA. The ESA data are based on the agree-

[5] Thus only 14 percent of the SSSI area is lowland grass or heath; but they account for 63 percent of area covered by management agreements. In contrast the uplands account for 65 percent of SSSI area but only 15 percent of land under agreements.

ments reached during the "first round" of those ESAs, starting in 1987. The main objectives of the Somerset Levels and Moors ESA is to conserve wetlands and their associated habitats. Producers can enter into two tiers of management agreement both requiring extensive management and grassland with tighter prescriptions for Tier 2 land. For agreeing to comply with these restrictions, initially for five years, Tier I producers receive £82 per hectare and Tier 2, £120. This ESA is 27,120 hectares of which 86 percent was eligible for management agreements and in 1991 the uptake was 9,640 hectares (41 percent) of which 8,110 hectares was in Tier 1 and 1,530 hectares in Tier 2. In 1989 the average rate of compensation payment for agreement land in this ESA was £104.4.

The Somerset Levels and Moors ESA contains 12 SSSI covering 5,920 hectares on 22 percent of the ESA. The objectives of these SSSI are similar to the ESA in protecting wetland habitat. In 1991 there was 170 SSSI management agreements covering 1,835 hectares and a sample of these is compared with the ESA. Data on the individually negotiated agreements for the SSSI were obtained from the NCC Land Agency Module and refer to 1989 and do not include any element of transactions cost.

The compensation cost of management agreements for a subset of SSSI is illustrated in Figure 2. The y intercept occurs at 1,846 ha: the area on which there are no management agreements. From then onwards the prices have been calculated as the average rates of compensation negotiated per hectare with the observations arranged in ascending order of value. The curve rises steeply as the first land comes under agreement, after that it rises slowly apart from one agreement reached at high compensation. The outlier, which may reflect superior negotiating skills, reduces the ability of the figure to portray the change in other levels of payments more clearly and has therefore been excluded in Figure 3, emphasizing that most agreements are between £100 and £220 per hectare, that is above the average level of ESA payments.

The existence of 1,846 hectares not under management agreement suggests that there has been no threat to this area or that NCC has secured its conservation through persuasion. However, a part of the area is un-

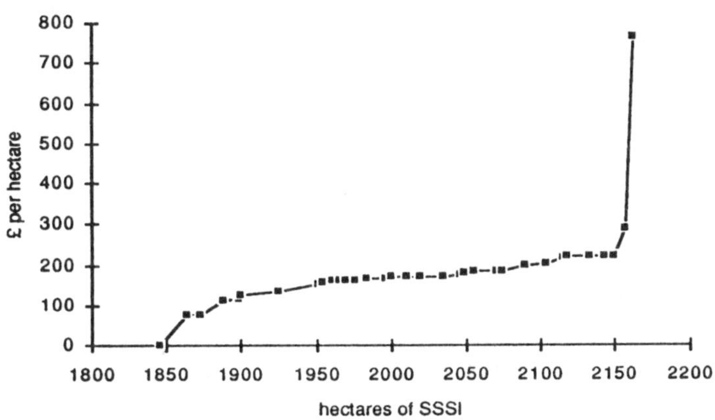

FIGURE 2
COST PER HECTARE OF SELECTED SOUTH WEST SSSI MANAGEMENT
(*Source*: NCC Land Agency Module [database], Peterborough, U.K.)

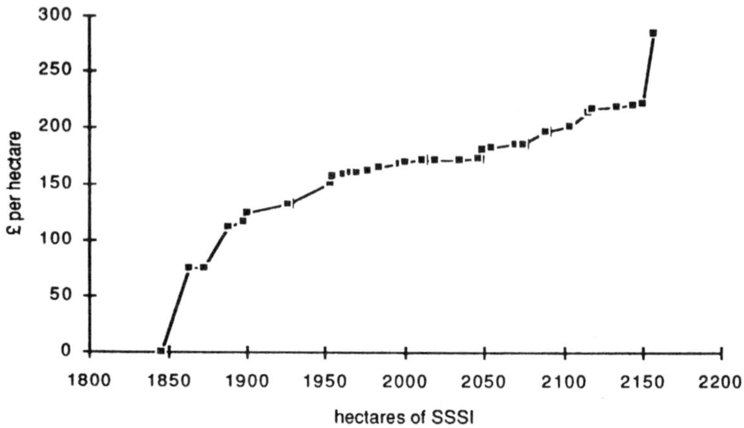

FIGURE 3
COST PER HECTARE OF SELECTED SOUTH WEST SSSI MANAGEMENT AGREEMENTS, EXCLUDING ONE OUTLIER
(*Source*: NCC Land Agency Module [database], Peterborough, U.K.)

der control of other conservation bodies such as the Royal Society for the Protection of Birds (RSPB) and Nature Conservation Trusts which would mainly provide similar public goods. So although compensation costs of protecting these SSSI average £26 per hectare, which compares favorably with ESA payments, it does not include payments by other agencies. The average cost of management agreements on land covered by agreement is £173 per hectare—higher than that for ESAs which averaged £84.2 in the Somerset levels and Moors up to 1989 (estimated from MAFF 1992a).

The second example was in the South East where the existence of a relatively large number of SSSI agreements enabled a supply curve to be constructed as illustrated in Figure 4. This shows an area of 16,573 hectares not under agreement, with an average payment per hectare of SSSI of £11 and of £228 per hectare of management agreement. The threats to these SSSI are mainly conversion of livestock grazing land to arable and loss of wet grassland. In terms of location this site is closest to the South Downs ESA. The objectives of this ESA are many and, although they include maintenance of wetlands, are predominantly concerned with chalk downland: so the public goods provided by each policy are not identical. In addition the management prescriptions for ESA Tier 1 are concerned with extensive use of grassland, attracting a payment of £35 per hectare and Tier 2, for arable conversion to livestock, £160 per hectare, averaging £129.2 per hectare in compensation cost in 1989 (MAFF 1992b). However, neither of these Tiers corresponds to specific threats in the SSSI, so that comparisons should perhaps be made between the SSSI and ESA such as the Broads or Brecklands offering payments ranging from £100 per hectare for Tier 1 to £200 per hectare for Tier 2 and even £300 per hectare for uncropped land. The average cost of ESA per hectare of the Broads ESA under agreement was £198.3 in 1989 MAFF (1992c).

The results of these habitat-specific comparisons are summarized in Table 3. As before, the low figure for SSSI is to be seen as a lower bound estimate, whilst the higher figure is the actual average per hectare of agreement on the South Downs. By contrast the ESA payments span the range available for different levels of participation.

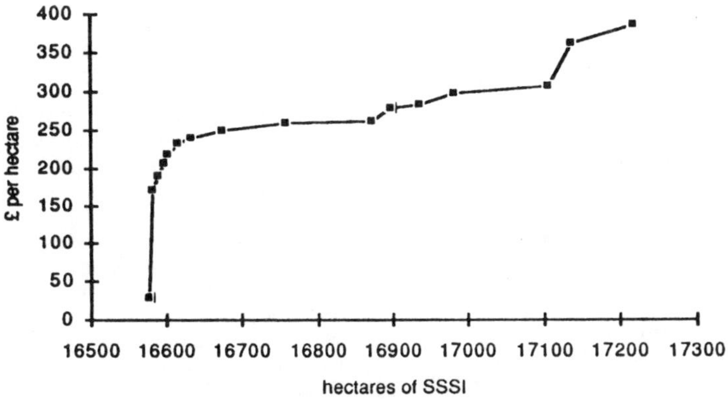

FIGURE 4
COST PER HECTARE OF SELECTED SOUTH EAST SSSI MANAGEMENT AGREEMENTS
(*Source*: NCC Land Agency Module [database], Peterborough, U.K.)

TABLE 3
AVERAGE COMPENSATION COSTS OF MANAGEMENT AGREEMENTS FOR SSSI AND ESAs: SOUTH WEST AND SOUTH EAST ENGLAND
(£ per hectare)

	South West	South East
SSSI		
Agreement Area	173	228
Whole Area	26	11
ESA	104	129

The conclusion from this comparison differs from that for aggregate data reported in Tables 1 and 2, in that the cost per hectare under management agreements is higher for SSSI. But moving the basis of comparison to cost per hectare "protected," the cost falls dramatically for SSSI. To that extent they are a cheaper means of protecting conservation sites in terms of financial costs per hectare. The inclusion of transactions costs from these estimates would partly bridge the gap between costs per hectare "protected."

More precision than this requires three conditions. First, that the SSSI compared would have a comparatively large share of their surface area under agreements, SSSI with comparatively high rates of compensation per hectare of agreement were selected (compare Tables 1 and 3). Second, the higher transactions costs for SSSI (excluded from Table 3) would raise their full cost towards the higher end of the range. Third, by contrast, the argument that all of the SSSI is protected by the mechanism would indicate the lower end of the range. A crude method of adjusting for these arguments would be to double the lower end of the range for SSSI and add 40 percent transactions costs (as argued in discussing Table 1) and to increase the ESA costs by 25 percent to allow for transactions costs at the reported average rate for all ESA (Table 2). This arithmetic would still leave SSSI costs per hectare lower than those of ESA, which is where they must remain until more accurate data on transactions costs are made available and better evidence is provided on the extent of under-protection of SSSI through existing policies.

V. CONCLUSIONS

Having estimated the above supply curves, we may then ask "What do they mean?" To specify the variables which would track the demand curve for these conservation goods might allow identification of an equilibrium for this market. This would be problematic not only because of the public nature of these goods, but also because of the difficulty of identifying the variables which determine demand. Whilst the level of payment of compensation will reflect the budgetary constraints binding the paying agency, there is no external (and probably no internal) way of isolating the variables which would translate the aggregate budget constraint into limits for particular activities. Some demand shifters, for example income, might be measurable but others are unknown and are likely to remain so, pending more work. Although demand side estimates of the value of ESAs have now been published (Willis, Garrod, and Saunders 1993), these estimates are difficult to compare with these supply functions because they measure expressed demand per person and not per hectare.

The low average cost of protection in SSSI, compared with ESA, is important in policy terms because the conservation agencies have been seeking to inject a "positive" aspect into their habitat management systems to escape from the "negative" image of PDOs. They have introduced standard or flat rate payment systems in some SSSI, which serves to reduce negotiation costs and loss of goodwill, by making the level of payments available evident before negotiations start. This development has achieved only a limited response (Whitby, Coggins, and Saunders 1990), and English Nature has now introduced policies to promote "Wildlife Enhancement" through payments for carrying out particular activities. At the same time, MAFF, on reaching the end of the first five years of its ESA designations, has renewed them and extended the area under such arrangements to nearly 2.4m hectares by 1994. The EU Regulation of 1992 has also doubled the maximum rate of contribution to compensation costs obtainable from EU funds, to as much as 50 percent of compensation paid. These expansionist adjustments, implying approbation of the ESA schemes so far in use, can now be seen to contradict the evidence of average costs and even the limited evidence of marginal costs presented here, though perhaps justified by wider political considerations.

Progress on this important question of how to secure public availability of conservation goods requires research on the costs of the various mechanisms available. This work must focus on the individual habitats being protected if different mechanisms are to be compared. Whether it will prove possible, using existing methodologies, to go further and measure the economic value of the benefits generated in a framework which allows comparison with the supply estimates presented here remains to be seen. Until appropriate benefit measures are available it will not be possible to estimate the relevant producer and consumer surpluses associated with these measures. Meanwhile we must content ourselves with the limited measurement of financial cost-effectiveness presented in this article. There is conformity between the national level and the two cases presented here; that ESAs are more expensive per hectare protected than SSSI.

Most economic attempts at environmental evaluation have worked through the demand side of the "market" and their publication has provoked a continuing debate. The estimates presented here, from the supply side, offer an alternative basis for measurement and go some way towards meeting the objections of critics in that they are based on actual transactions which have taken place: in the case of SSSI, individually negotiated transactions. The supply schedules therefore provide a potential check on current demand-based estimates. When the latter can be expressed as the demand for land under agreements, they will allow estimation of the equilibrium values.

Useful points to emerge from this analysis are the critical importance of recognizing that different assignments of property rights embodied in the two policies compared indicate different opportunity costs and levels of transactions costs as well as different equilibrium prices and, crucially, different levels of delivery of public goods. Comparison of measures must embody all such elements if they are to provide adequate policy advice.

Finally, the obvious major difference between individually negotiated agreements, which trace out an approximate supply curve for conservation goods, and ESAs, which pay each person delivering these goods a share of their full producer surplus, in addition to the supply price, has important implications for the relative size of transfer through the two systems. It seem unlikely that differences between them in terms of transactions cost will be sufficient to offset the difference in transfers, and SSSI therefore emerge as a "better buy," in terms of public expenditure, than ESAs. This conclusion awaits the delivery of better data, on both financial transfers and transactions costs, before it can be further tested.

References

Bromley, D. W. 1991. *Environment and Economy: Property Rights and Public Policy*. Oxford: Blackwell.

Colman, D., R. Crabtree, J. Froud, and L. O'Carroll. 1992. "Comparative Effectiveness of Conservation Mechanisms." Department of Agricultural Economics, University of Manchester.

Colman, D. R. 1994. "Ethics and Externalities: Agricultural Stewardship and Other Behaviors: Presidential Address." *Journal of Agricultural Economics* 45 (3):299–311.

Department of Environment (DoE). 1983. "Wildlife and Countryside Act 1981. Financial Guidelines." London: HMSO.

Felton, M. 1993. "Wildlife Enhancement Scheme Pilot Project Interim Review Report." Peterborough: English Nature.

Hodge, I. D., W. M. Adams, and N. A. D. Bourn. 1992. "The Cost of Conservation: Comparing Like with Like—A Comment on Brotherton." *Environment and Planning A* 24:1051–54.

Hooper, A. J. 1992. "Field Monitoring of Environmental Change in the Environmentally Sensitive Areas." In *Land Use Change: The Causes and Consequences*, ed. M. C. Whitby. London: HMSO.

Ministry of Agriculture Fisheries and Food (MAFF). 1990. "UK ESA Statistics." London: MAFF.

———. 1991. "Ministerial Information in MAFF (MINIM) 1990." London: MAFF.

———. 1992a. "The Somerset Levels and Moors: Environmentally Sensitive Area Report of Monitoring 1991." London: HMSO.

———. 1992b. "The South Downs Environmentally Sensitive Area Report of Monitoring 1991." London: HMSO.

———. 1992c. "The Broads: Environmentally Sensitive Area Report of Monitoring 1991." London: HMSO.

National Audit Office (NAO). 1992. "Inheritance Tax: Report by the Comptroller and Auditor General." London: HMSO.

———. 1994. "Protecting and Managing Sites of Special Scientific Interest in England." London: HMSO.

Oglethorpe, D. R., R. A. Sanderson, and J. R. O'Callaghan. 1995. "The Economic and Ecological Impact at the Farm Level of Adopting Pennine Dales Environmentally Sensitive Area (ESA) Grassland Management Prescriptions." *Journal of Environmental Planning and Management* 38:125–35.

Whitby, M. C., ed. 1994. *Incentives for Countryside Management: The Case of Environmentally Sensitive Areas*. Wallingford: CAB International.

———. ed. 1996. *The European Environment and CAP Reform: Policies and Prospects for Conservation*. Wallingford: CAB International.

Whitby, M. C., G. Coggins, and C. M. Saunders. 1990. "Alternative Payment Systems for Management Agreements." Peterborough: Nature Conservancy Council.

Whittaker, J. M., P. O'Sullivan, and J. P. McInerney. 1991. "An Economic Analysis of Management Agreements." In *Farming and the Countryside: An Economic Analysis of External Costs and Benefits*, ed. N. Hanley. Wallingford: CAB International.

Willis, K. G., G. D. Garrod, and C. M. Saunders. 1993. "Valuation of the South Downs and Somerset Levels Environmentally Sensitive Area Landscapes by the General Public; Summary Report." Centre for Rural Economy, University of Newcastle upon Tyne.

Withrington, D., and W. Jones. 1992. "The Enforcement of Conservation Legislation." In *Agriculture Conservation and Land Use: Law and Policy Issues for Rural Areas*, ed. W. Howarth and C. P. Rodgers. Cardiff: University of Wales Press.

B Conservation Management Priorities, Safe Minimum Standards and Opportunity Costs

[27]

Pergamon

S0959-3780(95)00017-8

Global Environmental Change, Vol. 6, No. 2, pp. 103–119, 1996

0959-3780/96 $15.00 + 0.00

Global biodiversity priorities

A cost-effectiveness index for investments

Dominic Moran, David Pearce and Anouk Wendelaar

Biodiversity investment priorities are a major concern for funding agencies and parties to the Convention on Biodiversity. We present an index to rank global investments which builds on previous methods by including cost-effectiveness criteria and considering the nature of threat and success in relation to priority nations. The cost-effective priority investment index (CEPII) is used to determine regional rankings for countries in the Asia–Pacific region, Sub-Saharan Africa and Central and South America and the Caribbean. Data limitations necessitate caution in the interpretation of priorities which, like other methods, are defined by subjective choice of success and threat surrogates. The index does, however, appear to have merit as a policy tool. Copyright © 1996 Elsevier Science Ltd

Dominic Moran and David Pearce are with the Centre for Social and Economic Research on the Global Environment, University College London and University of East Anglia, Norwich. Anouk Wendelaar is with Environmental Resources Management, London, UK.

[1]The total resources allocated to the Global Environment Facility (GEF) in its operational phase are US$2 billion over three years. Only part of this will be spent on biodiversity conservation.
[2]R L Pressey, C J Humphries, C R Margules, R I Vane-Wright and P H Williams, Beyond Opportunism: key principles for systematic reserve selection, *Trends in Ecological Evolution*, Vol 8 No 4, 1993,
continued on page 104

Although funding prospects are improving, there is a limited budget for global biodiversity conservation.[1] Hence some form of ranking device is needed to prioritize investments. Procedures suggested to date either omit or do not explicitly address important considerations, such as the cost of the intervention, or they do not propose an 'absolute' ranking system such as an index.[2]

We outline the development of single biodiversity investment 'cost-effectiveness' index which incorporates several relevant socio-economic factors and can be used to rank countries. The problem is approached using project ranking criteria; accounting for investment cost, and biodiversity benefit, as maximized through a representative 'currency' (such as species richness). The cost-effective priority investment index (CEPII), is derived by attenuating the resulting ranking with information on threat and the probability of a successful intervention. The resulting index is used to assess hypothetical investment potential in the Asia–Pacific, sub-Saharan Africa and Central and South America regions.

Implementing the Biodiversity Convention

In terms of guiding investment, the Biodiversity Convention is perhaps the least definitively prescriptive of all the global environmental conventions.[3] The framework agreement articulates numerous concerns, among which the contentious issues of sovereignty over genetic resources and biotechnological derivatives,[4] and the status of biodiversity as the 'common concern of humankind'.[5] There are however at least two factors complicating implementation. First, there is no precise baseline on which to formulate efficient policy. In the context of the Framework Convention on Climate Change and the Montreal Protocol, national energy plans permit some identification of an emissions baseline, which is normally not observable in the absence of similar national biodiversity planning. But species identification and the lack of a consensus diversity currency mean that few if any nations have a clear idea of a corresponding diversity baseline.[6] Second, there is no single appropriate technology to prevent biodiversity loss nor mitigate its consequences. In fact, the precautionary menu of indicative interventions (protected areas, biosphere reserves and vague appeals to sustainable use),[7] shows how poorly defined and completely habitat focused the conversation objective remains. The problem of where to conserve is part of the technology issue. From a global perspective it would seem reasonable to set priorities, which at the limit are a means of jointly implementing the global goal of maximum diversity.[8] But if each party sets conservation priorities based on differing criteria, the need for global consistency – the rationale for the Convention – may not be fulfilled. Negotiations to establish priority mechanisms have met resistance from some G77 nations and conservation organizations opposed on both financial and scientific grounds. Top-down approaches may be rational for allocating the budget of an organization such as the Global Environment Facility, but such aims need not be consistent with existing national priorities.

Cost-effectiveness and biodiversity investments

The conventional neoclassical economic approach to selecting investments is based in cost–benefit analysis (CBA). CBA assesses the net difference between investment costs and benefits over the life of the project and these are collapsed to a single money metric (present value) using an appropriate discount rate. Competing projects can be selected by ranking positive present values (or discounted benefit–cost ratios), as these are indicative of economically efficient investment opportunities. In effect the method provides a cost-effectiveness index.

The absence of observable monetary values for conservation benefits precludes a similar treatment of biodiversity. Dealing with global priority investments we are therefore in the realm of cost-effectiveness (CEA). A cost-effective ranking index is analogous to the conventional project cost–benefit ranking approach to projects outlined above. In contrast to CBA however, the absence of complete values for biodiversity[9] benefits means that the cost-effective approach is formulated in terms of available information on costs, and effectiveness defined in non-monetary terms. In other words, an index might be formulated in terms of any of a number of competing (non-monetary) measures describing biodiversity. Therefore an important issue is to select a best diversity maxim from several suggested definitions.

A cost-effective priority investment index (CEPII)

The generic form of the cost-effectiveness ratio developed below is:

$$\rho \frac{\Delta B}{C}$$

where ρ, $0 < \rho < 1$, is a composite threat/success indicator, C = cost and B = benefit, provides the basis of an index we subsequently develop.

On a suitable currency B

The conservation literature offers an array of measures for biological diversity covering, among others, taxonomic weighting criteria,[10] phylo-

continued from page 103
pp 124–128; E Dinerstein, E D Wikramanayake, 'Beyond hotspots: how to prioritise investments to conserve biodiversity in the Indo-Pacific region', *Conservation Biology* Vol 7, No 1, 1993, pp 53–65; T D Sisk, A E Launer, K R Switky and P R Erlich, 'Identifying Extinction Threats', in: *BioScience*, Vol 44, No 9, 1994, pp 592–604

[3] Relative to the unequivocal abatement or deposition targets arising from the Framework Convention on Climate Change (FCCC), the Montreal Protocol and the Second Sulphur Protocol

[4] Articles 15 and 19, UNEP 1992, Convention on Biological Diversity

[5] *Ibid*, (preamble)

[6] With obvious consequences for defining incremental cost as the difference between baseline action costs and cost of actions taken pursuant to the treaty

[7] Article 8, UNEP, *op cit*, Ref 4

[8] The distinction between this objective and a less exacting objective of species preservation should be made. How adequately the former objective is served by the latter is a debatable issue, see P Williams and K J Gaston, 'Measuring more of biodiversity: can higher-taxon richness predict wholesale species richness?' *Biological Conservation*, Vol 67, 1994, pp 211–217

[9] For a review of progress in the economic valuation of biological resources see D Pearce, and D Moran, The Economic Value of Biodiversity, Earthscan, London, 1994

[10] R I Vane-Wright, C J Humphries and P H Williams, 'What to Protect – Systematics and the Agony of Choice'. *Biological Conservation*, Vol 55, 1991, pp 235–254

geny based and related pattern and feature models,[11] phenotypic or morphological diversity,[12] organism or ecosystem functional diversity and redundancy.[13]

Although theoretically concise, such measures are data intensive. A more pragmatic approach has therefore focused research on surrogates, varying in complexity from centres of endemism,[14] use of higher taxon richness,[15] or the use of environmental (pattern) diversity linked by models to target characters or attributes.[16] Use of the first two in traditional priority setting,[17] is generally dictated by the availability of ground truthed information. Recent normative and surrogate approaches are to varying degrees influenced by several guiding principles defining the process of 'optimal' indicator selection. Most noteworthy are the characteristics of flexibility, irreplaceability and complementarity of relative areas, taxa, or any set of their attributes.[18] The complementarity concept is particularly germane to cost-effectiveness and is central to several recent algorithm based minimum set methods.[19] Further reference to cost-effectiveness is apparent in a parallel literature emphasizing the potential inefficiency arising from badly planned or default conservation strategies. Pressey and Tully[20] consider the cost of opportunistic or *ad hoc* reserve selection in Australia; while others highlight the disproportionate influence of political criteria,[21] or the adequacy of coverage afforded by reliance on one taxa within a favoured surrogate.[22] All approaches implicitly or explicitly highlight the difficulties of simultaneous optimization of economic and biological criteria.

Literature revisiting coarse species richness surrogates has emphasized the cost advantages of exploiting the potential richness predictive capacity of higher taxa richness,[23] and the dangers of assuming a nested relationship linking widely and rarely observed species.[24] Abstracting from this 'umbrella' function, practical considerations for adopting species richness include the use of taxonomic combinations – say one vertebrate one invertebrate – with contrasting habitat requirements and life history strategies;[25] or combinations of richness and endemism.[26] The global scale of the current analysis is limited to a similar scale, and accepting that all measures are value laden we make no unwarranted claims for the superiority of species richness. The measure B, is therefore represented by the sum of higher plants, mammals, birds, reptiles and amphibians, with data drawn from the World Conservation Monitoring Centre Biodiversity Data Sourcebook[27] (Table A1).

Note that the cost-effectiveness ratio represents the notional amount of biodiversity ΔB, conserved by an intervention. We would ideally like ΔB to be the change in the indicator with and without the intervention on a limited scale. While there may be possible proxies to measure 'biodiversity saved', we are generally restricted to a consideration of biodiversity currently in place. We normalize our favoured measures of richness and endemism to be on a km^2 basis.

Investment cost C

There are numerous cost considerations in area selection, though this single most limiting factor is frequently omitted in priority exercises. In minimum set type analyses, for example, a cost element implicit in a minimized [land] area numeraire is rarely considered in terms of its financial cost equivalent – perhaps that of area acquisition or the procurement of an easement. Biodiversity projects compete for limited

[11]M L Weitzman, 'What to preserve? An application of diversity theory to Crane conservation', *Quarterly Journal of Economics*, Vol 58, 1992, pp 157–183; A Solow, S Polasky and J Broadus, 'On the Measurement of Biological Diversity', *Journal of Environmental Economics and Management*, Vol 24, 1993, pp 60–68; D Faith, 'Phylogenetic pattern and the quantification of organismal biodiversity', Philosophical Transactions of the Royal Society of London. B Vol 345, 1994, pp 45–58

[12]P Williams, K Gaston and C J Humphries, 'Do conservationists and molecular biologists value differences between organisms in the same way?' *Biodiversity Letters*, Vol 2, 1995

[13]B H Walker, 'Biodiversity and Ecological Redundancy', *Conservation Biology*, Vol 6, No 1, 1992, pp 18–23

[14]ICBP = International Council for Bird Preservation, *Putting Biodiversity on the Map: Priority Areas for Global Conservation*, Birdlife International, Cambridge, 1992

[15]P Williams and K J Gaston, *op cit*, Ref 8

[16]D P Faith and P A Walker, 'Environmental diversity: on the best-possible use of surrogate data for assessing the relative biodiversity of sets of areas', submitted to *Biodiversity and Conservation*, 1995a; D P Faith and P A Walker, 'Hotspots and firestations: on the use of biotic and environmental data to estimate the relative biodiversity of different sets of areas', submitted to *Biodiversity Letters*, 1995b

[17]N Myers, 'Threatened biotas: "hot spots" in tropical forests', *The Environmentalist*, Vol 8, No 3, 1988, pp 187–208; N Myers, 'The biodiversity challenge: expanded hot-spots analysis', *The Environmentalist*, Vol 10, No 4, 1990, pp 243–256; R A Mittermeier, 'Primate diversity and the tropical forest: case studies from Brazil and Madagascar and the importance of the megadiversity countries', in: E O Wilson and Francis M Peters (eds) *Biodiversity*, National Academy Press, Washington, DC, 1988, pp 145–154; R A Mittermeier and T B Werner, 'Wealth of plants and animals unites "megadiversity" countries', *Tropicus*, Vol 4, No 1, 1990, pp 4–5; T D Sisk, *et al*, *op cit*, Ref 2; WCMC, *Priorities for Conserving Global Species Richness and Endemism*, WCMC Biodiversity Series No 3, Cambridge, 1994a

[18]R L Pressey, *et al*, *op cit*, Ref 2

[19]Complementarity based character selection involves a stepwise process of maximizing character representation in minimum sets of areas, see R L Pressey, H P Possingham and C R Margules, 'Optimality in reserve selection algorithms: when does it matter and how much?' submitted to *Biological Conservation*, 1995

[20]R L Pressey and S L Tully, 'The cost of *ad hoc* reservation: A case study in western New South Wales', *Australian Journal of Ecology*, Vol 19, 1994, pp 375–384

[21]A Metrick and M L Weitzman, *Patterns of Behaviour in Biodiversity Preservation*,

continued on page 106

funds and it makes sense to save the cheapest biodiversity first. The assumption of constant costs is clearly unrealistic and some assessment of how relative costs alter minimum set sequences is necessary.

For index purposes, C is ideally required to be the average unit cost per biodiversity intervention across countries. Since such detailed information is unavailable, international investment levels are taken as a best proxy to cost information for countries we wish to rank. Therefore we make the assumption that investment levels normalized per km^2 are indicative of costs to be covered. It is clear however that the figures we use may be influenced by other factors than cost, not least socio-political stability and a country's conservation track record which in itself will influence the effectiveness of foreign spending. We return later to cost interpretations and a consideration of alternative cost information.

Several investment series are available (accounting for overseas development assistance (ODA) investments by bilaterals, multilateral and NGOs). The proportion of world ODA currently spent on biological diversity is not known with certainty. Estimates are low,[28] and do not adequately reflect the recent emphasis on biodiversity issues[29] or the emerging role of Global Environment Facility (GEF) spending. Determining the actual use of biodiversity investment is not transparent. The primary cost data requirement for our index is international investment, given by bilateral, multilateral and NGO expenditure. The present analysis uses Abramovitz 1993 data (total US institutional spend for 1991)[30]. Or recently compiled data by WCMC (1994, unpublished) on total funding world-wide 1991–94 on biodiversity. The cost figure should be expressed as international expenditure, with attention paid to ensuring that the data are consistent across countries (same year/s, same donor classification).

Previous index approaches

Recalling that ΔB of Equation 1 is in non-monetary terms, the success/threat parameter ρ is best considered in the light of previous qualitative priority and descriptive approaches.

There are numerous limitations to the traditional qualitative focus of priorities on species rich or endemic countries or biogeographic units. First, lists produced by such exercises rarely provide a rationale for selecting among competing countries. Second, they do not refer to the costs of intervention, and as such they cannot provide absolute priority guidance to the investment decision maker. Third, those that mention threat or elements enhancing the probability of a successful intervention, normally fail to justify either as a priority ranking device. A high degree of threat for example is typically viewed as a reason to intervene – whereas, in fact, it may constitute a reason for non-intervention. Similar caveats can be made for aspects of a project's socio-economic environment (see for example Braatz *et al*[31]). However such factors rarely seem to be an integral part of a designation decision.

In an influential application for the Indo-Pacific region, Dinerstein and Wikramanayake[32] introduce the Conservation Potential/Threat Index (CPTI). The analysis actually fails to develop a single comprehensive index but does consider the extent of protected areas, forest habitats, species richness and deforestation rates. Cost information is used (represented by Abramovitz data and an unexplained reference to GEF funding), but is not related to benefit (species richness or ende-

continued from page 105

Policy Research Working Paper 1358, Policy Research Dept, World Bank: Washington, DC, 1994

[22] J R Prendergast, R M Quinn, J H Lawton, B C Eversham and D W Gibbons, 'Rare species, the coincidence of diversity hotspots and conservation strategies', *Nature*, Vol 365, 1993, 335–337

[23] Williams and Gaston, 1994, *op cit*, Ref 8

[24] P Williams, D Gibbons, C Margules, T Rebelo, C Humphries and B Pressey, 'Richness hotspots, rarity hotspots and complementary areas – a comparison of three area-selection methods for conserving diversity using British breed birds', submitted to Biological Conservation, 1995

[25] Sisk *et al*, *op cit*, Ref 2

[26] D Pomeroy, 'Centres of high biodiversity in Africa', *Conservation Biology*, Vol 7, No 4, 1993, pp 901–907

[27] WCMC, *Biodiversity Data Sourcebook*, WCMC Biodiversity Series No 1, Cambridge, 1994. For convenience Appendix Tables A1–A3 present sources for the Asia–Pacific region only. Data caveats for Latin America and sub-Saharan Africa are included as notes with their respective Tables 5 and 6

[28] UNEP has estimated that biodiversity investment is in the region of US$200 million a year and that multilateral agencies spend about US$75 million annually on biodiversity

[29] In 1989, for example, the US government spent US$23 million on biodiversity related projects, and in 1991/92 British government investments in such projects exceeded £37 million (WCMC 92)

[30] J N Abramovitz, 'Investing in Biological Diversity: US Research and Conservation Efforts in Developing Countries', World Resources Institute, Washington, DC, 1991; J N Abramovitz, 'Trends in Biodiversity Investments: US Based Funding for Research and Conservation in Developing Countries, 1987–1991'. Final Draft, July 1993 (in press)

[31] S Braatz, G Davis, S Shen and C Rees, *Conserving Biological Diversity: A Strategy for Protected Areas in the Asia-Pacific Region*, World Bank Technical Paper 193, World Bank, Washington, DC, 1992

[32] E Dinerstein and E D Wikramanayake, 'Beyond hotspots: how to prioritize investments to conserve biodiversity in the Indo-Pacific Region', *Conservation Biology*, Vol 7, No 1, 1993, pp 53–65

mism). Moreover the species index [benefit] is not combined with the potential/threat or cost information. The authors have used the cost (investment) data to interpret how well countries are financed, not how cost effective the investments have been. Bhutan is ranked at the top of category 1 in the cost analysis, which means it is well funded. The authors do not, however, state that it should therefore receive the highest priority. Neither do they allocate this priority to the countries receiving the lowest levels of funding – their slightly ambiguous conclusion is that high conservation potential exists in several countries which have received meagre support. We are given no clear indication as to whether existing high (or low) expenditure should warrant more expenditure or less.

Less ambitiously, Sisk *et al*[33] offer a set of low resolution databases to correlate anthropogenic threats (forest loss and population pressure), with species rich biota on a global basis. Indices derived using their 'Global Conservation Analysis Package' relate principally to threats which are not combined into any single measure. Instead, the authors classify critical cases – one assumes worthy of equal priority treatment – as those falling in the top 25% of any index. Again, although high index rankings coincide for several countries there is no unambiguous method to prioritize. Despite several notable omissions when comparing global with continental rankings, the authors claim the method to be as valid as any other currently available for setting global priorities.

The *ad hoc* nature of these conservation potential–threat studies is indicative of the difficulties encountered in reducing several data sets to a single comprehensive index number which objectively captures underlying interactions particularly those related to success and threat. Alternative – mainly geographical information system-based – approaches for amalgamating similar information have been suggested,[34] but their interpretation is no less problematic.

The cost-effective priority investment index (CEPII)

The suggested approach is flexible in the sense that it is amenable to adaptation using other surrogates of threat and or success as well as other forms of cost information.

Appendix 1 develops the CEPII combining the basic cost-effectiveness ratio with the attenuating factors describing threat and the probability of a successful intervention. We discuss these remaining basic and interrelated parts in turn.

The probability of success

Probabilities of successful intervention will depend on institutional characteristics, notably the commitment of the government, (ie its 'willingness to conserve'). Commitment could be reflected by an indicator which assesses the available infrastructure of a country – examples could be total km^2 of roads, staffing in protected areas or in natural resource related ministries. 'Ability' could also determine success. An 'ability' indicator might be the number of students graduating in biodiversity related disciplines in a particular country. Various other indicators are possible but most suffer from data problems. A simple indicator is area of 'protectable' land actually protected – the probability of success would therefore be given by the ratio of actual terrestrial protected area over total protectable area, per country.

[33] Sisk *et al*, *op cit*, Ref 2
[34] D M Olson and E Dinerstein, *Assessing the Conservation Potential and Degree of Threat Among Ecoregions of Latin America and the Caribbean: A Proposed Landscape Ecology Approach*, The World Bank, LATEN Dissemination Note 10, Washington, DC, 1994

On the basis of the above success indicator, a country with a high percentage of area protected would have a high probability of success. It should be noted however that a very high percentage of protected area – assuming none of which exists solely on paper – could indicate that a country has reached protection 'saturation' – there may be a maximum protection percentage beyond which a country will be reluctant to protect. Wilson[35] recommends that total global reserves should be expanded from 4.8%[36] to 10% of the land surface, to include as many undisturbed habitats as possible. To protect at least 10% of its total land area might therefore constitute a reasonable objective for a country. This somewhat abstracts from attenuating characteristics such as fragmentation and skewness in terms of biogeographical representativeness. We emphasize that such factors as well as size of remaining areas (protected or otherwise) matter because larger areas can withstand higher deforestation rates and may not require immediate interventions. Similarly it should be noted, however, that size of the intervention unit though difficult to incorporate into an index, will also bear on the probability of success.

There is a need to ensure congruence between index components used for surrogates for biodiversity, success and threat (see below). We use varying data sources to identify protected *forest* area as the basic unit of analysis for countries we rank. Concentration on forest biota is partly based on the assumption that a larger proportion of species diversity can best be safeguarded in forests. This proposition is debatable, though we do not pursue it here.[37]

The main reason for limiting the analysis to forests is data, and several problems were encountered. Forests are defined (FAO definition)[38] as including tropical rain forest, moist deciduous forest, dry deciduous forest, very dry forest, desert zone, and hill and montane zone, and excluding plantations. Species, protected areas and cost data are usually presented as totals across all habitats. As our analysis focuses on forests we would need species richness for forests, protected forest areas, and investment on forests for complete accuracy. Complete data for these categories are not available. An attempt has been made to give a more accurate figure for protected forests by subtracting protected wetlands from total protected area (see Appendix, Tables A2 and A3). Recent WCMC figures[39] for protected forest areas have been calculated by overlaying digitized maps of protected areas (IUCN management categories I–V) with those of forest cover and measuring the degree of overlap. However, the WCMC Biodiversity Map Library does not as yet include 100% cover of the protected areas for countries we rank and gives only partial validation for our figures.

The degree of threat

Consideration of threat is more problematic. A high degree of threat to biodiversity is commonly accepted as the rationale for intervention, and a basis for priority setting. However, the chances of success may well vary inversely with the threat. As such an appraisal of threat or vulnerability probability is highly relevant. In practice, much depends on the reasons for and imminence of, the threat. Thus, if it is population growth, the intervention is unlikely to lower the underlying threat. If it is institutional weakness (perhaps exemplified by low national expenditure on conservation), the intervention has a higher chance of succeeding. Threat analysis thus becomes essential.

[35]E O Wilson, *The Diversity of Life*, Penguin, London, 1992
[36]World Resources Institute, *World Resources 1992–93: Guide to the Global Environment*, Oxford University Press, Oxford, 1992
[37]For further discussion see Myers, *op cit*, Ref 17; P H Raven, Our Diminishing Tropical Forests, in E O Wilson (ed) *Biodiversity*, National Academy Press, Washington, DC, 1988
[38]FAO (1993). 'Forest Resources Assessment 1990: Tropical Countries', FAO Report, Food and Agricultural Organization of the United Nations, Rome. An alternative review of the problems involved in categorizing and detecting changes in forest cover can be found in R A Houghton, D S Lefkowitz and D L Skole, 'Changes in the landscape of Latin America between 1850 and 1985, I Progressive loss of forests', *Forest Ecology and Management*, Vol 38, 1991, pp 143–172
[39]WCMC, *op cit*, Ref 27

The most pressing anthropogenic threat[40] to biodiversity is the loss of natural habitat – mainly to arable cultivation, logging, settlement, poaching and pollution – resulting from human population growth. More proximate causes of erosion such as ill-defined land and resource rights and market and government failures have also been identified.[41] At the extreme, indicators of political turmoil and repression have been correlated with forest conversion in a cross-section of developing countries.[42]

Information on population growth and or density are the most accessible variables which have been positively correlated with habitat (ie forest) loss.[43] Deforestation rates themselves are also indicative of threat. Moreover we note that forest loss is the only type of habitat alteration adequately quantified on a global scale. Without making any claims for the preponderance of any particular species habitat relationship we adopt deforestation as a threat indicator. We recognize that rates are subject to a range of policy variables which are themselves subject to alteration, and that past deforestation may be a poor predictor of current rates. We consider the effects of using a population threat indicator in later analysis. Deforestation data derives from a number of sources, (Table A2).

An application of the CEPII

To facilitate comparison with the studies of Dinerstein and Wikramanayake[44] and Braatz *et al*[45] the mechanics of index derivation are illustrated with reference to data for the Asia-Pacific countries. Index rankings and referenced data sources for Central and South America and sub-Saharan Africa are presented in turn.

Patchiness of site level data dictates the national focus of the current analysis. This is also the relevant scale of decision making by national governments and foreign aid donors.[46] The three regions are however differentiated by broad macro-economic aggregates, which invariably impact, albeit historically, on the effort invested in biodiversity monitoring and conversation. Since historical performance provides the basis for the CEPII, it seems reasonable to segregate countries which have benefited from relatively high monitoring effort (SE Asia), from those where conservation investments have traditionally been low (sub-Saharan Africa). The regions can be differentiated on other grounds. Generally, South-East Asian countries are under severe threat from deforestation. The rate of forest clearance has roughly trebled in Asia since the early 1960s and is still rising. FAO figures confirm that South-East Asia along with Central America, has the world's fastest rates of tropical deforestation (FAO 1993). Both regions should be distinguished from sub-Saharan Africa in terms of current domestic funding and possibly the nature of threat.[47]

For all three regions, 'protectable' area has been specified as 'protectable' forest, area protected as protected forest, and unprotected area as unprotected forest. As previously stressed this is partly because of data availability, but also because diversity is strongly linked to forests. In addition, the severe deforestation threat makes an analysis of forested areas in two of three regions particularly pertinent.

For the Asia-Pacific region, countries considered in the analysis are Bangladesh, Bhutan, Cambodia, Fiji, India, Indonesia, (the Republic of) Korea, Lao PDR, Malaysia, Myanmar, Nepal, Pakistan, Papua New Guinea (PNG), the Philippines, the Solomon Islands, Sri Lanka, Thai-

[40]Although there are many 'natural' threats to biodiversity (such as floods or hurricanes), they occur more or less randomly and are impossible to plan for, making them unsuitable as indicators

[41]D W Pearce and J Warford, *World Without End*, Oxford University Press, New York, 1993

[42]R Deacon, 'Deforestation and the rule of law in a cross-section of countries', *Land Economics*, Vol 74, 1994, pp 417–430

[43]See K Brown and D W Pearce (eds), *The Causes of Tropical Deforestation*, UCL Press, London, 1994

[44]Dinerstein and Wikramanayake, *op cit*, Ref 2

[45]Braatz *et al*, *op cit*, Ref 31

[46]Notwithstanding a recent emphasis on cross boarder initiatives for species corridors and national parks (ie South Africa and Mozambique; GEPRENAF (Gestion Participative des Ressources Naturelles et de la Faune (Burkina Faso, Cote d'Ivoire and Mali)

[47]In the first instance the chosen threat surrogate for the CEPII is deforestation rates. Low or even negligible forest cover (in several countries), combined with high population growth mean that the latter is arguably a better threat surrogate in sub-Saharan Africa

land, Tonga, Vanuatu, Vietnam and Western Samoa. These are the same as those surveyed in Braatz *et al* (1992), except for Kiribati and the Maldives, which have been excluded from this analysis because they have negligible forest cover (Braatz *et al*). In Central and South America, omitted countries are French Guyana and several Caribbean small island states. For sub-Saharan Africa, data gaps exist for the Central African Republic, Sierra Leone and Togo.

Index calculation and country rankings

Data requirements for the CEPII are outlined in Appendix 1. Notwithstanding several inconsistencies (such as that for Pakistan reporting 18 550 km^2 of natural forest cover, and 36 550 km^2 under protection), they are sufficient for index calculation.

The threat ratio is calculated by dividing unprotected forest in 10-years time by unprotected forest now. We therefore need to assume that deforestation rates will remain constant over the next 10 years. The biodiversity indicator is species richness. In line with Dinerstein and Wikramanayake, the analysis is focused at the species level and assumes that all taxonomic groups have equal importance. Species richness is given by the total of certain well described taxa per country.

The index calculation and subsequent rankings are presented in Tables 1–4, which also vary cost and threat data series used in calculation. All the tables are based on the same index formulation (described in Appendix 1). Table 1 shows the mechanics of index calculation from raw data input using the deforestation rate as the threat and the Abramovitz

Table 1. Index calculation using (a) Abramovitz cost data and (b) deforestation rate as the threat

	(1) Total area km^2 A (Braatz)[a]	(2) Forested km^2 Af (FAO)[b]	(3) Protected forest km^2 Ap (Braatz)[a]	(4) Unprotected forest km^2 Au t (Af − Ap)	(5) Success ratio Ps Ap/Af	(6) Annual deforestation Rate	(7) Unprotected 10 yrs time Au t + n	(8) Threat ratio Au t + n/Au t	(9) Total SR B (WCMC)[c]	(10) SR per km^2 B/Af km^2	(11) Cost C/Af km^2 (Abram)[c]	(12) Index** (× 100)
Bangladesh	144 000	7 690	613	7 077	0.08	0.039	4 754	0.67	5 601	0.73	5.49	0.710
Bhutan	47 000	28 090	9 056	19 034	0.32	0.006	17 922	0.94	6 068	0.22	45.29	0.145
Cambodia	181 000	121 630	0	121 630	0.00	0.010	110 000	0.90	532	0.00	n/a	n/a
China	9 561 000	1 150 600	263 578	887 022	0.23	0.039	595 893	0.67	34 166	0.03	0.71	0.644
Fiji	18 000	8 100	53	8 047	0.01	0.001	7 967	0.99	1 746	0.22	0.00	n/a
India	3 288 000	517 290	122 401	394 889	0.24	0.006	371 825	0.94	17 881	0.03	2.10	0.367
Indonesia	1 905 000	1 095 490	163 309	932 181	0.15	0.010	843 048	0.90	25 315	0.02	2.98	0.105
Korea, Rep	99 000	48 000	7 510	40 490	0.16	0.000	40 490	1.00	2 978	0.05	n/a	n/a
Lao PDR	237 000	131 730	0	131 730	0.00	0.009	120 343	0.91	9 043	0.07	0.03	0.000
Malaysia	330 000	175 830	14 816	161 014	0.08	0.020	131 560	0.82	13 691	0.08	5.26	0.102
Myanmar	677 000	288 560	1 693	286 867	0.01	0.013	251 682	0.83	8 445	0.03	n/a	n/a
Nepal	141 000	50 230	10 999	39 231	0.22	0.010	35 480	0.90	7 835	0.16	13.84	0.225
Pakistan	796 000	18 550	35 170	(18 620)	1.90	0.034	(11 760)	0.71	5 725	0.31	3.45	12 001
Papua NG	463 000	360 000	290	359 710	0.00	0.003	349 063	0.97	12 796	0.04	1.04	0.003
Philippines	300 000	70 310	4 968	73 342	0.06	0.033	52 435	0.71	9 748	0.12	81.80	0.007
Solomon	27 000	24 400	0	24 400	0.00	0.000	24 400	1.00	3 454	0.14	6.15	0.000
Sri Lanka	66 000	17 460	7 071	10 389	0.40	0.014	9 023	0.87	3 704	0.21	14.79	0.505
Thailand	513 000	127 350	54 730	72 620	0.43	0.033	51 918	0.71	13 897	0.11	8.57	0.391
Tonga	700	500	0	500	0.00	n/a	500	1.00	509	1.02	232.00	0.000
Vanuatu	12 000	2 400	0	2 400	0.00	0.017	2 022	0.84	1 118	0.47	26.51	0.000
Vietnam	330 000	83 120	8 480	74 640	0.10	0.015	64 170	0.86	9 171	0.11	1.91	0.507
W Samoa	3 000	1 400	0	1 400	0.00	n/a	1 400	1.00	748	0.53	0.00	n/a

Notes: SR: species richness
B: biodiversity (benefit)
C: investment (cost)
n/a Not applicable
** The index is calculated by multiplying columns [(5) × (8) × (10)] and dividing the result by column (11)
(..) Denotes negative number
[a] Braatz *et al, op cit*, Ref 31
[b] FAO, *op cit*, Ref 38
[c] WCMC, *op cit*, Ref 27
[d] Jn Abramoritz, *Trends in Biodiversity Investments: US Bases Funding for Research and Conservation in Developing Countries 1987–1991* World Resources Institute, Washington, DC, 1993

Table 2. Countries ranked according to Table 1 index

Ranking	Index	Country	Reason for index n/a
1	12.001	Pakistan	
2	0.710	Bangladesh	
3	0.644	China	
4	0.507	Vietnam	
5	0.505	Sri Lanka	
6	0.391	Thailand	
7	0.367	India	
8	0.225	Nepal	
9	0.145	Bhutan	
10	0.105	Indonesia	
11	0.102	Malaysia	
12	0.007	Philippines	
13	0.003	Papua NG	
14	0.000	Lao PDR	
14	0.000	Vanuatu	
14	0.000	Solomon	
14	0.000	Tonga	
	n/a	Fiji	Zero cost
	n/a	W Samoa	Data unavailable – deforestation rate. Zero cost
	n/a	Cambodia	Data unavailable – cost
	n/a	Myanmar	Data unavailable – cost
	n/a	Korea, Rep	Data unavailable – cost

Table 3. Country index ranking using WCMC cost data

Ranking	Index	Country
1	6.292	Pakistan
2	0.844	Korea, Rep
3	0.190	Malaysia
4	0.047	Sri Lanka
5	0.030	Bhutan
6	0.026	Vietnam
7	0.014	Bangladesh
8	0.000	Lao PDR

cost data, which, although limited to US funding only, is more comprehensive in that costs for most countries are provided. A priority ranking derived from these data is shown in Table 2, which also highlights data gaps. Table 3 presents the index ranking derived using the WCMC (1994) cost data based on world-wide levels of investment. Table 4 considers the impact of using the population growth rate as the threat (combined with the Abramovitz cost data), as opposed to deforestation. Tables 5 and 6 present ranking for Latin America and sub-Saharan Africa both using WCMC cost data and deforestation rates.

Interpretation of the results

Asia–Pacific

Regional priorities (Table 2), are Pakistan, Bangladesh, China, Vietnam and Sri Lanka. Allowing for the observed inconsistencies in the Pakistan

Table 4. Country index ranking using population growth as threat

Ranking	Index	Country	Reason for index n/a
1	12.7674	Pakistan	
2	0.8491	China	
3	0.8126	Bangladesh	
4	0.5201	Sri Lanka	
5	0.4815	Vietnam	
6	0.4801	Thailand	
7	0.3215	India	
8	0.1958	Nepal	
9	0.1218	Bhutan	
10	0.1030	Malaysia	
11	0.0984	Indonesia	
12	0.0078	Philippines	
13	0.0022	Papua NG	
14	0.0000	Lao PDR	
14	0.0000	Vanuatu	
14	0.0000	Solomon	
14	0.0000	Tonga	
	n/a	Fiji	Zero cost
	n/a	W Samoa	Zero cost
	n/a	Cambodia	Data unavailable – cost
	n/a	Myanmar	Data unavailable – cost
	n/a	Korea, Rep	Data unavailable – cost

Table 5. CEPII index. Central/South America and the Caribbean. Calculation using (a) WCMC cost data and (b) deforestation rate as the threat

	(1) Total area km² A (WRI 94)[a]	(2) Forested km² Af (FAO 93)[b]	(3) Protected forest km² Ap (WCMC 94)[c]	(4) Unprotected forest km² Au t (Af − Ap)	(5) Success ratio Ps Ap/Af	(6) Annual deforestation Rate (FAO 93)[b]	(7) Unprot'd 10 yrs time Au t + n	(8) Threat ratio Au t + n/Au t	(9) Total SR B (WCMC 94)[c]	(10) SR per km² B/Af km²	(11) Cost C/Af km²	(12) Index** (× 100)
Bolivia	1084380	493170	62380	430790	0.12648782	0.012	361 799	0.88627693	19277	0.03908794	15.1469067	0.02892924
Brazil	8456510	5611070	67180	5543890	0.01197276	0.006	5 220 096	0.94159435	59214	0.01055307	3.28849934	0.00361775
Colombia	1038700	540640	42720	497920	0.07901746	0.007	464 143	0.93216434	54443	0.10070102	24.2675348	0.03056496
Ecuador	276840	119620	14110	105510	0.11795686	0.018	87 985	0.83390174	21999	0.18390737	78.147467	0.02314847
Guyana	196850	184160	580	183580	0.00314944	0.001	181 752	0.99004488	7558	0.0410404	9.74695917	0.0013129
Paraguay	397300	128590	1090	127500	0.00847655	0.027	96 970	0.76055111	8961	0.0696866	4.56489618	0.00984161
Peru	1280000	679060	20310	658750	0.02990899	0.004	532 369	0.96071237	20880	0.03074839	2.83480105	0.03116699
Suriname	156000	147680	4560	143120	0.03087757	0.001	141 695	0.99004488	6117	0.04142064	0.40628386	0.31166302
Venezuela	882050	456900	86450	370450	0.18920989	0.012	328 321	0.88627693	23132	0.05062815	1.18844386	0.71437564
Belize	22800	19960	2200	17760	0.11022044	0.002	17 408	0.98017904	3691	0.18491984	40.2805611	0.04959702
Costa Rica	51060	14280	4090	10190	0.28641457	0.029	7 592	0.7450618	13550	0.94887955	595.518207	0.03400192
Cuba	109820	17150	1390	15760	0.08104956	0.01	14 253	0.90438208	7038	0.41037901	29.4460641	0.10215521
Dominican Republic	48380	10770	276	10494	0.02562674	0.023	7 900	0.7527706	6071	0.56369545	217.084494	0.00500924
El Salvador	20720	1410	50	1360	0.03546099	0.022	1 089	0.80055016	3562	2.52624113	3.54609929	2.02238274
Guatemala	108430	42250	220	42030	0.0052071	0.017	35 407	0.84243263	9930	0.23502959	155.526627	0.0006629
Haiti	27560	230	40	190	0.17391304	0.048	116	0.61146202	5606	24.373913	2986.95652	0.08677567
Honduras	111890	46050	1100	44950	0.02388708	0.021	36 354	0.80877351	6745	0.14647123	35.4397394	0.00798457
Mexico	1908690	485860	10370	475490	0.0213436	0.018	417 170	0.87734727	28519	0.05869798	22.9613469	0.00478702
Nicaragua	118750	80130	21500	38630	0.35755862	0.019	31 887	0.82544867	8760	0.14568435	102.527856	0.04193806
Panama	75990	31170	8850	22320	0.28392685	0.019	18 424	0.82544867	11452	0.36740456	30.542188	0.28192977
Trinidad and Tobago	5130	1550	270	1280	0.17419355	0.021	1 035	0.80877351	2888	1.86322581	2836.77419	0.00925337
Jamaica	10830	239	20	219	0.08368201	0.072	104	0.47367423	3651	15.2761506	11552.3013	0.00524152

[a]WRI World Resources Institute, *World Resources 1993–1994: Guide to the Global Environment*, Oxford University Press, Oxford, 1994

[b]FAO, *op cit*, Ref 38

[c]WCMC, *op cit*, Ref 27

Notes: 1. Total (closed) forested area for El Salvador and Jamaican protected forest area both from sources cited in Sharma, *op cit*, Ref 52

2. Cost data for Surinam, El Salvador, Haiti and Panama from Abramovitz, *op cit*, Ref 30

3. Protected forest areas for the Dominican Republic, Haiti and Nicaragua are arbitrarily assumed to be approximately half total protected area, *World Resources 1994–95*, World Resources Institute

Table 6. CEPII index. African region. Calculation using (a) WCMC cost data and (b) deforestation rate as the threat

	(1) Total area km² A (WRI 94)[a]	(2) Forested km² Af (FAO 93)[b]	(3) Protected forest km² Ap (WCMC 94)[c]	(4) Unprotected forest km² Au t (Af − Ap)	(5) Success ratio Ps Ap/Af	(6) Annual deforestation Rate (FAO 93)[b]	(7) Unprotected 10 yrs time Au t + n	(8) Threat ratio Au t + n/Au t	(9) Total SR B (WCMC 94)[c]	(10) SR per km² B/Af km²	(11) Cost C/Af km²	(12) Index** (× 100)
Angola	1246700	230740	13000	217740	0.05634047	0.007	202 969	0.93216434	6411	0.02778452	1.49518939	0.09759322
Benin	11062	49470	1	49469	2.0214E-05	0.013	43 401	0.87734727	2813	0.05686275	0.00220336	0.04576915
Botswana	566730	142610	103000	39610	0.72224949	0.005	37 673	0.95111013	924	0.00647921	103.688381	0.0042925
B Faso	27380	44160	7860	36300	0.17798913	0.007	33 838	0.93216434	1703	0.03856431	108.627717	0.00589021
Burundi	25650	2330	230	2100	0.09871245	0.006	1 977	0.94159435	3205	1.37553648	2433.90558	0.00525296
Cameroon	465400	203500	11060	192440	0.05434889	0.006	181 200	0.94159435	9515	0.04675676	57.4791155	0.00416283
Chad	1259200	114340	33000	81340	0.28861291	0.007	75 322	0.93216434	2267	0.01982683	7.70509008	0.06922833
Congo	341500	198650	6600	192050	0.03322426	0.002	188 243	0.98017904	5128	0.02581425	223.483514	0.00037616
C d'Ivoire	318000	109040	5520	103520	0.05062362	0.01	93 622	0.90438208	4588	0.0420763	39.0040352	0.00493893
E Guinea	28050	18260	335	17925	0.01834611	0.004	17 221	0.96071237	3761	0.20596933	14.1840088	0.02559417
Ethiopia	1101000	141650	6223	135427	0.04393223	0.003	131 419	0.97040178	7707	0.05440875	22.4708789	0.01032247
Gabon	257670	182350	8910	173440	0.04886208	0.006	163 310	0.94159435	7477	0.04100356	4.41458733	0.04273333
Gambia	10000	970	20	950	0.02061856	0.008	877	0.92281941	1587	1.63608247	256.701031	1.3457E-09
Ghana	227540	95550	190	95360	0.00198849	0.013	83 664	0.87734727	4677	0.04894819	327.263213	2.6094E-05
Guinea	245860	66920	97	66823	0.00144949	0.012	59 224	0.88627693	3748	0.05600717	113.508667	6.3387E-05
G Bissau	23120	20210	0.01	20209.99	4.948E-07	0.008	18 650	0.92281941	1429	0.07070757	76.7936665	4.2043E-08
Kenya	569690	11870	1190	10680	0.10025274	0.006	10 056	0.94159435	8208	0.69149115	3778.18029	0.00172768
Liberia	96750	46330	930	45400	0.02007339	0.005	43 180	0.95111013	3074	0.0663501	0.50628103	0.2502081
Madagascar	581540	157820	2310	155510	0.01463693	0.008	143 508	0.92281941	10259	0.06500444	398.434926	0.00022037
Malawi	94080	34860	2460	32400	0.07056799	0.014	28 139	0.86849865	4798	0.13753626	62.8227194	0.01342743
Mali	122019	121440	3500	117940	0.02882082	0.008	108 837	0.92281941	2517	0.02072628	0.03046772	1.80927465
Mozambique	784090	173290	3750	169540	0.02164003	0.007	158 039	0.93216434	6616	0.03817878	0.86560102	0.08897223
Nambia	823290	125690	10000	115690	0.07955082	0.003	112 266	0.97040178	3995	0.03178455	21.0040576	0.01168325
Niger	1266700	25500	2380	23120	0.09333333	0	23 120	1	1791	0.07023529	0.26666667	2.45823529
Nigeria	910770	156340	3100	153240	0.01982858	0.007	142 845	0.93216434	6095	0.03898554	0.42314827	0.17029234
Rwanda	24670	1640	250	1390	0.15243902	0.003	1 349	0.97040178	3108	1.89512195	287.195122	0.09761304
Senegal	192530	75440	52	75388	0.00068929	0.007	70 274	0.93218434	2853	0.03781813	0.02120891	0.11457133
Sudan	2376000	429760	16000	413760	0.03723008	0.011	370 434	0.89528831	3108	0.00723194	6.91316083	0.00336422
Tanzania	886040	335550	1930	333620	0.00575175	0.012	295 680	0.88627693	11701	0.03487111	5.28386232	0.00336422
Uganda	199550	63460	690	62770	0.01087299	0.01	56 768	0.90438208	6935	0.10928144	126.946108	0.0008465
Zaire	2267600	1132750	51510	1081240	0.04547341	0.006	1 018 089	0.94159435	12604	0.0111269	1.61112337	0.02957106
Zambia	743390	323010	71200	251810	0.22042661	0.011	225 443	0.89528831	5797	0.01794681	8.64679112	0.04095994
Zimbabwe	386670	88970	2831	86159	0.03181971	0.007	80 296	0.93216434	5631	0.063291	41.2386198	0.00455225

[a]WRI World Resources Institute, *World Resources 1993–1994: Guide to the Global Environment*, Oxford University Press, 1994
[b]FAO, *op cit*, Ref 38
[c]WCMC, *op cit*, Ref 27
Notes:
1. Protected forest areas for Angola, Botswana, Burkina Faso, Chad, Malawi, Mali, Namibia, Niger and Sudan from sources cited in Sharma, *op cit*, Ref 52
2. Benin has certifiably zero protected forest area (and is covered 100% in the WCMC Biodiversity Map Library). Protected forest area is arbitrarily set to a small amount to avoid invalidating index calculation. Similarly, Guinea Bissau shows no recorded IUCN protected areas or forest reserves. Protected forest area is set to an arbitrarily small amount
3. For some countries there is no evidence of the overlap between forest and existing protected areas (IUCN categories). Ethiopia for example has 62 230 km² area protected but none is recorded as protected forest. We arbitrarily assume 10% is protected forest
4. Where species richness figures are not indicated for a taxonomic group but endemism figures are available, these have been included in total species richness instead. This should understate total species richness

data (an over-inflated success ratio),[48] the index is consistent in selecting two countries in the top five in all three versions, while both threat measures recommend the same top five countries in terms of biodiversity cost-effectiveness, albeit in a different order. Only the use of the incomplete WCMC cost data causes the index selection to include three other countries (Korea, Malaysia and Bhutan). This would seem to indicate that a decrease in forests is linked to an increase in population, as the rates are trending together within countries. Care should be taken in making such generalizations, but it is probably fair to say that both deforestation and population growth rate are acceptable surrogates for the threat of land conversion.

The rankings based on different costs cannot, strictly speaking, be compared. The WCMC data are not complete for all countries and the index is therefore not calculated for these. The WCMC cost data is also for a longer period – the costs have been annuitized and discounted for the period 1991 to 1993. Nevertheless, the analysis does show that rankings will differ when global cost data are used for the index instead of just US data.

Although the good result for Pakistan should be treated with caution, it can nevertheless be stated with a certain degree of confidence that

[48]It is worth noting, however, that Pakistan would still achieve the highest ranking even if its success ratio were only 0.2

Pakistan has a very good protection record and is therefore worthy of investment. Bangladesh achieves a high index figure essentially because of a high species richness indicator (second highest of all countries listed), and a relatively low cost figure (ie low level of investment). This result is interesting because it occurs despite severe deforestation. Bangladesh is almost completely deforested with less than 5% of the original forest cover remaining.[49] Looking at methods to ground truth index results for specific cases therefore becomes important. How accurate is the species richness indicator? Has it occurred because of sampling bias (ie collection in favoured areas), or is there genuinely a rich biodiversity resource? Are most of these species occurring in mangrove swamps or forests? As there is very little forest area left to work with in Bangladesh it may be ideal for investment if high biodiversity is concentrated in forests.

China receives a high ranking which is consistent with the Sisk *et al* and ICBP rankings. China's high index is attributable (mainly) to the low level of funding it receives. Of interest, however, is the fact that India is only ranked seventh in our index whereas it receives high priority in some of the other studies. Apart from cost (investment) India and China have very similar ratios except that China is more threatened. If China and India had the same cost (investment) then India would achieve the higher score, ie ranking. This is not the case however and therefore China does well because of cost.

Additional information pertaining to Vietnam is also useful in interpreting its index result. The remaining forests in Vietnam are very isolated, meaning that corridors are very important for biodiversity conservation. Investment might therefore focus on corridor provision.

Despite being quite expensive (high level of funding) and having medium species richness, Sri Lanka does well because of a high success ratio and low level of threat. It is worth noting, however, that population density is extremely high and projected to increase to 500 people per km^2 by the year 2125. This might constitute an important additional consideration in an investment decision.

Worthy of mention is that Indonesia and Malaysia, which traditionally score highly in most ranking procedures, do not do particularly well in this one (they are ranked 10th and 11th respectively). This would appear to be due to their relatively low levels of species richness – Indonesia however may be particularly disadvantaged here because of normalization. This issue requires further investigation.

Central and South America

Regional cost-effective priorities are El Salvador, Venezuela, Surinam, Panama and Cuba. The rankings for El Salvador, Surinam and Panama are derived from alternative cost figures[50] which are restricted to US investments and may therefore be understating true costs. Several megadiverse nations appear well funded. Interestingly, none of these countries are highlighted by the review of priority methods provided by Sisk *et al.*[51]

Sub-Saharan Africa

Regional priorities for sub-Saharan Africa coincide with some of the continent's poorest countries. In order these are: Niger, Mali, Liberia, Nigeria, Rwanda, Angola, Mozambique. Several data problems are

[49]M Beazley, *The Last Rainforests*, Mitchell, Beazley publishers, London, in association with IUCN, 1990

[50]Abramovitz, *op cit*, Ref 30

[51]Sisk *et al*, *op cit*, Ref 2, review the regional priority countries of 4 methods: (1) Regional hot-spots (Myers 1988, 1990 *op cit*, Ref 17); (2) Megadiversity countries (Mittermeier 1988 *op cit*, Ref 17); (3) Range-restricted birds (ICBP 1992, *op cit*, Ref 14); (4) their own areas of critical global or continental concern

noteworthy: cost figures for Mali, Niger and Liberia are again from Abramovitz. Protected forest area coverage for Mozambique is zero according to WCMC Biodiversity Map Library or 3750 km² according to source tables in Sharma.[52] Information on species richness and the extent of protected forest areas for several countries are derived from a number of sources (see Table 6). Of the seven countries listed, only Nigeria and Angola are considered as globally critical by Sisk *et al.*

Sensitivity analysis and cost revisited

From Table 1 it is evident that the index is working, although we now need to ascertain whether the index is well behaved in assigning ranks. Let us first of all consider an example of the index calculation for Bangladesh, before undertaking the sensitivity analysis.

From Table 1, for Bangladesh:

$$Ps = \underset{\text{(success ratio)}}{0.08} \times \underset{\text{(threat ratio)}}{0.67} = 0.0536$$

$$\text{CEPII} = \frac{0.0536 \times 0.73}{5.49} \times 100 = 0.710$$

The threat ratio signifies the percentage of unprotected forest remaining in Bangladesh in ten-years time. The higher this ratio, therefore, the lower the threat. The opposite is true for the success ratio – the higher the percentage of forest currently protected, the higher the probability of success. Other scenarios might be envisaged:

(1) An increase in threat. As threat increases, the index gets lower. Suppose the threat ratio for Bangladesh decreases from 0.67 to 0.50. *Ceteris paribus*, this results in an index of 0.53, which is lower than 0.710. This result would seem to bear out the proposition presented earlier in this paper and by others (see Faith and Walker 1995[53]) that it should not necessarily be a priority to invest where threat is highest. Threat is attenuated by probability of success and biodiversity, however. If both are very high for a particular country it may still prove cost-effective to invest.

(2) An increase in the probability of success. This leads to a higher index, *ceteris paribus*. As we would like the index to show that an investment should be more cost-effective if the probability of success is high, this is the desired result. The same is also true of biodiversity – the higher the species richness, the higher the index, *ceteris paribus*.

(3) A lower level of investment. Lower cost (investment) will lead to an increase in the index, *ceteris paribus*. This will therefore advantage those countries where conserving biodiversity is relatively 'cheap'. The biodiversity in place is there despite the threat and the low funding. Assuming that investment does reflect cost, we can choose however to interpret low investment (apparently cheap) countries differently. Low funding may for example signify that such countries are too expensive. Similarly the index result for highly funded countries could just as well be interpreted to signal cheap investments as showing that a country is too 'expensive', or alternatively that current high investment levels already occurring put the country beyond a critical saturation level after which costs begin to rise.

[52] N Sharma (ed), *Managing the World's Forests: Looking for balance between conservation and development*, Kendall/Hunt, 1992

[53] Faith and Walker, *op cit*, Ref 16

Care is needed in interpreting our cost data but the countries which have not received much funding to date – however the denominator is interpreted – do well in the index (relative to the other variables). Furthermore, whichever interpretation we choose, the data provide only a static picture and it should be noted that continued funding commitment is deemed to be extremely important in biodiversity conservation.[54] Once a country has received funding and built up infrastructure, it must continue to receive assistance – projects cannot be simply abandoned. These considerations should be kept in mind in any interpretation of the index rankings over time.

The foregoing discussion shows how cost data introduces potential ambiguity as well as being the differentiating factor in the priority rankings of Tables 1–4. Reliable cost data are likely to be difficult to come by in a consistent manner and even where investment cost information is available, a comprehensive treatment of costs should ideally also consider the role of *opportunity costs* in the threat element of the index.

Opportunity costs represent the returns from the next best alternative land use. These returns will vary over countries and regionally within countries, but are powerful indicators of imminent economic threat. The main competition to biodiversity conservation comes from the returns to agricultural production and rents from timber and mineral extraction. In most countries good geographically referenced data is generally available for all of these sectors and it is clear how the threat index may be influenced by the combination of distortionary/perverse domestic sectoral support policies and world market trends. Note that the use of opportunity costs introduces the complex task of achieving some correspondence between threatened forest areas and reliable data on returns.

Limitations of the index

In the face of accelerating habitat loss, the requirement for increased precision in biodiversity measurement and priority setting could well be a case of the best being the enemy of the good. Numerous other elements might legitimately be included.

Area or park size and fragmentation, which are two extremely important variables affecting successful biodiversity conservation, have not been incorporated. It has been shown[55] that the larger the protected area (park) and the more intact (no fragmentation), the higher the biodiversity (number of species). These factors are difficult to integrate into the index – ideally bigger parks should received a positive weighting. We have not attempted this here due to data limitations.

The index provides a snapshot, the only intertemporal assumption being that related to constant rates of deforestation. Funding availability, costs and domestic and international policy environments are all subject to unforeseen change which will not be reflected on the data for any particular ranking. Again, there is a premium on simplicity, but it is as well to consider some of these dynamics.

The index is mainly applicable to country data, as these are more readily available to compare biodiversity investment proposals between countries on the basis of cost-effectiveness. Regional specific data (ie within country) are not currently available for the indicators needed for the index. It will therefore be difficult to use the index to compare biodiversity projects within a particular country until such data are known. If national success, threat and biodiversity indicators were used

[54]Wilson, *op cit*, Ref 35
[55]*Ibid*

in the index for projects in the same country, no differential would be registered for these indicators; the resulting comparison would be based purely on the cost.

Conclusions

Although there is no single correct method for setting biodiversity conservation priorities there is scope for developing a harmonized methodology. Adequate priorities cannot be set with reference to scientific analysis alone. Social and economic factors – as the driving forces behind habitat loss – must be taken into account and the development of a unifying paradigm is clearly an area for cross-disciplinary research. The CEPII emphasizes cost-effectiveness which is consistent with maximizing conservation with a limited budget. In explicitly addressing cost, the index confronts the single most limiting factor and the one most notably absent from previous priority exercises. In addition to cost and biodiversity [benefit], the novel additions are the combined consideration of threat and the probability of success in index form. The methodology is no more than a suggestion of how the scope of priority assessment may be broadened to provide a more effective filter for national investments. While there are alternative methods for representing similar information the index approach offers distinct advantages of flexibility and convenience.

We caution against the overemphasis of the results presented here. Index rankings are relative and we suggest no absolute values to distinguish the performance of any one country relative to another. Furthermore, priorities can be set at global, national and local levels but data availability only allows us to assess at the level of analysis which happens to be furthest from where biodiversity actually resides. The resolution of this analysis provides an incomplete picture of the dynamics controlling biodiversity loss. The issue of whose priorities we are actually setting matters and global priorities can only be accurately interpreted accompanied by some understanding of local priorities, demands and unmet needs. Again, data requirements are a limiting source.

Appendix

Derivation of the Index

In order to calculate the index data are required for:

A: Total country area in km^2
Af: Forested area in km^2
Ap: Protected forest area in km^2
Au: Unprotected forest in km^2 (calculated as *Af* minus *Ap*)

This will provide sufficient data to calculate the probability of success. We now need,

k: either the deforestation rate or the population growth rate
B: biodiversity measure, species richness or endemism
C: cost given by world investment

The threat ratio, species and cost per km^2 of forested area, and consequently the index, can now be calculated.

Let the probability of a successful intervention be:

$$Ps = \frac{Ap_t}{A_t} * \frac{Au_{t+n}}{Au_t}$$

Where,

A_t = total protectable area at time t
Ap_t = area protected at time t
Au_{t+n} = unprotected (protectable) area in year $t+n$, and,
$0 < Ps < 1$. Now,

$$Au_{t+n} = Au_t(1-k)^n$$

where k represents the rate of growth of a selected threat, given by the rate of land conversion (estimated using either the deforestation rate or the population growth rate).

The term on the right-hand side measures existing (and past) commitment. The second term attenuates the first by the degree of threat.

The final index is:

$$E = \frac{Ps * \Delta B}{C}$$

$$E = \frac{\frac{Ap_t}{A_t} * (1-k)^n * \Delta B}{C}$$

where B = biodiversity measure and C = cost measure as discussed in the text.

This formula defines our index: the COST EFFECTIVE PRIORITY INVESTMENT INDEX (CEPII)

Table A1. Species richness

Country	Higher plants	Mammals	Birds	Reptiles	Amphibians	Total
Bangladesh	5000	109	354	119	19	5601
Bhutan	5468	109	448	19	24	6068
Cambodia	n/a	117	305	82	28	532
China	32200	394	1100	282	190	34166
Fiji	1628	4	87	25	2	1746
India	16000	317	969	389	206	17881
Indonesia	22500	515	1519	511	270	25315
Korea, Rep	2898	49	0	18	13	2978
Lao PDR	8286[a]	173	481	66	37	9043
Malaysia	12500	264	501	268	158	13691
Myanmar	7000	300	867	203	75	8445
Nepal	6973	167	629	80	36	7885
Pakistan	4938	151	476	143	17	5725
Papua NG	11544	242	578	249	183	12796
Philippines	8931	166	395	193	63	1748
Solomon	3172	47	163	57	15	3454
Sri Lanka	3214	86	221	144	39	3705
Thailand	12625	251	616	298	107	13897
Tonga	463	1	39	6	0	509
Vanuatu	1000	12	84	22	0	1118
Vietnam	8000[a]	273	638	180	80	9171
W Samoa	693	3	44	8	0	748

[a]All figures are from WCMC *op cit*, Ref 27 except for the LAO PDR and Vietnam bird figures which are from Dinerstein and Wikramanayake, *op cit*, Ref 2
n/a Data not available

Table A2. Forest areas and threat rates

Country	Total country area (square km)[a]	Natural forest cover (square km) (1990)[b]	Deforestation rate (%) (1981–90)	Population growth rate[e] (%) (1995–2000)
Bangladesh	144 000	7 690[c]	3.9[c]	2.6
Bhutan	47 000	28 090[c]	0.6[c]	2.3
Cambodia	181 000	121 630[c]	1.0[c]	1.8
China	9 561 000	1 150 600[d]	3.9[d]	1.2
Fiji	18 000	8 100[d]	0.1[d]	1.4
India	3 288 000	517 290[c]	0.6[c]	1.9
Indonesia	1 905 000	1 095 490[c]	1.0[c]	1.6
Korea, Rep	99 000	8 000[d]	0.0[d]	0.8
Lao PDR	237 000	131 730[c]	0.9[c]	2.6
Malaysia	330 000	175 830[c]	2.0[c]	1.9
Myanmar	677 000	288 560[c]	1.3[c]	2.0
Nepal	141 000	50 230[c]	1.0[c]	2.3
Pakistan	796 000	18 550[c]	3.4[c]	2.8
Papua NG	463 000	360 000[c]	0.3[c]	2.2
Philippines	300 000	78 310[c]	3.3[c]	2.1
Solomon	27 000	24 400[d]	0.0[d]	2.8
Sri Lanka	66 000	17 460[c]	1.4[c]	1.1
Thailand	513 000	127 350[c]	3.3[c]	1.3
Tonga	700	< 1 000[d]	n/a	1.4
Vanuatu	12 000	2 400[d]	1.7[d]	3.2
Vietnam	330 000	83 120[c]	1.5[c]	2.0
W Samoa	3 000	1 400[d]	n/a	1.0

[a]Braatz *et al*, *op cit*, Ref 31, Table A1 p 48 (using data from World Bank 1990 and WCMC 1992)
[b]FAO (1993) defines forest cover as including tropical rain forest, moist deciduous forest, dry deciduous forest, very dry forest, desert zone, and hill and montane zone
[c]FAO (1993) for the year 1990, Table 4b
[d]FAO (1988) as cited in Braatz *et al*, *op cit*, Ref 31
[e]WRI (1992) except for Tonga and Vanuatu, which are from Braatz *et al*, *op cit*, Ref 31, Table A1 p 48
n/a Data not available

Table A3. Protected areas

Country	Total protected area (square km)[a]	Protected wetlands (square km)[b]	Protected forest area (square km)[c]
Bangladesh	968	355	613
Bhutan	9 061	5	9 056
Cambodia[d]	0	20	0
China	283 578	20 000	263 578
Fiji	53	n/a	53
India	137 701	15 300	122 401
Indonesia	192 309	29 000	163 309
Korea, Rep	7 568	58	7 510
Lao PDR	0	0	0
Malaysia	14 880	64	14 816
Myanmar	1 733	40	1 693
Nepal	11 260	261	10 999
Pakistan	36 550	1 380	35 170
Papua NG	290	6 000	290
Philippines	5 729	761	4 968
Solomon	0	n/a	0
Sri Lanka	7 837	766	7 071
Thailand	55 140	410	54 730
Tonga	0	n/a	0
Vanuatu	0	n/a	0
Vietnam	8 975	495	8 480
W Samoa	0	n/a	0

[a]WCMC (1992) as cited in Braatz *et al*, *op cit*, Ref 31. Total protected areas for IUCN Categories I–V, Table A2 p 49
[b]Scott and Poole (1989) as cited in Braatz *et al*, *op cit*, Ref 31 (1992)
[c]The figure for protected forest areas is obtained by subtracting protected wetland areas from total protected areas
[d]There are inconsistencies in these data, highlighting the need for more accurate figures
n/a Data not available

ELSEVIER

Ecological Economics 22 (1997) 73–83

ECOLOGICAL ECONOMICS

ANALYSIS

Biodiversity, resilience and the control of ecological-economic systems: the case of fire-driven rangelands

Charles Perrings [a,*], Brian Walker [b]

[a] *University of York, York, UK*

[b] *Division of Wildlife and Ecology, CSIRO, Canberra, Australia*

Received 20 December 1992; accepted 31 December 1996

Abstract

The loss of resilience in systems characterised by multiple equilibria is indicated by a discontinuous change in the state of the system, or the transition from one locally stable state corresponding to a particular mix of species to another state corresponding to a different mix of species. The resilience of the system (the state of the range in this example) may be influenced by the management regime through its impact on the response to disturbances. The paper shows that optimal management of an event-driven system should be sensitive to the opportunities created by such events. © 1997 Elsevier Science B.V.

1. Introduction

A recent review of the problem of range conditions in Australia estimated that some 42% of arid and semi-arid lands were either severely or very severely degraded (Ludwig and Tongway, 1995a). Nor is the position any better elsewhere. In all cases range degradation is indicated by a change in the state of vegetation. This includes, for example, changes in the density and distribution of shrubs and perennial grasses in semi-arid grasslands, savannas and woodlands, together with changes in the patchiness of the landscape. Such changes represent discontinuous and sometimes irreversible switches between the multiple equilibria ('multiple stable states') of the systems. Because the characteristics and dynamics of such systems differ between states, some states may be preferred to others. Indeed, the problem of desertification derives from the fact that desertified states are economically less productive than — and so less preferred to — non-desertified states. Both shrub encroachment and the loss of perennial grasses have direct effects on the profitability

* Corresponding author.

PII S0921-8009(97)00565-X

of pastoral production, while the reduction of patchiness reduces the capacity of the system to use limited water and nutrients (Ludwig and Tongway, 1995b). This makes the transition between states an economically interesting problem.

The various states of a system can be thought of as basins of attraction around the system equilibria. The boundaries between basins mark the system thresholds. The transition between states may occur naturally as a consequence of climatic shift, for example. But it may also occur as a consequence of the management regime. The management regime can affect where a system is within the orbit of some attractor: i.e. where it is with respect to some equilibrium. It can also affect the nature of the equilibrium towards which it is tending. By determining where the system is with respect to a particular equilibrium, the management regime can influence its capacity to withstand the shocks that may cause it to flip from one state to another. As the capacity of a system to withstand shocks without change of state defines its resilience (sensu Holling, 1973), the management regime can influence the resilience of the system in any given state. For both reasons the state of the system is part of the choice set for decision-makers.

This paper considers the problem of the degradation of semi-arid rangelands as a problem in the management of complex non-linear systems. Its starting point is the cumulative evidence that grazing pressure in rangelands may lead to discontinuous and sometimes irreversible change in the state of the range (Noy-Meir, 1975; Noy-Meir and Seligman, 1979; Walker, 1988; Westoby et al., 1989; Friedel, 1991; Laycock, 1991). We are, however, concerned with the way in which grazing pressure interacts with other important effects. More particularly, we are concerned with the interaction between grazing pressure and fire. Research in Australia has shown that changes in the state of rangelands has usually been caused by grazing in conjunction with climate, fire and other factors (Hodgkinson and Harrington, 1991; Hodgkinson, 1991; 1992; Ludwig and Tongway, 1995b). This matches research results on woodland communities in East Africa which show that the interactions between climate, browsing, grazing and fire can have very similar effects (cf Dublin et al., 1990). In both cases a change in grazing pressure changes the probability of fire through its effects on fuel load and the establishment of woody plants. In this paper we consider the implications of this fact for the choice of stocking levels in fire-driven systems.

The paper is in five sections. Section two reviews the ecological links between climate, grazing and fire in semi-arid rangelands. Section three proposes an ecological-economic model that endogenises discontinuous change between states of the range. Section four considers the implications of the approach for ecosystem management. A final section offers some concluding remarks.

2. Fire, grazing pressure and multiple rangeland states

The dynamics of semi-arid rangelands share a number of common features

- the sensitivity of the system dynamics to 'rare' and extreme events including fire, flood and drought;
- the importance of the interaction between grazing pressure and extreme events in the transition between states; and sometimes
- the existence of multiple locally-stable states;

The dominant approach to these phenomena in range ecology is the 'state and transition' model due to Westoby et al. (1989). This assumes that there exists a number of rangeland states, each characterised by a different composition of species, and each associated with different properties at equilibrium. So, for example, a rangeland with viable seeds of woody plants in the soil would be defined to be in a different state to that same rangeland without seeds because of the different properties of their equilibria. The transition between states is thought to be a function of exogenous events, although the probability of transition for some given event varies with the way in which the system is managed. More particularly, whether a particular combination of events does or does not induce a change in the state of the system is held to depend on the forcing effect of herbivores, including livestock. Hence the 'model' may be summarised by a list of potential states and the transitions between them, together with the conditions that induce transition from one state to another.

The state and transition model is not, however, easy to formalise in a way that is tractable. Our approach is accordingly somewhat different, although we are concerned with the same phenomena. As in the state and transition model we acknowledge that the determinants of natural vegetation, exogenous natural shocks, and the forcing effects of grazing pressure are all strongly interactive. There is, for example, a strong correlation between soil nutrients, soil moisture and soil texture; soil texture determining the degree to which both soil moisture and soil nutrients are available to plants. Sandy soils have a low water holding capacity but a high proportion of soil moisture is available to plants. Heavy clays are the opposite. Plants adapted to sandy soils have a high ratio of underground to aboveground biomass. Plants adapted to clay soils are the opposite. One implication of this is that grazing or burning affects a smaller proportion of the biomass of plants on sandy than on clay soils. But equally, the sandier the soil the greater is the competitive advantage of deep-rooted woody plants over shallow-rooted grasses (Knoop and Walker, 1985).

The productivity of rangelands depends on the balance between palatable and unpalatable grasses available to herbivores, and between woody plants and grasses. The relation between these is strongly influenced by the patch structure of rangelands (cf Pickup, 1991; Tongway and Ludwig, 1989). Walker (1988) has conjectured that patchiness has two main implications, one positive and one negative. On the positive side, the high-production patches may provide higher quality food longer into the dry season than would occur in their absence. On the negative side, patchiness can be associated with a change in the composition of species in favour of woody plants, which reduces the economic productivity of the range. What is important here is that the process of species change is episodic and event-driven, rather than successional. In the absence of extreme events, net primary production varies in response to variation in rainfall, but it turns out that species composition remains essentially the same. Both the establishment and mortality of plants are functions of very particular conditions, which generally only occur in association with 'rare' events. So, for example, the establishment of cohorts of the perennial *Astrebla* grasses on the Mitchell grass plains of Australia is associated with particular phases of the El Niño Southern Oscillation, when good rains in late Autumn follow a year in which poor summer rains result in reduced vigour and abundance of the main competitor, *Danthonia* (Austin and Williams, 1988).

One implication of this—developed in the range ecology literature—is that the management of rangelands should be event-driven. It should be 'opportunistic'. For example, a combination of fuel load and weather conditions may provide an opportunity to change the state of vegetation by use of fire (Hodgkinson and Harrington, 1991). But the other side of the 'opportunistic' coin is that the probability that fire will regulate the species mix of the system depends on the effects of grazing pressure on fuel load. The level of grazing pressure affects fuel load through its impact on the distribution, abundance and physiological status of plants. In terms of climatic fluctuation, for example, grazing pressure directly influences the relations between topography, rainfall and erosion by increasing patchiness and so exaggerating both erosion and deposition (Stafford-Smith and Pickup, 1990). This polarises the effects of extreme climatic events. In source areas, it requires ever larger events to trigger a change. In sinks, on the other hand, irreversible episodic changes are prompted by ever smaller events (Walker, 1988).

It follows that an increase in grazing pressure implies a reduction in the proportion of palatable grasses and an increase in both unpalatable grasses and woody plants. There are various feedback effects involved. The feedback effect of special interest here is the effect of grazing pressure on fire. In many savanna rangelands, woody plants are controlled by a combination of periodic fire and the capacity of the grass layer to suppress woody seedlings (Knoop and Walker, 1985). In such cases an increase in grazing pressure both raises the probability that woody seedlings will successfully establish, and reduces the effectiveness of fire as a control. Once established, the growth of woody plants generates positive feedback effects. Reduced grass cover means that more water passes through to the sub-soil, taking nutrients with it and enhancing the competitive ability of plants which have more of their roots in deeper layers. In addition, reduced fuel load reduces the incidence of fire and hence the main control on the establishment of woody plants. The net effect is that the rangeland may be transformed from one state (characterised by one mix of species, and one level of productivity) to another state (characterised by another mix of species, and another level of productivity).

The impact of management decisions on fire varies depending on the institutional and other conditions governing the use of rangelands. In Australia, for example, fire has been suppressed by pastoralists as a direct goal of management (Ludwig and Tongway, 1995b). In Africa, fire has been suppressed as an incidental effect of grazing pressure. Higher stocking densities have reduced the build-up of fuel load, and so have inhibited fire. In both cases, this has reduced the role of fire in the regulation of shrub growth. The important point is that periodic fire is essential to the resilience of systems of this sort, and that disruption of the fire regime induces a loss of resilience, and a shift to a state with lowered grazing potential.

3. Optimal grazing in fire-driven rangelands

Following May (1977) considerable attention has been paid to modelling the dynamics of ecosystems with multiple equilibria or states. In the state and transition approach transition from one rangeland state to another is argued to be triggered by external shocks. But the management regime is thought to alter the sensitivity of the rangeland to such shocks. The limits of the local stability of any equilibrium define threshold values for livestock, vegetation, water flows and so on. If any of these variables are driven beyond such threshold values by the management regime, the system itself will switch from one equilibrium or state to another.

In semi-arid rangelands, the main shock to the system—fire—is not exogenous. Although fires may ignite spontaneously the probability of ignition is a function of fuel load, and this in turn is a function of management. We can capture this feature of the problem in a simple model. As in Perrings and Walker (1995) the optimal allocation of resources maximises some measure of welfare through choice of stocking density, constrained by the dynamics of density-dependent herd and vegetation growth. In this model we take account of both woody plant growth (and the competition between woody plants and grasses) and the effect of fire. The stocks comprise livestock, palatable grasses and woody plants. These are the state variables of the problem.

The evolution of these state variables is described below. All three are measured in terms of biomass, and their underlying growth functions are assumed to be logistic.

$$x_{t+1} - x_t = \alpha_t x_t(1 - \gamma x_t/g_t) - u_t \tag{1}$$

$$g_{t+1} - g_t = \beta_t g_t(1 - c_{gg}g_t/g_{\max} - c_{wg}w_t/w_{\max}) - \gamma_t x_t - b(g_t)g_t \tag{2}$$

$$w_{t+1} - w_t = \phi_t w_t(1 - c_{ww}w_t/w_{\max} - c_{gw}g_t/g_{\max}) - f(g_t)w_t \tag{3}$$

where x_t = livestock biomass at time t; g_t = grass biomass at time t; w_t = woody plant biomass at time t; c_{ij} = competition coefficients between grasses and woody plants; α_t = the growth rate of the herd at time t; u_t = herd offtake at time t; β_t = the rate of regeneration of grasses at time t; γ_t = the rate of depletion of grasses due to grazing at time t; $b(g_t)$ = the rate of depletion of grasses due to fire at time t; θ_t = the rate of regeneration of woody plants at time t; $f(g_t)$ = the rate of depletion of woody plants due to fire at time t.

The effects of fire are accordingly captured in the terms $b(g_t)g_t$ and $f(g_t)w_t$. Since the assumptions made about the unconstrained growth of the stocks are standard, it is these terms that distinguish the model. Indeed, the dynamics of the system depend critically on the form of the functions $b(.)$ and $f(.)$. To illustrate, consider the following very simple case. Suppose that there are no herbivores ($x = 0$) and that c_{gg} and c_{wg} have the values one and zero respectively. Suppose, too, that $b(g_t)$ is an exponentially increasing function of grass. That is, the proportion of grass destroyed in the event of fire increases exponentially with the stock of grass biomass (the fuel load). As the time interval, t, approaches zero, grass dynamics will then be as in Fig. 1.

In this simple case the impact of fire is shown by the difference $\beta_t g_t(1 - g_t/g_{\max}) - b(g_t)g_t$.

It will, first, reduce grass biomass at the equilibrium of the system (from $g_{\max}$ to $g_{\max}(b)$). This will also cause contraction of the stability domain of the system (measured by the distance from the origin to $g_{\max}(b)$. Other specifications of $b(g_t)$ will have different effects on the dynamics of the resource g_t. For example, any form of the function $b(g_t)$ which intersects the function $\beta_t g_t(1 - g_t/g_{\max})$ more than once will not only reduce the stability domain of the system: it will also fragment it. That is, the system may become characterised by multiple locally stable states (see Perrings and Dalmazzone, 1997).

Time is treated discretely to reflect the seasonality of rainfall and biomass production. In the normal case, the growth coefficients, α_t, β_t and θ_t may be expected to vary with rainfall (and so time). Fire, too, will have stochastic elements. For simplicity, the parameters are assumed to be constant in the example described in the next section, and it is assumed that fire will always occur if the fuel load exceeds a well-defined flashpoint. This is solely to make the underlying structure of the system's dynamics transparent. The economic problem is to maximise some index of welfare

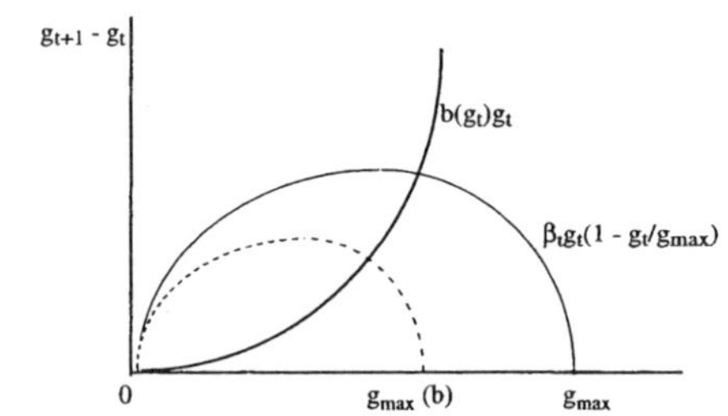

Fig. 1. Grass dynamics where destruction by fire is an exponentially increasing function of grass biomass.

from the use of x_t, g_t, w_t, through choice of the level of offtake, u_t, and subject to Eq. (1) Eq. (2) Eq. (3). For convenience, we assume that the appropriate index of welfare is a measure of net benefit or profit. Assuming a time horizon of T periods, the problem is accordingly to:

$$\max_{u_t} \sum_{t=0}^{T} \rho^t(p_t u_t - c_t x_t - r_t g_t) \tag{5}$$

where ρ = a discount factor; $\rho = [1/(1+\delta)]$ where δ is the discount rate; p_t = the market price of offtake; c_t = the net cost of maintenance of the herd; r_t = the access charge or price of the range.

All prices are per unit of biomass. This is subject to Eq. (1) Eq. (2) Eq. (3) and with

$$x_0, g_0, w_0 \quad \text{given.} \tag{6}$$

While $r_t g_t$, the value of the grazing resource, is shown as a cost in Eq. (5) similar to a grazing fee or land rent, it may also enter the objective function as a benefit. Since the state of the range may be expected to determine the market value of rangeland, an improvement in the quality of grass cover may be expected to offer a potential capital gain. Woody plants are assumed to be zero valued. In many cases woody plants will yield private benefits in the form of firewood or fencing material, and social benefits in the form of habitat, watershed protection and the like. Similarly, although the maintenance of the herd will always involve costs, in many cases these are partially or wholly offset by the benefits derived from draft power, non-meat products, status, insurance against adverse climatic conditions, and use as 'privileged currency' in social transactions (Perrings, 1991).

The first order necessary conditions for optimisation of this problem include Eq. (1), Eq. (2), Eq. (3) and Eq. (6), the maximum condition

$$0 = p_t - \rho\lambda_{t+1} \tag{7}$$

and the following conditions on the evolution of the shadow prices of livestock, grass and wood — respectively λ_t, ξ_t and μ_t:

$$\rho\lambda_{t+1} - \lambda_t = v_t - \rho\lambda_{t+1}[\alpha_t(1 - 2x_t/g_t) + m] + \rho\zeta_{t+1}\gamma_t \tag{8}$$

$$\rho\zeta_{t+1} - \zeta_t = r_t - \rho\lambda_{t+1}\alpha_t\gamma_t x_t^2/g_t^2 - \rho\zeta_{t+1}[\beta_t(1 - c_{gg}2g_t/g_{\max}) - (b_g g_t + b(.))] + \rho\mu_{t+1}(\phi_t c_{gw} w_t/g_{\max} + f_g w_t) \quad (9)$$

$$\rho\mu_{t+1} - \mu_t = -\rho\mu_{t+1}[\phi_t(1 - c_{ww}2w_t/w_{\max}) + f(.)] + \rho\zeta_{t+1}(\beta_t c_{wg} g_t/w_{\max}) \quad (10)$$

Since the Hamiltonian for this problem is linear in the control variable, we would expect a most rapid approach to the optimal path for the state variables. The decision-process is as follows. Given observations on the system at time t, the decision-maker calculates the optimal grazing pressure (which is just the optimal ratio between the state variables x_t and g_t) and adjusts the current grazing pressure to the optimal level through choice of offtake, u_t. We assume that the decision maker is capable of making the adjustment in one shot, although in most actual cases their capacity to make the adjustment may be constrained. Offtake may be either positive or negative: i.e. it may sometimes be optimal not just to sell livestock, but to restock the range. The optimal grazing pressure is obtained from the solution to the first order conditions. Specifically, denoting grazing pressure by

$$\psi_t \equiv x_t/g_t$$

the optimal grazing pressure at time t, ψ_t^* is a root of:

$$0 = \psi_t^2\alpha_t\gamma_t - \psi_t^2\alpha_t A_t + [A_t(\alpha_t + \delta - c_t/p_t)]\gamma_t^{-1} - r_t/p_t \quad (11)$$

where

$$A_t = \delta + [\beta_t(1 - c_{gg}2g_t/g_{\max}) - (b_g g_t + b(.))] - [(\beta_t c_{wg} g_t/w_{\max})(\phi_t c_{gw} w_t/g_{\max} + f_g w_t)] [\delta + \phi(1 - c_{ww}2w_t/w_{\max}) + f(.)]^{-1}$$

The solution requires, as one would expect, that the marginal benefit of the current level of grazing pressure equals the marginal cost of that level of grazing pressure, given both the set of relative prices that comprise the economic environment and the rate of discount. The optimal offtake is that level of offtake which adjusts the grazing pressure to its optimal value given current availability of graze and the current herd size.

A reduction in the availability of graze accordingly has two immediate effects. First, by tightening the density-dependent constraint on the rate of growth it induces an automatic reduction in the natural growth of the herd. Second, by increasing the grazing pressure associated with a herd of given size, it induces an increase in the optimal rate of offtake. Any fire event consequently has the same immediate effects. However, by simultaneously reducing the woody biomass on the range, it has a lagged but positive effect on the growth of grass. Since woody biomass is not, by assumption, reduced by livestock, fire is the only control on wood. Hence the optimal grazing pressure is such that there is sufficient build up in the fuel load to allow the system to burn at intervals. That is to say, the optimal grazing pressure will be lower in a system in which woody biomass competes with grass, than in a system involving grass alone. Of course, the outcome would be different if there were alternative means for controlling woody biomass, but this is not the case in most rangelands. To see the optimal management of a system with these characteristics, the next section reports a simple example.

4. Grazing strategies in fire-driven systems: an example

The resilience of fire-regulated systems depends on a process of 'creative destruction' in the sense of Holling (1973). There are numerous other examples of ecological systems whose resilience requires periodic renewal. Many economic systems have similar characteristics. The phenomenon of the business cycle (and longer 'waves') are examples of processes of renewal involving the 'creative destruction' of produced capital. In this case, the dynamics of the range depend on both the values of the competition, growth and depletion parameters and the form of the fire functions. To be able to say anything further we need to be precise about these functions. In general, the appropriate form of the fire function is an empirical question. In this section we assume that combustion depends only on fuel load, and that the fuel load consists of grass biomass only. It therefore increases with the rate of grass growth, and decreases with the level of grazing pressure. We assume that combustion occurs with probability one once the fuel load has reached a certain level. We further assume that any fire event removes a fixed proportion of both woody biomass and grass. The combustion point accordingly represents a discontinuity in the dynamics of both grass and woody biomass.

The fire functions for grass and woody biomass are assumed to have the following form:

$$b(g_t) = \sigma g_t[\mathrm{abs}((g_t - \gamma_t x_t)/g_f - 1) + ((g_t - \gamma_t x_t)/g_f - 1)] / 2[(g_t - \gamma_t x_t)/g_f - 1] \quad (12)$$

$$f(g_t) = \bar{\omega} w_t[\mathrm{abs}((g_t - \gamma_t x_t)/g_f - 1) + ((g_t - \gamma_t x_t)/g_f - 1)] / 2[(g_t - \gamma_t x_t)/g_f - 1] \quad (13)$$

where: g_f = the flashpoint (fuel load at which combustion occurs with probability one); σ = the proportion of grass biomass destroyed by fire; ω = the proportion of woody biomass destroyed by fire.

In both cases the switching term, $[\mathrm{abs}((g_t - \gamma_t x_t)/g_f - 1) + ((g_t - \gamma_t x_t)/g_f - 1)] / 2[(g_t - \gamma_t x_t)/g_f - 1]$, takes values of 0 or 1 depending on the magnitude of the current fuel load relative to the flashpoint. While this form of fire function is crude, it is sufficient to illustrate the role of grazing strategies in fire-driven systems. The greater the size of the herd and the greater the consumption requirements of each herbivore, the slower the build up of the fuel load, and so the less frequent fires will be. We emphasise that our concern in this paper is not to model the dynamics of a particular ecological system, but to explore the implications of interactions between grazing and fire for the management of a general class of ecological systems. Hence the parameters of the model have not been estimated. This is the subject of further work.

To get a sense of the impact of the management regime, it is useful first to consider the dynamics of the system in the absence of controlled offtake. To do this we suppose that the herd is subject to a constant rate of mortality:

$$u_t = \kappa x_t$$

where κ denotes the proportion of the herd that dies each period. It may be readily generalised to deal with problems of density-dependent mortality from external sources:

$$u_t = \kappa(x_t) x_t$$

but we do not consider this case here. To simplify further, we ignore the effect of rainfall on the value of ecological and other parameters of the model. Stochastic rainfall of the type that dominates in many of the semi-arid regions of Australia, for example, may be expected to show up in the model parameters. Given our concern with the impact of fire events, however, we abstract from the stochasticity of rainfall here. The initial and maximum values of the state variables, and the parameter values in this example are as follows:

$x_0 = 10$	$c_{gw} = 0.1$	$c_t = 0.2$
$g_0 = 100$	$\alpha_t = 0.3$	$r_t = 0.05$
$\omega_0 = 100$	$\beta_t = 3$	$g_f = 1.5\ g_{max}$
$g_{max} = 200$	$\gamma_t = 1$	$\sigma = 0{,}5$
$w_{max} = 200$	$\phi_t = 0.2$	$\omega = 0.4$
$c_{gg} = 1$	$\delta = 0.05$	$\kappa = 0.1,\ 0.2$
$c_{wg} = 0.5$	$p_t = 1$	
$c_{ww} = 1$		

We are interested on the effect of grazing on the fire regime and the composition of grass and woody biomass. This turns out to be highly sensitive to the herd mortality rate selected: the lower the herd mortality rate the lower the frequency of fire events. The reason for this is simply that lower herd mortality implies higher levels of grass consumption and so slows the build-up of fuel load. Fig. 2a,b shows the results of simulations over a thirty-year period assuming herd mortality rates of 10 and 20% respectively.

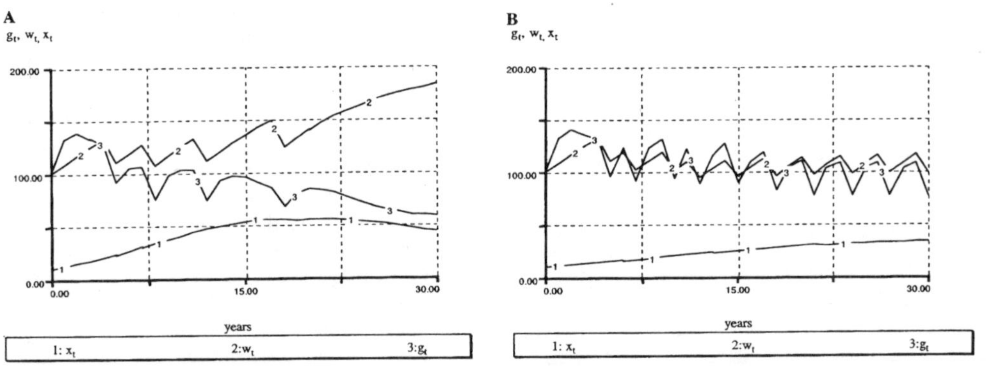

Fig. 2. (a) Grass, wood and herd dynamics in uncontrolled systems: herd mortality at 10%. (b) Grass, wood and herd dynamics in uncontrolled systems: herd mortality at 20%.

In the first case, because herd mortality is low relative to the natural rate of increase, the herd expands beyond the maximum sustainable size. Although the herd subsequently contracts as forage is depleted, the consumption of grass effectively squeezes fire out of the system. In typical rangelands this encourages woody seedlings to become established, further inhibiting the growth of grass. In the second case, the herd size converges to a sustainable level, and the system involves a mix of woody and grassy biomass regulated by fire events that occur every 4–5 years. It follows that a halving of the rate of herd mortality in this system is sufficient to cause the system to flip from a persistent state characterised by the both grass and woody plants, to one dominated by woody plants in which there is little room for livestock.

Now consider the optimal management of the same system. Recall that the decision process requires that the decision-maker is able to observe the current state of the range and to calculate the optimal grazing pressure associated with that state, where the optimal state is that which yields the highest net benefit under our various assumptions. The optimal level of offtake brings the current and the optimal level of grazing pressure into line with one another. It is worth repeating that we have assumed — for purposes of this exercise — that woody biomass has zero value in the system. Hence we might expect that the optimal strategy will support grass production at the expense of wood. It does. Furthermore, since income increases with offtake, we might also expect the optimal strategy to result in higher levels of offtake, and hence to increase the frequency of fire. It does.

Fig. 3 shows the time paths for the state and control variables in this problem — grasses/woody plant/livestock and offtake respectively. Offtake varies with the availability of grass, and this in turn depends on the fire regime and competition from woody plants. Eventually, the system settles into a pattern of fire-driven fluctuations in

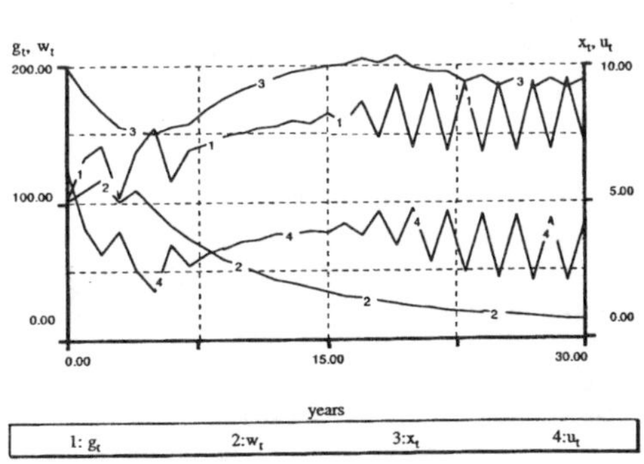

Fig. 3. Grass, wood and herd dynamics under an optimal offtake strategy.

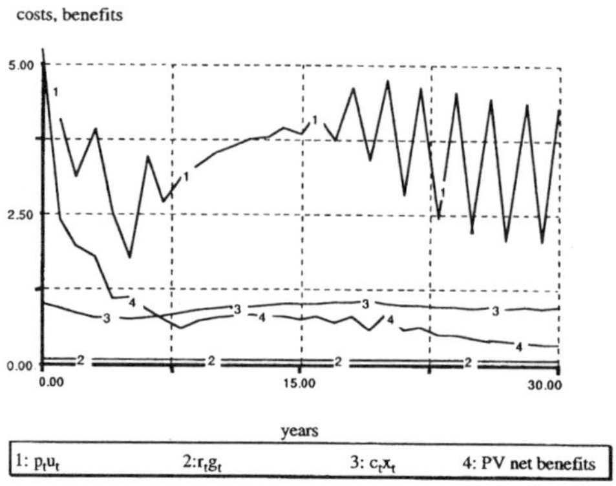

Fig. 4. Net benefits of range management under an optimal offtake strategy costs, benefits.

offtake, but the interim trajectory includes periods in which herd size is first reduced and then smoothly increased as the ratio between grass and woody biomass adjusts to the optimal level. The inclusion of stochastic effects does not, it turns out, change this underlying pattern.

Since the optimal level of offtake aligns current and optimal grazing strategies it responds to the loss of grass cover associated with fires. That is, herd size and offtake are dominated by the same cycle as are grasses and woody plants. This yields a flow of net benefits that has the same frequency (Fig. 4). Since grass cover is assumed to offer positive benefits to the decision-maker, aside from the provision of forage, there is an incentive to ensure that grass cover is threatened neither by the level of grazing pressure nor by the establishment of woody plants.

The outcome is sensitive to the choice of parameter values, and we have already made the point that they have not been estimated for any particular system. To illustrate, consider the effect of variation in two parameters: one ecological, the other economic. First, we take the effect of a change in the relative demands of livestock on the grazing resource. In this model the sensitivity of the resource to grazing pressure might be reflected by a change in any or all of the ecological parameters.

We choose to consider the consumption of grass per livestock unit — the parameter γ. The first effect of an increase in γ has already been discussed. As γ rises the frequency of fire decreases because of a reduction in the rate at which the fuel load builds up. Other effects have also been alluded to, including an increase in the rate of establishment of woody plants, an increase in the patchiness of the range with an attendant increase in run-off surfaces, etc. A doubling of γ with no change in any of the other parameters is indicated in Fig. 5. The dynamics of the system are still dominated by fire, but the frequency of these events falls, while the ratio of woody to grassy biomass increases. The initial adjustment in the balance between the state variables disappears, and the system moves very quickly into a pattern of sharp fluctuations in offtake. To illustrate the effect of a change in socio-economic or institutional conditions, consider the implications of an additional change in the cost of herd maintenance, c_i (see Fig. 6). The cost of herd maintenance is sensitive to a wide range of socio-economic conditions including security of land tenure, income and social and cultural factors.

The most noticeable effect is an increase in the optimal stock of grass, and a decrease in the optimal stock of wood and the size of the herd. The increase in grass is reflected in the frequency and severity of fires, and hence in the fluctuations in offtake.

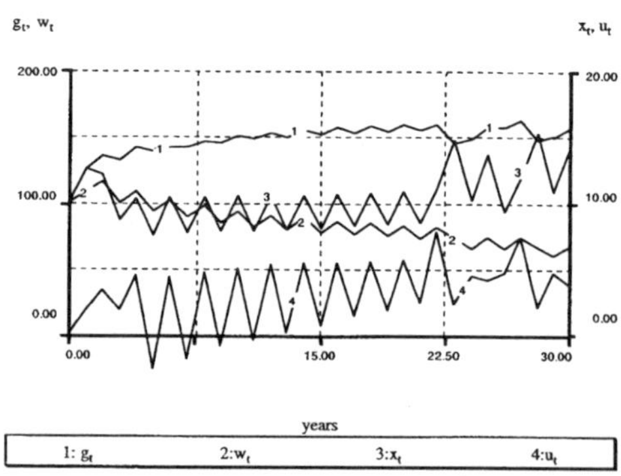

Fig. 5. Optimal offtake with heavy grazing pressure.

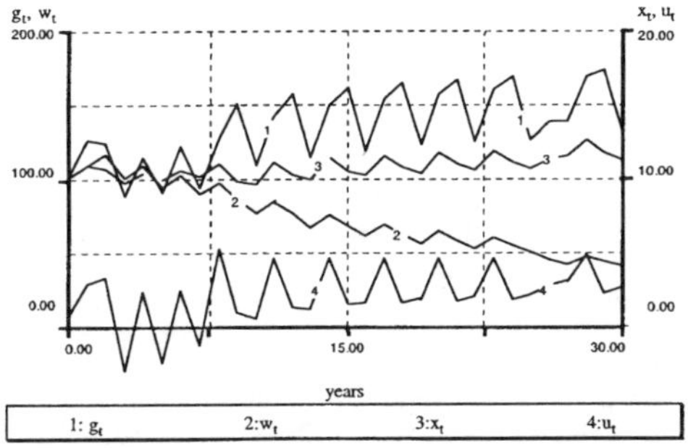

Fig. 6. Optimal offtake with heavy grazing pressure and high herd maintenance costs.

5. Discussion

We are concerned with the optimal management of systems whose dynamics reflect the sequential interaction between four system functions: exploitation, conservation, creative destruction and reorganisation (Holling et al., 1995). The first two functions represent, respectively, the colonisation of disturbed ecosystems, and the accumulation of energy and material in maturing systems. Creative destruction represents some abrupt change which releases both energy and material that have accumulated during the conservation phase. Reorganisation makes that released material and energy available for the next exploitive phase, either by reproducing an earlier pattern or by establishing a new pattern of development.

In particular, we are concerned with the class of systems in which fire is the vehicle of the creative destruction phase. Fire turns out to be particularly important in semi-arid rangelands where it not only releases nutrients for the development of grasses, but regulates the establishment of woody plants. The intensity and frequency of fire also determines which pattern of development the system will follow in the next phase. Rangelands are, as we have seen, characterised by multiple stable states. Whether a rangeland in one state persists or undergoes a more or less rapid transition to some other state depends on the intensity and frequency of fire.

Since grazing strategies affect both fire intensity and frequency, it is reasonable to consider how fire ought to be accommodated in the optimal control of rangelands. The empirical evidence suggests that grazing strategies more often than not fail to take account of the role of fire in regulating the establishment of woody plants. At this point we are concerned with an essentially methodological problem. While the range ecology literature has promoted the state and transition approach to range development, it is not immediately obvious what implications it has for modelling the choice of grazing strategies. The problem we have considered in this paper derives from the fact that fire allows for the discontinuous and persistent (if not irreversible) shift of the system between states characterised by a different mix of species. We consider whether the impact of fire on rangelands, and the choice of grazing strategy in the presence of fire, may be accommodated in models of optimal choice that have been developed for more 'well-behaved' cases. We have not, however, considered any particular rangeland, and the parameter set and simulations reported in the last section are designed merely to illustrate the outcome.

The optimal strategy in this case generates time paths for the state and control variables that are very different from the steady-state outcomes familiar from well-behaved systems satisfying the necessary and sufficient conditions for optimal control. The optimal levels of livestock, grasses and woody plants do not converge on any steady state values except where one or more of them is deleted from the system. Similarly, the optimal level of offtake varies with conditions elsewhere in the system — but may, in some cases, converge on a limit cycle. More importantly, both optimal and non-optimal paths may involve a change in the state of the system.

The model so far assumes a constant annual rainfall. In the discussion of Fig. 3 we observed that inclusion of stochastic effects (reflecting normal rainfall variation) does not change the underlying dynamics of the system. However, we recognise that changes in rangeland state are often brought about by the extremes of rainfall variation, as in the example of the Mitchell grass plains. Including such effects in the model will require explicit representation of the reaction of the state variables to extreme events.

Finally, while the focus of this paper and the empirical research to follow is fire-driven rangelands, it seems clear that they belong to a much wider class of systems in which resilience relies on a process of renewal involving creative destruction and reorganisation. All such systems share a number of characteristics. Hence their management requires a similar approach. At this stage we have not sought to develop a clear set of propositions relevant to the class as a whole, but that is a part of the research agenda to which the paper contributes.

References

Austin, M.P. and Williams, O., 1988. Influence of climate and community composition on the population demography of pasture species in semi-arid Australia. Vegetation, 77: 43–49.

Dublin, H.T., Sinclair, A.R.E. and McGlade, J., 1990. Elephants and fire as causes of multiple stable states in the Serengeti-Mara woodlands. J. Anim. Ecol., 59: 1147–1164.

Friedel, M.H., 1991. Range condition assessment and the concept of thresholds: a viewpoint. J. Range Management, 44: 422–426.

Hodgkinson, K.C. and Harrington, G.N., 1991. The case for prescribed burning to control shrubs in eastern semi-arid woodlands. Australian Rangelands J., 7: 64–74.

Hodgkinson, K.C., 1991. Identification of critical thresholds for opportunistic management of rangeland vegetation, Proceedings of the IVth Congres International des Terres de Parcours, Montpelier, France.

Hodgkinson, K.C., 1992. Elements of grazing strategies for perennial grass management in rangelands. Desertified Grasslands: Their Biology and Management, Linnean Society, London. 78–94.

Holling, C.S., 1973. Resilience and stability of ecological systems. Ann. Rev. Ecol. Syst., 4: 1–23.

Holling, C.S., Schindler, D.W., Walker, B.W. and Roughgarden, J., 1995 Biodiversity in the functioning of ecosystems: an ecological primer and synthesis. In Perrings C.A., K.-G. Mäler, C. Folke, C.S. Holling and B.-O. Jansson (eds.), Biodiversity Loss: Ecological and Economic Issues, Cambridge University Press, Cambridge: 44–83.

Knoop, W.T. and Walker, B.H., 1985. Interactions of woody and herbaceous vegetation in a Southern African savanna. J. Ecol., 73: 235–253.

Laycock, W.A., 1991. Stable states and thresholds of range condition on North American rangelands: a viewpoint. J. Range Management, 44: 427–433.

Ludwig, J.A. and Tongway, D.J., 1995a. Desertification in Australia: an eye to grass roots and landscapes. Environ. Monit. Assess., 37: 231–237.

Ludwig, J.A. and Tongway, D.J., 1995b. Spatial organisation of landscapes and its function in semi-arid woodlands, Australia. Landsc. Ecol. 10(1): 51–63.

May, R.M., 1977. Thresholds and breakpoints in ecosystems with multiple stable states. Nature, 269: 471–477.

Noy-Meir, I., 1975. Stability of grazing systems: an application of predator prey graphs. J. Ecol., 63: 459–481.

Noy-Meir, I. and Seligman, N.G., 1979. Management of semi-arid ecosystems in Israel. In: B.H. Walker, (Editor) Management of Semi-arid Ecosystems, pp. 113–160, Elsevier, Amsterdam.

Perrings, C. and Walker, B.H., 1995. Biodiversity loss and the economics of discontinuous change in semi-arid rangelands. In Perrings, C., Mäler, K.-G., Folke, C., Holling, C.S., Jansson, B.-O. (eds) Biological Diversity: Economic and Ecological Issues, Cambridge University Press: 190–210.

Perrings, C. and Dalmazzone, S., 1997. Resilience and stability in ecological economic systems. Working Papers in Environmental Economics and Environmental Management, EEEM, University of York.

Pickup, G., 1991. Event frequency and landscape stability on the floodplain systems of arid central Australia. Quart. Sci. Revs., 10: 463–473.

Stafford-Smith, D.M. and Pickup, G., 1990. Pattern and production in arid lands. Proc. Ecol. Soc. Austr., 16: 195–200.

Tongway, D.J. and Ludwig, J.A., 1989. Vegetation and soil patterning in semi-arid mulga lands of Eastern Australia. Austr. J. Ecol., 15: 23–34.

Walker, B.H., 1988. Autecology, synecology, climate and livestock as agents of rangelands dynamics. Austr. Range J., 10: 69–75.

Westoby, M., Walker, B. and Noy-Meir, I., 1989. Opportunistic management for rangelands not at equilibrium. J. Range Management, 42, 4: 266–274.

[29]

Biodiversity and Conservation 2, 168–181 (1993)

How useful are environmental safety standards in economics? – The example of safe minimum standards for protection of species

ANDREAS HOHL[1]* and CLEMENT A. TISDELL[2]‡
[1]*Institut für Landwirtschaftliche Betriebslehre, University of Hohenheim, Postfach 700562, W-7000 Stuttgart 70, Germany*
[2]*Department of Economics, The University of Queensland, Brisbane 4072, Australia*

Received 10 September 1992; accepted 7 October 1992

The use of environmental safety standards is being increasingly advocated, and economists are being increasingly required to adapt their policy models to take account of such standards, e.g. by the use of transferable environmental damage quotas. The use of these standards may be a consequence of a greater desire to achieve more sustainable development and preserve the natural wealth available to future generations. An important means to this goal is to preserve biodiversity and genetic variability. Safe minimum standards for conservation of wildlife have been suggested as a way to achieve this. Sometimes safe minimum standards for conservation of species are defined in terms of their minimum viable populations and their minimum habitat requirements. However, as a review of recent scientific literature indicates, these safe minimum standards are very uncertain.

Environmental safety standards have been integrated into economics in a variety of ways, e.g. via constraints to ensure sustainability at the macro-level, via minimum safety standards necessary to ensure survival of species. Despite the commendable support which safe minimum standards appear to give for nature conservation, they involve conceptual problems and uncertainties as outlined in this paper. In fact, safe minimum standards for the conservation of species do not exist – no standard ensures the continued existence of any species. This is not to deny that an improvement in standards can up to a point increase the probability of survival of a species for a specified period of time. It implies that minimum standards need to be cast in terms of those required to achieve a particular probability of survival of the targeted species for a specified period. It may, however, be impossible to achieve the goal specified, e.g. the probability of survival of a species aimed for, may be greater than can be achieved for any attainable standard. In general, the paper makes it clear that standards suggested by natural scientists and others need to be subjected to careful scrutiny before they can be used as a reliable basis for economic policy.

Keywords: environmental safety standards; conservation of species; probability of survival; safe minimum standards; game theory; social choice

Introduction

Since the early 1970s economists have taken a broader interest in environmental issues than previously. This change was largely due to natural scientists like Forrester (1971) and Meadows *et al.* (1972), who for the first time after the era of Malthus (1798) and

*To whom correspondence should be addressed.
‡Department of Economics, University of Queensland, St Lucia, QLD 4072, Australia.

Ricardo (1817) argued that there were 'limits to growth'. Fortunately not all of the early pessimistic scenarios, designed by the 'Club of Rome' and others, came into reality. However, undoubtedly there has always been a widespread view that there are natural limits to human development.

Economists have reacted in different ways to the new challenge (Tisdell, 1990a). The neoclassical mainstreamers argued that the price mechanism will act in favour of conservation and that human innovativeness is likely to overcome all difficulties. Therefore, they did not see any reason to change the growth orientated economic system. Others were less optimistic. Neo-Malthusians opted for a steady state economy (Daly, 1980) or even called for a reduction of the human population to a level sustainable by using renewable resources alone (Georgescu-Roegen, 1976).

The decisive question raised by these debates is still unanswered: how large can the human population become and how long can it be sustained with the given available set of resources? If the variables population size, resource stocks and resource-use intensity could be determined, we could predict the fate of mankind.

The objectives of this paper are not quite as ambitious. However, much of the incentive for defining environmental safety standards can be interpreted as attempts to contain an apparently uncontrolled and self-destroying human race.

Over the last 20 years it became obvious that environmental problems had been neglected and environmental costs had been externalized. Economists had to extend their theoretical horizon and consider a fourth production factor: the environment. In principle the English classical economists might however, have included this under land but in practice they did not emphasize environmental aspects of the economic problem. Today writers such as Pearce *et al.* (1989) do treat the environment as if it is an extra factor of production. Whereas the factors land, labour and capital are relatively homogeneous and arguably easy to quantify, the environment can only be taken into account in form of highly complex constraints. The heterogeneity and complexity of environmental influences demands an interdisciplinary approach. Today the findings of the natural or medical sciences have direct influences on economic modelling, no matter whether this relies on extended cost–benefit analysis or on cost efficiency analysis (a fixed standard approach).

Political decisions like where and how to construct an industrial plant that were formerly based almost entirely on economic considerations are influenced by a whole range of expert opinions from those expressed by geologists to those of toxicologists. Farmers, car drivers or garden owners have to observe ambient environmental quality standards for water, soil and air. Environmental safety standards define the framework for almost all enterprises. However, in many, if not in most cases, the environmental interdependencies are so complex that the scientific base of the safety standards is still incomplete or unsatisfactory. How should economists deal with this situation when political decision-makers and the public are demanding clear answers where there often are not more than crude rules of thumb?

After surveying different categories of environmental safety standards used in economics, this paper presents some of the special problems involved in defining and using safe minimum standards. It considers how precise these standards, as recommended by non-economic sciences, are and how well they can be integrated into economic theory. Finally, possible ways of dealing with political demands in relation to standards are discussed.

The new multiplicity of goals

Human welfare and inter-generational equity

Looking at classical definitions of economics (Samuelson, 1981), there is one inherent goal in all of them: the maximization of human welfare. Growing environmental problems did not lead to the abandonment of this goal but deprived it of its exclusivity. Subgoals or – depending on the philosophical view of the world – equally important goals emerged at its side.

The strong consensus, that human welfare should be maximized, has always been complemented by an equally strong dissent on how earthly wealth, however defined, should be won and distributed. One can see history as an endless chain of distributional conflicts, wiping out entire populations and peoples. Aspects of the distribution of wealth have been debated for many centuries and some have gained new importance with issues like poverty-caused environmental degradation being considered. However, there is a relatively new aspect of distribution: the distribution of wealth over time. Until recently, the time vector was not considered important. Current generations acted along similar lines to individuals guided by Adam Smith's 'invisible hand': what was in their interests was supposed to be good for future generations as well. As resources are known to be finite, this cannot be taken for granted any more. Many people feel a moral obligation to guarantee inter-generational equity. This change is very well characterized by the European Green Movement's slogan: 'We do not own the earth, we only have leased it from our children'.

Inter-generational equity can only be achieved if all development is sustainable or, in other words, if the services and qualities of the natural resource stock are maintained over time (Pearce and Turner, 1989). This implies the acceptance of two basic safety rules: renewable resources should not be exploited at a rate faster than natural regeneration occurs, and exhaustible resources should not be depleted at a rate faster than human inventiveness provides alternative technologies with substitute materials.

This is easily said but difficult to realize. Even if we assume an economically, socially and politically perfectly rational world with absolute control over resource utilisation, there are still unanswered questions. How fast does natural regeneration occur? What are the long-term effects of human resource exploitation on ecosystems? Decisive variables like growth and recruitment rates are governed by complex biological and ecological relationships: extinction thresholds, sigmoid growth curves, carrying capacities, and the like (Randall 1987). Production cycles can span several human generations. Despite ample research there is no available sustained management system for tropical rainforests (Fearnside, 1986; Keto et al., 1990; Keto and Scott, 1991). The management of the Queensland rainforests was not long ago presented as a model of sustainability to the world. The sustainable achievable yearly cut was estimated at 220 000 m^3, then 160 000 m^3 in the 1970s, in the 1980s it was revised down to 80 000 m^3 and later to 60 000 m^3, until the Australian Government decided to ban commercial logging in the rainforest altogether (Stocker *et al.*, 1977, Keto and Scott, 1991).

Prevention of pollution and preservation of the assimilative capacity of the environment.

Human welfare was in the past closely linked to the quantity of (material) goods available. This still holds true in many of the poor countries of the world. In the developed world, where all basic needs are satisfied, the growing side-effects of development have led to demands to replace quantity (of goods) by quality (of life).

One of the most obvious measures to improve the quality of life is to stop or reduce the release of toxic, pathogenic or otherwise deleterious substances into the environment. Catastrophes like Minamata, Bhopal or Chernobyl showed only the peak of the iceberg. Long-term developments like the lingering pollution of entire ecosystems may produce even more serious consequences. If the assimilative capacity of the environment is exceeded, even the output of such presumed harmless substances like carbon dioxide or methane may lead to global disaster.

Threshold values for substances considered deleterious to human health are among the best-known environmental safety standards. However, even the medical data, that appear so well founded, have their shortcomings. Grossly simplified, the traditional way to get a threshold value for a suspectedly harmful substance is to treat laboratory

Table 1. Options for pursuing genetic diversity in wild species

ALIVE
In situ
1. Single use wildland management
– Scientific reserves, strict nature reserves
2. Multiple use wildland management
– National parks
– Natural monuments
– Managed natural resources
– Protected landscapes
– Resource reserves
– Anthropological reserves
– Multiple use management areas
3. Areas other than wildlands
– Converted land managed for greater diversity
Ex situ
1. Whole organism
– Botanical gardens
– Zoos
– Culture collections (microorganisms)
2. Organism parts
– Sperm, egg and embryo banks
– Organ banks
– Seed and pollen banks
– Gene banks
– Tissue culture
DEAD
– Museums
– Herbaria

animals, normally rats, with growing doses of it, until 50% of the animals die. This is called the 'lethal dose 50 (LD_{50})'. The LD_{50}-value for the animals is multiplied with an estimated safety factor to get the 'acceptable daily intake (ADI) for humans'. The procedure to convert animal standards into human standards is today sufficiently refined to produce safe threshold values with a very high probability for the tested substance. However, no substance can be produced 100% pure. As there is still considerable danger coming from impurities, the effects of which may not be detected during the tests. The same accounts for reciprocal influences between various agents. Our environment is overloaded with myriads of potentially interactive substances, whose combined effects are largely unknown. Therefore, over time the recommended safety standards may prove not to be safe at all. Additionally, it has to be considered, that the probability of inadequate standards rises when the scientific threshold values have to be adapted to real world situations. The classical process of threshold value 'dilution' involves four steps: the scientifically recommended standard turns into the technically feasible, which is transformed into the economically bearable, which finally ends up as the politically enforceable.

Preservation of biodiversity and genetic variability within species

One of the greatest concerns in recent decades has been the irreversible loss of species and genetic varieties. The genetic base for biological production is narrowing. This may affect agriculture, forestry and the possiblities for medical advances. Not only the maintenance of existing production is at risk but the scope for developing new types of biological production and commodities based on living organisms is being reduced (Tisdell, 1990b). There are various possibilities to slow down genetic erosion (Goodland and Ledec, 1988). These are summarised in Table 1.

In order to preserve a species the decisive questions are:

- How many individuals are enough?
- How much habitat is enough?

The respective safety standards are the minimum viable population size and the minimum habitat size.

Defining minimum viable populations and minimum habitat sizes

According to Soulé (1987) the simple question 'How much is enough?' might be the quintessential issue in population biology. What are the minimum conditions to ensure a species' or population's long-term persistence and adaptation with a high degree of security in a given place?

When conservation efforts are made there are four key variables to be looked at: the persistence time, the degree of security, the population size and the habitat (or reserve) size.

It is impossible to define persistence time and the degree of security scientifically (Shaffer, 1987). Persistence cannot be forever, because extinction is a natural and inevitable process. Depending on the ethical point of view persistence time is assumed in population models ranging from the time span necessary to enable the next human generation to reconsider the species' future to evolutionary time dimensions. Assumptions on the degree of security require a value judgement too. The question 'How secure should the persistence of a given species be?' can only be answered by society.

Shaffer (1981, 1987) has shown that in order to estimate minimum viable population sizes and their respective area requirements the effects of various chance events on population persistence have to be considered:

(a) genetic uncertainty or random changes in genetic make-up due to the founder effect, genetic drift, or inbreeding, which alter the survival and reproductive probabilities of individuals;
(b) demographic uncertainty resulting from random events in the survival and reproduction of individuals;
(c) environmental uncertainty due to random, or at least unpredictable, changes in climate, food supply, and the populations of competitors, predators, parasites or pathogenic agents;
(d) natural catastrophes such as floods, fires, droughts or volcanic eruptions, which may occur at random intervals.

The genetic variability within a population or a species is a measure of its chances to persist in time. Mutation and immigration improve the genetic variability, whereas inbreeding, genetic drift and founder effects affect it adversely. The main obstacle to the short term survival of a population is inbreeding, caused by combination of deleterious genes (Gottschalk, 1978). The most important process endangering medium term survival is genetic drift, a change in gene frequencies caused by random recombination in meiosis (Crow and Kimura, 1970). A combination of both processes causes bottleneck or founder effects. They occur when a population is temporarily reduced in number or when a new population is founded by only a few individuals (Harris, 1984). To secure long term survival the genetic variability has to be big enough to enable the species to adapt to environmental changes, take part in the evolutionary process and develop new life forms.

Human knowledge of evolution is far too restricted to allow for complete genetic modelling. However, geneticists can provide rules of thumb for different time spans of persistence. As only the number of genetically active individuals in a population is important, the concept of effective population size is used (Harris, 1984). The number of non-reproductive individuals is subtracted from the actual population size and corrections such as those for unequal sex ratios are made. Considering genetic chance processes the effective population size to secure short-time survival should be at least 50 (Frankel and Soulé, 1981). For medium term survival it is estimated at 500 (Franklin, 1980), which will translate in an actual population size several times this number. To estimate minimum populations sizes for long-term persistence the effects of other random events have to be taken into account.

Schaffer (1987) suggested functional relationships for demographic, environmental and catastrophic uncertainty of the type shown in Fig. 1. Several important conclusions emerge from a perusal of these idealized functional forms. Average persistence times increase geometrically with increasing population sizes due to the effects of demographic uncertainty (Goodman, 1987). Therefore, demographic uncertainty will only pose an important hazard to relatively small populations (in the order of tens of hundreds). More troublesome is the relationship of average persistence times under the effects of environmental uncertainty. Persistence times increase linearly with increases in population size, which means that there is no critical population size that, once reached, ensures a

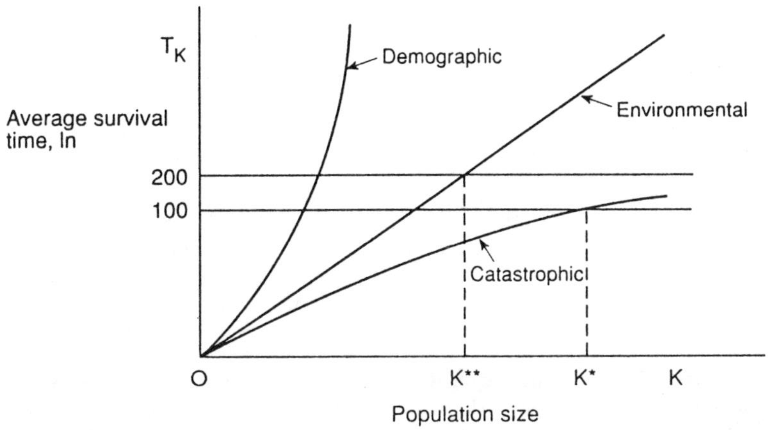

Figure 1. Functional forms of the relationship of the expected time to extinction, or average persistence time (T_k). To population size (K) for three classes of uncertainty. (Source: Shaffer, 1987)

high level of long term security. The most distressing function is the curve for catastrophic uncertainty. Here persistence times are believed to increase only with the natural logarithm of population size. Progressively increasing sizes of populations yield diminishing dividends in terms of increased persistence times. This curve is the theoretical confirmation of the inevitability of extinction (Shaffer, 1987). There is no curve for the relationship of persistence times to population size under the effects of genetic uncertainty because no clear relationship emerges from a consideration of genetic factors (Lande and Barrowclough, 1987). However, if environmental and catastrophic uncertainty effects are considered, genetic uncertainty will not always, perhaps not often, set the lower limit to acceptable population size.

Taking into account all the uncertainty effects it becomes obvious that there is no simple determination of minimum viable populations. Each case is different and requires a special viability analysis. For species with special social requirements, see e.g. Darling (1952). Therefore, Soulé (1987) only very reluctantly gives a rule of thumb: assuming 95% expectation of persistence, without loss of fitness, for several centuries, the minimum viable population of a vertebrate should be in the low thousands.

To calculate the minimum area requirements, the home ranges of the various species have to be known. A home range can be as little as 0.049 ha for the white footed vole or as much as 377 000 ha for the grizzly bear (Harris, 1984). The definition of home ranges is very time consuming, for this reason there are only data for a very limited number of species. It is assumed here that multiplication of the numbers for minimum viable populations and home ranges give the minimum area requirements. Edge and insularity effects, that are very important for the design of nature reserves (Harris, 1984; Soulé and Simberloff, 1986), are neglected. If the large predators, which can be considered as key species for entire biological communities, are looked at, the minimum area requirements are in the range of 1 000 000–100 000 000 ha (Shaffer, 1987; Soulé, 1987).

The safety standards suggested by scientists are highly inaccurate. The given numbers span a range of two or three orders of magnitude. The simple reason for this is the complexity of the biological world (Soulé, 1987).

Integrating environmental safety standards into economic theory

In traditional economics the state of the environment does not influence the calculation of economic optima. The Pareto optimality of allocation is independent of ecological states. Only since the beginning of the 1980s, with the publication of the World Conservation Strategy (IUCN, 1980) as a starting point, have sustainability aspects been considered. In order to achieve sustainability Pearce and Turner (1989) argue for an 'existence theorem': a guarantee that any economic optimum is associated with a stable ecological equilibrium.

There have been two alternative ways to approach this goal. Some analysts used an extended cost–benefit framework, utilising monetary valuations but also incorporating explicit recognition of uncertainty and irreversibilities. Others argued that there are principal limits to valuation and the assignment of probabilities. They adopted a fixed standard approach where macroenvironmental safety standards operate as binding constraints. A well known alternative concept of binding constraints is the safe minimum standard (SMS) that was first introduced by Ciriacy-Wantrup (1952) and later formalized by Bishop (1978). The SMS is a decision rule for problems involving irreversibility, very long time horizons and massive uncertainty like the extinction of plants and animals.

SMS has not been adequately tied to a theoretical model of social choice. Bishop's attempt to use game theory first seemed to prove successful. He used the Wald criteria (Wald, 1950) to minimise maximum losses (minimax) when making decisions under uncertainty. The minimax rule is an extremely pessimistic decision rule and is, therefore, biased towards conservation. The SMS approach makes preservation the preferred option unless it can be demonstrated that the social cost of preservation, i.e. the forgone development benefits, is unacceptably large (Pearce and Turner, 1989). In the game only two possible social states are assumed: in state 1, nothing unanticipated occurs in the future to create large social losses if the species ceases to exist (0); in state 2 the species turns out to be worth a large amount (Y) as, for example, a cure for cancer. The present value of net benefits from development is assumed to be x. The common base point is the intersection of 'Extinction' and 'State 1'. This leads to the loss matrix shown in Table 2.

Table 2. Bishop's original loss matrix

	State		
Strategy	1	2	Max loss
SMS	x	$x-Y$	x
Extinction	0	Y	Y

(Source: based on Bishop (1978), order of strategies exchanged)

Table 3. Revised loss matrix for the insurance game

	State		
Strategy	1	2	Max loss
SMS	0	0	0
Development	$L-B_d$	$-B_d$	$L-B_d$

(Source: based on Bishop and Ready (1990))

The original model contained a mistake. Bishop double counted the benefits from species preservation in state 2 of his loss matrix (it should be 'x' instead of '$x-Y$'). This has been criticized by Tisdell (1990b) and others and revised by Bishop himself (Bishop and Ready,1991). However, the double counting did not alter the original decision, as the revised matrix in Table 3 shows.

The situation in this so-called insurance game (because it insures against large future losses) is principally the same as in the first matrix. 'L' stands for 'large future losses', 'B_d' for 'development benefits'. Under the assumption that $L > B_d$, SMS is again chosen as the optimal strategy.

Bishop and Ready (1991) detected that only a slight change in the assumptions would reverse the conclusions. In a game called the lottery game, shown in Table 4, the disease is assumed to occur with certainty, but there is uncertainty about whether preservation of the species will lead to a cure. This game captures the argument that endangered species should be saved because they may lead to important new resources. Preserving the species is analogous to buying a lottery ticket. In state 1, the species does hold the cure, in state 2, it does not. The baseline for the game is no development and no cure.

The maximum possible loss under the SMS in the lottery game is 0, while under development it is $-B_d$, which is less than zero. The minimax strategy for the lottery game therefore is to choose development, and guarantee a gain of B_d, even if the extinction of the species and the losses to society associated with the disease are taken into account.

Bishop and Ready (1991) come to the conclusion that SMS cannot be motivated as the minimax loss solution to a two person game against nature. The situation is different when the Niehans–Savage rule of regret is considered for making decisions under uncertainty (Niehans, 1948; Savage, 1951). Using the minimax regret decision rule, the conslusion is that the SMS should be adopted for either kind of game. Individuals appear

Table 4. Loss matrix for the lottery game

	State		
Strategy	1	2	Max loss
SMS	$-L$	0	0
Development	$-B_d$	$-B_d$	$-B_d$

to consider regret in situations of massive uncertainty, rather than maximizing utility (Loomes and Sudgen, 1982). However, a general application is questionable. The Niehans–Savage rule is considered by decision theoreticians to be much weaker than the Wald criteria, because it does not fulfil the so-called 'mega-criteria' for decision rules (Borcherding, 1983), namely that the order, in which different alternatives are ranked by a decision rule should not be altered, if further alternatives are added or some of the existing alternatives are eliminated. This holds true for the minimax criteria but not for the minimax regret rule. There are still many open questions, and therefore it can be said, that currently no solid theoretical foundation exists for the use of SMS.

Additional limitations of the safe minimum standard as a guide to conservation of species

The safe minimum standard criterion as originally proposed by Ciriacy-Wantrup (1952) favours the conservation of species. However, closer examination of the requirements for applying the criterion reveal several considerable difficulties. While the criterion favours the conservation of as many species 'as possible', if available resources for the conservation of species are limited, it does not appear to provide a satisfactory guide to which species should be conserved (Tisdell, 1990b) given that some will be bound to disappear as a result of the economic choices made. It is unsatisfactory as a guide to determining which species should be conserved.

There is the further problem, as the above review indicates, that there appears to be no definite ecological–biological safe minimum standard for the conservation of any species either in terms of its population-size or its supply of habitat. The future survival of no species can be assured even though the probability of persistence of any species will rise as factors favourable to its survival are increased in availability.

As the standard of factors propitious to the survival of a species is increased, the probability of its survival for a specified period of time, say T_1, rises. Possibly the relationship tends to be of the logistic form shown in Fig. 2. There may be some standard below which the species will not survive (e.g. a standard of $S \leqslant S_0$) for a period T_1 but there may be no standard at which survival is assured. The probability of survival curve $P = \mathrm{F}(S)$ is not asymptotic to $p = 1$ but to a value of $p < 1$, e.g. $p = a$.

From Fig. 2, there is no standard which ensures the survival of the species for a period, T_1, so there is no safe minimum standard. However, if one wants to have a probability of 0.5 of survival of the species for a period of T_1 then the minimum necessary environmental standard is S_1. Similarly for other probabilities of continued survival, other minimum standards can be determined.

However, this raises the question of what is the status of the probability function $p = f(S)$. In practice, the function may be uncertain or even unknown. If it is uncertain, and if there is a probability distribution of probabilities, are the probabilities in Fig. 2 to be average or expected probabilities or the most likely (modal) probabilities? How should choice be affected by changes in the characteristics of the probability distribution of possibilities, e.g. an increased variance of the distribition? These issues are as yet unresolved.

It might also be noted that a minimum standard might be achieved in a number of different ways. For example, a number of different environmental combinations (e.g. combinations of habitat and population size of a species) may yield an 0.5 probability of the survival of a species for a period of T_1 or longer. As May (1992) points out determining the relationship between habitat provision and survivability of a species is

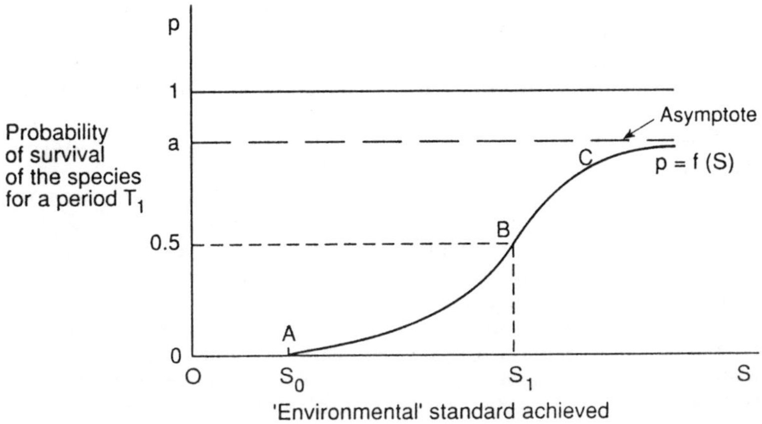

Figure 2. Relationship between (environmental) standard achieved and probability of survival of a species for a specific period of time (T_1).

extremely complicated. It does not depend merely on the size of habitat preserved. A common view expressed by ecologists is that the larger is a protected natural area the greater are the number of species likely to survive (MacArthur and Wilson, 1967). However, the relationship between the size of a protected area and the probability of survival of species is much more complicated than appears to be the case at first sight. Some of these limitations are reviewed by May (1992), e.g. complications raised by edge and patch effects.

In a world of climatic change, it also needs to be recognized that even very large protected areas may fail to save many species from extinction (see Goudie, 1990). Even large national parks and protected areas are like islands in a sea of human-dominated landscapes. Because human-dominated landscapes are an impediment to migration of many species, as climate changes, they may perish in their refuges because of their inability to migrate. The general problem has been discussed by Holdgate (1992). The drought during most of 1992 in Africa illustrates the problem. Hippopotami in the north-east of South Africa, in Kruger National Park, have been dying due to the drought. They are unable to move south. However, some have been captured and moved by humans to the south of South Africa in order to ensure their continued survival. Both natural corridors to facilitate movement of species and human assistance may become necessary for the survival of many species if global climate change occurs. Nevertheless despite such assistance, a number of species can be expected to disappear.

Conclusions

It has been shown that inaccuracies and uncertainties occur in the specification of minimum viable populations and minimum habitat sizes for survival of species. Also it was pointed out that the safe minimum standards have no sound theoretical foundation in game theory and further difficulties have been noted.

How should economists deal with the incompleteness of scientific knowledge? Should they ignore the limitations set by the environment on the grounds that much more research is needed and leave the blame to natural scientists? Or should they rather accept their part of responsibility to society and take a clear stand in the environmental debate? 'Administrators, policy makers, and managers have a right to ask for the bottom line . . . And biologists have the right and sometimes the obligation not to give an oversimplified, misleading answer to such a question' (Soulé, 1986).

A delicate balance has to be found between the obligation to consult decision makers and the obligation to disseminate only proved findings. However, decisions are necessary every day, and we have to keep in mind that in the political field non-action often only means that we approve the action of others. It may sound unsatisfactory but decisions on environmental issues still, for a considerable time to come, will have to be based on ethical considerations. As Bishop and Ready (1991) put it 'Society may choose to adopt the SMS not because it results from a rigorous model of social choice, but simply because individuals in the society feel that the SMS is "the right thing to do".' Without going any deeper into the value discussion it can be said that it does not necessarily take an ecocentric approach with sometimes quite cryptic intrinsic values to opt for the preservation of biodiversity. The anthropocentric, even the purely utilitarian approach should be enough to come to the same decision in a situation of irreversibility, very long time horizons and complete uncertainties (Tisdell, 1990b).

Scientists cannot wait for better safety values or better methods. The erosion of genetic variability, the degradation of the environment are taking place today. There are serious implications: if minimum viable population sizes of several thousands are accepted as necessary, and if long-term survival with a high degree of security is aspired to, it becomes quite obvious that in many countries there is already not enough space to preserve some of the bigger species in reserves alone. This means not only that international cooperation has to be enhanced, but also that other land uses, such as agriculture, have to be integrated into species conservation efforts. The latter should be easier in developed countries with their overproduction and their increasing demands for recreational areas. If inter-generational equity is thought to be a desirable goal, the dichotomy between developed and natural land has to be dissolved. In order to achieve sustainability we cannot restrict nature to parks, to natural monuments, and burn the earth or bury it under concrete elsewhere. Ecological interdependencies make interlinkages necessary: conservation networks must span from the most remote wildlands to the inner city parks. If the SMS, despite its theoretical shortcomings, is accepted as a model of social choice, it could alter the parties responsible for the burden of proof in the conservation versus development debate (Tisdell, 1990b). More generally, if the value of the 'precautionary principle' in relation to nature conservation is accepted whether on the basis of SMS or expected benefit, it suggests a change in the burden of proof as optimal. The developers would then have to prove in such a case that the preservation of species is the wrong choice because the forgone development benefits become unacceptably large.

More than ever policy makers ask for scientific advice and use scientific findings to achieve their goals. It is, therefore, of great political importance how environmental safety standards are defined and used.

References

Bishop, R.C. (1978) Economics of a safe minimum standard. *Amer. J. Agron. Econ.* **57**, 10–18.

Bishop, R.C. and Ready, R.C. (1991) Endangered species and the safe minimum standard. *Amer. J. Agron. Econ.* **73**, 309–12.

Borcherding, K. (1983) Entscheidungstheorie und Entscheidungs-hilfeverfahren für komplexe Entscheidungssituationen. In *Enzyklopädie per Psychologie – Methoden und Anwendungen in der Marktpsychologie* (M. Irle, ed.), Vol. 2, pp. 64–173. Göttingen: Holfgrefe.

Ciriacy-Wantrup, S.V. (1952) *Resource Conservation, Economics and Policies*. Berkeley: University of California Press.

Crow, J.F. and Kimura, M. (1970) *An Introduction to Population Genetics Theory*. New York: Harper and Row.

Daly, H. (1980) *Economics, Ecology and Ethics*. San Francisco: W.H. Freeman.

Darling, F.F. (1952) Social behaviour and survival. *Auk* **69**, 183–91.

Fearnside, P.M. (1986) *Human carrying capacity of the Brazilian rainforest*. New York: Columbia University Press.

Forrester, J.W. (1971) *World Dynamics*. Cambridge: Wright-Allen.

Frankel, O.H. and Soulé, M.E. (1981) *Conservation and Evolution*. Cambridge: Cambridge University Press.

Franklin, I.R. (1980) Evolutionary Change in small populations. In *Conservation Biology: an Evolutionary–Ecological Perspective* (M.E. Soulé and B.A. Wilcox, eds) pp. 135–50. Sunderland, MA: Sinauer Assoc.

Georgescue-Roegen, N. (1976) *Energy and Economic Myths, Institutional and Analytical Economic Essays*. New York: Pergamon Press.

Goodland R. and Ledec, G. (1988) *Wildlands, their Protection and Management in Economic Development*. Washington D.C.: The World Bank.

Gottschalk, W. (1978) *Allgemeine Genetik*. Stuttgart: Thieme.

Goudie, A. (1990) *The Human Impact on the Natural Environment*. Cambridge, MA: MIT Press.

Harris, L.D. (1984) *The Fragmented Forest – Island Biogeography Theory and the Preservation of Biotic Diversity*. Chicago: University of Chicago Press.

Holdgate, M. (1992) Protected areas and the future, The implications of change and the need for new policies. Paper presented to World Congress of National Parks and Protected Areas, Caracas, Venezuela, 10–21 February 1992.

IUCN (1980) *World Conservation Strategy*. Gland: IUCN, UNEP, WWF.

Keto, A.I. and Scott, K. (1991) Timber production from North Queensland rainforest. Paper presented at Colloquium, Sustainability in Natural Tropical Forest Management. Washington, D.C.: World Resources Institute.

Keto, A.I., Keto, A.I., Scott, K. and Olsen, M.F. (1990) *Sustainable Harvesting of tropical Rainforests, a Reassessment*. Brisbane: Rainforest Conservation Society.

Lande, R. and Barrowclough, G.F. (1987) Effective population size, genetic variation, and their use in population management. In *Viable Populations for Conservation* (M.E. Soulé, ed.) pp. 87–125. Cambridge: Cambridge University Press.

Loomes, G. and Sugden, R. (1982) Regret theory, an alternative theory of rational choice under uncertainty. *Econ. J.* **92**, 805–24.

MacArthur, R.H. and Wilson, E.O. (1967) *The Theory of Island Biogeography*. Princeton, New Jersey: Princeton University Press.

Malthus, T.R. (1798) *Essay on the Principle of Population*. London: MacMillan.

May, R.M. (1992) Ecological science and the management to protected areas. Paper presented in World Congress on National Parks and Protected Areas, Caracas, Venezuela, 10–21 February, 1992.

Meadows, D.L. (1972) *The Limits of Growth, a Report of the Club of Rome's Projection on the Predicament of Mankind*. New York: Universe Books.

Niehans, H. (1948) Zur Preisbildung bei ungewissen Erwartungen. *Schweizer Zeitschrift Volkswirtschaft Statistik* **84**, 433–56.

Pearce, D., Markandya, A. and Barbier, E.B. (1989) *Blueprint for a Green Economy*. London: Earthscan.

Pearce, D.W. and Turner, R.K. (1989) *Economics of Natural Resources and the Environment*. Baltimore: The John Hopkins University Press.

Randall, A. (1987) *Resource Economics*. New York: Wiley.

Ricardo, D. (1817) *On the Principles of Political Economy and Taxation*. London: John Murray.

Samuelson, P.A. (1980) *Economics*. New York: McGraw-Hill.

Savage, L.J. (1951) The theory of statistical decision. *J. Amer. Stat. Assoc.* **46**, 55–67.

Shaffer, M.L. (1981) Minimum population sizes for species conservation. *Bioscience* **31**, 131–4.

Shaffer, M.L. (1987) Minimum viable populations, coping with uncertainty. In *Viable Populations for Conservation* (M.E. Soulé, ed.) pp. 69–87. Cambridge: Cambridge University Press.

Soulé, M.E. (1986) Conservation biology and 'the real world'. In *Conservation Biology: Science of Scarcity and Diversity* (M.E. Soulé, ed.) pp. 1–12. Sunderland, MA: Sinauer Assoc.

Soulé, M.E. (1987) Where do we go from here? In *Viable Populations for Conservation* (Soulé, M.E. ed.) pp. 175–183. Cambridge: Cambridge University Press.

Soulé, M.E. and Simberloff, D. (1986) What do genetics and ecology tell us about the design of nature reserves? *Biol. Conserv.* **35**, 19–40.

Stocker, G.C., Gilmour, D.A. and Cassells, D.S. (1977) The future of our northern rain forests in the face of economic and political reality. Paper delivered at the Institute of Foresters of Australia Triennial Conference. Adelaide.

Tisdell, C. (1990a) The environment and economic welfare. *The University of Queensland Discussion Papers in Economics*, No. 42. Brisbane: University of Queensland.

Tisdell, C. (1990b) Economics and the debate about preservation of species, crop varieties and genetic diversity. *Ecol. Econ.* **2**, 77–90.

Wald, A. (1950) *Statistical Decision Functions*. New York: Wiley.

[30]

Report

Priya Shyamsundar and Randall Kramer

Biodiversity Conservation—At What Cost? A Study of Households in the Vicinity of Madagascar's Mantadia National Park

An important subset of the vast number of issues associated with the conservation of tropical forests are the costs and benefits incurred by local peoples. This paper presents the results of a study of the costs borne by households living near the Mantadia National Park, a protected area in eastern Madagascar. The villages in this region are subsistence economies that are based on swidden agriculture and forest product collection. Village communities have lost access to more than 800 ha of agricultural lands and to a significant amount of forest products as a result of the park. The net present value of costs to the average household because of protection is USD 419. The analysis undertaken also provides useful information about geographic variations in the costs borne.

Swidden agriculture in Eastern Madagascar. Photo: R. Kramer.

INTRODUCTION

Efforts to conserve tropical forests are challenged by a number of socioeconomic factors. An important subset of the complex issues associated with conservation are the costs and benefits incurred as a result of protection. Conservation results in a wide range of benefits and costs (1–3). These gains and losses accrue to a broad spectrum of people who vary by location, socioeconomic status, preferences, and time. The magnitude and distribution of benefits and costs can be a critical constraint to protection (4).

Local residents or peoples who live in the vicinity of protected areas are a particularly important group that is affected by conservation measures (5). Because of their geographic proximity, cultural and historical associations, and the likelihood that they will continue to live in the area, local residents are closely linked to protected areas in both time and space. Frequently, households make direct economic use of protected areas either by hunting and gathering forest products or by using these areas for agriculture (6, 7). The economic impact of conservation on local households is of vital consideration in evaluating the potential for successfully protecting a tract of tropical forests.

This paper presents the result of a study undertaken to assess the costs borne by local residents from the establishment of a protected area, the Mantadia National Park, in eastern Madagascar. The park's creation has resulted in households losing access to certain areas of land they previously used for swidden agriculture and for obtaining forest products. The economic implications of establishing the Mantadia National Park are evaluated by estimating the costs associated with losing access to these lands.

STUDY AREA

Madagascar is recognized to be one of the biologically richest countries in the world (8, 9). However, Madagascar is economically very poor. It is also rapidly losing its extraordinary natural wealth to important subsistence needs, particularly food production. Recognizing that Madagascar's unique natural heritage needs to be conserved, the Malagasy Government has begun establishing a system of parks and protected areas.

The Mantadia National Park is one of Madagascar's four national parks. It extends over an area of 9875 ha representative of the eastern rain forests. It contains the habitat for the indri, the largest known extant lemur in Madagascar (10). Though several villages lie within a short distance (1–8 km), there are no permanent human settlements within the Mantadia National Park. Most of the forested land in and around the park is under government ownership. However, villagers have traditionally used tracts of this land for a variety of purposes. Since the park was established, some villagers have been prosecuted for obtaining forest resources from within its boundaries.

A household survey conducted in 1991 provides several insights into the characteristics of the population living around the park. The average household size in the region is 4.6 persons. Most villages do not have direct access to medical facilities, running tap water, and electricity. Primary schools exist in several villages and the literacy rate in the region is relatively high at 52%. The average number of years of schooling per person, however, is only 2.36 years. The age distribution of the population surveyed indicates a high dependency ratio. The average age of the population surveyed is very low at 22 years, with 57% of the surveyed population being younger than 20 years and 37% being less than 10 years of age (11).

Villages in this region are subsistence economies based primarily on agriculture, and secondarily on the collection and consumption of different forest products. While there are two other important income generating activities undertaken in this region—timber extraction and mining, almost all households undertake *tavy*, a form of swidden agriculture commonly practiced in eastern Madagascar (12, 13). The average annual gross household income in this region is USD 279 (14). Of this income 54% comes from subsistence agriculture, 31% from the subsistence use of forest products, and the rest from other forms of labor.

Rice is the main staple. The average household grows 487 kgs of rice every year. Valued at 500 fmg (Malagasy francs) per kg, this amounts to an annual gross income of 243 500 fmg (USD 133) (15). The other important crops grown are manioc, bananas, coffee, taro, sweet potatoes, and different types of beans.

The villages in this region are also dependent on a number of different forest products. Fuel wood and a variety of construction wood are collected on a regular basis, numerous fish and animals are foraged for consumption, and several different types of grass are used for household purposes. For example, the average household collects 6164 kg of fuel wood in a year. The total value of the fuel wood collected is 25 249 100 fmg (USD 13 759) or USD 39 per household per year. Forest plants and herbs also serve as important sources of medicine.

In summary, people living in the vicinity of the Mantadia National Park are very poor and entirely dependent on the land and forests in the area. Thus, the establishment of the Mantadia National Park has profound implications for their lives and welfare.

THEORY AND METHODS

Welfare Implications of Establishing the Mantadia National Park

The establishment of the park has resulted in a change in the total area of the land that local households have access to for agricultural purposes and for obtaining forest products. For example, suppose initially, i.e., without the park, the average household had access to T ha of land. With the establishment of the park, the household loses access to ΔT ha ($\Delta T \geq 0$) of land. As long as the household is using optimal and positive quantities of land in its production activities, the loss in accessible land has a direct impact on household production, and thereby, on household income and consumption. The change in production and consumption behavior in turn affects household welfare.

The welfare loss to the average household can be indirectly established by estimating the profits the household forgoes by giving up lands within the park. With access to T ha of land, household profits, $\Pi(T)$, equal the net returns to the household from undertaking agriculture and collecting forest products in this land. The household maximizes profits subject to a number of production constraints.

$$\Pi(T) = p_f F + p_a A - p_k K - w(L_f + L_a)$$

subject to

$$F = F(l_f, K)$$
$$A = A(L_a, K)$$
$$L_a = L_o + L_h$$
$$L_f \geq 0,\ L_o \geq 0,\ L_h \geq 0,\ K \geq 0,\ T \geq 0,\ p_f \geq 0,$$

where F and A are vectors of forest products and agricultural goods that the household produces or collects; p_f and p_a are vectors that represent the value of forest products and agricultural goods to the household; K is a vector of capital goods used either in agriculture or in forest product collection; p_k reflects the cost of capital used; and, L_f and L_a represent labor used in forest product collection and agriculture. The wage rate is given by w. Labor used in agriculture is comprised of household labor (L_o) and hired labor (L_h). The household maximizes profits $\Pi(T)$ by choosing K, L_f, L_o, L_h, and T.

When the household loses ΔT ha of land as a result of the park, it forgoes annual profits, and, therefore, bears an annual cost $\Pi(\Delta T)$. Over time, the household bears a total cost equal to the sum of the discounted value (net present value) of profits foregone each year (16). This is given by:

$$C = \sum_{t}^{t'} \frac{\Pi_t(\Delta T)}{(1+i)^t}$$

where C refers to the net present value (NPV) of costs, $\Pi_t(\Delta T)$ refers to costs borne by the household in year t, t' refers to the last year of the household's planning horizon, and i refers to the discount rate. The NPV of costs is estimated using cash flow analysis.

The NPV of costs is a point estimate of the costs of establishing the park, i.e., C is obtained by assuming that annual foregone profits, $\Pi_t(\Delta T)$, are known and any changes in foregone profits can be predicted. However, annual profits may vary as a result of stochastic changes in agricultural and forest product yields and prices. This suggests that annual foregone profits are comprised of a known and a stochastic element, and can be defined as Π'_t:

$$\Pi'_t = \Pi_t(\Delta T;\ \theta_t)$$

where θ_t represents the random component of annual profits. Following the cash flow analysis, a risk analysis is undertaken to establish how the NPV of costs varies with stochastic variations in critical factors that affect cash flow.

Cash Flow Analysis: Assumptions and Methods

Cash flow analysis is an accounting tool used to evaluate the money value of the flow of resources to a group of people. In this study, three cash flows are developed to assess the value of the costs borne by local residents.

Based on aerial photographs and visits to the region, three geographical groups of villages around the Mantadia National Park were identified to be affected by the park's establishment. Group 1 includes the village of Vohibazaha, and Group 2 includes the *fokontony* (cluster of villages) of Volove. These villages lie to the east of the park. Group 3 consists of the southwestern *fokontony* of Falierana and Andasifahatelo. The first cash flow presents the costs of establishing the park to the average household in Group 1; the second cash flow reflects the costs to the average household in Group 2; and the third cash flow represents the costs to villagers in Group 3 (17).

The cash flows estimate the net present value of costs by evaluating the flow of foregone benefits from swidden agriculture and forest product collection to the average household and subtracting from it labor and capital costs. Each cash flow consists of two sets of estimates of foregone benefits: i) an initial year (1991) estimate of the net cash flow to the average village household from the harvesting of forest products and from swidden agriculture; and, ii) the final projected cash flow, which extends from 1991 to 2010 (18). This cash flow is based on the initial year's flow of benefits and costs, but also accounts for the impact of deforestation over time (19).

Aerial photographs indicate that Group 1 villagers have lost 208 ha of agricultural land, Group 2 villagers have lost 339 ha of agricultural land, and Group 3 villagers have lost 319 ha of agricultural land because of the park. The initial (1991) agricultural benefits foregone are based on these land-loss estimates, and are calculated to be a proportion of the total annual agricultural returns to the household (20). In the years that follow 1992–2010, net agricultural benefits foregone are assumed to increase at the rate of deforestation. The forest product benefits foregone are estimated by identifying products obtained mainly from primary forests and not from secondary forests (21).

Prices used in the three cash flows are farm-gate and local market prices. In most cases, the market prices used are from the market closest to the village under consideration (22).

Risk Analysis

Risk analysis is a formalized method for introducing uncertainty into project evaluation (23). It requires first that subjective or objective beliefs be formed about the likelihood of the occurrence of various uncertain events, i.e., probability distributions are assigned to stochastic variables such as prices. This information is then used to generate a probability distribution of the project's net present value or any other summary measure. Thus, risk analysis allows for the consideration of an entire range of possible project outcomes and presents information on the likelihood of the occurrence of each outcome.

In this study, three variables are considered stochastic: the price of rice, the price of coffee, and rice yields per ha. Rice is the most important crop in the region and variations in rice yields and rice prices are likely to significantly influence the NPV of foregone land use. Coffee is an important commercial crop, and its price is influenced by changes in the world market for coffee. Coffee is also the second most important crop in villages east of the park (in terms of gross income).

Using a Monte Carlo simulation approach, 1000 iterations are run on each cash flow. In each iteration a random sample from the assumed distributions of the stochastic variables is drawn and the net present value of costs computed. This results in a probability distribution of the net present value of costs.

DATA

The data for estimating costs were obtained from the household survey undertaken in 1991 and from a number of research visits to the region in 1991 and 1992. The survey was administered to 351 households in 17 villages in the vicinity of the Mantadia National Park. Preceding the survey was a rapid rural appraisal that identified the populations that were dependent on the park. A pre-survey was used to test the survey questionnaire (24).

The household survey provided information on land-use patterns, agricultural yields, forest outputs, forest product processing, and labor use. Market information, price information, and detailed village-level information were gathered through a series of open-ended unstructured interviews. Price data on agricultural and forest products are village level prices obtained from interviews with villagers, shop owners, household and village leaders, and from published reports. Price and quantity data were standardized to obtain the mean values used in the cash flows (25). Secondary information and statistics were obtained from different government agencies.

The primary data used to establish the costs to each group differ between the three cash flows. However, the secondary data used are common to all three cash flows. For example, agricultural labor-use data were unavailable at the household level and are common to all three cash flows. Parameter values, such as the discount rate and the foreign exchange rate, are also common across the cash flows.

Data on land-use patterns and areas cleared within the park were obtained by mapping 1991 aerial photographs of the park. The dot-grid method was used in making the calculations. The deforestation rate for forests in the park was estimated based on a study of the deforestation history of the region (26).

The data used in the risk analysis and the underlying assumptions are presented in Table 1. The mean values used for the three stochastic variables are the same as the values used to obtain the point estimates. The coefficients of variation used to estimate standard deviations, the correlation coefficient, and the probability distributions specified for each of the three stochastic variables are based on historical data (27). Historical data at the regional level were available for a 13 year period.

RESULTS AND DISCUSSION

Table 2 presents the net cash flow from forest products and agricultural products to the average household in the year 1991. The net cash flows represent the annual profits foregone by the average household as a result of the park.

The net cash flows presented in Table 2 are not equal to the annual total household income from agricultural production and forest product collection. Approximately 18% of the agricultural land used by the average household in Group 1 and 19% in Groups 2 and 3 lies within the park. Therefore, only 18–19% of annual agricultural income is estimated to be foregone as a result of the park. For household forest products 25% related income for Group 1 and 52% and 53% of forest product related income for Groups 2 and 3 is associated with the park —the remaining income is obtained from secondary forests near the villages.

Table 2 suggests that there are significant locational differences in the costs borne by households as a result of the park. The net annual income from forest products ranges from USD 8 for Group 1 to USD 57 for Group 3. The net annual agricultural income from the park ranges from USD 7 for Group 3 to USD 22 for Group 2.

The three groups also differ in the types of demands they make on the park. Groups 1 and 2 benefit from agricultural lands within the park more than they do from forest products. However, Group 3 benefits considerably more from forest product collection. This can be partially attributed to the share of forest products obtained from within the park. There is also a larger market for forest products in areas south of the park, which results in higher prices and in higher cash flow estimates.

Table 3 presents the net present value of the costs of establishing the Mantadia National Park to the average village household in each of the three groups of villages.

In all three geographical locations, the average household bears losses in use values from the establishment of the park. Cash flow analysis indicates that with a 11.5% discount rate and a 2% deforestation rate the net present value of costs per household from establishing the Mantadia National Park is: USD 240 for Group 1, USD 427 for Group 2, and USD 564 for Group 3. Thus, in light of the extreme poverty of the households living in the vicinity of the Mantadia National Park, the establishment of the park imposes a significant burden on the average household in the region.

Geographical variations exist in the costs of establishing the park. The NPV of costs per household differs within the eastern region (Groups 1 and 2), and between the east and the southwest (Group 3). On average, the establishment of the Mantadia National Park imposes much heavier costs on villages to the southwest of the park relative to villages to the east of the park. As before, this

Table 1. Risk analysis : Data and underlying assumptions.

Variables	Mean per household			Std Dev	Coefficient of variation	Correlation
	Group 1	Group 2	Group 3			
Paddy yields (kg)	566	554	400	319	0.21	0.68
Paddy prices (fmg)	500	500	50	75	0.66	0.68
Coffee prices (fmg)	800	800	900	267	0.56	–

USD - 1835 fmg

Table 2. Value of benefits foregone from forest product collection and agricultural production in 1991.

Village	Value of forest product benefits foregone by the average household in 1991 (USD)*			Value of agricultural benefits foregone by the average household in 1991 (USD)*		
	Cash in	Cash out	Net cash flow	Cash in	Cash out	Net cash flow
Group 1	9	1	8	26	11	15
Group 2	27	6	21	34	12	22
Group 3	69	12	57	24	17	7

* An exchange rate of 1835 fmg/1 USD has been used.

can be attributed partly to the higher dependence of the southwestern villages on the park and partly to the higher value of forest products in this region.

The choice of the discount rate has important implications for the estimated costs of establishing the Mantadia National Park. Estimates of the discount rate used by the households in the region are difficult to obtain and range from the official national bank rate of 11.5% to as high as 100% (28). Therefore, two different discount rates (11.5% and 25%) are used in this analysis. The lower discount rate of 11.5% results in an increase in the NPV of costs for each of the three groups of villages by approximately 50%.

Significantly, the choice of the rate of deforestation is not as important for understanding variations in estimated costs. The 2% deforestation rate reflects the rate of population growth in the region and the 4% rate reflects the historical rate of deforestation (11, 26). Using a 4% deforestation rate instead of 2%, changes the cost estimates by only 1 to 12 percentage points.

The results of the risk analysis from the Monte Carlo Simulations are presented in Table 4. Table 4 presents the mean value, minimum value, maximum value, range, standard deviation, and probability estimates of the net present value of costs for each sub-group.

Incorporating risk into the analysis allows us to consider how variations in key factors might affect the point estimates of NPV obtained from the cash-flow analysis. The expected mean NPV estimates range from USD 262 in Group 1 to USD 584 in Group 3. The estimated mean values differ by less than 10% from the point estimates obtained in the cash flow analysis. However, the risk analysis indicates that the net present value of costs lie within a wide range of values: from USD 10 for Group 1 to USD 4118 for Group 2. The high standard deviations also suggest that there is a great deal of variation in the probable NPV of costs—this is especially true of Group 2. Interestingly, the probability of the NPV of costs being greater than USD 500 is less than 15% for Group 1, but is as high as 30% for Group 2, and 55–60% for Group 3. Thus, not only is the mean NPV of costs higher for Groups 2 and 3 relative to Group 1, but the range of probable net present value of costs is also higher.

Table 3. The NPV of costs borne by the average household.

Village	Deforestation rate per year in %	NPV of costs per household 1991–2000 (1991 USD) i = 11.5%	i = 25%
Group 1	2	240	121
	4	273	129
Group 2	2	427	221
	4	473	232
Group 3	2	564	317
	4	577	320

Table 4. Risk analysis of the cash flows.

Categories	Group 1 (USD)*	Group 2 (USD)*	Group 3 (USD)*
Mean NPV of costs	262	461	584
Minimum	10	116	375
Maximum	1674	4118	2012
Standard deviation	212	313	186
Prob NPV > 0	100%	100%	100%
Prob NPV > Point estimate of NPV	> 40%	> 40%	> 40%
Prob NPV ≥ USD 500	≤ 15%	30%	55–60%

*The discount rate used is 11.5% and the deforestation rate used is 2%.

Table 5. Aggregate net present value of costs.

Village	Population	NPV of costs per household (USD)*	NPV of costs per capita (USD)*	Aggregate NPV of costs (USD)*	*Annualized costs per household (31) (USD)*
Vohibazaha (Group 1)	521	240	57	29 561	28
Andonaka	554	240	57	31 433	28
Volove (Group 2)	775	427	95	73 705	50
Falierana and Andasifahatelo (Group 3)	1554	564	110	170 892	66
All villages	3404	419	90	305 590	49

* A discount rate used is 11.5% and the deforestation rate used is 2%

AGGREGATION

Population aggregation is an important consideration in determining the total costs to local villagers as a result of establishing the Mantadia National Park. As already noted, the three cash flows developed correspond to the average household in Vohibazaha, Volove, and Falierana and Andasifahatelo. In establishing the total cost, the average NPV values are aggregated over the population of this set of villages (29).

Adjacent to the northern border of the park is set of villages that falls within the *fokontony* of Andonaka. Because of their proximity to the park, there is a high probability that the villagers in this region would use the lands within the park if they were free to do so. Therefore, the population of these villages is also included in aggregating costs. Population estimates for all the villages under consideration have been obtained from 1992 census data (30). Population figures and estimates of the aggregate NPV of costs are presented in Table 5.

As previously indicated in Table 3, cash flow analysis suggests that the average NPV of costs per household in Vohibazaha is USD 240, in Volove USD 427, and in Andasifahatelo and Falierana, USD 564. Furthermore, assuming that the costs borne by the populace of Andonaka are identical to those borne by the people in nearby Vohibazaha, a per household value of costs of USD 240 is obtained for households in Andonaka. Assuming that people outside of these areas do not use park lands, the aggregate net present value of the costs of establishing the Mantadia National Park to local peoples is USD 305 590 (32).

There are several interesting results that follow. The NPV of costs to the average household, across all the different sets of villages, is USD 419. Furthermore, the estimated annual cost borne by the average household is USD 49 (28). Thus, a total annual loss of USD 35 895 is incurred by local peoples as a result of the Mantadia National Park.

The aggregate NPV of costs is very sensitive to changes in the discount rate. If a 25% discount rate is chosen, the aggregate NPV declines to USD 164 949, which is almost half the cost estimated when a discount rate of 11.5% is assumed.

Finally, the aggregate NPV of costs is entirely dependent on the assumptions made about the populations affected by the park. It has been assumed that every household in the villages impacted by the park loses access to certain areas of land—either in the current period or in the near future. However, if it is assumed that only 50% of these households are impacted by the park, then the aggregate NPV of costs as a result of the Mantadia National Park is USD 152 792 with a 11.5% discount rate, and USD 82 474 with a 25% discount rate.

CONCLUSIONS

The cost analysis undertaken indicates that there are significant losses that local peoples bear as a result of the park. Households on average bear an annual (net) cost of USD 49 as a result of the park, which amounts to 18% of their total gross income in

1991. If all the households in the immediate vicinity of the Mantadia National Park are impacted by the park, the aggregate NPV of costs resulting from conservation is USD 305 590.

The risk analysis undertaken reinforces the results obtained from the cash flow analysis. There is at least a 40% chance that the costs are greater than the point estimates obtained from the cash-flow analysis. Risk analysis also indicates that there is a very wide range of values within which the costs can fall.

The value in undertaking this empirical exercise is to understand both the extent of the burden local residents have to bear as a result of conservation efforts, and to establish the relative differences in the burden borne. For policy decisions, the relative magnitude of the costs is perhaps more important than the absolute value of the costs. The analysis undertaken shows that costs vary according to geographical location. The largest burden is borne by villages that lie to the southwest of the park. Furthermore, different groups of villagers place different types of demands on the lands within the park. The value of forest products in the park is much higher for households in the southwestern region relative to households who live close to the northeastern borders of the park.

This paper has focused on the costs borne by local residents as a result of conservation. However, protection entails both costs and benefits. A full accounting of the benefits and costs of conservation would include amenity, non-use, and environmental values associated with the land and forests conserved. The household survey undertaken in this study did not suggest that households obtained any nonuse or amenity benefits from the park. However, there may be benefits from the environmental services provided by the forests in the park, such as flood control, that accrue to local communities.

The cost estimates presented in this paper represent a minimum value of the benefits required to make the park into an economically viable enterprise. The costs also represent the benefits required to make the park into a protected area that is accepted by local residents. Managers of integrated conservation and development projects in Madagascar are currently striving to find a balance between conservation and the satisfaction of local needs. The analysis presented in this paper shows, in very concrete terms, why local residents often resist conservation efforts. The magnitude of the costs suggests that the stakes are very high for local residents: if they are not actively involved in the planning of national parks and reserves, conservation efforts are unlikely to succeed (33).

References and Notes

1. Dixon, J.A. and Sherman, P.B. 1990. *Economics of Protected Areas: A New Look at Benefits and Costs*. Island Press, Washington DC, USA.
2. Dixon, J.A. and Sherman, P.B. 1991. Economics of Protected Areas. *Ambio* 20, 68–74.
3. Thomas, D.H.L., Ayache, F. and Hollis, G.E. 1991. Use and non-use values in the conservation of the Ichkeul National Park, Tunisia. *Environ. Conserv.* 18, 119–130.
4. van Schaik, C., Kramer, R.A., Salafsky, N. and Shyamsundar, P. 1993. *Biodiversity of Tropical Rainforests: Ecology and Economics of an Elusive Resource*. Center for Tropical Conservation Report, Duke University Center for Tropical Conservation, Durham, USA, 36 pp.
5. Wells, M., Brandon, K. and Hannah, L. 1992. *People and Parks. An Analysis of Projects Linking Protected Area Management with Local Communities*. World Bank, World Wildlife Fund and U.S.A.I.D., Washington DC, USA, 99 pp.
6. Lehmkuhl J.F., Upreti, R.K. and Sharma, U.R. 1988. National parks and local development: Grasses and people in Royal Chitwan National Park, Nepal. *Environ.Conserv.* 15, 143–148.
7. Infield, M. 1988. *Hunting, Trapping and Fishing in Villages within and on the Periphery of the Korup National Park*. Paper No. 6 of the Korup National Park Socio-economic Survey. Prepared by the World Wide Fund for Nature. Publication 3206/A9.6. WWF, East Sussex, England.
8. World Bank. 1988. Staff Appraisal Report: Democratic Republic of Madagascar. *Forests Management and Protection Project* (Report No. 6994-MAG), World Bank, Washington, DC, USA, 112 pp.
9. McNeely, J.A., Miller, K.R., Reid, W.V., Mittermeier, R.A. and Werner, T.B. 1990. *Conserving the World's Biological Diversity*. IUCN, WRI, CI, WWF-US, and the World Bank, Gland, Switzerland and Washington D.C., USA.
10. Harcourt, C. and Thornback J. 1990. *Lemurs of Madagascar and the Comoros. The International Union for the Conservation of Nature Red Book*. IUCN, Gland, Switzerland and Cambridge, U.K.
11. Shyamsundar, P. 1993. *Economic Implications of Tropical Forest Protection for Local Residents. The Case of the Mantadia National Park in Madagascar*. PhD Dissertation, Duke University, USA, 243 pp.
12. Dufresne, A. and Cloutier, A. 1989. *Plans the Gestion. Parc National Andasibe - Parc National Ankarafantsika*. Projet GPF, Volet Conservation et Protection (In French). International Union for the Conservation of Nature, Antananarivo, Madagascar.
13. Oxby, C. 1985. Forest farmers:The transformation of land use and society in eastern Madagascar. *Unasylva* 37, 42–51.
14. The exchange rate used is fmg 1835/1 USD.
15. The farm gate price of rice is used.
16. Two commonly-made simplifying assumptions have been adopted in order to keep the empirical analysis manageable. It is assumed that household production and consumption decisions are made sequentially and that the discounted sum of foregone annual profits equals the total value of the profits foregone if the household were maximizing a lifetime profit function.
17. Primary data on Andasifahatelo and Faliemana were unavailable. Therefore, in the third cash flow, data from respondents from villages to the south of the park are used as a proxy.
18. A 20 year project period was chosen because: i) a 20 year planning period for the park seems necessary to fully assess its impact; and ii) beyond 20 years discounted monetary benefits are so small that they do not add much to the net present value of costs.
19. The cash flows incorporate increases in agricultural benefits foregone because of likely deforestation over time. However, the cash flows are not adjusted for substitution activities likely to be undertaken over time. For example, households that no longer have access to hardwoods in the park (which are traditionally used for house construction), may replace these for timber species obtained in secondary forests. Lack of information on the direction and extent of substitution activities made it more appropriate to make the assumption of no substitution, rather than speculate about how substitution might occur. However, not incorporating substitution effects may mean that the cost estimates are biased upward.
20. Agricultural benefits foregone as a result of the park are obtained by considering the percentage of the total land used or controlled by the household that currently lies within the park boundaries.

$$\%\ \textit{Agricultural benefits foregone} = \frac{\textit{total agricultural land controlled within park}}{\textit{total agricultural land controlled}}$$

21. Because not all households understood where the park boundaries were, households were asked whether they obtained their forest products from primary or secondary forests. All products obtained from primary forests were assumed to have come from within the park.
22. Because there are relatively few buyers and sellers for several of forest products, market prices are to be treated as second best estimates of true marginal values.
23. Kramer, R. 1994. Advantages and limitations of benefit cost analysis for evaluating investments in natural disaster mitigation. *Invited paper for technical session: Economic aspects of disaster reduction for sustainable development*, World Conference on Natural Disaster Reduction, May 23–27, Yokohama, Japan.
24. The survey instrument was tested by conducting a pre-survey of 25 households (7% of the sample) in the village of Farahevitra.
25. The conversion factors used and their sources are available in Shyamsundar (11).
26. Green, G. M. 1993. *Remotely Sensed Determination of Deforestation History and Topographic Controls: Mantady National Park and Surroundings*. Unpublished Document, National Air and Space Museum, Washington DC,USA.
27. Probability distributions are specified for each of the three stochastic variables based on the distribution of historical data. The Kolmogorov-Smirnov (K-S) test was used to test the hypothesis that the empirical distributions fit a lognormal distribution. The null hypothesis could not be rejected at the 95% confidence level for any of the three chosen distributions.
28. The 11.5% discount rate is based on 1989 financial statistics on Madagascar published in *International Financial Statistics, Year Book 1991*, International Monetary Fund, USA.
29. Not every household in these villages uses lands within the park. However, their proximity to the park makes it highly likely that these households would use the park area if they were not forbidden to do so.
30. These data were obtained directly from government offices in Andasibe and Ambatavola.
31. The annual values have been calculated using the following equation:

$$\text{Annual costs} = \text{Estimated NPV} \cdot \left[\frac{i(1+i)^{20}}{(1+i)^{20}-1}\right]$$

32. People in other regions are unlikely to use lands within the park in the future. This is because the lands in the park that are accessible are land locked by villages that surround them.

33. Acknowledgement: We are grateful to the various people in Madagascar who patiently answered our questions and provided us with information about their lives. This research was undertaken with support from the Environment Department of the World Bank and the Biodiversity Support Program, a USAID funded consortium of the World Wildlife Fund, the Nature Conservancy and the World Resources Institute.
34. First submitted 27 June 1995. Accepted for publication after revision 17 April 1996.

**Periya Shyamsundar is a program officer with the World Environment and Resources Program at the John D. and Catherine T. MacArthur Foundation in Chicago.
Her address: 140 S. Dearborn Street, Suite 1100, Chicago, IL 60603, USA.**

Randall Kramer is an associate professor at the Nicholas School of the Environment, Duke University. His address: Nicholas School of the Environment, Duke University, Durham, NC 27708-0328, USA.

[31]

Opportunity Costs of Preserving Coastal Wetlands: A Case Study of a Recreational Housing Development

Sandra S. Batie and Carl C. Mabbs-Zeno

This article presents a case study of the development value of converting coastal wetlands[1] into recreational housing lots at Captain's Cove subdivision on Virginia's Eastern Shore. The development value estimates are valid measures of the opportunity cost of prohibiting recreational development in order to preserve Captain's Cove wetlands.

Such information can be a useful contribution to public decisions that must be made about protecting natural wetlands from conversion to other uses. For example, many coastal states and the federal government have policies to retain wetlands in their natural state. Most states that have wetlands programs base their wetlands management upon individual wetlands parcel permitting procedures. Because of this case-by-case nature of the permit programs, there is a use for economic value measurements of wetlands acres in altered and unaltered states.

The Commonwealth of Virginia, for example, manages coastal wetlands through citizen boards of unpaid citizens which review permits for wetlands development in local jurisdictions. Permits are to be issued for wetlands alteration only when "the anticipated public and private benefit of the proposed activity exceeds the anticipated public and private detriment" (Virginia Wetlands Act 1972). At the federal level, the Army Corps of Engineers issues permits under Section 404 of the Clean Water Act for wetlands alteration only when "the benefits of the proposed alteration outweigh the damage to the wetlands resource" (Office of Technology Assessment 1984).

However, the ecologic information necessary to develop monetary measures of the value of wetlands in their natural state and thereby to weigh the social costs of wetlands destruction[2] is lacking. In most cases, the linkages between the existence of a specific wetlands area and levels of environmental services are not well established. Furthermore, most of the previous research efforts to provide information on economic values are flawed by the use of inappropriate economic methods.[3]

Given the current inability to provide economic values of natural wetlands areas, one strategy for wetlands management would be the adoption of management guidelines which stress wetlands preservation, unless the expected value of foregone development is "unacceptably" large. This strategy is essentially that of maintenance of a safe minimum standard (SMS).[4] Of course, the determina-

Land Economics, Vol. 61, No. 1, February 1985
0023-7639/85/001-0001 $1.50/0

Respectively, professor of agricultural economics, Virginia Polytechnic Institute and State University, Blacksburg, and head of the Agricultural College, National University of Lesotho, Southern Africa. They thank Drs. Oral Capps, Thomas Johnson, Gary Lynne, and Leonard A. Shabman for their helpful comments.

This work is a result of research partly sponsored by NOAA, Office of Sea Grant, Department of Commerce, under Grant #04-6-158-44086 and 04-7-158-44086. The grant title is Alternative Management Strategies for Virginia's Coastal Wetlands. The U.S. Government is authorized to produce and distribute reprints for government purposes notwithstanding any copyright notation that may appear herein.

[1]"Wetlands" is defined for this paper as the vegetated coastal areas which are periodically flooded by tidal waters.

[2]References to wetlands "destruction" in this paper are not intended to imply any judgement on the value of wetlands development, rather it is used only to express the impact of development.

[3]See Shabman and Bertelsen (1979), Shabman and Batie (1980), Shabman and Batie (1983), and Batie and Shabman (1982) for an elaboration of this point.

[4]The safe minimum standard strategy (SMS) can be applied to conservation problems that involve long time

tion of an "unacceptably" large cost will be a matter for public debate. However, to facilitate this debate, economic analysts can identify the values of wetlands for development uses.

If the value of wetlands development were clarified, public decisions to deny wetlands development permits could be more easily defended. Specifically, lower development permits could be more easily defended. Specifically, lower development values make the argument for denial of a development permit more compelling (Shabman and Bertelsen 1979; Shabman et al. 1979). An important reason for concentrating on the estimation of development values is that market information is available—in contrast to that information needed for estimating preservation values. Thus, such research has the attractive attributes of a high probability of successful completion and considerable utility in public decisionmaking.

THE BENEFITS OF DEVELOPMENT

An analysis that estimates only the benefits of developing wetlands (i.e., the opportunity costs of preserving wetlands) is not without difficulty, however. Freeman has noted that changes in land value can be equated with changes in development benefits, if all other prices remain unchanged and if all profits and surpluses are eliminated or captured by land owners in the form of rents. The elimination of surpluses is more likely where there is strong competition for each type of parcel. However, observed differentials between improved and unimproved sites can be considered as the upper-bound estimate of the benefits of property enhancement even when economic surpluses remain (Freeman 1979). Such estimates should reflect the "with and without" conditions of the analysis. That is, the values of developing those building sites which are available within the accounting stance of the analysis, and which are alternative to the proposed wetlands development, influence estimation of the benefits of any particular wetlands conversion (Hitchens, Thampapillai, and Sinden 1978).

Figure 1 illustrates the importance of the identification of the "with and without" conditions. In Figure 1, the arrow (*a*) suggests that a newly developed wetlands area can be a factor input in a production process which yields a flow of final consumer services (Box II). For example, a filled wetlands may represent a location service when it allows a commercial or industrial operation to locate near a market, transportation facility, or raw material source. A developed wetlands may also be a direct input into production, as when a wetlands is drained and converted to agricultural production. Developed wetlands may also provide services utilized directly by final consumers when development allows location of residential and recreational home sites such as the case in Captain's Cove. These services, in turn, give rise to a change in economic surplus along linkage (*b*) in Figure 1. The economic benefit of the wetlands service (Box III) is dependent on alternative ways of providing similar service by either modified development (Box IV) or by development at an alternative site (Box V). Box IV represents a modified development plan for the wetlands area which may have higher development costs and/or a reduced service flow, but which reduces the loss of natural wetlands services. An example would be the construction of recreational homes on pilings above the wetlands rather than on a filled wetlands. Box V represents the possibility that a "similar" service flow could be realized from an alternative site; of course, the private wetlands owner may forego the returns on development if the alternative site is developed. An example can clarify this possibility. In coastal areas, a developer of recreational homes on a wetlands might be denied a permit. Those proposed homes might then be located at an alternative nonwetlands (fastlands) site.

With this perspective, Figure 1 can illustrate the proper basis for calculating the development benefits of a wetlands area. If there are no alternative sites and no technically feasible modified wetlands development

horizons, considerable uncertainty, and/or irreversibility. SMS is the maintenance of sufficient population and habitat to assure the survival of both—unless the social costs of doing so are unacceptably large (Bishop 1978).

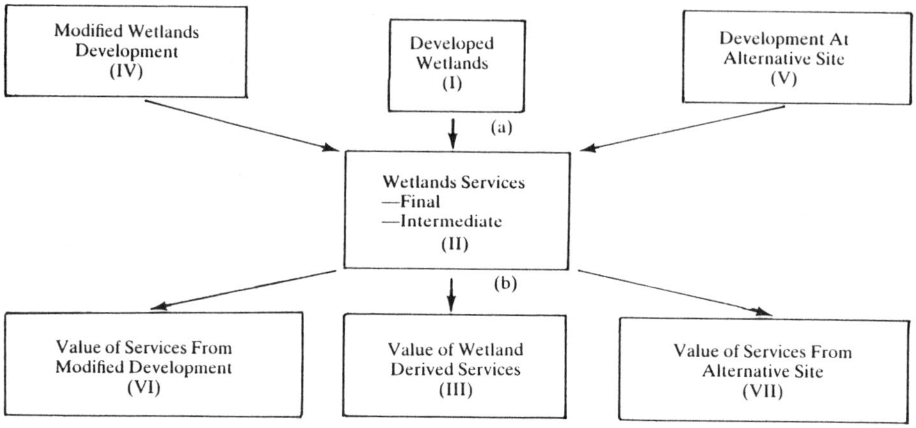

FIGURE 1.
DETERMINATION OF DEVELOPED WETLANDS VALUE

plans (i.e., Boxes VI and VII represent zero values), then the benefits from development of the site equal the change in economic surplus from the wetlands service (Box III). However, modified development or alternative site development may be feasible. If so, the modified development gives rise to values in Box VI, and the alternative site development gives rise to values in Box VII. In all cases, the value of wetlands development is the difference between the economic surpluses earned with developed wetlands and the economic surpluses which would have been earned without the wetlands development.

PROCEDURES

The concepts illustrated in Figure 1 were used to estimate benefits from development of wetlands in Captain's Cove subdivision in Accomack County, Virginia. In the development of this 1,500 acre subdivision, 150 acres of wetlands were filled or dredged to produce a marina, canals, and approximately 600 building lots of the total 4,877 lots. Thirteen percent of the site was left as wetlands and thirteen percent was converted to parks, canals, the marina, and two golf courses (U.S. Army Engineer District 1976).[5] Figure 2 provides a schematic of the subdivision and displays the location of filled wetlands and canals.

The estimation of the marginal returns to development of wetlands proceeded in several steps. First, a regression equation was estimated that relates lot characteristics to lot prices for a sample of lots at Captain's Cove. This was used to generate market prices for all wetlands lots. These prices were summed to obtain an estimate of gross returns from Captain's Cove wetlands development. Second, direct gross costs of developing wetlands were calculated using engineering estimates for converting wetlands to filled lots. Third, the net marginal returns from wetlands conversion were obtained. Because the development benefits will depend on the value of the

[5]Violation of permit regulations and denial of certain permits which were necessary to the overall plan of Captain's Cove have prevented completion of the housing development (Wood and Hill 1978). There is no evidence, however, that the subdividers were aware that their plans would fail due to the new and previously untested regulations, so this analysis accepts their behavior as rational in the risky enterprise of land speculation. The analysis presented here assumes the Captain's Cove's development was completed as planned.

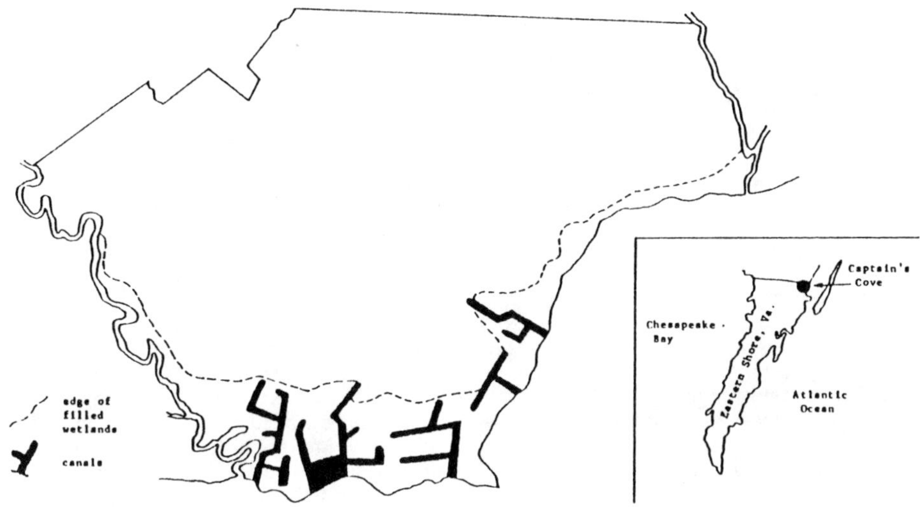

FIGURE 2.
CAPTAIN'S COVE SUBDIVISION, ACCOMACK COUNTY VIRGINIA

alternative development opportunities considered as well as the development services realized, these were calculated assuming two different alternative development scenarios: "fastland development alternatives do not exist" and "fastland development alternatives do exist."[6,7] Each scenario involved various amounts of wetlands destruction.

A third possible scenario, "modified development," in which the construction plans are modified so as to reduce the damage to any wetlands services from development of the wetlands (e.g., placing homes on pilings above the wetlands) was not considered in this case study. These are the values associated with Box VI, Figure 1.

The "no fastland development" scenario represents the case where, if the developers failed to develop the original planned amount of wetlands at Captain's Cove because of a permit denial, no offsetting development would take place. That is, development would not transfer to any other site. This is the same as estimating net values of development in Box III, Figure 1. The "fastland development" alternatives represents the case where the developers would have transferred development to a fastland site within the general area of the subdivision if their original development plans at Captain's Cove had been denied. This is equivalent to estimating the values associated with Box VII, Figure 1.

The estimated marginal returns to an additional increment of wetlands development included only those returns from land use that accrue to the developers of Captain's Cove and the developers of the alternative development sites. Net marginal returns to development were measured by subtracting the marginal private costs of development of the wetlands area from private income generated by the development. These estimates are the

[6]Marginal returns to wetlands destruction are the returns, as measured by willingness to pay, associated with an amount of acreage which is small relative to the total acreage in the state, which provides similar services. Since wetlands permits involve changes in land use for specific parcels rather than for all wetlands, the appropriate measure of value is marginal rather than average value of all wetlands. The use of marginal value assumes that wetlands sites provide similar services and that no individual site is unique.

[7]These calculations are equivalent to estimating a derived demand value for wetlands as one factor in the production of "filled building lots."

opportunity costs of preserving coastal wetlands for the marginal unit of wetlands.

Calculation of the private (developer) costs and returns for the case study are discussed below; the implications of the findings for administrators of public wetlands programs and for professionals interested in wetlands valuations are then identified.

GROSS RETURNS

The contribution of lot characteristics to lot prices for a sample of lots at Captain's Cove were estimated with a linear regression model. The estimated coefficients for the regression model were used to predict the selling price of any wetlands lot as a function of that lot's various attributes. Characteristics and sale prices for 980 lots (20% of the total number of lots at Captain's Cove) were used as observations (Brigham 1965; Diamond 1980). Most of the factors used in this regression model were measured as binary variables reflecting whether a specific lot possesses some attribute. Other included variables were the year when a lot was sold and the area of land in a lot (Table 1).[8]

It was hypothesized that buyers would have a positive willingness to pay for each of the features proxied by the binary variables and for larger lot size. The coefficient on time was hypothesized to represent those trends in demand that are due to time changes. These trends could include, for example, the effect of consumers becoming aware of the development's existence and being influenced by previous sales as well as the provision over time of some of Captain's Cove's development amenities (e.g., the yacht club and basin, the golf course). The effect of inflation was removed by expressing all prices used in the study in 1976 dollars.

Restricting analysis to a single subdivision allowed a focus on the variables which are important within a development site. For example, distance to cities and services is hypothesized as being important in affecting willingness to pay for recreational home lots. But, this effect does not vary within the present data set, and it is therefore contained in the intercept value. Also the restriction to a single subdivision means that net return to the developer is an accurate estimate of the development value of filling wetlands. The size of the subdivision is not so large as to result in a shift in the aggregate supply schedule for East Coast lots with water access; thus sales prices net of land and improvement costs measure development value at the margin.

Other attributes of lots which were considered but not included in the specification of the model reported in Table 1 are (1) whether the lot is adjacent to a park area, (2) whether the lot is adjacent to the yacht club, and (3) whether the lot is composed of filled wetlands. Lots next to the yacht club were grouped with lots next to canals in this model. Lots built on filled wetlands were hypothesized to be indistinguishable from lots on original fastland to purchasers. None of the coefficients on these variables was significant at the .7 level of confidence. Various functional forms were considered with respect to the time variable, but the linear model presented in Table 1 offered the highest level of significance for all included variables. (Numbers in parentheses are *t*-statistics. All coefficients and *t*-statistics are rounded to three significant digits.)

All variables included in the regression are significant when considered at a level of significance equal to .01. No indication of multicollinearity or heteroskedasticity problems were suggested by examination of supplementary statistics.

The coefficients in this regression may be interpreted as the marginal willingness of consumers to pay for a unit of the variable. For example, the coefficient on variable X_5 indicates a willingness to pay of $882 for ac-

[8]Consumer demand theory suggests that consumers' income should be included in this land price regression (Diamond 1980). If amenities of the wetlands lots are normal goods, the marginal value of the amenities will rise with income; thus, income is a "shifter" of the demand relationship. Income was not included, however, due to the lack of data. Because the primary use of the regression is prediction of lot price and because predictive power was reasonably good ($R^2 = .72$), it was felt this was adequate for the purpose of this study. The coefficients should be interpreted with caution, however, as they are most likely biased upwards from estimates holding income constant.

TABLE 1

REGRESSION OF CAPTAIN COVE'S RECREATIONAL LOT PRICES AS A FUNCTION OF LOT CHARACTERISTICS

$$Y = 4730 + 7410\,X_1 + 4180\,X_2 + 120\,X_3 + 0.157\,X_4 + 882\,X_5 - 1120\,X_6$$
$$\quad (14.5) \quad (37.3) \quad (28.8) \quad (9.83) \quad (7.07) \quad (4.96) \quad (3.14)$$

$R^2 = 0.719$

$n = 980$

where:

Y = price paid to subdivider in 1976 dollars;

X_1 = WATER = dummy variable with the value 0 for lots or 1 for waterfront lots;

X_2 = CANAL = dummy variable with the value 0 for lots not fronting on a canal or 1 for lots fronting a canal;

K_3 = TIME = quarter in which lot was sold beginning with the first quarter of 1969 equal to 1 (12 represents the last quarter of 1971);

X_4 = AREA = lot size in square feet (varied from 5,000 square feet to 31,300 square feet);

X_5 = SEWER = dummy variable with the value 0 for lots not on the sewer line or 1 for lots on the sewer line, and;

X_6 = WETLANDS = dummy variable with the value 0 for lots not adjacent to wetlands or 1 for lots adjacent to wetlands.

cess to sewer treatment, ceteris paribus. The intercept represents the value of a lot sold at the end of 1968 without waterfront, canal-front, or sewer, not on wetlands and with a lot size less than 5,000 square feet. The value is expressed in 1976 dollars and is equal to \$4,730. The intercept is interpreted as the approximate value of a lot of minimum size for building and possessing the other characteristics cited above.

The relative statistical importance of included variables in accounting for variation in the price of lots can be determined by examining the standard partial regression coefficients (Gujarati 1978). Whether a lot is adjacent to water or a canal[9] are the most important factors contributing to lot price, but all the factors included were found to be statistically significant.

To find the total gross private (developers') return from any development scenario, the price for each wetlands lot in the scenario, as determined by lot characteristics, is determined by using the regression equation. These prices are then summed.

GROSS PRIVATE COSTS

Private costs of development included the land costs and the costs of preparing the lot to meet specifications available in the environmental impact statement (U.S. Army Engineer District 1976).

Land at Captain's Cove was purchased by the developers as three separate parcels in three separate transactions. No difference in purchase price was specified for wetlands acreage as opposed to fastland. Rather, one price was paid for the entire subdivision. Therefore, to compute the cost of wetlands versus fastlands for this study, separate prices for these land types were estimated based on the differential in land price for fastlands and wetlands on Virginia's Eastern Shore. Based on a review of land transactions in the area, land prices for wetlands were assumed to have been 20% of fastland (nonwetlands) prices (Mabbs-Zeno 1980). In order to express the amounts paid in the three separate purchases in 1976 dollar terms, the real amounts actually paid were compounded at a real interest rate of 3% to 1976.[10] These totals were divided by the number of lots created on each type of land to yield estimates of land costs per lot: \$81.74 per wetlands lot and \$414.36 per fastland lot.

The cost of developing a purchased lot depends on both the undeveloped state and the developed state of the lot. The undeveloped state is defined for Captain's Cove according to whether a lot was initially wetlands or fastland and whether it is already leveed. Filled

[9]Some lots were on both water and canal.

[10]The real interest rate must be added to the inflation rate to find the approximate nominal interest rate. Thus, if inflation is 7% annually, a real interest rate of 3% would be observed at a nominal rate of 10%.

wetlands require levees to stabilize fill, but only lots along the border of the filled area actually have levees. The number and location of nearby filled lots, therefore, contribute to the cost of developing a wetlands lot.

The matrix of gross costs (inclusive of land costs) and returns for each form of undeveloped lot and for the major forms of developed lots is shown in Table 2. Lot improvement costs were estimated by a local construction firm. Also shown are the prices for developed lots derived from the recreation lot price regression. The cost of developing a wetlands lot which is not already protected by a levee into a canalfront lot, for example, is $9,439. Such a lot is predicted to sell for $17,268, based on the equation in Table 1. The net returns of developing a canalfront lot of $7,829 are a measure of economic surplus from the development.

MARGINAL NET RETURNS TO DEVELOPMENT

As Figure 1 illustrates, the development value of wetlands will depend on the alternatives available which will yield similar development services to the original plan. As noted earlier, it can be hypothesized that the developer had two possible types of scenarios from which to choose: no fastland development and fastland development. Costs were estimated for the most profitable alternative from each of these scenarios and for different amounts of wetlands development (from 1 to 150 acres). For a development plan which fills a given amount of wetlands, gross private costs were subtracted from gross returns to find net returns. The difference between net private returns for a given amount of wetlands destruction and net returns with one more acre of wetlands loss is the marginal net return of that acre.

Under the "no fastland development" scenario, denial of a permit to destroy, say, 58 acres in Captain's Cove would result in the developer applying for another permit for some lesser number of wetlands acres. However, the developer does not develop additional fastland lots to compensate for the denied wetlands acreage.

Calculation of net marginal returns under these conditions is the difference between direct costs and returns to the proposed 58 acres of wetlands conversion versus an alternative amount of 33 acres. This return is equal to $29,750 (Figure 3). Total development benefits for 58 acres of wetlands is $31,890,000; this is the total return for the entire subdivision including 58 acres of wetlands. This is equivalent to estimating the marginal and total values associated with Box III in Figure 1. Marginal returns peak between the 14th and the 20th converted wetland acre. This is because all possible shorefront lots have now been included. Stable returns of approximately $43,000 per acre occur above 58 acres of converted wetlands as all possible canalfront lots are included.

The "fastland development" scenario is

TABLE 2

AVERAGE COST AND REVENUE OF DEVELOPMENT AT CAPTAIN'S COVE (1976 DOLLARS, COMPOUNDED AT 3%)

	Undeveloped State of land	Developed State of Land		
		Shorefront	Canalfront	Interior
Cost[a]	Leveed Wetlands	$ 6,366	$ 6,073	$ 5,662
	Non-Leveed Wetlands	9,732	9,439	9,029
	Fastland	2,108	2,367	1,884
Revenue[b]	All	$22,108	$17,268	$11,004

[a]Costs were estimated by a local construction firm from specifications available in the environmental impact statement (U.S. Army Engineer District, 1976).

[b]Revenue is predicted by the regression of lot price on lot characteristics.

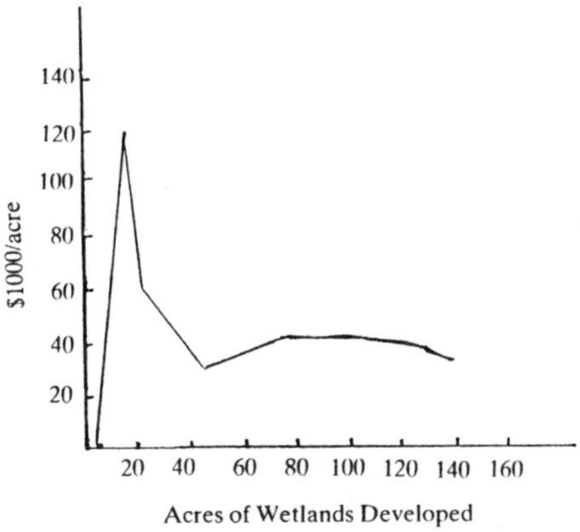

FIGURE 3
NET MARGINAL RETURNS TO DEVELOPMENT: NO FASTLAND ALTERNATIVE SCENARIO

the situation in which denial of a wetlands permit is compensated by a transfer of development of comparable recreational home lots to a fastland site. The marginal value of wetlands becomes positive after 10 acres when a fastland alternative is available. If a developer were to be denied a permit, for say, 147 acres of wetlands development, and develops instead 136 wetlands acres, the developer would compensate by adding an additional 11 acres to the fastland portion of the subdivision. If it is assumed that the lots added to the fastland portion of the subdivision are all interior lots, the marginal value of wetlands is $4,790—significantly less than the $37,040 calculated for the "no fastland alternative" for wetlands permits of 136 to 147 acres. This is equivalent to estimating the values of Box VII in Figure 1. The exact value will depend on the configuration of the ultimate subdivision and the number of lots which are shoreline, canal, or interior.

CONCLUSIONS

The state of hydrological and biological assessment of natural wetlands frequently is not of sufficient detail nor of proper design for use in economic valuation of specific wetlands sites. In contrast, economists are able to supply estimates of development value. Estimates of development benefits can provide a focus for an application of a "safe minimum standard" permit issuance rule. That is, when the permit could be denied (or modified to reduce wetlands destruction) when the development benefits foregone by doing so are not unacceptably large. Thus, the use of the development valuation framework for the provision of development value estimation appears to have considerable utility when data is lacking for a complete analysis of all social costs and benefits of natural resource use decisions.[12] For example the analysis of net marginal returns to wetlands, as illustrated at the Captain's Cove site, reveals several effects of allowing various levels of wetlands destruction. If it were reasonable to assume that the denial of a permit to develop Captain's Cove would mean no investment would be made, then the total returns foregone because of permit denial could be high. If a wetlands board focused on these returns, it is likely that the permit would be allowed unless very substantial negative externalities from development were identified.

In contrast, if a fastland alternative is available, a permitting board might restrict the development of wetlands acreage to a lesser amount when development values were not considered exceptionally large. In the Captain's Cove case, if the developers were effectively limited by a permit to less than 150 wetlands acres, they might still develop the Captain's Cove site at a lower profit level, but

[11]Ten acres is the number of acres at which the higher cost of development due to filling the wetlands just offsets the lower costs of putting in canals and a marina in wetlands. Hence, the marginal value to a developer is negative when a fastland alternative is available and when the number of acres under consideration is between zero and ten.

[12]The development benefit information is also pertinent to courts charged with assessing fines for noncompliance with wetlands permits. The developers of Captain's Cove were fined $4,500 for convictions on five counts of violating permitting laws (Wood and Hill 1978). It is clear from the net private returns calculated here that the magnitude of such a fine would not have provided a financial incentive to obey the law.

the pattern of roads, canals, and lots would have to be adjusted. An option available to a permitting board would be to allow the developer to dredge some wetlands in order to provide a marina in a region where water access is limited, for example.

References

Batie, Sandra S., and Shabman, Leonard A. 1982. "Estimating the Economic Value of Wetlands: Principles, Methods, and Limitations." *Coastal Zone Mgt. J.* 10(3):255–78.

Bishop, R. 1978. "Endangered Species and Uncertainty: The Economics of a Safe Minimum Standard." *Amer. J. of Agr. Econ.* 60:10–18.

Brigham, Eugene F. 1965. "The Determinants of Residential Land Use." *Land Economics* 41(Nov.):325–34.

Diamond, Douglas B., Jr. 1980. "The Relationship between Amenities and Urban Land Prices." *Land Economics* 56(Feb.):21–32.

Freeman, A. Myrick, III. 1979. *The Benefits of Environmental Improvement: Theory and Practice.* Baltimore: The John Hopkins University Press.

Gujarati, Damodar. 1978. *Basic Econometrics.* New York: McGraw-Hill.

Hitchens, M. T.; Thampapillai, D. J.; and Sinden, J. A. 1978. "The Opportunity Cost Criterion for Land Allocation." *Review of Marketing and Agricultural Economics* 46(Dec.):275–93.

Mabbs-Zeno, Carl C. 1980. "Institutional Purpose and the Management of Virginia's Coastal Wetlands." Ph.D. dissertation, Virginia Tech.

———, and Batie, Sandra S. 1980. "The Virginia Wetlands Act: A Case Study Comparing Academic and Applied Cost-Benefit Analysis." Presented at Southern Agricultural Economics Association annual meeting, Hot Springs, Arkansas, February 3–6.

Office of Technology Assessment, U.S. Congress. 1984. *Wetland's: Their Use and Regulation.* OTA-0-206. Washington, D.C., Mar.

Shabman, Leonard A., and Batie, Sandra S. 1983. "Socio-economic Values of Wetlands: A State-of-the-Art Review." Draft of final Report on Contract DACW39-82-M-4558, Waterways Experiment Station, Department of the Army, Engineer Corps, Vicksburg, Mississippi. 132 pp.

———, 1980. "Estimating the Economic Value of Coastal Wetlands: Conceptual Issues and Research Needs." In *Estuarine Prospectives.* ed. Victor S. Kennedy. New York: Academic Press.

———; and Mabbs-Zeno, Carl C. 1979. "The Economics of Wetlands Preservation in Virginia." *Journal of the Northeastern Agricultural Economics Council* 8(2):101–16.

Shabman, Leonard A., and Bertelsen, Michael K. 1979. "The Use of Development Value Estimates for Coastal Wetland Permit Decisions." *Land Economics* 55(May):213–22.

U.S. Army Engineer District. 1976. *Final Environmental Impact Statement: Captain's Cove Development.* Norfolk, Va.

Virginia Wetlands Act of 1972. Chapter 2-1, Title 62.1, Code of Virginia.

Wood, Lance D., and Hill, John R., Jr. 1978. "Wetlands Protection: The Regulatory Role of the U.S. Army Corps of Engineers." *Coastal Zone Management Journal* 4:371–407.

Part IV
Ethics

[32]

Valuing Wildlands

Holmes Rolston, III*

Valuing wildlands is complex. (1) In a philosophically oriented analysis, I distinguish seven meaning levels of value, individual preference, market price, individual good, social preference, social good, organismic, and ecosystemic, and itemize twelve types of value carried by wildlands, economic, life support, recreational, scientific, genetic diversity, aesthetic, cultural symbolization, historical, character building, therapeutic, religious, and intrinsic. (2) I criticize contingent valuation efforts to price these values. (3) I then propose an axiological model, which interrelates the multiple levels and types of value, and some principles for wildland management policy.

About 2 percent of the contiguous U.S. is wilderness (1.2 percent designated; 1 percent under study); 98 percent is developed, farmed, grazed, timbered, designated for multiple use. Another 2 percent might be suitable for wilderness or semiwild status—cut-over forests that have reverted to wilderness or areas as yet little developed. Decisions are being made about how to value these relict wildlands. Since they are almost entirely public lands, these are political decisions; but they are also taking place in the midst of a philosophical reassessment, coupled with ecological concerns, about how humans should value nature. They are political decisions entwined with reforming world views.

Since these are public land use decisions about wild nature, there is a tendency to think that the most useful principles and strategies are likely to be economic: that the nearest thing to an adequate theory of "resource use" is going to involve an estimate of benefits over costs in dollars; that wise use will be "efficient" use. Decisions ought to be democratic, since they are political and about public lands, but pitfalls in the democratic process are many. Those with political clout and savvy, those with concentrated high-order interests, a lot to gain or lose, outshout or outmanipulate the disorganized majority whose interests are diffuse and low-leveled. Organized small groups typically outact large latent groups; legislators react to pressure groups and defend their own interests. Agencies grow bureaucratic and sluggish; citizen preferences are difficult to register and aggregate; voters never have the options they prefer presented at the ballot box; and so on. One way

* Department of Philosophy, Colorado State University, Fort Collins, CO 80523. Rolston is associate editor of *Environmental Ethics*. This paper was written under Research Contract No. 28-K2-259, USDA Rocky Mountain Forest and Range Experiment Station. The author thanks Dennis Donnelly and J. Baird Callicott for stimulating criticism.

to minimize these pitfalls is to insist on a decision analysis which is more systematic, more scientific, which often means, more economic.

While it is widely recognized that some "amenity values" or "environmental values" are recalcitrant to quantification in dollars or other units, nevertheless, the effort to see how far cost-benefit analysis introduces some sense of order into an otherwise sprawling dispute over values continues. Legislators and government professionals are always sensitive to the charge of misusing public funds and resources, and if they can make economically based wildland decisions, they think that their decisions will then be as nearly scientific and democratic as can be. They believe that one sure route into human caring is by pricing, not in all cases, but routinely in matters of resource use. Some consideration of the just distribution of such costs and benefits will be needed, but that can come later. Initially, they say, an economic assessment of what these relict wildlands are worth is all that is required.

In the discussion to follow I challenge this strategy. My discussion is not a criticism of the utility of cost-benefit analysis, only an inquiry into whether the sorts of values to be protected on relict wildlands can be captured in economic terms. In section one I outline seven levels of value, applicable independently of wildland decisions, followed by twelve types of value associated with wildland use. In section two I question whether efforts to register these values in monetary terms succeed. In section three I suggest several philosophical principles which might improve wildland decisions. Section one is descriptive; section two is critical; section three is partisan; all sections are analytic.

I. A TAXONOMY OF VALUES

In the taxonomic levels and types narrated here, there is no reason to think of mutually exclusive categories, but there is every reason to think of identifiable dimensions of value that need to be factored into any complete analysis.

MEANING LEVELS OF VALUE

(1) *Value*$_{ip}$ = *individual preference value*. Value $_{ip}$ is what individuals prefer in contexts of choice. Valuing and its product, value, lie in the experience of interest satisfaction. In this sense valuing is subjective—valuing brings value into being within subject-owners, typically in their relationship to the world. Mere objects, including organisms that have no psychological life, no felt preferences, can have no individual preference value, though they may be resources for preferences. In our private case, for normal adults, what we ourselves prefer is reasonably evident from introspection coupled with action. The values$_{ip}$ of others can be known from verbal preferences and behaviors, although some situations compel discrepancies between attitude and action. Where there are constraints, it may be difficult to

ferret out true values. We also have to separate goals from failed performances, latent from manifest preferences, etc. Value_{ip} has seemed to many to be the motor force of all value, from which all the rest are derived.

(2) *Value_{mp} = market price*. Often an illuminating way out of the subjectivity of value_{ip} is to look to the market, which, though produced by value_{ip}, has empirical objectivity. Articles and services are regularly exchanged on the market, which in nonbarter societies invests them with a going price, a public, observable quantity, resulting from many individuals estimating the worth of having these commodities. Value_{mp} is a derivative of usefulness, rarity, labor, advertising, government regulation, etc., but in the end things are traded instrumentally to satisfy human interests, and their price must reflect preferences. If no one desired these things, the market would collapse. In wild nature, no monetary or barter economy exists, but humans, nevertheless, buy and sell natural things incessantly. They labor, trade, and own property, using nature for interest satisfactions, and this brings wild nature into the economic sphere.

(3) *Value_{ig} = individual good value*. If valuing is just preferring, one can hardly make mistakes about what one has valued, any more than one can be mistaken about having made a target of something, though mistakes abound with regard to whether what is preferred really brings satisfaction. We can make choices and purchases that are not in our best interest, not really. They bring momentary goods, or none at all, and soon leave us worse than before. Value_{ig} is what is in a person's interest, whether or not the individual chooses it. Preferences need to be constantly revised accordingly in terms of it. There is something raw about untutored preferences; individuals need education (even perfect knowledge) before they can competently say what (all) their goods are. Many things happen (such as the rain) which are of value whether we welcome them or not. Biological processes (photosynthesis) are vital even in our ignorance of them. Values_{ig} may not involve exercising preferences, much less marketing.

(4) *Value_{sp} = social preference value*. Through politics, ethics, religion, etc., individuals express a social will, often conflicting with some particular wills. Value_{sp} is what a society prefers when optionally allocating its time, resources, skills, energy, and money. Values characterize groups, not just individuals. Value_{sp}, a social trait, represents some amalgamating of value_{ip}, a psychological trait, though it is not clear how this does, or should, take place. Sociologists debate whether there is any social whole above the individual person-parts who compose it. Society itself has no center of experiences; only individuals do. Society is not capable of interest satisfactions; it enjoys no pleasures, suffers no pains. Such considerations lead many to claim that value_{sp} is some kind of fiction, a pragmatic operational concept (like a center of gravity) which is only apparent. Still, some social preferences seem to serve society at large, beyond the fact that, or regardless of whether, they satisfy individual interests. Social preferences, unless oppressive, seem to command more importance than particular individual preferences. At least they are relatively more enduring.

(5) *Value*$_{sg}$ = *social good value*. Society can err about what contributes to its well-being. Social choices too can be out of touch with reality. Further, given the prevailing pluralism, a negotiated consensus of values$_{sp}$ is likely not to be consistent. Even if only individuals have a well-being (society being an aggregate of related individuals), at least society functions and dysfunctions. The vague, beguiling slogan, "the greatest good for the greatest number," can mean "what most prefer on average" (value$_{sp}$) but usually means "what is on average functional in society" (value$_{sg}$). Part of the worth of a practice is whether it keeps society functioning smoothly, regardless of whether this agrees with the corporate will. The "open space" of the Indian Peaks Wilderness serves as a pressure release valve for the Denver-Boulder metropolitan area, and it might be more functional for Colorado society to maintain it so, even though the legislature voiced a preference, reflecting polls, that federal managers log the area for jobs and firewood.

(6) *Value*$_{or}$ = *organismic value*. Value is not just an economic, psychological, social, and political word, but also a biological one. Value$_{or}$ is what is good for an organism, and all preferences and goods of humans are really subsets of this more comprehensive notion. Various instrumental organic and environmental goods contribute to an organism's well-being, and that well-being is for the organism a telic end state, an intrinsic value, not always a felt preference. Survival value, lies at the core of evolutionary adaptation. Genetic information is of high organismic value, but has no necessary connection with sentience, experience, preference satisfactions, or markets. Wild creatures defend their lives as if they had goods of their own. An organism grows, repairs its wounds, resists death, and reproduces. Every genetic set is in this sense a (nonmoral) normative set, proposing what *ought to be* beyond what *is*. At this level, wild nature is a place of values prior to human decisions, and one thing the reforming world view asks is whether any concern for wild organismic value limits human decisions about land use.

(7) *Value*$_{es}$ = *ecosystemic value*. Like persons in society, organisms live in ecosystems, with a parallel between the good of the system and that of the individual. Persons are good in the roles they serve—mothers, wildlife biologists. Organisms fill niches and sustain a flourishing system, though not intentionally, perhaps unwillingly. Songbirds, which have intrinsic value$_{or}$ in themselves, have instrumental value regulating insect populations. Ecologists too (coached by sociologists?) have doubted whether ecosystems exist as anything over their component parts. Value$_{es}$ is even more convincingly a kind of fiction than value$_{sg}$, because in ecosystems there are no policy makers, no social wills, no goals. Though the ecosystem can seem more biologically real than the social system, it less evidently has a locus of value. Nevertheless, in some way that we poorly understand, some creativity (accidental or not) in the evolutionary ecosystem has formed all its organic species and processes, and we as humans are now much concerned with keeping these ecosystems running as they do. Some events can be better, some worse for the integrity of a biological community, as in the choice for a biodegradable pesticide.

II. CONTINGENT VALUATION OF WILDLANDS

Can the preceding array of value levels and types be reduced, wholly or in part, to economic terms, as a prerequisite for a cost-benefit analysis? J. V. Krutilla and A. C. Fisher noticed the difficulty and suggested some categories. "In confronting the need to evaluate preservation benefits, we find that there are a number of aspects of such benefits that we do not know how to estimate quantitatively. These are the value of natural environments that have remarkable qualities for scientific research; the value that individuals place on retaining an option when faced with actions having irreversible consequences; and the value that some individuals place on the knowledge of the mere existence of the gifts of nature, even when they feel certain they will never have or choose an opportunity to experience them *in situ*."[2]

Taking up the challenge, there have been a number of proposals to make economic estimates of various wildland values, typically (1) scenic value, (2) recreational value, (3) option value, (4) existence value, and (5) bequest value.[3] *Option value* is the value of retaining options that a set of natural entities, threatened with erosion or destruction by development, provides now and henceforth to a prospective user. *Existence value*, separable from option value, is the satisfaction to an individual of just knowing that the wild set exists, even if unvisited by himself (or others?). *Bequest value*, thought also separable, is the value that an individual places on bequeathing a wilderness to children and future generations.

---

[2] J. V. Krutilla and A. C. Fisher, *The Economics of Natural Environments* (Baltimore: Johns Hopkins University Press, 1975), p. 124; J. V. Krutilla, "Conservation Reconsidered," *American Economic Review* 57 (1967): 777–86.

[3] For contingent valuation of aesthetic, scenic visibility, see: Alan Randall, Berry C. Ives, and Clyde Eastman, "Bidding Games for Valuation of Aesthetic Environmental Improvements," *Journal of Environmental Economics and Management* 1 (1974): 132–49; Alan Randall, Berry C. Ives, and Clyde Eastman, *Benefits of Abating Aesthetic Environmental Damage from the Four Corners Power Plant, Fruitland, New Mexico*, New Mexico State University Agricultural Experiment Station Bulletin 618 (Las Cruces, 1974); David S. Brookshire, Berry C. Ives, and William D. Schulze, "The Valuation of Aesthetic Preferences," *Journal of Environmental Economics and Management* 3 (1976): 325–46; R. D. Rowe, Ralph C. d'Arge, and D. S. Brookshire, "An Experiment on the Economic Value of Visibility," *Journal of Environmental Economics and Management* 7 (1980): 1–19. For contingent valuation of wildlife experiences see: Judd Hammack and Gardner Mallard Brown, Jr., *Waterfowl and Wetlands: Toward Bioeconomic Analysis* (Washington, D.C.: Resources for the Future, 1974); David S. Brookshire, Alan Randall, and John R. Stoll, "Valuing Increments and Decrements in Natural Resource Service Flows," *American Journal of Agricultural Economics* 62 (1980): 478–88; David Brookshire, Alan Randall, et al., *Methodological Experiments in Valuing Wildlife Resources: Phase I Interm Report to the United States Fish and Wildlife Service*, 1977. For contingent valuation of option, existence, and bequest value in wilderness preservation see: Richard G. Walsh, John B. Loomis, and Richard A. Gillman, "Valuing Option, Existence, and Bequest Demands for Wilderness," *Land Economics* 60 (1984): 14–29. For a sensitive overview, cautiously advocating contingent valuation, see: William D. Schulze, Ralph C. d'Arge, and David S. Brookshire, "Valuing Environmental Commodities: Some Recent Experiments," *Land Economics* 57 (1981): 151–72. See also George L. Peterson and Alan Randall, eds., *Valuation of Wildland Benefits* (Boulder, Colo.: Westview Press, 1984).

The most plausible of the five to be measured economically is recreational value. Recreation is sold elsewhere on the market and can perhaps be shadow priced. Though wilderness experience is unsold, it costs something—gas, meals, lodging enroute, fishing licenses, gear (?), campground fees. Access costs are typically $10–15 per day. Does this represent the value of the visit? Difficulties arise. Wilderness is a free good provided by Mother Nature, like fresh air and sunshine. Wilderness is a nonmarket service provided by governments, like the military and public schools. The wilderness experience is an atypical recreational experience and what the user actually spends poorly reflects its value.

To overcome this difficulty, contingent valuation is proposed. We can discover what users would be willing to pay. "The applicable rule to follow . . . is to use that procedure which appears to provide the best measure or expression of willingness to pay by the actual consumer of the recreation good or service."[4] "Techniques of benefit estimation have been developed sufficiently to make benefit-cost analysis fully applicable to appraisal of recreation alternatives on public forests. Willingness of users to pay is the appropriate way to measure benefits, and can be estimated from either the participation behavior of recreation users or by surveying a sample of participants to learn their preferences."[5] From what the users (think they) are willing to pay, we now subtract access costs for the net benefit supplied by the government, the "consumer's surplus," a value$_{ip}$ received but not paid for.[6] Aggregated over multiple visits and visitors, the total renders objective the otherwise subjective (though real) wildland recreational value, which can be compared with timber sales.

This is imaginary money, which the users pay no one and which also does not remain in their pockets. Still it might provide a reliable estimate, provided the user has a congenial context of judgment. But on wildlands there is little reason to ask willingness to pay except under the shadow of protecting wildlands from alternative uses, and in this context the answer becomes a competitive bid. Rather ill-defined, it is something akin to a proposal that we take from them what they have had by tradition, public benefit, or right, and then ask what they are willing to pay to defend it. The precise status of any "right" to recreation, scenic views, open space, wildlife encounters, religious experiences (all integrated with recreation)

---

[4] U. S. Water Resources Council, "Principles and Strategies for Planning Water and Related Land Resources," *Federal Register* 38, no. 174 (10 September 1973): 24, 778–866, citation on p. 24,804; U. S. Water Resources Council, *Procedures for Evaluation of National Economic Development (NED) Benefits and Costs in Water Resources Planning*, Subpart K, *Recreation, Federal Register* 44, no. 242, (14 December 1979): 72, 950–65.

[5] J. F. Dwyer and M. D. Bowes, "Benefit-Cost Analysis for Appraisal of Recreation Alternatives," *Journal of Forestry* 77, no. 3 (1979): 145–47, with comment by A. A. Dyer and J. G. Hof, 147–48, and vigorous discussion by others, *Journal of Forestry* 78, no. 1 (1980): 21–28.

[6] Consumer's surplus is "the maximum a consumer will pay for a given amount of a good, less the amount he actually pays" (E. J. Mishan, *Cost-Benefit Analysis: An Introduction* [New York: Praeger Publishers, 1971], p. 31).

on public lands is obscure, and, of course, developers also have "rights" on "lands of many uses." But citizens have long been told that these are "your national forests" (as the signs say on entry), which they are encouraged to use, managed by public servants whom (directly or indirectly) they have elected. They are wholly unaccustomed to paying to keep developers out; their presumptions are the other way round: that the developers who gain (and not the losers) ought to bear the full cost of what they remove from the public till.

The better question might be what the users would accept in payment to give up their recreation. But now we are asking a quite unfamiliar question. Although it does presume a willing seller rather than a defender of a threatened good, the user is made the owner of a good—a unit of recreation (integrated with historical, cultural symbolization, or life support values)—that he is ill-adept at selling. It introduces a host of concerns (examined below) about whether a citizen can entertain the thought of selling, corporately with others, a public good for private gain, *x* dollars. The upshot is that in these surveys economists are unsure whether to ask willingness to pay or to accept payment. Even if they have theories about what should be the starting reference point, their survey respondents do not operate out of these theories.

Clever respondents might bid or sell high to save the wilderness. (Or should they bid low to prevent entrance fees?) They could misrepresent to gain the outcomes they value. However, strategic scheming seems beyond most respondents, and can (in part) be checked against by alternative questions that put all on a limited budget. A better grounded fear is of nonstrategic pricing. If they think their estimates are not going to set any policy, they will not care what they say, especially if busy or puzzled. They must not simply be preferring but helping to project what will take place. The incentive for being responsible is the same as that for bias. Even if earnest, one is being asked to price what one has little clear strategy for pricing. This is especially true in the package of up to a dozen goods, some of which (re-creation beyond recreation, aesthetic experience, religious experience, amateur scientific study, character building) make it difficult to differentiate recreational value in isolation from the rest.

The difficulties are compounded if we try to price aesthetic value. (Is this a subset of, or something else than, recreational value?) "This is a national park, to be affected by a proposed coal-fired power plant nearby. How much additional entrance fee would you pay to have for thirty days a year (of relevant atmospheric conditions) a scenic visibility of 25, 50, 75 miles? What would you sell such views for?" This supposes that an individual in private trade might gain or sell a public good such as air quality or scenic vistas, more or less murky. It asks him to translate a good which is collective and nonmarket (scenery) into a good which is private and market (consumer dollars). It asks an individual to presume the right to sell (not only on his part but implicating others) what he thinks of as a nonexclusive, nonrival public good. "This wilderness has coal under it. How much would you pay (taxes, utility hikes?) not to lose *x* encounters with elk, coyotes, brown creepers?" (Is this an aesthetic value or respect for value$_{or}$ mixed with value$_{es}$?) These questions suppose a voluntary exchange (power for scenery, dollars per wildlife encounter), but in fact the exchange will be individually involuntary; it will be an agency decision levied upon all. The sets of persons who lose the vistas and the wildlife are really quite different from the persons who gain the power or the operators who profit.

The frequency of "infinity," "zero," or "not for sale" responses indicates that for many (up to half) the question is in one way or another off the map.[7] Persons expect to pay for some good produced, an artifact or service, some material that has been or can be labored over. But here on wildlands there is no "producer" or "supplier" whom it seems right to compensate.[8] No one has labored over anything, and why should anyone pay or be paid? In a way, the "victim-must-pay" inquiry is like paying to prevent the theft of a good one thought he had by democratic process or by gift of nature. It is something like paying for protection on the streets of Chicago, hardly a free market. It forces the user to be a consumer and compromising pragmatist in a decision context in which he ought to be a citizen, esthete, philosopher, and ethicist.

The format allows no account of what the recreator thinks in the fuller dimensions of his person about the larger benefits of public wildlands, free goods provided by nature and preserved by the government, about income constraints, the inequitable distribution of wealth, or the best things in life being free. The respondent is blinded to such considerations as the need for diverse options in recreation, rights of minority users, future trends, needs of future generations, intrinsic natural values, the psychological or social desirability or the quality of various forms of recreation. He must skip all historical, therapeutic, or cultural symbolization values. Already a victim who must estimate his willingness to pay to defend something he owns, the citizen is further victimized by the narrowed context into which he must squeeze his preferences. Only dollar answers count.

If we try to press beyond recreational and scenic values to option, existence, and

[7] See I. M. Gordon and J. L. Knetsch, "Consumer's Surplus Measures and the Evaluation of Resources," *Land Economics* 55 (1979): 1–10; Randall, Ives, and Eastman, *Benefits*, pp. 19, 24; Shultze, d'Arge, and Brookshire, "Valuing Environmental Commodities," p. 166; Rowe, d'Arge, and Brookshire, "Experiment," p. 9.

[8] Subsidized public recreation is not at issue here. On-site wildland users may be enjoying benefits which have maintenance costs (upkeep of roads, trails, game supervision, ranger salaries) and which taxpayers at large are bearing. Questions about whether on-site recreators should bear more of the actual costs are plausibly subject to willingness-to-pay evaluation, although the lack of a realistic market context troubles answers here too. The unspecified mix of recreational with other benefits preserved on public lands (life support, genetic diversity), which taxpapers do enjoy, complicates the issue. But should-users-bear-the-maintenance-cost questions are conceptually different from willingness to pay to prevent economic development, pollution, species extinctions, scenic despoilations, which impinge on recreational opportunities.

bequest values, the problems rapidly grow worse. While some economists caution that we do not know how to price these values, others suppose that here too willingness to pay is the appropriate measure. A Colorado household is willing to pay, on average, about $4.04 annual option value, $4.87 annual existence value, and $5.01 annual bequest value, for a total of $13.92 annually for 1.2 million acres of wilderness. This sum rises to $18.75 for 2.6 million acres ($5.44 option value, $6.56 existence value, $6.75 bequest value). For 5 million acres, the total willingness to pay is $25.30 ($7.34 option value, $8.86 existence value, $9.10 bequest value). Multiplied by the number of Colorado households, we calculate the benefits to the state.[9] But why not use willingness to sell here? Perhaps because no one could seriously imagine that compensation would be paid the sellers. In any case, what have such numbers measured?

Such categories as existence, option, and bequest values promise to package up a fuzzy assortment (roughly the range of types 2–12, minus some recreation), but as values grow intangible, social and ecosystemic, the individual's capacity to price them becomes progressively poorer. All the problems met with in pricing recreational or scenic value return with a vengeance. As consumers, respondents are being asked to express their convictions in dollars about things over which they have no market experience, and dollars are, after all, units that have their everyday home in markets. Even as citizens voting in referendums they are accustomed to answer "Is-it-worth-it-in-dollars?" kinds of questions that have to do with the purchase of market materials and services (bombers, library books, real estate, building construction, man-hours of police time). They are unaccustomed to citizen "purchases" of genetic diversity, wildlife encounters, scenic vistas, or wild cultural symbols, especially on lands that the citizens already own and in situations where no money changes hands.[10]

The respondent has never operated in any market vaguely resembling these kinds of goods. How do we price, for instance, "the value that some individuals place on the knowledge of the mere existence of the gifts of nature, even when they feel certain that they will never have or choose an opportunity to experience them

[9] Walsh, Loomis, and Gillman, "Valuing Option, Existence, and Bequest Demands," p. 25.

[10] Contingent valuation needs to be distinguished from citizens' referendums. In April 1981 Larimer County, Colorado, citizens voted for a half cent sales tax over six months to purchase the Soderberg Ranch containing Horsetooth Mountain, on the skyline of Fort Collins. Following public debate, a local jurisdiction voted to use mutual coercion to buy something on the market, which they did not own, in order to preserve wild recreational, aesthetic, and cultural embodiment values. There is nothing hypothetical and nonmarket here, and no victim-must-pay-for-what-he-already-owns overtones. They were buying a working ranch, with the owner using the proceeds to relocate, competing with offers from real estate developers. They were not separately bidding, but had a corporate target figure, translated into a tax, on which they were voting yes or no. That is a citizen choice using votes to allocate dollars. By contrast wildlands are already public lands, and citizens (if they so regard themselves as respondents) are being asked about a hypothetical willingness to pay to protect what they already own, with nothing traded that bears any relationship to actual markets.

in situ"?[11] Since biological diversity, a value we are only learning to appreciate, is not something on the market, how can one price it? Does one include it somewhere within option, existence, or bequest values, or is it omitted in such a survey? Might it be that for $3.00 per person Americans will sell the whooping crane into extinction? Most respondents do not know how to resolve questions of extinction even within their own hierarchy of personal needs, much less to price these for social policy. Does it matter whether the respondent is ignorant about the intricacies of the ecosystem he judges, whether the estimator of scientific value is a scientist? Even allowing, as economists often do, that some environmental values remain uncaught, how do we know which ones are left out and not confusedly stated somewhere in the dollar amounts for option, existence, and bequest values?

It might first seem to help to cast the whole question into a citizens' rather than a consumers' orientation. Respondents are cautioned that they must not think of their replies as consumer purchases, despite the fact that their dollar figures go into cost-benefit equations, and even though a frequent economist's term for the value so captured is "consumers' surplus." They must think of them more like citizen votes, using dollars as a kind of proxy, indicating what they think citizens ought to pay for wildland benefits, comparably to the way they pay for law courts or military protection. The survey is a citizen's device only partly to cooperate with but more to counter the real market. Willingness to pay levied taxes will have more citizen orientation than individual entrance fees, purchasing a day's recreation, or utility hikes, consumer-style transactions. But the payment of extra taxes does nothing to eliminate the victim-must-pay distastefulness of the procedure.

Willingness to pay becomes a kind of game, used not so much to imagine anything on a consumers' market as to elicit citizen choices, preferences stated in monetary amounts in order that these can be weighed into the equations, competing with alternate uses which do involve marketed goods. Despite appearances, we are not asking a "What's-it-worth-to-you-personally?" question, where each bids dollars depending on tastes and purchasing power. We are asking, "What's it worth to citizens on average, yourself included, in tax dollars, if all pay fairly proportioned amounts, and if all have the same nonconsumptive access to these public benefits that we cannot expect the market to supply?" In terms of the value level analysis, we are not asking a value_{mp} question, despite the dollar appearance of the answers, nor even a value_{ip} or value_{ig} question, but a value_{sp} and value_{sg} question.

[11] Those who value the mere existence of wildlands, though they never expect to visit them, are often called "off-site users," a term revealing the utilitarian, anthropocentric coordinates of this value-mapping system, made worse by contingent valuations treating them as off-site consumers. Such categories lie in an alien reference frame from that of those trying to report convictions about intrinsic values in nature. "Off-site users" grotesquely illustrates the extent to which a non-negotiable paradigm ("resource use") can be twisted to accommodate anomalous phenomena. It exactly reverses what those so categorized would say of themselves: that wildlands are valuable when left alone, apart from questions of human use.

The difficulty now is that the question is not what it appears, and even if the respondent comes to understand the pretending involved, he has no rules by which he can translate $value_{sg}$ into $value_{mp}$, none by which he can integrate value types 2–12 on a scale commensurate with the market value of wildland products. The more he operates as a citizen, the more the privatized form of the question is remote from what he is really trying to indicate, and the less his capacity to do any pricing. He may also feel that he is being forced to play this game, in the sense that no answer eliminates his opinion, and wonder whether zero bids or infinity demands (or others that the interviewer considers out of range) will be eliminated on grounds of noncooperation. If a respondent states a huge sum (a recreator who personally places high religious value on wilderness experiences, a citizen who wishes to protect the rights of those who do, or an Earth citizen who holistically values ecosystems), will this make it into the cost-benefit equations, or be tossed out as a monkey-wrench answer? The citizen ought not in principle be asked to couple sufficient money with his nonmarket policy preferences; and when he is asked this, he does not in practice reliably know how to answer.

The respondent has no idea how to do any calculations; yet on the basis of his guesstimates, economists do metric calculations, overly refining what are really raw data. All this number crunching creates the illusion of mathematical exactitude covering up what were, to begin with, iffy replies in a cramped hypothetical context. Nor is the use of the respondent's behavior to correct verbal misjudgments of willingness to pay reliable, because behavior is already infected with the inequities in the prevailing distribution of wealth. Meanwhile, it will take considerable intellectual subtlety for the respondent to understand the differences between willingness to pay and willingness to accept payment, between consumer and citizen uses of dollars, between option, existence, and bequest values, between hypothetical and actual markets, to say nothing of reliably attaching dollar amounts to these issues. It is fortunate that most visitors to wildernesses, where these surveys are usually conducted, are college educated!

The wilderness purist thinks the procedure is profane; the ethicist protests that justice is wholly overlooked. How we treat the environment is not always a matter of economics; it is sometimes a matter of conscience. If the psychiatrist Jung is right, some of our emotions toward the land rise out of the collective unconscious; yet we are asked to price them in a few minutes. In psychology, advocates of values clarification argue that we have not identified a person's values until we find options he or she has chosen (1) freely (2) from genuine alternatives (3) after reflection, understanding the outcomes. The person must (4) remain satisfied with the choices, (5) be willing to advocate them publicly, and (6) act upon them (7) repeatedly. Not one of these criteria is satisfied here.

The admission-price-to-nature question could hardly have been asked in any earlier century or in a nonindustrial country, being the product of an economy biased toward production maximizing dollars. The mentality of the methodology by which we seek a solution is what created the problem, decimating the wilds. "Where will you put your dollars?" is as theory-laden as "Where is your center of gravity?" But unlike theories in physics, here we have a value$_{mp}$-laden theory that already purports to know what counts as costs and benefits. What we want to know here (or ought to want to know) are the citizens' *convictions* about goods in nature and their appropriate response to them, their public conscience, and we only confuse those we interview when we ask about their *desire* to pay. The question is about principles, not pocketbooks. They are being asked if an ideology is for sale. Desire-to-pay questions elsewhere in life are kept carefully separate from what we believe to be noble or ignoble behavior. The cash question is incompatible with answers of breadth. To insist on pricing is to insist on a category mistake ("the gold standard") where attitudes are inchoate.[12]

Advocates and even critics can throw up their hands. They can say that the technique is well intended, that there is nothing better, and that we cannot avoid running an economy. Some dollar values are better than none at all, for otherwise these intangible values get lost in the midst of pressures for economic use. Especially as the technique has been lately refined, the results significantly aid wilderness preservation. Environmentalists can fight fire with fire and prevent the burning of their wildlands on the altar of progress. Perhaps. But (to change the figure) a philosopher hates to fly in a conceptually flawed airplane, even though for the time being it is pragmatically flying in the right direction. If the model is a muddle but we use it anyway, being unwilling to face its fallacious reductionism, we are disinclined to admit the complexities of the real world and to look for truth in likelier directions.

III. WILD DECISION RULES

We do not want to play the game by the old rules, but to rewrite the rules. The point of my narration of levels and types of value (section one) was to display a richer value spectrum than we have reason to believe can be caught by economic valuation (section two). Can we say anything positively (section three) to order decisions? These principles will not constitute a procedural set. Like a compass, they will orient general directions of travel. Specific paths will have to be figured out locally. The rules will only begin to map some presently wild terrain. But they will be more dependable for discovering and protecting wildland values than economic reductions, which as we just found normally lead us astray.

(1) *Use an axiological model.* A diagrammatic schema (figure 1) should guide policy decisions on remnant U.S. wildlands in the contiguous United States. It is applied elsewhere (abroad, Alaska, outside environmental ethics) at your own risk. Above the line is a humanistic sequence, below it is a naturalistic sequence.

[12] M. Sagoff, "At the Shrine of Our Lady of Fatima, *or* Why Political Questions Are Not All Economic," *Arizona Law Review* 23, no. 4 (1981):1283–98.

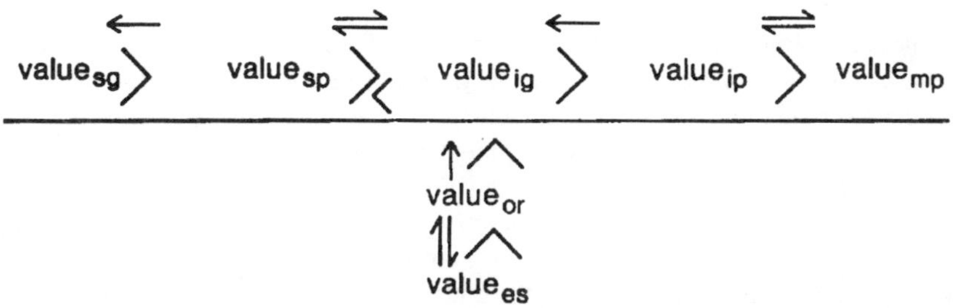

Fig. 1. An ordering of levels of value.

Social goods ordinarily override (>) social preferences, although the latter routinely produce (←) such goods. In turn, this combination overrides individual goods, though these feed and determine social preferences, which reciprocally also promote (⇌) individual goods. A caveat, the small wedge checking the larger one, (≷) specifies that some individual goods, few but crucial, veto some social preferences. Murder is not justified to obtain wilderness, even supposing society preferred this. Individual goods trump those individual preferences which commonly produce them. Individual preferences are what produce (→) market prices, and the existence of a market produces in turn (←) the satisfaction of individual preferences so that there is a two-way arrow (⇌).

On remaining wildlands, all this valuing overrides market value. The social drama is superimposed on the underlying ecosystem, and the diagram suggests the shrunken wild domain. Organismic value, as a populational form of life (e.g., endangered species, not individuals) ordinarily trumps what (usually minor) social or personal goods compete with it, though organisms ordinarily contribute social values. Pileated woodpeckers are good in themselves and also enrich human experience. Ecosystemic value is more basic than the life-form level, though biosystems are naturally selected to interfit (⇅) the two. Both contribute to serial human goods.

The diagram yields *prima facie* rules. For exceptions, the burden of proof is on the dissenter. On remnant wildlands the odds are that the sacrifice of wild values will not both contribute to long-term human goods and be justified. The wilder and rarer the land, the lower the probability that any consumptive economic use will override other values. Virgin forests will stay intact. The demand for wilderness is increasing quite as rapidly as the demand for timber; although the latter can be satisfied on private lands, the former cannot.

I am advocating a kind of *maxi-min* principle following the ratio of continental domestication to wildness so that a minimum level (2 percent, 4 percent) of wildland values (intrinsic and human instrumental) is maximized (at 98–2, 96–4 odds), opposing a *maxi-max* principle (maximum consumption increasing from 96 percent to 98 percent to 100 percent our available acreage to raise our already high standards of living). That two to four percent surfeit is a "consumers' surplus" we do not need. It is a few more pounds on already fat people, who need the rigorous leanness that disciplined exposure to wilderness can give. Considering probable economic productivity, rather than acreage, there is no reason to think that on average U.S. citizens would be even one percent better off if all wildlands were sacrificed. From this viewpoint the odds are 99–1 that the sacrifice of wildness makes sense in order to achieve economic gain at the cost of values 2–12 lost.

The areas richest in resources have long since been domesticated and sacrificing the remaining wildlands is scraping the bottom of the barrel, a matter of diminishing marginal returns (notwithstanding newly developed technologies), although from other points of view the destruction of noneconomic value would be enormous. Public lands, often left over from the public domain in the West or reacquired after marginal use in the East, tend not to be economically productive in agriculture, timber, or minerals.

(2) *Maximize noncompeting value types*. The twelve types of natural value are incommensurable, but are also largely noncompeting in all cases but the first. Although they cannot be commensurated, they do not have to be. There is no translation unit by which $A + B - C = D$. Genetic diversity (indicated by species counts, low extinction rates) cannot meaningfully be added to, or subtracted from, recreational benefits (visitor days). But we can easily add together and simultaneously enjoy multiple noncompeting incommensurables. Historical, character-building, or therapeutic uses of wildlands will seldom interfere with scientific and religious uses, and nowhere are science and religion more congenial than here. The cultural value carried by eagles requires the preservation of a flourishing wild population. Aesthetic uses (measured on preference scales) do not upset the life support value (indexed by energy flow?) in an ecosystem. So we are not forced to prefer only one value. The commensurability of values is, thus far, a pseudoproblem, because these nonconsumptive values reinforce and need not be traded off against each other.[13] They do not use up wilderness in the ordinary resource sense at all, making it into something different from what it spontaneously was.

---

[13] Some counterexamples: Sioux and Cheyenne Indians lost a legal bid to halt recreational development at Bear Butte State Park in the Black Hills, a conflict of religion with recreational value; other kinds of recreation would have been nonrival. In a Maine referendum on moose hunting, the aesthetic, recreational, cultural symbolization (the state animal) values of environmentalists, as well as their respect for intrinsic organismic value, clashed with the differing recreational, aesthetic, character-building, and historical values of the hunters. In Florida, fishermen sought, unsuccessfully, to have the park service relax restrictions on fishing in the Everglades, restrictions protecting the endangered crocodile. The Defenders of Wildlife protested regulations which allow the twilight shooting of game birds, on grounds that low visibility makes it difficult to distinguish common from endangered species. These involve clashes of forms of recreational value with scientific, genetic diversity, organismic, ecosystemic, and alternate forms of recreational value. But such conflicts are minimal beside economic conflicts on wildlands. Though they are not handled by the rules proposed here, willingness-to-pay surveys are even more irrelevant.

Only economic uses tend to consume wildlands. The question faced here is not whether the past economic reduction of the continent was justified, but whether it is enough. How much more of the goods (values$_{ip, sg}$) we already amply have (fiber, timber, energy) can we obtain by consuming the surviving wildlands, and do we want these as economic benefits (value$_{mp}$) by trading away the nonconsumptive set of values? Although this seems to pit market value against some or all of values 2–12, for a wider public it only pits a little more extracted from the 2 percent wild (the 4 percent half wild) of what we have extracted already from the 98 percent (96 percent) in impressive amounts. Yet this "more" could as easily and efficiently be extracted from nonwild lands, which are far and away already the richest ones economically. If we ask who are the chief beneficiaries (a few operators, perhaps more workers) and who are the trickle-down beneficiaries (the public), we may find that this gains a few concentrated, short-term values$_{ip, ig}$ (profits, jobs perhaps unobtainable elsewhere) traded against values$_{sg}$ and diluted but extremely widespread, long-term value$_{ip}$ (recreational benefits, religious experiences). Do we have enough consumption? If more is needed, need it come by sacrificing wildlands?

On remaining wildlands, economic use is typically consumptive, and that gives preferential treatment to one class of users and uses (value type 1) at the cost of depriving other users of their alternate uses (value types 2–12). The nearest that policy can come to nonpreferential treatment is by nonconsumptive use. Everyone can use wildlands, but no one ought be allowed to use them up. Sometimes disparagingly called a "lockup" use, that use treats all users alike, even though (indeed, just because) the would-be economic user is prohibited from "takeover" use. "Lockup" only prevents "takeover." It is more likely to result in gradually unlocking an equal distribution of benefits of the other types over time. It simultaneously protects organismic and ecosystemic value, leaving wildlands in a spontaneous natural state.

If basic needs were at stake, or if the continent were not already 98 percent developed and developable, this might be unjustly prohibitive. But it is not unjust on the last 2 percent, where only more of already abundantly possessed goods is at stake. While any economic use is somewhat consumptive, since it sacrifices primordial wildness to some degree, some economic uses are more consumptive than others. A clear-cut takes the forest out of other uses for a century, and the second-growth forest is never the equal of the virgin forest. A cut by selection leaves half the original wildness and may only halfway destroy values 2–12.

Beyond this, however, values of the sort here defended tend not to aggregate additively, even when noncompeting. They integrate into a gestalt in which a calculus is irrelevant. A single value change may throw the whole pattern into a different light. Colored floodlights on Old Faithful, illuminating it for night viewing, would add tourist revenue, and yet break the picture of wildness for most park visitors. A small-scale, out-of-sight logging operation returns economic benefits; while the noise of the chainsaw carried around the valley may subtract only a little from the solitude of some wilderness travelers, it destroys it entirely for others. Various thresholds of perception and value experience are crossed, which are not incorporated in aggregating procedures.

(3) *Keep remaining public wildlands off the market.* The military, police, courts, schools, museums, churches, scientific societies, historical parks—all cost money and have budgets, but are not businesses expected to produce income in any cost-efficient sense, not even when they capture (by fees, admission charges, or contributions) part of the value of their services. Though we sometimes decide how much they are worth in dollars, their purposes are to produce nonmonetary values. Wildness costs something to preserve, though only the minutest fraction of what these other social activities cost. Costs are largely opportunity costs forgone, so far as these cannot be achieved on the 98 percent of the continent on the market.

Even on semiwild lands, it is a category mistake to compare market efficiency in a Weyerhaeuser timber tract with a national forest. Where national forests are quasi-market operations, and little more, they should be sold or leased to private entrepreneurs, who will operate them better. On truly "national" forests we protect other values, and here market-style questions are awkward because what we want on wildlands is what the market never sells—a hiking trail, a trout stream, a scenic view, a wilderness experience. No one invests in land to lease it for bird watching or butterfly collecting. That is economic nonsense. The purpose of public wildlands is to provide benefits that we cannot expect on the market.

The point of a value analysis should not be to translate all or as much as possible of values 2–12 or levels 1–7 into an economic common denominator, but to display the wide spectrum of types and levels and give decision makers and citizens a strengthened persuasion where these really do (or do not), in aggregate or gestalt, beat economic considerations on the fractional remaining wildlands. We want a policy to protect these value dimensions not because they are covert economic values, but because they are not economic values at all.

(4) *Do not use remaining wildlands nonrenewably or consumptively to satisfy the basic needs of a minority in society.* The way to feed the hungry is by a redistribution of produce from lands in private hands, not by further exploitation of the fractional public wildlands. Very few will be put permanently to work by wildland exploitation, and these can surely as well be employed elsewhere in the enormous American economy. In a free economy, left otherwise unchanged, it is impossible to assure that the benefits from sacrificing wildlands will go to the poor; indeed, they probably will not. What the disadvantaged (or the middle or upper class) think about trading wildland values to help the disadvantaged becomes irrelevant when considered apart from whether in the prevailing economy this transfer can reliably be expected to take place.

Nor will the individual economic gains of the poor ($value_{ig, mp}$) overcome the social losses ($value_{sg}$) in value types 2–12. Poverty problems should be solved where they arise: in the mainstream economic sector, not on wildlands. To pit the trivial pleasures of an elite (a few fit, wealthy backpackers) against the needs of the many (starving in the ghettos) is confused. While such a choice may not be spurious in underdeveloped nations, it is in the United States, to which this analysis is applied. Even elsewhere, unjust social structures will often prevent goods obtained by sacrificing wildlands from benefiting the poor in any long-term way. It is everywhere futile to sacrifice wildlands to benefit the disadvantaged unless the social structures are just enough to make it probable that this transfer will take place. Failing that, the issue is a smokescreen which merely protects vested interests; it sacrifices almost every kind of value type and level only to delay needed social reforms, keeping in place a social disvalue. The basic needs of all can be met, and would already be, if the system were just. Social injustice, condoned, does not justify destroying natural values, as yet unappreciated.

(5) *Increase options*. By this we increase our possibilities to actualize preferences ($value_{ip}$) and so increase freedom and the quality of life. Most Americans are oversupplied with market artifacts and undersupplied with sites for experiencing pristine nature. They live in an urban or rural environment 99 percent of their time, in a wild environment 1 percent or less. We should, therefore, manage wildlands to meet needs that are unmet and unmeetable elsewhere (desires for a temporary exit from society or to see bighorn sheep). This is more true when these needs are intense, even in a minority, and if society's meeting these needs involves doing little or nothing. Also, there is truth in the adage that one should manage for the specialized user (the fly fisherman, the backpacker) rather than the general user (the picnicker, car camper), since it is likely that the interests of the former will be keener, resulting in higher quality experiences for which there are no substitutes. Economic uses that consume wildlands destroy our liberties here. Although society as a whole should increase all sorts of options, in remnant wildland decisions we increase options in wildness alike for ourselves and future generations. The pluralist model of "multiple use" here means such multiple uses as are noncompeting and nonconsumptive. Competitive and consumptive use is provided for elsewhere in the market sector. All this optimizes social diversity.

(6) *Make explicit the latent value judgments in quantitative models*. The numbers look hard—11,176 visitor days, $4,175,000 timber sales, willingness to pay $2.32 per person, discount rate 6.7 percent, preference 7.3 (scale of 10) in age group 25–35, with 12+ years of education. Although some will say that quantification makes values explicit, remember that the numbers are no harder than the theories out of which they come, as limited as the concepts which generate them. Unless the theories and concepts are explored and remain visible in the discussion,

the numbers can deceive. All numbers in science are theory-laden, in environmental and social science often laden with soft theories. When used to persuade, they arise from value-laden theories. They contain large empirical margins of error and value judgments that need to be made manifest. To take visitor days as a value indicator concentrates on the value$_{ip}$ level, value type 3, and leaves unanalyzed all other levels and types. To use any discount rate prefers the present over future generations.

Quantitative techniques, when ineptly or mechanically used, can obscure important value relations, and even when used at their best can never substitute for judgment, intuition, scope of understanding, and verbal assessment. Only the latter skills can suggest, at the start, what values are worth quantifying and how to try to quantify them, and in the end, only the latter can interpret the numbers that emerge. Dollar signs give the impression that a host of problems have been overcome interrelating value$_{sg}$ to value$_{mp}$. The "number values," nevertheless, are meaningless except in the context of an interpreted narrative of values. Numbers may disguise, rather than expose, value judgments. Indeed, the numbers are sometimes little more than tropes, used for the sake of giving life or emphasis to an idea. They are used in a different sense from the way numbers are usually used, not to count empirical things exactly at all, but only symbolically to stand for values when they are felt to be real but the amount present, though important, is unverifiable. We should distrust any numbers for which there is only one indicator, since there is no cross-checking and the cramped value judgments are less easy to expose.

In the end, wildland decisions are not a data-driven process; rather the data is caged by a value-driven theory. The data seldom changes anyone's mind, but is gathered and selected to justify positions already held, and ignored or reinterpreted if it is in conflict with favored positions. We should decide first about the latent ideology, only secondarily about the number analysis.

(7) *Protect minority interests, especially where this is nonconsumptive and requires doing nothing*. Consumptive minority interests, especially if they require expensive action (building a road to keep a local mill going) should not override nonconsumptive majority interests. On the other hand, nonconsumptive minority interests (mountaineering, bird watching) which can be satisfied by doing almost nothing are cheap and easy to protect and are noncompeting between values 2–12. The protection of such minority interests is a long-standing majoritarian American value. Nothing is actually taken away from the majority in protecting these interests, although opportunity costs may be forgone. These may be significant to a few, but are not likely to be great to the community as a whole.

Opportunity costs forgone (value$_{mp}$) make sense as real social costs (value$_{sg}$) only when these opportunities are available only on the site in question. From the perspective of value$_{sg}$, the timber lost on a wilderness site is no irretrievable disvalue if by better management techniques an equal amount might have been grown elsewhere, although managers are too inefficient or uninterested to attempt this.[14] Similarly, in assessing the costs of the Cranberry Wilderness in West Virginia at $30.77 per visitor day, it makes little sense to count as by far the largest factor $223,609 per year in opportunity costs of coal left underground and unmined, since there is a three-hundred-year supply of coal elsewhere in the Appalachian coal strata, and no reason to think the Cranberry coal should be mined soon or needs to be in order to meet national needs.[15] Nor should one forget, when development takes place and these opportunity costs become dollar benefits instead, that a spectrum of other opportunity costs (values 2–12) appears, opportunities lost perhaps forever.

Any delay, moreover, brings opportunity to see whether even the majority do not gain in values 2–12 more than they lose in opportunity costs. We gain benefits for the minority and the benefits of waiting for the majority. This rule protects against a danger in what otherwise seems so democratic—settling things by referendum. One purpose of government is to see that the fully considered will of the majority is done, where there is no injustice; another is to see that this is not imposed on the minority unjustly. A strategy for insuring this might be, for instance, to prefer nonconsumptive minority interests over consumptive simple majority interests and, furthermore, to require a two-thirds majority, on grounds that consumptive use of relict wildlands requires more caution and a quite unequivocal demonstration that this is in fact the considered majority view—that it is a use worth imposing at irreversible sacrifice on the minority and with irreversible loss of option to the majority.

Policy here should favor nonconsumptive minority interests over consumptive minority interests, since those who want to "lock up" the land delay but do not irrevocably destroy the option of developing it, while developers who "use up" the land irrevocably erode the original set of wild values carried by it.

(8) *Do not underestimate diffused values*. Although lots of persons may be hurt by a decision, they are not hurt very much (a diminished vacation vista), while a few (the mill operators who clearcut) have a lot to gain. We can expect the latter to lobby full force; the former, however, will be disorganized, slow to realize what is at stake, and ill-represented in the public participation process. Yet, the aggregate loss of value—now to the majority—can be much larger than the gain to the minority. The scenery is marred for a century. Even when the amenity values are accurately represented, they are intangible.

---

[14] Julie F. Gorte and W. Wendell Fletcher, analysts for the Office of Technology Assessment, find that simple management practices (such as thinning) on only 30–40% of the U.S. commercial forestland base could double the current harvest levels, and at the current levels Americans already consume more wood than people in any other nation. "Technology, Timber and the Future," *Renewable Resources Journal* 2, nos. 2 & 3 (Autumn 1983 & Winter 1984): 16–19.

[15] R. W. Guldin, "Predicting Costs of Eastern National Forest Wilderness," *Journal of Leisure Research* 13 (1981): 112–28.

Sometimes the agenda can determine the result. The hard values, considered first, are thoroughly assessed; the soft values, held in abeyance to be considered last, are only loosely (and wearily) thrown in, after the hard values are already massively in place. If the agenda is the other way round, with as much attention given to specifying the soft values, and only afterward considering whether economic benefits override these, the results could be different. The path by which the choice is made can bias the choice.

All this tends to stunt the attention given to soft values, that is, to underrepresent the widespread, low-level intangibles and overestimate the concentrated, intense tangibles. Even the use of willingness to pay, well-intended though it is, routinely underappraises the soft values, for reasons we have examined. Moreover, one needs to identify the changes in value distribution over different populations, always an ethically relevant consideration even beyond aggregates, to avoid unjustly benefiting the few while burdening the many.

Some contend that a preservationist policy imposes concentrated costs (the few, local timber companies who lose a lot) and diffuse benefits (the public, who gain slight recreational benefits). Such policy is rarely implemented not only because it is difficult to maintain legislatively, but even more because it is ethically unsatisfactory. The better gestalt is to turn the picture around and say that a development policy, at least on relict wildlands, permits concentrated benefits (the timber sellers and buyers) and diffuse costs (intangibles over a larger public). The latter is the more logical and ethical gestalt, because the public, which is the present owner and has its benefits of wildness taken away, is the loser. The timber operators, who do not now own the timber but who buy it for profit, are the gainers.

Furthermore, if the timber companies do not pay the full costs, or if they destroy what money cannot buy, then they (and their customers) are subsidized at the public loss, a situation which is even more ethically unsatisfactory. The ideal in a democracy is for the majority to have their well-considered way, provided there is no injustice to the minority, and provided that there is a full counting of the production and distribution costs and benefits. The concept of opportunity costs forgone should not be applied to minority would-be exploiters of the land, who are not now its owners, but to the present owners, the public. If this public uses wildlands as one kind of resource and it cannot simultaneously use them as another kind of resource, it must consider its opportunity costs forgone, especially if such opportunities cannot be had elsewhere. But opportunity costs forgone need not apply to consumptive opportunists on public lands. That is why we should not pit concentrated costs against diffuse benefits, but turn the issue around and pit concentrated benefits against diffuse costs.

(9) *Recognize that wildland decisions must be one place where the model (myth?) of the perpetually expanding economy is broken.* Four hundred and fifty years ago Europeans began to enter what they naively called an empty continent. Abetted by the industrial revolution in the last two hundred years and the explosion

of science and technology in the last hundred, the American economy has been on a growth trip unprecedented in the history of the world. A national tradition conflicts with the preservation of wildland values. Many people cry, "More!" Wildland managers, aside from the natural values they positively protect, negatively have to say, "Enough!" It will always be possible to increase the GNP a bit by sacrificing wildlands. But now is the time, here is the place, to draw the line at the boundaries of the wildland remnants and say: this far, but no farther with the expanding economy.

Expand it, if you must, on nonwild lands. It is futile to sacrifice the relict wildlands and to confront the collapsing growth myth only after they are gone. To confront it now saves so much in values 2–12. This policy also helps us anticipate the steady-state economy and not tumble into it by default and catastrophe. We have no reason to think that the last fraction exploited will leave us any nearer to satisfied consumer desires in a system designed ever to escalate those desires. Americans are already rich and need to learn when enough is enough. In this sense, wildland decisions are not peripheral "recreational" matters but frontline challenges to a governing paradigm. Wildland managers are not simply supplying values additionally to those generated in the domestic economy; they are confronting the slowdown, and, at wildland boundaries, the breakdown of a traditional economy in favor of noneconomic values. Wildland decisions are rewriting history, terminating and reevaluating the transcontinental growth trip. At the wilderness boundary, we should post a sign: "Enter the wilderness. Abandon the GNP rat race here. Learn to be wilderness rich."

In this context one frequently hears proposals that developers and preservationists ought to meet each other halfway. Compromise is frequently a necessity and often moral in policy decisions, but there is no logic by which fairness is always meeting each other halfway or by which conflicting values are usually optimized by compromise. There is even less reason to believe this in wildland disputes, where 98 percent of the continent is developed, 2 percent remains wild, and developers propose to preservationists that they should meet halfway over the remaining 2 percent.

(10) *Expect wildland decisions to awaken previously latent and newly emerging values*. Environmental values are among those things until recently taken for granted, naively appreciated, or unappreciated. Not until developers threatened the mountain on the skyline did we realize what it meant to us. Not until they proposed to drain the marsh could we say that we would rather have it left alone, and even now it is difficult to articulate why. We never miss the water until the well runs dry. We learn what is at stake only when we learn that it is at stake. We awaken to goods when their opposites threaten, or awaken to inconsistencies in our own value sets when we cannot have our cake and eat it too. In these matters we can hardly expect a congressional mandate ever to be as groundbreaking as it ought to be, though we can reasonably expect it to express social preferences over

economic interests. One cannot get anything through Congress that is very complex or controversial. The earliest growth in value awareness comes somewhere back in the grassroots, but decision models need to help it along. Deciders should find the trend, not the mean; they should lead, not just follow. They should set principles, not just sum preferences.

I close with a factual claim and a plea. A trend approximating this axiological model (above the solid line) and these rules can be found in congressional legislation over the last twenty years (the National Environmental Policy Act; the Endangered Species Act, the various Wilderness Acts, the Wild and Scenic Rivers Act), usually expressed vaguely as a desire to protect "environmental," "ecological," or "amenity" values from economic usurpation. The Federal Land Policy and Management Act of 1976 "declares that it is the policy of the United States that" there be "a combination of balanced and diverse resource uses . . . without permanent impairment of the productivity of the land and the quality of the environment with consideration being given to the relative values of the resources and not necessarily to the combination of uses that will give the greatest economic return."[16]

What is novel here is adding the understory of natural value, which, while not explicit in the legislation, is permitted by and consistent with it, and more ethically advanced (nonanthropocentric) than can yet be expected of congressional legislation. What we still need is a kind of emancipation proclamation for the wildness that remains, which can be issued with the full assurance that the benefits to the emancipators will outweigh their costs, with these to be added to the benefits to the emancipated (as was true with the Emancipation Proclamation of 1863). This is a call for humans to respect the plenitude of being, once so vast and now so quickly vanishing, which surrounds us in the wild world.

Can humans genuinely gain by exploiting the fractional wilds that remain? What does it profit to gain the whole world, only to lose it—to gain it economically, only to lose it scientifically, aesthetically, recreationally, religiously, as a wonderland of natural history, as a realm of integral wildness that transcends and supports us—and perhaps even to lose some of our soul in the trade-off?

---

[16] *Federal Land Policy and Management Act of 1976* (Public Law 94–579), Secs. 102, 103. (90 STAT. 2743) There is similar language in the *Multiple Use Sustained Yield Act of 1960* (Public Law 86-517), Sec. 4(a). (74 STAT. 215)

[33]

ELSEVIER

Ecological Economics 14 (1995) 113–127

ECOLOGICAL ECONOMICS

Analysis

Evaluating ecosystem states: Two competing paradigms

Bryan G. Norton

School of Public Policy, Georgia Institute of Technology, Atlanta, GA 30332, USA

Received 1 October 1993; accepted 16 August 1994

Keywords: Ecosystem valuation; (paradigms); Environmental management; Sustainability; Environmental risk decision square

1. Introduction: environmental policy and analysis: a shared crisis

Environmental management faces a crisis: legislation, high-level committees, and international agreements all urge that natural systems be managed to protect their health and systematic integrity; [1] but these widely employed analogies have not yet yielded a consensus in operational and specific management directives. Inaction may not be an option, however, because observable trends in the quality of the environment are forcing upon us the conclusion that traditional approaches to resource management are inadequate to the problems of today and the next century. Thus, while there is a growing consensus among ecologists, managers, and environmentalists that we must supplement traditional concerns for human health with more ecosystem-oriented management criteria, this consensus has not led to major changes in management or environmental decision-making.

Unfortunately, this crisis in management has an analogue in environmental policy analysis, the very place where one might first search for new directions for management. While there exists no shortage of arguments showing why we need more comprehensive, systematic management objectives, there is a foundational disagreement among policy analysts regarding how environmental objectives should be formulated, discussed, and analyzed. I believe these foundational disagreements can best be understood as "extraparadigmatic" disagreements. A "paradigm" is here defined as a constellation of concepts, values, and assumptions, as well as accepted practices, that give unity to a scientific discipline (Kuhn, 1970). The purpose of this paper is to compare alternative approaches to valuing natural resources over time and to show that the disagreements that separate mainstream, neoclassical environmental economists from the growing insurgency of the "ecological economists" amounts to an extraparadigmatic disagreement. Once this is established, it will be possible to say something about the conduct and logic of extraparadigmatic dis-

[1] For example, the Clean Water Act of 1972 specifically calls for management of water bodies to protect their biological integrity. The Great Lakes Treaty is based firmly on a commitment to protect the integrity of the Great Lakes Ecosystems. In general, there have been many proposals to institute whole ecosystem management in diverse situations. The common element of all of these initiatives is emphasis on monitoring and protecting characteristics of whole ecosystems.

Elsevier Science B.V.
SSDI 0921-8009(95)00024-0

agreements, and take beginning steps toward characterizing models that would combine current economic reasoning with other informational sources within a more comprehensive decision framework.

The multifaceted problem of intertemporal comparisons of values turns on a cluster of conceptually related foundational questions (Norton, 1994). These unresolved questions have blocked progress in developing a comprehensive and interdisciplinarily adequate account of environmental valuation. This interrelatedness can be observed by noting the highly connected nature of the opposed answers of mainstream and ecological economists to three related questions about time preference.

1. . How should we *measure and compare values that are experienced at different times*?
2. . How should we place a value on the *risk of irreversible loss* of a natural feature or productive ecological process?
3. . How should we evaluate changes in the *scale* of an economy vis-a-vis its ecological and physical context? Do increases in throughput in the economy, which often result from economic growth, significantly impact the ability of future generations to achieve an equal or better life than is possible in the present? If so, how should we address the resulting questions of intergenerational equity in the use of resources?

I will argue that answers to questions in this cluster define an important "fault line" in the intellectual bedrock underlying environmental policy discussion and debate. While considerable disagreement and confusion exists on both sides of the fault line, I will argue that positions on these three questions are sufficiently inter-locked so that it is not inaccurate to argue that there are *two competing paradigms* for the discussion of intergenerational impacts. I will show how answers to these questions lead in turn to differing analyses of the idea of "sustainability". I introduce the device of "risk decision squares" as *maps* of the evaluative space in which environmental decisions are made. This neutral vocabulary exhibits the assumptions, conceptual commitments, and principles of economics and ecology. Ecological economics represents a new paradigm because it introduces a two-pier process of policy evaluation. Two-tier systems of analysis are characterized by their representation of some decisions as essentially "economic" in nature, but other decisions as requiring alternative methods of evaluation, such as consideration of equity among generations. These two-tier systems of analysis apply economic/utilitarian analysis within broad constraints set by long-term, physically described constraints.

2. Definitions of sustainability locate a paradigmatic split in policy models

In a series of lectures and papers, Robert Solow (1974,1986,1992,1993) has defended the view that "sustainability" can be fully defined, characterized, and measured within the neoclassical theory that shapes the mainstream economic tradition of resource analysis. Solow's basic idea is that the obligation to sustainability "is an obligation to conduct ourselves so that we leave to the future the option or the capacity to be as well off as we are". He doubts that "one can be more precise than that." A central implication of Solow's view is that, while to talk about sustainability is "not empty ... there is no specific object that the goal of sustainability, the obligation of sustainability, requires us to leave untouched." Solow correctly acknowledges that the plausibility of this account of sustainability is essentially connected to the principle that no resource is irreplaceable, that every natural resource has an adequate substitute. In the jargon of economists, all resources are "fungible". Given this complex of essentially related ideas, Solow also, and validly, draws the conclusion that the problem of obligations to the future can be understood as simply the task of defining a rational and intergenerationally equitable *investment policy*. Sustainability, within this complex of principles and assumptions, is a matter of balancing consumption with adequate investment so that the future faces a nondeclining stock of total capital. Note that, on this view, monetary capital, labor, and natural resources are interchangeable elements of capital. Within this set of definitions, the future cannot fault us as long as we leave the next generation as able to fulfill their needs and desires as we have been in our generation. This nifty simplification of the intergenerational problem therefore succeeds, if it succeeds at all, on the heroic principle of intersubstitutability of resources because the intergenerational fungibility of resources follows from that foundational and constitutive principle. See Freeman (1986) for a brief explanation of the importance of fungibility in the economists' system of analysis.

Ecological economists have challenged the heart of this complex of assumptions and economic models. In particular, they have questioned Solow's principle that all forms of capital can be aggregated together and compared across generations. They argue that certain elements, relationships, or processes of nature represent irreplaceable resources, and that these resources constitute a scientifically separable and normatively significant category of capital—natural capital. This position directly contradicts Solow's central conclusion that sustainability is achieved, provided simply that the *total* stock of capital is not declining. We therefore face what is described by Daly and Cobb (1989) as the clash between "strong" and "weak" senses of sustainability. If there is to be an "accounting" and a judgment of the moral acceptability of one generation's bequest to all subsequent ones, on this view, there must be physically defined constraints on alteration of physical and ecological processes, which are apparently not associated in any measurable way with impacts on human welfare. The call of ecological economists for a structured bequest package cuts deeply against the grain of the mainstream paradigm because the acceptance of physically characterized constraints as a part of the definition of sustainability demands inputs of information that are not directly characterizable in the language of individual preferences. What is at stake, in particular, is what counts as relevant data for computing intergenerational balances—mainstream economists opt for a more highly aggregated data base with which to compute intergenerational fairness than would ecologists and ecological economists. The definition of sustainability is therefore practically important because it determines what type of information will be relevant in decisions affecting future generations. The conflict between paradigms also manifests itself in competition for research funding.

3. Competing paradigms

The growth of the new field of ecological economics, the phenomenon of ecologists becoming more active in policy formation, and the expanding number of calls by political leaders for a more integrated and ecosystemic approach to environmental policy (see, for example, Gore, 1992) are all symptoms of an extraparadigmatic debate regarding environmental goals and objectives and, especially, of the types of information that are crucial for forming rational policy in the new millennium. In extraparadigmatic debates, there is always danger of the two sides talking past each other because, lacking a shared paradigm, communication is difficult. It is hoped that clear recognition of the extraparadigmatic nature of ecological economics, and tentative steps toward a more encompassing set of concepts, may increase communication and hasten the development of a new, more inclusive paradigm. The ideal outcome would be a set of integrative models of environmental problems in which information from multiple disciplines is integrated into a rational, long-term approach to environmental management.

I begin by sketching competing answers of mainstream and ecological economists to our three questions above, noting how the three answers are inter-related, and how crucial assumptions determine that positions on the three questions stand and fall together.

A. Mainstream economists have generally dealt with the first question, the question of intertemporal preference, by the device of discounting (see, for example, Lind et al., 1982). *Homo economicus* "discounts" values across time, preferring to enjoy benefits sooner rather than later, and to delay losses. Economic analyses therefore discount values that will be experienced in the future; in business, costs or benefits expected in future years are discounted at a rate roughly equivalent to the rate of real interest. But at any significant positive discount rate, such as one a business might use to calculate the present value of future returns on investments, economic calculations imply that we care very little for even the next generation, and not at all for distant ones. This viewpoint is consonant with Solow's simplification of intertemporal ethics. The value of all future outcomes and options depends on the willingness of present consumers to pay for those outcomes and to hold open those options. This result follows from the intertemporal fungibility assumption that all human values can be expressed within a single temporal scale, the present. Some economists, noting that significant public goods will be destroyed if all decisions are made on a "private" discount rate, have propounded the concept of a "social discount rate", a slower rate that applies to investments in public goods. But discounting, while undoubtedly a useful method in many shorter-term contexts, has been criticized and at least one leader in the field has conluded that "searching for the 'correct' social rate of discount is searching for a will o' the wisp" (Page, 1988).

Ecological economists, by contrast, emphasize the importance of scalar issues. The hierarchical approach takes as its basic assumption that smaller-scale subsystems change at a more rapid pace than do the larger super-systems that form their ecological and physical context. According to one version of this approach, different valuational systems apply at different physical scales, with temporal policy horizons set by the social values impacted by the policies under consideration (Page, 1977; Norton and Ulanowicz, 1992).

As ecological economists attempt to address these contextual and scalar issues that arise in addressing intertemporal changes in the ability of nature to produce a flow of goods and services, they have found Solow's simplification a barrier to modelling the full complexity of long-term environmental problems. Highly aggregated data is too coarse to retain information about impacts of human activities that affect the patchy, multiscalar landscape on many different scales in space and time. Fungibility, which is essential if we are to aggregate these time scales together into the present, implies information about the temporal and spatial scale of human impacts is lost. Solow's fungibility assumption is therefore rejected by ecological economists, and with it they reject the entire paradigm represented by Solow's definitions and simplifications of intergenerational obligations.

Just as intersubstitutability is a unifying principle in Solow's system, the ecological economists' position is unified by the belief that healthy ecological systems—defined as multiscalar patchworks of inter-related processes—are essential elements of any "fair" bequest package. The future cannot be compensated by technologies or financial/cultural capital as substitutes for creative and productive systems, according to ecological economists; they believe that the bequest we offer future generations must be structured, not unstructured. Further departing from Solow, the health of these large systems will be measured with physical descriptors rather than economic criteria. By abandoning Solow's project of reducing all intertemporal environmental values to present preferences, ecological economists have apparently exiled themselves from the discourse of mainstream environmental economics.

B. Mainstream environmental economists have recognized the importance of irreversibilities in decision making (Krutilla, 1967; Krutilla and Fisher, 1975; Smith, 1980; Fisher and Krutilla, 1985; Solow, 1993), arguing that significant public goods will be lost if questions of preservation are left to unguided markets. The highly regarded "Natural Areas Program" at Resources for the Future, for example, was based on the premise that natural resource scarcities will not hinder economic productivity, but that there is a danger of destroying "amenities" such as parks and natural wonders because individuals might have economic incentive to develop and irreversibly change outstanding natural features which should be protected as public goods. Note that this dual accounting system for private and public goods goes nicely with the dual conceptualization of private and social discount rates. If Page and others are correct that there exists no "correct" social discount rate, this approach lacks clear operationalization. Solow explicitly follows Krutilla in his description of our obligations to set aside natural wonders that are scarce. Solow incorporates this general approach into his program, arguing that there will be unique and irreplaceable natural features that we should set aside to protect "unspoiled nature as a component of well-being". He explains: "It is perfectly okay, it is perfectly logical and rational, to argue for the preservation of a particular species or the preservation of a particular landscape. But it has to be done on its own, for its own sake, because this landscape is intrinsically what we want...not under the heading of sustainability" (Solow, 1993). But this is not a surprising position: the conclusion that irreversible changes affect only amenities is conceptually equivalent to the Solow conclusion that future generations cannot complain as long as they are left with the capacity to be as well off as we are, in general.

C. Having seen that the mainstream economists' answer to questions 1 and 2, and the ecological economists' rejection of them, both turn on their oppositions regarding fungibility across resources and across time. It has now become evident that their positions similarly entail opposed positions on scalar issues. We have seen that Solow, following in the tradition of Krutilla and other writers on the valuation of natural areas, treats the protection of natural areas and natural features of the landscape as bounded decisions regarding how to use—develop or preserve—specific natural areas. Compare two examples. First, assume someone today proposes damming the Yosemite valley to provide hydropower and irrigation water to jump-start the sagging California economy. Second, suppose someone suggests that, since corrective measures to "Save the (Chesapeake) Bay" are so expensive, that the Chesapeake be "sacrificed" to development. Notice, first, that the scale of these two examples is importantly different. The Chesapeake consists of a bay stem plus a complex of tributaries each of which are comparable in scale to the Merced River in Yosemite Valley. Because of the greater scale and complexity of the Chesapeake system, it is not implausible to refer to it as a "whole ecosystem", whereas the Yosemite Valley is more plausibly thought of as an element in a larger system.

Perhaps because the Yosemite Valley is encompassed within a National Park which has a long-established record of the willingness-to-pay of recreationists and others, it seems well within the capabilities of an economic analysis to produce plausible, and very high, use and non-use values that derive from protecting the valley. Given this established commercial and semicommercial productivity, and equally plausible measures of "existence" or "intrinsic value" enhanced by uniqueness or at least rarity, it is easy to see how the general outlines of an "economic" analysis of the decision to dam Yosemite might be constructed. But in the larger-scale case of the Chesapeake Bay, a collapse of the large-scale ecological system is more difficult to interpret in terms of losses to individual human welfare. This is not to say that there would be no impact on human welfare if the Bay is "written off", but rather to say that consumers and analysts would find it much harder to identify and classify as either "use-values" or "existence values", the values they derive and expect to derive from the complex, beautiful, and productive system. Because the Chesapeake Bay is the keystone resource in the region, a failure of the Bay system would have so many systemic effects on the broader ecology and economy of the region that calculation and aggregation of welfare measures —benefits and costs to individuals—are apparently beyond computation. Because the change is a change in a large-scale ecosystem process/function, one would have to calculate the outcomes of interacting series of changes in multiple layers of ecological processes over a whole landscape. That is, while the distinctive nature and well-documented history of use at Yosemite makes it a reasonable "unit" of economic analysis as a "rare" amenity, it seems much less likely that an over-all collapse of the health of the Chesapeake Bay could be associated with measurable welfare impacts on individuals. But it is change at this ecosystem level that is the apparent concern of advocates of the ecosystem risk and ecosystem health paradigms. As more and more environmental policy is formulated in terms of ecosystem-level impacts, the question arises whether the mainstream economic paradigm can forge any rational connection between the ecological processes that form the "production function" for ecological services, and associate these with units of individual welfare.

Ecological economists have opposed the fungibility hypothesis on the grounds that, while small-scale alterations of particular resource-and service-bearing ecosystems may be overcome by improvements in technology and know-how, at some scale these alterations affect the health and integrity of the functioning system. Ecological economists believe that ecosystems are not equilibrium systems, but rather dynamic systems whose essence is to be self-organizing and self-creative. This emphasis on large-scale dynamics and ecosystem organization causes the ecologists to expect systems to behave in nonlinear ways and considers the possibility that this behavior can be induced by gradual increases in the scale of changes to the landscape. Ecological economists cannot accept fungibility because they see the structure and processes of physical systems as vulnerable to incremental human choices, and they fear that, at some scale, the destruction of ecological systems and processes will result in direct impacts on human welfare. But they do not believe that this collapse can be correlated with present-value estimates of future utility because the analysis is carried out in a nonequilibrium system rather than an equilibrium system. It apparently follows that the differences between ecological economists and mainstream economists represents a serious conceptual disagreement, a disagreement closely related to the principle of irreversibility they respectively employ. Relying upon the principle of substitutability of resources, mainstream economists have concluded that resource scarcity has not and will not hamper future growth of the economy. Reduction of resource stocks and degradation of resource-producing systems are not expected to affect productivity. Therefore, the problem of irreversibility is reduced to the problem of protecting the *quality* of future life (Krutilla, 1967; Solow, 1993). This conclusion reduces the limits on economic growth to the "project" scale. It is not clear that many such "projects" of protection, designed to protect amenity and quality-of-life values of the future, add up to protection for whole ecosystems. On the contrary, it seems unlikely that large-scale ecological systems and habitats for far-ranging animals can be protected unless we manage the entire landscape, including privately used and exploited lands as well as nature preserves (Harris, 1984).

What is distinctive and important about recent calls for ecosystem management is a focus on larger-scale dynamics across whole landscapes and longer periods of time. The idea of ecosystem health/integrity, which attributes both descriptive and normative characteristics to varying ecosystem states, challenges the mainstream paradigm to develop an adequate method for associating changes in ecosystem states with changes in human welfare.

Indeed, it can be said that, so far, the mainstream economic paradigm has not been hospitable to values that emerge on the ecosystem level. Even when environmental managers such as officials at EPA wish to act to protect these whole-system characteristics they are stymied because there exist no accepted means by which to measure benefits derived from whole ecosystems, and they have little hope of justifying programs directed to this purpose in terms acceptable to auditors at the Office of Management and Budget. For example, in a recent National Academy of Sciences study, which attempted to weigh the relative costs of prevention versus accommodation to global climate change, zero value was assigned to the damages to natural systems. The panel justified this exclusion by stating that no adequate methods for ecosystem valuation exist (NAS, 1992). Again, this sort of exclusion—and protestations against it—are characteristic of extraparadigmatic debate.

A highly connected system of assumptions can often appear inflexible in the face of new problems and unexpected results. But it is also important not to conclude too quickly that a highly connected paradigm cannot account for new problems or data—the strength of a paradigm is measured by the ability of its practitioners to devise new theoretical and methodological means by which to encompass new information into the conceptual framework of the paradigm. It is not here claimed that the mainstream paradigm cannot, ever, develop a more comprehensive methodology for dealing with time preference and irreversibility. Indeed, it would be very interesting to see attempts to design contingent valuation studies to see just how much consumers and citizens are willing to pay to protect large-scale ecological systems and the landscapes they support. Or, some new methodology for assessing values of ecosystem-level changes may emerge. In the meantime, I think it is fair to say on the basis of the arguments developed in this part that mainstream economists and their opponents among ecological economists exhibit disagreements that are extra-paradigmatic.

Although it is perhaps too early to call the ecological economists' approach to intertemporal time preference a "new paradigm", I think it is no exaggeration to say that the ecological economists are uncomfortable with the tools offered by mainstream economics to estimate the values of protecting large, landscape-level ecological systems, and that their attempts to specify certain physical features of ecological systems as natural capital signals an intention to create an alternate paradigm that includes values and concerns that cannot be characterized in the mainstream paradigm.

4. A neutral conceptual geography

What is lacking is a broader, but neutral, vocabulary and discourse in which these problems can be addressed. It is helpful to envision the intellectual terrain as represented in Fig. 1, which represents issues separating ecological and mainstream economists in neutral terms (Norton, 1992). This "environmental risk decision square", plots the variables of reversibility against magnitude of impacts, defining a decision space on which various policy decisions can be located, depending on types of possible risks.

The decision square of Fig. 1, as outlined here, represents the logical space in which environmental policy decisions are made and therefore provides a neutral space in which we can define possible positions/assumptions/basic principles regarding the subject matter of the field we seek to understand: the overlap of economics and ecology, especially as they pertain to "policy".[2] The risk decision space can represent in neutral terms the three issues stated above that separate mainstream from ecological economists. It will

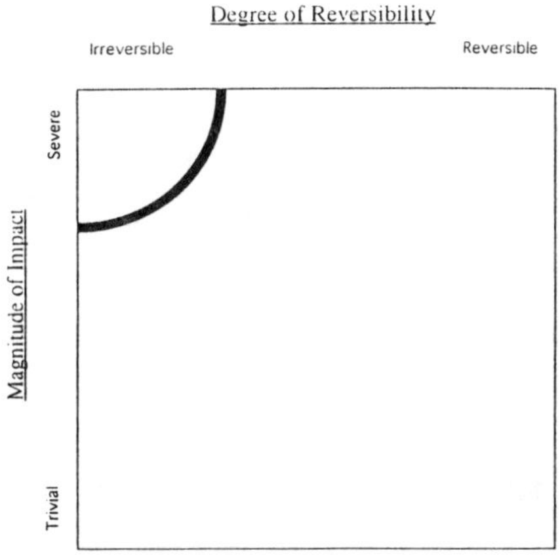

Fig. 1. Neutral version.

[2] It should be noted that the model as here presented is incomplete as a *decision* model because it does not incorporate a consideration of uncertainty/probability. Presumably, the obligation to act to avoid a risk to the future depends also on the degree of probability attached to the likelihood of the risk, given current knowledge, that a negative outcome will occur. That is, if the present faces two risks, the negative outcomes of which would be equally harmful to an equal number of people, we assume that there is a stronger obligation to act to avert a risk that has a 40 percent chance of negative outcome than to avert a risk that has a 0.05 likelihood of occurring. It may be possible to model probability in a third dimension, but work in this area is too speculative to guide policy at this time. For simplicity we will use only versions that plot decisions in two dimensions. In effect, that means we are assuming that in comparing two risks the risks carry equal levels of likelihood of negative outcomes. This means that the model must be used in conjunction with the judgment of managers, especially regarding the likelihood of various impacts.

therefore be an important metascientific tool for examining the conceptual commitments of various approaches to defining sustainability.

We can represent what we said, above—that economists and ecologists employ a different "paradigm"—by recognizing that ecological economists would, at the expense of being unable to aggregate fully across the whole space, draw distinctions in *types of risks*, breaking the risk decision space into distinct regions where different considerations and criteria apply. They therefore differ from mainstream economists who assume that all types of risks can be expressed as dollar values which vary gradually and at the margins, representing their decision to treat all risks and benefits in "fungible" terms.

For mainstream economists, then, options for preference satisfaction, and accordingly for future welfare, are not prejudiced by irreversible changes in ecological systems or other alterations of the physical world—the decision square therefore need not specify a time scale for reversibility. The horizontal axis can be understood simply as the degree of substitutability of a new resource for any damaged resource, with the goal being to ascertain the correct dollar figures a consumer would be willing to accept for the destruction of a resource that will require development of a suitable substitute. This simplified version of the risk decision square can be represented as Fig. 2 (Crosson and Toman, 1991). [3]

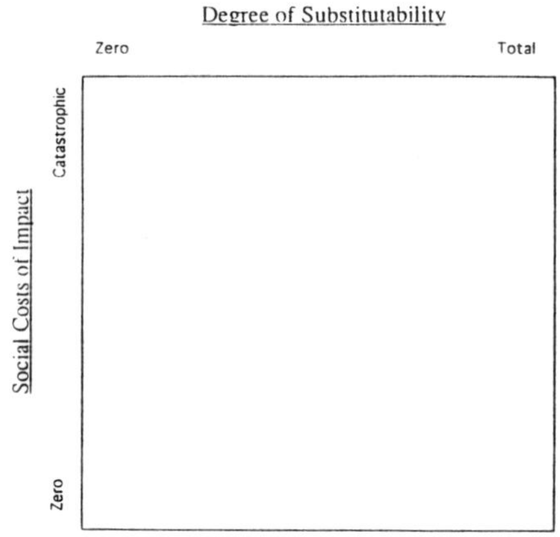

Fig. 2. Economist's version.

But this version of the decision space breaks conceptual neutrality because it treats reversibility as having no temporal dimension, and as a matter of increments at the margin of a system at equilibrium. It amends the traditional methods of mainstream economics only by recognizing that there may be decisions which, because of the potential scale of their impacts and because of the potentially large cost of "substituting" other means to attain the same level of welfare, may require special attention. With this amendment, the mainstream economic paradigm is made more sensitive to problems of scale and of irreversibility, but it maintains its commitment to a unitary decision space, positing no essential difference in the type of risks involved as one moves into the northwest corner of the space.

For mainstream economists, in other words, all risks remain of the fungible and compensable type. For the ecological economist, on the other hand, some risks can be understood as fungible and compensable, as aggregatable with other costs and benefits, and others fall in the "red zone" and must be understood as involving a more fundamental (moral?) obligation to the future. The risk square, and the question whether the space is continuous or not, dramatize the difference in the attitude of mainstream and ecological economists regarding intertemporal values. The technique of discounting is adequate to represent decisions made throughout a continuous decision space, whereas a multi-dimensional method of analysis is necessary if, as ecologists believe, information regarding temporal and spatial scale must be represented in decision processes. We have not, however, violated our policy of concep-

[3] While this version of the decision square models the mainstream economic viewpoint, it should be noted that Toman (1992) goes on to develop an interesting and useful distinction between the context of ordinary economic thinking and the very different context of decision making that would be encountered in, for example, a constitutional convention.

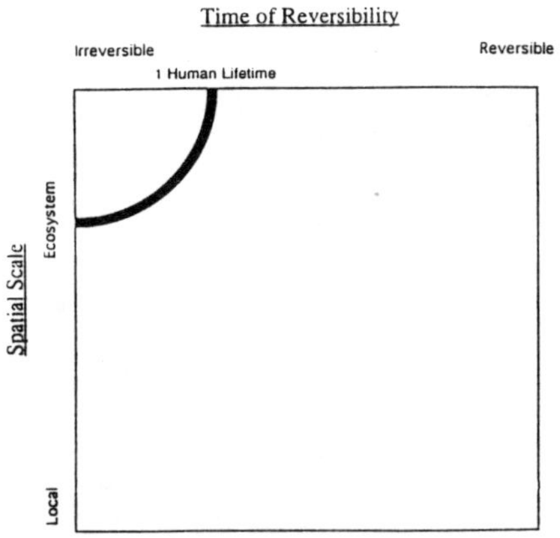

Fig. 3. Ecologist's version.

tual neutrality simply by adopting Fig. 1, because the doctrinaire economist who says that all decisions will be decided on costs and benefits as measured in markets can represent that position by asserting that the "red area" is empty. The economists' conceptualization thereby represents the limiting case in which there are zero cases of nonfungible resources.

Having introduced Fig. 2 as representing the economists' decision space by assuming reversibility can be treated as substitutability, we can elaborate Fig. 1 in a more ecological direction by incorporating the working principles of ecological economics, creating a more complex decision space which is represented here as Fig. 3. First we can note that ecological economists, purely by virtue of their emphasis on ecological systems and processes, will focus on the temporal aspects of change. For them, irreversibility is not just interpreted as an abstract concept of substitutability of one resource for another, as measured against units of welfare available to consumers—it will include the more concrete parameter of reversal of impacts in ecological time and space. The horizontal axis can therefore be calibrated as a measure of how long, given a particular impact or disturbance, it would take for natural processes to reverse that impact. The horizontal axis therefore locates decisions and policies which incur certain risks according to the restoration time necessary to repair damage if negative impacts occur as a result of that decision or policy. The risk decision space therefore represents the range of positions separating economists, ecologists, and ecological economists, with version 1 representing a neutral space, and 2 and 3 representing the risk decision space as it would be interpreted, respectively, by economists confident of substitutability among resources, and by ecologists who are not.

The ecological viewpoint, here, can be elaborated by introduction of "hierarchy theory (Allen and Starr, 1982; O'Neil et al., 1986; Norton, 1990,1991) which has as its assumption that larger systems of nature change more slowly than do the smaller subsystems that compose them. We therefore assume that an ecologically adequate paradigm of management will be multiscalar, and that spatial and temporal scales of management systems will exhibit a positive correlation, with spatially limited subsystems changing at a faster pace and according to a different dynamic than do the larger systems that form their environment. Now, relying on hierarchy theory and its principle that larger, super-systems change more slowly than do smaller subsystems, we can correlate the vertical axis with spatial scale, creating a grid that will represent decisions as located on the space. Risks which threaten harm that is prevalent over a large geographical area and irreversible for a very long time will be clustered in the upper left of the square, suggesting that decisions in this region entail questions of intergenerational equity. We expect, given our model, that decisions in this area may be governed by moral constraints. Decisions that risk small-scale and quickly reversible impacts are not significant on the ecological scale and fall in the area governed by economic reasoning. Applying reasoning such as this, Norton and Ulanowicz (1992), for example, have argued that since a commitment to sustain biodiversity is consensually understood to be a commitment to do so for many generations of humans (at least 150 years), we can conclude that the focus of biodiversity policy should be landscape-level ecosystems which normally change on a different temporal scale, an ecological scale that is slow relative to changes in economic behavior. If we can isolate the dynamics driving local economic opportunities from the dynamic supporting biodiversity, it may be possible to encourage both and avoid policy gridlock such as has occurred over the spotted owl in Northwest old-growth forests.

Human economics, which is paced to individual decision making, would on this perspective be understood as describing values that emerge in a subsystem of an encompassing physical system. Individual economic values are expressed on short scales and, provided they do not collectively add up to a trend that is significant on the larger ecosystem scale, these can be governed by individual free choice. In this sphere, economic analysis (perhaps corrected for considerations of interpersonal equity) accurately models individual decision making. When individual choices cumulatively impact large systems in an irreversible way, morally based considerations of equity across generations come into play. Good management, therefore, involves identifying and protecting processes crucial to the complex structure of the ecological system, which is to say that good management allows economic freedom, provided choices of individuals are damped out at a smaller scale and do not threaten to introduce ecosystem-level change that is irreversible. We can therefore define decisions that have impacts within two distinct spatio-temporal dynamics—individual choices affecting small subcomponents of a system that are reversible in one human lifetime, and decisions which threaten to create change that affects whole ecosystems. If these large-scale changes cannot be reversed within a single human lifetime, then they must be treated as decisions affecting intergenerational equity. The classification system corresponding to this ecological paradigm can be represented by the sorting diagram in Fig. 4. Sustainability of whole-system processes and the structure necessary to continue them is therefore one of those social decision areas that are a matter of intergenerational equity.

Given these contrasts we can, following Page (1977), begin to characterize a "two-tier" or "hybrid" approach to environmental decision making as an alternative to the unidimensional decision processes of mainstream economics. This position would see the decision space faced by environmental managers as split into regions, with the corner of the square that is characterized by major negative outcomes that are irreversible as representing an area of risk where values do not vary continuously with consumptive values. It could then be argued that moral strictures apply in these areas. This area of the decision space differs in kind from the regions where reversibility is high, cost is low, or both. In these latter regions, we will be inclined to accept the usefulness of economic methods on the assumption that, in these decisions, the future cannot fault us if we compensate them for destructive impacts of our activities on natural systems with increased technological know-how, monetary capital, etc. In these regions, in other words, decisions are regarded as fungible.

The central issue facing the policy analytic community can now be formulated quite simply: Are there any environmental decisions that are located in the red area, and thereby governed by

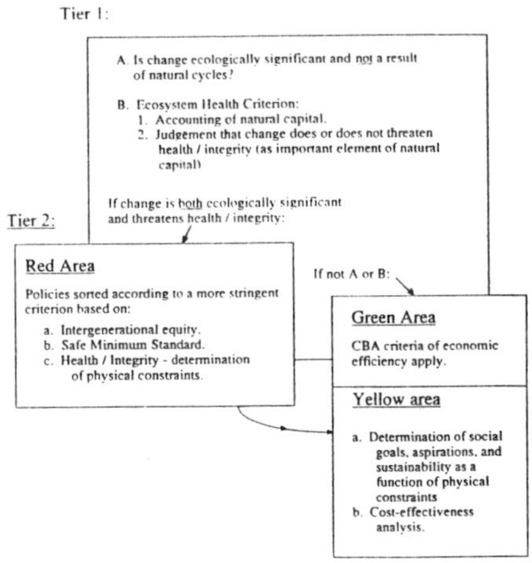

Fig. 4. Sorting diagram based on risk decision square.

moral rather than economic criteria? If there are, the goal should be to determine how many decisions, and precisely which decisions, comprise these two broad categories, recognizing that introduction of constraints is always under a burden of proof to establish moral strictures because constraints on individual choice, invoked by placing some resource use decision within the red area, are in principle limitations on the freedom of consumers to pursue their welfare as they see fit.

5. The extra-paradigmatic debate analysed

It would be comforting if the disagreements associated with these three issues were amenable to empirical confirmation or disconfirmation because then we could conclude that more scientific study would resolve the dispute between mainstream economists and ecologists. But these principles are, I think, too abstract to be directly supportable, or refutable, by empirical evidence. This is true because, as Kuhn (1970) and others before him have shown (Duhem, 1906; Quine, 1953,1960), observation is theory-bound and because, as noted above, the central assumptions that serve to constitute a scientific discipline are accepted in clusters. The problem in extraparadigmatic disagreements is that there exists no shared conceptual basis, no shared assumptions, and no consensually accepted methodology according to which intellectual and policy agreements can be submitted to empirical resolution. It will be useful to discuss some of the factors that are important in understanding extraparadigmatic disagreements.

Speaking informally, the tendency of practitioners of different disciplines to talk past each other is typical of disputes that involve several scientific disciplines. These seemingly insoluble differences are exacerbated when two disciplines have both theoretical differences *and* differences regarding the role their respective disciplines should have in governance and in society. In this situation, intellectual debate often deteriorates into turf wars as competing disciplines jostle to gain more territory and more grant funding. More formally, the failures of communication involved can be characterized by distinguishing between scientific disagreements that take place within a paradigm, accepting its basic assumptions and principles, and those disagreements that arise among scientists, perhaps from different disciplines or adherents of a competing paradigm, when there are no accepted disciplinary principles and assumptions to which all disputants can appeal.

As originally stated in Kuhn (1962), the dichotomy between "normal" science (science undertaken within an accepted and largely unquestioned paradigm) and "revolutionary" science (science as it goes through a periodic and sometimes stormy transition to a new paradigm) was surely overdrawn. Even Kuhn himself has regretted some of the passages in which he overemphasized the nonrational elements that affect revolutionary science, and has since rejected interpretations of the revolutionary process as subjective (Kuhn, 1970). Nevertheless, I believe (and I think that most contemporary philosophers of science would agree) that Kuhn's argument at least established that there is a useful, if vague and messy-to-describe, difference in the "logic" of science in two different situations. In situations where important empirical results are being achieved within an accepted paradigm, and all practitioners can agree what results are important and why, then science is comparatively "normalized"—these conditions are conducive to further experimental work. [4] When these conditions do not hold, debate enters an extraparadigmatic ("revolutionary") phase.

According to convincing arguments showing that "crucial experiments" are impossible in science—which has led to important revisions in our understanding of hypothesis-testing and of the positivist approach to science—every experiment is based on "background assumptions". If an experiment does not result in the results "expected" given an experimental hypothesis, it is possible to either reject the hypothesis or to adjust the background assumptions (Quine, 1960). Practically, this implies that if these background assumptions are widely accepted within a field, research questions can be formulated precisely, tested, and results can lead to further research. The shared assumptions train our attention upon outcomes of experiments and each result points toward new experiments. When a paradigm is under challenge, however, it may be impossible to find a sufficiently firm conceptual basis on which to define a hypothesis that will be both testable and yield results that all the disciplinary inquirers can accept as important. In this situation, the logic of scientific inquiry shifts. Many assumptions that could be built upon to form important hypotheses during normal science are now questioned. This is a time at which to loosen the intellectual hinges, to try many different models, and engage in disciplinary crossovers to increase communication across disciplines. Many hypotheses will ideally be floated and experimental work becomes more exploratory and hypothetical rather than precise and predictive. It is probably healthy, from the view of advancing science, that disciplines cycle through periods in this way because creativity of all types apparently results from a mixture of order and randomness.

One more cautionary note is necessary. It is important, in talking about a "paradigm" to be very clear as to what discipline, or level of resolution, one is pinpointing. Kuhn tended to use the term very loosely, applying it to disciplines, schools of thought within disciplines, and sometimes to little more than groups of advocates for conceptual innovations of any magnitude (Toulmin, 1972). I therefore feel justified in using the term to refer two somewhat different, but also similar aspects of the total picture of science and management. First, I use the term narrowly to apply to accepted principles, etc. which govern professional activities within an established scientific discipline such as economics, ecology, or medicine. But I am also here extending the term to apply to the broader policy analysis and implementation community, treating this broader, multidisciplinary, and activist community as seeking a more inclusive paradigm that includes a scientific, a normative, and a political/economic component, hopefully in some reasonable balance. The search for a comprehensive approach to environmental management, then, represents a search on a broader level for an action-oriented paradigm that encompasses disciplinary paradigms in such a way as to integrate information from the natural and social sciences. Given these two uses, it will be possible to discuss, for example, the role of the disciplinary paradigms of economics and ecology, within the encompassing paradigm for environmental policies and management.

Having noted above that the logic of extra-paradigmatic disagreements differs from intra-paradigmatic inquiry, we can now note that the internal logical structure of paradigms may also differ significantly. The mainstream economic paradigm is reductionistic in the sense that most economists favor explanations and data that can be expressed in terms of a single measure—dollars as representing units of individual welfare. Their system of value measurement, because it is monistic in this sense, is "highly connected"—all of the statements of the discipline can be comprehended within a single measure and evaluated under a central principle.

In general, it is considered an advantage for a discipline to agree upon a small number of principles from which more particular consequences follow. The degree to which a discipline has reduced its concepts and belief structure to a few theoretical principles measures the "connectedness" of the system. Connectedness is valued because it represents theoretical elegance and reduces the likelihood of conflict among principles of action. Solow's proposed understanding of sustainability has the distinct advantage that it

[4] Note: This condition must be described as a *sociological condition* of a group of people, the disciplinary practitioners, rather than as a logical distinction between two types of discourse. Attempts to formally and logically specify a sharp logical difference between "empirical questions" and "linguistic/conceptual questions" have apparently failed and are likely impossible to achieve without arbitrary fiat (Quine, 1953, 1960) Clear recognition that science is done by scientists who inevitably make judgments not reducible to relevant observations and deductions therefrom undermines any attempt to provide a logical characterization of a single rational response to any given body of data. Nevertheless, I think the distinction between paradigmatic and extra-paradigmatic disagreements is a useful one in the sociology of science.

comprehends questions of sustainability within the highly connected and sparse theoretical assumptions of neoclassical economics. As a result of this high degree of connectedness, Solow's approach allows us to measure sustainability by measuring present preferences of consumers, providing a single metric (present dollars) as a measure of welfare across all generations. Using this approach to interpret and measure values has the tremendous advantage that, to the extent price-values represent patterns of real behavior reflecting individual's willingness-to-pay, those values are concrete in the sense that they represent that individuals' judgment of the value of a thing *in competition with other goods and services*. These features of economic analysis of public values combine to provide an extraordinarily powerful set of analytic tools for the analysis of public values.

While connectedness is a desirable characteristic when one is communicating within a paradigm, systems that are highly connected can appear inflexible in the face of new situations and problems. This characteristic of a discipline can be understood if we distinguish between (a) the "intuitive subject matter" of a discipline and (b) the "reach" of its methods. In clinical psychology, for example, antisocial behaviors, self-destructive behaviors, and perhaps also the moods that lead to them constitute events that require attention of psychologists and form the intuitive subject matter of the discipline. It would be considered a criticism of a paradigm in this field if it could not classify, describe, and explain behaviors in these categories—its methodological principles, we would say, are inadequate to account for the intuitive subject matter of psychology. But advocates of a paradigm that is accused of insufficient reach may, rather than yield to criticism, answer by denying that some phenomena that are referred to in general conversation are "real", arguing that they are actually bogus entitities that are the ontological fallout, the theoretical dross, of failed paradigms. They believe, on this basis, that it is in fact an advantage of their paradigm that it makes no reference to these entities. Behavioral psychologists, for example, would argue that it is an advantage of their paradigm that they make no reference to unobservable mental states such as moods.

To return to the case at hand, advocates of the mainstream can either deny that there are important connections between states of ecological systems and the state of welfare of human populations in the present or distant future, or they can agree that there are impacts on welfare that predictably result from changes in ecosystem-level functioning and attempt to devise methods and means to describe and quantify this connection. In the first case, mainstream economists are denying that effects on ecosystems are real elements in the subject matter of environmental values; in the latter case, they admit that ecosystem changes affect their measures of human welfare, and accept the burden to show how those changes can be expressed as dollar figures.

Two points follow. First, extra-paradigmatic disagreements often go in circles because there is neither agreement on basic principles nor on the scope of the true subject matter of the discipline. To put the point in philosophical terminology, methodological disputes become the manifestations of "ontological" disagreements—differences in language and methods of measurement reflect differences in beliefs about what are really the constituents of the existing world. Second, sparseness of conceptual foundations has disadvantages as well as advantages. The burden of proof can shift against a highly connected paradigm if it repeatedly fails to encompass within its traditional, narrowly defined, methods an element of its subject matter that is widely accepted as important.

6. Conclusion

Ecological economists, for their part, advocate a more structured bequest package containing an appropriate mix of natural and human capital. They do not believe that a set of "sustainable" policies can be specified without essential reference to the physical states of the large, functioning ecological systems that provide an ongoing stream of ecological services. The movement to introduce concepts such as ecosystem health and integrity into policy analysis and implementation is best understood as a movement to define a structured and appropriate bequest from the current to future generations, especially as that package affects the functional characteristics of whole ecological systems. This expanded paradigm implies a multi-tiered decision process, one that seeks protection of both economic and ecological resources. While little has been said here to resolve this dispute, it is hoped that the characterization of the debate as an extraparadigmatic one, and the development of a neutral vocabulary, will further the goal—central to the Ecosystem Valuation Forum—of increased communication and a more fruitful understanding of the current crisis in policy analysis and formation.

Let me end on a final, conciliatory note, and state one plea. The idea that extraparadigmatic debates must be decided in nonrational forums unless the disagreements can be reduced to solving a conceptual or empirical "puzzle" within an unquestioningly adopted set of assumptions is based on a (useful) myth. Interdisciplinary cooperation can broaden debates in policy development if it is given a chance, and that chance is the creation of interdisciplinary panels that focus on problems and on finding solutions. These forums, in which experts from different disciplines pledge to work together, are capable of creating rational dialogue and developing a more interdisciplinary language. It is therefore very important to continue research into the theoretical questions isolated by the discussions in the first phase of the Ecosystem Valuation Forum, and to do them in conjunction with real case studies. To go to case studies only at this time would be to cut the theoretical head off the more organismic and dynamical models that could emerge from an integration of economics with ecology and the physical sciences. The point, after all, about all the discussion of "paradigms" is that if we are more aware of them and create dialogue across them, new, more inclusive conceptual frameworks will emerge. Most of the great scientific advances in recent years have come from disciplinary cross-overs and interdisciplinary teams (the most publicly noted case being the discovery of the double helix structure of DNA molecules. There is no reason to think that a scientific approach to the environment, which is the study of a complex, multilayered systems that provide the context of our lives, should be resolved within narrow disciplinary boundaries.

Acknowledgements

Research for this paper was supported by the Army Environmental Policy Institute. The author thanks V. Kerry Smith, Michael Toman, William Anderson, and Claire Huppertz Miller for careful review and comments on an earlier version. R. Gordon Dailey, Jr. assisted in the research and graphics preparation.

References

Allen, T.F.H. and Starr, T.B., 1982. Hierarchy: Perspectives for Ecological Complexity. Chicago: University of Chicago Press.

Crosson, P. and Toman, M., 1991. Economics and Sustainability: Balancing Trade-offs and Imperatives., Resources for the Future Discussion Paper ENR-9105, Jan. 1991.

Daly, H. and Cobb, J., 1989. For the Common Good. Boston: Beacon Pre

Duhem, P., 1906. La Théorie Physique: Son Objet et Sa Structure. Paris: Chevalier and Rivière.

Fisher, A.C. and Krutilla, J.V., 1985. Economics of nature preservation. In: A.V. Kneese and J.L. Sweeney (Editors), Handbook of Natural Resource and Energy Economics, Vol. I. Amsterdam: North-Holland Press.

Freeman, III, A.M., 1986. The ethical basis of the economic view of the environment. In: People, Penquins and Plastic Trees, edited by Donald VanDeveer and Christine Pierce. Belmont, CA: Wadsworth Publishing Company.

Gore, A., 1992. Earth in the Balance. Boston, MA: Houghton Mifflin Company.

Harris, L., 1984. The Fragmented Forest. Chicago: The University of Chicago Press.

Krutilla, J.V., 1967. Conservation Reconsidered. Am. Econ. Rev., 57: 777–786.

Krutilla, J.V. and Fisher, A.C., 1975. The Economics of Natural Resources: Studies in the Valuation of Commodity and Amenity Resources. Baltimore: John Hopkins University Press.

Kuhn, T.S., 1962, 1970. The Structure of Scientific Revolutions. Chicago: University of Chicago Press.

Lind, R.C. et al., 1982. *Discounting for Time and Risk in Energy Policy*. Washington, D.C., Resources for the Future.

NAS (National Academy of Sciences), Panel on Policy Implications of Greenhouse Warming, 1992. Policy Implications of Global Warming. Washington DC.

Norton, B.G., 1990. Context and hierarchy in Aldo Leopold's theory of environmental management. Ecol. Econ., 2: 119–127.

Norton, B.G., 1991. Toward Unity Among Environmentalists. New York: Oxford University Press.

Norton, B.G., 1992. Sustainability, human welfare, and ecosystem healthy. Environ. Values, 1: 97–111.

Norton, B.G., 1994. Future generations, obligations to. In: Encyclopedia of Bioethics, 2nd edn. New York: Macmillan.

Norton, B.G. and Ulanowicz, R.E., 1992. Scale and biodiversity policy: a hierarchical approach. Ambio, 21: 244–249.

O'Neil, R.V., et al., 1986. A Hierarchical Concept of Ecosystems. Princeton, NJ: Princeton University Press.

Page, T., 1977. Conservation and Economic Efficiency. Baltimore: The Johns Hopkins University Press.

Page, T., 1988. Intergenerational Equity and the Social Rate of Discount. Washington, DC: Resources for the Future.

Quine, W.V.O., 1953. From A Logical Point of View. Cambridge: Harvard University Press.

Quine, W.V.O., 1960. Word and Object. Cambridge, MA: MIT University Press.

Smith, V.K., 1980. The evaluation of natural resource adequacy: elusive quest or frontier of economic analysis. Land Econ., 56: 257–298.

Solow, R.M., 1974. The economics of resources or the resources of economics. Am. Econ. Rev. Proc., 64: 1–14.

Solow, R.M., 1986. On the intergenerational allocation of natural resources. Scand. J. Econ., 88: 141–149.

Solow, R.M., 1992. An Almost Practical Step toward Sustainability. Invited lecture on the occasion of the fortieth anniversary of Resources for the Future, Washington, DC, October 8, 1992.

Solow, R.M., 1993. Sustainability: an economist's perspective. In: Robert and Nancy Dorfman (Editors), Economics of the Environment: Selected Readings. New York: W.W. Norton and Company.

Toman, M.A., 1992. The difficulty in defining sustainability. Resources, 61 (Winter): 3–6.

Toulmin, S., 1972. Human Understanding, Vol. I. Princeton, NJ: Princeton University Press.

[34]

ELSEVIER

Ecological Economics 15 (1995) 141–144

ECOLOGICAL ECONOMICS

Forum

Harmony and pathology

Talbot Page

Department of Economics, Box B, Brown University, Providence, RI 02912, USA

Keywords: Harmony model; Pathology model

Arrow et al. (1995) have made a real contribution in identifying and developing a consensus among ecologists [1] and economists, a contribution much needed for improving environmental policy [2]. The main elements can be summarized:

(1) The environment is a collection of interlocking and nested systems. The economy is another collection of interlocking and nested systems. The two sets of systems are themselves interlocking and nested, with the economy nested inside the environment (inputs to the economy come from the environment and waste products go back into the environment).

(2) The environment provides many of our essential life support systems. Yet the equilibria that maintain these life support systems are sometimes fragile and can be undermined by economic activities. We need to pay closer attention to the links between economic activities and intervene when necessary to insure against the possible adverse impacts on environmental systems.

(3) Institutions and institutional incentive structures shape economic behavior. When important or essential environmental systems are threatened by economic activities, institutional incentive structures should be modified to insure sustainable enviromental and economic systems.

The three points fit together to provide a constructive policy framework as well as a long-term research agenda. Already, economists view markets as systems, systems ecology is well established, and many of the mathematical models of the two fields are similar, sometimes identical [3]. Virtually all ecologists, as well as most economists, understand that we are utterly dependent on environmental life support systems. And by now most environmental economists focus on institutional design as the basis for policy development, and so do game theorists and economists studying mechanism design.

In this note, I will say a little more about the second point, on the links between the environment and the economy. On this point economists and ecologist often talk at cross purposes and the divergence is at a fundamental, philosophical level, below the surface where it is not clearly articulated. While economists understand in principle that we are dependent on the life support systems of the environment (air, water, soil, minerals, forests, energy, waste sinks, etc.), they tend to be less troubled by this

[1] For brevity I will use "ecologist" to refer to a much wider group of biologists and other scientists.

[2] The authors' stated goal was to "establish a substantive dialogue ... to gauge whether an interdisciplinary consensus exists ... and to determine what can be said about ... policy." (Arrow et al., 1995, footnote 1).

[3] For example, prisoners' dilemmas have been used in both ecology and economics, and Bayesian games to describe the behavior of spiders and Coasean bargaining.

Elsevier Science B.V.
SSDI 0921-8009(95)00071-2

dependence than ecologists. To many economists, there is a natural harmony between what is good for the economy and what is good for the environment. I suspect that most ecologists think this view is wrong-headed and dangerously complaisant.

The belief in a natural harmony may explain why there has been so little attention to environmental problems when formulating economic policy. Conventionally, many economists take income growth to be good for the economy, and many conclude that income growth will, in the long run, be good for the environment. Beckerman (1995) and other economists have long argued that environmental quality is a luxury good. This means that as incomes go up, the demand for environmental quality goes up disproportionally more. Grossman and Krueger recently found some empirical support for the idea that environmental quality is a luxury good, with the qualification that, starting with very low income, as income grows environmental quality first declines, then as income grows further, environmental quality grows disproportionally more. The argument easily extends: "If you observe harmful environmental effects from our economic policies, don't worry, these are just transitional effects. In the long run, environmental quality will improve. Economic growth is the best remedy for this awkward transitional period of environmental deterioration."

To this argument authors diplomatically say, "Yes, ... but." They agree on the plausibility of the claim, but then in just a few paragraphs point out its limitations and cite some contrary evidence. Ecologists may wonder how an idea so limited in application became so widely accepted and broadly applied, with (to them) its obviously falsely reassuring consequences for policy.

Part of the answer, I think, has to do with how deeply embedded is the idea of harmony in economics and other social sciences. We see it in population studies, where there is another "U-shaped" relation. The empirical claim of demographic transition is that starting with low incomes and economic development, there will be low rates of population increase; with initially increasing economic growth and development population growth will increase; but with further increasing economic growth and development the rates will fall again (to environmentally sustainable levels). "Development is the best contraceptive," as the slogan goes. If world population is growing rapidly, don't worry, this is just an awkward transitional period. To many ecologists this is another falsely reassuring policy implication, that we need do nothing special to stabilize population.

The Laffer curve, another "U-shaped" relation, posits that with very low tax rates there will be little government revenue, with increasing tax rates the government revenue will increase, but with further increasing tax rates the government revenue will decrease, as economic activity is choked off by the taxes. While it may be a stretch to think of the Laffer curve in terms of an idea of harmony between what is good for the individual and what is good for the larger society, consider Laffer's argument: Taxes are already too high, and what is bad for the individual (high tax rates) is also bad for society (low government revenues and deficits). Doing what is good for the individual (lowering tax rates) achieves what is good for society (more economic growth and more revenues with less burdensome taxes).

A more clear-cut example of the idea of a harmony between the smaller parts and the larger system is *Scarcity and Growth* by Barnett and Morse (1961). In this book is the argument that as natural resource depletion proceeds people will anticipate future scarcities, prices will be bid up, incentives will be created for developing new technologies and substitutes, and the resource base will be renewed. Natural competition among the individual players in the market will take care of the larger environment. (I know this argument pretty well, since I made the counter-argument that these market forces were *tendencies* and there was no guarantee that the new technologies would come on line in time to bail us out, and I felt the full force of Barnett's conviction.)

Another example of the idea of harmony is Coase's policy prescription for controlling environmental harms: define and protect individual property rights and let the individuals bargain. If the bargaining costs are kept low enough, the results will be close to efficient and good for society as a whole.

The idea of harmony in economic systems underlies the competitive model in economics and goes back to Adam Smith's invisible hand. The idea is presently in a strong resurgence under the names of structural adjustment, liberalization, Thatcherism, and the Contract with America. Its policy prescriptions are lower taxes, less regulation, freer markets and free trade. This movement has brought the largest economic policy shift, in the last decade, in the World Bank's and the International Monetary Fund's economic policies toward developing countries and has brought large policy shifts in the developed world as well.

Ecologists, on the other hand, take harmony less for granted and often see the possibilities of pathology. When the rabies virus "chose" to reproduce by exiting from the dog's mouth and facilitated the process by making the dog crazy, biting, and frothing at the mouth, this was not in the dog's interest. The report of this Forum does not take for granted a natural harmony between the environment and the economy. How can an interdisciplinary consensus and productive collaboration evolve if economists are so committed to the idea of harmony and the ecologists so haunted by its lack?

I have drawn the picture too starkly. If we look we will find the idea of harmony in ecology as well. In the controversy between group selection and individual selection, the advocates of evolution by group selection believed that there is a natural harmony between what is good for the individual and what is good for the species. Those who believed in individual selection (for example, Dawkins in the *Selfish Gene*) believed that there is often a conflict between the interests of the individual and the interests of the species.

Correspondingly, in economics we find the idea of pathology as a well-established tradition, though not so deeply embedded as the harmony of the competitive model. It is well accepted that there are strong economic incentives for individual crimes, but crime is not in the interest of society; analogously, economic incentives leading to pollution do not lead to the social good. Scott Gordon (1954) wrote a classic paper describing how the incentives of individual fisherman conflict with the well-being of fishermen as a whole, and his model predicted the currently observed collapse of fisheries around the world. His analysis is fundamentally the same as Pigou's (Pigou, 1920) analysis of road congestion. Pigou wrote in the early 20th century and pioneered the concept of externality, which characterizes the all too possible pathology between the interests of the individual actor and the larger society.

The fact that Hardin (1968) rediscovered the concept of externality in his "Tragedy of the Commons," apparently independently of Pigou, Gordon, and other economists, is another instance of a growing consensus among ecologists and economists. Although ecologists and economists criticize Hardin's classic paper on various grounds, the key issue here is his rediscovery of the concept of externality and how one individual can impose costs on others. Pigou's view of externality was roundly criticized by Coase (1990), yet Pigou's analysis and policy perspective remains the dominant one in textbooks of general economics as well as specialized ones on environmental and resource economics. Much of non-cooperative game theory can be reinterpreted in terms of externality.

Thus the parallel between economics and ecology is deeper than it first appears. In both there are strong commitments to the idea of harmony; in both there is a longstanding controversy between the adherents of harmony and those of pathology [4]; in both the idea of pathology appears to have come later as a challenge and (partial) replacement to the idea of harmony (at least in evolutionary biology and much of economics). The main difference appears to be that in ecology the ideas of pathology and harmony came more simultaneously and the idea of pathology was more widely accepted and applied than in economics. In economics greater recognition of the possibilities of pathology has come with the shift in perspective from the competitive model to a game theoretic one. In ecology, game theory has also led to a greater emphasis on the possibilities of pathology.

When we recognize that both perspectives, of harmony and pathology, are in both ecology and economics, we get closer to the possibility of interdisciplinary consensus and collaboration. The next step seems obvious—after moving away from an either/or type of debate, where either harmony is the true relationship or pathology is—is to identify

[4] Even *Scarcity and Growth*, which is widely viewed as a book of resource optimism, carries the controversy within it. Morse, who wrote much of the chapters on institutions, was more cautious than Barnett and believed that with increasing pressure from scarcity and growing complexity institutions might fail to develop sufficiently effective incentive structures.

conditions leading toward harmony and conditions leading to pathology, in both fields. I think this step will reveal remarkable overlaps. In the two fields the conditions may turn out to be the same or analogous, in many cases. If so, we will truly be on the road to consensus and collaboration, and before long the two fields may be viewed as one larger field as was envisioned by Costanza and others several years ago.

References

Arrow, K., Bolin, B., Costanza, R., Dasgupta, P., Folke, C., Holling, C.S., Jansson, B.-O., Levin, S., Mäler, K.-G. Perrings, C. and Pimentel, D., 1995. Economic growth, carrying capacity, and the environment. Science, 268: 520–521.

Barnett, H. and Morse, C., 1961. Scarcity and Growth. Johns Hopkins Press, Baltimore, MD.

Beckerman, W., 1995. Two Cheers for an Affluent Society.

Coase, R., 1990. Notes on the problem of social cost. In: The Firm, the Market, and the Law. The University of Chicago Press, Chicago, IL.

Gordon, S., 1954. The economic theory of a common-property resource: the fishery. J. Polit. Econ., April, Sections I, II, III.

Hardin, G., 1968. The tragedy of the commons. Science, 168, December 13.

Pigou, A.C., 1920. The Economics of Welfare. MacMillan and Co., London.

[35]

Article Alan Randall

The Value of Biodiversity

Given the present threat to biodiversity worldwide, the article considers various philosophical approaches to the value of biodiversity. A clear case can be made, on instrumentalist and utilitarian grounds, that humans have good reason to place a high value on preserving biodiversity. Many conservationists and philosophers find such reasoning unsatisfactory, perhaps because it would leave biodiversity unprotected in the event that people develop technologies that substitute for naturally-occurring genetic material or find substitutes for the amenity value of natural environments. I then consider non-consequentialist philosophical approaches that define right action as that which respects moral duties or, alternatively, rights. Like the utilitarian alternative, these theories provide no assurance that biodiversity will prevail over other valid concerns unless one asserts at the outset that preservation of biodiversity is a first principle that trumps all others. Without according first-principle status or pre-eminent value to biodiversity, it is still possible to develop appealing philosophical theories in which biodiversity counts. It is reasonable to argue, starting from either moral or utilitarian principles, that in the absence of overriding constraints policy may be decided on the basis of benefits and costs. However, it is reasonable also to insist that a safe minimum standard of protection for biodiversity is an appropriate constraint to place on the policy-decision process. I conclude with some reasons for cautious optimism that benefit cost analysis can be applied successfully to biodiversity issues.

The evidence seems strong that a mass extinction of species is underway and that humankind bears most of the blame. This issue entered public consciousness in the early 1960s, when Rachel Carson (1) drew attention to the damage caused by pesticides to insects, birds and aquatic life. Since then, this concern has been generalized progressively, from particular species, to species in general, to the integrity of the diversity of ecosystems on this planet. In the United States, attempts to enforce the endangered species legislation highlighted the conflict between habitat protection and other human goals.

Many of the world's richest ecosystems and many of the world's poorest people can be found in the tropics. So, it is not surprising that the understandable human aspiration for improved living conditions frequently clashes with biodiversity objectives in the tropics. Several international development agencies and the World Bank have announced policies to protect and promote biodiversity. As the concern with loss of species becomes more generalized, institutionalized, and internationalized, the confrontations between biodiversity and other human goals will become even more wrenching.

It is desirable that we find ways to live in greater harmony with nature, but it is unlikely that the problem can be solved completely by discovering and enhancing complementarities between people and the environment. We will need also to develop a rationale for choosing, when conflicts between biodiversity and other human goals cannot be circumvented.

REASONS FOR CONCERN ABOUT BIODIVERSITY

There is little doubt that we are presently embarking on a mass extinction that may well be worse than any since the last major ice age. This time, humans bear most of the blame and have it within their collective power to sharply reduce its severity if decisive action can be taken.

The fact that biodiversity, worldwide, is under threat does not establish sufficient reason for concern. It is necessary to show also that biodiversity is worth caring about.

To many people, the most immediate reasons for caring about biodiversity are *instrumental* and *utilitarian*. A diversity of species in a variety of viable ecosystems serves as an instrument for people seeking to satisfy their needs and preferences. Many of the instrumental services that nature provides for people are obvious: food and fiber from domesticated plants and animals that were bred and selected from wild ancestors; and chemicals and pharmaceutical products with biotic origins.

Some attempts to develop a rationale for preserving biodiversity focus on these kinds of services. The argument proceeds as follows. There are many instances in which biotic resources have proven valuable to people and, together, the total of these services is enormous. However, the majority of species on this earth are yet to be catalogued and systematically evaluated for their commercial potential. But, since many of those species we know have proven useful, it is reasonable to expect that many of the presently unknown species will turn out to be useful, too (2). In addition, it is reasonable to expect continued technological progress, although we cannot predict its direction (3). So, new uses may be discovered for known species not currently thought useful. By standard statistical notions, the usefulness of many species under present technologies suggests a positive probability that literally any species, known or unknown, will eventually prove useful. Thus, we should approach the potential loss of any species with the presumption that its expected value to humans is positive.

This appeal to current and expected usefulness as commercial raw material, perhaps the crudest of the utilitarian arguments, has been the basis of many economists' cases for preservation.

The knowledge arguments—that species represent a store of genetic information for future use—have been extended to include ethnobiological knowledge. Ecosystems and indigenous human cultures co-evolved in considerable harmony, it is conjectured. This implies that disruption of indigenous cultures by colonists from elsewhere threatens destruction not only of the ecosystem but also the folk knowledge that enhances its value (4–6).

In the 1970s, the rapidly growing field of environmental economics established that people have not only indirect demands for natural systems as sources of raw materials, but also direct demands as amenities. Amenity values include use values and existence values. Use values include aesthetic and recreation values and, in this latter connection, it is suggestive to observe that travel is now the fastest growing industry worldwide and adventure travel is the fastest growing segment of the travel industry. Existence value arises from human satisfaction from simply knowing that some desirable thing or state of affairs exists.

The instrumental utilitarian argument for preserving biodiversity recognizes not only demands but also production possibilities.

Natural ecosystems serve as effective assimilators of wastes. In this way, wetlands help purify water, and forests assimilate greenhouse gases and help restore the oxygen balance. Natural ecosystems contribute to water and air quality objectives.

More generally, ecosystems are complex and fragile, and it is a tenet of ecology that everything has its place in the broader scheme of things. Thus, there is a presumption that species are not only useful directly as suppliers of raw materials and amenities but also indirectly for their contribution to ecosystem support. Thus, species that have no conceivable value in providing raw materials and amenities (if there are any) could still be valued for the ecosystem support services they provide to species that are more directly valued.

The ultimate instrumentalist argument—that all species must be preserved because the loss of any would initiate processes leading inexorably to the collapse of the whole ecosystem—seems clearly false. In its place, the Ehrlichs (7) have introduced, by analogy, a statistical argument. Their much-cited rivet popper justifies his continued removal, one-by-one, of rivets from airplane wings by reasoning that the practice must be safe since no planes have been lost yet. There is an obvious logical fallacy in the rivet popper's claim: each successive inconsequential loss of a rivet does not serve to confirm the low probability of the practice causing a crash. Rather, since the initial numbers of rivets is finite and the number needed for safe operation is smaller yet, each inconsequential rivet removal increases the probability that the next one will be disastrous. So, by analogy, each particular extinction increases the probability that the next one will unravel the whole ecosystem.

Taken together, these various considerations amount to a convincing argument, in the instrumentalist utilitarian tradition, that preserving biodiversity should be a serious consideration for a society of rational human beings pursuing what are essentially homocentric goals. At the very least, biotic resources are resources to be allocated carefully, rather than squandered or merely wasted. They are to be valued for their amenity services as well as their usefulness as raw materials; use and existence values count; the concept of value encompasses but goes beyond commercial values: ecosystem and environmental support services are recognized; and, where ignorance is rampant, statistical arguments are used to infer positive expected values. Right action, it follows, is whatever maximizes value including the value produced by biotic resources.

One could argue that mainstream economics is one version of the homocentric, instrumentalist, utilitarian system of thought. Hence, the general approach to valuing biodiversity developed above is hospitable to mainstream economic thinking.

The major alternative approaches define right action in terms of respect for rights and/or moral duties. Duty-based moral theories attempt to identify the moral obligations that humans have toward other living things and the morally correct actions these obligations entail. Moral obligations may be asserted as first principles. Conflicts among duties must be resolved by reasoning; if possible, the ranking of duties must be deduced from first principles. Ehrenfeld (8), for example, asserts that human beings have overriding moral obligations toward biodiversity.

A contractarian approach argues that arrangements are justified if they respect the rights of all the affected parties. In duty-based reasoning, "rights" is often used to mean moral claims; one respects rights (in this sense) by observing moral obligations. In contractarian theories, rights are enforceable claims. Change occurs when all affected parties, endowed with enforceable rights, consent to it; without consent, the *status quo* prevails. While consent justifies change, the lack of consent for change is insufficient to justify the *status quo*.The starting point must itself be justified directly, typically by arguing that it was chosen by voluntary agreement among all concerned or, more likely, that it could have emerged from voluntary agreement under ideal conditions such as those posited by Rawls (9).

It seems that the contractarian approach is a non-starter in serious discussion of the applied ethics of biodiversity. The idea of a constitutional convention among all kinds of living things to establish a just starting point, with subsequent adjustments justified by the voluntary and informed consent of all parties, seems too far-fetched. On the other hand, contractarian thought experiments might prove useful, in the Rawlsian tradition, as a way to deduce a set of moral principles for human behavior with respect to the biota.

To summarize this section, then, the reasons for caring about biodiversity are that it is under serious threat *and* at least one of the following: i) the consequences of failing to take action will be unpleasant, ii) humans have a moral obligation not to permit the destruction of biodiversity, and iii) for humans to permit continued loss of biodiversity would violate the rights of the non-human biota under a just constitution.

CAN "THE SLICK TERRAIN" BE AVOIDED?

I suspect Ehrenfeld (8) speaks for many conservationists, who "would like to see (conservation) find a sound footing outside the slick terrain of the economists and their philosophical allies." For these conservationists, the homocentric, instrumentalist and utilitarian rationale for conservation is slick terrain, in that it provides a rationale for valuing biodiversity but its value is always relative to the values attached to other things.

Some of the conservationists and their philosophical allies have provided clues as to what is needed, to develop a rationale for biodiversity that avoids the slick terrain. Such a rationale should not depend on:

- the human utility function. Many conservationists suspect that any rationale for preservation that depends on human preferences provides less than iron-clad guarantees. Preferences for biodiversity may wane, or simply be overwhelmed by competing claims on behalf of other things that also give us pleasure.
- instrumentalist arguments. Arguments that biodiversity is valued not so much for its own sake but because it serves as an instrument for various human purposes—e.g., as a storehouse of genetic information for agriculture and medicine—could be undermined by technological change. Ehrenfeld (8, p. 213) claims that pharmaceutical researchers do not in fact tramp through the jungles searching for exotic species with medicinal prospects; a strategy of computer modeling, organic synthesis of promising molecular structures, and screening of the resultant synthetic compounds is more efficient. It is not yet clear whether the emerging technologies of bioengineering will enhance the value of naturally occurring genetic material or merely substitute for it.
- "divide and conquer strategies." Norton (10) argues that we should refuse even to try to answer questions about the value of the biodiversity losses associated with particular policy proposals, considered one by one. The piecewise, or "divide and conquer," valuation strategy would inevitably trivialize the aggregate losses from human actions. The whole of the losses far exceeds the sum of its estimated pieces. Lovejoy (11) makes a similar argument. Norton's solution is to deny the validity of any quantified values for piecemeal losses of biodiversity, while insisting that the value of biodiversity *in toto* "is the value of everything there is" (p. 205).
- trade-offs. Norton (10, p. 204) complains that the act of asking about the value of biodiversity is itself a measure of the unique arrogance of humankind. Ehrenfeld (8) while citing the Old Testament, offers an argument that perhaps owes more to the concept of hubris in Greek tragedy: "Assigning value to that which we do not own and whose purpose we cannot understand except in the most superficial ways is the ultimate in presumptuous folly" (p. 216). I interpret their arguments as not being opposed to the idea that biodiversity is valued in any absolute sense. Rather, what is being opposed is the idea that biodiversity can be valued in a relative sense; that one can compare its value to that of other good things and adjust society's production and consumption patterns for the better by making marginal trade-offs.

In textbook economics, the optimal combination of competing goods is almost always a mix; solutions where one consumes only one of the goods are treated as special cases and presumed to be fairly rare. The argument against trade-offs in the case of biodiversity must presumably argue that biodiversity is to be maximized regardless of other values. To make such an argument, it would be necessary to depart from the customary assumptions that production possibilities and preferences are smooth and continuous, or to impose some additional constraint on the choices that can be made.

The kinds of rationale for biodiversity that would be developed by mainstream economists and homocentric utilitarian philosophers would surely fail the "slick terrain" test in every way: human preferences would count, instrumentalist arguments would have some force, choices would be conceptualized as typically piecemeal rather than all-or-none (especially about something so global as biodiversity), and the interaction of technical possibilities and human preferences would often lead to trade-offs (although by no means always in favor of material goods at the expense of biodiversity). It is interesting to consider whether alternative philosophical approaches can avoid the slick terrain.

First, the homocentric, utilitarian approaches are a subset of a broader group of consequentialist theories that claim, loosely speaking, that the rightness of an action should be judged by the goodness of its consequences. Consequentialist theories do not have to be homocentric: Bentham was in principle open to the idea that the utility of the animals could count; he just could not visualize any obvious way of incorporating it into the utilitarian calculus (12, p. 311). More generally, utilitarianism evaluates all consequences in terms of their contribution toward preference satisfaction, whereas consequentialism is open to other ways of evaluating consequences.

Clearly, a consequentialist could accord preservation of biodiversity the status of the pre-eminent value. Then all proposed actions would be evaluated in terms of their effects on biodiversity, and those with the most favorable consequences for biodiversity would be chosen. Other objectives would be subordinated to biodiversity. Such a consequentialist scheme would avoid the slick terrain, but only at the cost of subordinating other worthy objectives such as enhancing the life prospects of the very worst-off people. Of course, according these other objectives the status of pre-eminent values too, would return us to the slick terrain. The primacy of biodiversity cannot be assured in a clash of pre-eminent values.

Ehrenfeld (8) offers a solution to the "slick terrain" problem: "If conservation is to succeed, the public must come to understand the inherent wrongness of the destruction of biological diversity" (p. 215). Clearly, Ehrenfeld's is a duty-based approach: right action is that which respects the moral obligation of human beings to preserve biodiversity.

When several considerations have moral force (cannibalism is morally evil, while self-preservation is morally worthy), clashes among them (under what conditions, if any, would self-preservation justify cannibalism?) can be resolved only via deduction from higher moral principles. The slick terrain can be avoided only by asserting that preservation of biodiversity is a trump among moral principles, one that defeats all others. Without such an assertion, Ehrenfeld's "inherent wrongness" does not solve his problem. Surely, many would argue that enhancing the life prospects of the worst-off people has moral force at least as powerful as that of protecting biodiversity. Again, biodiversity is on slick terrain.

It seems to be a general conclusion that, unless conserving biodiversity can be accorded either first-principle or pre-eminent value status, the case for it will always be circumstantial: i.e., dependent on the technical possibilities and the strength of competing duty-based or consequential claims. Norton (10) and Ehrenfeld (8) claimed the high ground by resisting all the circumstantial approaches as mere manifestations of the moral repugnancy of homocentrism. But, their victory is empty, since it depends on first-principle or pre-eminent value status for biodiversity, and such status is unlikely to survive scrutiny given the powerful appeal of many other candidates.

TOWARD A LESS-SLICK TERRAIN FOR BIODIVERSITY

We can do more than deny the viability of a first-principle case for biodiversity. It is possible to work, more affirmatively, toward constructing a strong but defensible circumstantial case.

First, consider a duty-based approach. Assume that preserving biodiversity and enhancing the life prospects of the worst-off people are both moral goods. However, the claims of humans trump those of nonhumans. From these moral principles, it can be deduced that humans should make some, but not unlimited, sacrifices for biodiversity. This result endorses the basic idea of a *safe minimum standard* (SMS): a sufficient area of habitat should be preserved to ensure the survival of each unique species, sub-species, or ecosystem, unless the costs of doing so are intolerably high (2, 13). One could add an additional moral premise—marginal increments in the welfare of people count for less in the case of the already well-off than for those who are currently immiserated—and the implication is that one could identify situations in which the well-off have an obligation to subsidize the SMS in impoverished places. Obviously, specific applications remain debatable: just how seriously should we take each particular threat of extinction? Just how high a cost of meeting the SMS would be intolerable in each case? Nevertheless, the general decision rule seems fairly robust, and it is derived from principles that enjoy fairly wide support.

Alternatively, consider a utilitarian approach. The benefit cost (BC) approach applies classical utilitarian principles with just one nod toward the value system of mainstream economics. The basic value data in BC analyses are those of the market-willingness to pay (WTP) for desired changes and willingness to accept (WTA) compensation for changes that are not desired—and as such reflect not only the preferences but also the endowments of the valuer. The rule that benefits and costs be aggregated anonymously—i.e., without regard to the identity and welfare status of the gainers and losers—is merely an economic operationalization of Bentham's classic rule of the greatest good for the greatest number.

As humans comprehend the technology of natural systems and how it limits the performance of anthropogenic technology, this understanding is reflected in a valid BC analysis. As human preferences extend to the amenity and existence services provided by diverse ecosystems, the valuations that emerge are fully reflected in a valid BC analysis.

One of the more persistent arguments against the BC approach—and many alternative expressions of utilitarianism—is that preferences may be myopic. How can we be assured that the lure of immediate gratification will not induce us to make decisions that will surely have very unpleasant consequences later on? Elster (14) has shown that "binding" behavior—Ulysses bound himself to the mast in advance to prevent himself from doing what he was quite sure he would do in the heat of the moment, i.e., steer his ship into the rocky waters separating it from the sirens—is consistent with both rational behavior and utilitarianism. Thus, one logically coherent utilitarian strategy would be to make policy choices on the basis of benefits and costs, but subject always to the constraint that actions we are reasonably sure we (or future generations of people we care about) will regret are forbidden. Biodiversity issues may be decided by consulting a BC analysis but subject to a safe minimum standard or similar constraint (15). Net benefits are maximized because benefits are good consequences, and the constraints are imposed because the consequences of not satisfying them are terrible.

Incidentally, one may arrive at a similar policy stance via duty-based reasoning. My colleague, the philosopher Donald C. Hubin (16) suggests the following. It is hard to conceive of a plausible moral theory that does not, in the absence of overriding concerns, give a good deal of weight to the satisfaction of human preferences. Thus, we should take seriously a rule that policy issues be decided on the basis of benefits and costs, but always subject to constraints identified by moral reasoning. Net benefits are maximized because human preference satisfaction is morally worthy, and the constraints are imposed because they ensure that higher moral goods can trump preference satisfaction in the event of conflict.

If we accept that satisfaction of human preferences is at least a consideration, and that BC analysis can provide a method of systematically accounting preference satisfaction, then estimated benefits and costs are at least a consideration when making choices about biodiversity. If environmental economists could be satisfied with claiming no more than this, and if conservationists could discipline themselves to eschew the grandstanding strategy of denouncing any and all moral theories that do not accord first-principle or pre-eminent value status to biodiversity, a harmonious collaboration ought to be possible.

THE PROSPECTS FOR BENEFIT EVALUATION OF BIODIVERSITY

A sound argument for considering empirical estimates of benefits and costs when deciding policy with respect to biodiversity requires that these empirical estimates provide information, as opposed to misinformation or noise. In other words, the argument for considering the output of BC analysis depends on the possibility of producing estimates that reflect the underlying WTP and/or WTA with sufficient accuracy and precision that the results are informative rather than misleading or merely confusing. Those whose main line of attack is on the moral underpinnings of BC analysis in the context of biodiversity (8, 10, 17) tend also to embellish their case with assertions about the inherent nonquantifiability of the benefits and costs of preserving biodiversity or the arbitrariness of the estimate obtained using the available methods. In contrast, Randall (18) takes a relatively optimistic view of the feasibility of generating informative BC estimates.

There has been substantial progress in all aspects of environmental benefit estimation in the last quarter-century. The theory has been developed, consistently with the economic theory of welfare change measurement. Explicit concerns of environmental economics have been addressed: the benefits of complex policies and the relationship between total value and the value of policy components (19); the distinction between use values and existence values (20); use value under uncertainty (21, 22); a total value framework for benefit estimation (23); the concept of quasi-option value to account for situations in which current decisions might forestall the opportunity to benefit from knowledge emerging in the future (3); and development of the theory and methods for estimating values of nonrival, nonexclusive, and nonmarketed environmental services and amenities, including contingent valuation (24–26) and methods that use market observations for weak complements (27, 28). A very substantial body of empirical estimates of environmental benefits has been assembled. The practice of estimating environmental benefits has become routine in the United States policy process, encouraged by legislation, Presidential Executive Orders, and the Office of Management and Budget.

There is no denying that challenging problems remain in environmental benefit estimation. However, there seems little doubt that environmental benefit estimation is a progressing research program (29) that has already achieved important successes.

Biodiversity is an especially challenging topic for environmental benefit estimation. In order to evaluate policies pertaining to biodiversity, it is necessary to develop scenarios that describe baseline conditions and project them into the future, and do the same for with-policy conditions. Yet scientific data bases and the understanding of the basic natural science relationships are often very incomplete. BC analysis will reflect this scientific uncertainty and ignorance (but, to be fair, so will any other process by which people try to come to terms with the policy issues surrounding biodiversity).

The basic economic data for BC analysis, WTP and WTA, are likely to be volatile, as new information is added to the very sparse information base that most people have about biodiversity issues. However, this volatility reflects the basic reality of decision processes that start with limited information and revise expectations as new information becomes available (30) and surely applies not only to WTP and WTA but also to any other measure of public preferences about biodiversity policy.

For several reasons,—the relative importance of existence values, the importance of uncertainty and therefore of *ex ante* value (22, 23), and the relative difficulty of identifying serviceable weak complements—contingent valuation seems destined to play a major role in benefit estimation for biodiversity. While contingent valuation is gaining respectability, it remains a controversial method (26). Nevertheless, a number of studies have successfully estimated existence or preservation values with contingent valuation (31–35). The recent development of referendum methods for contingent valuation (25, 26) will be especially important for evaluating biodiversity. These methods appeal to the public perception of public goods, such as biodiversity; and they rely on scenarios built around political rather than market institutions and should, therefore, be more readily adaptable to international contexts.

While contingent valuation will likely be the method most widely used in evaluating biodiversity benefits, there may be important opportunities to use weak complementary methods. The growing importance of adventure travel to, and vacation and retirement homes near, exotic environments suggests a limited but nontrivial role for travel cost and hedonic price methods.

In summary, biodiversity presents some of the most difficult challenges for the rapidly-developing theory and methods of environmental BC analysis. Nevertheless, there are promising prospects for successful empirical applications.

CONCLUSIONS

Let me attempt to summarize the argument with a few succinct statements.

Some conservationists and philosophers are contemptuous of any theory of value that treats the value of biodiversity as relative to the values of other concerns. Yet, it seems that any coherent duty-based or consequentialist theory will do that, unless the theory asserts that preservation of biodiversity is a first principle or a pre-eminent value that trumps all others. And there seems no overpowering reason to accord trump status to biodiversity.

Without asserting first-principle or pre-eminent value status for preserving biodiversity, duty-based and consequentialist theories can be developed in which biodiversity counts.

A sound argument can be made that human preference satisfaction counts morally. Thus, benefits and costs (as environmental economists conceptualize them) count in duty-based theories as well as consequentialist theories of value. This suggests that benefits and costs have a place in a more complete theory of the value of biodiversity.

The idea that benefits and costs count in a more complete theory of biodiversity does not exclude other moral or consequential considerations. One admissable theory would be that policy should implement the strategy with the greatest net benefits, subject to the safe minimum standard (SMS) constraint. The SMS constraint could be derived from duty-based or from consequentialist principles.

Real and considerable difficulties exist in obtaining reliable empirical estimates of the benefits and costs of preserving biodiversity. Such estimates are likely to be volatile, reflecting the reality that the current information base is small and new information may change perceptions dramatically. Nevertheless, and in full awareness of the challenges, I believe the effort could be made and the results should be taken seriously.

References and Notes

1. Carson, R. 1962. *Silent Spring*. Houghton Mifflin. Boston.
2. Bishop, R.C. 1978. Economics of endangered species. *Am. J. Agricultural Economics 60*:10-18.
3. Fisher, A.C. and Hanemann, W.M. 1986. Option value and the extinction of species. In: *Advances in Applied Microeconomic*. Smith, V.K. (ed.) JAI Press, Greenwich, Vol. 4, p. 169-190.
4. Southgate, D. 1988. *Efficient Management of Biologically Diverse Tropical Forests*. IIED/UCL, London Environmental Economics Centre, 19 p.
5. Norgaard, R.B. 1984. Coevolutionary development potential. *Land Economics 60*:160-173.
6. Norgaard, R.B. 1988. The rise of the global exchange economy and the loss of biological diversity. In: *Biodiversity*. Wilson, E.O. (ed.) National Academy Press, Washington, p. 206-211.
7. Ehrlich, P.R. and Ehrlich, A. 1981. *Extinction*. Random House, New York, 305 p.
8. Ehrenfeld, D. 1988. Why put a value on biodiversity. In: *Biodiversity*. Wilson, E.O. (ed.) National Academy Press, Washington, p. 212-216.
9. Rawls, J. 1971. *A Theory of Justice*. Harvard University Press, Cambridge, MA.
10. Norton, B. 1988. Commodity, amenity, and morality: the limits of quantification in valuing biodiversity. In: *Biodiversity*. Wilson, E.O. (ed.) National Academy Press, Washington, p. 200-205.
11. Lovejoy, T. 1986. Species leave the ark one by one. In: *The Preservation of Species*. Norton, B. (ed.). Princeton University Press, Princeton, NJ, p. 13-27.
12. Bentham, J. 1970. *An Introduction to the Principles of Morals and Legislation*. Hafner, Darien, CT.
13. Ciriacy-Wantrup, S. von. 1968. *Resource Conservation: Economics and Policies*, (3rd ed.) University of California Division of Agricultural Sciences, Berkeley.
14. Elster, J. 1979. *Ulysses and the Sirens*. Cambridge University Press, Cambridge, UK.
15. A referee questions the compatibility of utility maximization and the SMS. Formally, the referee claims, the SMS violates the underlying expected utility theory, presumably because the SMS implies discontinuous preferences. To the contrary, given the enormous uncertainties that plague biodiversity issues and the high cost of information, a rational utility maximizer with continuous preferences may well choose to

invoke a simple decision heuristic such as the SMS. Elster extends the rational utilitarian argument by showing that, when the pursuit of immediate gratification would have disasterous consequences for one's own utility later, binding behavior is rational. Further, there is really not much at stake in the referee's question, because mixed strategies are permissable in moral thinking, and one such strategy — maximize net benefits, but subject to a SMS constraint — is not without appeal. Thus, my response makes three claims. Binding is consistent, first, with an expanded view of expected utility maximization; second, with rational utilitarian behavior; and, third, with coherent moral thinking more generally.

16. Hubin, D.C., Department of Philosophy, The Ohio State University, Columbus, OH, personal communication.
17. Sagoff, M. 1988. *The Economy of the Earth*. Cambridge University Press, New York.
18. Randall, A. 1988. What mainstream economists have to say about the value of biodiversity. In: *Biodiversity*. Wilson, E.O. (ed.) National Academy Press, Washington, p. 217–223.
19. Hoehn, J.P. and A. Randall. 1989. Too many proposals pass the benefit cost test. *American Economic Review 79*:544–551.
20. Randall, A. and Stoll, J.R. 1983. Existence value in a total valuation framework. In: *Managing Air Quality and Scenic Resources at National Parks and Wilderness Aeas*.Rowe, R.D. and Chestnut, L.G. (ed.) Westview Press, Boulder, p. 265–274. 21. Graham, D.A. 1981. Cost-benefit analysis under uncertainty. *American Economic Review 71*:7 15–725.
22. Smith, V.K. 1987. Nonuse values in benefit cost analysis. *Southern Economic Journal 54*: 19–26.
23. Randall, A. 1989. Nonuse benefits. In:*Measuring the Demand for Environmental Improvement*.Braden, J.B. and Kolstad, C.D. (ed.). ILENR/RE-EA-89/18. University of Illinois, Urbana, 10.1-10.19.
24. Randall. A., Ives, B.C., and Eastman, C. 1974. Bidding games for valuation of aesthetic environmental improvements. *Journal of Environmental Economics and Management 1*: 132–149.
25. Hoehn, J.P. and Randall, A. 1987. A satisfactory benefit cost indicator from contingent valuation. *Journal of Environmental Economics and Management 14*: 226–247.
26. Mitchell, R.C. and Carson, R.T. *Using Surveys to Value Public Goods: The Contingent Valuation Method*. Resources for the Future. Washington, 463 p.
27. Maler, K-G. 1974. *Environmental Economics: A Theoretical Inquiry*. Johns Hopkins University Press. Baltimore.
28. Bradford, D.F. and Hildebrandt, G.G. Observable preferences for public goods. *Journal of Public Economics 8*: 111–131.
29. Lakatos, I. 1970. Falsification and the methodology of scientific research programmes. In: *Criticism and the Growth of Knowledge*.Lakatos, I. and Musgrave, A. (ed.). Cambridge University Press, London.
30. Viscusi, W.K. 1985. Are individuals Bayesian decision makers? *American Economic Review 75*: 381–385.
31. Bennett, J.W. 1984. Using direct questioning to value existence benefits of preserved natural areas. *Australian Journal of Agricultural Economics 28*: 136–152.
32. Bowker, J.M. and Stoll, J.R. 1988. Use of dichotomous choice nonmarket methods to value the whooping crane resource. *American Journal of Agricultural Economics 70*: 372–381.
33. Brookshire, D.S., Eubanks,L.S. and Randall, A. 1983. Estimating option prices and existence values for wildlife resources. *Land Economics 59*: 1–15.
34. Majid, I., Sinden, J.A. and Randall, A. 1983. Benefit evaluation of increments to existing systems of public facilities. *Land Economics 59*: 377–392.
35. Stoll, J.R. and Johnson, L.A. 1984. Concepts of value, nonmarket valuation, and the case of the whooping crane. *Transactions of the Forty-ninth North American Wildlife and Natural Resources Conference 49*: 382–393.

36. The author gratefully acknowledges support from the National Sciences Foundation under Grant No. BBS-8710153, and the comments of two referees and the substantive suggestions of . Hubin, D. C and Farmer, M. C.
37. First submitted April 11, 1990, accepted for publication after revision September 12, 1990.

Alan Randall has been professor of Resource Economics and Environmental Policy at Ohio State University since 1985. He was previously professor of Agricultural Economics, University of Kentucky, Lexington. Professor Randall has published widely in welfare economics, benefit-cost analysis, valuation of environmental goods and, in recent years, the philosophical issues raised by the increasing use of benefit-cost analysis in environmental policy. His address: Dept. of Agricultural Economics and Rural Sociology, The Ohio State University, 2120 Fyffe Road, Columbus, OH 43210, USA.

Name Index